The Behavior of
the Laboratory Rat

THE BEHAVIOR OF
THE LABORATORY RAT

A Handbook with Tests

Edited by

IAN Q. WHISHAW
BRYAN KOLB

Department of Psychology and Neuroscience
Canadian Centre for Behavioural Neuroscience

OXFORD
UNIVERSITY PRESS

2005

OXFORD
UNIVERSITY PRESS

Oxford   New York
Auckland   Bangkok   Buenos Aires   Cape Town   Chennai
Dar es Salaam   Delhi   Hong Kong   Istanbul   Karachi   Kolkata
Kuala Lumpur   Madrid   Melbourne   Mexico City   Mumbai   Nairobi
São Paulo   Shanghai   Taipei   Tokyo   Toronto

Published by Oxford University Press, Inc.
198 Madison Avenue, New York, New York, 10016
www.oup.com

Oxford is a registered trademark of Oxford University Press

Library of Congress Cataloging-in-Publication Data
Whishaw, Ian Q., 1939–
The behavior of the laboratory rat : a handbook with tests /
edited by Ian Q. Whishaw, Bryan Kolb.
p. cm.   ISBN 0-19-516285-4
1. Rats—Behavior. 2. Rats as laboratory animals.
I. Kolb, Bryan, 1947– II. Title.
QL737.R666W52 2004   616'.02733—dc22   2004041514

9 8 7 6 5 4 3 2 1

Printed in the United States of America
on acid-free paper

To Samuel Anthony (Tony) Barnett (1915–2003)

Behavioral neuroscientists know S.A. Barnett best for his book *The Rat: A Study in Behavior*, first published in 1963. He was the author of over 150 papers (the last of which is the second chapter of this volume) and nine books. He was a graduate of Oxford University in Zoology. He advised the British Government on plague during WWII and was the Chair of Zoology in Glasgow from 1971 to 1983. Tony was an enthusiastic broadcaster, contributing for years to the BBC series *Occam's Razor*.

Preface

The principal function of the nervous system is to produce behavior. Thus, the ultimate goal of most behavioral work with laboratory animals in neuroscience is to understand how molecular events in the nervous system come to produce behavior and, as a corollary, how changes in molecular events produce differences in behavior. Understanding these issues offers hope for understanding the nature of the human mind, which some may argue is the fundamental question in neuroscience. But perhaps even more important is that understanding brain–behavior relationships offers a way to find treatments for dysfunctions of behavior, whether they are in the province of neurology or psychiatry. Advances in molecular and cellular neuroscience have been dramatic over the past two decades, but most of these advances have been independent of an understanding of how they relate to behavior. This is changing. Neuroscientists oriented toward molecular research are increasingly looking to the ultimate function of the phenomenon that they have been studying—behavior. For the majority of behavioral studies, this means studying the behavior of the laboratory rat.

This book has three objectives. Our first objective is to present an introduction of rat behavior to neuroscience students. In choosing the rat as the subject species, we made the assumption that this species will remain, as it has in the past, the primary subject used the laboratory investigations of behavior. Our second objective is to describe the organization and complexity of rat behavior. The major theme emerging from many lines of research on rat behavior is that understanding the rules of behavioral organization will be central in understanding the structural basis of behavior. Our third objective is to update, as much as is possible, previous compendiums of rat behavior. Behavioral neuroscience continues to be a diverse field of research in which there remain many competing experimental methods and hypotheses, and we believe that collectively, the chapters of this volume reflect that diversity.

As we have noted, advances in the field of neuroscience have kindled an interest in behavior by researchers in many of its subdisciplines. For many of these researchers, behavior may previously have seemed unrelated to their studies and of little direct interest. Many researchers with primary training in diverse fields such as genetics or biochemistry are now looking at brain–behavior questions for the first time and, like all fields, the literature can be bewildering to the novice. We therefore asked the authors of the chapters of this book to imagine that students from another discipline were coming to them with questions related to how they could incorporate behavior into their research programs. For example, we asked them to imagine a student coming from such areas as medicine, chemistry, or genetics who had no special training in behavior but now saw behavior as relevant and necessary to research questions in which they were interested. The challenge we presented was, "Could they summarize their field of expertise so that the novice student would gain an understanding of the questions, methods, and potential findings in that line of research?" We also asked them to make their summary brief, to emphasize methodology, and to minimize as much as possible literature reviews. Our expectation is that a novice student would read their chapter as a first introduction to the study of rat behavior. This introduction would then guide the student as he or her gained further expertise in the practical application of that information and in study of the larger body of literature.

In our view, the problem that we posed is not fictive. During the past few years, we received large numbers of telephone calls, e-mails, and queries regarding behavior from people with diverse backgrounds in neuroscience. We have also been asked to speak at meetings about behavior to scientists who we could not have imagined would have been interested in behavior or in the research that we do. The tone of this interest is best represented by the comments of an acquaintance, a molecular biologist, who stated: "I could not have imagined that this psycho mumbo-jumbo was going to be important, but now I see it as the only game in town." We realize that in answer to such interest, it is simply not good enough to say, "We could look at that for you," or "Perhaps you should collaborate with an expert on that behavior." Rather, it is likely that behavioral methods will become a part of many lines of research, so the challenge for behaviorists is to make their science accessible to other investigators. Indeed, we have encountered the problem in reverse as we have added more molecular techniques to our behavioral studies. Accordingly, we asked the authors of the chapters in this book to make their research accessible to new students of behavior.

Of course, this book is about rat behavior. The rat, *Rattus norvegicus*, was the first species to be domesticated for the purpose of scientific research. In the 100 years since the rat was first introduced to the laboratory, it has generated an incredibly large body of literature. It has been the primary subject with which psychological theories have been tested; it has been the primary subject for the study of behavioral pharmacology; and it has been the primary subject for the investigation of brain chemistry, anatomy, and physiology. One reason for the popularity of the rat is that

it is a behavioral generalist. Rats are found in virtually every ecological niche on earth, and they have proved to be adaptable and successful in all of them. For the purposes of answering behavioral questions, we assert that as a behavioral generalist, the rat will remain the primary species for continued investigations into the organization of brain and behavior and the structural basis of behavior. In being a generalist, the rat is very much like the human, a species with which it is commensal. It is likely that the genes, neural structure, and behavior of generalists have properties that are similar. This is why the rat continues to be the primary model used to study a wide range of questions related to human behavior and health.

With the development of genetic engineering, a line of inquiry that uses the mouse as the prime vertebrate model, we should address the question of whether the mouse might not have been a better species about which to compile a behavioral book. The laboratory mouse is the laboratory rat's closest domestic relative, but we do not think that the two species are so similar that one can be substituted for the other. This is especially so with respect to behavior. Both species have been used for behavioral research for approximately 100 years, and each has apparently found its laboratory niche. There is little doubt that for many questions related to motor functions, regulatory functions, and especially cognitive functions, the rat has been the species of choice. We think that it will continue to be so. In contrast, mice will likely continue to be the subject of choice in genetic studies in neuroscience, and presumably the study of many behavior genetic questions will retain the mouse as the primary laboratory subject. However, the behavioral study of the mouse in genetics is fundamentally different from the primary questions addressed in behavioral studies in the rat and must be the subject of a separate volume.

Two excellent, and still relevant, books have been previously devoted to rat behavior.

Norman L. Munn's 1950 "Handbook of Psychological Research on the Rat: An Introduction to Animal Psychology" is directed to many of the same questions as is the present book. It describes general activity, unlearned behavior, sensory processes, learning, social behavior, and rat models of neuropsychiatric diseases. It also emphasizes methods of study. S. A. Barnett's 1963 book, "The Rat: A Study in Behavior," covers much of the same ground but with more emphasis on the ethogram, or profile of behavior, of the wild rat. In what way is the present book different from these predecessors? We think that the primary advancement in understanding rat behavior is the emergence of understanding how rat behavior is organized. For example, rat grooming, play and aggression, exploration, cognition, and other activities are organized with both fixed and open syntax. The understanding of this organization provides new avenues for the investigation of genetic, neural, and hormonal regulation of behavior. This organization has also led to the development of computer-based behavioral analysis systems that aid in using behavior as an assay for other scientific manipulations.

Although this book consists of 43 chapters on different aspects of rat behavior, and thus is comprehensive, it is not exhaustive. Our major difficulty in editing the book was in insisting that authors substantially shorten their chapters to make the book manageable as a single volume. Indeed, we could have doubled the number of chapters without covering every aspect of rat behavior, but we believe that the selection of chapters presented here provide more than adequate grist for an introduction to the study of the rat in behavioral brain research.

We express a special thanks to all of the authors who generously contributed time to write a chapter for this book. We also express our thanks to Fiona Stevens of Oxford University Press, who approached us and persuaded us to compile this handbook and gave us the liberty of selecting a structure of our own choosing.

Lethbridge, Alberta, Canada   I. Q. W.
                                                                     B. K.

SUGGESTED READINGS

Barnett SA (1963) The rat: a study in behavior. Chicago and London: The university of Chicago Press.

Munn NL (1950) Handbook of psychological research on the rat: an introduction to animal psychology. Boston: The Riverside Press Cambridge.

Contents

Contributors

JEFFREY R. ALBERTS
Department of Psychology
Indiana University
Bloomington, Indiana

J. WAYNE ALDRIDGE
Departments of Neurology and Psychology
University of Michigan
Ann Arbor, Michigan

HYMIE ANISMAN
Institute of Neurosciences
Carelton University
Ottawa, Ontario, Canada

MICHAEL C. ANTLE
Department of Psychology
Columbia University
New York, New York

BERNARD W. BALLEINE
Department of Psychology and the Brain Research
   Institute
University of California, Los Angeles
Los Angeles, California

S. ANTHONY BARNETT*
Aranda, Australia

JILL B. BECKER
Department of Psychology
Reproductive Sciences Program and Neurosciences
   Program
University of Michigan
Ann Arbor, Michigan

YOAV BENJAMINI
Department of Zoology
Tel Aviv University
Tel Aviv, Israel

D. CAROLINE BLANCHARD
Department of Neurobiology
University of Hawaii
Honolulu, Hawaii

*Deceased.

ROBERT J. BLANCHARD
Department of Neurobiology
University of Hawaii
Honolulu, Hawaii

MARK S. BLUMBERG
Department of Psychology
Indiana University
Bloomington, Indiana

STEVE L. BRITTON
Functional Genomics Laboratory
Medical College of Ohio
Toledo, Ohio

RICHARD BROWN
Department of Psychology
Dalhousie University
Halifax, Nova Scotia, Canada

RUSSELL W. BROWN
Department of Psychology
East Tennessee State University
Johnson City, Tennessee

MICHELE R. BRUMLEY
Department of Psychology
University of Iowa
Iowa City, IA

BAUKE BUWALDA
Department of Animal Physiology
University of Groningen
Haren, The Netherlands

SAMUEL W. CADDEN
The Dental School
University of Dundee
Dundee, Scotland

JOHN K. CHAPIN
Department of Physiology and Pharmacology
SUNY Downstate Medical Center
Brooklyn, New York

PETER G. CLIFTON
Department of Psychology
University of Sussex
Brighton, United Kingdom

SIETSE F. DE BOER
Department of Animal Physiology
University of Groningen
Haren, The Netherlands

ROBERT M. DOUGLAS
Centre for Macular Research
Department of Ophthalmology and Visual Sciences
University of British Columbia
Vancouver, British Columbia, Canada

ANNA DVORKIN
Department of Zoology
George S. Wise Faculty of Life Sciences
Tel Aviv University
Tel Aviv, Israel

RICHARD H. DYCK
Department of Psychology
Department of Cell Biology and Anatomy
University of Calgary
Calgary, Alberta, Canada

DAVID EILAM
Department of Zoology
Tel Aviv University
Tel Aviv, Israel

MICHAEL S. FANSELOW
Department of Psychology
University of California, Los Angeles
Los Angeles, California

ALISON S. FLEMING
Department of Psychology
University of Toronto at Missassauga
Missassauga, Ontario, Canada

BENNETT G. GALEF, JR.
Department of Psychology
McMaster University
Hamilton, Ontario, Canada

ROBBIN L. GIBB
Canadian Centre for Behavioural Neuroscience
Department of Psychology and Neuroscience
University of Lethbridge
Lethbridge, Alberta, Canada

ILAN GOLANI
Department of Zoology
George S. Wise Faculty of Life Sciences
Tel Aviv University
Tel Aviv, Israel

LINDA HERMER-VAZQUEZ
Department of Physiology and Pharmacology
SUNY Downstate Medical Center
Brooklyn, New York

RAYMOND HERMER-VAZQUEZ
Department of Physiology and Pharmacology
SUNY Downstate Medical Center
Brooklyn, New York

ANDREW N. IWANIUK
Department of Psychology
University of Alberta
Edmonton, Alberta, Canada

WILLIAM J. JENKINS
Department of Psychology
Reproductive Sciences Program and Neurosciences
   Program
University of Michigan
Ann Arbor, Michigan

NERI KAFKAFI
Maryland Psychiatry Research Center
University of Maryland
College Park, Maryland

LAUREN GERARD KOCH
Functional Genomics Laboratory
Medical College of Ohio
Toledo, Ohio

BRYAN KOLB
Canadian Centre for Behavioural Neuroscience
Department of Psychology and Neuroscience
University of Lethbridge
Lethbridge, Alberta, Canada

JAAP M. KOOLHAAS
Department of Animal Physiology
University of Groningen
Haren, The Netherlands

ALEXANDER W. KUSNECOV
Department of Psychology
Biopsychology and Behavioral Neuroscience Program
Rutgers, The State University of New Jersey
Piscataway, New Jersey

DANIEL LE BARS
Institut National de la Santé et de la Recherche
    Médicale (INSERM)
Paris, France

DINA LIPKIND
Department of Zoology
George S. Wise Faculty of Life Sciences
Tel Aviv University
Tel Aviv, Israel

VEDRAN LOVIC
Department of Psychology
University of Toronto at Missassauga
Missassauga, Ontario, Canada

GERLINDE A. METZ
Canadian Centre for Behavioral Neuroscience
Department of Psychology and Neuroscience
University of Lethbridge
Lethbridge, Alberta, Canada

KLAUS A. MICZEK
Department of Psychology, Psychiatry, Pharmacology,
    and Neuroscience
Tufts University
Medford, Massachusetts

RALPH E. MISTLBERGER
Department of Psychology
Simon Fraser University
Burnaby, British Columbia, Canada

GUY MITTLEMAN
Psychology Department
University of Memphis
Memphis, Tennessee

GILLIAN MUIR
Biomedical Sciences
Western College of Veterinary Medicine
University of Saskatchewan
Saskatoon, Saskatchewan, Canada

DAVE G. MUMBY
Department of Psychology
Concordia University
Montreal, Quebec, Canada

SERGIO M. PELLIS
Canadian Centre for Behavioral Neuroscience
Department of Psychology and Neuroscience
University of Lethbridge
Lethbridge, Alberta, Canada

VIVIEN C. PELLIS
Canadian Centre for Behavioral Neuroscience
Department of Psychology and Neuroscience
University of Lethbridge
Lethbridge, Alberta, Canada

JOHN J.P. PINEL
Department of Psychology
University of British Columbia
Vancouver, British Columbia, Canada

BRUNO POUCET
Laboratoire de Neurobiology de la Cognition,
    UMR 6155
CNRS—Universite Aix-Marseille I
Marseille, France

GLEN T. PRUSKY
Canadian Centre for Behavioural Neuroscience
Department of Psychology and Neuroscience
University of Lethbridge
Lethbridge, Alberta, Canada

STEPHANIE L. REES
Department of Psychology
University of Toronto at Missassauga
Missassauga, Ontario, Canada

SCOTT R. ROBINSON
Department of Psychology
University of Iowa
Iowa City, Iowa

NEIL E. ROWLAND
Department of Psychology
University of Florida
Gainesville, Florida

EVELYN SATINOFF
Department of Psychology
University of Delaware
Newark, Delaware

ETIENNE SAVE
Laboratoire de Neurobiology de la Cognition,
    UMR 6155
CNRS—Universite Aix-Marseille I
Marseille, France

TIM SCHALLERT
Department of Psychology
University of Texas at Austin
Austin, Texas

HEATHER SCHELLINCK
Department of Psychology
Dalhousie University
Halifax, Nova Scotia, Canada

BURTON SLOTNICK
Department of Psychology
University of South Florida
Tampa, Florida

GRETA SOKOLOFF
Department of Psychology
Indiana University
Bloomington, Indiana

ALAN C. SPECTOR
Department of Psychology
University of Florida
Gainesville, Florida

ROBERT J. SUTHERLAND
Canadian Centre for Behavioural Neuroscience
University of Lethbridge
Lethbridge, Alberta, Canada

HENRY SZECHTMAN
Department of Psychiatry and Behavioural
    Neurosciences
McMaster University
Hamilton, Ontario, Canada

MATTHEW R. TINSLEY
Department of Psychology
University of California, Los Angeles
Los Angeles, California

DALLAS TREIT
Department of Psychology
University of Alberta
Edmonton, Alberta, Canada

DOUGLAS G. WALLACE
Canadian Centre for Behavioural Neuroscience
Department of Psychology and Neuroscience
University of Lethbridge
Lethbridge, Alberta, Canada

IAN Q. WHISHAW
Canadian Centre for Behavioural Neuroscience
Department of Psychology and Neuroscience
University of Lethbridge
Lethbridge, Alberta, Canada

MARTIN T. WOODLEE
Department of Psychology
University of Texas at Austin
Austin, Texas

Natural History

Evolution

ANDREW N. IWANIUK

1

ON THE ORIGIN OF
Rattus norvegicus

The Norway, or "laboratory," rat (*Rattus norvegicus*) has been used in behavioral, neural, physiological, and other forms of research for more than a century. The evolutionary history of this species is often dismissed as unimportant in psychological and biomedical research because the aim is not to understand evolutionary biology but rather to use the rat as a model system to investigate a specific aspect of organismal biology. It is not our intention to critique these experiments because they are integral to our understanding of animal behavior, anatomy, molecular biology, and physiology. It is, however, important to acknowledge that the rat did not evolve in a vacuum and that the morphological, physiological, and behavioral changes imposed by "domestication" are still a result of the evolutionary process.

This chapter addresses the evolution of the laboratory rat from the origins of rodents in general to the speciation of the genus *Rattus*. This is not meant to be a complete review of all the taxonomy and phylogenetic history of *Rattus* and higher-level taxonomic ranks, because discussions of this are provided elsewhere (Carleton and Musser, 1984; Luckett and Hartenberger, 1985; Musser and Carleton, 1993; Nowak, 1999). Instead, I provide a summary of the evolutionary events that led to *R. norvegicus*. Because palaeontology, taxonomy, and phylogenetics are intimately related to one another, this chapter is organized in terms of the taxonomy of *R. norvegicus* (Table 1–1). Evolutionary relationships and palaeontological history are discussed with reference to other groups of the same taxonomic rank. For example, the order Rodentia is placed in the context of other mammalian orders. By summarizing the evolutionary history of *R. norvegicus*, we aim to provide a basic understanding of how the species has evolved that may be instructive in interpreting the results of behavioral experimentation and/or comparative analyses.

ORDER RODENTIA

To understand the evolution of *R. norvegicus*, it is necessary to begin with the history of the rodents in general and their relationship to other mammalian taxa. The order Rodentia is the most abundant of all of the mammalian orders, numbering close to 2000 species. Rodents are found on every continent, except Antarctica, and account for almost half of all placental mammals. They are readily distinguishable from other mammals by an array of morphological features (Luckett and Hartenberger, 1993, 1985), the most prominent of which is their distinctive dental morphology. Rodent incisors are large, unrooted, and persistently growing teeth with enamel only on the upper surface that maintains a beveled cutting edge. The surface morphology of the molars is also distinctive, and the jaw structure exhibits adaptations that allow considerable movement during grinding (Hand, 1984).

Table 1–1. Taxonomy of the Genus *Rattus*

Class Mammalia
  Order Rodentia
    Superfamily Muroidea
    Family Muridae
      Genus *Rattus*

Species		
	adustus	*annandalei*
	*argentiventer**	*baluensis*
	bontanus	*burrus*
	colletti	*elaphinus*
	enganus	*everetti*
	*exulans**	*feliceus*
	foramineus	*fuscipes*
	giluwensis	*hainaldi*
	hoffmanni	*hoogerwerfi*
	jobiensis	*koopmani*
	korinchi	*leucopus*
	*losea**	*lugens*
	lutreolus	*maclearii*[†]
	marmosurus	*mindorensis*
	mollicomulus	*montanus*
	mordax	*morotaiensis*
	nativitatis[†]	*nitidus**
	*norvegicus**	*novaeguineae*
	osgoodi	*palmarum*
	pelurus	*praetor*
	ranjiniae	*rattus**
	sanila	*sikkimensis*
	simalurensis	*sordidus*
	steini	*stoicus*
	tanezumi	*tawitawiensis*
	timorensis	*tiomanicus**
	tunneyi	*turkestanicus*
	villosissimus	*xanthurus*

Note: The taxonomy is taken from Guy and Musser (1993).
*Commensal species.
[†]Species that recently became extinct.

Despite some broad similarities in morphology, rodents are a morphologically and behaviorally diverse order. They span a range of locomotor behaviors that include gliding, climbing, swimming, underground digging, hopping, and running. Not only do rodents exhibit this range of locomotor behaviors; in most instances, they have evolved independently many times. For example, subterranean locomotion has evolved at least three times (Muroidea, Geomyoidea, Bathyergoidea). Similarly, there is a broad range of social systems from uniparental monogamy/polygamy to complex, multimale/multifemale societies have also evolved independently many times.

This behavioral diversity belies the fact that the rodents form a monophyletic group. That is, all rodents share a common ancestry that is not shared with nonrodent species. Although the issue of rodent monophyly was questioned by molecular studies of the guinea pig (*Cavia porcellus*) (Graur et al., 1991; D'Erchia et al., 1996), more recent studies agree that rodents are monophyletic (Adkins et al., 2001; Madsen et al., 2001; Murphy et al., 2001a,b; Huchon et al., 2002). There is, however, some debate regarding the position of Rodentia relative to other mammalian orders in phylogenetic trees.

Traditionally, the order Lagomorpha (hares, rabbits, and pikas) is considered to be most closely related to rodents (i.e., a sister-group) based on their morphological similarity (Shoshani and McKenna, 1998) (Fig. 1–1A). The rodents and lagomorphs together comprise a clade termed Glires. Morphological similarities also link the elephant shrews (order Macroscelidea) as the sister-group to the Glires (Fig. 1–1A). Molecular studies have demonstrated markedly different relationships between many traditional mammalian orders, but they all agree that rodents and lagomorphs should be placed together (Huchon et al., 2001; Madsen et al., 2001; Murphy et al., 2001a,b) (Fig. 1–1B). The broadly based studies also agree in placing the Glires as sister-group to a clade containing primates (Madsen et al., 2001; Murphy et al., 2001a,b).

In terms of dating the origin of the Rodentia, the superorder Glires likely originated between 64 and 104 million years ago (mya) (Archibald et al., 2001; Murphy et al., 2001a). The diversification of rodents is therefore estimated at 65 mya at the earliest on the basis of both palaeontological (Alroy, 1999; Archibald et al., 2001) and molecular (Bromham et al., 1999; Foote et al., 1999; Eizirik et al., 2001) data. Thus, the predecessors of rodents may have coexisted with dinosaurs, but true rodents did not evolve until after the Cretaceous–Tertiary Period boundary (i.e., <65 mya),

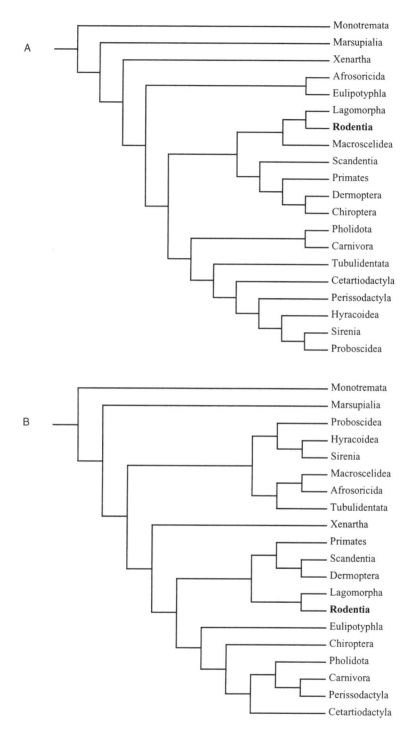

Figure 1–1. Two alternative phylogenetic hypotheses of the interordinal relationships of all mammals. Contrary to "traditional" taxonomies, the Insectivora is broken up into two orders: Afrosoricida and Eulipotyphla. Also, the Cetacea and Artiodactyla are merged into the singular order Cetartiodactyla. The first tree (A) is based on morphological characters and places rodents as a sister-group to the lagomorphs (Shoshani and McKenna, 1998). The second tree (B) differs in many of the interordinal relationships but retains the sister-group relationship between Rodentia and Lagomorpha (Murphy et al., 2001a). Although the monotremes (Monotremata) were not sampled in the analysis presented by Murphy et al. (2001a), they are included here as the basal group to provide consistency between the two trees.

contrary to some estimates (Hedges et al., 1996; Kumar and Hedges, 1998).

SUPERFAMILY MUROIDEA/ FAMILY MURIDAE

The superfamily Muroidea is the most diverse rodent taxon, with more than 1300 species (Musser and Carleton, 1993). Within this superfamily, there is consensus that only one family is represented by extant species: the Muridae (Musser and Carleton, 1993). This family includes the "true" rats (*Rattus*) as well as numerous genera of mice (e.g., spiny mice, deer mice), voles, lemmings, gerbils, and hamsters. The diversity in life history and morphology within murids is almost as great as that of rodents as a whole, with a wide span of locomotor modes, social systems, and ecology. Murids are generally distinguished from other rodent families by their molar morphology (Carleton and Musser, 1984), but they also possess a "primitive" middle ear structure (Lavocat and Parent, 1985) and several unique developmental features (Luckett, 1985).

The kangaroo-rats (superfamily Dipodoidea) are generally considered to be the sister-group to the Muridae (Fig. 1–2b). This is based on a number of shared morphological features (Luckett and Hartenberger, 1985) as well as molecular evidence (Nedbal et al., 1996; Adkins et al., 2001). More recently, Huchon et al. (2002) demonstrated a possible relationship between the Geomyoidea (pocket gophers) and the Muridae, using the most comprehensive sampling of rodent species in molecular phylogenetics thus far (Fig. 1–2a). Although this relationship was strongly supported in their analyses, several alternative arrangements could not be discounted. Further investigation of this arrangement is therefore warranted as this could result in several taxonomic changes at the superfamily level of rodents.

The diversification of the rodent superfamilies and families is estimated to have occurred in late Palaeocene to early Eocene (≈55 mya) (Hartenberger, 1998). A basal lineage leading to the Muridae branched off at this point, with "modern" murids apparent in the middle Eocene (36.5 to 49 mya). The oldest murid discovered thus far, the hamster-like *Cricetodon*, was found in Mongolia and has been dated as late Eocene (34.2 to 36.5 mya) (Li and Ting, 1983, in Carleton and Musser, 1984). Fossil taxa more closely resembling *Rattus* do not appear to occur for another 20 million years (see later).

SUBFAMILY Murinae

Within the Muridae, the subfamily Murinae is the most speciose lineage. There are more than 500 recognized species within the Murinae that span more than 120 genera (Musser and Carleton, 1993; Nowak, 1999). The subfamily Murinae encompasses a broad range of species of diverse ecological and morphological forms that include several specialized genera that resemble species from other subfamilies, families, and orders. For example, *Hydromys* resemble aquatic shrews; *Echiothrix*, elephant shrews; *Komodomys*, gerbils; *Crateromys*, squirrels; and *Nesokia*, gophers. Other genera are more "typically" murine, such as *Rattus* and the laboratory or house, mouse (*Mus musculus*). Other murid rodents, such as hamsters, gerbils, and voles, are divided into several other subfamilies (Fig. 1–3).

In terms of the relationships between the murines and other murid subfamilies, there is general agreement that Gerbillinae is the sister-group to Murinae on the basis of both morphological (Flynn et al., 1985) and molecular (Dubois et al., 1999; Michaux et al., 2001) evidence (Fig. 1–3). The position of several other subfamilies has, until recently, remained controversial (Chevret et al., 1993; Dubois et al., 1999; Michaux et al., 2001).

There is some debate over what represents the first murine (Flynn et al., 1985). Some authors claim that it is the middle Miocene (15

Figure 1–2. These two phylogenetic trees presented illustrate two alternative topologies of the interrelationships between rodent families. (A) The maximum likelihood tree derived from reconstructions of three nucleotide sequences (Huchon et al., 2002). Here, the Muridae (shown in bold) forms a clade with the pocket gophers (Geomyoidea) as a sister-group. (B) In contrast, this second tree represents the one of several topologies depicted in Adkins et al. (2001). Although differences were present in the interfamilial relationships between some trees, all of them agree with placing the jerboas and kangaroo rats (Dipodoidea) as sister-group to the Muridae (shown in bold). The other families and superfamilies include squirrels (Sciuridae), mountain beaver (*Aplodontia rufa*) (Aplodontidae), dormice (Gliroidea), scaly-tailed flying squirrels (Anomaluridae), Springhare (*Pedetes capensis*) (Pedetidae), beavers (Castoridae), gundis (Ctenodactylidae), cane rats (Thryonomyoidea), African mole rats (Bathyergoidea), Old World porcupines (Hystricoidea), New World porcupines (Erethizontoidea), chinchillas (Chinchillidae), degus (Octodontidae), and guinea pigs and cavies (Cavioidea).

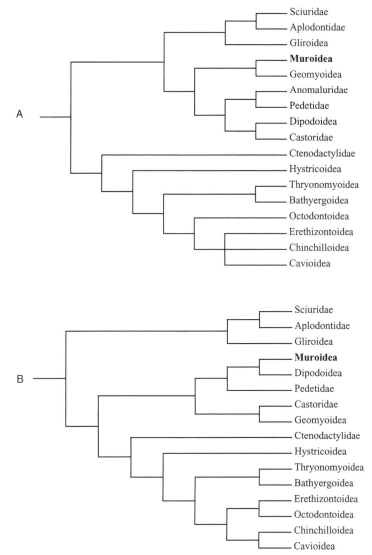

to 16 mya) *Antemus* from Thailand (Jacobs, 1977; Hand, 1984; Jaeger et al., 1986). The inclusion of *Antemus* in the Murinae has, however, been questioned because of morphological similarities with other subfamilies. Therefore, others consider *Progonomys* to be the earliest murine (Flynn et al., 1985) at a more recent age of 11 to 12 mya. Due to the fragmentary material recovered thus far, no resolution has been posited for the affinity of *Antemus*, but an origin of 15 to 16 mya agrees with molecular estimates (see later).

The point at which the murines diverged from the other subfamilies is difficult to pin-point, as they appear to have first evolved in central Asia, where mammalian palaeontology is still in its early stages. Based on the available evidence, a divergence time of between 16 and 23.8 mya (i.e., late Oligocene to early Miocene) seems likely (Hartenberger, 1998; Tong and Jaeger, 1993). This is supported by molecular estimates of 17.9 to 20.8 mya (Michaux et al. 2001). In contrast, the hamsters (Cricetinae) are known from middle Eocene deposits (Hartenberger, 1998) dating between 36.5 and 49 mya. Thus, the murines underwent an explosive radiation relatively late in murid rodent evolution.

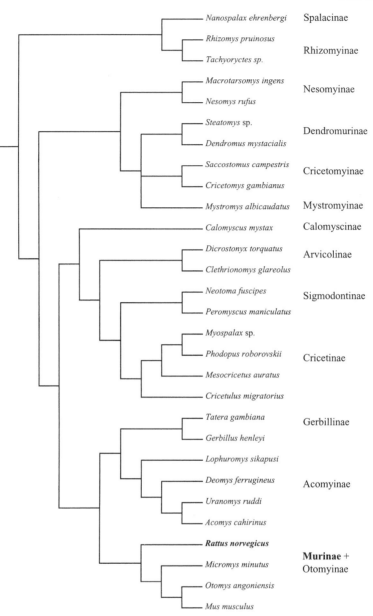

Figure 1–3. This phylogenetic tree depicts the interrelationships between subfamilies within the Muridae based on a combined analysis of LCAT and vWF genes (Michaux et al., 2001). Note that the spiny mice (Acomyinae) and gerbils (Gerbillinae) form a sister-group relationship with a clade composed of the Murine (in bold) and the vlei rats (Otomyinae). The other subfamilies include hamsters (Cricetinae), voles and lemmings (Arvicolinae), blind mole rats (Spalacinae), bamboo rats (Rhizomyinae), New World murids (e.g., deer mice, wood rats, muskrat) (Sigmodontinae), mouse-like hamsters (Calomyscinae), pouched mice and rats (Cricetomyinae), and several diverse African clades (Nesomyinae, Dendromurinae, and Mystromyinae).

GENUS *Rattus*

The status of the genus *Rattus* has undergone numerous changes since its first use by Fischer in 1803 (Musser and Carleton, 1993). Several genera that were once considered to belong to the genus *Rattus* have been further subdivided into other genera, such as *Praomys*, *Mastomys*, and *Apomys* (Nowak, 1999). There is, in fact, still some debate regarding how to define the genus itself (Carleton and Musser, 1984; Musser and Holden, 1991; Musser and Carleton, 1993). Despite these problems, *Rattus* can generally be distinguished from other murine genera by long body fur, sparsely haired tails with overlapping scales, and stout skulls with relatively large auditory bullae and prominent coronoid processes (Watts and Aplin, 1981). These characters are by no means definitive, however, as they are also present in other genera. A precise description of characters that typify the genus is currently wanting.

Coinciding with the uncertainty of the boundaries of the genus *Rattus* are uncertainties of the relationships between murine genera. A number of studies have depicted phylogenetic relationships within the Murinae (Robinson et al., 1997; Martin et al., 2000; Suzuki et al., 2000; Michaux et al., 2001), but Watts and Baverstock (1995) provided the most comprehensive number of species. In their analysis, Watts and Baverstock (1995) found that four biogeographical clades could be recognized: southeast Asian, Australasian, New Guinean, and African (Fig. 1–4). Based on their analyses, they suggested that the first murines arose 20 mya, leading to the basal members of the lineage. This is supported by fossil evidence that estimated the divergence time between *Rattus* and *Mus* at approximately 12 mya (Jaeger et al., 1986). At 8 mya, the African, southeast Asian, and Australasian clades underwent a rapid speciation event, resulting in "bushy" phylogenetic trees and the aforementioned problems of delimiting generic boundaries. The age at which *Rattus* diverged from other members of the Asian clade is uncertain, but based on Watts and Baverstock's (1995) phylogeny, it would have occurred within the past 8 million years (see Fig. 1–4). This estimate agrees with the earliest *Rattus* fossil being recorded from the late Pliocene (<3 mya) of China (Xue, 1981, in Jaeger et al., 1986). Such a recent origin of *Rattus* suggests that their occurrence in Australasia occurred well after other murines first colonized the area 4 to 6 mya (Hand, 1984).

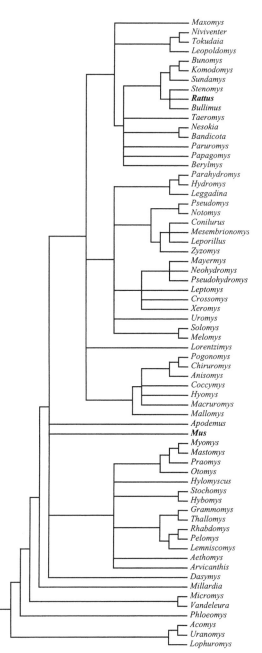

Figure 1–4. This phylogenetic tree depicts intergeneric relationships within the Murinae based on microcomplement fixation of albumin (Watts and Baverstock, 1995). Both *Rattus* and *Mus* are shown in bold, and the date of divergence between these two genera is approximately 12 million years (Jaeger et al., 1986). Note that although *Acomys*, *Lophuromys*, and *Uranomys* were included in the murine phylogeny of Watts and Baverstock (1995), more recent studies have shown that they form a monophyletic clade separate from the Murinae (Dubois et al., 1999; Michaux et al., 2001) (see Fig. 1–3).

Rattus norvegicus

Following this pattern of most speciose lineages, the species *R. norvegicus* is a member of
the most speciose genus of murid rodents
with up to 50 species being recognized (Table
1–1) (Musser and Carleton, 1993; Nowak,
1999). As discussed earlier, the status of this
genus is uncertain, but in general they have a
similar appearance to *R. norvegicus*. In Australia, *R. norvegicus* differs from native *Rattus*
primarily by their longer tails and larger skulls,
but also there are differences in tail and foot
pad coloration, a tapered snout, and coarse fur
(Watts and Aplin, 1981).

Another feature that can aid in distinguishing *R. norvegicus* from other *Rattus*
species is that it occurs in urban and other disturbed environments. This is not, however, a
unique feature of *R. norvegicus* as several
Rattus species are commensal. That is, they
are found primarily close to human habitation
and prefer this habitat to others. Some of these
species are well known pests in a variety of locales, such as the Polynesian Rat or Kiore (*R.
exulans*), Black Rat (*R. rattus*), and *R. norvegicus*. The other species (see Table 1–1) are less
well known by Western researchers but are
no less destructive (e.g., Leung et al., 1999).

Some of the remaining *Rattus* species can
also be found in close proximity to humans
but generally occur in more remote habitats.
For example, the Australian Bush Rat (*R.
fuscipes*) is often found in suburban areas
(Watts and Aplin, 1981; Menkhorst, 1995;
Strahan, 1998) and the Long-haired Rat
(*R. villossissimus*) can overwhelm towns and
farms when in plague proportions (Watts and
Aplin, 1981; Strahan, 1998). These are not considered to be commensal species, however, as
they do not depend on human-altered environments. In fact, most *Rattus* species prefer
undisturbed areas in environments that include arid plains, rain forests, coastal heath
and subalpine regions.

Structurally, there are few differences in
behavior between *R. norvegicus* and other

Rattus species (Begg and Nelson, 1977; Barnett et al., 1982; Beeman 2002). In general,
R. norvegicus tend to be more aggressive in
their social interactions than other *Rattus*
species that have been examined (Barnett et
al., 1982), but this is true only for "wild"
norvegicus as laboratory strains tend be more
docile (see Chapters 2 and 3). Behaviorally,
all *Rattus* are best described as generalist
species that are flexible to environmental
perturbations. It is perhaps this flexibility that
has enabled their success as introduced
species. It should be noted, however, that
this behavioral flexibility is not unlimited and
that some species are susceptible to extreme
environmental changes. For example, two
species of *Rattus* native to Christmas Island
(*macleari* and *nativitatis*) have become extinct
since European settlement (Nowak, 1999). In
addition, two species are currently listed as
endangered (*baluensis* and *enganus*) and 13
more as vulnerable (International Union for
Conservation of Nature and Natural Resources, 2002). Thus, even the seemingly lim-
itless behavioral adaptability of *Rattus* species
appears to have its limits.

In terms of their interspecific relationships, relatively little is known. Musser and
Carleton (1993) described five groupings of
Rattus species: (1) *norvegicus* (1 species), (2) a
R. rattus group composed of various Asian
species (21 species), (3) a native Australian
group (6 species), (4) a native New Guinea
group (8 species), and (5) a Sulawesi/Phillip-
ines Islands group (5 species). The remaining
10 species that Musser and Carleton (1993) include in *Rattus* are described as possessing unresolved phylogenetic affinities. Of particular
interest is that *norvegicus* is sufficiently distinctive to warrant its own monotypic group.
The distinctiveness of *norvegicus* is further supported by several molecular studies (Chan,
1977; Chan et al., 1979; Baverstock et al., 1986;
Verneau et al., 1997; Suzuki et al., 2000;
Dubois et al., 2002; but see Pasteur et al., 1982)
(Fig. 1–5). How *R. norvegicus* is related to other
Rattus species and divergence dates between

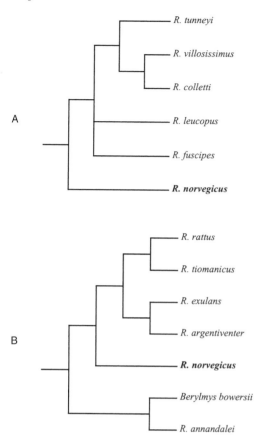

Figure 1–5. These two phylogenetic trees depict the relationships between *Rattus* species. (A) This tree is derived from Baverstock et al. (1986) and includes the endemic Australian species and *R. norvegicus* (shown in bold). (B) This second tree is derived from Chan et al. (1979). *Berylmys bowersii* is included because *Rattus annandalei* was shown to be more closely related to it than other *Rattus* species.

WHAT DOES IT ALL MEAN?

Given what is known about the evolutionary history *R. norvegicus*, how can this information be used to understand the biology and behavior of *R. norvegicus*? Cross-species analyses of anatomy, behavior, and ecology can yield profound insights into the ultimate mechanisms underlying species phenotypes as well as their possible functions. For example, the comparative approach has been instrumental in understanding the evolution of behavior (Martins, 1996; Lee, 1999) and the nervous system (Butler and Hodos, 1996).

From a comparative perspective, information on evolutionary history can be used to examine the evolution of the behavior of *R. norvegicus*. One example is provided by a comparative analysis of social play behavior in muroid rodents (Pellis and Iwaniuk, 1999). *R. norvegicus* possesses a highly complex form of social play (see Chapter 28) compared with other rodents. There are a number of reasons that complex play has evolved in *R. norvegicus*, such as evolutionary history, social system, ecology, and others. Using a phylogeny of 13 murids, Pellis and Iwaniuk (1999) analyzed several features of murid rodent social play and assessed the relative importance of phylogeny and social system on the evolution of these features. Contrary to many other behaviors that are closely linked with phylogenetic history, the complexity of social play did not conform to any observed pattern across the phylogeny. Further analyses indicated that the degree of sociality was positively correlated with the complexity of social play. Thus, the more social a murid, the more complex is its play. Last, using play traits to construct their own "phylogeny," Pellis and Iwaniuk (1999) demonstrated that the play of *R. norvegicus* most closely resembled that of the Golden Hamster (*Mesocricetus auratus*) rather than other murines (see Fig. 1–3). This means not only that play is variable within muroids but also that convergent forms of play can and have evolved in divergent rodent lineages.

the different species has remained largely unexplored.

R. norvegicus itself is presumed to have originated in Asia, but its long association with humans makes a more precise location difficult to determine. Two suggested points of origin are the steppes north of the Caspian Sea (Matthews, in press) and northern China (Musser and Carleton, 1993). From this general area, it spread throughout Europe in the mid-1700s. By the late 1800s, *norvegicus*, and *R. rattus*, had spread across much of North America as well as successfully invading innumerable ocean ports around the world.

The implications that this has for theories concerning the evolution of play in murids and other mammals are far-reaching and are still being investigated (e.g., Iwaniuk et al., 2002; Pellis and Iwaniuk, 2004).

Without a phylogeny to use with phylogenetically based statistical methods, the study of Pellis and Iwaniuk (1999) would have been superficial. In fact, further interspecific comparisons of the behavior of murids and murines could yield similar insight into the behavioral evolution of *R. norvegicus*. Comparable data are not, however, available for many other rodents. This curtails our ability to determine whether observations made on the behavior and biology of *R. norvegicus* are representative of rodents or, for that matter, other mammals. For example, are many of the behaviors of *R. norvegicus* typical of all *Rattus* species or specific to *R. norvegicus*? Thus, our understanding of *R. norvegicus* behavior would be greatly enlightened by further studies on other murines and *Rattus* species.

CONCLUSION

The evolution of the laboratory rat is characterized by a series of explosive radiations that occurred within relatively narrow time frames at the level of order, family, subfamily, and genus. As a result of these evolutionary radiations, the phylogenetic relationships between many taxa remain unresolved. Similarly, the fossil record is patchy and the first appearance of many taxa is debatable due to multiple instances of convergent evolution of morphology throughout the evolutionary history of rodents. The recent use of more comprehensive species sampling and multiple genes is painting a clearer picture of rodent evolution, but there are many unanswered questions. In particular, there are many uncertainties regarding the origin and relationships between and within the genus *Rattus* that could aid in delineating the boundaries of

this speciose genus. Furthermore, future research into the radiation of the genus *Rattus* could yield insight into the general features of radiative evolution as well as the potential for commensalism to aid in dispersal and speciation. From the perspective of understanding the behavior of *R. norvegicus*, such research will yield insight into how and why some behaviors and not others have evolved in *R. norvegicus*.

ACKNOWLEDGMENTS

I would like to thank the editors, Ian Whishaw and Bryan Kolb, for inviting my contribution to this volume and for their editorial assistance and Karen Dean for reviewing earlier drafts of this chapter. Financial support was provided by a Monash University Postgraduate Publications Award to the author.

REFERENCES

Adkins RM, Gelke EL, Rowe D, Honeycutt RL (2001) Molecular phylogeny and divergence time estimates for major rodent groups: evidence from multiple genes. Molecular Biology and Evolution 18:777–791.

Alroy J (1999) The fossil record of North American mammals: evidence for a Paleocene evolutionary radiation. Systematic Biology 48:107–118.

Archibald JD, Averianov AO, Ekdale EG (2001) Late Cretaceous relatives of rabbits, rodents, and other extant eutherian mammals. Nature 414:62–65.

Barnett SA, Fox IA, Hocking WE (1982) Some social postures of five species of *Rattus*. Australian Journal of Zoology 30:581–601.

Baverstock PR, Adams M, Watts CHS (1986) Biochemical differentiation among karyotypic forms of Australian *Rattus*. Genetica 71:11–22.

Begg RJ and Nelson JE (1977) The agonistic behaviour of *Rattus villosissimus*. Australian Journal of Zoology 25:291–327.

Beeman C (2002) An analysis of skilled forelimb movements in the Bush Rat (*Rattus fuscipes*). Unpublished honours thesis, Monash University, Clayton.

Bronham L, Phillips MJ, Penny D (1999) Growing up with dinosaurs: molecular dates and the mammalian radiation. Trends in Ecology and Evolution 14:113–118.

Butler AB and Hodos W (1996) Comparative vertebrate neuroanatomy: evolution and adaptation. New York: Wiley-Liss.

Carleton MD (1984) Introduction to rodents. In: Order and families of recent mammals of the world (Anderson S and Jones JK Jr, eds.), pp. 255–265. New York: John Wiley & Sons.

Carleton MD and Musser GG (1984) Muroid rodents. In: Order and families of recent mammals of the world (Anderson S and Jones JK Jr, eds.), pp. 289–380. New York: John Wiley & Sons.

Chan KL (1977) Enzyme polymorphism in Malayan rats of the subgenus *Rattus*. Biochemical Systematics and Ecology 5:161–168.

Chan KL, Dhaliwal SS, Yong HS (1979) Protein variation and systematics of three subgenera of Malayan rats (Rodentia: Muridae, genus *Rattus* Fischer). Comparative Biochemistry and Physiology B 64: 329–337.

Chevret P, Denys C, Jaeger J-J, Michaux J, Catzeflis FM (1993) Molecular evidence that the spiny mouse (*Acomys*) is more closely related to gerbils (Gerbillinae) than to true mice (Murinae). Proceedings of the National Academy of Sciences U.S.A. 90:3433–3436.

D'Erchia AM, Gissi C, Pesole G, Saccone C, Arnason U (1996) The guinea-pig is not a rodent. Nature 381: 597–600.

Dubois J-YF, Catzeflis FM, Beintema JJ (1999) The phylogenetic position of "Acomyinae" (Rodentia, Mammalia) as sister group of Murinae + Gerbillinae clade: evidence from the nuclear ribonuclease gene. Molecular Phylogenetics and Evolution 13:181–192.

Dubois J-YF, Jekel PA, Mulder PPMFA, Bussink AP, Catzeflis FM, Carsana A, Beintema JJ (2002) Pancreatic-type ribonuclease 1 gene duplications in rat species. Journal of Molecular Evolution 55:522–533.

Eizirik E, Murphy WJ, O'Brien SJ (2001) Molecular dating and biogeography of the early placental mammal radiation. Journal of Heredity 92:212–219.

Felsenstein J (1985) Phylogenies and the comparative method. The American Naturalist 126:1–25.

Flynn LJ, Jacbos LL, Lindsay EH (1985) Problems in muroid phylogeny: relationship to other rodents and origin of major groups. In: Evolutionary relationships among rodents: a multidisciplinary analysis (Luckett WP and Hartenberger J-L, eds.), pp. 589–618. New York: Plenum Press.

Foote M, Hunter JP, Janis CM, Sepkoski JJ Jr (1999) Evolutionary and preservational constraints on origins of biologic groups: divergence times of eutherian mammals. Science 283:1310–1314.

Graur D, Hide WA, Li W-H (1991) Is the guinea-pig a rodent? Nature 351:649–652.

Hand S (1984) Australia's oldest rodents: master mariners from Malaysia. In: Vertebrate zoogeography and evolution in Australasia (Archer M and Clayton G, eds.), pp. 905–912. Sydney, NSW: Hesperian Press.

Hartenberger J-L (1998) Description de la radiation des Rodentia (Mammalia) du Paléocène supérieur au Miocène; incidences phylogénétiques. Comptes Rendus de l'Academie Sciences, Paris, Sciences de la Terre et des Planètes 326:439–444.

Hartenberger J-L (1996) Les débuts de la radiation adaptative des Rodentia (Mammalia). Comptes Rendus de l'Academie Sciences, Paris, Sciences de la Terre et des Planètes 323:631–637.

Harvey PH and Pagel MD (1991) The comparative method in evolutionary biology. Oxford: Oxford University Press.

Hedges SB, Parker PH, Sibley CG, Kumar S (1996) Continental breakup and the ordinal classification of birds and mammals. Nature 381:226–229.

Huchon D, Catzeflis FM, Douzery EJP (2000) Variance of molecular datings, evolution of rodents and the phylogenetic affinities between Ctenodactylidae and Hystricognathi. Proceedings of the Royal Society of London, Series B 267:393–402.

Huchon D, Madsen O, Sibbald MJJB, Ament K, Stanhope MJ, Catzeflis F, de Jong WW, Douzery EJP (2002) Rodent phylogeny and a timescale for the evolution of Glires: evidence from an extensive taxon sampling using three nuclear genes. Molecular Biology and Evolution 19:1053–1065.

International Union for Conservation of Nature and Natural Resources (2002) 2002 IUCN red list of threatened species. Available at: http://www.iucn.org. Acessed June 5, 2003.

Iwaniuk AN, Nelson JE, Pellis SM (2001) Do big-brained animals play more? comparative analyses of play and relative brain size in mammals. Journal of Comparative Psychology 115:29–41.

Jacobs LL (1977) A new genus of murid rodent from the Miocene of Pakistan and comments on the origin of the Muridae. Paleobios 25:1–11.

Jaeger J-J, Tong H, Buffetaut E (1986) The age of *Mus-Rattus* divergence: paleontological data compared with the molecular clock. Comptes Rendus de l'Academie Science, Paris, Sciences de la Terre et des Planètes 302:917–922.

Kumar S and Hedges SB (1998) A molecular timescale for vertebrate evolution. Nature 392:917–920.

Lavocat R and Parent J-P (1985) Phylogenetic analysis of middle ear features in fossil and living rodents. In: Evolutionary relationships among rodents: a multidisciplinary analysis (Luckett WP and Hartenberger J-L, eds.), pp. 333–354. New York: Plenum Press.

Lee PC (ed.) (1999) Comparative primate socioecology. Cambridge: Cambridge University Press.

Leung LKP, Singleton GR, Sudarmaji R (1999) Ecologically-based population management for the rice-field rat in Indonesia. In: Ecologically-based management of rodent pests (Singleton GR, Hinds LA, Leirs H, Zhang Z, eds.), pp. 305–318. Canberra: Australian Centre for International Agricultural Research.

Luckett WP and Hartenberger J-L (1993) Monophyly or polyphyly of the order Rodentia: possible conflict between morphological and molecular interpretations. Journal of Mammalian Evolution 1:127–147.

Luckett WP and Hartenberger J-L (1985) Evolutionary relationships among rodents: comments and conclusions. In: Evolutionary relationships among rodents: a multidisciplinary analysis (Luckett WP and Hartenberger J-L, eds.), pp. 685–712. New York: Plenum Press.

Madsen P, Scally M, Douady CJ, Kao DJ, DeBry RW, Adkins R, Amrine H, Stanhope MJ, de Jong WW, Springer MS (2001) Parallel adaptive radiations in two major clades of placental mammals. Nature 409:610–614.

Martin Y, Gerlach G, Schlotterer C, Meyer A (2000) Molecular phylogeny of European muroid rodents based on complete cytochrome *b* sequences. Molecular Phylogenetics and Evolution 16:37–47.

Martins EP (ed.) (1996) Phylogenies and the comparative method in animal behavior. Oxford: Oxford University Press.

Mathews F (2004) *Rattus norvegicus*. Mammalian Species, in press.

Menkhorst PW (1995) Mammals of Victoria: distribution, ecology and conservation. Melbourne: Oxford University Press.

Michaux J, Reyes A, Catzeflis F (2001) Evolutionary history of the most speciose mammals: molecular phylogeny of muroid rodents. Molecular Biology and Evolution 18:2017–2031.

Murphy WJ, Eizirik E, Johnson WE, Zhang YP, Ryder OA, O'Brien SJ (2001a) Molecular phylogenetics and the origins of placental mammals. Nature 409:614–618.

Murphy WJ, Eizirik E, O'Brien SJ, Madsen O, Scally M, Douady CJ, Teeling E, Ryder OA, Stanhop MJ, de Jong WW, Springer MS (2001b) Resolution of the early placental mammal radiation using Bayesian phylogenetics. Science 294:2348–2351.

Musser GG and Carleton MD (1993) Family Muridae. In: Mammal species of the world: a taxonomic and geographic reference, 2nd edition (Wilson DE and Reader DM, eds.), pp. 501–756. Washington, DC: Smithsonian Institution Press.

Musser GG and Holden ME (1991) Sulawesi rodents (Muridae, Murinae): morphological and geographical boundaries of species in the *Rattus hoffmanni* group and a new species from Pulau Peleng, Malay Archipelago. Bulletin of the American Museum of Natural History 206:322–413.

Novacek MJ (1992) Fossils, topologies, missing data, and the higher level phylogeny of eutherian mammals. Systematic Biology 41:58–73.

Nowak RM (1999) Walker's mammals of the world, 6th edition. Baltimore: Johns Hopkins University Press.

Pasteur N, Worms J, Tohari M, Iskandar D (1982) Genetic differentiation in Indonesian and French rats of the subgenus *Rattus*. Biochemical Systematics and Ecology 10:191–196.

Pellis SM and Iwaniuk AN (2004) Evolving a playful brain: a levels of control approach. International Journal of Comparative Psychology, in press.

Pellis SM and Iwaniuk AN (1999) The roles of phylogeny and sociality in the evolution of social play in muroid rodents. Animal Behaviour 58:361–373.

Robinson M, Catzeflis F, Briolay J, Mouchiroud D (1997) Molecular phylogeny of rodents, with special emphasis on murids: evidence from the nuclear gene LCAT. Molecular Phylogenetics and Evolution 8:423–434.

Shoshani J and McKenna MC (1998) Higher taxonomic relationships among extant mammals based on morphology, with selected comparisons of results from molecular data. Molecular Phylogenetics and Evolution 9:572–584.

Strahan R (1998) Mammals of Australia. Sydney, NSW: Reed Books/New Holland Press.

Suzkui H, Tsuchiya K, Takezaki N (2000) A molecular phylogenetic framework for the Ryukyu endemic rodents *Tokudaia osimensis* and *Diplothrix legata*. Molecular Phylogenetics and Evolution 15:15–24.

Tong H and Jaeger J-J (1993) Muroid rodents from the Middle Miocene Fort Ternan locality (Kenya) and their contribution to the phylogeny of muroids. Paleontographica Abteilung A 229:51–73.

Verneau O, Catzeflis F, Furano AV (1997) Determination of the evolutionary relationships in *Rattus* senso lato (Rodentia: Muridae) using L1 (LINE-1) amplification events. Journal of Molecular Evolution 45:424–436.

Watts CHS and Aplin HJ (1981) Rodents of Australia. Sydney, NSW: Angus & Robertson Press.

Watts CHS and Baverstock PR (1995) Evolution in the Murinae (Rodentia) assessed by microcomplement fixation of albumin. Australian Journal of Zoology 43:105–118.

Ecology

S. ANTHONY BARNETT

2

Among mammals, the miscalled "Norway rat" (*Rattus norvegicus*), known also as the "brown rat" although it is often gray but sometimes black, is rivaled as a pest only by the house mouse (*Mus domesticus* vel *musculus*) and the "ship" or "black" rat (*Rattus rattus*), which as a rule is not black. *R. norvegicus* also displays peculiar features that make it of special interest to both ethologists and experimental psychologists. This chapter provides an overview of the ecology of the wild *R. norvegicus*.

ENVIRONMENTS: RATS AS COMMENSALS

The hordes of rats that are present in some human communities are due to our growing and storing of great concentrations of food, construction of shelter in buildings and drains and at the edges of fields, and killing of the carnivorous species that prey on small mammals. Moreover, we often kill rats only when they are numerous. The resulting slaughter may be impressive, but the likely result is a surviving population that can breed at a high rate.

In agricultural land and gardens and on the banks of canals and streams (in which Norway rats readily swim), they make extensive burrows that form irregular systems of branching and conjoining passages in which they breed.

Norways climb well, although not as well as *R. rattus*. Until the 18th century in northern Europe, *R. rattus* was without competitors, but when Norways invaded from the east, they largely replaced the black rat. Buildings in cities often housed *R. rattus* in the attics and Norway rats on the lower floors. Norways are also numerous in large sewers; in those of at least one American city, they are said to be the main food of a population of feral alligators. However, urban environments will harbor few rats if they have well-maintained buildings, a modern sewer system, and a high standard of hygiene.

Rarely, Norways live independent of people, such as on seashores.

Seeking shelter is a feature of rat behavior. When food can be carried, it is eaten in a nest or under cover. It may also be hoarded for consumption later.

In burrows and elsewhere, rats build nests of straw or of fragments of textile material or paper dexterously shredded and arranged. In addition to nesting material, items sometimes carried to the nest include fragments of wood, stones, and cakes of soap. Nest building is enhanced by cold.

Females rear their litters in nests that they defend against intruders. The movements of a female retrieving strayed young resemble those of a rat hoarding objects.

EXPLORATION, NEOPHILIA, AND NEOPHOBIA

In stable conditions, members of established colonies move, usually at night, on beaten tracks between nests and sources of food, water, and bedding. They commonly run with their vibrissae and fur in contact with a vertical surface. Rubbing of the fur leaves a smear from skin secretions; the odor of the smear is thought to attract other rats.

Norways are also highly exploratory and range widely from settled pathways. Movement in the living space (home range) can be analyzed under three categories. (1) Movements may be a search for food, water, nesting material, or shelter; movements then cease when the objective is reached. (2) Small mammals such as Norway rats regularly patrol the whole of their home range; places that have not been recently visited are likely to be approached first. (3) Exploratory movements, especially of new places, may continue when all basic needs seem to be satisfied. The second and third types of movement are examples of *neophilia*, or the tendency to approach what is unfamiliar.

The three aspects of movement can be distinguished experimentally. Figure 2–1 shows a "plus maze" in which a wild Norway may live for many days. Suppose a rat, already familiar with the maze, is confined in the nest box, without food, for some hours, and the arms of the maze are then opened. The rat soon moves out and, during the next few hours, spends much time eating and drinking; each meal, however, is followed by a brief patrol of all parts of the environment, including those that contain no food or other reward.

If different palatable foods are offered in the arms, the rat makes a meal of one; then, as it patrols the maze, it also samples the other foods. *Sampling* is an aspect of neophilia. It is appropriate for an omnivore, and, as we see later, it is important for food selection.

A rat may also be allowed to become familiar with a maze with access to only three arms. If the fourth arm is then opened and so gives access to a new place, it is quickly explored. Such neophilic behavior is typical of small mammals.

Wild Norways are, however, atypical in their response to strange objects. A common experience for farmers and warehousekeepers is to lay traps or bait on runways and to find that the Norways then give up their usual nightly visits, as if aware of danger. This *neophobia* is a response to a discrepancy. A novel object *in a familiar place* is avoided. It is the principal explanation of the popular reputation of rats for intelligence. If, in a plus maze, a strange object is put in an arm that has already been explored, farther entry into the arm is delayed. Similar avoidance has been systematically observed among rats in free populations.

A decline in food consumption may be used as an index of the avoidance of a new object. However, the automatic recoil, as shown by wild Norways, is an indiscriminate avoidance of *any* new object in a familiar place, edible or inedible, dangerous or harmless, usually at a distance. It is not a response to a strange odor or taste.

In their response to new objects, the domestic varieties differ greatly from the wild type. Kept in small cages and offered food in a strange container, domestic rats soon approach the container and continue to eat as before. Wild Norways, in identical conditions, may remain at the back of the cage and stop eating for days.

Other species of Muridae that are dependent on human communities are neophobic, but the independent species of the genus, *Rattus*, so far studied, is not. Hence neophobia is probably a result of natural selection among rats that have lived with our ancestors at least since a settled agriculture began.

The neophobic response is, however, not irrevocably fixed ("innate" or "instinctive"). Wild Norways infesting a landfill, in which everything is, in effect, likely to be a new object, are hardly neophobic. In their unstable environment, they habituate to incessant change and so develop a novel response to novelty.

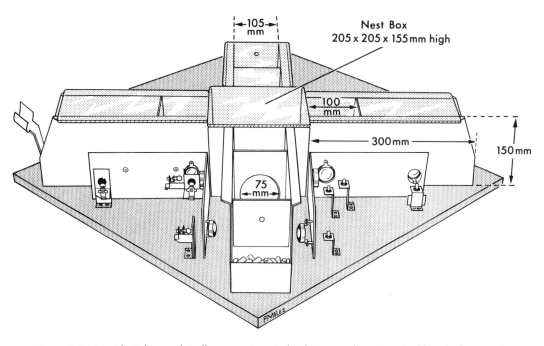

Figure 2–1. A residential maze that allows experiments that last many days. An animal lives in the central box. Food, water, nesting material, or other objects can be offered at the ends of the arms. Visits to the arms and duration of stay are recorded and analyzed with a computer.

FOOD

MEALS AND ENERGY INTAKE

Like other rodents, Norways are equipped to cope with hard foods such as seeds and nuts; they can even gnaw through lead pipes. Typically, a small, hard object, such as a wheat grain, is held in the forefeet while it is eaten. Individuality, however, is shown in feeding patterns; one rat, eating flour, may bury its nose in the food, while another sits, dexterously scooping the food into its mouth with one paw; yet another may use both forepaws.

In a stable environment, wild rats eat in the darkness, a few grams at a time, at regular intervals. In the short term, adult Norways adjust these meals to maintain a steady weight, but they grow slowly throughout most of their adult lives, up to about 700 grams in the most favorable conditions.

The mealtimes of rats can be altered by training. If food is made available at only one time in each 24-hour period, rats regularly assemble at that time. In a colony of wild Norways, individual mealtimes may also be influenced by dominance relationships: the oldest or heaviest rats have priority at food sources.

FOOD SELECTION

Wild Norways with access to a variety of foods may seem to be indiscriminate. They may prey on small birds or mammals; some eat snails; and those that live on the banks of streams may eat fish. Wild Norways that feed largely on wheat grains quickly accept newly available alternative foods, such as cabbage leaves or raw meat.

The selection of foods is influenced by flavor, odor or texture. Rats favor sweet mixtures, whether they contain sugar or saccharin. Wild rats also prefer finely divided wheatmeal to whole grains. Some edible oils, such as arachis, added to grains increase acceptance, but butyric acid or aniseed oil is a deterrent. (Some domestic Norways readily accept both.)

The most notable influences on feeding are related to the internal effects of foods. A prominent feature of selection by Norway rats is their ability to acquire aversions. The preceding section describes the avoidance of novelty (neophobia), which is typical of the species. Another, very different kind of withdrawal, shared with other mammals, arises from adverse, individual experience.

A pile of unfamiliar food on a rats' runway is at first avoided: it is a "new object" (the neophobic response). The delay is greater if the food is in an unfamiliar container. However, eventually the food is sampled. After this, the food is again avoided, as if "fear" and "curiosity" (or hunger) are opposed; the former gradually gives place to unhesitating consumption.

If, however, the food contains a poison, the small amount first taken may cause illness but not be lethal. For a time, the animal then stops eating. When it resumes eating, as a rule it refuses the toxic mixture—it has become "bait shy." It may even reject constituents of the mixture with a distinct taste, such as sugar. The survival value of neophobia, combined with acquired aversions to poisoned foods, has been observed in many field experiments. These two features make infestations of Norways (and of *R. rattus*) difficult to manage.

For experimental psychologists, such aversions have two unexpected features. (1) In conventional experiments on learning, habits are usually acquired slowly and require many trials, but aversions arise from a single experience. (2) In laboratory experiments, the interval between a stimulus and the animal's response is usually brief, but the interval between ingesting a poison and the development of illness may be several hours. It is an example of *learning after a long delay*. Evidently, when illness is involved, an immediate impact is not needed for learning. This has been confirmed in many experiments on laboratory rats.

Acquiring an aversion is the obverse of the ability to choose favorable foods (*dietary self-selection*), which also has been extensively analyzed in experiments on domestic Norways. The importance of sampling foods is evident when rats are offered a cafeteria-type array of food with several mixtures of different nutritional values. (This resembles conditions in freedom. The chow used in laboratories is a complete diet but is quite unnatural.) Nutritionally deficient animals tend to choose unfamiliar foods and so make finding the best one more likely.

SOCIAL INTERACTIONS

All accounts of the social lives of animals use expressions derived from human social action. Examples are *status system* (or *dominance hierarchy*), *dominance* and *subordinacy, courtship*, and *territory*. Such uses are inevitable but incur the danger of *anthropomorphism*, that is, describing the animals as if they were human. The account that follows tries to avoid this hazard.

SOCIAL INFLUENCES ON FEEDING

Nineteenth-century writings contain a persistent and popular story of rat warning others about the dangers of poison bait. Most such tales are preposterous. Nonetheless, it is reasonable to ask whether a rat's feeding behavior is affected by encounters with other rats. In this aspect of social behavior, domestic rats resemble wild Norways. Seemingly, the domestic varieties retain remarkable abilities that are presumably crucial for the survival of wild rats in freedom.

Wild Norways are "central place foragers." When they return home after feeding, they smell of the food that they ate. Experiments have tested the effects of such odors. One question was: Would another rat, given a choice, be influenced *against* the food if the first rat were ill? No such effect has been found. Food aversions, it seems, are not acquired by detecting illness in others.

Young rats can, however, be drawn away from toxic foods by the behavior of adults. When young rats become independent, they tend to go where other rats are already feeding. Such local enhancement may lead wild Norways, in freedom, to eat nourishing rather than harmful food.

In contrast to aversions, positive preferences are socially influenced. The choice of food of a wild or a laboratory Norway rat can be changed by the postprandial odor of another. Hence, both wild and laboratory Norways can transmit information about food to their neighbors.

SOCIAL RELATIONSHIPS

In social interactions other than feeding, wild Norways differ greatly from the domestic varieties. All *domestic* Norway rats—white, hooded, even brown or black—are, in ordinary laboratory conditions, peaceful. They represent a typical result of domestication, because they can be caged and moved around with little regard to their companions. (In special conditions, individuals of some strains can, however, be induced to interact quite violently.)

The absence of structured social interactions among laboratory rats once led to a strongly held belief that rats have little social life. (Until they saw the contrast on film, some experimenters refused to accept that the social interactions of domestic Norways differ greatly from those of the wild type.)

Male wild Norways, trapped from crowds, are sometimes found to be scarred, evidently due to bites by other rats. They have therefore been suspected of continuous strife. Yet, if several adult males are put together in a large cage or enclosure, with plenty of food, water, and nest sites, they play, grow, and look sleek. The postures they adopt include those described later as accompanying conflict but are harmless. In such conditions, no individual has the opportunity to establish a territory.

When, however, wild male Norways meet as strangers, one of which is resident in its living space, the outcome is different. Initially, one is likely to crawl under the other (Fig. 2–2). Perhaps this distinctive performance deters attack. Another usually peaceful act is grooming (strictly, allogrooming); while huddling, one rat nibbles at the fur of another. These interactions may be put under the heading of "social sedation," in contrast to "social stress."

An enigmatic act, seen during meetings between males, has been called the *threat posture* (TP) (Fig. 2–3). "Threat," however, implies an intention to punish or hurt, but no convincing means exists of deciding what a rat *intends*. In the TP, the back is arched, the legs are fully extended, the hairs are erect, and the head is usually turned toward the opponent. The posture often precedes an attack and sometimes also occurs after it, or two rats may be in the TP at the same time.

The most striking interaction is *attack* (Fig. 2–4), in which one male leaps at another with rapid adductions of the forelimbs (visible

Figure 2–2. During encounters, one rat often crawls under another. (From Barnett, The rat: A study in behavior. Drawn by Gabriel Donald from a photograph.)

Figure 2–3. The "threat posture," often seen during encounters by male Norways, is common to all of the species of *Rattus* studied. (From Barnett, The rat: a study in behavior. Drawn by Gabriel Donald from a photograph.)

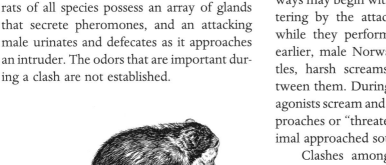

Figure 2–5. This posture, although it occurs during clashes and is called *boxing*, is nonviolent. (From Barnett, The rat: a study in behavior. Drawn by Gabriel Donald from a photograph.)

only on sped-up cinefilm). Sometimes it is accompanied by a brief bite. Rarely, the attacker bites and holds on.

Attacks are not relentless but are often interrupted: the agonists turn to "boxing" (Fig. 2–5) or to other positions, especially the TP. During these intervals, if the agonists have separated, an attacked rat may approach the attacker.

Interactions do not depend only on vision and contact. Rats regularly sniff other rats, and objects are often also marked with urine. Wild rats of all species possess an array of glands that secrete pheromones, and an attacking male urinates and defecates as it approaches an intruder. The odors that are important during a clash are not established.

Figure 2–4. Attack by a male Norway. (From Barnett, The rat: a study in behavior. Drawn by Gabriel Donald from a photograph.)

Scent marking by mammals was at first assumed to be a defense of a territory—an instance of a common presumption that social behavior is predominantly combative. Later, many scent marks have been shown to be attractive. Some observers, however, believe that wild Norways attack other males only if they possess a strange odor.

Social interactions are accompanied by sounds. A hostile encounter between Norways may begin with percussion—tooth chattering by the attacker. During encounters, while they perform the postures described earlier, male Norways also utter pure whistles, harsh screams, and intermediates between them. During attack and boxing, both agonists scream and whistle, but when one approaches or "threatens" another, only the animal approached sounds off.

Clashes among wild Norways can be quantitatively analyzed by introducing a strange male into the living space of another. For maximum effect, the resident should have female companionship. Males might therefore be supposed to fight for females. However, this is not so: in small colonies, a female Norway in estrus is followed by several males, which take turns. Females not in estrus are ignored; no pairs are formed.

In experiments, a female need not be present during the actual encounter. Some have been staged with a single resident male, while the female was shut away. The visitor then typically makes the first approach, but lively action is nearly always begun by the resident: the newcomer fends off the attacker or runs away. The encounters therefore hardly rate as fights, for they are one-sided.

The encounters fall in the category of territorial behavior. A *territory* is a region, occupied by an individual, a family, or a larger group, from which other members of the species are excluded. (Some ethologists reserve the term for a *defended* region.) It requires learning, by each individual, about the environment and companions or neighbors. Attacks by a resident on intruders are examples of territorial defense. Another example, already mentioned, is defense of a nest by a female with a litter.

SOCIAL STATUS AND ENIGMATIC DEATH

When adult males are introduced to groups of wild Norways of both sexes, no group action occurs. Three kinds of adult male emerge. "Alpha" males are always large and move about freely and initiate attacks on intruders; small rats do not overcome much larger ones. Others, the "beta" males, adapt themselves to an inferior role; they keep away from the alphas but feed well and gain weight. They attack intruders only if the alphas are first removed; they may then adopt the status of an alpha. Both alphas and betas also appear among Australian longhaired rats (*R. villosissimus*).

Last are the "omega" males. Although attacks are intermittent and brief, after a day or two under attack some rats decline in weight, move slowly, and eat hesitantly with a drooping posture and bedraggled appearance; eventually, they die. Similar debility can result from encounters between black and longhaired rats. In human terms, the omegas seem dejected or seriously depressed. This condition has been fancifully likened to "voodoo

death," in which a person dies after being cursed by a witch doctor.

Deaths during clashes have, less implausibly, been attributed to killing by other rats. However, in several hundred closely observed encounters, many quite vigorous, between wild rats of five species, no rat was *killed* by another rat. An unwounded omega rat was likely to collapse and die, sometimes quickly but more often after hours or days of increasing debility. Unexplained deaths have also been reported in free populations and in conditions in which some rats could reach food only by running the gauntlet of others nesting near the food.

This baffling phenomenon has been called "death of unknown origin," or DUO, by analogy with the physician's "pyrexia of unknown origin," or PUO.

"SOCIAL STRESS"

The search for the causes of DUO has involved experiments on several species of *Rattus* and, in particular, on tree shrews (*Tupaia belangeri*), which also display the three kinds of individuals named here alpha, beta, and omega.

Findings from early studies of DUO suggested that both attacked and attacking rats were adapting to stressful conditions. During conflict, the adrenal glands enlarge. The adrenal response helps to prepare an animal for exertion. Correspondingly, the blood sugar level of attacked wild Norways is high. Collapse is not due to hypoglycaemia.

Observations on wild rodent populations, however, led to conjectures on the involvement of infection, especially of the kidneys. And, in controlled conditions, glomerulonephritis was found in rats that were socially stressed. If, then, a bacterial infection contributes to DUO, it should be possible to prevent it; in experiments, the antibiotic neoterramycin prevented death among socially stressed rats.

Attention was therefore directed to resistance to pathogens. The flaring of infections during social stress suggests immune suppression. During the response to stressors, glucocorticoids secreted by the adrenal cortex interact with the immune system. In severely adverse conditions, these hormones can diminish the immune response.

Hence an immunological account of DUO seems possible. One hypothesis concerns cytokines, soluble proteins secreted by lymphocytes and other cells during infection. Cytokines have an anomalous "side effect"; they can induce loss of appetite and lethargy, and so may contribute to the collapse of omegas.

If so, they provide an example of a response that is usually adaptive but can be fatal. Here, therefore, is an anomaly. In biology it is usual to ask of a trait: How does it contribute to survival? Yet to ask this of DUO sounds ludicrous. Can death have survival value? The question, however, implies an assumption that is often made but is incorrect: that every feature of an organism is a direct result of natural selection, and therefore aids (or has aided) survival or breeding. Many traits are, however, only indirect consequences of natural selection: they go incidentally with advantageous features and do not themselves contribute to survival. Darwin called this phenomenon *correlated variation*.

DUO therefore has five major peculiarities. (1) Death may occur without wounding or other obvious cause. (2) Most of the physiological changes found during the collapse of omegas resist the effects of stressors and should therefore prevent death. (3) The postures of lethal encounters resemble those seen also during harmless interactions ("play"). (4) Betas evidently possess special features, not yet identified, that enable them to adapt to attack. (5) DUO has no obvious explanation in terms of survival value.

ETHOLOGICAL QUESTIONS

The social interactions of wild rats also raise questions concerning social signals and aggression.

The "Social Signal"

Like genetics and experimental psychology, social ethology began with simple concepts. Standard signals, said to be "innate," were identified for each species studied and called "releasers"; each represented a distinct state of the signaler and reliably evoked a similarly "fixed action pattern" or standard response from another.

The social signals of rats (like those of many other animals) do not correspond to this digital concept. They form a fluctuating complex that evokes a similar diversity of responses. Some ethologists call such encounters "negotiations"; indeed, like many political and financial transactions, they are difficult to interpret.

"Aggression" and "Drive"

Another difficulty arises because the response of a resident rat to an intruder is often called aggression.

Calling defense "aggressive" is an instance of the widespread custom of putting diverse activities, such as hunting and bird song, under this one heading. In ordinary speech, "aggression" signifies ungoverned violence or unprovoked assault intended to cause injury. However, defense of an occupied area is not unprovoked, nor is it ungoverned.

Further, animals are sometimes held to have an aggressive drive that compels them to attack members of their own species. The drive may said to build up internally if it is not expressed. But territorial dense occurs only in clearly defined circumstances. An adult male wild rat deprived of encounters with strange males does not turn on members of its own group as substitutes.

Each of the many activities called aggressive requires separate analysis. In such analysis, exemplified by the preceding account, imagined drives do not help; the use of the blanket term "aggression" obscures the special features of the behavior under study.

BREEDING AND POPULATIONS

The elaborate social interactions of wild Norways allow them to exist in large and densely crowded populations. The conditions that encourage the growth of the populations were described earlier, but they do not tell us what limits population growth. Possible checks on numbers include, shortage of food or water, lack of shelter, predators, pathogens, and negative social interactions.

If a population is isolated and in constant conditions, increase may follow the S-shaped "logistic" curve. Increase is slow at first, becomes rapid, and then slows and stops. After many rats have been killed by poisoning, some populations of Norways seem to follow this ideal curve. Those reduced by only about 50% breed at a high rate and quickly restore their numbers; others, brought to about 10%, recover slowly. (These findings have implications for pest management.)

The decline at the top of the curve suggests an adverse influence, of which the effect becomes greater in proportion to population density. If such density-related factors could be revealed, we could learn much about how populations are regulated. Sometimes, indeed, altering a single feature, such as food supply or shelter, can have a distinct effect on numbers. Often, however, no single factor can be identified as crucial. Nor do populations remain constant when they have reached an apparent maximum. The logistic curve is notoriously a model to which few natural populations conform.

Crowding, measured by numbers to an area or volume, is influenced by territorial behavior. When, however, the food supply is lavish, population densities are reached much above those in other environments. Territorial spacing seems then to be, up to a point, in abeyance. Here, evidently, is an instance of density-related factors interacting.

Populations are continually influenced by the conditions we offer them. Reduced cover and shortage of nesting material make constructing nests and rearing young more difficult. They also alter behavior. Prominent features of groups of Norways are contact, crowding, and fecundity; hence, a complete picture should include the balance between social sedation and social stress. However, the favorable effects of crowding, or at least of contact, have been little acknowledged.

To understand what regulates populations, we need to know not only about the rats' food, shelter, and diseases but also about how these interact and how they influence their social lives.

DIVERSITY

The Norway rat, even the domestic Norway, is sometimes called "the rat." In fact, however, about 300 species of the genus *Rattus* have been described. Of these, much is known about the polymorphic black rat. This species, which has been very successful as a commensal of *Homo sapiens*, can be tamed, yet it has never been domesticated. (It even sometimes throws up an occasional albino mutant.)

Another notable species is the rice rat (*Rattus argentiventer*), which has been described as the most important rodent pest in southeast Asia. It not only eats large amounts of rice but also cuts down the rice plants at the base, after the fields have been drained. Yet its biology remains not fully known.

Not all "rats" are members of the genus *Rattus*. Indian mole rats (*Bandicota bengalensis*) have a large distribution in southeast Asia. Dense populations live in the godowns of Calcutta and other Indian cities where, early in the twentieth century, they replaced *R. rattus*; they are also a conspicuous pest in rice fields.

We have no explanation of the geography of these species, nor any full account of their interactions. They do, however, remind us that the Norway, despite its prominence, is only one species among many, all of which present unsolved problems.

FURTHER READING

Barnett SA (1975) The rat: a study in behavior, 2nd edition. Chicago: University of Chicago Press.

Barnett SA (2001) The story of rats. Sydney: Allen & Unwin.

Galef BG (1996) Social enhancement of food preferences. In: Social learning in animals (CM Heyes and BG Galef, eds.) pp. 49–64. San Diego: Academic Press.

Holst D v (1998) The concept of stress and its relevance for animal behavior. Advances in the Study of Behavior 27:1–131.

Singleton GR (ed.) (1999) Ecologically-based management of rodent pests. Canberra: ACIAR.

Strains

LAUREN GERARD KOCH AND STEVEN L. BRITTON

3

An understanding of the development of inbred strains relates directly to the use of experimental design for probing at cause and effect at all levels of biological organization.

THEORETICAL BASIS OF INBRED STRAINS

HARDY-WEINBERG PRINCIPLE

The rudimentary ideas of the Hardy-Weinberg principle are a good starting point for understanding the development and use of inbred animals. The Hardy-Weinberg principle is a hypothetical representation of genotypic frequencies that will occur in randomly mating populations. This principle has use because it simplifies to a multinomial expression that describes the genetic content of a population (Falconer and Mackay, 1996a).

For an entire population, there are numerous alleles (alternate forms of the same gene) for most genetic loci. For example, the A locus could have five variants in a population designated as A_1, A_2, A_3, A_4, and A_5. Each individual within a population, however, has only two alleles at each locus and these can be either identical (homozygotic, e.g., A_1A_1 or A_2A_2, etc.) or different (heterozygotic, e.g., A_1A_2 or A_3A_4).

To understand the Hardy-Weinberg principle, consider a large population that contains only two alleles (A_1 and A_2) for a specific gene. The frequency of allele A_1 is the proportion of all A alleles in the population that are of the A_1 type. This includes individuals who are homozygous (A_1A_1) and half of the alleles in individuals who are heterozygous (A_1A_2). The frequency of allele A_2 is the proportion of all A alleles in the population that are of the A_2 type. Let p equal the frequency of all A_1 alleles and q equal the frequency of all A_2 alleles such that in fractional representation, $p + q = 1$.

Because each gamete contains only one of the A variants, each A locus has the probability of being represented in progeny proportional to its allelic frequency represented in the parental population if mating is random. Thus, as alleles recombine from the parental gene pool via fertilization, there is p probability of an individual acquiring an A_1 allele and $p \times p = p^2$ probability of acquiring an A_1A_1 combination. Likewise, there is $q = 1 - p$ chance of acquiring an A_2 allele and q^2 probability of both variants being of the A_2 type (A_2A_2 genotype). Extending this logic, there is pq chance that the first allele is A_1 and the second allele is A_2. The chance of the first allele being A_2 and the second allele being A_1 is qp. Therefore, the combined chance of obtaining a heterozygote (A_1A_2 or A_2A_1) for the A locus is $(p \times q) + (q \times p) = 2pq$.

These ideas form the basis of the Hardy-Weinberg principle that describes a theoretical relationship for gene frequencies and genotype frequencies in a population. By this principle, picking an individual randomly from a population is tantamount to picking two genes at random from the entire gene pool of a population. As from above, the probability of obtaining A_1A_1 is p^2, the probability of A_1A_2 is $2pq$, and the probability of A_2A_2 is q^2. The sum of all the fractional probabilities accounts for all of the A_1 and A_2 variants as given by the Hardy-Weinberg principle in its usual form:

$$p^2 + 2pq + q^2 = 1 \qquad \text{(Eq. 1)}$$

Not obvious from first inspection, the Hardy-Weinberg equation models only for populations in which gene pool frequencies do not change. The Hardy-Weinberg principle operates with the assumption that the following five conditions are met:

1. The population contains an infinite number of individuals.
2. Genotypes do not influence mate choice.
3. No mutation or natural selection occurs.
4. No migration in or out of the population occurs (closed population).
5. An equal evolutionary fitness exists among the individuals.

A population that conforms to these conditions is declared to be in *Hardy-Weinberg equilibrium*. This principle thus provides a theoretical standard for analysis of conditions in which changes do occur, such as with inbreeding. *Inbreeding* is the mating between individuals related closely enough to produce a nonrandom distribution of genotypes. The consequence of inbreeding is to increase the frequency of homozygous genotypes (A_1A_1 and A_2A_2) and thus decrease the frequency of heterozygous genotypes (A_1A_2). Stated another way, inbreeding represents deviation from Hardy-Weinberg proportions ($p^2 + 2pq + q^2$) by reduction of heterozygous genotypes, with no attendant loss of individual allelic frequencies.

DEFINING AN INBRED STRAIN

In diploid organisms, sister-brother (full-sib) mating represents the closest form of inbreeding. At the most, two parents can have four different allelic variants and thus produce four different genotypes among the progeny. The probability that the sibs have the same genotype is 1:4. The magnitude of inbreeding can be estimated from the relative loss of heterozygosity across generations and is termed the *coefficient of inbreeding* (F). The measure of the coefficient of inbreeding ranges from 0 to 1 and reflects deviation from Hardy-Weinberg proportions (Fig. 3–1). Consider a baseline population that was initially in Hardy-Weinberg equilibrium such that the frequency of heterozygous genotypes was $2pq$. The coefficient of inbreeding (F) after t generations of inbreeding is estimated as

$$F = (2pq)_i - (2pq)_c / (2pq)_i \qquad \text{(Eq. 2)}$$

where $(2pq)_i$ is the initial frequency of heterozygotes at baseline, and $(2pq)_c$ is the current frequency of heterozygotes after t generations of inbreeding.

When $F = 0$, the genotype frequencies are not different from Hardy-Weinberg proportions. When $F = 1$, inbreeding is com-

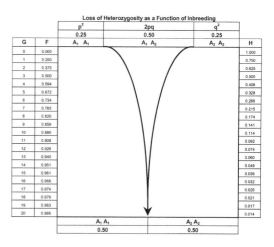

Figure 3–1. Idealized graphic depicting loss of heterozygosity with 20 generations of sister-brother inbreeding. Initial conditions presume two kinds of alleles (A_1 and A_2) available at locus A. Twenty-five percent are present in the A_1A_1 homozygotic form; 50%, as heterozygotes (A_1A_2); and 25%, in the A_2A_2 homozygotic form. The coefficient of inbreeding (F) increases and the remaining heterozygosity (H) decreases with each generation.

plete, no heterozygous genotypes are present, and the population contains only homozygous genotypes in p (A_1A_1) and q (A_2A_2) frequencies.

F can also be expressed as $1 - H$, where H is the frequency of heterozygotes. Therefore, $H = 1$ in a population that is in Hardy-Weinberg equilibrium. The reduction in heterozygosity with inbreeding via full-sib mating can be calculated from this recursion relation (Hedrick, 2000):

$$H_{t+2} = \left(\frac{1}{2} H_{t+1} \right) + \left(\frac{1}{4} H_t \right) \quad \text{(Eq. 3)}$$

where H_t = the heterozygosity at the tth generation.

So, consider the start of inbreeding with a closed base population that is in equilibrium ($H_0 = 1$). From the above recurrence relation, the heterozygosity would decrease with each subsequent generation (H_t) in a sequence $H_0 = 1$, $H_1 = 0.75$, $H_2 = 0.625$, $H_3 = 0.5$, $H_4 = 0.406$, $H_5 = 0.328$, or given by the ratios 2:2, 3:4, 5:8, 8:16, 13:32, 21:64, etc, approaching zero as a limit. Conveniently, but perhaps not apparent from initial inspection, the denominators double and the numerators for the declines in H_t follow a Fibonacci sequence (Atela et al., 2002) where each subsequent number is the sum of the previous two numbers (Crow, 1986). From this sequence, the remaining H_t at any generation can be calculated from solution of the above sequence that, as shown in Figure 3–2, follows an exponential relationship: $H_t = 0.944e^{-0.2117x}$.

Likewise, the coefficient of inbreeding in the tth generation of full-sib mating can also be expressed as a recursion relationship (Hedrick, 2000):

$$F_{t+2} = \left(\frac{1}{4} \right) + \left(\frac{1}{2} F_{t+1} \right) + \left(\frac{1}{4} F_t \right) \quad \text{(Eq. 4)}$$

An *inbred strain* is defined as one that has been sister-brother mated for at least 20 generations as agreed on by The International

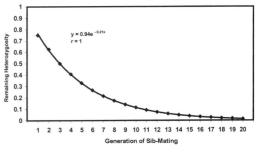

Figure 3–2. On average, decreases in heterozygosity with each generation of sister-brother mating follow an exponential decay. The numerator of this decay is represented by a Fibonacci sequence. In addition to breeding schemes, many natural phenomena, such at spiral growth patterns in plants and seashells, follow this pattern (Atela et al., 2002).

Committee on Standardization Nomenclature for Mice (Silver, 1995). As shown (see Figs. 3–1 and 3–2), 20 generations of inbreeding produce a population of animals that have on average 1.4% remaining heterozygosity and are thus about 98.6% homozygotic (Hartl and Clark, 1988).

VARIATION

Traits can be divided into either mendelian or quantitative. Mendelian traits are those for which a genetic difference at a single locus is sufficient to cause a difference in phenotypic expression of a given character. Quantitative traits do not manifest as discrete phenotypes in populations but distribute with continuous variation. Continuous variation is the result of the variable presence and expression of many genes (i.e., polygenic) within a population as they interact with the environment. Most traits of physiological, morphological, clinical, or behavioral interest are of the quantitative type, and their inherent complexity enhances the importance of well-defined inbred strains.

Two statements about the components of variation summarize the usefulness of inbred strains: (1) environmental factors account for within-strain variation, and (2) genetic factors account for between-strain variation. Thus, a common starting point is to assess both the within- and between-strain variations in a panel of inbred strains for a trait of interest.

HERITABILITY

The major assumption of quantitative genetics is that variation from the mean value of a trait is caused by the additive influence of genetics plus environment (Falconer and Mackay, 1996b) and can be expressed as

$$V_P = V_G + V_E \qquad \text{(Eq. 5)}$$

This equation implies that variance in phenotype (V_P) can be partitioned into genotypic variance (V_G) and environmental variance (V_E). The measure most widely used as a descriptor of the contribution of genetic factors is heritability, and two types can be considered. Broad sense heritability (H^2) is the sum of all genetic factors that influence the phenotype at the population level and is given by the expression $H^2 = V_G/V_P$. Broad sense heritability is also called the *degree of genetic determination*. In theory, the phenotypic variance can be estimated if one of the two components (G or E) is eliminated. Experimentally, genetic variance can be more closely controlled from a highly inbred line with identical genotypes. The major assumption with this approach is that the environmental variance is similar between inbred strains and that the genotypes respond similarly to the environment. This is problematic because of gene–environment interactions. Indeed, some strains have been found to be more variable and sensitive to environmental differences for a given trait (Crabbe et al., 1999).

The more useful and commonly referred to narrow sense heritability (h^2) or simply heritability is expressed as $h^2 = V_A/V_P$, where V_A is defined as the additive genetic variance. Additive genetic variance is that portion of genetic variance that causes offspring to resemble their parents and is considered the variance associated with the average effects of substituting one allele for another. Estimates of heritability range from 0.0 to 1.0. If $h^2 = 0$, there are no genetic contributions to phenotypic variance. If $h^2 = 1$, all phenotypic variation can be accounted for by genetic factors. (Note that the symbol h^2 is itself heritability, and not the square of the term in the arithmetic sense, a use chosen by Sewall Wright [1921].)

Two approaches can be used to estimate h^2 for a given trait. The first uses information on the resemblance between relatives and is easier conceptually but more difficult experimentally. In a widely heterogenic population (outbred), it is more likely for relatives to possess the same allelic variants for genes. That is, the offspring from parents high for a trait would also be expected to be high for a trait. In contrast, offspring from parents that demonstrate low for a trait would more likely be low for a trait. These ideas form the basis for using the regression of mean offspring values on the mean value of the parents (mid-parent value) as an estimate of h^2. If the trait is inherited additively with complete fidelity, such that the values of offspring are highly similar to the parents, then the slope of the regression line (h^2) equals 1. In contrast, if no additive similarity exists between parent and offspring, $h^2 = 0$.

As a second approach, h^2 can also be estimated for measures of a trait from a panel of inbred strains and is based on two assumptions related to the properties of inbreds. First, individuals within each inbred strain are similar genetically. Trait variation within a strain is thus attributed to the environmental variance such that an estimate of V_E can be obtained. Second, variation between strains derives from genetic differences and can be used to estimate V_A. The critical factor is that each inbred strain represents almost exclusively homozygous genotypes.

As indicated earlier, narrow sense heritability (h^2) is estimated from the ratio of additive genetic variance (V_A) to phenotypic variance (V_P) (Falconer and Mackay, 1996b), with the phenotypic variance being the sum of the V_A and environmental variance (V_E):

$$h^2 = \frac{V_A}{V_A + V_E} \quad \text{(Eq. 6)}$$

V_A and V_E can be estimated by partitioning Equation 6 into the between-group variance component (ΣB^2) and the within-group variance component (ΣW^2) from analysis of variance using these expressions:

$$\sum B^2 = \frac{(MSB - MSW)}{n} \quad \text{(Eq. 7)}$$

$$\sum W^2 = MSW \quad \text{(Eq. 8)}$$

where MSB is the mean square between strains, MSW is the mean square within strains, and n is the number of animals in each strain. ΣB^2 equals $2V_A$, and V_E is approximated by ΣW^2. Combining Equations 6, 7, and 8 yields an estimate of h^2 (Hegmann and Possidente, 1981):

$$h^2 = \frac{\frac{1}{2}\left[\frac{MSB - MSW}{n}\right]}{\frac{1}{2}\left[\frac{MSB - MSW}{n}\right] + MSW} \quad \text{(Eq. 9)}$$

The relationship between strain differences and V_A as they relate to the effect of inbreeding on character variance are presented by Crow and Kimura (1970) and summarized by Hegmann and Possidente (1981). In brief, V_A within an inbred line [$V_A(i)$] increases relative to the V_A in a randomly mating population [$V_A(r)$] as the coefficient of inbreeding (F) increases, as given by

$$V_A(i) = V_A(r)(1 + F) \quad \text{(Eq. 10)}$$

As a result, if F approaches 1, the estimate of variance among inbred strains should be twice the V_A for a trait in a randomly bred population from which the inbred strains were de-

rived. This explains why the estimate of V_A in Equation 9 is divided by 2.

If the number of rats is not equal for all strains, a weighted average value for n can be calculated as suggested by Sokal and Rohlf (1981):

$$\text{Weighted } n = \frac{1}{a - 1}\left[\sum^a n_i - \left(\frac{\sum^a n_i^2}{\sum^a n_i}\right)\right] \quad \text{(Eq. 11)}$$

where a is the number of strains, and n_i is the number of rats in each strain.

DEVELOPMENT OF MODELS

INBRED STRAINS DERIVED FROM SELECTED LINES

A very broad idea emerges from consideration of the above fundamentals on genetic variance and heritability. Based on R. A. Fisher's 1930 theorem of natural selection, traits peripherally associated with evolutionary fitness, such as morphology, behavior, and complex physiology, demonstrate more additive genetic variance (i.e., h^2) than do traits essential to fitness because of less pressure from natural selection. This generalization is consistent with the demonstration of success in artificial selection for traits such as blood pressure (Knudsen et al., 1970; Yamori et al., 1972) and aerobic capacity (Koch and Britton, 2001) in rats.

Many of the currently available inbred rat strains were not developed from lines first selectively breeding for a trait but were simply inbred to increase genetic uniformity. As a consequence, evaluation of a panel of inbred strains may yield only minimal between-strain variance for a desired trait. In this case, two-way artificial selection can be used to create low and high lines widely different for a trait from which inbred strains can be subsequently produced.

AVAILABILITY OF GENETICALLY HETEROGENEOUS RATS

The development of the widely heterogeneous outbred stock of N:NIH rats from the Animal Resource Center of the National Institutes of Health (NIH) is a major resource for the creation of artificially selected lines. These rats are available from the NIH and are somewhat ideal as a founder population. This genetically segregating stock of rats originated from the intentional crossbreeding of eight disparate inbred strains of independent origin (AxC 9935 Irish [ACI], Brown Norway [BN], Buffalo [BUF], Fischer 344 [F344], Marshall 520 [M520], Maudsley reactive [MR], Wistar-Kyoto [WKY], and Inbred Wistar [WN]) by Hansen and Spuhler in 1979 [1984]. (See also Rat Genome Database, http://www.rgd.mcw.edu/strains) The outbred stock is managed such that the gene frequency remains stabilized, the inbreeding is minimized within the closed colony, and the genetic variability remains substantial. Each rat from the N:NIH stock is a genetic admixture of the eight founder inbred strains and in theory can be a useful source for genetic analysis of behavioral traits (Mott et al., 2000).

GENERAL APPROACH TO SELECTION

Selective breeding begins by measuring the trait of interest in a large founder population that has wide genetic heterogeneity. The contrasting lines for the trait (low and high) are started by breeding rats that demonstrate the extreme values of the founder population. Then, at each subsequent generation, progeny are phenotyped and selected as the "best" for the trait and bred to create the next generation. This process is repeated until the change in the population mean produced by selection (selection response) plateaus, which typically indicates exhaustion of additive genetic variance for the trait (Fig. 3–6). The degree of heterozygosity in a selected population can be increased above that random bred by making the contributions from each family more equal; this can be accomplished by taking the "best" female and male from each mating and using them as parents in the next generation (within-family selection). Within-family selection coupled with a systematic rotational breeding design keeps mating between relatives at the minimum to maintain the rate of inbreeding (ΔF) per generation just less than 1% if at least 13 families are used in both the low and high selected lines. ΔF = $1/(4N)$, where N = number of individual parents in each line at each generation (Falconer and Mackay, 1996a).

SELECTION ON THE GENETIC COMPONENT

In general, the more information that is available per each individual about the trait, the more accurate is the selection. For example, if five repeat measures of running capacity were available, it would seem logical to select on the average of these five estimates of running capacity (Nicholas, 1987). Despite this, we took a different approach when we started large-scale selection for aerobic treadmill running capacity in 1998 (Koch and Britton, 2001). For each rat, the single best value of five trials was deemed as the measure most closely associated with the genetic component of running capacity. This idea of estimating the genetic component from the best trial rather than the average of all trials has two origins. (1) The environment can have an infinite negative influence on capacity (i.e., a detrimental environment can take the capacity to zero). Factors such as subtle differences in daily housing or handling conditions could cause a genetically superior rat to perform below its maximal ability for a given trial. (2) However, the environment can have only a finite positive influence on a trait. That is, a favorable environment cannot cause a rat to exceed values above its genetically determined upper limit of capacity. Thus, we reasoned the rat's best trial of five comes closest to the genetically determined upper limit of its capacity. Although we used only the best day for selection, it should be noted that testing across 5 days matches the average estrus cycle for female rats and thus eliminates this as a variable.

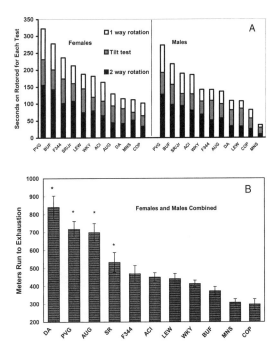

EXPERIMENTAL USE OF INBREDS

Inbred strains of animals have been a core substrate in the analysis of complex phenotypes and remain as one of the basic tools for progress in functional genomics. The major value of inbred strains emanates from the close genetic uniformity that facilitates phenotyping, genotyping, and the opportunity for multiple investigators to repeatedly evaluate the same genotype.

Our laboratory group has scanned panels of inbreds for measures of physical capacity that include aerobic endurance treadmill running capacity, sensorimotor capacity, and strength (Barbato et al., 1998; Biesiadecki et al., 1998; Koch and Britton, 2003).

Figure 3–3A shows the degree of variation in sensorimotor ability among 11 inbred strains of rats (Biesiadecki et al., 1998). Sensorimotor capacity was estimated on the basis of three separate tests. Rats from each strain were tested to determine how long each could remain on (1) a rotating cylinder as the velocity of rotation increased every 5 seconds (one-direction rotation test), (2) a rotating cylinder that reversed direction every 5 seconds and increased velocity every 10 seconds (two-direction rotation test), and (3) a platform that was tilted 2° every 5 seconds from 22° to 47° (tilt test). The distribution among the strains for each test was continuous and normally distributed. On all three tests, the Black hooded PVG strain was consistently the highest ranking strain (for both males and females), whereas the Copenhagen (COP) and Milan Normotensive Strain (MNS) strains were consistently the lowest ranking strains. The large

Figure 3–3. (A) Variation in sensorimotor capacity based on three tests of motor ability (one-way rotation, tilt test, and two-way rotation) measured as timed performance (seconds) in a panel of 11 inbred strains. The PVG strain demonstrated the greatest ability, whereas rats of the MNS and COP strains represented strains with the lowest performance on all three tests. (B) For aerobic running capacity (distance run to exhaustion in meters), the DA and COP displayed the greatest difference. (Adapted from Biesiadecki et al. (1998) and Barbato et al. (1998.)

differential in sensorimotor performance between PVG and MNS or COP strains suggests that these strains can be used as contrasting models of sensorimotor capacity.

The same panel of 11 inbred strains was tested for aerobic treadmill endurance running capacity to exhaustion as estimated from duration of the run, distance run, and vertical work performed to take into account variation in body weight (Barbato et al., 1998). The COP rats were the lowest performers and the DA rats were the highest performers on all estimates of running. The wide divide in performance between COP and DA strains of rats represents identification of genetic substrate for exploration of aerobic endurance capacity and related traits such as economy of running, oxygen transport pathways, and heart function.

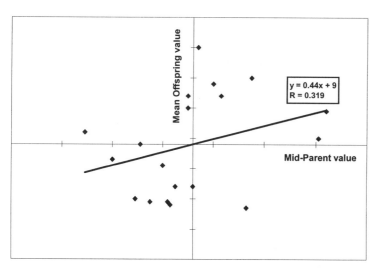

Figure 3–4. Regression of mean offspring on mid-parent value for one-way rotation test for sensorimotor capacity. Each point represents the mean value of one set of parents (along x-axis) and mean value of their offspring (along y-axis). Axes intersect at the mean value for all parents and all offspring and are marked at 10 second intervals. The slope of the line estimates the heritability (h^2) and has a value of 0.44. (Data from Koch and Britton [2003].)

Figure 3–4 displays an estimate of heritability for generalized sensorimotor capacity (in N:NIH stock) as estimated from time on a rotorod. The mean of the offspring from each of 19 families was regressed on the mid-parent value of each family and yielded a slope of 0.44. In the panel of 11 inbred rats that were tested, h^2 for sensorimotor capacity as estimated by comparing the between- and within-strain variances averaged 0.39 in females and 0.48 in males (Koch and Britton, 2003). These results demonstrate a wide phenotypic variation and a heritable component to sensorimotor capacity sufficient for success in the development of contrasting rat genetic models by divergent artificial selection.

COSEGREGATION ANALYSIS

Contrasting inbred strains can be characterized at organ, tissue, cell, and molecular levels of organization to identify physiological, behavioral and morphological differences that may be responsible for variations in the trait; these are termed "likely determinant phenotypes" (Jacob and Kwitek, 2002). The identification of the genes responsible for a given phenotype using divergent inbred strains is based on two widely held principles of biology: (1) genes cause traits, and not vice versa, and (2) genes that cause a given trait will remain associated

with that trait and other genes will segregate randomly relative to the trait.

Understanding the use of inbreds in cosegregation studies is one path of genetic analysis that has theoretical and heuristic value. Like individuals, two inbred strains can demonstrate wide variation for a given trait. The central idea is to follow the association of genes or downstream gene products (such as messenger RNA, protein, or subordinate physiological and biochemical traits) with values of the phenotype in a segregating population. A segregating population is one in which alleles recombine randomly to yield new genotypes in mating crosses. This approach works because loci and pathways causative of a given trait will remain associated with that trait and other genes and nonassociated traits will segregate randomly. The most informative segregating population is created from two sequential crosses of inbred strains. The first cross is between two different inbred parental strains (P_1 and P_2) that differ significantly for a trait. The assumption is that the two strains have contrasting allelic variants that dictate the phenotypic difference. These original contrasting strains are often referred to as the "low strain" and "high strain" to indicate the directional difference in trait measure. The $P_1 \times P_2$ cross yields an F_1 (first filial) population composed of close to identical heterozygotes. A subsequent $F_1 \times F_1$ cross yields the desired segregating F_2 (second filial) population in which allelic variants recombine randomly, producing a variation of genotypes and thus a variation in the distribution of phenotypic trait values (Fig. 3–5).

Following a single locus that distributes via discrete mendelian methods allows one to understand the usefulness of cosegregation when extended to analysis of a polygenic trait. Consider two allelic variants (A_1 and A_2) that produce sufficient variation that the homozygotic (A_1A_1 and A_2A_2) and the heterozygotic (A_1A_2) genotypes can be distinguished by phenotype. Assign the low strain (P_1) as the homozygotic genotype A_1A_1 and the high strain (P_2) as the homozygotic genotype A_2A_2. Crossing $P_1 \times P_2$ produces only A_1A_2 heterozygotes in the F_1 population. These crosses can be followed using Punnett squares:

Gametes from Low Strain (P_1)

		A_1	A_1	
Gametes	A_2	A_1A_2	A_1A_2	(F$_1$ Population)
from High				
Strain (P$_2$)	A_2	A_1A_2	A_1A_2	

An intercross between F_1 heterozygotes produces a segregating population that yields all possible genotypes in the ratio of 1:2:1 (1 A_1A_1, 2 A_1A_2, 1 A_2A_2):

Gametes from F$_1$ Male

		A_1	A_2	
Gametes	A_1	A_1A_2	A_1A_2	
from F$_1$				(F$_2$ Population)
Female	A_2	A_1A_2	A_2A_2	

Thus, for a single locus with a large effect, segregation of the phenotype can be followed in an F_2 population. That is, phenotypic expression of the homozygotes (A_1A_1 and A_2A_2) separates to express the low and high values while the heterozygotes demonstrate intermediate values. Note the simple genotype associated with creation of an F_2 population originated from two inbred strains. For each locus, only three allelic combinations are possible and can be followed for association between genotype and phenotype as shown in Figure 3–5. Genetic markers are identifiable physical locations on a chromosome (loci) whose inheritance can be followed similarly as described for

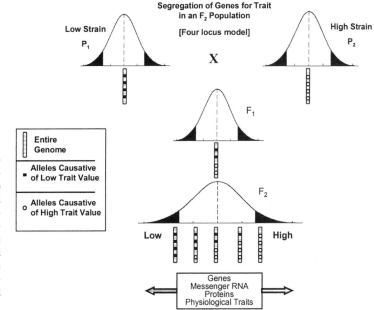

Figure 3–5. The biological determinants of the differences in trait between two inbred strains (P$_1$ and P$_2$) can be evaluated via cosegregation analysis. Two sequential crosses produce an F$_2$ population in which alleles recombine randomly and can be used to determine which genes and downstream products cosegregate with the distribution of the phenotype. (Adapted from Britton and Koch [2001].)

Mendelian traits and form the basis for genomic scans (see Jacob and Kwitek, 2002).

THE BENEFITS OF INBREDS DERIVED FROM SELECTED LINES

Four properties make inbred lines originally created from two-way artificial selection useful (Fig. 3–6). First, as indicated earlier, models can be created for traits for which there exists only minimal variance between the already available inbred strains. Second, because the selection process often carries the phenotypic means of the low and high lines beyond the range of the extremes of the founder population (Falconer and Mackay, 1996a), the signal measurements of the trait can be made to differ substantially between the lines. Third, by maintaining a high level of heterozygosity at each generation during

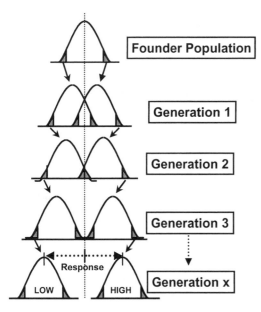

Figure 3–6. An idealized selective breeding paradigm. Selective breeding begins by measuring a trait in a large group of widely heterogenous animals from which selections of choice breeders are made to concentrate alleles expressing extreme values of the phenotype. At each generation, progeny are phenotyped, selected as the "best," and bred to create the next generation. This process is repeated until the change in population mean (selection response) has plateaued or a desired difference is attained between the selected lines.

selection, the main complement of contrasting alleles causative of trait difference will be concentrated in both the divergent lines and the subsequently developed inbred strains. Fourth, selection across many generations interprets into inadvertent selection for insensitivity to subtle changes in environment. This lack of influence by environmental effects in selected strains can be of great benefit to the reproducibility of the phenotype. Inbred strains that differ markedly for a trait but did not originate from selected lines often demonstrate wide variation in phenotype in response to similar controlled experimental conditions (Crabbe et al., 1999).

THE FUNDAMENTAL CHALLENGE FOR ALL MODELS

Although we focused here on genetic models, the development and use of any kind of animal model are associated with subtle problems, especially for models of complex diseases. First, physical and chemical maneuvers applied to mimic disease conditions such as ligation of cerebral arteries (stroke), intracerebral injection of 6-hydroxydopamine (Parkinson's disease), or administration of streptozocin (diabetes mellitus) more accurately reflect response to injury rather than emulating the progression of disease. Second, although it seems that disease models derived from selection would be a highly informative alternative, direct selection for a disease is also problematic. Selection would be based on currently known measurable traits of the disease and not the full array of underlying mechanisms. The issue is that traits are not mechanisms, such that selection based on measures thought to characterize a disease will leave components either unrepresented or weighted inappropriately. Third, the problem in selecting on a disease trait is amplified because it appears that diseases emerge not as discrete events but as biological complexes with coordinately regulated gene clusters, such as the pathogenic cascades represented by Parkinson's disease

(Tieu et al., 2003) and metabolic syndrome X (Lopez-Candales, 2001).

To address this challenge, we currently hypothesize that clinically relevant models can emerge from two-way artificial selection for high-order, complex, physiological traits such as sensorimotor function. We predict that selection for the extremes of fundamental trait variation will create two populations: (1) a low line that displays low physiological function and disease-related traits and (2) a high line that displays high physiological function, no disease-like traits, and resistance to the development of disease.

This hypothesis derives from our highly speculative ideas about the association between our evolutionary history and oxygen (Britton, 2003). The underlying proposal is that 2 billion years of evolution in an oxygen environment has determined that oxygen metabolism occupies a central feature of our biology (DesMarias, 2000). It appears that evolution followed the increased free energy transfer afforded by the widened redox potential when oxygen is the final electron acceptor in oxidation reactions (Baldwin and Krebs, 1981). Obligatory for the use of oxygen in energy transfer pathways was the simultaneous co-evolution of enzymes that detoxify the reactive oxygen species that are byproducts of oxidation reactions. Thus, the pathways that mediate both oxidation reactions and oxygen detoxification reactions constitute a large part of our biology (Young and Woodside, 2001; Myers et al., 2002). Our extension is that essentially all disease will resolve at the molecular level into problems associated with oxygen utilization. For example, it has been shown that an increased expression of genes involved in oxidative phosphorylation contributes to human diabetes (Mootha et al., 2003) and that increased production of reactive oxygen species is related to the cellular dysfunction and biochemical alterations present in Parkinson's disease (Tieu et al., 2003). Development and study of well-defined animal models will be useful in resolving mechanisms of cellular injury and death.

ACKNOWLEDGMENTS

This work was supported by grants from the U.S. Public Health Service, National Institutes of Health (Heart, Lung and Blood Institute grant HL-64270 and the National Center for Research Resources grant RR-17718) to S.L.B. and L.G.K.

REFERENCES

Atela P, Gole C, Hotton S (2002) A dynamical system for plant pattern formation: a rigorous analysis. Journal of Nonlinear Science 12:641–676.

Baldwin JE and Krebs H (1981) The evolution of metabolic cycles. Nature 291:381–382.

Barbato JC, Koch LG, Darvish A, Cicila GT, Metting PJ, Britton SL (1998) Spectrum of aerobic endurance running performance in eleven inbred strains of rats. Journal of Applied Physiology 85:530–536.

Biesiadecki BJ, Brand PH, Koch LG, Metting PJ, Britton SL (1998) Phenotypic variation in sensorimotor performance among eleven inbred rat strains. American Journal of Physiology 276:R1383–R1389.

Britton SL (2003) Is there an answer? IUBMB Life 55:429–430.

Britton SL and Koch LG (2001) Animal genetic models for complex traits of physical capacity. Exercise and Sport Sciences Reviews 29:7–14.

Crabbe JC, Wahlsten D, Dudek BD (1999) Genetics of mouse behavior: interactions with laboratory environment. Science 284:1670–1672.

Crow JF (1986) Basic concepts in population, quantitative and evolutionary genetics. New York: W.H. Freeman & Company.

Crow JF and Kimura M (1970) An introduction to population genetics theory. New York: Harper and Row.

DesMarias DJ (2000) When did photosynthesis emerge on earth? Science 289:1703–1705.

Falconer DS and Mackay TFC (1996a) Introduction to quantitative genetics, 4th edition. Essex, England: Addison Wesley Longman, Ltd.

Falconer DS and Mackay TFC (1996b) Heritability. In: Introduction to quantitative genetics, pp. 160–183. Essex, England: Addison Wesley Longman Limited.

Fisher RA (1930) The genetical theory of natural selection. Oxford, England: Clarendon Press.

Hansen C and Spuhler K (1984) Development of the National Institutes of Health genetically heterogeneous rat stock. Alcoholism, Clinical and Experimental Research 8:477–479.

Hartl DL and Clark AG (1988) Principles of population genetics, 2nd edition. Sunderland, MA: Sinauer Associates, Inc.

Hedrick PW (2000) Quantitative traits and evolution. In:

Genetics of populations, pp. 445–500. Sudbury, MA: Jones and Bartlett Publishers.

Hegmann JP and Possidente B (1981) Estimating genetic correlations from inbred strains. Behavior Genetics 11:103–114.

Jacob HJ and Kwitek AE (2002) Rat genetics: attaching physiology and pharmacology to the genome. Review. Nature Review Genetics 3:33–42.

Knudsen K, Dahl LK, Thompson K, Iwai J, Leith G (1970) Effects of chronic salt ingestion: inheritance of hypertension in the rat. Journal of Experimental Medicine 132:976–1000.

Koch LG and Britton SL (2001) Artificial selection for intrinsic aerobic endurance running capacity in rats. Physiological Genomics 5:45–52.

Koch LG and Britton SL (2003) Genetic component of sensorimotor capacity in rats. Physiological Genomics 13:241–247.

Lopez-Candales A (2001) Metabolic syndrome X: a comprehensive review of the pathophysiology and recommended therapy (review). Journal of Medicine 32:283–300.

Mootha VK, Lindgren CM, Eriksson KF, Subramanian A, Sihag S, Lehar J, Puigserver P, Carlsson E, Ridderstrale M, Laurila E, Houstis N, Daly MJ, Patterson N, Mesirov JP, Golub TR, Tamayo P, Spiegelman B, Lander ES, Hirschhorn JN, Altshuler D, Groop LC (2003) PGC-1alpha-responsive genes involved in oxidative phosphorylation are coordinately downregulated in human diabetes. Nature Genetics 34:267–273.

Mott R, Talbot CJ, Turri MG, Collins AC, Flint J (2000) A method for fine mapping quantitative trait loci in outbred animal stocks. Proceedings of the National Academy of Sciences of the United States of America 97:12649–12654.

Myers J, Prakash M, Froelicher V, Do D, Partington S, Atwood JE (2002) Exercise capacity and mortality among men referred for exercise testing. New England Journal of Medicine 346:793–801.

Nicholas FW (1987) Veterinary genetics. New York: Oxford University Press.

Silver LM (1995) Mouse genetics: concepts and applications. New York: Oxford University.

Sokal RR and Rohlf FJ (1981) Biometry: the principles and practice of statistics in biological research, 2nd edition. San Francisco: W.H. Freeman and Company.

Tieu K, Ischiropoulos H, Przedborski S (2003) Nitric oxide and reactive oxygen species in Parkinson's disease. IUBMB Life 55:329–335.

Wright S (1921) Systems of mating. Genetics 6:111–178.

Yamori Y, Ooshima A, Okamoto K (1972) Genetic factors involved in spontaneous hypertension in rats: an analysis of F_2 segregation generation. Japanese Circulation Journal 36:561–568.

Young IS and Woodside JV (2001) Antioxidants in health and disease. Journal of Clinical Pathology 54:176–186.

Individual Differences

GUY MITTLEMAN

4

Intrinsic to any discussion of human behavior is the notion that people display vast individual differences in behavior in response to the same situation. Knowledge of individual differences has fueled a variety of disciplines, including personality and social psychology, and individual differences in the behavior of human beings have long been recognized as a tool in the investigation of the neuropsychophysiological bases of a variety of diseases. The purpose of this chapter is to suggest that rats, like humans, despite going through a genetic bottleneck when domesticated, display profound individual differences that have relevance for understanding a variety of normal and pathological conditions.

Vast individual differences in the behavioral response to drugs of abuse as well as subsequent patterns of self-administration have been observed in humans. As stated by O'Brien et al. (1986), "Some addicts go for months or years using heroin or cocaine only on weekends before becoming a daily (addicted) user. Others report that they had such an intense positive response that they became addicted with the first dose. Similar variation in initial response may also be observed in animals." These individual differences in the reinforcing effects of addictive drugs (de Wit et al., 1986) are considered by many clinicians to be a primary factor in the vulnerability to addiction shown by some individuals (O'Brein et al., 1986). Although much research has been devoted to determining the substrates of addiction (for review, see Koob and Bloom, 1988), there have been relatively few investigations into the mechanisms that underlie individual differences.

GENESIS OF THE CONCEPT OF INDIVIDUAL DIFFERENCES IN RATS

Elliot Valenstein was the first to suggest that rats displayed consistent individual differences. He observed that in response to electrical stimulation of the lateral hypothalamus (ESLH), rats displayed a variety of different behaviors, including eating, drinking, gnawing, hoarding, grooming, aggression, retrieval of young, and male copulatory behavior (for a review, see Valenstein, 1975). Some animals consistently responded to the electrical stimulation by eating or drinking, whereas others displayed only an increase in locomotor activity. The explanation that these differences in the behavioral response to ESLH could be attributed to corresponding differences in electrode placement within the lateral hypothalamus was eliminated. Even with indistinguishable electrode placements and experimental procedures, different rats showed very different responses to brain stimulation (Cox and Valenstein, 1969; Valenstein et al., 1970). Moreover, rats with electrodes implanted at different hypothalamic sites showed a strong tendency to display the same behavior from both electrodes (Valenstein et al., 1970). Furthermore, Wise (1971) tested rats for evoked eating and drinking using moveable hypothalamic electrodes. Animals that displayed an evoked consummatory response both ate and drank, and continued to do so as electrodes were advanced as much as 1.5 mm in a dorsal-to-ventral direction through the hypothalamus. Within the limits of this positive area, movement of the electrode had little effect on current threshold for evoking eating and drinking. Animals that displayed no response failed to eat or drink at any stimulated site. Wise concluded that the variability between animals reflected individual response tendencies rather than differences in the site of stimulation. Bachus and Valenstein (1979) destroyed hypothalamic cells and fibers surrounding stimulating electrodes that evoked drinking in Long-Evans rats. Although the larger lesions extended up to 3.0 mm from the center of the electrode and higher current levels were required to excite more distal neuronal elements, all animals continued to drink when stimulated. Thus, stable characteristics of animals, and not the precise neuroanatomical locus of the electrode, accounted for the response to hypothalamic stimulation. It was also demonstrated that there were strain differences (Long-Evans versus Sprague-Dawley) in the likelihood of evoking consummatory behavior with ESLH, suggesting a genetic link (Mittleman and Valenstein, 1981). Valenstein (1969) concluded that this evidence provided justification for postulating that individual animals had a "prepotent" and intrinsic tendency to respond in a characteristic manner to ESLH.

INDIVIDUAL DIFFERENCES AND RELATED PSYCHOLOGICAL AND NEUROPHYSIOLOGICAL CHARACTERISTICS

Because eating and drinking elicited by ESLH are forms of nonregulatory ingestive behavior, it seemed logical to compare behaviors elicited by ESLH with those occurring in another situation that elicited nonhomeostatic consumption. One of the best-known methods for eliciting nonregulatory ingestion is the schedule-induced polydipsia paradigm (Falk, 1961). When rodents, for example, are food but not water deprived and given intermittent presentations of small amounts of food, many develop excessive fluid consumption, called schedule-induced polydipsia (SIP). Such excessive drinking has been reported in many animal species and occurs under a variety of schedules of reinforcement (for reviews, see Wallace and Singer, 1976; Roper, 1981; Wetherington, 1982). It has been suggested that SIP is an example of the larger category of adjunctive behaviors that occur in situations where strongly motivated appetitive or consummatory behaviors are interrupted or thwarted and probably shares characteristics in common with displacement behaviors that occur in more natural settings (Tinbergen, 1952; Falk, 1966, 1969, 1971).

Mittleman and Valenstein (1984) implanted 42 adult, male Long-Evans rats with hypothalamic electrodes and screened them for eating and drinking elicited by ESLH. Twenty-four animals initially ate or drank (ESLH-positive) in response to stimulation, whereas 18 (ESLH-negative) did not. When tested in SIP, ESLH-positive animals rapidly acquired drinking, which increased from 2 ml on day 1 to more than 11 ml on day 10, whereas ESLH-negative rats hardly drank at all on day 1, and consumption increased to only 4 ml by day 10. This study shows that individual differences in response to ESLH are predictive of differences in SIP. Thus, the behavioral differences observed in ESLH are not unique to this paradigm but are representative of a more global characteristic. Predictability extended to cognitive ability and "emotionality." That is, SIP-positive animals more rapidly learned an active avoidance response and showed less freezing when confronted with an aggressive resident rat in the resident–intruder paradigm (Dantzer et al., 1988).

The individual differences observed in these paradigms also have been associated with biological differences in the response properties of mesolimbic dopamine systems (e.g., Antelman and Szechtman, 1975; Robbins and Koob, 1980; Wallace et al., 1983; Fibiger and Phillips, 1986). Individual differences in the behavioral response to amphetamine were first investigated as an indirect means of assessing dopaminergic activity. Long-Evans rats that readily engage in SIP consistently show an enhanced response to a single injection of D-amphetamine (Mittleman and Valenstein, 1985). In addition, in response to repeated injections of this drug, SIP-positive rats (Fig. 4–1, left, solid line), in comparison to negative animals (dashed line), showed more rapid behavioral sensitization of overall stereotypy (Mittleman et al., 1986). Because dopaminergic activity is the primary substrate of stereotyped behavior (Creese and Iversen, 1975), these results suggested differences in dopaminergic neural systems in SIP-positive and -negative rats. Neurochemical evidence for such a difference was confirmed using an indirect measure of dopamine turnover (see Fig. 4–1, right). Foot shock was used to increase dopamine utilization in SIP-positive (solid bar) and -negative (open bar) Long-

Evans rats. Clear differences between the groups emerged. Positive rats showed significantly greater increases in dopamine turnover in the striatum and nucleus accumbens than did SIP-negative (open bar) animals, as assessed by ratios of 3,4-Dihydroxyphenylacetic acid to dopamine and high-performance liquid chromatography–with electrochemical detection (Mittleman et al., 1986). That Long-Evans rats have significantly different individual response profiles to single and repeated injections of amphetamine as well as associated regional differences in dopamine metabolism was confirmed by Segal and Kuczenski (1987).

In agreement with these lines of research, more recent investigations have confirmed the general concept that individual differences in the unconditioned response to psychomotor stimulants are related to the response properties of dopaminergic systems. Sabeti et al. (2003) reported that, in outbred Sprague-Dawley rats, individual differences in the locomotor response to cocaine are directly related to corresponding differences in the rate of dopamine clearance in the nucleus accumbens. The mechanism underlying this indi-

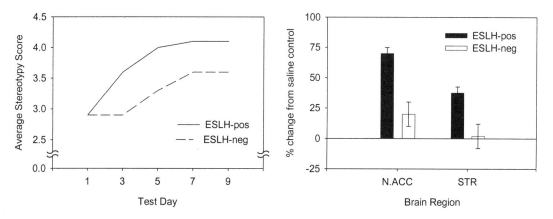

Figure 4–1. (*Left*) The development of amphetamine-induced stereotypy in SIP-positive (*solid line*) and -negative rats (*dashed line*). Both groups received an injection of 3.0 mg/kg D-amphetamine sulfate once every 3 days for 27 days (a total of nine injections). Overall stereotypy was rated on an 8 point scale. The average rating of the 2 hour test session is shown. (*Right*) SIP-positive (*solid bar*) and -negative (*open bar*) rats received 20 minutes of intermittent footshock (1.3 mA for 0.5 second every 15 seconds). Dopamine and dihydroxyphenylacetic acid (DOPAC) levels in the nucleus accumbens and striatum were determined, and the DOPAC/dopamine ratio was used as an indirect measure of dopamine utilization. Results are expressed as the percentage increase over an unshocked control group composed of both positive and negative animals. (Modified from Mittleman et al. [1986].)

vidual difference was differential inhibition of nucleus accumbens dopamine transporters.

Because individual differences in ESLH, SIP, and the behavioral responsiveness to psychostimulants can be linked to corresponding neurochemical differences in forebrain dopamine, it was predicted that this relationship would extend to individual differences in addictive liability. This possibility was confirmed by Piazza and colleagues using an animal model of intravenous drug self-administration (Weeks, 1962; Schuster and Thompson, 1969). Piazza et al. (1993) tested male Sprague-Dawley rats for spontaneous locomotion in a novel environment. Animals could be divided into "high" and "low" responders based on the amount of spontaneous locomotion exhibited when tested in a circular corridor for 2 hours (Piazza et al., 1989). Animals were then implanted with intracardiac catheters and allowed to self-administer a low dose of D-amphetamine sulfate (10 μg/20 μl/inj). Behavioral differences in self-administration were highly correlated with differences in activity levels (Piazza et al.,

1989). Figure 4–2 (left) shows the significant differences in the number of drug requests (as indicated by nose pokes into the "active" hole) made by rats that showed high and low levels of locomotion in response to a novel environment. Also shown is the number of responses made into the "inactive" hole by these two groups. Responding in the inactive hole was quite similar, suggesting that these two groups differed specifically in the number of drug requests, which minimized the role of nonspecific activational effects of amphetamine in the observed difference in drug requests.

The rats were then gradually food deprived to 85% of their baseline weight and tested for SIP using standard procedures. As indicated (see Fig. 4–2, right), animals that were predisposed to self-administer amphetamine (SA+) acquired SIP significantly more rapidly than animals that did not self-administer (SA−). This figure also shows a subgroup of SA− rats that were tested for an additional 5 days in the self-administration paradigm, dur-

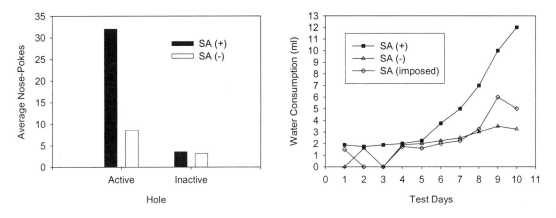

Figure 4–2. (*Left*) Self-administration testing was conducted in test chambers with a hole (2 cm above the floor) in each of the two short walls. Each hole was monitored by an infrared photocell beam such that nose pokes in one of the holes (defined as "Active") turned on the infusion pump for 2 seconds, injecting 10 μg of D-amphetamine sulfate into the animal's venous system. Nose pokes in the other hole (defined as "Inactive") had no effect. During each 30 minute session, the number of nose pokes in both the active and the inactive holes was recorded for animals that were predisposed to self-administer amphetamine (SA+) and those that did not (SA−). (*Right*) Rats were given 10 daily 30 minute SIP tests in standard operant cages equipped with a drinking tube connected to a water-filled graduated cylinder. After each test, the total amount of water consumed was recorded. Filled squares designate SA+ animals (those that acquired SA); open triangles indicate D-amphetamine–negative rats. Also shown (open diamonds) is the performance of SA (imposed) rats that were yoked to SA+ animals so that they received the same amount and timing of amphetamine infusions. (Modified from Piazza et al. [1993].)

ing which drug requests from a randomly selected SA+ rat were "played back" in the order that the rat received the amount and timing of amphetamine administrations. These SA− *imposed* animals were then tested in the SIP paradigm to determine if the amount of prior amphetamine exposure was a factor that influenced responsiveness in SIP. As shown (see Fig. 4–2, right), these animals did not differ from the SA− animals, suggesting that prior drug exposure was not an important factor influencing the development of individual differences in SIP.

Taken together, a number of possible conclusions are suggested by these results. First, they suggest a relationship between individual differences in the behavioral response to activating or "arousing" conditions such as SIP or ESLH and profound individual differences in forebrain dopamine systems as demonstrated by behavioral, psychopharmacological and neurochemical means. Rats that readily engaged in excessive drinking during SIP showed significantly greater behavioral responsiveness to single or repeated injections of D-amphetamine. These drug-induced differences in behavior were related to significantly greater stress-induced neurochemical responsiveness of the mesolimbic dopamine system. Second, these results further indicated that the consistent behavioral differences observed in SIP are related to individual differences in the predisposition to self-administer the psychoactive compound amphetamine. This was demonstrated by the high level of concordance between amphetamine SA and SIP. This result is particularly important because it provides further evidence that these individual differences in responsiveness are part of a more global characteristic. From the results cited earlier, it appears that this "global" characteristic may include cognitive, emotional, pharmacological, and neurochemical components.

These results support the notion that consistent individual differences in behavior can be used as a means for investigating the neuro-

physiological and psychological underpinnings of behavior, but they do not, by themselves, indicate that the investigation of individual differences in rats has relevance toward understanding the individual differences in vulnerability to addiction observed in humans. Nevertheless, a national survey by the Substance Abuse and Mental Health Services Administration (SAMHSA) (2003, p. 55) indicates that in 2002, approximately 9.4% of the population of the United States was classified as substance abusers or substance dependent, based on the criteria specified in the *Diagnostic and Statistical Manual of Mental Disorders*, 4th edition (American Psychiatric Association, 1994).

INDIVIDUAL DIFFERENCES IN RATS HAVE FACE VALIDITY WITH RESPECT TO HUMAN INDIVIDUAL DIFFERENCES AND CANNOT BE EXPLAINED BY PROXIMAL EXPERIENTIAL FACTORS

In an effort to determine if the observed individual differences in self-administration we had observed in rats were similar to the vast individual differences seen in humans, we investigated the acquisition of amphetamine self-administration in naïve rats. The goals of this experiment were to (1) document individual differences in the acquisition of drug taking, (2) simultaneously determine any dose preferences during the acquisition process, and (3) investigate any experiential factors that might explain the development of individual differences in amphetamine self-administration.

Male Long-Evans rats (Harlan; n = 207; age range, 50 to 125 days) were maintained on food and water ad libitum throughout testing. They were initially tested for locomotion in a novel environment using the methods of Piazza et al. (1993). The animals were then implanted with a catheter and tested in experimental chambers in which one curved wall contained a row of five 2.5 cm round holes (a central hole and two on each side) illuminated by a yellow LED. A nose poke response into each hole was detected by an infrared photobeam. A house light located near the ceiling of the experimental chamber signaled the availability of drug. Drug self-administration contingencies were programmed such that an animal received a different drug dose depending on the hole where a nose poke was made. Each injection was followed by a 30 second time-out period, during which the house light and the lights within each hole were extinguished and any additional nose poke responses were recorded but had no programmed consequences. Drug concentration and pump delivery rate (2.0 μl/s) were kept constant throughout an experimental session, whereas unit dose of drug per injection was controlled by varying the duration of pump action (i.e., volume of injected solution). In this experiment animals could sample five doses (0.0, 0.018, 0.032, 0.056, and 0.10 mg/kg per injection) by making a nose poke response into the five different stimulus "holes."

Rats had 23 hours of daily access to the drug for a total of 10 days. The animals' first experimental session was composed of a sampling component and a choice component, whereas subsequent experimental sessions were composed of the choice component only. During the sampling component, rats had to request each drug dose before access to all doses was permitted. Provided that the five doses were sampled before the end of the experimental session, the choice component began. During this period, the programmed experimental contingencies were identical to those described for the sampling component except that all five drug doses were continuously available after an injection-produced time-out. The dose associated with each hole was counterbalanced using a Latin square to reduce any response bias. During the first 5 days of testing, animals responded on a fixed ratio 1 (FR 1) schedule. This was changed to an FR 2 schedule for the final 5 test days to determine if the animals were willing to expend more effort to receive the drug.

A total of 146 animals completed the experiment. Figure 4–3 (top) describes the range of individual differences in amphetamine self-administration. Average daily drug taking in individual rats ranged from a mean of 0.08 to 28.2 mg/kg. Because of this vast range of responding, we arbitrarily divided the group into thirds using the standard deviation. Low responders were defined as those animals self-administering the mean amount or less (range, 0.81 to 2.02 mg/kg per day; n = 116, or 80% of the population). Middle responders ranged from the mean to 1 SD above the mean (range, 2.03 to 5.90 mg/kg per day; n = 15, or 10% of the population), and high responders were those animals that self-

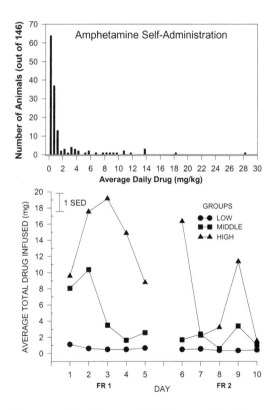

Figure 4–3. (*Top*) The sample distribution of the 146 rats tested for 10 days of D-amphetamine self-administration. Average daily drug infused ranged from 0.81 to more than 28 mg/kg per day. (*Bottom*) Self-administration behavior of the low, middle, and high groups (see text for definition). Moderate to high levels of amphetamine self-administration were associated with a "binge and crash" pattern. The switch to an FR 2 schedule reduced responding in the high and middle groups.

administered amounts greater than 1 SD above the mean (range, 5.91 to 28.2 mg/kg per day; n = 15, or 10% of the population). Analyses indicated that the groups differed significantly from one another and that responding in the high and middle groups was significantly reduced by the switch to an FR 2 schedule. As shown in Figure 4–3 (bottom), both of these groups showed a characteristic "binge and crash" pattern of responding for drug.

Significant dose preferences that were group dependent appeared on test day 1. As indicated in Figure 4–4, rats in the high group showed the clearest preference. These animals preferred the middle dose (0.032 mg/kg per injection) over the 0.0 control dose. Low responders took significantly more of the middle (0.032 mg/kg per injection) and high (0.10 mg/kg per injection) doses in comparison to the control dosage. Surprisingly, rats in the middle group failed to develop a dose preference. When considered over the 10 test days (Fig. 4–5), dose preferences remained consistent only in the high-responder group. These animals self-administered the middle dose preferentially throughout testing. Perhaps because of the lower levels of self-administration

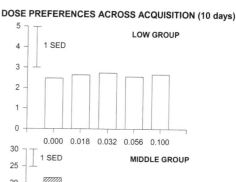
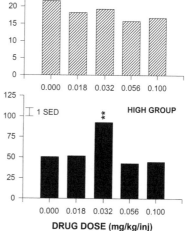

Figure 4–5. Dose preferences of the low, middle, and high groups over the 10 days of testing. Although the low group showed an initial preference, it was not maintained throughout testing. The middle group never exhibited a consistent preference over the 10 days; only the high group maintained a consistent dose preference throughout testing. **$p < .01$ difference from the 0.0 mg/kg per injection dose. SED, standard error of the difference in means.

and corresponding reduced experience with the drug, animals in the middle and low groups failed to show a significant dose preference when considered over the 10-day testing interval.

We also wanted to determine if a variety of factors were related to the vast individual differences in amphetamine self-administration. Because age is a significant factor in human abuse (SAMHSA, 2003), it was included as a factor. It also seemed reasonable that differences in self-administration might be a simple outgrowth of differences in activity. Thus, more active animals might be more likely to nose poke for drug. Activity in a novel circu-

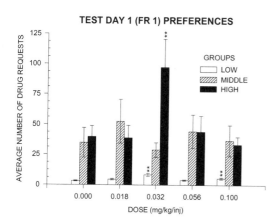

Figure 4–4. Dose preferences of the low, middle, and high groups when first given the opportunity to self-administer amphetamine on test day 1. Both the low and high groups rapidly developed a significant preference for the 0.032 (mg/kg per injection) dose. In contrast, the middle group did not show an initial dose preference. **$p < .01$ difference from the 0.0 mg/kg per injection dose.

lar corridor and activity in the test chamber (day 1) before the initial infusion were included for these reasons. The first drug dose selected was included as a representative experiential variable that might influence subsequent abuse. Because there are experiential or environmental influences on the likelihood of drug taking (Everitt et al., 2001) we reasoned that self-administration might be influenced by the first dose selected. For example, it seemed possible that the likelihood of self-administration would be reduced if the first dose selected had no effects (0.0 mg/kg per injection) or was potentially aversive to a naïve animal (0.10 mg/kg per injection). Finally, to determine if characteristics of initial drug taking predicted individual differences in self-administration, the latency to the first drug request and total amount of drug infused on day 1 were also included.

Table 4–1 presents the results of these analyses. Age was significantly negatively correlated with individual differences in amphetamine self-administration, in that, much like humans, larger amounts of drug taking were associated with younger ages. Differences in self-administration were not a byproduct of general activity. Neither activity in the circular corridors nor activity in the test chambers was related to the amount of self-administration. Initial experience in self-administration also failed to significantly influence subsequent self-administration, as the first dose selected was unrelated to the amount of drug infused. Not surprisingly, initial behavior related to self-administration was predictive of individual differences in drug taking. The latency to commence self-administration was negatively correlated with drug taking in that rats with shorter latencies to self-administer took more amphetamine. Self-administration on day 1 also significantly predicted drug taking over the 10-day test interval.

Considered together, these results indicate that, similar to human individual differences in vulnerability to addiction, about 10% of the rats showed high levels of abuse. These differences could be categorized along two dimensions: the amount of drug requested and in terms of dose preferences. These individual differences in amphetamine self-administration were related to age of the animal and initial self-administration but not to differences in activity or initial contact with the drug. Two conclusions are suggested. First, rats appear to be good models for studying human vulnerability to addiction. They show a wide range of individual differences in drug taking and, with repeated drug contact, a similar percentage go on to show substantial abuse. Second, it seems most parsimonious to attribute these observed individual differences to some sort of innate biological predisposition that varies with age.

Table 4–1. Summary of the Relationship between Predictor and Performance Variables

Predictor Variables	Average Total Drug Infused per Day*
Age (days)	−.2106
	.0107
	146
Activity counts in circular corridor	.1132
	.1739
	146
Activity before first drug request (day 1)	−.0382
	.6672
	129
First drug dose selected (day 1)	.0318
	.7034
	146
Latency to first drug request (day 1)	−.1634
	.0487
	146
Total drug infused (day 1)	.5602
	.0000
	146

*Values given as Pearson correlations, two-tailed P values, and No. of observations.

CONCLUSIONS

Variability in the response to drugs, lesions, or various other experimental manipulations are frequently observed in animals, but they are typically acknowledged as simply representing experimental error along with the natural range of variation occurring between individuals of the same species. Although this traditional attitude does have merit, results of the investigations presented in this chapter suggest that much within-experiment variability consists of individual differences. As illustrated in this chapter, exploiting these individual differences has a potentially large payoff in terms of modeling the behavioral and physiological differences observed between humans as well as providing a means of exploring the factors that control the expression of such differences. In addition, because behavioral and neurochemical differences are preexisting, it is unnecessary to induce differences by using artificial means such as lesions, pharmacological agents, or radical changes in the environment. Thus, using an individual differences approach, the behavioral and neurochemical sequelae of these differences may be investigated in intact organisms. It should be noted that individual differences are being identified in many different tasks. A current search on PubMed using the keywords "rat," "individual differences," and "behavior" returned more than 300 articles.

ACKNOWLEDGMENTS

The author wishes to thank the following for their assistance with some of the experiments reported in this chapter: including Carrie L. Van Brunt, Rachel Chase, Mary Houts, Pat Le Duc, Paul Rushing, Peter Pierre, and Paul Skjoldagger. The self-administration experiment was supported by NIDA grant 1R29DA07517.

Dr. William Marks, a cognitive psychologist and long-time supporter of the Neuroscience Program at the University of Memphis, died unexpectedly while this chapter was being written. This work is dedicated to his memory.

REFERENCES

American Psychiatric Association (1994) Diagnostic and statistical manual of mental disorders, 4th edition. Washington, DC: American Psychiatric Association.

Antelman SM and Szechtman H (1975) Tail pinch induces eating in sated rats which appears to depend on nigrostriatal dopamine. Science 189:731–733.

Bachus SE and Valenstein ES (1979) Individual behavioral responses to hypothalamic stimulation persist despite destruction of tissue surrounding the electrode tip. Physiology and Behavior 23:421–426.

Cox VC and Valenstein ES (1969) Distribution of hypothalamic sites yielding stimulus-bound behavior. Brain Behavior and Evolution 2:359–376.

Creese I and Iversen SD (1975) The pharmacological and anatomical substrates of the amphetamine response in the rat. Brain Research 83:419–436.

Dantzer R, Terlouw C, Tazi A, Koolhaas JM, Bohus B, Koob GF, Le Moal M (1988) The propensity for schedule-induced polydipsia is related to differences in conditioned avoidance behaviour and in defense reactions in a defeat test. Physiology and Behavior 43:269–273.

De Wit H, Uhlenhuth EH, Johanson CE (1986) Individual differences in the reinforcing and subjective effects of amphetamine and diazepam. Drug and Alcohol Dependence 16:314–360.

Everitt BJ, Diskinson A, Robbins TW (2001) The neuropsychological basis of addictive behavior. Brain Research Brain Research Reviews 36:129–138.

Falk JL (1961) Production of polydipsia in normal rats by an intermittent food schedule. Science 133:195–196.

Falk JL (1966) The motivational properties of schedule-induced polydipsia. Journal of the Experimental Analysis of Behavior 9:19–25.

Falk JL (1969) Conditions producing psychogenic polydipsia in animals. Annals of the New York Academy of Science 157:569–593.

Falk JL (1971) The nature and determinants of adjunctive behavior. Physiology and Behavior 6:577–588.

Fibiger, HC, Phillips AG, (1986) Reward, motivation and cognition: psychobiology of meso-telencephalic dopamine systems. In: Handbook of physiology: higher neural functions (Bloom FE and Plum F, eds.), pp. 647–675. Washington, DC: American Physiological Society.

Koob GF and Bloom FE (1988) Cellular and molecular basis of drug dependence. Science 242:715–723.

Mittleman G and Valenstein ES (1981) Strain differences in eating and drinking evoked by electrical stimulation of the hypothalamus. Physiology and Behavior 26:371–378.

Mittleman G and Valenstein ES (1984) Ingestive behavior evoked by hypothalamic stimulation and schedule-induced polydipsia are related. Science 224:415–417.

Mittleman G, Castaneda E, Robinson TE, Valenstein ES (1986) The propensity for non-regulatory ingestive behavior is related to differences in dopamine systems: behavioral and biochemical evidence. Behavioral Neuroscience 100:213–220.

Mittleman G and Valenstein ES (1985) Individual differences in non-regulatory ingestive behavior and catecholamine systems. Brain Research 348:112–117.

O'Brien CP, Ehrman, RN, Terns JN (1986) Classical conditioning in human opioid dependence. In: Behavioral analysis of drug dependence (Goldeberg SR and Stolerman IP, eds.), pp. 329–338. London: Academic Press.

Piazza PV, Mittleman G, Deminiere JM, Le Moal M, Simon H (1993) Relationship between schedule-induced polydipsia and amphetamine self-administration: individual differences and the role of experience. Behavioural Brain Research 55:185–194.

Piazza PV, Demeniere JM, Le Moal M, Simon H (1989) Factors that predict individual vulnerability to amphetamine self-administration. Science 245:1511–1513.

Robbins TW and Koob GF (1980) Selective disruption of displacement behavior by lesions of the mesolimbic dopamine system. Nature (London) 285:409–412.

Roper TJ (1981) What is meant by the term "schedule-induced" and how general is schedule induction? Animal Learning and Behavior 9:433–440.

Sabeti J, Gerhardt GA, Zahniser NR (2003) Individual differences in cocaine-induced locomotor sensitization in low and high cocaine locomotor-responding rats are associated with differential inhibition of dopamine clearance in nucleus accumbens. Journal of Pharmacology and Experimental Therapeutics 305:180–190.

Schuster CR and Thompson T (1969) Self-administration and behavioral dependence on drugs. Annual Review of Pharmacology 9:483–502.

Segal DS and Kuczenski R (1987) Individual differences in responsiveness to single and repeated amphetamine administration: behavioral characteristics and neurochemical correlates. Journal of Pharmacology and Experimental Therapeutics 242:917–926.

Substance Abuse and Mental Health Services Administration (2003) Results from the 2002 National Survey on Drug Use: National Findings. Rockville, MD: Office of Applied Studies, NHSDA Series H-22, DHHS publication No. SMA 03-3836.

Tinbergen N (1952) "Derived" activities: their causation, biological significance, origin, and emancipation during evolution. Quarterly Review of Biology 27:1–32.

Valenstein ES (1969) Behavior elicited by hypothalamic stimulation: a prepotency hypothesis. Brain Behavior and Evolution 2:295–316.

Valenstein ES (1975) Brain stimulation and behavior control. In: Nebraska Symposium on Motivation (Cole JK, Sonderegger TB, eds.), pp. 251–292. Lincoln: University of Nebraska Press.

Valenstein ES, Cox VC, Kakolewski JW (1970) A reexamination of the role of the hypothalamus in motivation. Psychological Review 77:16–31.

Wallace M and Singer G (1976) Schedule induced behavior: a review of its generality, determinants and pharmacological data. Pharmacology Biochemistry and Behavior 5:483–490.

Wallace M, Singer G, Finlay J, Gibson S (1983) The effects of 6-OHDA lesions of the nucleus accumbens septum on schedule-induced drinking, wheel running and corticosterone levels in the rat. Pharmacology Biochemistry and Behavior 18:129–136.

Weeks JR (1962) Experimental morphine addiction: method for automatic intravenous injections in unrestrained rats. Science 138:143–144.

Wetherington CL (1982) Is adjunctive behavior a third class of behavior? Neuroscience and Biobehavioral Reviews 6:329–350.

Wise RA (1971) Individual differences in the effects of hypothalamic stimulation: the role of stimulation locus. Physiology and Behavior 6:569–572.

Sensory Systems

Vision

GLEN T. PRUSKY AND ROBERT M. DOUGLAS

5

In some ways the rat is an odd choice as a laboratory animal. For one, the rat is nocturnal, whereas most laboratories operate on a diurnal schedule with testing done in relatively bright light. Second, the rat evolved in a crowded ground environment of forest underbrush, wetlands, grassy fields, and underground burrows. Such places are as rich in olfactory, auditory, and tactile information as they are in visual information. Laboratory conditions are the opposite: open, uncrowded, and clean. Testing boxes and rooms thus provide an unnatural visual environment and a minimum of other sensory information. The consequences for vision and visually guided behavior are not obvious, and the situation has not been helped by the limited experimental data about rat vision. Despite an early start by Lashley and others, the psychophysical analysis of rat vision has lagged as vision researchers focused their efforts on larger mammals like cats and primates. This relative state of ignorance has lead to two widespread but common misconceptions that we wish to correct in this chapter. First, many vision scientists believe the rat is almost blind and thus is a poor choice as an experimental model of vision. In fact, the rat has a typical mammalian visual system that functions quite well, and rats can learn demanding visual tasks rivaling those used in primates. Second, those interested in animal cognition often assume that the rat sees what a human sees, or that all rats see equally. Differences between stimuli may be obvious to the experimenter, but they may not be so for a rat, let alone for different strains

of rats. Furthermore, even if visual information is available in an experimental setting and would be used by a primate in a task like reaching, the visual information may not necessarily be used (e.g., Whishaw and Tolmie, 1989). In an effort to place rat vision in a sensible experimental setting, we discuss in this chapter some of the methodologies used to measure rat vision, the known visual capabilities of rats, and the implications of rat vision for those researchers who use rats in vision-based experiments.

EXPERIMENTAL METHODS FOR MEASURING RAT VISION

Lashley's jumping stand may have been the first method used to quantify rat vision (Lashley, 1930; Seymoure and Juraska, 1997), and it is still used to a limited extent. Y-mazes (Seymoure and Juraska, 1997), conditioned aversion (Dean, 1978), and operant tasks (Keller et al., 2000; Jacobs et al., 2001) have all been used with some success, but in general, these methods require a considerable amount of time to train and test rats, which probably accounts for their limited popularity. Some experimenters have also used a modification of the Morris water task, in which rats learn to swim to a platform that is raised above the water's surface (Morris et al., 1982). However, viewing distances are hard to control in this situation, making quantitative measurements nearly impossible, and the task confuses visual detection with visual acuity (this issue is dealt with in more detail later in the chapter). On the other hand, swimming to an exit platform and using distant visual cues seem to come naturally to the rat and prompted us to develop the *visual water task* (Prusky et al., 2000) to psychophysically measure visual discrimination thresholds. As shown in Figure 5–1, a

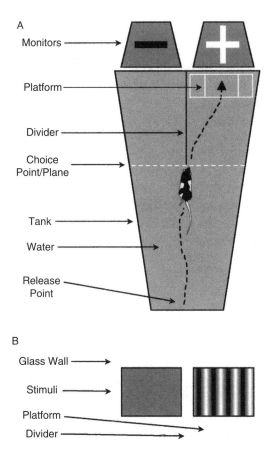

Figure 5–1. Visual water task. (A) Top view. Apparatus consists of a trapezoidal tank containing water with a midline barrier creating a Y-maze. Two monitors face into the arms of the maze and display either a + (reinforced) or a − (nonreinforced) stimulus. A platform is always submerged below the + stimulus regardless of its left/right position. Rats are released into the pool at the narrow end and then swim to the divider. They inspect each picture from this vantage and then choose to swim toward one of them. If they select the + stimulus, they are rewarded quickly with escape from the water and the trial is scored as correct. If they choose the − stimulus, they are obligated to swim until they find the escape platform on the opposite side, and the trial is scored as incorrect. The path of a rat executing a correct response is illustrated. (B) Front view of apparatus and visual stimuli configured for typical testing of visual acuity; gray Vs grating.

trapezoidal-shaped tank is made into a Y-maze with a central divider and with computer monitors placed behind a glass wall at the end of each arm. A platform is submerged below a positive stimulus displayed on one of the two monitors. Once rats are taught that the images on the screens are clues to the location of the platform, the animals stop at the end of the divider and inspect both screens before making their choices. At no time have we had to explicitly reinforce this behavior; rats spontaneously stop and compare the screens before taking a chance on going the wrong way. The water aids in dispersing odor trails and focuses the rat's attention on the computer monitors without generating a great deal of stress. On land, rats have a rich array of sensory inputs to consider, but when in water, rats seem to know that vision is the best modality to use and that the visual cues will be some distance away. Besides the task exploiting ecologically relevant behavior, the use of computer-generated stimuli is also advantageous. The computer monitors allow stimuli to be presented that are impossible to produce with printed cards; we have used stimuli that consist of moving patterns and varying contrasts with a wide range of values. In addition, computer control permits the automatic interleaving of stimuli and animals, which greatly increases throughput in the laboratory. Perhaps the major advantage of the *visual water task* is that the training and testing of visual thresholds can be completed much faster than with the other methods cited above.

THE RAT EYE

The small eye of the rat is reasonably efficient at gathering light, but it has relatively poor optics. The rat retina appears homogeneous and lacks a fovea or area centralis. The retina contains both rods and cones, with the proportion of cones being about 1% (LaVail, 1976). The cone system is often overlooked in rats;

however, almost all behavioral testing reported in rats has taken place in photopic (laboratory lighting) conditions in which the rods are not functioning. Rats are typical mammalian dichromats with short-wavelength cones comprising about 10% of the total cone population (Szel and Rohlich, 1992). There is a twist, however: the short-wavelength cone is most sensitive in the ultraviolet, with the peak at 359 nm (Fig. 5–2A). The mid-wavelength cone has peak sensitivity at 510 nm, and there is little sensitivity beyond 650 nm. The evolutionary significance of the ultraviolet sensitivity is not known, but in a series of elegant experiments, Jacobs et al. (2001) have shown that rats can make color discriminations using their two cone systems.

SPATIAL VISION

Acuity is the most common measure of vision and is the smallest spatial pattern that can be resolved. Distinguishing between two small dots and one large one, for example, depends on having adequate acuity. Most pigmented rats have acuities of 1.0 to 1.1 cycles per degree (c/d) when measured with vertical or oblique gratings (Dean, 1978; Burch and Jacobs, 1979; Keller et al., 2000; Prusky et al.,

2000). With horizontally oriented gratings, their acuity is 1.4 c/d (Bowden et al, 2002). This is much lower than normal human visual acuity (20:20) of 30 c/d. However, humans have a highly specialized fovea, and the rat acuity is not greatly dissimilar from that of human peripheral retina. In addition, much of the world that interests a rat will be viewed at close distances and they likely see these objects in considerable detail (see Figs. 5–3 and 5–4 for examples).

Humans can detect single elements smaller than the acuity limit if they have sufficiently high contrast. For example, we can see stars even though they subtend angles smaller than 1 minute of arc, which corresponds to 30 c/d. The brightness is averaged over the larger area and the element is seen as blurred, covering the wider area with a lower contrast. Humans also have good contrast sensitivity (>100, corresponding to <1% difference between light and dark), so we can detect the presence of very small single elements. Although rats have an inverted U-shaped contrast sensitivity curve characteristic of all vertebrates, the peak sensitivity is lower than that of humans (Keller et al., 2000). As shown in Figure 5–2B, rats have a peak sensitivity of about 25, meaning they can detect a 4% difference in a pattern of 0.1 to 0.2 c/d.

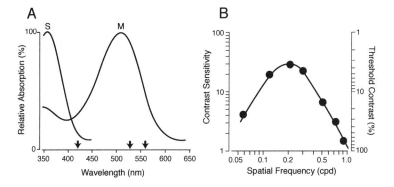

Figure 5–2. (A) Relative sensitivities of the two cone pigments in the rat. The short-wavelength pigment (S) has a sensitivity that peaks in the ultraviolet range. This is quite different from the human short-wavelength pigment (*leftmost arrow*). The more abundant medium wavelength (M) pigment has a sensitivity more similar to that of human M and L wavelength pigments. (Redrawn from Jacobs et al. [2001]). (B) Contrast sensitivity for Long-Evans rats. Contrast sensitivity is the reciprocal of the threshold contrast (*right side*) at each spatial frequency. Peak sensitivity is about 0.2 c/d, and the acuity cutoff is just above 1.0 c/d.

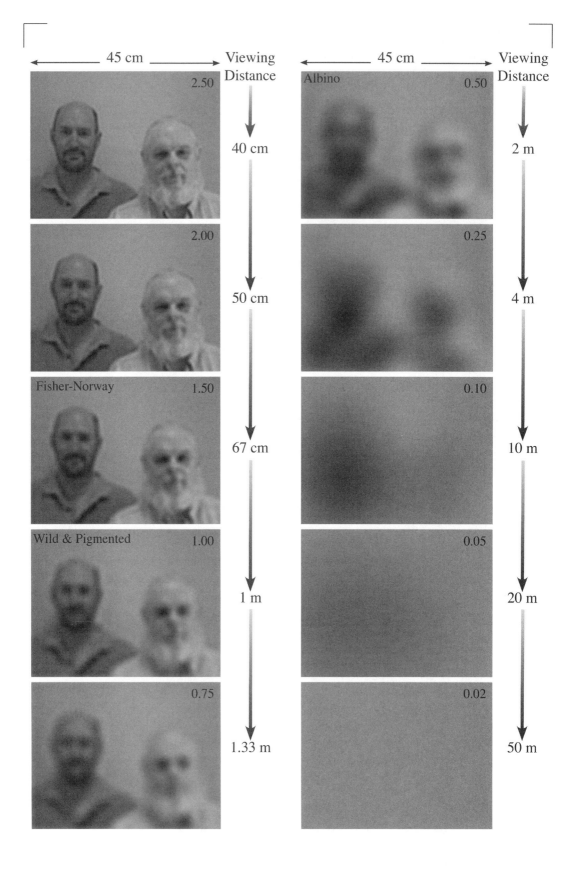

45 cm   Viewing Distance

45 cm   Viewing Distance

2.50   40 cm

2.00   50 cm

Fisher-Norway   1.50   67 cm

Wild & Pigmented   1.00   1 m

0.75   1.33 m

Albino   0.50   2 m

0.25   4 m

0.10   10 m

0.05   20 m

0.02   50 m

Figure 5–3. It is difficult to judge whether a rat can see a visual cue in a behavioral task because the visibility of the cue depends on its size, its distance from the rat, and the acuity of the rat. This series of 10 images of the authors (G.P. on left; R.D. on right; width of image is 45 cm) can be used in two ways. First the images model what animals with different acuities would see at a distance of 1.0 m. The acuities (in c/d) are shown in the top right corner of each image. The images have been filtered to emulate different acuities by removing all spatial frequencies above the frequency listed. Frequencies near the acuity limit are modeled as an attenuation. Second, the series of images can also be used to gauge what a normal wild or pigmented rat would see at different distances (viewing distances are shown at the right of each image). For animals with other acuities, the figure can be used in the same way by simply multiplying the distances by the acuities. For example, an albino rat with an acuity of 0.5 c/d will see at 50 cm what a normal rat sees at 1.0 m. Conversely, a Fisher-Norway rat with an acuity of 1.5 c/d can be seen at 1.5 m and see the same as a normal rat at 1.0 m.

Although rats have relatively low-resolution retinas, subsequent neural processing appears to make excellent use of the information provided by the retina. For example, we have found that rats can discriminate gratings that differ in orientation by less than 3° (Bowden et al., 2002). We have also obtained evidence for rat hyperacuity in a Vernier acuity task. Hyperacuity in orientation and Vernier discriminations is thought to reflect the specialized processing of visual cortical circuitry. We also have evidence that rats can detect motion coherence in a field of moving dots; a capability that in primates is thought to be due primarily to extrastriate processing (Neve et al., 2002). The coherence thresholds for rats

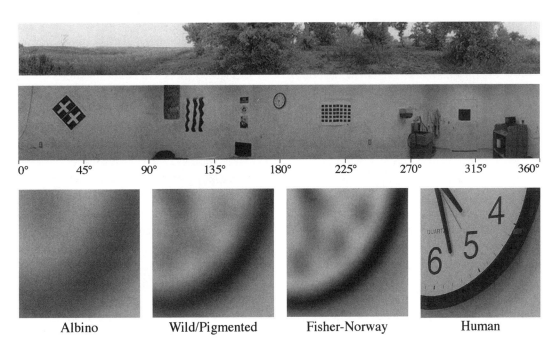

Figure 5–4. The visual environment for rats in the laboratory is often radically different from what they would experience in nature. The top picture is a 360 degree panorama of an outdoor setting where rats might live. The picture has been filtered as in Figure 5–3 to model that spatial frequencies that a wild rat can see. Note that despite its low acuity, when one considers the large visual field, there is a considerable amount of visual information available to the animal. The middle panel is a 360 degree panorama of a Morris water maze testing room at the Canadian Centre for Behavioural Neuroscience, filtered as above and taken from the center of the pool. Although the large, high-contrast features of the cues that have been placed on the walls are visible to the rat, many fine details are not. For example, there is a clock on the wall in the middle of the panel that is visible as an oval, but the interior details are blurred. The degree of blurring of the clock for different strains of rats (Albino, wild/pigmented, Fisher-Norway) and humans (30 c/d; 20/20 Snellen acuity) is presented in the bottom row of pictures. The degree of visual detail available to any rat is much less than what a human experimenter could see, but there is also significant variability between rat strains. Also note the reduced contrast of the heavy black rim for the albino rat, which will make even large objects like the clock less useful as navigation cues for these animals.

in this task are higher than those of humans, possibly reflecting the much greater cortical processing possible in the human brain, but it is still remarkable that rats can even do the task and detect motion coherence when as few as 25% of the dots in a display move in one direction. Besides having this "global" motion perception system, rats also have a more localized system for detecting the motion of individual dots. The optimal dot displacement of 2° for the rat is close to that of 1.0° for humans (Braddick, 1980).

In summary, the rat visual system is specialized for low-light and low-resolution vision. However, it has a typical mammalian contrast sensitivity function, dichromatic color vision, and a visual cortical system capable of precise spatial and temporal analysis. In addition, various visual thresholds can be psychophysically quantified in the visual water task as efficiently, or more efficiently, than those thresholds can be generated in primates.

IMPLICATIONS OF RAT VISION FOR INVESTIGATORS

The laboratory rat has a reputation among many vision scientists as an animal with poor vision. As a consequence, the rat has not been widely adopted as an animal model for investigating the structure and function of the mammalian visual system. In some ways, this reputation is deserved; the acuity and contrast sensitivity of the laboratory rat are lower than that of other popular models, such as cats and ferrets, and the rat does not have trichromatic color vision, a fovea, or the elaborate functional divisions of the lateral geniculate nucleus, striate, and extrastriate cortex that do many primates, including humans. These facts alone, however, do not fully explain why rats as models of mammalian vision have not been widely embraced, because the rat model has many other merits that would be of great value for studying the nature of mammalian vision. Instead, the lack of simple behavioral

techniques to psychophysically quantify rat vision has probably curtailed studies of vision more than anything else. The recent development of the visual water task to readily measure rat spatial vision, motion sensitivity, orientation sensitivity, and so on will undoubtedly increase the popularity of rats as fundamental models of mammalian vision.

In contrast to the relative minor use of rats in specific vision-based studies, rats are commonly used as subjects in visuobehavioral tasks to investigate the neural substrates of mammalian cognitive function. Although most would agree that reduced visual competence in a rat should negatively influence the measurement of its cognitive function in these tasks, little attention has been paid to this issue. Again, this is likely due to the historical lack of simple and fast behavioral tasks to measure vision in control experiments. Inbreeding and transgenic manipulations in rats have created many strains with desirable traits for experimental brain research. Among these are strains with mutations that affect the visual system specifically and, as such, provide valuable models of visual system diseases or offer experimental advantages for studying the structure and function of the visual system. Some mutations, such as albinism, however, can negatively affect the visual system, whereas unique genetic combinations in some strains may augment normal visual function.

These and other characteristics of the laboratory rat visual system have major implications for experimental studies of the mechanisms of mammalian vision and for using rats as models for experimental brain research in general. Here we provide three examples of such implications from our work.

EFFECTS ON VISION OF RAT DOMESTICATION AND INBREEDING

To assess the effects of rat domestication and inbreeding on visual function, we measured the visual acuity of six different strains (inbred, outbred, and albino) of laboratory rats and compared their acuity with that of wild rats. We found no significant effect of inbreeding or outbreeding on acuity; however, we did find that all albino strains had acuity measurements about half those of wild rats.

Albinism produces a number of structural abnormalities in the visual system, including neuroretinal abnormalities resulting from nonpigmented retinal pigment epithelium (Jeffery, 1998), abnormal decussation of retinal ganglion cell axons at the optic chiasm (Lund et al., 1974), and abnormal interhemispheric connections of the visual cortex (Abel and Olavarria, 1996). The most likely explanations for reduced visual acuity in albino rats are that excessive light scattering within the retina (Abadi et al., 1990) make the albino eye a rather poor image-forming device and that light-induced retinal degeneration (Birch, 1977) results in poor spatial sampling. It is also possible that deficient central visual processing plays a role in the acuity deficits of albino rats. The absence of melanin or the melanin-related agent (Rice et al., 1999) responsible for anomalous axonal decussation at the optic chiasm in albinos (Jeffrey, 1997) may also produce errors of interhemispheric connectivity of the visual cortex (Abel and Olavarria, 1996) and result in anomalous visual cortical processing. Given the numerous visual deficits in albino rat strains, it is almost certainly wise to avoid using them in behavioral experiments that depend on vision.

Surprisingly, our study also found that one pigmented strain, Fisher Norway, had a significantly higher acuity than that of wild rats, as well as other pigmented and albino strains tested: their grating threshold was approximately 50% higher than that of other pigmented strains and 150% higher than that of the albino strains. The most likely explanation for this finding is that a genetic difference accounts for their enhanced acuity. Fisher-Norway rats are an F_1 cross between an inbred Fisher 344 female and an inbred Brown Norway male. Although we did not measure the acuity of the Brown Norway strain in this

study, we did measure the acuity of Fisher 344 animals and found their acuity to be about 0.5 c/d. It is possible that the Brown Norway strain possesses acuity higher than other pigmented strains in our study, and that the Fisher-Norway animals owe their high acuity to genes present in the Brown Norway genome. The superior visual acuity of the Fisher-Norway strain also raises the possibility that the acuity of 1.0 c/d we measured for wild rats is lower than that of native *Rattus norvegicus* or that there is substantial heterogeneity in the visual acuity of wild rats.

It is also possible that Fisher 344 and Brown Norway strains carry alleles that are deleterious for high-resolution vision within the strains, but the unique combination of genes in Fisher 344/Brown Norway heterozygotes results in alleles that are beneficial to visual acuity. This may produce animals with enlarged eyes, smaller receptive fields in the visual cortex, or other structural changes that could lead to higher-resolution vision. As a consequence, the Fisher-Norway strain may be an attractive model for studying the mechanisms of visual perception in nocturnal rodents.

Figure 5–3 graphically illustrates how the perception of a normal pigmented rat varies as a function of distance. The figure can also be used to compare the relative visual perception of rat strains with different acuities.

A recent study has shed some light on the implications of variation in rat visual acuity for performance in the Morris water task (Harker and Whishaw, 2002). When combined with the results of our study, the data indicate that visual acuity alone does not predict performance on place or matching-to-place versions of the Morris water task (Harker and Whishaw, 2002). For example, despite having superior (Fisher-Norway) or equal visual acuity (Dark Agouti) relative to the Long-Evans strain, both of these strains were relatively impaired. Even among albino strains whose visual acuity did not differ in our study, Harker and Whishaw (2002) reported that there are significant strain differences in performance. These data do not rule out the possibility that visual acuity can influence rat behavioral performance in vision-dependent tasks, because all of the albino strains with poor visual acuity (Sprague-Dawley, Wistar, Fisher 344) in Harker and Whishaw's (2002) study were impaired relative to the Long-Evans (pigmented) strain. Therefore, it is likely that rat strains vary in a number of brain functions, including visual function, and this variation can contribute to differences in performance on complex behavioral tasks.

SPATIAL LEARNING

The initial experiments that used the Morris water task (Morris et al., 1982) to measure place learning and memory anticipated that interpretations of the data could be confounded by abnormal visual, motor, or motivational function. One configuration of the task that was developed to control for these noncognitive factors, including reduced visual function, was the *cued-platform task*. The rationale of this task was that if animals could not learn to swim directly to a visible platform in the pool, they probably had noncognitive, possibly visual, impairments and should be excluded from further study. Conversely, it was reasoned that animals that could swim directly to the cued platform were relatively free of these deficits, and therefore, their performance in a place configuration of the water maze most likely reflected their cognitive ability to learn and remember the platform location in space.

Recently, however, we performed a study that challenges the ability of the cued-platform task to detect animals with visual deficits sufficient to affect place learning (Prusky et al., 2000c). First, we binocularly deprived rats during the critical period for visual plasticity, which reduced their adult visual acuity by about 30%. We then measured the place learning of these animals in the Morris water task and found that they had a significant impairment in learning

the task. These same animals, however, were not impaired in their ability to locate a cued platform. These data indicate that place learning deficits should be expected in animals with 30% to 100% reductions in visual acuity (blind animals are known to be impaired in place learning). However, screening animals with 30% reduced acuity is not possible using a typical cued platform version of the Morris water task.

The major limitation of the cued-platform task for identifying animals with visual deficits is that the task does not accurately measure the visual function required for accurate place learning. For example, any image can be analyzed as if it is formed by a set of sine waves of different spatial frequencies, and the receptive field organization of mammalian retinal and cortical cells appears to measure the magnitude and location of the different spatial frequency components. The ability to identify cues and locate a place in visual space is limited by sensitivity to the highest spatial frequencies. In a typical cued platform task, the single platform cue appears dark against a white pool wall. The single object can be decomposed into many frequencies, any of which can be used to locate the platform. That is, in the absence of distractors or multiple visual cues that must be discriminated, detection of a single large, high-contrast cue may be possible with vision consisting of only the lowest spatial frequency detectors. Moreover, performance in the cued platform task is not an accurate measure of visual acuity because the cue is usually too large and there is little control of the viewing distance. Even from the farthest viewing position in a 1.5 m pool, a 10 cm cue will subtend about 5° of visual angle, corresponding to a grating acuity of 0.1 c/d, an order of magnitude lower than the 1.0 c/d that normal rats can see.

Considering the relatively low visual acuity of normal rats ($\approx$1.0 c/d), it is also possible that apparent cognitive impairments in the Morris water task could be the result of insufficient visual cueing in the test room. The overall size, contrast, illumination, and stability of the cues should be considered as well as their relationship to the animal as it swims about the pool; stable distal cues should subtend more than $1°$ of visual angle after considering the viewing positions from within the pool. As a guide to experimenters in selecting visual cues, Figure 5–3 shows what a normal rat can see of a well-known object, human heads, at a variety of distances. Figure 5–4 models what rats see in a typical laboratory setting for the Morris water maze.

The visual competence of rats is pertinent to the interpretation of data from water maze studies or from other visual based behavioral tasks where visual function may be compromised. For example, a study by Lindner et al. (1997) reported that blind rats performed better than atropine-treated animals in place learning, but the groups could not be differentiated in a cued platform task. Other studies have correlated water maze performance with the loss of photoreceptors in aging Sprague-Dawley rats (e.g., Osteen et al., 1995). It is possible that visual deficits contributed to deficient place learning in these studies, however, the only way this uncertainty can be resolved is if the visual function of atropine-treated animals or animals with retinal degeneration is measured independently.

VISUAL PLASTICITY

A critical period for visual plasticity early in life in which the nature of visual experience shapes the structure and function of the developing visual system was first delineated by Wiesel and Hubel (1970). Although this demonstration in cats, and later in monkeys, provided a model for how abnormal visual experience in developing humans can lead to amblyopia, the lack of simple psychophysical measures of cat and monkey vision meant that most studies of visual cortical plasticity that followed used electrophysiological measures of visual function, such as cortical ocular dominance. This is surprising because the best clinical measure of human amblyopia is visual

perception, and the relationship between ocular dominance and vision is correlational, not causal (Murphy and Mitchell, 1987). Cats and monkeys also have a number of limitations that are absent in the rat. For example, in the rat there is easy access to the whole of visual cortex, the cortex is flat, there are large stereotaxically defined monocular and binocular regions of primary visual cortex, and an abundance of biochemical and molecular tools are available to investigate cellular function in the rat. In addition, the rat shows a developmental ocular dominance plasticity that is fundamentally the same as that in other mammals (Fifkova, 1968; Fagiolini et al., 1994). We used the visual water task to demonstrate that impoverishment of the visual environment during the rat's critical period leads to permanent amblyopia (Prusky et al., 2000b). Ocular dominance experiments in cats and primates have virtually taken for granted that interocular competition is the mechanism of developmental visual plasticity. Our behavioral results showing large deficits from binocular deprivation suggest that this simple story is incomplete and that there are additional processes that contribute to the development of amblyopia. The rat may also be one of the best species in which to study these factors: it has the advantages listed earlier; in addition, there are many transgenic and inbred strains available, rats are easily housed and handled, rats have precisely tuned visual cortical receptive fields (Girman et al., 1999), and we now have the ability to quickly measure its visual function. In fact, the rat may be the best choice to intergrate the results of many diverse methodologies to study the mechanisms of how visual experience is translated into mature visual function.

CONCLUSIONS

The development of the visual water task has enabled researchers to rapidly and accurately quantify the vision of rats. Rats have a typically mammalian visual system specialized for low-resolution vision, and many of the visual capabilities of rats are surprisingly good. Therefore, the rat visual system is a very good model for studying the mechanisms of mammalian vision. In addition, it is important to consider the unique visual capabilities of rats and strain differences in rat vision when designing behavioral experiments that depend on vision.

REFERENCES

Abadi R, Dickinson CM, Pascal E, Papas E (1990) Retinal image quality in albinos. A review. Ophthalmic Paediatrics and Genetics 11:171–176.

Abel PL and Olavarria JF (1996) The callosal pattern in striate cortex is more patchy in monocularly enucleated albino than pigmented rats. Neuroscience Letters 204:169–172.

Bowden WF, Douglas RM, Prusky GT (2002) Horizontal bias in rat visual acuity. Program No. 260.18. Abstract Viewer/Itinerary Planner. Washington, DC: Society for Neuroscience. Online.

Braddick OJ (1980) Low-level and high-level processes in apparent motion. Philosophical Transactions of the Royal Society of London. Series B Biological Sciences 290:137–151.

Birch D and Jacobs GH (1977) Effects of constant illumination on vision in the albino rat. Physiology and Behavior 19:255–259.

Birch D and Jacobs GH (1979) Spatial contrast sensitivity in albino and pigmented rats. Vision Research 19:933–937.

Dean P (1978) Visual acuity in hooded rats: effects of superior collicular or posterior neocortical lesions. Brain Research 156:17–31.

Fagiolini M, Pizzorusso T, Berardi N, Domenici L, Maffei L (1994) Functional postnatal development of the rat primary visual cortex and the role of visual experience: dark rearing and monocular deprivation. Vision Research 34:709–720.

Fifkova E (1968) Changes in the visual cortex of rats after unilateral deprivation. Nature 220:379–381.

Girman SV, Sauve Y, Lund RD (1999) Receptive field properties of single neurons in rat. primary visual cortex. Journal of Neurophysiology 82:301–311.

Harker KT and Whishaw IQ (2002) Impaired spatial performance in rats with retrosplenial lesions: importance of the spatial problem and the rat strain in identifying lesion effects in a swimming pool. Journal of Neuroscience 22:1155–1164.

Hughes A (1977) The refractive state of the rat eye. Vision Research 17:927–939.

Jacobs GH, Fenwick JA, Williams GA (2001) Cone-based vision of rats for ultraviolet and visible lights. Journal of Experimental Biology 204(Pt 14):2439–2446.

Jeffery G (1997) The albino retina: an abnormality that provides insight into normal retinal development. Trends in Neurosciences 20:165–169.

Jeffery G (1998) The retinal pigment epithelium as a developmental regulator of the neural retina. Eye 12:499–503.

Keller J, Strasburger H, Cerutti DT, Sabel BA (2000) Assessing spatial vision-automated measurement of the contrast-sensitivity function in the hooded rat. Journal of Neuroscience Methods 97:103–110.

Lashley KS (1930) The mechanism of vision: I. A method for rapid analysis of pattern vision in the rat. Journal of General Psychology 37:453–460.

LaVail MM (1976) Survival of some photoreceptors in albino rats following long-term exposure to continuous light. Investigative Ophthalmology and Visual Science 15:64–70.

Lindner MD, Plone MA, Schallert T, Emerich DF (1997) Blind rats are not profoundly impaired in the reference memory Morris water maze and cannot be clearly discriminated from rats with cognitive deficits in the cued platform task. Cognitive Brain Research 5:329–333.

Lund RD, Lund JS, Wise RP (1974) The organization of the retinal projections to the dorsal lateral geniculate nucleus in pigmented and albino rats. Journal of Comparative Neurology 58:383–403.

Morris RG, Garrud P, Rawlins JN, O'Keefe J (1982) Place navigation impaired in rats with hippocampal lesions. Nature 297:681–683.

Murphy KM and Mitchell DE (1987) Reduced visual acuity in both eyes of monocularly deprived kittens following a short or long period of reverse occlusion. Journal of Neurosciences 7:1526–1536.

Neve AR, Prusky GT, Douglas RM (2002) Perception of motion coherence in rats. Program No. 353.17. Abstract Viewer/Itinerary Planner. Washington, DC: Society for Neuroscience, 2002. Online.

O'Steen WK, Spencer RL, Bare DJ, McEwen BS (1995) Analysis of severe photoreceptor loss and Morris water maze performance in aged rats. Behavioral Brain Research 68:151–158.

Prusky GT, West PWR, Douglas RM (2000a) Behavioral assessment of visual acuity in mice and rats. Vision Research 40:2201–2209.

Prusky GT, West PWR, Douglas RM (2000b) Experience-dependent plasticity of visual acuity in rats. European Journal of Neuroscience 12:3781–3786.

Prusky GT, West PWR, Douglas RM (2000c) Reduced visual acuity impairs place but not cued learning in the Morris water task. Behavioral Brain Research 116:135–140.

Rice DS, Goldowitz D, Williams RW, Hamre K, Johnso PT, Tan SS, Reese BE (1999) Extrinsic modulation of retinal ganglion cell projections: analysis of the albino mutation in pigmented mosaic mice. Developmental Biology 21:41–56.

Seymoure P and Juraska JM (1997) Vernier and grating acuity in adult hooded rats: the influence of sex. Behavioral Neuroscience 111:792–800.

Szel A and Rohlich P (1992) Two cone types of rat retina detected by antivisual pigment antibodies. Experimental Eye Research 55:47–52.

Whishaw IQ and Tomie JA (1989) Olfaction directs skilled forelimb reaching in the rat. Behavioral Brain Research 32:11–21.

Wiesel TN and Hubel DH (1970) The period of susceptibility to the physiological effects of unilateral eye closure in kittens. Journal of Physiology (London) 206:419–436.

Somatosensation

6

LINDA HERMER-VAZQUEZ, RAYMOND HERMER-VAZQUEZ, AND JOHN K. CHAPIN

In many ways, the rat somatosensory system, at both the sensory periphery and throughout the central nervous system (CNS), is homologous to that found in other mammals, including primates. Conserved elements of this system include multiple types of cutaneous, proprioceptive, nociceptive, and thermal receptors; somatosensory afferents from the spinal cord including the dorsal column pathways, the spinothalamic and spinoreticular tracts, and the spinocervical tract; and ascension through the dorsal column nuclei, multiple thalamic nuclei including the ventral posterior lateral nucleus (VPL) (for the nonvibrissal body surface) and the ventral posterior medial nucleus (VPM) (for vibrissal information) and at least two neocortical somatosensory areas (SI and SII; Krubitzer, 1995; Paxinos, 1995). Likewise, systems for sensorimotor integration and learning, such as the basal ganglia and cerebellum, have been conserved across rats and other mammals, as have the main motor output pathways (the corticospinal, rubrospinal, vestibulospinal, and reticulospinal tracts). Additionally, there is extensive homology across rats and other mammals in terms of cytoarchitecture, cytochemistry, and the roles played by different neurotransmitters and neuromodulators (Paxinos, 1995; Aboitiz, 2001). These facts underscore the general point that in many ways rats are a good mammalian exemplar.

Nevertheless, rats are adapted to a dark, cluttered environment of underground burrows and densely wooded terrain, usually close to water, making the sensory world of the rat quite different from that of most primates, especially humans. The rat has its own foveal somatosensory system: its whiskers, perioral areas, and forepaws, as well as its keen olfactory system. As the rat moves through a new environment, it constantly whisks and sniff objects directly in front of it and uses the somatosensory receptors on the tip of its nose, in many cases, to determine where to next place its forepaw (L. Hermer-Vazquez and R. Hermer-Vazquez, unpublished data). Then it gingerly places its forepaw on the object's surface where the nose had just been, as it continues to "map" the new object using somatosensory and olfactory information. Thus, the rat relies on information that is integrated from snout and forepaw sensory receptors.

Correspondingly, the somatosensory cortical areas devoted to representation of the nose, whiskers, mouth, forepaw skin, and long, whisker-like hairs on the underside of the rat's wrists ("sinus hairs," as explained later) are greatly enlarged relative to the sensory representations of other body parts, and the receptive fields for these foveal body regions are much more sharply defined (Fig. 6–1a and b) (Chapin and Lin, 1984). Moreover,

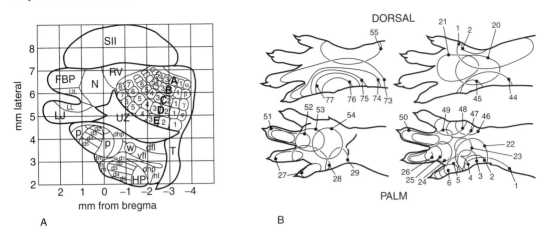

Figure 6–1. (*A*) Map of the cutaneous representation of the rat's body in S1 cortex. Note the specific cortical regions devoted to cutaneous maps of the forepaw (fp) and digits (d2-d5) and thumb (t), wrist sinus hair (w), and mystacial vibrissaw (A-E, 1-8), which together comprise the rat's somatosensory fovea. T, trunk; hl, hindlimb; HP, hindpaw; dhp, dorsal hindpaw; d1-d5, hindpaw digits from 1 to 5; hm, hindlimb muscle; vfl, ventral forelimb; dfl, dorsal forelimb; w, wrist sinus hairs; dfp, dorsal forepaw; d2-d5, forepaw digits from 2 to 5; t, thumb; uz, zone that was unresponsive during mapping; A-E, 1-8, rows (from dorsal to ventral) and numbers (from caudal to rostral) of facial whiskers; RV, rostral small vibrissae, N, nose; FBP, frontobuccal pads; UL, upper lip; LL, lower lip; LJ, lower jaw. (*B*) Receptive field centers for isolated single units in SI for the palmar surface versus the dorsal hand. Note the finer representation of the skin on the palm in SI.

the rat has two primary cortical representations of the forepaw, and one of them, now referred to as "caudal M1," contains overlapping somatosensory inputs to layer 4 and motor outputs descending from layer 5. There are also separate cortical representations for the rodent-specific somatosensory peripheral organs referred to earlier. The mystacial vibrissae are represented as separate granular aggregates (in the cortex, "barrel fields") in caudolateral SI. Cells in this region are extremely responsive to whisker manipulation but much less so to the manipulation of the underlying skin and fur adjacent to the whiskers. In contrast, the more rostral cortical representation of the perioral surface has large receptive fields extending into the peripheral cortical whisker zone, which are much more responsive to stimulation of the skin and fur. Finally, the rat has another distinctive somatosensory apparatus: the sinus hairs on its ventral wrists. These long, tiny hairs are represented cortically by their own

granular aggregates, similar to the granular aggregates that comprise the barrel fields for the whiskers. Also as with the whiskers, the skin and fur adjacent to the wrist sinus hairs are represented in a more classic manner by the larger caudal and rostral forepaw–forelimb receptive fields (Chapin and Lin, 1984).

In this chapter, we describe five principles for how sensorimotor behaviors are learned and performed by rats, based on new findings from neuroscience. We focus mainly on nonvibrissal somatosensory processing but discuss examples from the whisker-tactile system or from other sensory modalities when they illustrate a point well. Taking the necessary biological constraints into account, studies of rat somatosensory behavior are pioneering for behavioral neuroscience. Rats also make excellent subjects for experiments in that they eagerly perform movements repetitively and stereotypically, minimizing variation in how each trial is executed.

PRINCIPLE I: ANALYZED SOMATOSENSORY FEEDBACK INFORMATION IS CONSTANTLY INFLUENCING THE ASCENDING SOMATOSENSORY DATA STREAM IN RATS

Both "bottom-up" stimulus feature-based information and relatively more processed "top-down" information interact to produce responses of cells in intermediate levels of the somatosensory system such as the somatosensory thalamus. This has been repeatedly demonstrated in studies of the somatosensory receptive fields of thalamic and cortical neurons, which have made clear that both ascending sensory information from the periphery and descending corticofugal projections determine somatosensory receptive field structure. For example, Shin and Chapin (1990) found that motor cortical stimulation decreased the response time in thalamic VPL neurons to mechanical stimulation of the forepaw. Extending that line of reasoning, Krupa et al. (1999) determined the short-latency and long-latency receptive fields of thalamic VPM neurons, and then tested them again under (1) muscimol inactivation of SI cortex or (2) muscimol SI inactivation combined with lidocaine inactivation of ascending somatosensory input. Under each type of reversible chemical blockade, immediate receptive field reorganization occurred. Moreover, the findings suggested that GABAergic corticofugal feedback appeared to suppress short-latency responses caused mainly by ascending influences in many thalamic neurons, whereas corticofugal glutamatergic excitatory influences appeared to be necessary for many thalamic cells to exhibit long-latency responses. Results in rats using other sensorimotor protocols as well support the general notion that receptive fields are created by multiple ascending and descending influences and are held under a dynamic tension that allows immediate reorganization after any change in these inputs. Thus, top-down, cognitive information as well as bottom-up data streams—which, crucially, are not always predicted by the top-down processes—can be taken into account in the rat's analysis of its somatosensory world.

PRINCIPLE II: RATS ARE CONSTANTLY EVALUATING INFORMATION ACROSS MULTIPLE TIMESCALES TO MORE ACCURATELY PREDICT WHAT WILL HAPPEN IN THEIR WORLD

Consistent with the preceding data on descending influences on ascending sensory inflow, it is widely agreed that rats are constantly reevaluating the past at different timescales and combining the resulting knowledge with their current perceptions to predict the future at different timescales (Llinas, 2001). That is, they use their intellect as well as their small size and nocturnal nature to elude predators. The fact that these evaluative processes are occurring on multiple timescales complicates one's interpretation of rat behavior.

Many older models of rat information processing are based on the simplifying assumption that tasks are learned and maintained as a simple function of the number of training trials up through reaching an asymptotic performance level and degraded as a simple function of lack of practice or interference by new memories (Baddeley, 1992). This view suggested that time flows forward linearly, in a regular manner, throughout task acquisition and performance; that is, the rat is using its experience on the current trial to shape its performance on the next trial, until some maximum level is reached and maintained. For example, rats in a recent study of ours (Hermer-Vazquez et al., in press) were trained daily on the skilled reach-to-grasp-food task (Whishaw and Pellis, 1990) described earlier. On day 1 of training, the sample of rats grasped the target correctly on 27% of trials and improved daily and linearly in their performance until day 6, when their success rate began to asymptote at approximately 68% correct. Consistent with this view of task learning, it was shown that the rat caudal digit-wrist motor cortex, which contains an overlapping somatosensory representation of the forepaw (Chapin and Lin, 1984), expands and grows new synapses during the learning of this task, whereas no similar changes were detected in the (more strictly motor) rostral M1 (Kleim et al., 2002). Moreover, the degree of long-term potentiation in rat M1 synapses has been found to correlate with reaching skill acquisition (Rioult-Pedotti et al., 2000).

Nevertheless, other data suggest that the information processing relevant to task learning and performance flows in a more complex manner, as rats attempt to better understand their past to better predict the future. These processes occur at multiple timescales throughout different levels of the nervous system. For example, at the somatosensory periphery, slowly adapting versus rapidly adapting cutaneous receptors in the rat glabrous skin (Paxinos, 1995) allow simultaneous perception of different aspects of the tactile world (Johnson, 2001). At the same time, even at very low levels of the somatosensory–motor interface, multiple feedback processes are occurring. For example, the spinal stretch reflex illustrates how proprioceptive feedback is constantly influencing muscle tension. Modification of neural processing by "evaluative" feedback is seen at longer timescales as well; for example, corticothalamic feedback projections modify subcortical sensory processing, as described in the section on Principle I (e.g., Krupa et al., 1999). And at still longer time-scales, such as hours, considerable data now suggest that after rats spend a day whisking, sniffing, and manipulating objects in a novel environment, the rats reevaluate their sensorimotor performance on task from the prior day during slow-wave sleep, developing better internal models for those tasks as evidenced by the fact that performance often improves after a complete sleep cycle with no additional overt practice (Lee and Wilson, 2000; Poe et al., 2000). All of these processes,

occurring at distinctive timescales, help the rat use its past to more accurately predict its future.

PRINCIPLE III: INFORMATION FROM MULTIPLE SPATIAL SCALES IS PROCESSED SIMULTANEOUSLY IN THE RAT

Three or four decades ago, many researchers advocated the view that mammalian sensory systems represent their sensory surfaces topographically and with high resolution and that the high-resolution, body-centered information gained at the sensory surface was gradually transformed at higher levels of the nervous system by cells with progressively larger receptive fields into object-centered representations. It is now widely recognized that information is processed at both fine and broad spatial scales simultaneously. For instance, at the somatosensory surface, type 1 cutaneous receptors process mechanical somatosensory data with high spatial resolution, and type II receptors have larger, less well-defined receptive fields (Vallbo and Johansson, 1984). The principle of simultaneous processing of sensory data at multiple spatial scales holds at higher levels of the nervous system as well.

For instance, in many cases, different microcircuits in the CNS—defined by their cell types, neurochemistry, and connectivity—process information at their own, distinctive spatial scales. It is now known that in all thalamic nuclei, including the sensory relay nuclei, a matrix of calbindin-immunoreactive cells projects diffusely throughout the cortex, regardless of sensory topography or even sensory domain (Jones, 2001). In contrast, the precisely projecting, topographically ordered cells more classically associated with thalamic relay efferents stain positively for parvalbumin. Thus, for example, the relay nucleus for body somatosensory afferents, VPL, contains both parvalbumin-positive, precisely projecting, somatotopically organized afferents to layer 4 of cortical area SI, with correspondingly small re-

ceptive fields, and much more widely projecting, calbindin-staining cells whose axons target
the superficial cortical layers in multiple sensory modalities. These diffusely projecting cells
play a critical role in coordinating activity
across brain regions, particularly in view of evidence that many layer IV corticothalamic feedback neurons target these transcortically projecting matrix cells (Jones, 2002).

Likewise, at the primary somatosensory
cortical level, the forepaw and perioral areas are
represented by cells with small as well as cells
with large receptive fields (Chapin and Lin,
1984), just as cells in the septa between precisely
mapped cortical barrel fields have much larger
receptive fields (Brecht and Sakmann, 2002). Rat
somatosensory cortical and thalamic processing
therefore shares spatial-related features with
others subcortical entities such as the basal ganglia, which contains topographically more ordered regions that process information with
high spatial resolution (striosomes), as well
more diffusely organized region (matrix cells)
(Brown et al., 2002), and the cerebellum, where
distant regions of the body surface are adjacently
located in the cerebellar "mosaic" (Bower and
Woolston, 1983). Thus, either rat behavioral
tasks need to be extremely well controlled if
they are to be based primarily on the hypothesis of high-resolution, topographic processing
(e.g., by localizing perception and motion to a
small and isolated portion of the body) or the
experimenter's hypotheses need to account for
processing occurring at multiple spatial scales.

PRINCIPLE IV: RAT SENSORY AND MOTOR PROCESSING ARE CONSTANTLY INFLUENCING ONE ANOTHER

Both older and newer neuroanatomical and
neurophysiological data indicate that all levels
of the neuraxis, including the spinal cord, "ascending" sensory information, and "descending" motor information, influence each other.
For example, it has been known for many years
that at the level of spinal reflexes, proprioceptive sensory information about the stretch state
of a muscle feeds back to regulate muscle tension in rats and other mammals (Kandel et al.,
2000). Newer data demonstrate that sensorimotor interplay occurs at higher levels of the
rat neural system as well. For instance, outputs
from the rat's basal ganglia, once thought to be
a motor-learning structure, project to layer 1 of
its somatosensory cortex and other primary
neocortical areas (McFarland and Haber, 2002),
likely influencing the processing of all cortical
cells with dendrites extending into layer 1. Data
such as these indicate that sensory and motor
areas are continuously interacting via multiple,
parallel looping structures. Likewise, all "sensory" thalamic "relay" nuclei, including the VPL
(processing somatosensory information), the
medial geniculate nucleus (MGN) (processing
auditory information), and the lateral geniculate
nucleus (LGN) (processing visual data), receive
axon collaterals of layer 5 efferents from M1 and
the premotor cortex (Guillery and Sherman,
2002). A fact that is not widely appreciated is
that these axon collaterals constitute a larger
portion of inputs to the thalamic relay nuclei
than do the ascending sensory fibers! Thus, descending motor commands appear able to
modulate or even drive thalamic sensory processing. These facts are among many more
pieces of anatomical and physiological data suggesting that sensory and motor processing are
constantly modulating each other.

The constant interaction of sensory and
motor datastreams is well illustrated by rat olfactory as well as somatosensory behaviors.
For example, the strength of an odor modulates sniffing intensity, even on the first sniff
of the odor (Johnson et al., 2003). Furthermore, this modulation of sniffing intensity occurs so rapidly that it is thought that cortical
processing cannot be involved; the modulation must take place at brain stem or spinal
levels (Johnson et al., 2003). In our studies of
rats performing an olfactory-driven, reach-to-
grasp food task (Whishaw and Pellis, 1990),
the presence or absence of a food-related odor

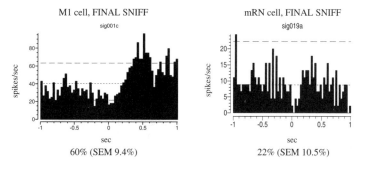

Figure 6–2. Perievent histograms for representative single cells in the rat caudal primary motor cortex (M1) and magnocellular red nucleus (mRN), centered around the final sniff of the food pellet before lifting the paw to initiate reaching. In each graph, the center horizontal line depicts the cell's firing rate, and the lines above and below it show 2 SDs from the mean (i.e., statistical significance of rate modulation). It can be seen that at the final sniff moment, each cell's firing rate is significantly depressed. The binwidth for these perievent histograms is 25 milliseconds. The percentages below each graph show the proportion of single units recorded in each area that displayed such significant modulation on final sniff.

determines whether the rat will lift its paw and guides the spatial accuracy of the reach (Hermer-Vazquez et al., unpublished data). Indeed, rats appear to sniff the target just before lifting the paw to gain the initial spatial coordinates for the impending reach and then take several more "update sniffs" during the overt arm-movement phases of the task. These behavioral observations have a corresponding neurophysiology. For instance, while recording single units from the digit-wrist area of caudal M1 and the magnocellular red nucleus as rats perform this task, we have found that many neurons in both areas that code the overt arm-movement phases are strongly modulated by olfactory information

taken in during the final sniff of the food target just before lifting the paw (Hermer-Vazquez et al., in press) (Fig. 6–2).

Somatosensory processing also appears to constantly guide the rats' reaching maneuvers. We have found that many rat M1 units are particularly active during phases of the reaching task in which somatosensory information is being evaluated, such as lifting the paw off the ground, brushing the paw against the shelf on which the food target rests on the way to the target, and contacting the food pellet itself (Hermer-Vazquez et al., in press) (Fig. 6–3). The cells whose firing rates increase as the rat's paw hits the shelf are likely responding to movement of the sinus hairs of the ventral wrist,

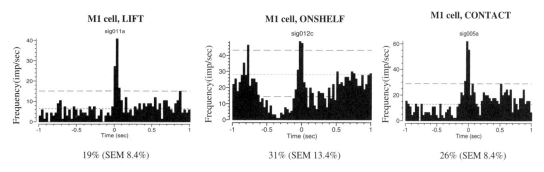

Figure 6–3. Perievent histograms for single cells recorded from the rat caudal M1 centered around each of three reaching-task events in which the cutaneous inputs to the paw change and are presumably evaluated by M1 cells: lifting of the paw off the ground (*left*), the paw making contact with the shelf (*middle*), and the paw making contact with the food pellet (*right*). Other elements of the graphs, such as the binwidth, are the same as in Figure 6–2.

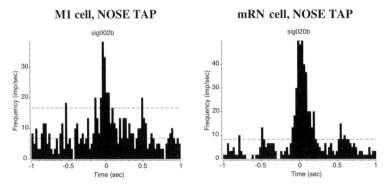

Figure 6–4. Perievent histograms for single cells recorded from the rat caudal M1 and mRN, with upper-body motor fields, that also respond strongly to cutaneous inflow from the perioral region. Other elements of the graphs, such as the binwidth, are the same as for Figures 6–2 and 6–3.

and the large percentage of cortical "on-shelf" cells suggests that this phase of the task is a major calibration point for the final approach to the target.

The role of somatosensory processing in motor cells' activity is especially clear once the task is well learned. When reaching tasks are first being learned, the latencies of spike rate increases in S1 and M1 are relatively consistent with the "data in, cognitive transformations performed, data out" task model, in that S1 units peak in their firing several tens of milliseconds before the M1 cells' peaking. In one study, the inadequacy of that model was strikingly illustrated. As animals became more proficient at the task, roughly one third of all recorded S1 units developed much longer latencies, consistent with "motor" processing, and roughly one-third of all recorded M1 cells developed early, "somatosensory" latencies (J. Chapin, unpublished data). Consistent with this fact, we have also found that many neurons in both the rat's caudal M1 and its magnocellular red nucleus respond as strongly, if not more strongly, to sensory input from the body regions corresponding to their motor fields. For example, Figure 6–4 shows representative cells from M1 and the magnocellular red nucleus (mRN) with an upper-body motor field responding to mechanical taps on the ratís nose. The sharp increase in firing rate shown by the perievent histograms at the moment of the tap, as with the other data presented earlier, supports the view that task-related sensory and motor information dynamically and continuously interact even in structures such as the rat's red nucleus that were classically thought to be "motor."

PRINCIPLE V: RAT BEHAVIORS APPEAR TO BE ORGANIZED INTO SURVIVAL-RELATED REPERTOIRES THAT CAN BE ADAPTED TO NOVEL CIRCUMSTANCES

Increasing evidence suggests that cortical, subcortical, and spinal motor circuits are organized according to synergistic activity of groups of muscles, or perhaps even according to the whole, complex movements they produce, rather than being organized somatotopically or musculotopically (Graziano et al., 2002). As a minimum, it is clear that individual cells can code for synergies of muscles rather than the movements of single muscles or joints (d'Avella et al., 2003). For example, we have found that during skilled reaching, red nucleus cells appear to code for combined limb movements and postural shifts (Hermer-Vazquez et al., in press). Moreover, new findings with primates as well as rats suggest that whole, survival-related movements, such as moving the hand toward the mouth and opening the jaw, defensively blocking objects from hitting the face, or manipulating objects in a frontal "workspace," are coded by motor-related circuits from the spinal level (Strick, 2002) to higher motor cortical regions (Graziano et al., 2002)

Rat behavioral research has produced results consistent with this view of how movement control is organized in the brain. For instance, using the reach-to-grasp food task described earlier, Metz and Whishaw (1996) found that rats did not adjust their grip size for food pellets of varying diameters. On the basis of these findings, they argued that manual movements in rats are organized in terms of stereotypic movements. Our observations of how rats learn the standard reach-to-grasp task are also consistent with this view. For instance, on the first few training trials rats often reach for the food pellet, and then, even if they succeed in grasping it, they do not retract the pellet toward their mouth, open their mouth, and place the pellet inside. Rather, after barely starting to retract their paw, they drop the pellet and let their arm go limp temporarily. Then, after several more trials they gradually pull the pellet closer and closer to their mouth. Similarly, on many initial trials rats fail to contact and grasp the target, and do not even extend their paw forward a sufficient distance, but still they open their mouth and retract their pelletless hand (L. Hermer-Vazquez and R. Hermer-Vazquez, unpublished data). Thus, a crucial part of learning to perform the whole maneuver appears to be learning to conjoin two stereotypic movements: extending the paw outward and grasping the object, and then pulling it toward and inside the mouth. We have evidence that in the rat M1, the neural control for skilled reaching and for the "reachlike" phase of locomotion, in which the forepaw is lifted off the ground, projected forward, and placed back down, is similar: In both cases, M1 cells preferentially encode the lift and paw-down phases of the movements (Hermer-Vazquez et al., in press). These findings suggest that the interjoint timing required for one type of movement can be gradually shifted over the course of learning so that it produces a different, although related, movement.

The view that rats learn sophisticated motor behaviors by shaping a preestablished repertoire to new circumstances has dramatic implications for one's task analysis. When the rat is learning a new and difficult motor task, for instance, instead of learning a completely novel and lengthy series of joint torques and angles, the animal may start with a subset of its hard-wired movements and then combine and subtly adapt them to the current circumstances.

FINAL COMMENTS

Using recent findings from neuroscience, we have begun to sketch a new view of the psychological information processing that occurs as rats learn and execute a new somatosensory-motor task. This new view includes the facts that the rat's mind is continuously processing aspects of the task at multiple temporal and spatial scales, that sensory and motor processing are highly fused, and that whole, stereotypic movements appear to be represented in the rat's brain and mind. We believe that if researchers in the areas of rat perception, cognition, and behavior base their task designs and task analyses on these principles, it will facilitate developing a more accurate psychological understanding of the rat's performance. For example, if a researcher hypothesizes that rats will learn a new task by adapting a set of partly known preexisting motor behaviors, it will greatly simplify his or her understanding of how to kinematically code the evolving motor sequence. However, researchers still must validate their task analyses by testing their hypotheses about how the animals perform their tasks, because rats do have distinctive somatosensory and other adaptations whose deployment may not always be obvious to the experimenter.

REFERENCES

Aboitiz F (2001) The origin of isocortical development. Trends in Neurosciences 24:202–203.
Baddeley A (1992) Human memory: theory and practice. Boston: Allyn and Bacon.

Bower JM and Woolston DC (1983) Congruence of spatial organization of tactile projections to granule cell and Purkinje cell layers of cerebellar hemispheres of the albino rat: vertical organization of cerebellar cortex. Journal of Neurophysiology 49:745–766.

Brecht M and Sakmann B (2002) Dynamic representation of whisker deflection by synaptic potentials in spiny stellate and pyramidal cells in the barrels and septa of layer 4 rat somatosensory cortex. Journal of Physiology 543(pt 1):49–70.

Brown LL, Feldman SM, Smith DM, Cavanaugh JR, Ackermann RF, and Graybiel AM (2002) Differential metabolic activity in the striosome and matrix compartments of the rat striatum during natural behaviors. Journal of Neuroscience 22:305–314.

Chapin JK and Lin C-S (1984) Mapping the body representation in the SI cortex of anesthetized and awake rats. Journal of Comparative Neurology 229:199–213.

D'Avella A, Saltiel P, and Bizzi E (2003) Combinations of muscle synergies in the construction of a natural motor behavior. Nature Neuroscience 6:300–308.

Graziano MS, Taylor CS, Moore T, and Cooke DF (2002) The cortical control of movement revisited. Neuron 36:349–362.

Guillery RW and Sherman SM (2002) Thalamic relay functions and their role in corticocortical communication: generalizations from the visual system. Neuron 33:163–175.

Hermer-Vazquez L, Hermer-Vazquez R, Moxon KA, Kuo K-S, Viau V, Zhan Y, and Chapin JK (2003) Distinct temporal activity patterns in the rat M1 and magnocellular red nucleus during skilled versus unskilled movement. Behavioural Brain Research, in press.

Hermer-Vazquez L, Hermer-Vazquez R, and Chapin JK (2003) Olfactomotor coupling prior to skilled, olfactory-driven reaching. PNAS, under review.

Iwaniuk AN and Whishaw IQ (2000) On the origin of skilled forelimb movements. Trends in Neuroscience 23:372–376.

Johnson BN, Mainland JD, and Sobel N (2003) Rapid olfactory processing implicates subcortical control of an olfactomotor system. Journal of Neurophysiology 90:1084–1094.

Johnson KO (2001) The roles and functions of cutaneous mechanoreceptors. Current Opinion in Neurobiology 11:455–461.

Jones EG (2001) The thalamic matrix and thalamocortical synchrony. Trends in Neuroscience 24:595–601.

Kandel ER, Schwartz JH, and Jessell TM (eds.) (2000) Principles of neural science, 4th edition. New York: McGraw-Hill, Health Professions Division.

Kleim JA, Barbay S, Cooper NR, Hogg TM, Reidel CN, Remple MS, Nudo RJ (2002) Motor learning-dependent synaptogenesis is localized to functionally reorganized motor cortex. Neurobiology of Learning and Memory 77:63–77.

Krubitzer L (1995) The organization of neocortex in mammals: are species differences really so different? Trends in Neuroscience 18:408–417.

Krupa DJ, Ghazanfar AA, Nicolelis MA (1999) Immediate thalamic sensory plasticity depends on corticothalamic feedback. Proceedings of the National Academy of Sciences 96:8200–8205.

Lee AK and Wilson MA (2002) Memory of sequential experience in the hippocampus during slow wave sleep. Neuron 36:1183–1194.

Llinás RR (2001) I of the vortex: from neurons to self. Cambridge, MA: MIT Press.

McFarland NR and Haber SN (2002) Thalamic relay nuclei of the basal ganglia form both reciprocal and nonreciprocal cortical connections, linking multiple frontal cortical areas. Journal of Neuroscience 22: 8117–8132.

Metz GA and Whishaw IQ (1996) Skilled reaching an action pattern: stability in rat (Rattus norvegicus) grasping movements as a function of changing food pellet size. Behavioural Brain Research 116:111–122.

Paxinos G (ed.) (1995) The rat nervous system. San Diego: Academic Press.

Poe GR, Nitz DA, McNaughton BL, Barnes CA (2000) Experience-dependent phase-reversal of hippocampal neuron firing during REM sleep. Brain Research 855:176–180.

Rioult-Pedotti MS, Friedman D, Donoghue JP (2000) Learning-induced LTP in neocortex. Science 290: 533–536.

Shin HC and Chapin JK (1990) Mapping the effects of motor cortex stimulation on somatosensory relay neurons in the rat thalamus: direct responses and afferent modulation. Brain Research Bulletin 24:257–265.

Strick L (2002) Stimulating research on motor cortex. Nature Neuroscience 5:714–715.

Vallbo AB and Johansson RS (1984) Properties of cutaneous mechanoreceptors in the human hand related to touch sensation. Human Neurobiology 3:3–14.

Whishaw IQ and Pellis S (1990) The structure of skilled forelimb reaching in the rat: a proximally driven movement with a single distal rotatory component. Behavioural Brain Research 41:49–59.

Pain

7

DANIEL LE BARS AND SAMUEL W. CADDEN

Sherrington (1906) introduced the concept of "nociception" (from the Latin *nocere*, "to harm"). Within that concept, "nociceptive" stimuli are those that threaten the integrity of the body and/or directly activate a collection of discrete sensory organs or nerves known as *nociceptors*. These stimuli set off a varied but limited repertoire of somatic and vegetative reflex and behavioral responses that are associated with the perception of pain. Sherrington also described pain as "the psychical adjunct of a . . . protective reflex," underlining that pain triggers reactions and induces learned avoidance behaviours that may decrease whatever is causing it. As a result, pain may limit the (potentially) damaging consequences, and it is the behaviors associated with this end point that define many "pain responses" in rats.

The complexity of pain is underlined by the definition of pain adopted by the International Association for the Study of Pain (IASP): "an unpleasant sensory and emotional experience associated with actual or potential tissue damage or described in terms of such damage." The painful experience is more than a sensory experience that discriminates the intensity, location, and duration of a stimulus; it is also characterized by an emotional aversive state, which pushes one to action (motivation). This emotion is a fundamental and inextricable part of the painful experience and not a reaction to the sensory aspect. Therefore, the pain gets our attention, interferes with activity, and mobilizes strategies for defense. Zimmermann (1986) reinterpreted the

IASP definition of pain so that it could be applied to animals: "an aversive sensory experience caused by actual or potential injury that elicits progressive motor and vegetative reactions, results in learned avoidance behaviour, and may modify species specific behaviour, including social behaviour."

The purpose of models in the rat is to model human pain. Thus, there is a need for these models to reflect the different categories of human pain, such as physiological pain (nociceptive pain), inflammatory pain, and neuropathic pain (neuropathic: pertaining to disease of the nervous system). The first two types usually occur after injury and are often associated with each other. During an inflammatory episode, the threshold for pain is lowered so that (1) innocuous physical contact may become painful (allodynia) and/or (2) a nociceptive stimulus is perceived as more intense pain than usual (hyperalgesia). Although these pains can exceed the duration of the stimulus, they usually disappear after the related injury has healed.

By contrast, neuropathic pain results from an injury or a pathological transformation of the somaesthesic system that evolves an abnormal and unsuitable mode of functioning. In addition to the usual symptoms of inflammatory pain, there are continuous or paroxysmal "spontaneous" pains (e.g., sensations of "electrical" discharges), pains stemming from insensible regions (e.g., the paradoxical "anesthesia dolorosa"), paraesthesia (tingling, pricking, dullness), dysaesthesia (very unpleasant, but not painful, sensations), and sometimes sympathetic disturbances. There is also the very paradoxical situation usually known as *phantom pain* that occurs after deafferentation (e.g., avulsion of the brachial plexus or amputation of a limb) and thus in the absence of a nociceptive stimulus or nociceptors. Taken together, these symptoms are described by patients as "strange" or sometimes even as not being pain but "worse than pain." It is unclear whether these all have corollaries in the rat.

SENSATION AND REACTION

It is fairly obvious that in the context of rat models of pain, psychological pain is difficult to monitor. The absence of verbal communication in rats is undoubtedly an obstacle to the evaluation of pain. Nevertheless, most pain responses displayed by humans can be objectively measured behaviorally, and very similar behavioral responses can be measured in rats in response to similar stimuli. There are circumstances when there can be little doubt that a rat is feeling pain, such as when it is responding to stimuli with vocal responses. On the other hand, it is far more difficult to certify that at a given moment a rat feels no pain because it is presenting no typical physical signs or overt behaviors. This is particularly so given that we know that immobility or prostration is sometimes the only response accompanying pain. The question of pain in rats can be approached only with anthropomorphic references, although differences probably do exist by comparison with humans, notably with respect to certain cerebral structures (Bateson, 1991).

By contrast with the polymorphic nature of the pain that is described as a sensation in humans, that in rats can be estimated only by examining their reactions. This is essentially the same difficulty as is faced by the pediatrician, the geriatrician, or the psychiatrist when dealing with patients incapable of expressing themselves verbally. In those cases, too, the

semeiology is not unequivocal. It has to be taken in context and placed in an inventory, as its meaning will differ depending on the degree of maturation (or degradation) of the nervous system.

The study of behavioral reactions provides the only indicator of the perceived, disagreeable sensation resulting from a stimulus that would be algogenic (pain producing) if experienced by a human. But it must never be forgotten that these responses are often not specific: for example, escape can result from any disagreeable stimulus whether it is nociceptive or not. In addition, it should be remembered that the existence of a reaction is not necessarily evidence of a concomitant sensation (Hardy et al., 1952).

The observed reactions in the rat cover a wide spectrum ranging from the most elementary reflexes to far more integrated behaviors (e.g., escape, avoidance). In almost every case, it is a motor response that is monitored. By contrast, vegetative responses are considered only occasionally.

MODELS OF CHRONIC PAIN

There are two types of chronic pain models in the rat—models of rats with induced arthritis and models of rats with lesions of the central or peripheral nervous systems.

The injection of complete Freund's adjuvant in the rat brings about a severe general malaise associated with ankylosing spondylitis (Butler, 1989). In other models, the arthritis is limited to one joint.

Rat models of neuropathies are based on a total or partial deafferentation of an area of the body. The cutting of several neighboring dorsal roots triggers, within 2 to 3 weeks, a type of self-mutilation "autotomy " (Dong, 1989; Kauppila, 1998). As with anesthesia dolorosa in humans, this behavior is thought to be elicited by the pain evoked by deafferentation. However, such an interpretation is disputed by some because the introduction of a female into the cage of an operated rat is sufficient to eliminate the autotomy behavior.

There are three types of rat models of neuropathic pain triggered by partial deafferentation (Seltzer, 1995; Bennett, 2001). The chronic constriction injury model is produced by a loose ligature around the rat sciatic nerve, resulting in a loss of myelinated fibers but— for the main part—the survival of C-fibers (Bennett and Xie, 1988). The partial nerve transection model is produced by partial section of the rat sciatic nerve (Seltzer et al., 1990). The spinal nerve transection model is produced by the section of the L5-6 spinal dorsal roots, with the hindpaw remaining partially innervated through the L4 root (Kim and Chung, 1992). In addition, there are some models that were designed to replicate known anatomoclinical entities in humans (e.g., the diabetic neuropathy produced by streptozocin and neuropathies elicited by antineoplastic drugs such as Taxol (paclitaxel), vincristine, or cisplatin). All of these models are characterized by evidence of allodynia and hyperalgesia when stimuli are applied to the affected zone. With the notable exception of the observation of autotomy, purely behavioral approaches to spontaneous pain in the rat are rare. The same tests are used to evaluate stimulus-evoked pain in normal rats and in rats with chronic pain. The main difference results from the fact that the tests are applied to a body and a nervous system with a different history; however, the triggered behavior and the other measured variables are the same.

BEHAVIORAL REACTIONS

There are two main types of reaction to a nociceptive stimulus: (1) those organized by centers that are relatively "low" within the hierarchy of the central nervous system and (2) more complex ones organized by higher centers in the central nervous system.

The former can be elicited in decerebrate animals and were termed "pseudo-affective reflexes" by Sherrington (1906). They include (1) basic motor responses (withdrawal, jumping, contractures, etc.), (2) neurovegetative reactions, generally in the context of Selye's "alarm reaction," with an increase in sympathetic tone (tachycardia, arterial hypertension, hyperpnea, mydriasis, etc.), and (3) vocalization. The latter categories include conditioned motor responses that result from a period of learning and sometimes can be very rapid. Behavioral reactions (escape, distrust of objects responsible for painful experiences, avoidance, aggression, etc.) or modifications of behavior (social, food, sexual, sleep, etc.) are often observed. It must be noted, however, that even if active motor reactions are frequent, passive responses are seen just as often in animals, such as immobility, which allows the animal to preserve a painless posture. Furthermore, motor atonia is a general response to illness, regardless of whether the condition is painful.

INPUT AND OUTPUT: THE STIMULUS AND THE RESPONSE

Behavioral tests in the rat need to be appropriate. The stimuli have to be quantifiable, reproducible, and noninvasive (Beecher, 1957; Lineberry, 1981). All nociceptive stimuli can be defined by a number of different parameters: (1) the physical nature of the stimulus, (2) the site of application of the stimulus, and (3) the history of this site of stimulation.

PHYSICAL NATURE OF THE STIMULUS

Regardless of whether a stimulus is electrical, thermal, mechanical, or chemical, it is essential that three of its parameters are controlled: the intensity, the duration, and the surface area to which it is applied. These three parameters determine the "global quantity of nociceptive information" that will be generated and carried toward the central nervous system by the peripheral nervous system.

SITE OF APPLICATION OF THE STIMULUS

Clinical pain can originate from somatic, visceral, articular, or musculotendinous tissues. In nociceptive tests, stimuli are usually applied to cutaneous and, to a lesser extent, visceral structures. We know that some areas of skin can have a specific, particular function. For example, the rat tail, a structure used in many nociceptive tests, is an essential organ for thermoregulation and balance (see Chapters 12 and 21).

PREVIOUS HISTORY OF THE STIMULATED SITE

Tests for acute pain involve healthy tissues and, occasionally, acutely inflamed tissues (of a few days' standing at most). Tests for chronic pain relate to rheumatic or neuropathic conditions that last for longer periods of time (from weeks up to months).

Because the application of the stimulus must not produce lesions, one often defines a limit for how long the animal should be exposed to the stimulus (the "cut-off time"). This limit is absolutely necessary when the intensity of the stimulus is increasing. Furthermore, the repeated application of a stimulus can sensitize peripheral receptors and/or produce central sensitization.

REQUIREMENTS FOR BEHAVIORAL MODELS OF NOCICEPTION

Ideally, a behavioral model for nociception in the rat possesses the following characteristics (Lineberry, 1981; Vierck and Cooper, 1984; Ramabadran and Bansinath, 1986; Hammond, 1989; Watkins, 1989; Tjølsen and Hole, 1997; Le Bars et al., 2001; Berge, 2002).

SPECIFICITY

The stimulus must be nociceptive ("input specificity"). Although this is common sense,

it is not always easy to confirm that it is being achieved. For example, the appearance of a flexion reflex does not inevitably mean that the stimulus is nociceptive or that it is a nociceptive flexion reflex. Indeed, flexion reflexes are not triggered exclusively by nociceptive stimuli (Schomburg, 1997). It must be possible in the behavioral model to differentiate responses to nociceptive stimuli from responses to non-nociceptive stimuli. In other words, the quantified response must be exclusively or preferentially triggered by nociceptive stimuli ("output specificity"). In this respect, one must remember that some innate and acquired behaviors can be triggered by aversive stimuli that are not nociceptive/painful.

SENSITIVITY

It must be possible to quantify the response and for this variable to be correlated with stimulus intensity within a reasonable range (from the pain threshold to the pain tolerance threshold). In other words, the quantified response must be appropriate for a given type of stimulus and monotonically related to its intensity. The model must be sensitive to manipulations, notably pharmacological, that would reduce the nociceptive behavior in a specific fashion.

VALIDITY

The model must allow the differentiation of nonspecific behavioral changes (e.g., in motility, attention, etc.) from those triggered by the nociceptive stimulus itself. In other words, the response being monitored must not be contaminated by simultaneous perturbations related to other functions, notably if these have been introduced by a pharmacological agent. The test validity (i.e., the degree to which the test actually measures what it purports to measure) is undoubtedly one of the most difficult issues to determine (Berge, 2002; Hansson, 2003).

RELIABILITY

Consistency of scores must be obtained when animals are retested with an identical test or equivalent form of the test. In this context, the repeated application of the stimulus must not produce lesions.

REPRODUCIBILITY

Results obtained with a test must be reproducible not only within the same laboratory but also between different laboratories.

Before describing tests that endeavor to meet these requirements, it is worth noting that they can be divided into two categories: those measuring a threshold and those measuring supraliminal responses. However, both categories permit the study of only one point on the stimulus–response curve, be it the threshold or an arbitrary point farther up the curve. As a result, they allow only a rough appreciation of the gain of the process (Tjølsen and Hole, 1997).

TESTS BASED ON THE MEASUREMENT OF REACTION TIME FOR ESCAPE BEHAVIOR

Escape tests are based mainly on the application of thermal stimuli to the skin. Heat constitutes a relatively selective stimulus for nociceptors, and radiant heat has the advantage of not producing a concomitant tactile stimulus. Nevertheless, heating is progressive and results in thermoreceptors being activated before nociceptors are recruited. Just as there is this sequence of activation of thermoreceptors and then nociceptors, there is a sequence of a hot sensation followed by a painful one. As a result, the possibility that the same stimulus is successively a conditioning and a conditioned stimulus cannot be ruled out. In addition, one has to address the question of the meaning that can be ascribed to measurements of reaction time when a stimulus is gradually increasing in intensity (Le Bars et al., 2001).

TAIL-FLICK TEST

There are two variants of the tail-flick test. One consists of applying radiant heat to a small surface of the tail. The other involves submersing the tail in water at a predetermined temperature. Both provoke a vigorous withdrawal movement of the tail (d'Amour and Smith, 1941). It is the reaction time of this movement that is recorded. The "tail-flick" is a spinal reflex as witnessed by the persistence of at least its shorter-latency form after section or cold block of upper parts of the spinal cord.

PAW WITHDRAWAL TEST

This test is comparable to the tail-flick test but offers the advantages that (1) it does not involve the preeminent organ of thermoregulation in rats (the tail) and (2) it can be applied to freely moving animals (Hargreaves et al., 1988). However, the latter is also a disadvantage in that the position of the leg becomes a factor of variability, because the background level of activity in the flexor muscles varies with the position of the animal.

HOT PLATE TEST

This test consists of introducing a rat into an open-ended cylindrical space with a floor consisting of a heated metallic plate (Woolfe and MacDonald, 1944). Two behavioral components can be measured by their reaction times: paw licking and jumping. Both are considered to be supraspinally integrated responses. However, licking the forepaw is a typical behavior for heat dissipation (Roberts and Mooney, 1974). Although such behavior is relatively stereotyped in the mouse, it is more complex in the rat, which sniffs, licks its forepaws, licks its hindpaws, straightens up, stamps its feet, starts and stops washing itself,

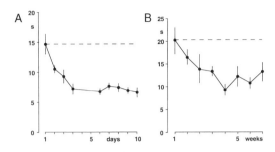

Figure 7–1. Learning phenomena in the hot plate test. In the experiments, measurements were made of the delay between after the animal had been put on the hot plate and before the animal began to lick the paws. (A) Sandkühler et al. (1996) repeated the test daily in Sprague-Dawley rats. (B) Lai et al. (1982) repeated the test each week on a Wistar strain of rat. It can be seen that in both cases, four or five tests were sufficient to almost halve the reaction time. (Modified from Sandkühler et al. [1996] and Lai et al. [1982].)

etc. (Espejo and Mir, 1993). Furthermore, this test is very susceptible to learning, which makes it delicate to interpret (Fig. 7–1).

TEST BASED ON THE MEASUREMENT OF THRESHOLD FOR ESCAPE BEHAVIOR

Threshold tests are based on the application of mechanical or electrical stimuli to the skin or an internal organ.

APPLICATION OF INCREASING PRESSURE

An increasing pressure is applied to a small area on the hindpaw and interrupted when the threshold is reached (Green et al., 1951). This produces successive responses of reflex withdrawal of the paw, a more complex movement whereby the rat tries to release its trapped limb, a struggle, and, finally, a vocal reaction. Although the first of these is undoubtedly a proper spinal reflex, the latter two clearly involve supraspinal structures. With the aim of improving the sensitivity of this test, Randall and Selitto (1957) proposed comparing thresholds observed on a healthy paw and an inflamed paw.

APPLICATION OF CALIBRATED PRESSURE

Skin pressure tests involve applying a fiber of a given diameter to the skin (Handwerker and Brune, 1987). Pressure is applied until the fiber bends. The use of a range of fiber diameters (variously called Semmes-Weinstein fibers or von Frey hairs) makes it possible to determine the threshold for evoking a response in the animal (e.g., a flexion reflex). This test is a prized tool in models of neuropathic pain (e.g., Kim and Chung, 1992). A technical difficulty with this approach relates to the sensitivity of such fibers to humidity (Möller et al., 1998).

DISTENTION OF HOLLOW ORGANS

Distention tests include colorectal distention by means of an inflatable balloon that produces avoidance behavior, reflex activities in the abdominal muscles, and quantifiable vegetative responses (increased arterial pressure, tachycardia) (Ness and Gebhart; 1988). A development of this model involves first inflaming the colon by the administration of chemical agents. Distention of the vagina or uterus has also been used in female rats (Berkley et al., 1995).

APPLICATION OF ELECTRICAL STIMULI

The application of electrical stimuli has the advantages of being quantifiable, reproducible, and noninvasive and of producing synchronized afferent signals. However, intense electrical stimuli nondifferentially excite all peripheral fibers, including large-diameter fibers that are not directly implicated in nociception as well as fine $A\delta$ and C fibers, which mediate thermoreceptive as well as nociceptive information.

Electrical stimuli of gradually increasing intensities can be delivered in trains through subcutaneous electrodes in the tail of the rat (Carroll and Lim, 1960; Levine et al., 1984). The gradually increasing currents produce, in succession, reflex movement of the tail, vocalization at the time of stimulation, and, finally, vocalization, which continues beyond the period of stimulation (a vocalization afterdischarge). These responses are organized on a hierarchical basis; they depend on the different levels of integration of the nociceptive signal in the central nervous system: the spinal cord, the brain stem, and the thalamus/rhinencephalon. The last of these can reflect affective and motivational aspects of pain behavior (Borszcz, 1995a).

Electrical stimuli can also be applied in the form of single, short-duration pulses. These elicit, in succession, twitching, escape behavior, vocalization, and biting of the electrodes. Again, these responses are hierarchically organized, with the last one being the most coordinated. They also depend on different levels of integration of the nociceptive signal (Charpentier, 1968).

In an attempt to overcome the drawback that intense electrical stimuli excite non-nociceptive as well as nociceptive nerves (see earlier), some investigators have stimulated tissues in which they believe all the afferent nerve fibers are nociceptive. Most commonly, the dental pulp has been used for this purpose. Contrary to commonly held belief, however, it is not certain that all of the afferent fibers in the dental pulp are nociceptive, although the proportion of them that are may be greater than that in other tissues (Le Bars et al., 2001). In addition, the anatomical arrangement of the dental tissues in the rat is such that it is difficult to apply electrical stimuli that excite pulpal nerves without exciting (non-nociceptive) nerves in the surrounding tissues (e.g., Hayashi, 1980; Jiffry, 1981). Some workers have shown that the exclusive activation of rat pulpal nerves is possible provided adequate care is taken (e.g., Rajaona et al., 1986, Myslinski and Matthews, 1987), but it seems that this has not always been done. Two types of response have been monitored in such models: the dysynaptic jaw opening reflex (e.g., Vassel et al., 1986), which is organized within the trigeminal structures of the brain stem in a fashion similar to spinal

reflexes elsewhere in the body (e.g., Sumino, 1971), and the appearance of more complex reactions such as scratching, head movements, and vocalization (e.g., Rajaona et al., 1986), which involve coordination at higher centers.

TESTS BASED ON THE OBSERVATION OF BEHAVIOR

The main types of these tests involve using an intradermal or intraperitoneal injection of an irritant, algogenic, chemical agent as the nociceptive stimulus. In this section, we also consider the description of complex vocal patterns elicited by electrical stimuli.

INTRADERMAL INJECTIONS OF IRRITANT AGENTS

The most commonly used substance for intradermal injections is formalin (the formalin test). When it is injected into the dorsal surface of the rat forepaw, formalin provokes a "painful" behavior that can be assessed on a four-level scale related to the posture of the injected paw (see Fig. 7–2): 0 indicates normal posture; 1, paw remaining on the ground but not supporting the animal; 2, paw clearly being raised; and 3, paw being licked, nibbled, or shaken (Dubuisson and Dennis, 1977). An initial phase can be observed about 3 minutes after the injection; then, after a quiescent period, there is a second phase between minutes 20 and 30. The first phase results essentially from the direct stimulation of nociceptors, whereas the second phase involves a period of sensitization during which inflammatory phenomena occur.

INTRAPERITONEAL INJECTIONS OF IRRITANT AGENTS

The intraperitoneal administration of agents that irritate serous membranes provokes a very stereotyped behavior, which is characterized by

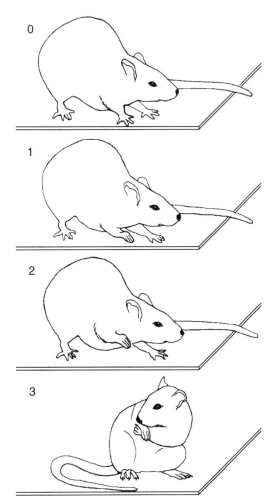

Figure 7–2. Behavior triggered by an intradermal injection of formalin into the right forepaw (see text). (Modified from Dubuisson and Dennis [1977].)

abdominal contractions, movements of the body as a whole (particularly of the hindpaws), twisting of dorsoabdominal muscles, reduction in motor activity, and motor incoordination (writhing behavior). These "behaviors" are all considered to be reflexes (Hammond, 1989).

STIMULATION OF HOLLOW ORGANS

Tests that involve the injection of algogenic substances directly into hollow organs are used as models for visceral pain. Administration of formalin into the rat colon can produce a complex, biphasic type of "pain" behavior involving an initial phase of body stretching and contraction of either the flanks or the whole body and a second phase that predominantly involves abdominal licking and nibbling (Miampamba et al., 1994). Similarly, a number of models have been developed for bladder or uterine pain whereby reflexes and/or more complex behaviors have been observed after the administration of irritants into the organ (e.g., McMahon and Abel, 1987; Pandita et al., 1997; Wesselmann et al., 1998).

Giamberardino et al. (1995) studied the behavior produced by the surgical introduction of dental cement, to mimic a calculus, into the ureter. This produced something akin to episodes of writhing behavior over a 4-day period. A concomitant hyperalgesia in the abdominal muscles provided evidence of visceromuscular convergence. To the best of our knowledge, this is the only animal model of "referred" pain.

VOCALIZATION ELICITED BY ELECTRICAL SHOCKS

Complex vocal patterns can be produced by electrical stimuli of short duration (Jourdan et al., 1995). Three types of emissions have been identified (Fig. 7–3).

1. Two distinct "peeps," the energies of which are distributed across a wide range of audible frequencies without a defined structure, occur. The first peep results from activation of relatively rapidly conducting Aδ fibers, and the second peep results from activation of slowly conducting C fibers.
2. "Chatters" are characterized by formants composed of a fundamental frequency and its harmonics; these constitute a very elaborate response, the physical characteristics of which are similar to human words.
3. Ultrasonic emissions, inaudible to humans and made up of a fundamental frequency, without harmonics, between 20 and 35 kHz, occur with mild modulations.

The characteristics of the first two peeps emitted by the rat are reminiscent of the phenomenon of "double pain" observed in humans after a brief, sharp nociceptive stimulus. The

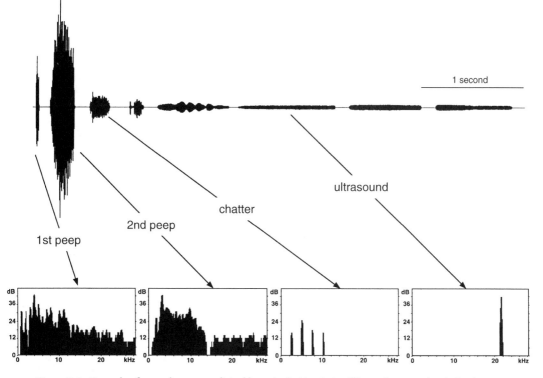

Figure 7–3. Example of a vocal response elicited by a single 20 mA, 2 millisecond square electrical pulse applied to the base of the tail of a rat. (*Top*) Recording during the 6 seconds after the stimulus. This response consisted of eight components separated by silences: two peeps, two chatters, and four ultrasonic emissions, as defined by the corresponding spectrograms shown in the insets below. These spectrograms revealed that the first two peeps corresponded to strong emissions within a wide spectrum of frequencies. By contrast, the two subsequent chatters were pure, slightly modulated sounds with a fundamental frequency and corresponding harmonics. The last four components were pure ultrasound with a single, 21.7 kHz, frequency. (Modified from Jourdan et al. [1995].)

physiological meanings of the other components of the response are more difficult to understand. In line with Pavlovian conditioning, the "chatters" can be triggered by a light signal (Borszcz, 1995b). The ultrasonic emissions may reflect the affective state and the degree of anxiety of the animal, because they can also be recorded in other experimental situations that generate fear or stress (Sales and Pye, 1974; Haney and Miczek, 1994). In addition, they are sensitive to anxiolytic drugs (Tonoue et al., 1987; Cuomo et al., 1992).

FURTHER CONSIDERATIONS AND CONCLUSION

Behavioral tests of nociception are not entirely satisfactory (Le Bars et al., 2001). Their first weakness lies in the difficulty of controlling the stimuli used to trigger a nociceptive reaction. More important, even when the physical parameters of the external stimuli are well controlled, that does not necessarily result in an equally well-controlled effective stimulus. Indeed, the stimulus that affectively activates the peripheral nociceptors is also influenced by the physiological state of the targeted tissue. This state is determined by the vegetative system, mainly through thermoregulation, systemic arterial blood pressure, and local vasomotor tone. For example, variation of local temperature may confound many experimental protocols, not only those using heat as a stimulus (Tjølsen and Hole, 1997). The second great weakness of these models lies in the nature of the dependent variable, which is usually the determination of the threshold for a reaction. Most of the models do not allow the study of stimulus–response relationships although these are really an essential element of sensory physiology. Furthermore, it is often not the threshold itself that is measured but rather a response time to a stimulus of increasing intensity. Because when the skin is subjected to a constant source of radiation, the temperature increases with the square root of time, such an approach is highly questionable when radiant heat is used.

These considerations invite prudence in the interpretation of results obtained using tests of nociception in the rat. It is worth stressing that these considerations also relate to rat models of chronic pain insofar as the tests applied in these models are the same ones, or almost the same ones, as those described for acute pain. Faced with the simplistic and reductionist approaches available, one is inclined to conclude that valid behavioral approaches of pain should be promoted in animals, notably in the rat.

REFERENCES

Bateson P (1991) Assessment of pain in animals. Animal Behavior 42:827–839.

Beecher HK (1957) The measurement of pain: prototype for the quantitative study of subjective responses. Pharmacological Reviews 9:59–209.

Bennet GJ (2001) Animal models of pain. In: Methods in pain research (Kruger L, ed.), pp. 67–91. Boca Raton: CRC Press.

Bennett GJ and Xie YK (1988) A peripheral mononeuropathy in rat that produces disorders of pain sensation like those seen in man. Pain 33:87–107.

Berge OG (2002) Reliability and validity in animal pain modeling. In: Pain and brain (Tjølsen A and Berge OG, eds.), pp. 77–89. Bergen: University of Bergen.

Berkley KJ, Wood E, Scofield SL, Little M (1995) Behavioral responses to uterine or vaginal distension in the rat. Pain 61:121–131.

Borszcz GS (1995a) Increases in vocalisation and motor reflex thresholds are influenced by the site of morphine microinjection: comparisons following administration into the periaqueductal gray, ventral

medulla, and spinal subarachnoid space. Behavioral Neuroscience 109:502–522.

Borszcz GS (1995b) Pavlovian conditional vocalizations of the rat: a model system for analyzing the fear of pain. Behavioral Neuroscience 109:648–662.

Butler SH (1989) Animal models and the assessment of chronic pain: critique of the arthritic rat model. In: Issues in pain measurement, advances in pain research and therapy, vol 12 (Chapman CR and Loeser JD, eds.), pp. 473–479. New York: Raven Press.

Carroll MN and Lim RK (1960) Observations of the neuropharmacology of morphine-like analgesia. Archives Internationales de Pharmacodynamie et de Thérapie 125:383–403.

Charpentier J (1968) Methods for evaluating analgesics in laboratory animals. In: Pain: proceedings of the International Symposium on Pain organised by the Laboratory of Psychophysiology (Soulairac A, Cahn J, Charpentier J, eds.), pp. 171–200. New York: Academic Press.

Cuomo V, Cagiano R, De Salvia MA, Mazzoccoli M, Persichella M, Renna G (1992) Ultrasonic vocalisation as an indicator of emotional state during active avoidance learning in rats. Life Sciences 50:1049–1055.

D'Amour FE and Smith DL (1941) A method for determining loss of pain sensation. Journal of Pharmacology and Experimental Therapeutics 72:74–79.

Dong WK (1989) Is autotomy a valid measure of chronic pain? In: Issues in pain measurement, advances in pain research and therapy, vol 12 (Chapman CR and Loeser JD, eds.), pp. 463–472. New York: Raven Press.

Dubuisson D and Dennis SG (1977) The formalin test: a quantitative study of the analgesic effects of morphine, meperidine and brain stem stimulation in rats and cats. Pain 4:161–174.

Espejo EF and Mir D (1993) Structure of the rat's behaviour in the hot plate test. Behavioural Brain Research 56:171–176.

Green AF, Young PA, Godfrey EI (1951) A comparison of heat and pressure analgesiometric methods in rats. British Journal of Pharmacology 6:572–585.

Hansson P (2003) Difficulties in stratifying neuropathic pain by mechanisms. European Journal of Pain 7:353–357.

Hammond DL (1989) Inference of pain and its modulation from simple behaviors. In: Issues in pain measurement, advances in pain research and therapy, vol 12 (Chapman CR and Loeser JD, eds.), pp. 69–91. New York: Raven Press.

Handwerker HO and Brune K (1987) Deutschsprachige Klassiker der Schmerzforschung [Classical German contributions to pain research]. Hassfurt: Tagblatt-Druckerei KG.

Haney M and Miczek KA (1994) Ultrasounds emitted by female rats during agonistic interactions: effects of morphine and naltrexone. Psychopharmacology (Berlin) 114:441–448.

Hardy JD, Wolff HG, Goodell H (1952) Pain sensation and reaction. Baltimore: Williams & Wilkins.

Hargreaves K, Dubner R, Brown F, Flores C, Joris J (1988) A new and sensitive method for measuring thermal nociception in cutaneous hyperalgesia. Pain 32:77–88.

Hayashi H (1980) A problem in electrical stimulation of incisor tooth pulp in rats. Exp Neurology 67:438–441.

Jiffry MT (1981) Afferent innervation of the rat incisor pulp. Experimental Neurology 73:209–218.

Jourdan D, Ardid D, Chapuy E, Eschalier A, Le Bars D (1995) Audible and ultrasonic vocalisation elicited by single electrical nociceptive stimuli to the tail in the rat. Pain 63:237–249.

Kauppila T (1998) Correlation between autotomy-behavior and current theories of neuropathic pain. Neuroscience and Biobehavioral Reviews 23:111–129.

Kim SH and Chung JM (1992) An experimental model for peripheral neuropathy produced by segmental spinal nerve ligation in the rat. Pain 50:355–363.

Kontinen VK and Meert TF (2003) Predictive validity of neuropathic pain models in pharmacological studies with a behavioral outcome in the rat: a sytematic review. In: Proceedings of the 10th World Congress on Pain, Progress in Pain Research and Management, vol 24 (Dostrovsky JO and Carr DB, eds.), pp. 489–498. Seattle: IASP Press.

Lai YY and Chan SHH (1982) Shortened pain response time following repeated algesiometric test in rats. Physiology and Behavior 28:1111–1113.

Le Bars D, Gozariu M, Cadden SW (2001) Animal models of nociception. Pharmacological Reviews 53:597–652.

Levine JD, Feldmesser M, Tecott L, Gordon NC,, Izdebski K (1984) Pain induced vocalization in the rat and its modification by pharmacological agents. Brain Research 296:121–127.

Lineberry CG (1981) Laboratory animals in pain research. In: Methods in animal experimentation, vol 6, pp. 237–311. New York: Academic Press.

McMahon SB and Abel C (1987) A model for the study of visceral pain states: chronic inflammation of the chronic decerebrate rat urinary bladder by irritant chemicals. Pain 28:109–127.

Miampamba M, Chery-Croze S, Gorry F, Berger F, Chayvialle JA (1994) Inflammation of the colonic wall induced by formalin as a model of acute visceral pain. Pain 57:327–334.

Möller KA, Johansson B, Berge OG (1998) Assessing mechanical allodynia in the rat paw with a new electronic algometer. Journal of Neuroscience Methods 84:41–47.

Myslinski N and Matthews B (1987) Intrapulpal nerve stimulation in the rat. Journal of Neuroscience Methods 22:73–78.

Ness TJ and Gebhart GF (1988) Colorectal distension as a noxious visceral stimulus. Physiologic and pharmacologic characterization of pseudoaffective reflexes in the rat. Brain Research 450:153–169.

Pandita RK, Persson K, Andersson KE (1997) Capsaicin-induced bladder overactivity and nociceptive behaviour in conscious rats: Involvement of spinal nitric oxide. Journal of the Autonomic Nervous System 67:184–191.

Rajaona J, Dallel R, Woda A (1986) Is electrical stimulation of the rat incisor an appropriate experimental nociceptive stimulation? Experimental Neurology 93:291–299.

Ramabadran K and Bansinath M (1986) A critical analysis of the experimental evaluation of nociceptive reactions in animals. Pharmaceutical Research 3:263–270.

Randall LO and Selitto JJ (1957) A method for measurement of analgesic activity on inflamed tissue. Archives Internationales de Pharmacodynamie et de Thérapie 111:409–419.

Roberts WW and Mooney RD (1974) Brain areas controlling thermoregulatory grooming, prone extension, locomotion, and tail vasodilation in rats. Journal of Comparative Physiology and Psychology 86:470–480.

Sales GD and Pye D (1974) Ultrasonic communication by animals. London: Chapman and Hall.

Sandkühler J, Treier AC, Liu XG, Ohnimus M (1996) The massive expression of c-fos protein in spinal dorsal horn neurons is not followed by long-term changes in spinal nociception. Neuroscience 73:657–666.

Schomburg ED (1997) Restrictions on the interpretation of spinal reflex modulation in pain and analgesia research. Pain Forum 6:101–109.

Seltzer Z (1995) The relevance of animal neuropathy models for chronic pain in humans. Seminars in Neurosciences 7:211–219.

Seltzer Z, Dubner R, Shir Y (1990) A novel behavioral model of neuropathic pain disorders produced in rats by partial sciatic nerve injury. Pain 43:205–218.

Sherrington CS (1906) The integrative action of the nervous system. New York: C. Scribner's Sons.

Sumino R (1971) Central pathways involved in the jaw opening reflex in the cat. In: Oral-facial sensory and motor mechanisms (Dubner R and Kawamura Y, eds.). New York: Appleton-Century-Croft.

Tjølsen A and Hole K (1997) Animal models of analgesia. In: Pharmacology of pain (Dickenson AH and Besson JM, eds.), pp. 1–20. Berlin: Springer-Verlag.

Tonoue T, Iwasawa H, Naito H (1987) Diazepam and endorphine independently inhibit ultrasonic distress calls in rats. European Journal of Pharmacology 142:133–136.

Vassel A, Pajot J, Aigouy L, Rajaona J, Woda A. (1986) Effects, in the rat, of various stressing procedures on the jaw-opening reflex induced by tooth-pulp stimulation. Archives of Oral Biology 31:159–163.

Vierck CJ and Cooper BY (1984) Guideline for assessing pain reactions and pain modulation in laboratory animal subjects. In: Advances in pain research and therapy, vol 6 (Kruger L and Liebeskind JC, eds.), pp. 305–322. New York: Raven Press.

Watkins LR (1989) Algesiometry in laboratory and man: current concepts and future directions. In: Issues in pain measurement, advances in pain research and therapy, vol 12 (Chapman CR and Loeser JD, eds.), pp. 249–265. New York: Raven Press.

Wesselmann U, Czakanski PP, Affaitati G, Giamberardino MA (1998) Uterine inflammation as a noxious visceral stimulus: behavioral characterization in the rat. Neuroscience Letters 246:73–76.

Woolfe G and MacDonald Al (1944) The evaluation of the analgesic action of pethidine hydrochloride (Demerol). Journal of Pharmacology and Experimental Therapeutics 80:300–307.

Zimmermann M (1986) Behavioural investigations of pain in animals. In: Assessing pain in farm animals (Duncan IJH and Molony Y, eds.), pp. 16–29. Brussels: Office for Official Publications of the European Communities.

Vibrissae

8

RICHARD H. DYCK

The brain provides all animals with the means to detect, interpret, and act on specific classes of sensory stimuli. Rats, like other animals, are equipped with sophisticated arrays of sensory receptors that are especially well suited to allow them to negotiate their particular environments. Rats have evolved in complex and diverse environments that are rich in olfactory, gustatory, auditory, visual, and tactual information. However, because rats are nocturnal, by necessity they must be especially reliant on nonvisual stimuli. One of the most conspicuous of sensory receptor arrays, which are readily apparent when one observes the behaving rat, are the long, specialized hairs found on the face and nose commonly called tactile hairs, sinus hairs, whiskers, or vibrissae.

Rats, like all mammals (except for humans and a few nonprimate species), possess vibrissae that are essential for their survival. Like common pelagic fur or hair, vibrissae are rigid columns of dead epidermal cells that are deeply imbedded in epidermal hair follicles. What distinguishes vibrissae from pelagic hairs are their distinct structure, length, sensory innervation, motor control, and, most important, their function. Although pelagic hair is distributed widely across the skin, vibrissae are predominant on the eyebrows, cheeks, lips, and chin with small groups sometimes appearing in other areas like the abdomen or flexor surface of the wrist (Pocock, 1914; Sokolov and Kulikov, 1987). Although each of these groups of tactile hairs is important in guiding various aspects of the rat's behavior, the remainder of this chapter is concerned with the most widely studied, those located on the rat's upper jaw, the so-called mystacial (moustache) vibrissae.

Despite the apparent ecological and anatomical salience of tactual information provided by the vibrissae, there is a surprising paucity of studies that assess their role in guiding the rat's behavior (see Gustafson and Felbain-Keramidas, 1977, for the most recent review). This is particularly true compared with the volumes of published studies that deal with the anatomy and physiology of the rodent vibrissal somatosensory system. This chapter provides a review of the anatomical characteristics of the rat's vibrissa sensory apparatus and its growth and dynamic functional characteristics. This information is essential because knowledge of the use of the vibrissa sensory apparatus and the constraints under which it operates allows an understanding of its role in the generation of behavior in laboratory and natural settings.

So large and particular distribution of an exquisitely sensible nerve, it is reasonable to suppose, must be for the purpose of some sensible function.

—Broughton, 1823

STRUCTURE AND GROWTH

The rat provides a very useful model to understand the mechanisms underlying the development, structure, and function of the mammalian somatosensory system. In the vibrissal system, the exquisitely structured topographical organization of the periphery is maintained in the one-to-one functional representation of the vibrissae in the brain stem barrelettes, ventrobasal thalamic barreloids, and barrels within layer 4 of the somatosensory vibrissa cortex (Jones and Diamond, 1995). The usefulness of this system is exemplified by the ability to unambiguously identify these functional compartments using simple anatomical stains (for example, see Fig. 8–3). In addition, because the vibrissae are continuously replaced and spontaneously regrow when trimmed or plucked, this model system allows one to easily manipulate the sensory input at the level of the peripheral receptor and to perform subsequent analyses using the same peripheral array. Moreover, this system is amenable to within-animal designs whereby experimental and control manupulations occur in the same animal. This section is intended to briefly provide the reader with a basic understanding of several of these parameters that should prove useful in the design of experimental methods for behavioral studies using the rat vibrissa system.

VIBRISSA FOLLICLE

Vibrissae are distinguished from other kinds of hair by connective tissue capsules that form around the follicles. The capsule of each vibrissa is composed of spongy cavernous tissue in the lower half and an open ring sinus in the upper portion, with both cavities filled with blood (Vincent, 1913). In addition to providing metabolic requirements to the follicular tissues and nerves, the turgid blood sinus is believed to be essential for amplifying vibratory information from the vibrissae that activates the sensory nerve endings in the follicle.

VIBRISSAL ORGANIZATION

The mystacial vibrissae are arrayed in a bilaterally symmetrical, stereotypic manner on the rat's cheek and upper lip (Fig. 8–1). The caudal array, consisting of about 35 vibrissae, are distributed on the mystacial pad in five well-

defined caudorostrally oriented rows that are referred to, from dorsal to ventral, as row A to E (Fig. 8–2). The vibrissae are also arranged in dorsoventrally oriented arcs numbered 1 to 7 from caudal to rostral. The four longest, caudalmost vibrissae are not aligned within rows and are therefore referred to as straddlers (α to δ; see Fig. 8–2). Collectively, because of their relative length, this group has also been referred to as the macrovibrissae. Vibrissae within the same arc are approximately the same length, and they decrease in length, successively, caudorostrally. The straddlers are 45 to 60 mm in length, with vibrissae in arcs 1 through 4 projecting 40 to 44, 33 to 35, 23 to 25, and 11 to 16 mm, respectively (Ibrahim and Wright, 1975).

The more rostral, furry buccal pad consists of 40 to 70 vibrissae arranged in five rows that line the rat's upper lip (see Figs. 8–1 and 8–2). Because of their shorter length (<7 mm), this group of vibrissae has also been referred to as microvibrissae (Brecht et al., 1997).

SENSORY INNERVATION

The mystacial vibrissae are innervated by the infraorbital branch of the maxillary division of the sensory portion of the trigeminal nerve

Figure 8–1. More than 70 specialized tactile hairs project from each side of the rat's upper jaw. The caudal-most array consists of approximately 35 macrovibrissae, whereas the rostral-most array lining the rat's upper lip consists of 40 to 70 microvibrissae.

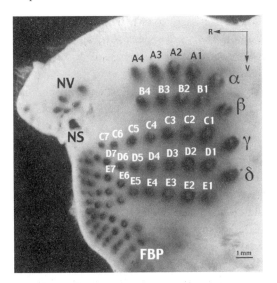

Figure 8–2. The organization of the rat's mystacial vibrissae on the skin of the rat's upper jaw. A–E correspond to the five rows of larger vibrissae located on the mystacial pad. The vibrissae are also organized in the dorsal/ventral plane into seven arcs (1–7). The longest, caudal-most vibrissae lie between rows and are referred to as "straddlers" (α–δ). The furry buccal pad (FBP) is densely populated with normal pelagic fur as well as vibrissae; however, only vibrissae follicles contain blood sinuses, which are made visible in this figure by staining with xylene. NV, nasal vibrissae; NS, nostril; R, rostral; V, ventral. (Figure freely available online from Barrels Web: http://www.neurobio.pitt.edu/barrels/pics.htm; courtesy of S. Haidarliu and E. Ahissar.)

tinct advantage of the vibrissal sensory system is that the vibrissae are continuously replaced throughout the animals life and they effectively regenerate after they are plucked or trimmed (Oliver, 1966), thereby providing the laboratory researcher with a significant degree of control over experimental parameters.

Vibrissae are present when rats are born and reach their adult length in the second postnatal month (Ibrahim and Wright, 1975). The short, rostral vibrissae grow at a rate of about 0.5 mm/day, while the longest vibrissae grow at a rate of 1.5 to 2 mm/day. Regardless of their position on the face, all vibrissae require approximately 4 weeks to reach their maximum length, after which a new vibrissa appears from the same follicle approximately 1 week later. When it achieves a length of one-half to three-fourths of its final length, the old vibrissa falls out and its function is taken over by the new whisker (Ibrahim and Wright, 1975). Trimming the old vibrissa at

(nerve V). The nerve is divided into fascicles, each of which innervates a row of vibrissae (Dörfl, 1985), indicating that vibrissae are functionally more closely affiliated within rows than they are between rows (Simons, 1983, 1985). Each vibrissa is exclusively innervated by as many as 200 large, myelinated sensory nerve endings that originate from cell bodies in the trigeminal ganglion (Vincent, 1913; Zucker and Welker, 1969), thereby differentiating these sensitive tactile structures from normal hair, which is sparsely innervated.

GROWTH AND REPLACEMENT

It is common in behavioral and neural science to establish the role of a particular structure or system by assessing the functional consequences of its removal or inactivation. A dis-

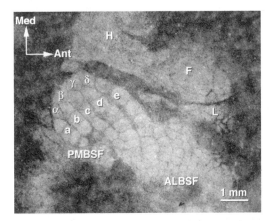

Figure 8–3. Functional compartments (barrels) in layer 4 of the somatosensory cortex show a one-to-one correspondence to individual vibrissae on the contralateral face. These compartments are readily identifiable using simple histological methods, in this case staining for synaptic zinc (see Land and Akhtar, 1999 for methods), which exemplifies the utility of the vibrissal sensory system in empirical studies assessing the relationship between brain structure and brain function. PMBSF, the posteromedial barrel subfield, corresponds to the cortical representation of the mystacial macrovibrissae. ALBSF, the anterolateral barrel subfield, corresponds to the cortical representation of the microvibrissae. H, hindlimb; F, forelimb; L, lower lip.

or above the level of the skin does not affect the rate of its growth or that of the new whisker. However, when a whisker is plucked from the follicle, a new whisker begins to grow immediately thereafter (Oliver, 1966).

In total, the life cycle of an individual vibrissa is approximately 2 months.

MOTOR ASPECTS OF VIBRISSAE FUNCTION

In all mammals, tactile sensitivity is enhanced by the relative movement of an object across the receptor periphery. By the second postnatal week of life (Welker, 1964), the rat's mystacial vibrissae are active during exploratory and discriminative behaviors, with individual whiskers serving as elements in a receptive array, scanning across object surfaces.

Three distinct behavioral states have been described in relation to whisking behavior. The first is *quiet behavior*, where whisker movement is absent while the animals are standing or sitting still (Fanselow and Nicolelis, 1999). The second, referred to as *whisker-twitching behavior*, also occurs during immobility but involves small-amplitude rhythmic movements at a rate of 7 to 12 Hz (Semba and Komisaruk, 1984). Finally, during active exploration, the whiskers are rhythmically swept forward (*protraction*) and backward (*retraction*) (Welker, 1964) at a frequency of 5 to 9 Hz (Carvell and Simons, 1990; Berg and Kleinfeld, 2003), with these movements finely coordinated with sniffing and discrete head movements (Welker, 1964). A whisking cycle lasts around 120 milliseconds, with two-thirds of it involving whisker protraction. Although the potential exists for individual caudal vibrissae to move independently of one another (Sachdev et al., 2002), for the most part, all mystacial vibrissae are observed to move as a single unit, with bilateral symmetry (Vincent, 1912). Protractions are produced by contractions of muscles that form a sling

around the base of each whisker follicle. Retraction is a passive process, largely mediated by the viscoelastic properties of facial tissues (Dörfl, 1982; Carvell et al., 1991), but it can be actively assisted by muscles that move the underlying mystacial pad (Berg and Kleinfeld, 2003).

Whisking movements provide a means for actively sensing the proximate environment directly in front of the rat. Because most sensory receptors respond preferentially to changes in the sensed signal, active movement of the sensory organs enhances these changes, even when the environment is stationary. Whisking also facilitates hyperacuity, a process that enables this sensory system to achieve a higher effective resolution than is allowed by the peripheral receptor size or spacing.

The muscles that control individual vibrissae and mystacial pad movements are innervated by the buccal branch of the facial motor nerve (nerve VII) (Dörfl, 1985). This nerve can be readily transected without affecting the sensory innervation of the vibrissae (Semba and Egger, 1986). This paradigm has been proved useful for assessment of the role of whisking in the tactile function of vibrissae (Krupa et al., 2001) and in the development of the vibrissa sensory system (Nicolelis et al., 1996). The results of these studies indicate that the ability of rats to discriminate objects at high resolution, by using only their vibrissae, is dependent on intact whisking behavior. Furthermore, these studies have shown that the development of normal tactile perception is dependent on intact whisking behavior during early stages of development. As such, whisking performs the critically important function of creating motion between the vibrissa hair shaft and objects that it contacts. Rats whisk as they probe their immediate environment for the presence of objects, obstacles, or food. This carefully regulated motor action is, therefore, linked intimately to the acquisition and processing of ecologically important tactile sensory information.

METHODS USED TO ANALYZE
THE KINEMATICS OF WHISKING

The most effective means of observing and measuring the dynamic activity of vibrissae during exploratory whisking behaviors is by using high-speed recording techniques. Early, low-resolution analyses of vibrissae movements were described by recording the rat with motion picture film as it was engaged in exploratory behavior (Welker, 1964). Videographic analysis replaced film as photographic technologies advanced, however, both the spatial and temporal resolutions of videography are low, and the quantitative analysis of these data is extremely time consuming (Carvell and Simons, 1990). On-line tracking of individual whisker movements using optoelectronic and piezo-electric methods, which have very high spatial (26 μm) and temporal (1 millisecond) resolution, have provided significant advances in the kinematic analysis of vibrissae movements (Bermejo et al., 1998; Bermejo and Zeigler, 2000; Bermejo et al., 2002). However, behavioral analysis is limited by the fact that the head of the rat must be fixed to isolate movements of the vibrissae alone. The continued development of high-speed digital video cameras now permits monitoring of whisker displacement in freely moving animals along two dimensions (from the side and overhead) at a rate of as high as 1000 frames per second using a shutter speed of 200 microseconds (Hartmann et al., 2003). This level of resolution allows the analysis of whisker movements in behaving animals at frequencies far above typical whisking ranges, including resonant frequencies, which are believed to be required for texture discrimination (Hartmann et al., 2003; Neimark et al., 2003).

SENSORY ASPECTS OF
VIBRISSAE FUNCTION

The important role of vibrissal sensation in guiding rat behavior was first systematically investigated in rats by Vincent (1912), who as-

sessed their ability to learn and navigate a maze after discrete sensory manipulations. The strategy that she and most of her successors have used was to establish the normal function of vibrissae by assessing changes in behavior that followed vibrissal removal. Moreover, it was readily apparent to her that rats do not use the vibrissal sense in exclusion of other sensory systems. It is, therefore, necessary to exclude the potential contribution of the other senses in guiding behavior to assess the particular role of vibrissae. By systematically removing the vibrissae, the eyes, and the olfactory bulbs individually or in combination, Vincent was able to conclude that the vibrissae are "delicate tactile organs, which function in equilibrium, locomotion, and the discrimination of surfaces in distinct ways . . ." (1913, p. 69). Most of her conclusions have been confirmed and reaffirmed over the years by the few stalwart researchers who have also taken up this area of study. Many of the methods have been refined or enhanced by technological advances, and new details regarding the role of vibrissae have emerged, but the conclusions are essentially the same—the rat uses its vibrissae to determine the position, size, texture, and shape (identity) of objects that it encounters in its environment. The remainder of this chapter details the most relevant of these studies that have been applied in an attempt to define the role of sensory processing by the vibrissal system in these functions.

OBJECT POSITION

The deflection of vibrissae in the sensory array, individually or in combination, encodes the spatial location or position of objects, vertically and horizontally, in vibrissal space. A head-orientating movement seen in response to active palpation of the vibrissae with a fine wooden dowel is a simple and effective way of assessing general sensitivity of the vibrissal system. Another simple task, the forelimb placing task, takes advantage of the instinctual response that a rat makes, when suspended by the torso with the limbs hanging freely, when the vibrissae touch a surface. Animals with an intact and functioning vibrissal sensory system reach with the forelimb ipsilateral to the stimulated vibrissae and place it on the surface of the object. Animals with lesions of the whisker to cortex sensory pathway show an impaired response or none at all.

DISTANCE DETECTION

Rats can determine their distance from an object relying solely on information provided by the whiskers. This ability is exemplified by a simple task that requires a rat to traverse two elevated platforms that are separated by a variable sized gap to obtain a food reward (Hutson and Masterton, 1986). In the gap-crossing task, it is necessary for the animals to be blinded, blindfolded, or tested in the dark, to restrict the salient cues to that solely provided by vibrissae sensation. The rats are trained to cross a gap that is widened at 1-cm intervals until they can no longer step across it. At this point the animals must stretch their

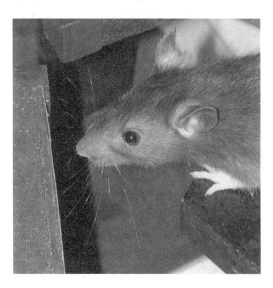

Figure 8–4. Rats use their vibrissae to scan the proximate environment for objects and to determine their position, size, texture, and shape. Here, a rat whisks its vibrissae across a textured stimulus to make the appropriate discrimination in a forced choice test and then jumps to the correct platform to receive a food reward.

bodies over the gap and extend their vibrissae to make contact with the reward platform and then to use this information to accurately guide their leap. A "feel-before-jumping" behavior must be reinforced in the first few days of training, by widening the runway gap to 60 cm. The rats cannot jump this distance, and if they try, they fall more than 32 cm to the benchtop. Trained rats with an intact vibrissae system can reliably span gaps of 16 cm. This distance is reduced to that spannable by its nose ($\approx$13 cm) when all of the vibrissae are removed (Hutson and Masterton, 1986), or the anatomical pathway from the periphery to the cortical representation of the vibrissae is damaged (Hutson and Masterton, 1986; Jenkinson and Glickstein, 2000). Removal of subsets of the vibrissae reduces the spannable gap distance to that sensed by the remaining vibrissae.

DETERMINATION OF OBJECT SIZE

A novel and ingenious behavioral task that has been used to determine the rat's ability to discriminate the size of an aperture using only their vibrissae was described by Krupa et al. (2001). The apparatus consists of a large reward area connected to a small discrimination chamber by a small passage. Animals are trained to discriminate the size of an aperture in the discrimination chamber, using only their large facial vibrissae, and to indicate their choice (wide or narrow) by poking their nose into either one of two spatially distinct sensors in the reward chamber. The animals are put on a water-restriction schedule so that correct responses can be reinforced by a water reward. The difference between wide and narrow apertures is systematically reduced to sensitively assess the rat's ability to undertake fine-grain distance detection. Rats were found to be able to discriminate between small differences in aperture width (3 mm) after 30 training sessions. By trimming the large mystacial vibrissae or inactivating the barrel cortex, the authors were able to establish that the ability to discriminate distances was dependent on an intact sensory pathway from the periphery to the cortex. However, abolition of active whisking by transection of the facial nerve had no deleterious effect, indicating that movement of the body through space provides sufficient mechanical activation of the appropriate vibrissae to make aperture size discriminations.

TEXTURE DISCRIMINATION

We explore the texture of an object by moving our fingers across it. Rats scan their environment and objects within it by moving their vibrissae. The presence of an object or ridges, grooves, and irregularities across its surface are transduced by deflections of the vibrissae shaft into signals from the sensory receptors in the follicle.

Vibrissa-based texture discrimination in rats has been studied using several means, but the simplest and easiest task was developed almost 100 years ago (Richardson, 1909), with variations on the design still effectively used today (Hutson and Masterton, 1986; Guic-Robles et al., 1989; Carvell and Simons, 1990; Prigg et al., 2002). In their basic form, tests of texture discrimination force the rat to make a choice between two stimuli varying only in the degree of roughness. The apparatus consists of a start platform and two choice platforms separated by an adjustable gap. During each trial, rats are required to stretch across the gap from the start platform to palpate the surface of a stimulus that is attached to the front of the choice platforms (as in Fig. 8–4). The length of the gap is adjusted so that the stimulus can be palpated only by the mystacial vibrissae and the animals are forced to make a choice by jumping to the correct platform. Rats are fitted with blindfolds (Carvell and Simons, 1990), binocularly occluded (Guic-Robles et al., 1989), or tested in the dark (with infrared light) to restrict the sensorium to only vibrissal sensation. If they make a correct choice, the food-restricted animals are availed with a food reward that is concealed behind a door on the correct platform. Rats trained in this task have been shown to reliably detect differences in surface textures as small as 30 μm spaced at 90 μm intervals, at a level comparable to primates using their fingertips (Carvell and Simons, 1990). When their whiskers are trimmed, accuracy falls to chance levels with even the coarsest discriminanda (Guic-Robles et al., 1989).

OBJECT RECOGNITION

Lashley (1950) first suggested that vibrissal sensation might endow the rat with the ability of form recognition. Although this hypothesis has often been restated, direct empirical confirmation of this contention was absent until very recently. In 1997, Brecht et al. described an object discrimination task that required sighted or blind rats to discriminate between an array of cookies of varying sizes and geometric forms using their vibrissae. Cookies were either sweetened or embittered with caffeine, which is odorless, to train the desired discrimination. Testing of sighted animals was performed under infrared illumination or in total darkness. The target cookie (small, sweet triangle) was presented in random positions among 15 distractors in a 4 × 4 array. By observing whisker movements and behavioral changes after selective whisker removal, the authors were able to doubly dissociate two distinct vibrissal systems. They determined that the longer, laterally oriented macrovibrissae were critically involved in spatial tasks, such as distance detection, but were not necessary for object recognition. On the other hand, the microvibrissae were found to be essential for object recognition but not for spatial tasks.

VIBRISSA SYSTEM PLASTICITY

The behavioral tasks that are described in this chapter provide the means with which to determine the functional capabilities of the rat's vibrissa sensory system. Behavioral studies applied to the vibrissal system are vastly outnumbered by those detailing its anatomy and physiology. The amalgamation of these areas and their application to complex behavioral processes such as perceptual learning and memory (Harris et al., 1999; Harris and Diamond, 2000), or the role of vibrissal sensation in activity- and experience-dependent modifications of central neuronal networks, during development and in adulthood (see Fox, 2002, for a review), is a large and growing field and holds promise for keeping behavioral scientists busy for decades to come.

CONCLUSIONS

Rats can use their mystacial vibrissae to perform a variety of tactile discriminations and behaviors. Using behavioral tasks, the vibrissal array has been demonstrated to function as an active, skin-like receptive surface used for fine-grained texture discriminations. In addition, the vibrissae can be likened to a retina-like sensor that uses the large, peripheral macrovibrissae to detect coarse components of the sensory field, such as object location and distance, while the microvibrissae are analogous to the retinal fovea, involved in fine-grained resolution required for the accurate discrimination of objects. Because of the one-to-one correspondence of a vibrissa with its cognate cortical barrel (Fig. 8–3), the rat vibrissal system provides an excellent system for studies designed to understand the relationship between sensory function and its central nervous system representation.

BIBLIOGRAPHY

Berg RW and Kleinfeld D (2003) Rhythmic whisking by rat: retraction as well as protraction of the vibrissae is under active muscular control. Journal of Neurophysiology 89:104–117.

Bermejo R and Zeigler HP (2000) "Real-time" monitoring of vibrissa contacts during rodent whisking. Somatosensory and Motor Research 17:373–377.

Bermejo R, Houben D, Zeigler HP (1998) Optoelectronic monitoring of individual whisker movements in rats. Journal of Neuroscience Methods 83:89–96.

Bermejo R, Vyas A, Zeigler HP (2002) Topography of rodent whisking—I. Two-dimensional monitoring of whisker movements. Somatosensory and Motor Research 19:341–346.

Brecht M, Preilowski B, Merzenich MM (1997) Functional architecture of the mystacial vibrissae. Behavioral Brain Research 84:81–97.

Broughton SD (1823) On the use of whiskers in feline and other animals. London Medical and Physical Journal 49:397–398.

Carvell GE and Simons DJ (1990) Biometric analyses of vibrissal tactile discrimination in the rat. Journal of Neuroscience 10:2638–2648.

Carvell GE, Simons DJ, Lichtenstein SH, Bryant P (1991) Electromyographic activity of mystacial pad musculature during whisking behavior in the rat. Somatosensory and Motor Research 8:159–164.

Dörfl J (1982) The musculature of the mystacial vibrissae of the white mouse. Journal of Anatomy 135:147–154.

Dörfl J (1985) The innervation of the mystacial region of the white mouse. Journal of Anatomy 142:173–184.

Fanselow EE and Nicolelis MA (1999) Behavioral modulation of tactile responses in the rat somatosensory system. Journal of Neuroscience 19:7603–7616.

Fox K (2002) Anatomical pathways and molecular mechanisms for plasticity in the barrel cortex. Neuroscience 111:799–814.

Guic-Robles E, Valdivieso C, Guajardo G (1989) Rats can learn a roughness discrimination using only their vibrissal system. Behavioral Brain Research 31:285–289.

Gustafson JW and Felbain-Keramidas SL (1977) Behavioral and neural approaches to the function of the mystacial vibrissae. Psychological Bulletin 84:477–488.

Harris JA and Diamond ME (2000) Ipsilateral and contralateral transfer of tactile learning. Neuroreport 11:263–266.

Harris JA, Petersen RS, Diamond ME (1999) Distribution of tactile learning and its neural basis. Proceedings of the National Academy of Science, USA 96:7587–7591.

Hartmann MJ, Johnson NJ, Blythe Towel R, Assad C (2003) Mechanical characteristics of rat vibrissae: resonant frequencies and damping in isolated whiskers and in the awake behaving animal. Journal of Neuroscience 23:6510–6519.

Hutson KA and Masterton RB (1986) The sensory contribution of a single vibrissa's cortical barrel. Journal of Neurophysiology 56:1196–1223.

Ibrahim L and Wright EA (1975) The growth of rats and mice vibrissae under normal and some abnormal conditions. Journal of Embryology and Experimental Morphology 33:831–844.

Jenkinson EW and Glickstein M (2000) Whiskers, barrels, and cortical efferent pathways in gap crossing by rats. Journal of Neurophysiology 84:1781–1789.

Jones EG and Diamond IT (eds.) (1995) The barrel cortex of rodents. New York: Plenum Press.

Krupa DJ, Matell MS, Brisben AJ, Oliveira LM, Nicolelis MA (2001) Behavioral properties of the trigeminal somatosensory system in rats performing whisker-dependent tactile discriminations. Journal of Neuroscience 21:5752–5763.

Land PW and Akhtar ND (1999) Experience-dependent alteration of synaptic zinc in rat somatosensory barrel cortex. Somatosensory and Motor Research 16:139–150.

Lashley K (1950) Personal communication. In: Handbook of psychological research on the rat (Munn NL, ed.). New York: Houghton-Mifflin.

Neimark MA, Andermann ML, Hopfield JJ, Moore CI (2003) Vibrissa resonance as a transductino mechanism for tactile encoding. Journal of Neuroscience 23:6499–6509.

Nicolelis MA, De Oliveira LM, Lin RC, Chapin JK (1996) Active tactile exploration influences the functional maturation of the somatosensory system. Journal of Neurophysiology 75:2192–2196.

Oliver RF (1966) Histological studies of whisker regeneration in the hooded rat. Journal of Embryology and Experimental Morphology 16:231–244.

Pocock RI (1914) On the facial vibrissae in the mammalia. Proceedings of the Zoological Society of London 889–912.

Prigg T, Goldreich D, Carvell GE, Simons DJ (2002) Texture discrimination and unit recordings in the rat whisker/barrel system. Physiology and Behavior 77:671–675.

Richardson F (1909) A study of sensory control in the rat. Psychology Reviews Monograph Supplement 12:1–124.

Sachdev RN, Sato T, Ebner FF (2002) Divergent movement of adjacent whiskers. Journal of Neurophysiology 87:1440–1448.

Semba K and Komisaruk BR (1984) Neural substrates of two different rhythmical vibrissal movements in the rat. Neuroscience 12:761–774.

Semba K and Egger MD (1986) The facial "motor" nerve of the rat: control of vibrissal movement and examination of motor and sensory components. Journal of Comparative Neurology 247:144–158.

Simons DJ (1983) Multi-whisker stimulation and its effects on vibrissa units in rat SmI barrel cortex. Brain Research 276:178–182.

Simons DJ (1985) Temporal and spatial integration in the rat SI vibrissa cortex. Journal of Neurophysiology 54:615–635.

Sokolov VE and Kulikov VF (1987) The structure and function of the vibrissal apparaus in some rodents. Mammalia 51:125–138.

Vincent SB (1912) The functions of the vibrissae in the behavior of the white rat. Behavior Monographs 1:7–81.

Vincent SB (1913) The tactile hair of the white rat. Journal of Comparative Neurology 23:1–36.

Welker WI (1964) Analysis of sniffing of the albino rat. Behaviour 12:223–244.

Zucker E and Welker WI (1969) Coding of somatic sensory input by vibrissae neurons in the rat's trigeminal ganglion. Brain Research 12:138–156.

Olfaction

9

BURTON SLOTNICK, HEATHER SCHELLINCK, AND RICHARD BROWN

RATS ARE MACROSMATIC

Mammals that have a well-developed olfactory system and a keen sense of smell and that largely depend on olfactory cues for operating in their environment are classified as macrosmatic (Moulton, 1967; Rouquier et al., 2000). This group comprises virtually all mammals except primates. Although some macrosmatic mammals, such as bats and chinchillas, have other specialized sensory systems for detecting prey or avoiding predators, for the most part, their single most important exteroceptive sense is smell. In the rat and other rodents, the olfactory epithelium is complex and densely packed with olfactory sensory neurons, the olfactory bulbs are large relative to the rest of the forebrain, and projections from the olfactory bulb not only terminate in essentially all of the cortex below the rhinal fissure, a cortical field that is probably the largest sensory cortex in the rodent brain, but also extend into the amygdala and hypothalamus. All of these anatomical features are much reduced in the brain of primates that, in contrast to rodents, have a well-developed visual system. To borrow a phrase from Freud, anatomy, in this regard, is destiny and virtually every aspect of rodent life is guided by and often largely dependent on the sense of smell.

The extent to which smell dominates the life of the rat may not be fully appreciated by investigators who use primarily nonolfactory stimuli in studies of learning. Olfaction is the first sensory system to become functional, and even prenatal or perinatal exposure to odorants can affect later postnatal behavior (e.g., Pederson and Blass, 1982; Smotherman, 1982; Hepper, 1990; Hudson, 1993; Abate et al., 2002). Infant rodents are easily conditioned to approach or avoid odors (Sullivan and Wilson, 1991) and quickly learn to prefer the familiar odors of their mother and nesting material and avoid novel odors, and odors associated with foodstuffs very early in life can have long-term effects on food preference. As adults, rats communicate largely by odors; they leave odor marks and trails that identify their age, sex, territory, and dominance status (Rainey, 1956; Brown, 1985; Galef and Buckley, 1996). These odor signals are released from specialized scent glands and are in urine and feces and even in the animal's breath. Indeed, a considerable literature attests to the influence of breath odors in social transmission of food preferences (see Chapter 34). Perhaps more familiar to most investigators are the "Bruce effect," the "Whitten effect," and the "Vandenberg effect," phenomena primarily demonstrated in mice, that reveal dramatic changes in behavior and reproductive status produced by exposure to a conspecific's odor stimuli. A number of other behavioral effects largely or entirely dependent on olfactory cues, including the communication of dominance and sexual status, nesting behavior, kin recognition, and, in the neonate, nipple attachment and suckling, are not graced by the names of investigators but provide additional evidence for a dominant role of olfaction in rodents.

Brown (1979) classified these various signals into two categories: those that reflect the emotional status or level or arousal of the individual and those that identify the individual. Emotive odors are those produced or released only in special circumstances such as sexual arousal, maternal behavior, or fear. Identifier odors are those produced by the body's normal metabolic processes that are stable over time. The latter may be influenced by dietary conditions (Schellinck et al., 1997) but appear largely determined by genetic factors. A major effort in contemporary olfactory research has focused on determining both the genetic origin and the chemical expression of these olfactory cues. Behavioral, genetic, and biochemical analyses have revealed that both the major histocompatibiity complex (Boyse et al., 1991; Schellinck et al., 1995; Schellinck and Brown, 1999; Schaefer et al., 2002; Beauchamp and Yamazaki, 2003) and the major urinary proteins (Brennan et al., 1999; Hurst et al., 2001; Nevison et al., 2003) appear to be the source of a unique chemical signature that can be learned and remembered by rodents.

Recently, a quite different line of olfactory studies emerged that were driven by two sets of findings. First, when provided with odor cues, rats showed a remarkable ability to learn simple and complex discrimination tasks (Dusek and Eichenbaum, 1997; Slotnick, 2001). Second, a series of molecular biological and anatomical studies identified olfactory receptor genes and the organizational principles governing the projections of olfactory sensory neurons onto the olfactory bulb (Buck, 1996; Mombaerts et al., 1996). These later findings provided the basis not only for a now widely accepted view of how odors are coded by the brain but also the generation of a variety of gene-targeted mice in which selected features of the olfactory system are altered. Behavioral studies are being designed to exploit these findings, to assess potential changes in olfaction in genetically modified mice (e.g., Zufall and Munger, 2001), and to test hypotheses generated by advances in the molecular biol-

ogy of the olfactory system. Such investigations require the development of sophisticated methods for generating and controlling odor stimuli together with behavioral tests for psychophysical analyses of olfactory function.

SPECIAL PROBLEMS IN CONTROLLING THE STIMULUS

The generation, control, and measurement of odors present the investigator with special problems, which are not encountered in dealing with stimuli for other modalities (Dravnieks, 1972, 1975). For mammals, odor stimuli are gases, generated by vaporization from some odorant substance. Most natural odors come from a complex source whose components are generally difficult to identify but vaporize at different rates, resulting in a stimulus whose constituents vary over time. The contribution of each component of this complex will be a function of individual masses and vapor pressure of each, the partial pressure of the component in the odorant mixture, the substrate on which it is deposited, relative humidity, and, of course, the extent to which each component changes with changes in temperature. Unless controlled, the vapor itself will diffuse into the atmosphere, perhaps form "odor plumes" carried by prevailing air currents or simply become increasingly less concentrated as distance from the point source increases.

Unfortunately, there are no simple devices to measure odor concentration or identify the components of a vapor or, for psychophysical tests, to generate "pure" odor stimuli or, for that matter, nonodorous stimuli. Even tubing systems designed to control vapor flow must contend with a variety of problems, including determining optimal flow parameters, potential contamination by component materials, and adsorption of odorant molecules on tubing walls. A further complication, and often the primary topic of a research endeavor, is the biological constraints on the effectiveness of vapors in stimulating olfactory receptor neurons. Thus, sensory neurons may be relatively unresponsive to the major component of a vapor complex but be exquisitely sensitive to quite minor components.

BEHAVIORAL METHODS: DIFFUSION OF ODORS FROM A POINT SOURCE

Olfactory studies have used a very wide variety of methods and these lend themselves to a number of potential classificatory schemes. Because of the unique problems associated with controlling odor stimuli, we consider separately methods used in those studies that exert little or no control of the stimulus, allowing the odor to passively diffuse from a point source, and those that use olfactometric devices in an attempt to generate, control, and deliver defined quantities of the stimulus.

Behavioral methods used in early studies of olfaction in rats were relatively simple and, for the most part, the effects obtained were quite strong and did not require either precise control of the stimulus or control of stimulus sampling by the subject. Such methods continue to be used in studies examining learning and memory in rats. Nonetheless, the often idiosyncratic methods and limited control of the stimulus used in such studies virtually preclude replication from one laboratory to another or meaningful comparisons among their outcomes. A description of the some of these tests and their potential limitations follows.

HABITUATION TESTS

These tests are based on the observation that successive presentations of the same stimulus odor will result in a decrease in investigatory behavior, that is, habituation. Then, when a different odor stimulus is presented to the same subject, the habituated response will reoccur or will be "dishabituated." This behavioral test has been used to determine if the second stimulus is perceived as different from

first. The tests are of two types: the habituation-discrimination test and the habituation-dishabituation test.

In the habituation-discrimination test, one odor is presented a number of times, usually in a small odor pot or on a filter paper, and the response to it is measured; then, the initial stimulus and a second novel odor are presented simultaneously. If the rat investigates the novel odor more often than the original stimulus, one can conclude that it is able to discriminate between the two odors. Thus, the test provides a simple assessment of odor memory. The time between the first and second phase of the test varies depending on the nature of the research question. The test has been used frequently to assess memory for the odors of conspecifics. In some instances, rather than presenting odors, the whole animal is presented. This form of the test is often referred to as the social recognition test (Bhutta et al., 2001).

The procedure in the habituation-dishabituation *test* is similar to that just described except that only one odor is presented at a time in both phases of the task. The habituation phase involves repeated presentations of one odor, and in the dishabituation phase, samples of a second odor are presented. In some instances, the test begins with a no-odor-adaptation period during which the subject is put in the test chamber and presented with the odor vehicle several times (Brown, 1988; Schellinck et al. 1995). Sundberg et al. (1982) devised a standard approach to scoring the behavioral response of rats during initial investigation and habituation of the odors. This method has been used extensively to provide evidence that congenic strains of rats that differ only in the genes of the major histocompatibility complex produce discriminably different urine odors.

The principal benefit of both of these tests arises from the simple design, minimal need for equipment, and speed and ease of testing. An advantage of the habituation-dishabituation task is that it eliminates the problems associated with creating a mixture of test odors. Nonetheless, both forms of the test are somewhat time consuming, require continuous observation of the subjects and the measurement of a not always well-defined "investigative behavior." Moreover, a null outcome is open to several interpretations. When dishabituation occurs, it seems clear that the two odors can be discriminated. The failure of the test stimulus to produce dishabituation could reflect a failure to discriminate, a disinterest in the test odor, or that the test odor is discriminable but not sufficiently different to produce a clear dishabituation effect.

UNCONDITIONED PREFERENCE TESTS

These tests are analogous to two bottle preference tests used in studies of taste. Quite simply, two or more odors are placed in a test arena, and the frequency and/or duration of a subject's approach to and investigation of each odor is recorded. Usually, the test chamber includes a neutral area in which the animal is contained before exposing it to the odors. To reduce the possibility that the subject's responses to the odors will be limited because of either a neophobic response or generalized investigation, it is appropriate to include a habituation period to the test chamber before introducing the odor stimuli (Schellinck et al., 1995). The experimenter generally records the time the subject spends investigating each odor and thus should be blind with regard to experimental conditions. If possible, the experiment should be videotaped or a computerized tracking system used so that rater reliability can be assessed. Although there is an increasing trend toward testing animals in their home cages, this is not recommended because rodents tend to sniff indiscriminately at all novel objects so presented.

Preference tests can be also be used with neonatal and juvenile rats. Because of their limited motor abilities, a simple two-choice apparatus is used for testing pups. Often, the pups are placed on a mesh floor and the odors diffuse through the mesh from containers placed below. To avoid the recording of false-positive results because of chance movements, a middle neutral zone can be incorporated into the test apparatus. Differences in defining what behavior constitutes a choice makes it difficult to compare results among studies. For example, does turning the head toward the odor constitute a preference, or must half of the subject's body extend into the preference zone?

Except when very strong preference effects are obtained (e.g., Kavaliers and Ossenkopp, 2001), it is often unclear how the results from such tests should be interpreted. Greater investigation of one odor may reflect an absolute or a relative preference for one of the test odors, the relative novelty of each odor, a difference in detectability of the odors, or possibly an aversion to the least-sampled odor (Brown and Wilner, 1983; Amiri et al., 1998). A failure to display a preference may result from both odors being equally attractive or aversive. A variety of factors that influence activity, including the subject's circadian rhythm and the ambient temperature, may also affect test outcomes. The use of preference tests also has a number of serious shortcomings regarding the control of odor dispersion. When two or more odors are presented simultaneously, their vapors may mix, consequently increasing the difficulty in discriminating between them. Moreover, because the odors will reach the subject over time regardless of whether the animal investigates them, there is a possibility that an animal need not approach an odor directly to show a preference. It is also possible that the rat may habituate to the odor before showing a measurable response.

SIMPLE ASSOCIATIVE LEARNING TESTS

The static presentation of odors has also been used in associative learning paradigms with both young and adult rats. In these tasks, one odor is presented with reinforcement and a second odor is presented without any associated reward. A number of daily sessions are usually run in a quasi-randomized order for a specified number of days. To assess the success of the task, preference tests may follow odor conditioning. If an animal investigates or spends more time in the vicinity of the odor in the absence of reinforcement, this is considered to be an indication of the effectiveness of the conditioning procedure. Johanson and Teicher (1980) paired an odor with oral infusions of milk in a warm environment to create a conditioned odor preference in neonatal rats. Sullivan, Leon, and colleagues (Sullivan et al., 1991, 1994; Johnson and Leon, 1996) used a similar conditioning paradigm to examine the neurobiology of olfactory learning. They exposed neonatal rats to an odor paired with tactile stimulation (i.e., stroking with a sable brush) that simulates maternal contact. A Y-maze preference test was used to determine if the conditioning was successful. Schellinck and colleagues developed a simple discrimination task to assess odor learning in adult mice and rats (Brennan et al., 1998; Fairless and Schellinck, 2001; Forestell et al., 2001; Schellinck et al., 2001b). The animal is presented with a pot in which finely cut wood chips are diffused with an odor from a lower level. One odor is paired with sugar reinforcement buried in the woodchips (CS+ stimulus) and a second odor is presented alone (CS− stimulus). In a subsequent preference test, the rodent is presented with both odors but no sugar. The animals only need to be food restricted before the preference test, and the use of digging rather than sniffing provides an objective and easily scored measure of preference. In the preference test, the subjects dig consistently and almost exclusively in the woodchips containing the CS+ odor. Mice rapidly learn the task and remember the discrimination for at least 90 days (Schellinck et al., 2001a and unpublished observations). Studies to assess the long-term memory of rats using this task are yet to be completed. The appetitive learning paradigms described here

are easy to set up and are useful for assessing the neurobiological basis of learning and memory (Wilson and Sullivan, 1994; Forestell et al., 1999, 2002). Given the extraordinary robustness of the effect, the test is not particularly useful for assessing higher cognitive functions.

Mazes were among the first tests used to determine if rodents could learn to make a discrimination between the correct or reinforced stimulus and the incorrect or unreinforced stimulus (Bowers and Alexander, 1967). They continue to be popular. For example, the radial arm and more simple mazes have been used to examine odor memory and, learning (Staubli et al., 1986; Reid and Morris, 1992; Steigerwald and Miller, 1997) and, in mice, odor-cued maze learning has been used in an extensive series of studies of volatile signals of the major histocompatibility complex (Singer et al., 1997; Yamazaki et al., 1979; Beauchamp and Yamazaki, 2003). As discussed by Stevens (1975), a number of precautions should be observed in using mazes to study odor detection. The reinforcement itself should not be a cue as rats have been shown to discriminate between the odor of a 45 mg food pellet and clean air in a T-maze (Southall and Long, 1969). An enclosed start box should be used to prevent the experimenter from unintentionally cueing the subject in any fashion, and the experimenter should be unaware of the location of the correct stimulus. Rats have been shown to track odors of themselves and other rats (Wallace et al., 2002), so it is essential that the arms of the maze are cleaned between trials and experimental sessions. Clearly, unless special precautions are taken, mazes are not well suited for studies of odor learning. Further, it may be difficult to compare results from maze studies and those that use more rigorous methods. Thus, the failure of Reid and Morris (1992) to replicate aspects of olfactory stimulus control originally demonstrated using an olfactometer (Slotnick and Katz, 1974; Nigrosh et al., 1975) was probably due, in part, to shortcomings in using a maze to study complex olfactory learning.

Rats are quite adept at digging in a substrate to obtain a food reward, and various forms of digging tasks are often used to assess olfaction in both mice and rats (Berger-Sweeney et al., 1998; Zagreda et al., 1999; Mihalick et al., 2000; Schellinck et al., 2001b). The paradigm requires that a food-restricted animal find food reinforcement by digging in a substrate of sand, gravel, or bedding material for a food particle. Finding buried food has been widely used as a simple measure of whether some experimental procedure has disrupted or eliminated the sense of smell (Alberts and Galef, 1971; Hendricks et al., 1994; Genter et al., 1996). Despite their popularity and simplicity, these tests have the various shortcomings common to those using uncontrolled diffusion of odors from a point source and, in addition, have proved to be a poor predictor of anosmia (Xu and Slotnick, 1999; Slotnick et al., 2000a).

Examples of better and more interesting uses of digging behavior as a measure of odor detection and discrimination are tests used by Eichenbaum and associates. Rats are trained to dig in a container of odorized sand for a food reward and later tested for their ability to detect which of several containers has the odor target. Variants of these test methods have been used to demonstrate complex odor-based learning in rats in which correct responding is dependent on the configuration of the stimuli and not simply on associations between individual stimuli and reinforcement (e.g., Dusek and Eichenbaum, 1997; Fortin et al., 2002; Van Elzakker et al., 2003).

Conditioned odor aversion (COA) and odor-cued taste avoidance are two other associative learning methods for assessing odor detection and discrimination. Procedures in producing a COA are analogous to those used in the better known conditioned taste aversion paradigm. What is perhaps less well known is that odors can be effective conditioned stimuli for aversion learning and, when presented as intraoral stimuli, can support single-trial and long-delay learning (Slotnick et al., 1997). Indeed, COA has been demonstrated in rats of all ages, including those in the prenatal and early postnatal stages of life (Rudy and Cheatle, 1979, 1983; Smotherman, 1982; Smith et al., 1993).

Odor-cued taste avoidance learning requires only that rats accustomed to drinking from a spout be exposed on the training trial to a bitter and odorous solution. Learning occurs in only one or two trials, and in subsequent tests, rats sniff at the drinking spout but do not sample the liquid if the odor is detected (Darling and Slotnick, 1994). These tasks have an advantage over food-finding tasks that have not been fully exploited: both are easy to implement, allow reasonably good control of the odor stimulus, and support single-trial learning and long-term memory for odors.

OLFACTOMETRY

CONTROL OF THE STIMULUS

Olfactometers are devices that generate an odor whose concentration, flow parameters, and temporal parameters can be specified, controlled, and varied. When combined with operant conditioning, olfactometers provide powerful tools for training rats to attend to and differentially respond to selected features of the stimulus. Although the term *olfactometer* is often applied to any device that allows some control over odor flow, including mazes equipped with fans for directing air currents, those meeting these criteria are based largely on the design principles first described by Tucker (1963) and Moulton and Marshall (1976). These devices can generate odors either by a series of air dilutions of an odorant saturated vapor or from the headspace of an odorant dissolved in an odorless liquid. The latter method is by far the easiest to instrument and has the advantage that multiple channels, each with its own odorant solution, can be incorporated.

Olfactometers for use with rats have evolved from relatively crude devices (Williams and Slotnick, 1970) or complex multicomponent devices (Bennett, 1968; Sakellaris, 1972) to simpler and computer-controlled designs that are easily cleaned and maintained. Olfactometers and training methods designed for research with rodents have been described in detail (Slotnick and Schellinck, 2002). The multiple-channel liquid dilution system (Bodyak and Slotnick, 2000) is relatively easy to construct and provides the control needed for studies of odor sampling behavior, simple odor discrimination, discrimination between odor mixtures, absolute and intensity difference detection thresholds for monomolecular odors, odor masking, odor memory, odor quality identification, and other odor sensory tasks that demand reasonably precise presentation of the stimulus. As described by Slotnick and Schellinck (2002), the ease with which the multiple-channel liquid dilution can be cleaned, together with the use of pinch valves and disposable tubing and odor saturator containers, has greatly minimized odor contamination of components, a problem that plagues most olfactometer designs.

ODOR CONTROL OF BEHAVIOR

In a series of published and unpublished studies, we assessed a variety of operant training procedures and methods for presenting the stimulus. Initially, rats were trained in a wind tunnel and were essentially immersed in the stimulus on each trial (Slotnick and Nigrosh, 1974). When it became apparent that rats were easily trained to sample odor stimuli, we used a simple Plexiglas chamber fitted with a vertically oriented glass tube for presenting odors. Traditional operant shaping procedures were used to train the rat to sample an odor by inserting its snout into an opening in the side of the tube. Snout insertions, detected with a photobeam, initiated a trial in which odorized air was added to a constant stream of clean air for 1 to 2 seconds. A discrete trials, go, no-go procedure was used in which a designated response made after sampling the

S+ stimulus was rewarded with water. A response made after sampling the S− stimulus was neither reinforced nor punished. The response "manipulandum" was a stainless steel water delivery tube located outside of the odor-sampling tube. Rats were required to sample the stimulus (i.e., keep their snout in the odor sampling tube) for some minimum time (generally 0.15 second) before responding and shorter samples resulted in immediately aborting the trial and repeating it on the next trial. The designated response was making a fixed number of licks (generally 10 or more) on the water delivery tube.

As described later, this training procedure was remarkably effective in gaining stimulus control of behavior. Other training methods, including using symmetrical reinforcement with the go/no-go procedure or using two water delivery tubes, each associated with a different odor, proved far more troublesome and less efficient. We also rejected the use of the more traditional free operant discrimination procedure because the extended exposure to the stimulus required to determine a response rate might produce adaptation. In an effort to minimize adaptation effects we initially used a 60- or 90-second intertrial interval. It quickly became apparent that a much shorter intertrial interval was preferred by rats, and they worked far more efficiently with the minimum interval we thought was needed for the clean air stream to clear the odor-sampling tube between trials (≈5 seconds). With these parameters, well-motivated rats completed a trial, on average, every 12 to 15 seconds and could maintain essentially perfect performance with suprathreshold stimuli over hundreds of trials. The number of trials that could be given in a session was limited only by satiation effects and, with reinforcement volume of 0.04 ml, 400 to 600 trials could be run in daily sessions. Such multiple trials sessions are particularly useful for psychophysical studies.

Despite the fact that responding on S− trials (false alarms) was not punished, rats quickly acquired odor detection and discrimination tasks, and it was clear that nonreinforcement after responding was sufficiently aversive to support response inhibition. Errors on S+ trials (misses) seldom occurred and, in general, more than 95% of errors were false alarms. Thus, acquisition functions were largely a function of learning to inhibit responding on S− trials.

Acquisition of simple odor discrimination tasks could be remarkably rapid (Nigrosh et al., 1975). For example, after extended training on S+-only trials, subsequent acquisition was essentially errorless; rats did not respond or made only one or two responses to the S− stimulus when it was abruptly introduced. Even without initial training on the S+ stimulus, rats almost always achieved near perfect performance on simple two-odor discrimination tasks within 20 to 60 training trials. Additional evidence for strong stimulus control by odors was that odor stimuli easily overshadowed auditory or visual stimuli but that even extended training on visual stimuli failed to disrupt acquisition of an odor discrimination. Further, when rats were trained equally well on both auditory and olfactory discriminations, responses in stimulus competition tests (e.g., pairing the S− auditory stimulus with the S+ odor stimulus) were almost entirely determined by the sign value of the odor stimulus (Nigrosh et al., 1975).

After training on several odor discrimination tasks, rats became quite sophisticated observers; when new stimuli were used, they carefully sampled the odor on each trial and quickly learned to inhibit responding to the stimulus not associated with reward. We exploited this finding in a series of learning-set studies designed to determine if rats, like primates, could acquire a win-stay/lose-shift response strategy when presented with a series of odor discrimination tasks. In both sequential discrimination reversal tasks and when given a series of novel two-odor discrimination tasks, the number of errors to a criterion of 90% correct responding in a block of 20 tri-

als rapidly decreased; by the end of training, most rats achieved criterion performance in the first block of trials, and some made no errors at all (Slotnick and Katz, 1974; Nigrosh et al., 1975; Slotnick, 1984). This level of performance was all the more remarkable because it was achieved after training on only 10 to 20 problems. Thus, on learning set tasks, rats trained with odors performed at least as well as did primates trained with visual stimuli and, like primates, demonstrated the acquisition of a response strategy—in Harlow's (1949) terms, they learned to learn. In response to criticisms of this work by Reid and Morris (1992), Slotnick et al. (2000b) fully replicated these learning set outcomes using a different olfactometer apparatus and a variety of test procedures.

In related studies, it was shown that rats readily acquired both matching-to-sample and non–matching-to-sample tasks when odors were used (Otto and Eichenbaum, 1992; Lu et al., 1993) and even showed evidence of learning to learn to match to sample (Lu et al., 1993). Their capacity to acquire many different odor discriminations and to remember odors was challenged by tasks that required discrimination among eight odors presented in random order within a training session (Slotnick et al., 1991). Not only did rats quickly sort out which of these odors were associated with reinforcement and which were not, but also, in repeated sessions using novel sets of eight odors, their performance rapidly improved, and by the seventh or eighth such set, most needed only three to five exposures to each odor to achieve criterion performance. When tested on a reversal task using odors from a set in the middle of the series, they made many errors, thus demonstrating memory for this earlier set of odors.

OLFACTORY PSYCHOPHYSICS

Determination of absolute detection threshold and other measures of odor sensitivity requires the precise control of the stimulus conferred by olfactometers. Psychophysical studies with rats have generally used either a modified staircase psychophysical procedure (Youngentob et al., 1997) or, more commonly, a modification of the descending method of limits procedure in which rats receive extensive training in detecting some initial suprathreshold concentrations of the stimulus. *Concentration*, defined as the percent air or liquid dilution of the odorant source, is then decreased in quarter-log or half-log steps, and training at each step is continued until the animal fails to reach some criterion level (generally 75% or 80% correct responding in a block of 20 or 40 trials) in a fixed number of trials. As might be expected, lower thresholds are obtained using small changes in odorant concentration and by overtraining at each concentration. Specifying the precise molecular concentration of the stimulus is fraught with a variety of technical and methodological problems (Dravineks, 1975) and often requires assumptions about vapor saturation levels and other factors that are, at best, difficult to confirm. Fortunately, however, only the most demanding sensory studies require precise measures of the physical stimulus and, for most behavioral studies, the relative change in sensitivity produced by some experimental manipulation provides a perfectly acceptable end point. Thus, for example, Apfelbach et al. (1991) showed the extent to which odor sensitivity varied with both age and density of olfactory receptor neurons, and Slotnick and Schoonover (1993) determined that transection of the lateral olfactory tract in the rat decreased odor sensitivity to amyl acetate by approximately 2.2 orders of magnitude.

Olfactometers lend themselves to other useful measures of odor sensitivity, including intensity difference threshold (Slotnick and Ptak, 1977; Slotnick and Schoonover, 1993) and odor masking (Laing et al., 1989). In the latter case, the rat may be required to discriminate between two odors, A and a combination of A and B. In subsequent sessions, the proportion of B in the mixture is gradu-

ally reduced until the discrimination can no longer be made. Generally, at that point, the concentration of B alone is well above detection threshold but is masked by the presence of the A stimulus. Variants of this discrimination task have been used (e.g. Laing et al., 1989; Lu and Slotnick, 1998; Dhong et al., 1999; Slotnick and Bisulco, 2003) and the test could easily be extended to examine discrimination of more complex mixtures of odors.

ODOR QUALITY PERCEPTION

Perhaps one of the most challenging tasks in experimental studies of olfaction is assessing odor quality perception—determining the perceived similarity between odors and whether some manipulation produces a change in the perceptual quality of an odor. Such issues are particularly relevant now because contemporary molecular biological and anatomical studies provide an empirical basis for several theories of odor coding (Xu et al., 2000). A challenge is to develop methods to test these theories at the level of behavior. The traditional procedure, and probably still the gold standard for measuring sensory quality perception in animals, is to determine a stimulus generalization gradient. But it is unclear whether this method can be applied in the olfactory modality because odors differ from one another along numerous dimensions and it is likely that some, as yet unidentified, combination of these dimensions are the determinants of odor quality. Alternative methods for assessing odor quality perception in rats have been reported, but none are completely satisfactory. One unique approach is that of Youngentob and his associates (1990, 1991). Rats were trained to sample an odor from a central port and then traverse a runway associated with that odor for a water reward. Five odors were used, and each odor signaled which of five runways had to be selected. It was assumed that errors in response choices would reflect the extent to which rats confused one odor with another and, hence, provide a measure of the perceptual similarity between odors. However, in practice it was found that rats did not learn the task easily but that once it was acquired, few errors were made, even between structurally similar odors. A quite different approach is that of Slotnick and Bodyak (2002), who trained rats to discriminate among structurally homologous odors and unrelated odors and later tested their memory for the positive and negative assignments of these odors. It was assumed that more errors in the memory test would be made on odors that were similar perceptually. The memory test was performed under extinction and rats had no feedback for correct or incorrect responding. Control rats in this study had near-perfect memory scores for both the structurally homologous and unrelated odors. Rats with olfactory bulb lesions performed almost as well and, hence, failed to show any marked effect of bulbar lesions on odor quality perception. Although this particular test proved useful for assessing potential changes in perception resulting from brain lesions, the excellent performance of control rats in the Slotnick and Bodyak study (2002) and in the Youngentob et al. studies (1990, 1991) indicate that different and more sensitive methods may be needed to index odor quality perception in rats.

OLFACTION, ANOSMIA, AND OTHER CHEMICAL SENSES

Olfactory sensory neurons are not the only sensory neurons that respond to vapor stimuli. Macrosmatic mammals possess a well-developed accessory olfactory system and sensory neurons in the vomeronasal organ may have quite low thresholds for odors (Leinders-Zufall et al., 2000). Indeed, determining whether the response to an odor is mediated by the main or accessory olfactory system (or both) is nontrivial and a matter of considerable interest in studies of pheromonal stimuli. There

are several experimental methods for disassociating the two systems: the vomeronasal organ can be removed without damage to the main olfactory epithelium (Wysocki et al., 1991) or the vomeronasal nerves, which are compact and travel along the medial aspect of the olfactory bulbs, can be transected (Fleming et al., 1992). It is far more difficult to eliminate the main olfactory system while leaving the accessory system intact. The (main) olfactory epithelium is extensive and an integral part of the respiratory system and it would be difficult to completely ablate it without serious complications. Olfactory nerves extend in a diffuse manner through the cribriform plate, making their transection problematical. Costanzo and colleagues (Costanzo, 1985; Yee and Costanzo, 1995) have, however, described the use of a specially constructed knife to transect these fibers, although that transection may also interrupt the accessory olfactory nerve. Some caustic agents and toxins may destroy olfactory sensory neurons but leave the accessory olfactory system intact (Setzer and Slotnick, 1998; Slotnick et al., 2000a), but confirmation of these outcomes requires careful histological analyses and effects may be critically dose dependent.

Zinc sulfate is a caustic metallic salt that destroys epithelial tissue, and syringing the nasal vault with this agent has been used in numerous studies in an attempt to produce anosmia in the rat. However, as generally practiced, the procedure is only partly effective in destroying the olfactory epithelium, and both anatomical studies and sensitive olfactometric tests reveal residual unaffected areas of the epithelium as well as considerable residual olfactory function in zinc sulfate–treated rats (Slotnick and Gutman, 1977; Slotnick et al., 2000a). The only completely reliable method for producing a frank anosmia (complete loss of smell) in rodents is surgical removal of the olfactory bulbs. Complete removal of the bulbs is essential because olfactory function may be mediated by even small remnants of olfactory bulb tissue (Lu and Slotnick, 1998). The surgery is somewhat difficult because the posterior aspect of the olfactory bulbs lies under the frontal pole cortex and some care is required to excise the bulbs completely without producing significant ancillary damage to surrounding structures. The lesions will necessarily invade the anterior olfactory nucleus whose anterior pole extends into the lateral aspect of the bulb. Completely olfactory bulbectomized adult rats have no deficits in performing the operant olfactometer task but respond at chance over hundreds of trials on a simple odor detection problem even when tested with relatively high concentrations of an odor (Slotnick and Schoonover, 1992). The emphasis here is on olfactory bulbectomy in the adult rat because there is evidence for regeneration and functional projections to the forebrain in the olfactory bulbectomized neonatal rat (Hendricks et al., 1994).

A potential confound in many studies of olfaction arises from cues that may be mediated by trigeminal receptors in the nasal vault or cornea, and sensory receptors in the trachea that can respond to certain vapors. Of these, a contribution of corneal receptors and those in the trachea can be ruled out in detecting odors if olfactory bulbectomy results in anosmia. However, the nasociliary branch of the opthalamic nerve provides sensory input to part of the mucous membrane of the nasal cavity. It travels just lateral to the olfactory bulb in the rat and is invariably transected when the bulb is removed. Thus, olfactory bulbectomized rats will also have reduced input from trigeminal receptors. Trigeminal receptors respond to irritants and, although they generally have a much higher threshold than olfactory sensory receptors, can respond to high concentrations of many commonly used odorants (Laska et al., 1997). Perhaps the simplest way to minimize or eliminate a potential contribution of these nonolfactory receptors in an olfactory study is to use odorant concentrations that are judged as nonirritating by human observers or, better, concentrations that are well below the known threshold for these receptors. Trigeminal and respiratory thresholds have been established for many commonly used odorants (Nielsen et al., 1984; Silver et al., 1986; Silver, 1992; Schaper, 1993; Cometto-Muniz et al., 2002).

CONCLUSIONS

Rats and other rodents live in an olfactory world and any attempt to understand rodent biology must take into account the importance of olfaction for social behavior, feeding, learning, and orientation in the environment. The role of odors in the control of rodent behavior has long been a primary topic in ethologically oriented studies of rat behavior, and the relatively simple tests used in these studies have served to demonstrate the influence and importance of odors. However, recent advances in odor control of learning and in the molecular biology of olfaction have required the use of more sophisticated test procedures and better control and understanding of the stimulus. No single test will serve the myriad facets of olfactory research but all olfactory tests must take into account the unique features of and problems in dealing with vapor state stimuli and controlling the attentive or sampling behavior of the subject.

REFERENCES

Abate P, Varlinskaya EI, Cheslock SJ, Spear NE, Molina JC (2002) Neonatal activation of alcohol-related prenatal memories: impact on the first suckling response. Alcohol Clinical and Experimental Research 26:1512–1522.

Alberts JR, Galef BG Jr (1971) Acute anosmia in the rat: a behavioral test of a peripherally-induced olfactory deficit. Physiology and Behavior 6:619–621.

Amiri L, Dark T, Noce KM, Kirstein CL (1998) Odor preferences in neonatal and weanling rats. Developmental Psychobiology 33:157–162.

Apfelbach R, Russ D, Slotnick B (1991) Olfactory development and sensitivity to odors in rats. Chemical Senses 16:209–218.

Beauchamp GK, Yamazaki K. (2003) Chemical signalling

in mice. Biochemistry Society Tranactions 31(Pt 1): 147–151.

Bennett MH (1968) The role of the anterior commissure in olfaction. Physiology and Behavior 3:507–515.

Berger-Sweeney J, Libbey M, Arters J, Junagadhwalla M, Hohmann CF (1998) Neonatal monoaminergic depletion in mice (Mus musculus) improves performance of a novel odor discrimination task. Behavioral Neuroscience 112:1318–1326.

Bhutta AT, Rovnaghi C, Simpson PM, Gossett JM, Scalzo FM, Anand KJ (2001) Interactions of inflammatory pain and morphine in infant rats: long-term behavioral effects. Physiology and Behavior 73:51–58.

Bodyak N and Slotnick B (2000) Performance of mice in an automated olfactometer: odor detection, discrimination and odor memory. Chemical Senses 24:637–645.

Bowers JM and Alexander BK (1967) Mice: individual recognition by olfactory cues. Science 158:1208–1210.

Boyse EA, Beauchamp GK, Bard J, Yamazaki K (1991) Behavior and the major histocompatibility complex of the mouse. In: Psychoneuroimmunology (Ader R and Felten D, eds.), pp. 831–846. San Diego: Academic Press.

Brennan PA, Schellinck HM, De La Riva C, Kendrick KM, and Keverne KB (1998) Changes in neurotransmitter release in the main olfactory bulb following an olfactory conditioning procedure in mice. Neuroscience 87:583.

Brennan PA, Schellinck HM, Keverne EB (1999) Patterns of expression of the immediate-early gene egr-1 in the accessory olfactory bulb of female mice exposed to pheromonal constituents of male urine. Neuroscience 90:1463–1470.

Brown RE (1979) Mammalian social odors: a critical review. Advances in the Study of Behavior 10:10–162.

Brown RE (1988) Individual odors of rats are discriminable independently of changes in gonadal hormone levels. Physiology and Behavior 43:359–363.

Brown RE and Mcdonald DW (1985) Social odours in mammals. Oxford: Oxford University Press.

Brown RE and Willner JA (1983) Establishing an "affective scale" for odor preferences of infant rats. Behavioral and Neural Biology 38:251–256.

Buck LB (1996) Information coding in the vertebrate olfactory system. Annual Review of Neuroscience 19:517–544.

Cometto-Muniz JE, Cain WS, Abraham MH, Gola JM (2002) Psychometric functions for the olfactory and trigeminal detectability of butyl acetate and toluene. Journal of Applied Toxicology 22:25–30.

Costanzo RM (1985) Neural regeneration and functional reconnection following olfactory nerve transection in hamster. Brain Research 361:258–266.

Darling FMC and Slotnick BM (1994) Odor-cued taste avoidance: A simple and efficient method for assessing olfaction in rats. Physiology and Behavior 55:817–822.

Dhong HJ, Chung SK, Doty RL (1999) Estrogen protects against 3-methylindole-induced olfactory loss. Brain Research 824:312–315.

Dravnieks A (1972) Odor measurement. Environmental Letters 3:81–100.

Dravnieks A (1975) Instrumental aspects of olfactometry. In: Methods in olfactory research (Moulton DG, Turk A, Johnston JW, eds.). New York: Academic Press.

Dusek JA and Eichenbaum HB (1997) The hippocampus and memory for orderly stimulus relations. Proceedings of the National Academy of Science 94:7109–7114.

Fairless DS and Schellinck HM (2001) Assessing the effects of 192-saporin lesions on rat olfactory learning and long term olfactory memory using a modified odour preference task. Canadian Conference on Brain and Behavioural Science, Laval University, Quebec City.

Fleming AS, Gavarth K, Sarker J (1992) Effects of transections to the vomeronasal nerves or to the main olfactory bulbs on the initiation and long-term retention of maternal behavior in primiparous rats. Behavioural and Neural Biology 57:177–188.

Forestell CA, Schellinck HM, Drumont S, Lolordo VM (2001) Effect of food restriction on acquisition and expression of a conditioned odor discrimination in mice. Physiology and Behavior 72:559–566.

Forestell CA, Schellinck HM, Lolordo VM, Brown RE, Wilkinson M (1999) Olfactory conditioning and immediate early gene expression in mice. Chemical Senses 24:618.

Fortin NJ, Agster KL, Eichenbaum HB (2002) Critical role of the hippocampus in memory for sequences of events. Nature Neuroscience 5:458–462.

Galef BG Jr and Buckley LL (1996) Use of foraging trails by Norway rats. Animal Behaviour 51:765–771.

Genter MB, Owens DM, Carlone HB, Crofton KM (1996) Characterization of olfactory deficits in the rat following administration of 2,6-dichlorobenzonitrile (dichlobenil), 3,3′-iminodipropionitrile, or methimazole. Fundamentals of Applied Toxicology 29:71–77.

Harlow H (1949) The formation of learning sets. Psychological Review 56:51–65.

Hendricks KR, Knott JN, Lee ME, Gooden MD, Evers SM, Westrum LE (1994) Recovery of olfactory behavior. I. Recovery after a complete olfactory bulb lesion correlates with patterns of olfactory nerve penetration. Brain Research 648:121–133.

Hepper PG (1990) Foetal olfaction. In: Chemical Signals

in Vertebrates (Macdonald DW, Muller-Schwarze D, Natynczuk SE, eds.), pp. 282–286. Oxford: Oxford University Press.

Hudson R (1993) Olfactory imprinting. Current Opinions in Neurobiology 3:548–552.

Hurst JL, Payne CE, Nevison CM, Marie AD, Humphries RE, Robertson DH, Cavaggioni A, Beynon RJ (2001) Individual recognition in mice mediated by major urinary proteins. Nature 414: 631–634.

Johanson IB and Teicher MH (1980) Classical conditioning of an odor preference in three-day old rats. Behavioral and Neural Biology 29:132.

Johnson BA and Leon M (1996) Spatial distribution of [14C]2-deoxyglucose uptake in the glomerular layer of the rat olfactory bulb following early odor preference learning. The Journal of Comparative Neurology 376:557–566.

Kavaliers M and Ossenkopp KP (2001) Corticosterone rapidly reduces male odor preferences in female mice. Neuroreport 12:2999–3002.

Laing DL, Panhuber H, Slotnick B (1989) Odor masking in the rat. Physiology and Behavior 45:689–694.

Laska M, Distel H, Hudson R (1997) Trigeminal perception of odorant quality in congenitally anosmic subjects. Chemical Senses 22:447–456.

Leinders-Zufall T, Lane AP, Puche AC, Ma W, Novotny MV, Shipley MT, Zufall F (2000) Ultrasensitive pheromone detection by mammalian vomeronasal neurons. Nature 405:792–796.

Lu XM and Slotnick BM (1998) Olfaction in rats with extensive lesions of the olfactory bulbs: implications for odor coding. Neuroscience 84:849–866.

Lu XM, Slotnick BM, Silberberg AM (1993) Odor matching and odor memory in the rat. Physiology and Behavior 53:795–804.

Mihalick SM, Langlois JC, Krienke JD, Dube WV (2000) An olfactory discrimination procedure for mice. Journal of the Experimental Analysis of Behavior 73:305–318.

Mombaerts P, Wang F, Dulac C, Chai SK, Nemes A, Mendelsohn M, Edmondson J, Axel R (1996) Visualizing an olfactory sensory map. Cell 87:675–686.

Moulton DG (1967) Olfaction in mammals. American Zoologist 7:421–429.

Moulton DG and Marshall DA (1976) The performance of dogs in detecting alpha-ionone in the vapor phase. Journal of Comparative Physiology 110:287–306.

Nevison CM, Armstrong S, Beynon RJ, Humphries RE, Hurst JL (2003) The ownership signature in mouse scent marks is involatile. Proceedings of the Royal Society of London Series B Biological Sciences 270:1957–1963.

Nielsen GD, Bakbo JC, Holst E (1984) Sensory irritation

and pulmonary irritation by airborne allyl acetate, allyl alcohol, and allyl ether compared to acrolein. Acta Pharmacologica Toxicology (Copenhagen) 54:292–298.

Nigrosh B, Slotnick BM, Nevin JA (1975) Reversal learning and olfactory stimulus control in rats. Journal of Comparative and Physiological Psychology 80:285–294.

Otto T and Eichenbaum H (1992) Complementary roles of the orbital prefrontal cortex and the perirhinal-entorhinal cortices in an odor-guided delayed-nonmatching-to-sample task. Behavioral Neuroscience 106:762–775.

Pedersen PE and Blass EM (1982) Prenatal and postnatal determinants of the first suckling episode in albino rats. Developmental Psychobiology 15:349–355.

Reid IC and Morris RGM (1992) Smells are no surer: rapid improvement in olfactory discrimination is not due to the acquisition of a learning set. Proceedings of the Royal Society of London Series B Biological Sciences 247:137–143.

Rouquier S, Blancher A, Giorgi D (2000) The olfactory receptor gene repertoire in primates and mouse: evidence for reduction of the functional fraction in primates. Proceedings of the National Academy of Sciences 97:2870–2874.

Rudy JW and Cheatle MD (1979) Ontogeny of associative learning: acquisition of odor aversions by neonatal rats. In: Ontogeny of learning and memory (Spear NE and Campbell BA, eds.). New Jersey: Hillsdale.

Rudy JW and Cheatle MD (1983) Odor-aversion learning by rats following LiCl exposure: ontogenetic influences. Developmental Psychobiology 16:13–22.

Sakellaris PC (1972) Olfactory thresholds in normal and adrenalectomized rats. Physiology and Behavior 9:495–501.

Schaefer ML, Yamazaki K, Osada K, Restrepo D, Beauchamp GK (2002) Olfactory fingerprints for major histocompatibility complex-determined body odors, II: relationship among odor maps, genetics, odor composition, and behavior. Journal of Neuroscience 22:9513–9521.

Schaper M (1993) Development of a database for sensory irritants and its use in establishing occupational exposure limits. American Industrial Hygiene Association Journal 54:488–544.

Schellinck HM and Brown RE (1999) Searching for the source of urinary odours of individuality in rodents. In: Advances in Chemical Signals in Vertebrates (Johnson RE, Muller-Schwarze D, Sorenson PW, eds). 267. New York: Kluwer Academic/Plenum Publishers.

Schellinck HM, Forestell CA, Dill P, Lolordo VM, Brown RE (2001a) The development of a simple associative olfactory test of learning and memory. In: Chemical Signals in Vertebrates 9 (Marchlewskw Koj A, Lepri JJ, Müller-Schwarze D, ed.). New York: Kluwer Academic Publishers.

Schellinck HM, Forestell C, Lolordo V (2001b) A simple and reliable test of olfactory learning and memory in mice. Chemical Senses, 26:663–672.

Schellinck HM, Rooney E, Brown RE (1995) Odors of individuality of germfree mice are not discriminated by rats in a habituation-dishabituation procedure. Physiology and Behavior 57:1005–1008.

Schellinck HM, Slotnick BM, Brown RE (1997) Odors of individuality originating from the major histocompatibility complex are masked by diet cues in the urine of rats. Animal Learning and Behavior 25: 193–199.

Setzer AK and Slotnick BM (1998) Disruption of axonal transport from olfactory epithelium by 3-methylindole. Physiology and Behavior 65:479–487.

Silver WL (1992) Neural and pharmacological basis for nasal irritation. Annuals of the New York Academy of Science 641:152–163.

Silver WL, Mason JR, Adams MA, Smeraski CA (1986) Nasal trigeminal chemoreception: responses to n-aliphatic alcohols. Brain Research 376:221–229.

Singer AG, Beauchamp GK, Yamazaki K (1997) Volatile signals of the major histocompatibility complex in male mouse urine. Proceedings of the National Academy of Science 94:2210–2214.

Slotnick BM (2001) Animal cognition and the rat olfactory system. Trends in Cognitive Sciences 5:216–222.

Slotnick BM (1984) Olfactory stimulus control in the rat. Chemical Senses 9:157–165.

Slotnick BM and Bisulco S (2003) Detection and discrimination of carvone enantiomers in rats with olfactory bulb lesions. Neuroscience 121:451–457.

Slotnick BM and Bodyak N (2002) Odor discrimination and odor quality perception in rats with disruptions of connections between the olfactory epithelium and the olfactory bulb. Journal of Neuroscience 22:4205–4216.

Slotnick BM, Glover P, Bodyak N (2000a) Does intranasal application of zinc sulfate produce anosmia in the rat? Behavioral Neuroscience 114:814–829.

Slotnick BM and Gutman L (1977) Evaluation of intranasal zinc sulfate treatment on olfactory discrimination in rats. Journal of Comparative and Physiological Psychology 91:942–950.

Slotnick BM, Hanford S, Hodos W (2000b) Can rats acquire an olfactory learning set? Journal of Experimental Psychology: Animal Behavior Processes 26:399–415.

Slotnick BM and Katz H (1974) Olfactory learning-set formation in rats. Science 185:796–798.

Slotnick BM, Kufera A, Silberberg AM (1991) Olfactory

learning and odor memory in the rat. Physiology and Behavior 50:555–561.

Slotnick BM and Nigrosh BJ (1974) Olfactory stimulus control evaluated in a small animal olfactometer. Perceptual and Motor Skills 39:583–597.

Slotnick BM and Ptak J (1977) Olfactory intensity difference thresholds in rats and humans. Physiology and Behavior 19:795–802.

Slotnick BM and Schellinck H (2002) Methods in olfactory research with rodents. In: Frontiers and methods in chemosenses (Simon SA and Nicolelis M, eds.), pp. 21–61. New York: CRC Press.

Slotnick BM and Schoonover FW (1984) Olfactory thresholds in normal and unilaterally bulbectomized rats. Chemical Senses 9:325–340.

Slotnick BM and Schoonover FW (1992) Olfactory pathways and the sense of smell. Neuroscience and Biobehavioral Reviews 16:453–472.

Slotnick BM and Schoonover FW (1993) Olfactory sensitivity of rats with transection of the lateral olfactory tract. Brain Research 616:132–137.

Slotnick BM, Westbrook F, Darling FMC (1997) What the rat's nose tells the rat's mouth: long delay aversion conditioning with aqueous odors and potentiation of taste by odors. Animal Learning and Behavior 25:357–369.

Smith FJ, Charnock DJ, Westbrook RF (1983) Odor-aversion learning in neonate rat pups: the role of duration of exposure to an odor. Behavioral and Neural Biology 37:284–301.

Smotherman WP (1982) Odor aversion learning by the rat fetus. Physiology and Behavior 29:769–771.

Southall PF and Long CJ (1969) Odor cues in a maze discrimination. Psychonomic Science 16:126–127.

Staubli U, Fraser D, Kessler M, Lynch G (1986) Studies on retrograde and anterograde amnesia of olfactory memory after denervation of the hippocampus by entorhinal cortex lesions. Behavioral and Neural Biology 46:432–444.

Steigerwald ES and Miller MW (1997) Performance by adult rats in sensory-mediated radial arm maze tasks is not impaired and may be transiently enhanced by chronic exposure to ethanol. Alcohol Clinical and Experimental Research 21:1553–1559.

Stevens DA (1975) Laboratory methods for obtaining olfactory discrimination in rodents. In: Methods in olfactory research (Moulton DG, Turk A, Johnston AW Jr, eds.). New York: Academic Press.

Sullivan RM, McGaugh JL, Leon M (1991) Norepinephrine-induced plasticity and one-trial olfactory learning in neonatal rats. Developmental Brain Research 60:219–228.

Sullivan R and Wilson DA (1991) Neural correlates of conditioned odor avoidance in infant rats. Behavioral Neuroscience 105:307–312.

Sullivan RM, Wilson DA, Lemon C, Gerhardt GA (1994) Bilateral 6-OHDA lesions of the locus coeruleus impair associative olfactory learning in newborn rats. Brain Research 643:306–309.

Sundberg H, Doving K, Novikov S, Ursin H (1982) A method for studying responses and habituation to odors in rats. Behavioural and Neural Biology 34:113–119.

Tucker D (1963) Physical variables in the olfactory stimulation process. Journal of General Physiology 46:453–489.

Van Elzakker M, O'Reilly RC, Rudy JW (2003) Transitivity, flexibility, conjunctive representations and the hippocampus. I. An empirical analysis. Hippocampus 13:292.

Wallace DG, Gorny B, Whishaw IQ (2002) Rats can track odors, other rats, and themselves: implications for the study of spatial behavior. Behavioural Brain Research 131:185–192.

Williams J and Slotnick BM (1970) A multiple-choice airstream design for olfactory discrimination training of small animals. Behavioral Research Methods and Instrumentation 2:195–197.

Wilson DA and Sullivan RM (1994) Neurobiology of associative learning in the neonate: early olfactory learning. Behavioral and Neural Biology 61:1.

Wysocki CJ, Kruczek M, Wysocki LM, Lepri JJ (1991) Activation of reproduction in nulliparous and primiparous voles is blocked by vomeronasal organ removal. Biology of Reproduction 45:611–616.

Xu F, Greer CA, Shepherd GM (2000) Odor maps in the olfactory bulb. Journal of Comparative Neurology 422:489–495.

Xu W and Slotnick BM (1999) Olfaction and peripheral olfactory connections in methimazole-treated rats. Behavioural Brain Research 102:41–50.

Yamazaki K, Yamaguchi M, Baranoski L, Bard J, Boyse EA, Thomas L (1979) Recognition among mice: evidence from the use of a Y-maze differentially scented by congenic mice of different histocompatibility types. The Journal of Experimental Medicine 150:755–760.

Yee KK and Costanzo RM (1995) Restoration of olfactory mediated behavior after olfactory bulb deafferentation. Physiology and Behavior 58:959–968.

Youngentob SL, Markert LM, Hill TW, Matyas EP, Mozell MM (1991) Odorant identification in rats: an update. Physiology and Behavior 49:1293–1296.

Youngentob SL, Markert LM, Mozell MM, Hornung DE (1990) A method for establishing a five odorant identification confusion matrix task in rats. Physiology and Behavior 47:1053–1059.

Youngentob SL, Schwob JE, Sheehe PR, Youngentob LM (1997) Odorant threshold following methyl bromide-induced lesions of the olfactory epithelium. Physiology and Behavior 62:1241–1252.

Zagreda L, Goodman J, Druin DP, McDonald D, Diamond A (1999) Cognitive deficits in a genetic mouse model of the most common biochemical cause of human mental retardation. Journal of Neuroscience 19:6175–6182.

Zufall F and Munger SD (2001) From odor and pheromone transduction to the organization of the sense of smell. Trends in Neuroscience 24:191–193.

Taste

10

ALAN C. SPECTOR

The taste world of the common laboratory rat (*Rattus norvegicus*) appears to be remarkably similar to that of the human. Rats and humans qualitatively categorize similarities and differences among taste compounds in comparable ways. The rat's robust approach and avoidance responses to taste compounds, as well as its innate oromotor reflexes to chemical stimuli, parallel the hedonic reactions of humans. Perhaps that is why rats have successfully earned a living as pests in urban dwellings. To be sure, there are some interesting exceptions;—for example, aspartame does not appear to be "sweet" to rats—but on the whole the similarities between rats and humans far outweigh the differences in regard to taste perception. Thus, from the perspective of the final output of gustatory system, the rat serves as an excellent and economical animal model of human taste and affect.

Although the output of the gustatory system appears to be similar between rats and humans, the underlying anatomy has some notable differences. In nonhuman primates (and presumably humans), the second-order taste neurons of the nucleus of the solitary tract (NST) project directly to the gustatory zone of the thalamus, but in rats the taste-related NST output reaches the thalamus through an obligate synapse in select subdivisions of the parabrachial nucleus (PBN). In rats, the gustatory zone of the PBN not only contributes to the thalamocortical pathway but also sends fibers to ventral forebrain, including the amygdala, substantia innominata, and the hypothalamus, all of which are structures associated with motivated behaviors (Norgren, 1995). In nonhuman primates, these ventral forebrain structures receive their taste input directly from gustatory cortical areas such as the insula, operculum, and orbitofrontal cortex (Pritchard, 1991). Although the functional significance of these anatomical differences remains unclear, the rat anatomy offers some experimental opportunities to selectively manipulate the flow of ascending gustatory information in an effort to discern the functional organization of the gustatory system and to gain insight into related processes including motivation and affect, and learning and memory.

Taste function can be heuristically divided into three domains (Spector, 2000). First, taste stimuli possess qualitative signatures that allow the animal to identify them. Second, taste stimuli have motivational properties in that they can promote or discourage ingestion or be neutral. The motivational domain can be further subdivided into processes associated with the procurement of the stimulus (appetitive behavior) and processes involved in the reflex-like oromotor actions triggered by the contact of the stimulus with taste receptors (consummatory behavior). Third, some taste stimuli can trigger physiological reflexes such as salivation that help facilitate digestion and assimilation of ingested substances. Depending on the experimental procedure, taste stimulus–induced responses can be classified into one or more of the above domains, and it is important to recognize that a given procedure might focus more on one function than on another. With that in mind, the purpose of this chapter is to review the methodological and conceptual issues associated with some common behavioral procedures used to assess taste function. I focus on laboratory rodents, but many of the principles can be generalized to other species.

STIMULUS PREPARATION

Chemical stimuli should be prepared as precisely as possible. It is best to use reagent-grade chemicals dissolved in a reasonably pure source of water with minimal contamination. Unfortunately, contaminants could come from sources beyond the experimenter's control such as the actual purity of the chemical purchased. Nevertheless, excellent water distillers and reverse osmosis deionizers and filtration cartridges are available. Whenever possible, it is best to prepare the solutions fresh. There are, however, situations in which it might be advisable to prepare a solution somewhat in advance of its use. For example, some organic compounds undergo mutorotation over time between anomeric forms until equilibrium is reached. Potentially this could cause variability in responsiveness if a compound were used before the equilibrium was achieved, assuming that the different anomers had significantly different physiological properties. If solutions are prepared and stored under refrigeration, they should be allowed to reach room temperature before use. If compounds are light sensitive, solutions can be kept in glassware and bottles that are wrapped in aluminum foil to help minimize light exposure.

In the literature, there are two common conventions used to define the concentration of a solution: percentage weight by volume (%w/v) and molarity (M). Researchers who study feeding prefer the former because metabolizable carbohydrate solutions that are equivalent in their percentage weight by vol-

ume concentration have an identical caloric density. For researchers interested in taste processes, however, it is preferable to represent concentration in terms of molarity because isomolar solutions have the same number of molecules per unit volume (see Pfaffmann et al., 1954). For example, to say that a 10% fructose solution is "sweeter" than a 10% sucrose solution would not be meaningful from a taste perspective because the former has about twice the number of molecules per unit volume.

Most taste research is conducted with stimuli presented in liquid phase, because it is difficult to uniformly mix solid chemicals into food and to be certain of the concentration that is reaching taste receptors. Under some circumstances, spillage of powdered food can be a problem. To overcome these practical problems, investigators have dissolved chemical stimuli into gelatin solutions and then let it solidify (e.g., Rowland et al., 2003). For the remaining portion of this chapter, I discuss procedures involving liquid stimuli, although other forms of stimulus presentation are available.

Experimenters should be cautious in their assumptions about the relative intensity of individual qualitative components of mixtures of chemical compounds. Within limits, the sense of taste appears to be more *analytic* than it is *synthetic*. In other words, when chemical compounds of different taste qualities are mixed in solution, observers can report the individual qualitative components in the mixture, at least with binary and ternary combinations (see Smith and Theodore, 1984; Laing et al., 2002; Frank et al., 2003). This is like the recognition of notes in a musical chord and contrasts with the synthetic nature of color vision in which observers are unable to detect the wavelength components of combined light sources (e.g., white light arising from the equal mixture of color primaries). Despite the analytical character of taste sensations, caution should be exercised regarding assumptions about the sensory properties of mixtures (Laing et al., 2002; Schifferstein, 2003). Some compounds can interact in mixtures and affect receptor processes in the peripheral gustatory system in unexpected ways. In such cases, the whole is not the sum of the parts. For example, in hamsters, a weak concentration of quinine hydrochloride (a "bitter" tasting compound to humans) that does not itself lead to stimulation of the chorda tympani nerve when it is placed on the anterior tongue will nonetheless significantly suppress the responses of this nerve to sucrose when the two compounds are presented in mixture (Formaker et al., 1997). In some cases, mixture suppression could be mediated in the central nervous system (Lawless, 1979; Travers and Smith, 1984; Vogt and Smith, 1993). Synergies can also occur. Responses to monosodium glutamate, as assessed by chorda tympani nerve recordings in rodents or intracellular Ca^{2+} concentration changes in heterologous expression systems for taste receptor proteins, are enhanced when the amino acid salt is mixed with the $5'$-purine nucleotide inosine monophosphate (e.g., Yamamoto et al., 1991; Li et al., 2002). Various types of mixture interactions have been identified both electrophysiologically and psychophysically.

INTAKE TESTS

The most common measure used to assess responsiveness to taste stimuli is the intake test. Various procedural incarnations of such tests can be found in the literature, but they all involve the measurement of the amount consumed from one or more bottles containing chemical solutions. The duration of a test can last from a few minutes to days. In the two-bottle version, one bottle contains a taste stimulus and the other bottle contains either water or a different taste stimulus. These tests are simple and do not involve specialized equipment, and no extensive training of animals or laboratory personnel is required.

METHODOLOGICAL CONSIDERATIONS

When more than one bottle is available, stimulus sampling and position preferences become issues. In a short-term test, animals may never sample one of the alternatives. Sampling can be encouraged by first presenting one solution and then presenting the other solutions. Once all of the stimuli have been sampled, they can be presented simultaneously (Nachman, 1962). This assumes that animals have been trained on a restricted fluid-access schedule and are "primed" to ingest solutions when they are presented. In both short-term and long-term tests, animals might prefer ingesting from one bottle over another on the basis of its position. Researchers have controlled for this by switching the positions of the bottles halfway through the test. This can be made difficult by light/dark phase changes, but a 48-hour test avoids the problem. If there are three or more bottles, this strategy becomes complicated and preferences actually vary with the number of bottles presented (Tordoff and Bachmanov, 2003). Perhaps this is why researchers rarely use more than two bottles. The order in which various concentrations of a stimulus are presented can also influence preference and aversion (e.g., Flynn and Grill, 1988; Fregly and Rowland, 1992).

CONCEPTUAL AND INTERPRETIVE CONSIDERATIONS

The advantage of the intake test is its relative ease of use compared with other, more involved procedures (see later). A drawback of this method is its difficulty in dissociating the effects of postingestive stimulation from those related to taste. For example, it is known that rodents display an inverted-U–shaped concentration-intake function for several types of compounds, including sugar solutions. It is thought that the descending limb of such functions is caused by the stimulation of postingestive receptor systems (Davis and

Levine, 1977). Indeed, if the ingested contents are allowed to drain out of an open gastric or esophageal cannula (i.e., sham drinking), intake monotonically increases with concentration (Mook, 1963; Geary and Smith, 1985). In another example, Rabe and Corbit (1973) were able to recapitulate the inverted-U–shaped concentration-intake function for NaCl derived from a 1 hour one-bottle test in rats by orally presenting a single low concentration of NaCl while simultaneously infusing varying concentrations of NaCl directly into the stomach. This does not show that the taste of NaCl cannot also produce a inverted-U–shaped concentration-response function; rats sham drinking NaCl show similar curves to normal-drinking rats. The Rabe and Corbit (1973) study does show, however, that taste is not necessary.

To the extent that the amount ingested can be attributed to taste stimulation, intake tests rely on the hedonic characteristics of the chemical stimuli to drive the behavior. Consequently, these tests do not necessarily reveal much about the perceived qualitative identity of taste stimuli (with certain paradigmatic exceptions such as sodium appetite; see Nachman, 1962). In other words, discriminable stimuli can lead to the same degree of preference or aversion in an intake test. Moreover, a lack of preference or aversion does not necessarily mean that the stimulus is undetectable; it simply means that the taste compound is hedonically neutral. For example, in one experiment, C57BL/6J (B6) mice generated a rather flat concentration-preference function up to approximately 0.1 M NaCl derived from 48-hour two-bottle tests (NaCl versus water; ascending concentration series) and then displayed progressive degrees of avoidance as the concentration was raised further (Eylam and Spector, 2002). However, these same animals were able to clearly detect the low concentrations of NaCl as indicated by their performance on a conditioned signal discrimination task in an operant response–based task (see later). Interestingly, the epithelial sodium channel blocker amiloride, which in-

terferes with one of the sodium taste transduction pathways, had no effect on the preference–aversion function for NaCl but shifted the sensitivity curve measured in the operant task by close to 1 order of magnitude (Eylam and Spector, 2002).

OROMOTOR AND SOMATIC TASTE REACTIVITY

When chemical stimuli contact taste receptors, they can elicit stereotypical oromotor and somatic responses from the animal that vary in a concentration-dependent fashion (Grill and Norgren, 1978a; Grill and Berridge, 1985; Grill et al., 1987). These responses are referred to as taste reactivity and fall into one of two classes: ingestive responses and aversive responses. Ingestive responses are elicited by normally preferred stimuli such as sucrose and include tongue protrusions, lateral tongue, and mouth movements. Aversive responses are elicited by normally avoided stimuli such as quinine and include gapes, chin rubs, forelimb flails, and head shakes. Ingestive responses are normally accompanied by consumption of the stimulus, whereas aversive responses are normally accompanied by fluid ejection from the oral cavity (Fig. 10–1).

METHODOLOGICAL CONSIDERATIONS

Taste reactivity behavior is best quantified when the stimulus is delivered directly into the oral cavity under experimenter control. In rodents, an intraoral cannula can be implanted chronically in anesthetized animals; there are a variety of surgical methods to accomplish this (Grill and Norgren, 1978a; Hall, 1979; Grill et al., 1987; Spector et al, 1988).

After recovery from intraoral surgery (2 weeks), the animal is habituated to the test environment, which usually consists of a small Plexiglas arena, the wall of which sits on posts that elevate it about 1 cm above the floor. A mirror is positioned on an angle below the

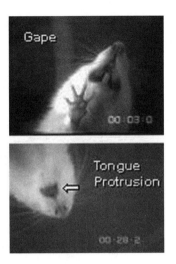

Figure 10–1. Ventral views of two characteristic taste-elicited oromotor responses. (*Top*) The gape, hallmark of aversive taste reactivity, is characterized by a large opening of the mouth accompanied by retraction of its corners, usually exposing the incisors. (*Bottom*) The tongue protrusion, hallmark of ingestive taste reactivity, is characterized by extension of the tongue across the plane of the incisors (*arrow* points to still-frame image of protruding tongue).

transparent floor. A length of PE-160 tubing, that has a small piece of PE-100 tubing attached to one end and flanged by heat, is connected to the cannula (the heat-flanged end), and the other end of it is attached to a commutator on the ceiling of the chamber. In turn, another length of PE-160 tubing connects the commutator to a hyperdermic needle, which in turn is attached to a syringe anchored in a syringe pump. Fluid is pushed through the tubing to eliminate the dead space before attachment to the rat. When the infusion is scheduled to start, the pump is activated; once the fluid reaches the oral cavity, there is a reflexive series of mouth movements on which the timer is started.

Some investigators attempt to quantify taste reactivity during real time, but it is best to videotape the session for later detailed analysis. The video camera is aimed at the mirror and focused on the ventral view of the mouth, and the operator merely follows the animal as it moves in the chamber. Keeping the chamber relatively small (e.g., ≈24 cm diameter arena) limits the range of locomotion. The tape is analyzed for different types of taste reactivity behavior. Some responses are distinct brief events that can be counted (e.g., tongue protrusions); others are longer and are timed (e.g., passive drip). A description of the different types of taste reactivity behavior that have been scored can be found elsewhere (Grill and Berridge, 1985; Grill et al., 1987; Spector et al., 1988).

Unfortunately, there is a certain degree of subjectivity in any scoring method that is inescapable. For example, sometimes gapes are difficult to distinguish from large mouth movements, and tongue protrusions can be difficult to distinguish from small mouth movements depending on the amplitude of the response. Accordingly, the scoring can be conducted by a trained observer who is "blind" to the experimental treatment of the animal. To circumvent the vulnerability of scoring videotape, some investigators have implanted electromyographic electrodes in the tongue and jaw musculature through which they can measure signature waveforms representing certain types of oromotor responses (e.g., Kaplan and Grill, 1989; Chen and Travers, 2003).

CONCEPTUAL AND INTERPRETIVE ISSUES

There is a growing literature on the theoretical meaning of taste-elicited oromotor and somatic behaviors (Grill and Berridge, 1985; Breslin et al., 1992; Parker, 1995; Berridge, 1996). Taste reactivity experiments are cumbersome to conduct, and analysis is tedious. Thus, the investigator should have strong justification for taking this methodological path. That said, there are some very strong rationales for the use of this procedure. It is perhaps the best way to quantify purely consummatory behavior. In neural preparations that do not voluntarily eat and drink, such as the chronic supracollicular decerebrate rat, it is the only way to behaviorally assess taste responsiveness (Grill and Norgren, 1978b). Even in intact animals, the taste reactivity procedure provides a way to circumvent the floor effects associated with intake and licking tests when animals no longer approach the drinking spout. Moreover, animals can be tested in a nondeprived state, and very small volumes of fluid stimuli can be used and immediate responses measured, minimizing the contribution of postingestive factors. Taste reactivity can also be modified by learning, as is strikingly obvious when rats progressively change their initial ingestive responses to 30 second sucrose infusions delivered every 5 minutes after an LiCl injection (which causes visceral

malaise), to an aversive profile of behavior (Breslin et al., 1992) (Fig. 10–2).

BRIEF-ACCESS TASTE TEST

The brief-access taste test is designed to assess unconditioned licking responses to taste stimuli during very short duration trials (Young and Trafton, 1964; Davis, 1973; Smith et al., 1992; Markison et al., 2000; Glendinning et al., 2002; Spector, 2003). Just as the taste reactivity procedure involves the measurement of immediate responses to small volumes of

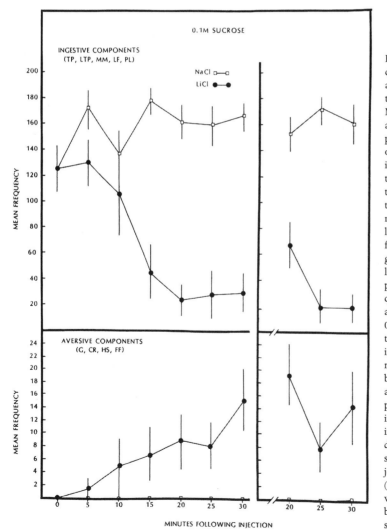

Figure 10–2. (*Left*) Mean ($\pm$ SE) frequency of ingestive responses (*top*) and aversive responses (*bottom*) to intraoral infusions (0.5 ml/30 s) of 0.1 M sucrose presented immediately and every 5 minutes after an intraperitoneal injection of LiCl (n = 6) or NaCl (n = 8) in rats. The LiCl-injected rats begin to change their taste reactivity profile from ingestive to aversive, presumably as the LiCl takes effect. Naïve rats that begin to receive sucrose after a 20 minute delay will not display an aversive profile of responding (not shown), suggesting that the curves shown in the left represent a rapid conditioning process. (*Right*) Mean ($\pm$ SE) frequency of ingestive responses (*top*) and aversive responses (*bottom*) to 0.1 M sucrose during a subsequent test 4 days later in the same rats. The infusions were given 20, 25, and 30 minutes after injection of LiCl in both groups. The rats that received a lithium injection 4 days earlier display an aversive profile of responding, whereas the rats that were originally injected with NaCl are still displaying an ingestive profile of responding, although they were injected with LiCl 20 minutes earlier. (Reprinted from Spector et al. [1988] with permission. Copyright © 1988 by the American Psychological Association.)

taste stimuli, the brief-access taste procedure, pioneered by P. T. Young and his students (e.g., Young and Trafton, 1964), has some of the same methodological attributes but reflects the contribution of both appetitive and consummatory behavior. This procedure requires the use of an automated stimulus delivery and lick monitoring system, commonly referred to as a *gustometer* (e.g., Slotnick, 1982; Spector et al., 1990; Thaw and Smith, 1992; Reilly et al., 1994). Some of these devices are commercially available, and others have been custom built. All are capable of delivering multiple taste solutions during a session and recording licking responses from the stimulus-access drinking spout.

METHODOLOGICAL CONSIDERATIONS

In general, each solution from the test array is presented for a very brief trial on the order of seconds ($\approx$5 to 30 seconds). Normally, the trial timer starts with the first lick. Solution presentations are randomized without replacement within blocks of trials. Consequently, satiation and fatigue factors are uniformly distributed across the stimuli as the session progresses. The shorter the trial duration, the more trials that are initiated by the animal before satiation occurs. On the other hand, if the trial duration is too brief, then the range of potential variation in lick rate is curtailed. In the case of normally preferred stimuli, animals can be tested in a nondeprived state, although they are usually trained initially to sample the stimuli when water-deprived. In the case of normally avoided stimuli, however, animals must be tested while on a water-restriction schedule, pitting the drive to rehydrate against the aversiveness of the solution. Often times, the concentration of a single chemical solution is varied. The resulting concentration–response curves are very orderly and sigmoidal regardless of whether the stimulus is preferred or avoided. Because this technique is automated and involves very little animal training, a reasonably

large number of animals can be tested in a single day. For example, with 30 minute sessions and 5 gustometers, more than 70 animals could potentially be tested. Hence, this procedure shows promise for high-throughput taste screening in mutaginized rodents (see Glendinning et al., 2002).

CONCEPTUAL AND INTERPRETIVE ISSUES

The basic premise of the brief-access taste test is that the global lick rate (differentiated from the local lick rate, which is simply the reciprocal of the modal fundamental interlick interval [ILI]) during a given trial is a measure of the affective potency of the stimulus. Depending on the experimental manipulation, however, there can be factors other than taste that can affect licking. For example, a sucrose concentration–response function would be virtually flat if tested in a water-deprived rat. Indeed, if one animal is more "thirsty" than another, this could also influence the degree of licking even with normally avoided stimuli. This latter possibility can easily be controlled by merely forming a ratio of stimulus licking to water licking. Thus, differences in drive state or motor faculty can be factored out. When testing normally preferred stimuli in nondeprived animals, this latter statistical manipulation is not meaningful and in fact is inadvisable to use, because minor changes in water licking can have a disproportional effect on the ratio. One way investigators can adjust for potential differences in general rate of licking across animals tested in nondeprived conditions is to measure the ILI during a water test under conditions of water restriction and then use the reciprocal of that value to determine the maximum licks possible in a trial. Thus, the lick rate to each stimulus can be scaled by the maximum possible lick rate for each subject individually (Glendinning et al., 2002). Although it depends on motivational properties of the taste stimulus to drive responsiveness, the brief-access taste test is one of the few methods available to assess perceived intensity in the suprathreshold domain and is becoming a popular alternative to the two-bottle preference test.

TASTE STIMULI AS CONDITIONED CUES

All of the procedures discussed up to this point assess taste function in the affective domain. Yet it is possible for two taste solutions to generate identical values in these measures, because the two stimuli are hedonically equivalent under the test conditions, but nonetheless be qualitatively discriminable. Investigators can make inferences about the qualitative properties of chemical solutions by using the taste stimuli as conditioned cues. Procedures in which rats are trained to discriminate between two or more taste stimuli provide information about the perceived qualitative differences (or lack thereof) of chemical compounds. In other cases, rats can be trained to respond in a specific fashion to a single taste stimulus, and the degree to which the animal generalizes its responses to other chemical solutions can provide information on the perceived qualitative similarity of the stimuli.

METHODOLOGICAL CONSIDERATIONS

In classical conditioning procedures, taste compounds can serve as conditioned stimuli (CSs). A common example of this approach is the application of the conditioned taste aversion paradigm. It is well established that when animals ingest a novel taste stimulus followed by aversive visceral consequences, usually caused by the controlled administration of an emetic such as LiCl (although rats cannot vomit), they will subsequently avoid ingesting the substance (Riley and Tuck, 1985). This procedure has been exploited to make inferences about the perceived taste quality and intensity of chemical solutions (Nachman, 1963; Tapper and Halpern, 1968; Nowlis et al., 1980;

Spector and Grill, 1988). For example, ingestion of 0.3 M NaCl (CS) could be followed by LiCl injection, and on future occasions the degree of avoidance to a variety of test stimuli could be evaluated in comparison to that observed in unconditioned control animals. The ratio of *conditioned* avoidance of a test solution relative to the CS (i.e., 0.30 M NaCl) is taken as an index of similarity between the two stimuli. Such procedures have been used with one- or two-bottle intake tests as the primary measure of avoidance or with gustometers (e.g., Spector et al., 1990), which can measure lick rates to a variety of stimuli delivered in a controlled fashion during trials distributed within a single session (i.e., brief-access taste test).

In operant conditioning procedures, taste stimuli can be used as discriminative cues, which signal the opportunity for reinforcer delivery contingent on the execution of a specific response. These procedures require the use of a gustometer. We have successfully used a two-response stimulus discrimination procedure in which animals are trained to make one response (e.g., press right-hand lever) in the presence of a given stimulus (e.g., NaCl) and another response (e.g., press left-hand lever) in the presence of a different stimulus (e.g., water). Correct responses are rewarded with small volumes of water. The animals are tested while water-deprived, which promotes stimulus sampling and potentiates the reinforcing efficacy of water (St. John et al., 1997). We have used this procedure to measure detection thresholds (Fig. 10–3) and taste quality discrimination (Fig. 10–4) (Spector et al., 1996; St. John et al., 1997; St. John and Spector, 1998; Geran and Spector, 2000; Kopka and Spector, 2001; Geran et al., 2002; Spector and Kopka, 2002). Other operant conditioning procedures also have been successfully used in conjunction with different types of gustometers to measure taste sensitivity, discrimination, and generalization (see Spector, 2003, for a review).

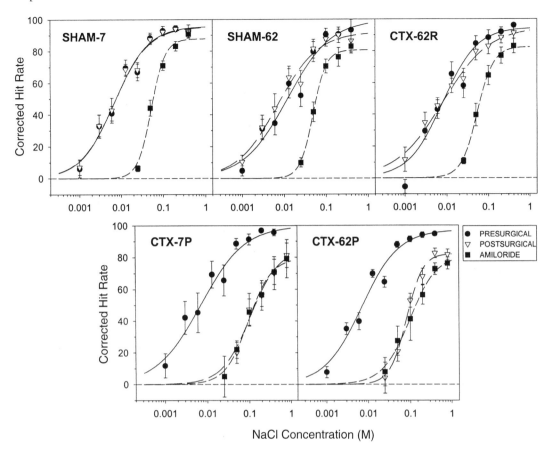

Figure 10–3. Mean (± SE) percentage of trials with correct responses when stimulus was present, corrected for the false alarm rate (referred to as Corrected Hit Rate) in a two-response operant procedure in which five groups of rats were trained to press one lever when water was presented and to press a different lever when NaCl was presented. Note that when the chorda tympani nerve (CT) was transected and prevented from regenerating (CTX-7P and CTX-62P), sensitivity to NaCl was profoundly disrupted relative to sham-operated rats (SHAM-7 and SHAM-62). In contrast, when the CT regenerated (CTX-62R), taste sensitivity to NaCl recovered completely and the performance-disrupting effect of stimulus adulteration with the epithelial sodium channel blocker amiloride also returned to normal. These results highlight the importance of the signals in the CT, which innervates only about 13% of the total taste buds, and the importance of epithelial sodium channels in taste receptor cells in maintaining sensitivity to NaCl. It is noteworthy that, historically, two-bottle tests have not revealed any major effects of CT transection on NaCl taste. (Reprinted from Kopka and Spector [2001] with permission. Copyright © 2001 by the American Psychological Association.)

CONCEPTUAL AND INTERPRETIVE CONSIDERATIONS

The use of taste stimuli as conditioned cues to assess gustatory function in the sensory/discriminative domain falls under the rubric of classic animal psychophysics and, as such, engenders standard interpretive issues (Blough and Blough, 1977; Berkley and Stebbins, 1990; Spector, 2003). First, it is important to demonstrate and maintain stimulus control of behavior. In the case of detection threshold determinations or multiple tests for avoidance after taste aversion conditioning, some extinction can occur. In the former, this can happen as a result of the presentation of too many subliminal concentrations, but it is easily discerned by attending to the false alarm rate (proportion of water trials in which the animal reported the stimulus was present) and

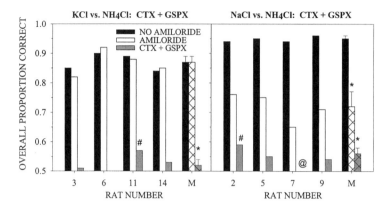

Figure 10–4. Overall proportion of correct responses collapsed across stimuli, concentrations, and sessions on an operant conditioning procedure in which rats were trained to discriminate between KCl and NH$_4$Cl (*left*) or NaCl and NH$_4$Cl (*right*) before (*solid bars*) and after (*gray bars*) combined bilateral transection of chorda tympani nerve and greater superficial petrosal nerve (CTX + GSPX). The rats were also presurgically tested during sessions (AMILORIDE) in which all of the stimuli were mixed in the epithelial sodium channel blocker amiloride hydrochloride (100 μM). Concentration of the stimuli was varied to render intensity an irrelevant cue. Note that amiloride treatment only affects performance on the NaCl-versus-NH$_4$Cl discrimination, likely because of its effect on sodium taste. Also note that transection of the gustatory branches of the seventh cranial nerve virtually eliminates the discrimination, despite that close to 70% of the total oral taste buds remain. Interestingly, transection of the glossopharyngeal nerve, which denervates close to 60% of the total taste buds, does not have any effect on this taste discrimination or others, supporting the hypothesis that the gustatory nerves in the rat are functionally specialized (see St. John and Spector [1998]). (Reprinted from Geran et al. [2002] with permission.)

to the performance on clearly detectable concentrations. The inclusion of the latter can help maintain stimulus control. In taste aversion conditioning, in which intake tests are used, some extinction could occur as stimulus testing progresses across days; thus, it is important to include CS probe trials to ensure that the aversion is still strong.

With gustometers, extraneous cues associated with stimulus delivery can guide an animal's behavior despite the best efforts to minimize them. At the end of the experiment, we routinely fill all of the fluid reservoirs in the gustometer with water and arbitrarily assign them as the "right-lever" or "left-lever" stimulus and examine whether the animals can perform the discrimination task without the presence of the chemical cue.

It is important to also recognize that chemical stimuli can potentially interact with the olfactory and trigeminal systems. For example, as the concentration of sucrose increases, so does its viscosity, and there is evidence that rats can smell sucrose at concentrations above 0.03 M (Rhinehart-Doty et al., 1994). Thus, if a manipulation in the gustatory system fails to alter conditioned responses to a taste stimulus, it might be due to nongustatory influences. In some experiments, the contribution of olfactory or trigeminal cues can be ruled out based on the profile of results (see Fig. 10–4), but in other cases the possibility remains an interpretive caveat.

When assessing differences and similarities in taste quality among compounds, the perceived intensity of the stimuli must be considered. Animals can use intensity cues to discriminate among stimuli. Likewise, generalization can take place based on either the intensity or the quality of a test compound relative to the training stimulus. Thus, the failure of a test stimulus to generalize to a training stimulus may not necessarily mean that the two compounds have different taste qualities but might merely reflect that the test compound is relatively weaker. So, it is advisable to vary the concentration of test stimuli to help dismiss this possibility. Alternatively, each test stimulus could also serve as a training stimulus in different experimental groups; if generalization between two stimuli takes place asymmetrically (i.e., when one stimulus is trained and the other is tested but not vice versa), this would raise the concern that the relative intensity of the taste solutions was influencing the results.

Finally, in most procedures involving the use of taste stimuli as conditioned cues, the animals are placed on food or water restriction schedules to provide the motivation to sample the stimuli and perform responses. Although this is an effective means of generating substantial behavior, such restriction schedules have physiological consequences that could affect the gustatory system. There are examples of hormonal manipulations affecting taste-responsive neurons and leading to upregulation of transduction-related ion channels in taste receptor cells (Giza and Scott, 1987; Giza et al., 1990; Herness, 1992; Gilbertson et al., 1993; Nakamura and Norgren, 1995; Tamura and Norgren, 1997; Lin et al., 1999). Thus, a given profile of results may depend on the physiological state of the animal during behavioral testing.

FINAL REMARKS

This chapter has provided a review of the major methodological and interpretive issues associated with some common behavioral techniques applied in the study of taste processes in laboratory rats. Because of the limited scope of the chapter, other useful procedures (e.g., taste contrast procedures, progressive ratio, and other reinforcement schedules involving taste stimuli as reinforcers) were not discussed, but some of the issues addressed

here would be pertinent to these other techniques. In closing, it is important to recognize that a given manipulation (e.g., genetic, anatomical, and pharmacological) could potentially affect one taste function while leaving another unaltered. Accordingly, a variety of procedures should be used by investigators to comprehensively assess the impact of experimental treatments on taste function in nonhuman animals.

ACKNOWLEDGMENTS

The author would like to thank Shachar Eylam and Laura C. Geran for providing constructive comments on an earlier version of this chapter. Part of the work presented here was supported by a grant from the National Institute on Deafness and Other Communications Disorders (R01-DC01628).

REFERENCES

Berkley MA, Stebbins WC (1990) Comparative perception: basic mechanisms. New York: John Wiley & Sons.

Berridge KC (1996) Food reward: brain substrates of wanting and liking. Neuroscience and Biobehavioral Reviews 20:1–25.

Blough D and Blough P (1977) Animal psychophysics. In: Handbook of operant behavior (Honig WK and Straddon JER, eds.), pp. 514–539. Englewood Cliffs, NJ: Prentice-Hall, Inc.

Breslin PA, Spector AC, Grill HJ (1992) A quantitative comparison of taste reactivity behaviors to sucrose before and after lithium chloride pairings: a unidimensional account of palatability. Behavioral Neuroscience 106:820–836.

Brosvic GM and Slotnick BM (1986) Absolute and intensity-difference taste thresholds in the rat: evaluation of an automated multi-channel gustometer. Physiology and Behavior 38:711–717.

Chen Z and Travers JB (2003) Inactivation of amino acid receptors in medullary reticular formation modulates and suppresses ingestion and rejection responses in the awake rat. American Journal of Physiology: Regulatory, Integrative and Comparative Physiology 285:R68–R83.

Davis JD (1973) The effectiveness of some sugars in stimulating licking behavior in the rat. Physiology and Behavior 11:39–45.

Davis JD and Levine MW (1977) A model for the control of ingestion. Psychological Review 84:379–412.

Eylam S and Spector AC (2002) The effect of amiloride on operantly conditioned performance in an NaCl taste detection task and NaCl preference in C57BL/ 6J mice. Behavioral Neuroscience 116:149–159.

Flynn FW and Grill HJ (1988) Intraoral intake and taste reactivity responses elicited by sucrose and sodium chloride in chronic decerebrate rats. Behavioral Neuroscience 102:934–941.

Formaker BK, MacKinnon BI, Hettinger TP, Frank ME (1997) Opponent effects of quinine and sucrose on single fiber taste responses of the chorda tympani nerve. Brain Research 772:239–242.

Frank ME, Formaker BK, Hettinger TP (2003) Taste responses to mixtures: analytic processing of quality. Behavioral Neuroscience 117:228–235.

Fregly MJ and Rowland NE (1992) Comparison of preference thresholds for NaCl solution in rats of the Sprague-Dawley and Long-Evans strains. Physiology and Behavior 51:915–918.

Geary N and Smith GP (1985) Pimozide decreases the positive reinforcing effect of sham fed sucrose in the rat. Pharmacology, Biochemistry and Behavior 22: 787–790.

Geran LC, Garcea M, Spector AC (2002) Transecting the gustatory branches of the facial nerve impairs NH_4Cl vs. KCl discrimination in rats. American Journal of Physiology: Regulatory, Integrative and Comparative Physiology 283:R739–R747.

Geran LC and Spector AC (2000) Sodium taste detectability in rats is independent of anion size: the psychophysical characteristics of the transcellular sodium taste transduction pathway. Behavioral Neuroscience 114:1229–1238.

Gilbertson TA, Roper SD, Kinnamon SC (1993) Proton currents through amiloride-sensitive Na+ channels in isolated hamster taste cells: enhancement by vasopressin and cAMP. Neuron 10:931–942.

Giza BK and Scott TR (1987) Intravenous insulin infusions in rats decrease gustatory-evoked responses to sugars. American Journal of Physiology: Regulatory, Integrative and Comparative Physiology 252:R994–R1002.

Giza BK, Scott TR, Antonucci RF (1990) Effect of cholecystokinin on taste responsiveness in rats. American Journal of Physiology: Regulatory, Integrative and Comparative Physiology 258:R1371–R1379.

Glendinning JI, Gresack J, Spector AC (2002) A high-throughput screening procedure for identifying mice with aberrant taste and oromotor function. Chemical Senses 27:461–474.

Grill HJ and Berridge KC (1985) Taste reactivity as a measure of the neural control of palatability. In: Progress in psychobiology and physiological psychology, vol. 11 (Epstein, AN and Sprague J, eds.), pp. 1–61. New York: Academic Press.

Grill HJ and Norgren R (1978a) The taste reactivity test. I. Mimetic responses to gustatory stimuli in neurologically normal rats. Brain Research 143:263–279.

Grill HJ and Norgren R (1978b) The taste reactivity test. II. Mimetic responses to gustatory stimuli in chronic thalamic and chronic decerebrate rats. Brain Research 143:281–297.

Grill HJ, Spector AC, Schwartz GJ, Kaplan JM, Flynn FW (1987) Evaluating taste effects on ingestive behavior. In: Techniques in the behavioral and neural sciences, vol 1: feeding and drinking (Toates F and Rowland N, eds.), pp. 151–188. Amsterdam: Elsevier.

Hall WG (1979) The ontogeny of feeding in rats: I. ingestive and behavioral responses to oral infusions. Journal of Comparative and Physiological Psychology 93:977–1000.

Herness MS (1992) Aldosterone increases the amiloride-sensitivity of the rat gustatory neural response to NaCl. Comparative Biochemistry and Physiology [A] 103:269–273.

Kaplan JM and Grill HJ (1989) Swallowing during ongoing fluid ingestion in the rat. Brain Research 499:63–80.

Kopka SL and Spector AC (2001) Functional recovery of taste sensitivity to sodium chloride depends on regeneration of the chorda tympani nerve after transection in the rat. Behavioral Neuroscience 115: 1073–1085.

Laing DG, Link C, Jinks AL, Hutchinson I (2002) The limited capacity of humans to identify the components of taste mixtures and taste-odour mixtures. Perception 31:617–635.

Lawless HT (1979) Evidence for neural inhibition in bittersweet taste mixtures. Journal of Comparative and Physiological Psychology 93:538–547.

Li XD, Staszewski L, Xu H, Durick K, Zoller M, Adler E (2002) Human receptors for sweet and umami taste. Proceedings of the National Academy of Sciences of the United States of America 99:4692–4696.

Lin WH, Finger TE, Rossier BC, Kinnamon SC (1999) Epithelial Na+ channel subunits in rat taste cells: localization and regulation by aldosterone. Journal of Comparative Neurology 405:406–420.

Markison S, Gietzen DW, Spector AC (2000) Essential amino acid deficiency enhances long-term intake but not short-term licking of the required nutrient. Journal of Nutrition 129:1604–1612.

Mook DG (1963) Oral and postingestional determinants of the intake of various solutions in rats with esophageal fistulas. Journal of Comparative and Physiological Psychology 56:645–659.

Nachman M (1962) Taste preferences for sodium salts in adrenalectomized rats. Journal of Comparative and Physiological Psychology 55:1124–1129.

Nachman M (1963) Learned aversion to the taste of

lithium chloride and generalization to other salts. Journal of Comparative and Physiological Psychology 56:343–349.

Nakamura K and Norgren R (1995) Sodium-deficient diet reduces gustatory activity in the nucleus of the solitary tract of behaving rats. American Journal of Physiology: Regulatory, Integrative and Comparative Physiology 269:R647–R661.

Norgren R (1995) Gustatory system. In: The rat nervous system (Paxinos G, ed.), pp. 751–771. Sydney: Academic Press.

Nowlis GH, Frank ME, Pfaffmann C (1980) Specificity of acquired aversions to taste qualities in hamsters and rat. Journal of Comparative and Physiological Psychology 94:932–942.

O'Keefe GB, Schumm J, Smith JC (1994) Loss of sensitivity to low concentrations of NaCl following bilateral chorda tympani nerve sections in rats. Chemical Senses 19:169–184.

Parker LA (1995) Rewarding drugs produce taste avoidance, but not taste aversion. Neuroscience and Biobehavioral Reviews 19:143–151.

Pfaffmann C, Young PT, Dethier VG, Richter CP, Stellar E (1954) The preparation of solutions for research in chemoreception and food acceptance. Journal of Comparative and Physiological Psychology 47:93–96.

Pritchard TC (1991) The primate gustatory system. In: Smell and taste in health and disease (Getchell TV, Doty RL, Bartoshuk LM, and Snow JB Jr., eds.), pp. 109–125. New York: Raven Press.

Rabe EF and Corbit JD (1973) Postingestional control of sodium chloride solution drinking in the rat. Journal of Comparative and Physiological Psychology 84:268–274.

Reilly S, Norgren R, Pritchard TC (1994) A new gustometer for testing taste discrimination in the monkey. Physiology and Behavior 55:401–406.

Rhinehart-Doty JA, Schumm J, Smith JC, Smith GP (1994) A non-taste cue of sucrose in short-term taste tests in rats. Chemical Senses 19:425–431.

Riley AL and Tuck DL (1985) Conditioned food aversions: a bibliography. Annals of the New York Academy of Sciences 443:381–437.

Rowland NE, Robertson K, Green DJ (2003) Effect of repeated administration of dexfenfluramine on feeding and brain Fos in mice. Physiology and Behavior 78:295–301.

Schifferstein HNJ (2003) Human perception of taste mixtures. In: Handbook of olfaction and gustation (Doty RL, ed.), pp. 805–822. New York: Marcel Dekker.

Smith DV and Theodore RM (1984) Conditioned taste aversions: generalization to taste mixtures. Physiology and Behavior 32:983–989.

Smith JC, Davis JD, O'Keefe GB (1992) Lack of an order effect in brief contact taste tests with closely spaced test trials. Physiology and Behavior 52:1107–1111.

Spector AC (2000) Linking gustatory neurobiology to behavior in vertebrates. Neuroscience and Biobehavioral Reviews 24:391–416.

Spector AC (2003) Psychophysical evaluation of taste function in non-human mammals. In: Handbook of olfaction and gustation (Doty RL, ed.), pp. 861–879. New York: Marcel Dekker.

Spector AC, Andrews-Labenski J, Letterio FC (1990) A new gustometer for psychophysical taste testing in the rat. Physiology and Behavior 47:795–803.

Spector AC, Breslin P, Grill HJ (1988) Taste reactivity as a dependent measure of the rapid formation of conditioned taste aversion: a tool for the neural analysis of taste-visceral associations. Behavioral Neuroscience 102:942–952.

Spector AC and Grill HJ (1988) Differences in the taste quality of maltose and sucrose in rats: issues involving the generalization of conditioned taste aversions. Chemical Senses 13:95–113.

Spector AC, Guagliardo NA, St. John SJ (1996) Amiloride disrupts NaCl versus KCl discrimination performance: implications for salt taste coding in rats. Journal of Neuroscience 16:8115–8122.

Spector AC and Kopka SL (2002) Rats fail to discriminate quinine from denatonium: implications for the neural coding of bitter-tasting compounds. Journal of Neuroscience 22:1937–1941.

St. John SJ, Markison S, Guagliardo N, Hackenberg TD, Spector AC (1997) Chorda tympani transection and selective desalivation differentially disrupt two-lever salt discrimination performance in rats. Behavioral Neuroscience 111:450–459.

St. John SJ and Spector AC (1998) Behavioral discrimination between quinine and KCl is dependent on input from the seventh cranial nerve: implications for the functional roles of the gustatory nerves in rats. Journal of Neuroscience 18:4353–4362.

Tamura R and Norgren R (1997) Repeated sodium depletion affects gustatory neural responses in the nucleus of the solitary tract of rats. American Journal of Physiology: Regulatory, Integrative and Comparative Physiology 273:R1381–R1391.

Tapper DN and Halpern BP (1968) Taste stimuli: a behavioral categorization. Science 161:708–710.

Thaw AK and Smith JC (1992) Conditioned suppression as a method of detecting taste thresholds in the rat. Chemical Senses 17:211–223.

Tordoff MG and Bachmanov AA (2003) Mouse taste preference tests: why only two bottles? Chemical Senses 28:315–324.

Travers SP and Smith DV (1984) Responsiveness of neurons in the hamster parabrachial nuclei to taste mixtures. Journal of General Physiology 84:221–250.

Vogt MB and Smith DV (1993) Responses of single hamster parabrachial neurons to binary taste mixtures: mutual suppression between sucrose and QHCl. Journal of Neurophysiology 69:658–668.

Yamamoto T, Matsuo R, Kiyomitsu Y, Kitamura R (1988) Taste effects of "umami" substances in ham-

sters as studied by electrophysiological and conditioned taste aversion techniques. Brain Research 451:147–162.

Young PT and Trafton CL (1964) Activity contour maps as related to preference in four gustatory stimulus areas of the rat. Journal of Comparative and Physiological Psychology 58:68–75.

Motor Systems

III

Posture

11

SERGIO M. PELLIS AND VIVIEN C. PELLIS

We typically think of behavior as involving movement, be it eating, walking, or copulating. But to move effectively, an animal must support its body and make fine postural adjustments to protect the body's stability (Martin, 1967). One important lesson to be learned from this recognition of the importance of postural support for the genesis of movement is that some forms of lack of movement, or immobility, should still be considered, and studied, as behavior.

With regard to postural support, there are a variety of defensive mechanisms used to maintain or regain stability (Monnier, 1970). Close inspection of postural support responses reveals them to be complex phenomena, guided by all suitable sensory systems—vestibular, tactile, proprioceptive, and visual (Magnus, 1924). In addition, postural responses are themselves composed of independent motor output modules (or programs) that can be disassociated during development and in pathology (Pellis, 1996).

Studies with rats have been central to the understanding of postural support mechanisms. Thus, examples from this research on rats are used to illustrate the importance of looking carefully at states of immobility, identifying the sensory controls guiding postural responses, and fractionating such responses into their independent modules. For the latter, one type of postural support response, righting, is examined in detail to illustrate the need for using highly specific tests to evaluate these independent modules.

IMMOBILITY

When unconscious, such as during sleep, a rat may appear to have a flaccid body tone. This is a state in which not only is the animal unresponsive to its surroundings but also its body seems unprepared to deal with environmental contingencies. There are drug-induced states of immobility that are dramatically different from this image of inertness. Both high doses of morphine, an opioid agonist, and haloperidol, a dopamine antagonist, produce states of immobility. Although in both cases, the treated rats appear inert, they in fact adopt bodily postures of preparedness for action but not the same kind of action in both (De Ryck et al., 1980; De Ryck and Teitelbaum, 1983).

Haloperidol induces a state in which a rat will actively maintain static, stable equilibrium but will not spontaneously move. This state of readiness to act in order to defend its stable position can be seen even when the rat is left standing by itself on a tabletop (Fig. 11–1A). The limbs are splayed out, and the body is raised off the ground. If challenged by being pushed to one side, it will resist that displacement by shifting its body weight in the direction of the oncoming force (see later).

In contrast, morphine induces a state of immobility in which the limbs appear frozen in a step-cycle (Fig. 11–1B). If it is pushed abruptly, the rat will run forward for a few steps before again becoming immobile, but if it is pushed gently, it can be rolled over onto its side. That is, in morphine-induced immobility, the pos-

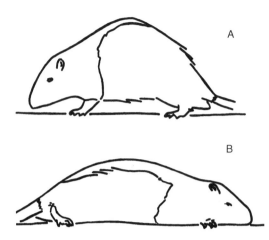

Figure 11–1. Immobility induced by different drugs produces states that can be very different with regard to the postural mechanisms available to the animal. In haloperidol-induced immobility, rats have all of their postural support mechanisms intact, so that when left standing on a tabletop, the head is held off the ground, the body has a curved configuration, and the limbs are pillar-like (A). In contrast, in morphine-induced immobility, the head and body are not supported off the ground, and the rat's limbs are positioned as if in the middle of a step cycle; this is especially evident in the visible hind foot (B). (Adapted from Pellis et al. [1986].)

BRACING

Taking protective actions against an horizontally displacing force is a quintessential measure to maintain an upright, stable orientation. Shifting one's body weight into the direction of the displacing force can be seen in dynamic situations as well, such as when one is running fast in a tight circle (Gambaryan, 1974). Naturally occurring shifts in body weight, which buttress the body against a displacing force, can be evaluated in various tests of skilled action or locomotion (Whishaw et al., 1994; Miklyaeva et al., 1995).

Bracing responses in their purest form are most readily analyzed in rats with dopamine blockade or depletion. For example, if a displacing force is applied to the side of such a rat, it leans laterally toward the displacing force. Similarly, if it is pushed forward in a "wheelbarrow" configuration (i.e., when it is held elevated by its rump), it pushes backward with its forepaws rather than step in the direction of the displacement. An undrugged rat typically steps away or steps forward in these tests (Schallert et al., 1979; Pellis et al., 1985).

A simple way to test bracing is to place the rat on an horizontal platform with a mildly rough surface to allow it to grip and to then raise its tail end so that gravity begins pushing the animal forward. An intact rat typically responds with a positive geotaxis response and turns to face up or slowly walks down as the slope increases. In contrast, a cataleptic rat pushes its body back (Fig. 11–2A) and so brace against the displacing force (Crozier and Pincus, 1926; Morrissey et al., 1989; Field et al., 2000).

If the board is tilted further, so that the rat can no longer resist the downward force of gravity, it begins to slide forward. At the moment its postural stability is lost, the rat explosively jumps forward. On landing, it again becomes immobile. This test shows that the forward jump is strictly a defensive response to a challenge to the rat's postural stability and not a self-initiated forward move-

tural support mechanisms are suppressed but the locomotor mechanisms are not, whereas in haloperidol-induced immobility, the locomotor mechanisms are suppressed but the postural support mechanisms are not. Combined treatment with haloperidol and morphine suppresses both the postural and the locomotor mechanisms (Pellis et al., 1986).

Immobility, then, is not just inertness but may involve states with or without postural support. To properly assess the effects of different experimental treatments, any immobility state that is created needs to be carefully evaluated so as to discern the behavioral capabilities available to the subject. An inadvertent advantage arising from haloperidol-induced immobility (or catalepsy) is that such dopamine-blocked rats afford the opportunity to study postural support responses independently of locomotor, exploratory, and other movement systems (Teitelbaum et al., 1982). Two postural responses, bracing and righting, are described in some detail.

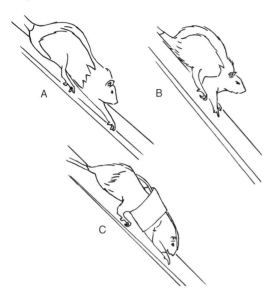

Figure 11–2. Bracing in a cataleptic rat can be tested by placing it on an horizontal board, with a mildly rough surface, which allows it to grip. If the tail end of the board is then gradually tilted, the rat will begin to brace by pushing its limbs forward, shifting its body weight backwards (A). When the rat can no longer resist the downward force of gravity, it begins to slide forward. As the rat slides forward, it raises its head, by dorsiflexing its neck (B). If the dorsiflexion of the neck is prevented, by attaching a tablespoon onto the rat's back with a harness so that its head is then capped by the concave bowl of the spoon, the rat will not jump forward (C). (Adapted from Morrissey et al. [1989] and Teitelbaum and Pellis [1992], respectively.)

ment. That the jumping response in cataleptic rats reflects postural responses disconnected from other behavior systems is illustrated by the two triggers that are necessary for it to occur.

The initial trigger for jumping is the rat feeling the loss of stability in its feet. If small rods are placed on the board so that the rat can grasp them with its feet, it is able to maintain its stability, even when the board approaches a slope of 70° or more, as compared to a typical slope of 50° to 60°. Such stability can be further exaggerated by using a wire mesh, which provides even greater support (Morrissey et al., 1989).

Once the rat feels its feet slide, a second trigger is released, when the head is raised by dorsiflexing the neck (see Fig. 11–2B). This

dorsiflexion appears to be necessary to trigger the forward thrust of the rat's back legs. If a tablespoon is attached onto a rat's back with a harness, so that the rat's head is capped by the concave bowl of the spoon and thus is prevented from dorsiflexing (see Fig. 11–2C), then the cataleptic rat will not jump forward. Indeed, as the rat is sliding down, it can be observed to be making small upward head movements; however, the presence of the spoon prevents them from being of sufficient magnitude to trigger the pushing of its hind legs (Teitelbaum and Pellis, 1992).

When the slope is steep, undrugged rats occasionally jump forward. But if the platform were placed so that its front is flush with the edge of a table, a forward jump would lead the rat to land somewhere off the table. In this situation, an undrugged rat orients to one of the sides of the board and then jumps onto the surface of the table. In contrast, the drugged rat jumps forward and off the table (with a soft cushion being provided as a landing pad) (Morrissey et al., 1989). That is, the drugged rat fails to use visual cues to modify its jump. The failure to incorporate vision in organizing protective postural responses is common to various behavioral tests when using cataleptic rats (Pellis et al., 1987). Furthermore, depending on the test used, when marshalling a postural defense, cataleptic rats, unlike undrugged ones, may give precedence to tactile and proprioceptive information over vestibular information (Pellis et al., 1985; Cordover et al., 1993).

As can be seen from this description of bracing, postural mechanisms can be studied in a manner that allows the sensory devices that regulate their expression to be identified. Another type of postural response, righting, is used to illustrate the motor diversity of these mechanisms.

RIGHTING

Like many other animals, rats, when sleeping, will "voluntarily" relinquish a stable, upright position. As rat pups suckle beneath an upright, standing mother, to gain access to a nipple, pups typically roll over onto their backs (Eilam and Smotherman, 1998). Similarly, during fighting, rats may roll over onto their backs so as to defend themselves against attacks from their opponents (Pellis and Pellis, 1987). In contrast to these behaviors, in righting, a standing position is regained from a recumbent one.

As demonstrated by Magnus (1926), righting can be elicited by multiple sensory inputs, each one capable of independently triggering righting. He thus labeled each form of righting by the type of sensory input involved as well as by the body part that initiated the righting movement. These different types of righting mature independently, and so can be disassociated developmentally, just as they can be with some forms of brain damage and sensory deactivation. Furthermore, there are specific rules of organization that give one form of righting dominance over the others in particular functional contexts (Pellis, 1996).

Righting can occur from a static position, such as when a rat is placed supine on the ground, and from a dynamic position, such as when a rat falls, in a supine position, through the air. In either case, different righting responses can be triggered by tactile/proprioceptive, vestibular, and visual inputs. Unlike some other animals, such as cats, in rats, vision can modulate the timing of dynamic righting, but it cannot initiate such righting (Pellis, 1996). Just as each form of righting involves its own distinctive sensory input and motor output, so each requires a particular testing paradigm to exclude the activation of the other forms of righting.

TESTS OF RIGHTING

In some cases, given that more than one sensory system can initiate the righting of the same part of the body (Magnus, 1926), either some test paradigm that restricts input from the competing sensory systems must be used,

or the sensory systems must be directly blocked by some physiological manipulation. In the following tests, emphasis is placed on disentangling the types of sensory inputs without the need to eliminate those systems. This approach makes it easier for researchers to evaluate test subjects for many forms of righting without the chronic removal of sensory systems that may interfere with general motor function, such occurs after labyrinthectomy (Chen et al., 1986). Of great advantage is videotaping sequences of righting using high shutter speeds, to allow frame-by-frame inspection of the movements performed.

TACTILE/PROPRIOCEPTIVE

Contact with the surface of the skin of its flanks or back provides a rat with information regarding its recumbent position, which allows the initiation of righting. Based on this sensory input, one of three forms of righting can be triggered from a static, recumbent position. When a rat is in contact with the ground but is falling from a stable position, some combination of tactile and proprioceptive information can trigger righting in this dynamic context.

Trigeminal Righting

When the dorsum or the flanks of a rat's head are in contact with the ground, tactile information via the trigeminal nerve triggers righting of the head by a rotation of the neck. If the rat is unrestrained, the righting can proceed cephalocaudally. This was first described by Troiani et al. (1981) in the guinea pig.

In rats, trigeminal stimulation induces rotation of the head, and as the head is rotated, it characteristically maintains firm contact with the substrate. The only other form of righting capable of eliciting rotation of the head is that triggered by vestibular information. In rats, vestibularly triggered head rotation is a transient feature, present only in early development; later vestibular input triggers rotation by the shoulders (Pellis et al., 1991).

Therefore, when placing a rat on the ground, in whether the supine or laterally recumbent position, the experimenter should firmly apply pressure to the exposed part of its body. Then, the rat's head should be released after it is pushed down to ensure that it has contact with the ground; this reveals whether its trigeminal righting program is functional.

Body Tactile Righting

Contact with the surface of either the dorsum or the flank triggers the rat's righting response. However, there are two distinct righting programs that can be elicited in this manner. One involves a rotation that begins at the shoulders (body-on-head), and the other begins with a rotation of the hindquarters (body-on-body) (Magnus, 1926). To test these forms of righting, the rat can be placed on one of its flanks, with the experimenter's hand placed over and then pressed down onto its exposed flank. To avoid triggering trigeminal righting, the rat's head should not make contact with the experimenter's hand, and the other side of its head should be left to jut out over the edge of the table's surface.

Under normal circumstances, body-on-head righting is dominant over body-on-body righting. Therefore, to test for the presence of body-on-body righting, body-on-head righting has to be inactivated. To do this, the experimenter can place one hand over the rat's shoulders and the other over its pelvis, and then release the hand over the pelvis. If the body-on-body program is present, the rat's hindquarters rotate to prone. Evaluating the presence of body-on-head righting, however, is more difficult because shoulder rotation can also be triggered vestibularly. In such a case, a labyrinthectomy is necessary to negate the influence of the vestibular information (Chen et al., 1986). Even so, if prior testing of vestibular righting (see later) had revealed that such righting were not present, then the subject does not need to undergo labyrinthectomy to test for body-on-head righting.

Early in development, body tactile forms of righting involve the rat pushing with its limbs rather than using axial rotation (Pellis et al., 1991). If an animal reverts to pushing with its limbs to right, it may indicate brain damage (Martens et al., 1996). Therefore, when testing the rat from the laterally recumbent position, both the body-on-head and body-on-body forms of righting can be evaluated by the experimenter for whether they demonstrate the adult-typical axial rotation pattern or have regressed to the more primitive form of righting involving limb action. The distinction between these two forms can be readily seen on videotaped sequences of righting. When righting with axial rotation, the rat tucks the paw nearest to the ground close to its body; this prevents the paw from obstructing the body's rotation to prone (Fig. 11–3A). In contrast, when righting using limb action, the rat places the paw closest to the ground beneath its body; the paw is thus in a position to flip the rat's body to prone by pushing against the ground (Fig. 11–3B).

Tactile/Proprioceptive Dynamic Righting

A distinctive, dynamic form of righting is triggered in the rat when it falls while in contact

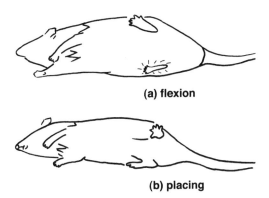

(a) flexion

(b) placing

Figure 11–3. When righting with axial rotation, the rat tucks the paw that is nearest to the ground close to its body (a). This prevents the paw from obstructing its body's rotation to prone. When righting using limb action, the rat places the paw that is closest to the ground beneath its body (b). The paw is now in a position to flip the rat's body to prone by pushing against the ground. (Adapted from Pellis et al. [1989b].)

with the ground. This form of righting matures later in development than do the static forms of tactile righting (Pellis and Pellis, 1994) and can best be tested by holding the rat in a bipedal position, so that it stands on its hindlimbs. The rat is held by the experimenter from the back under its arms and is then pulled back, onto the ground. If its righting is intact, as the rat begins to fall, it should rotate to face the direction in which it is falling, so that by the time it lands, it does so in a prone position. The rotation typically begins in the rat's shoulders and progresses caudally. As with body-on-head righting, this form of righting can also be triggered vestibularly, and so either pretesting for vestibular function (see later) or labyrinthectomy is needed to evaluate fully whether it is intact.

VESTIBULAR

There are three distinctive ways in which vestibular input can affect righting. Two involve the rat righting its head from a static position based on otolithic function, and the third involves the rat righting itself when falling based on semicircular function (Monnier, 1970).

Static Asymmetrical
When held by the experimenter, laterally in the air, so that both flanks of its body have contact but its head is left untouched, the rat's head rotates toward the ground, so that it is prone relative to gravity. That is, from a situation of asymmetrical vestibular information, the head turns to a normal prone orientation that has equal vestibular input on both sides of the head. In this lateral orientation, it is the otoliths on the side of the rat's head that face the ground that provide the stimulus for righting.

A developmental dimension to this form of righting may also be used to detect any regression to a more primitive functioning. In early development, when this form of righting first appears, a rat's head rises upward so its snout points skyward and then its head rotates toward the ground. However, at later

stages of maturity, the rat's head rotates directly to prone (Pellis et al., 1991).

Static Symmetrical
When the rat is held by the base of the tail and is lifted into the air so that its head is pointing downward, the rat's initial response is to dorsiflex its head to bring it toward a prone orientation relative to the ground. As the rat begins to crawl up one of its flanks very quickly after the initial dorsiflexion of its head, it is useful to videotape this test. Such a confounding response can be blocked by pretreatment with a dopamine antagonist such as haloperidol (see earlier). In the absence of vestibular function, the rat's head rapidly ventroflexes (Pellis et al., 1991b).

Dynamic
The classic way to test this response is to drop the rat from a height onto a soft cushion. The rat should be picked up by the experimenter with one hand under its shoulders and lifted off the ground and held by the dorsum of the pelvic girdle with the other hand. In this position, the rat should then be raised to the desired height. When the experimenter feels the rat to be relaxed, in that it is no longer squirming in the experimenter's grip, the experimenter should rapidly swing away his or her hands, releasing the rat to fall. Within about 30 to 60 milliseconds, an intact rat begins to right in the manner described earlier, so that on landing it is fully prone (Pellis et al., 1991a). In the absence of vestibular input, righting does not occur, and in cases of partial loss of vestibular function, righting may be delayed or incomplete (Chen et al., 1986; Wallace et al., 2002).

Although some researchers have used mechanical devices for holding animals in an attempt to standardize the hold-and-drop procedure (Warkentin and Carmichael, 1939; Schonfelder, 1984), we have found that there is no substitute for holding the animal in one's hands, as the experimenter is then able to use tactile cues to ensure that the rat is dropped only when fully relaxed (see also Crimieux et al., 1984). Indeed, if the rat struggles or tenses before its release, its righting can be initially inhibited or, due to its struggling, these movements may interfere with the progression of smooth, cephalocaudal axial rotation.

VISUAL

As noted, vision does not trigger righting in rats, but it can modulate its onset. Modulation can best be evaluated in the righting-while-falling-in-the-air paradigm. By alternating trials at two different heights (e.g., 60 cm and 30 cm), it can be determined whether the difference in height can influence the onset of the rat's righting. Using videotapes of these trials, the latency of onset can be counted in terms of the number of frames that have elapsed before the first evidence of shoulder rotation is seen. The difference between the two heights can be expressed in milliseconds. Typically, there is a 30 millisecond difference between these two heights; rats begin righting sooner when dropped from the lower height (Pellis et al., 1989a; Pellis et al., 1991c). If vision is blocked, rats initiate air righting at the same latency, regardless of the height of the drop (Pellis et al., 1989a, 1996).

CONCLUSION

Postural support is not something that can be left for kinesiologists and physiologists to deal with; it is integral to the production of movement, and so organized sequences of behavior (Martin, 1967). In early development, postural concerns have a more obvious impact on behavior (Pellis and Pellis, 1997). Even in adulthood, the limitations imposed by the capacity to modify postural support may account for species differences in the types of movement strategies used and in the bodily configurations adopted (Berridge, 1990; Pellis, 1997). Yet the analysis of postural support mechanisms is rarely incorporated into analyses of movement (for rare exceptions, see Whishaw et al., 1994; Miklyaeva et al., 1995).

As can be seen from this short, selective review, the postural support system of rats is quite complex, with even seemingly small subcomponents (e.g., righting) requiring multiple techniques to be fully assessed. Thus, to fully comprehend rats' behavior, their postural support mechanisms need to be understood and that understanding must be incorporated into studies of their movements.

BIBLIOGRAPHY

Berridge KC (1990) Comparative fine structure of action: rules of form and sequence in the grooming patterns of six rodent species. Behaviour 113:21–56.

Chen Y-C, Pellis SM, Sirkin DW, Potegal M, Teitelbaum P (1986) Bandage backfall: labyrinthine and non-labyrinthine components. Physiology and Behavior 37:805–814.

Cordover AJ, Pellis SM, Teitelbaum P (1993) Haloperidol exaggerates proprioceptive-tactile support reflexes and diminishes vestibular dominance over them. Behavioural Brain Research 56:197–201.

Cremieux J, Veraart C, Wanet MC (1984) Development of the air righting reflex in cats visually deprived since birth. Experimental Brain Research 54:564–566.

Crozier WJ and Pincus G (1926) The geotropic conduct of young rats. Journal of Genetic Physiology 10:257–269.

De Ryck M and Teitelbaum P (1983) Morphine versus haloperidol catalepsy in the rat: an electromyographic analysis of postural support mechanisms. Experimental Neurology 79:54–76.

De Ryck M, Schallert T, Teitelbaum P (1980) Morphine versus haloperidol catalepsy in the rat: a behavioral analysis of postural mechanisms. Brain Research 201:143–172.

Eilam D and Smotherman WP (1998) How the neonatal rat gets to the nipple: common motor modules and their involvement in the expression of early motor behavior. Developmental Psychobiology 32:57–66.

Field EF, Whishaw IQ, Pellis SM (2000) Sex differences in catalepsy: evidence for hormone-dependent postural mechanisms in haloperidol-treated rats. Behavioural Brain Research 109:207–212.

Gambaryan PP (1974) How animals run. New York: Wiley.

Magnus R (1924) Körperstellung. Berlin: Springer.

Magnus R (1926) On the co-operation and interference of reflexes from other sense organs with those of the labyrinths. Laryngoscope 36:701–712.

Martin JP (1967) The basal ganglia and posture. London: Pitman Medical Publishing.

Martens DJ, Whishaw IQ, Miklyaeva EI, Pellis SM (1996) Spatio-temporal impairments in limb and body movements during righting in an hemiparkinsonian rat analogue: relevance to axial apraxia humans. Brain Research 733:253–262.

Miklyaeva EI, Martens DJ, Whishaw IQ (1995) Impairments and compensatory adjustments in spontaneous movement after unilateral dopamine-depletion in rats. Brain Research 681:23–40.

Monnier M (1970) Functions of the nervous system: volume II, motor and sensorimotor functions. Amsterdam: Elsevier.

Morrissey TK, Pellis SM, Pellis VC, Teitelbaum P (1989) Seemingly paradoxical jumping in cataleptic haloperidol-treated rats is triggered by postural instability. Behavioural Brain Research 35:195–207.

Pellis SM (1996) Righting and the modular organization of motor programs. In: Measuring movement and locomotion: from invertebrates to humans (Ossenkopp K-P, Kavaliers M, Sanberg P, eds.), pp. 116–133. Austin, TX: R.G. Landes Company.

Pellis SM (1997) Targets and tactics: the analysis of moment-to-moment decision making in animal combat. Aggressive Behavior 23:107–129.

Pellis SM and Pellis VC (1987) Play-fighting differs from serious fighting in both targets of attack and tactics of fighting in the laboratory rat Rattus norvegicus. Aggressive Behavior 13:227–242.

Pellis SM and Pellis VC (1994) The development of righting when falling from a bipedal standing posture: evidence for the disassociation of dynamic and static righting reflexes in rats. Physiology and Behavior 56:659–663.

Pellis SM and Pellis VC (1997) The prejuvenile onset of play fighting in laboratory rats (Rattus norvegicus). Developmental Psychobiology 31:193–205.

Pellis SM, Chen Y-C, Teitelbaum P (1985) Fractionation of the cataleptic bracing response in rats. Physiology and Behavior 34:815–823.

Pellis SM, Pellis VC, Teitelbaum P (1987) 'Axial apraxia' in labyrinthectomized lateral hypothalamic-damaged rats. Neuroscience Letters 82:217–220.

Pellis SM, Pellis VC, Teitelbaum P (1991a) Air righting without the cervical righting reflex in adult rats. Behavioural Brain Research 45:185–188.

Pellis SM, Pellis VC, Teitelbaum P (1991b) Labyrinthine and other supraspinal inhibitory controls over head-and-body ventroflexion. Behavioural Brain Research 46:99–102.

Pellis SM, Pellis VC, Whishaw IQ (1996) Visual modulation of air righting involves calculation of time-to-impact, but does not require the detection of the looming stimulus of the approaching ground. Behavioural Brain Research 74:207–211.

Pellis SM, Whishaw IQ, Pellis VC (1991c) Visual modulation of the vestibularly-triggered air-righting in rats involves the superior colliculus. Behavioural Brain Research 46:151–156.

Pellis SM, de la Cruz F, Pellis VC, Teitelbaum P (1986) Morphine subtracts subcomponents of haloperidol-isolated postural support reflexes to reveal gradients of their recovery. Behavioural Neuroscience 100:631–646.

Pellis SM, Pellis VC, Morrissey TK, Teitelbaum P (1989a) Visual modulation of vestibularly triggered air-righting in the rat. Behavioural Brain Research 35:23–26.

Pellis SM, Pellis VC, O'Brien DP, de la Cruz F, Teitelbaum P (1987) Pharmacological subtraction of the sensory controls over grasping in rats. Physiology and Behavior 39:127–133.

Pellis SM, Pellis VC, Chen Y-C, Barzci S, Teitelbaum P (1989b) Recovery from axial apraxia in the lateral hypothalamic labyrinthectomized rat reveals three elements of contact righting: cephalocaudal dominance, axial rotation, and distal limb action. Behavioural Brain Research 35:241–251.

Pellis VC, Pellis SM, Teitelbaum P (1991) A descriptive analysis of the postnatal development of contact-righting in rats (Rattus norvegicus). Developmental Psychobiology 24:237–263.

Schallert T, De Ryck M, Whishaw IQ, Ramirez VD, Teitelbaum P (1979) Excessive bracing reactions and their control by atropine and L-DOPA in an animal analog of Parkinsonism. Experimental Neurology 64:33–43.

Schonfelder J (1984) The development of air-righting reflex in postnatal growing rabbits. Behavioural Brain Research 11:213–221.

Teitelbaum P, Szechtman H, Sirkin DW, Golani I (1982) Dimensions of movement, movement subsystems and local reflexes in the dopaminergic systems underlying exploratory locomotion. In: Behavioral models and the analysis of drug action (Spiegelstein MY and Levy A, ed.), pp. 357–385. Amsterdam: Elsevier.

Teitelbaum P and Pellis SM (1992) Towards a synthetic physiological psychology. Psychological Science 3:4–20.

Troiani D, Petrosini L, Passani F (1981) Trigeminal contribution to the head righting reflex. Physiology and Behavior 27:157–160.

Wallace DG, Hines DJ, Pellis SM, Whishaw IQ (2002) Vestibular information is required for dead reckoning in the rat. Journal of Neuroscience 22:10009–10017.

Warkentin J and Carmichael L (1939) A study of the development of the air-righting reflex in cats and rabbits. Journal of Genetic Psychology 55:67–80.

Whishaw IQ, Gorny B, Tran-Nguyen LTL, Castaneda E, Miklyaeva EI, Pellis SM (1994) Making two movements at once: impairments of movement, posture, and their integration underlie the adult skilled reaching deficit of neonatally dopamine-depleted rats. Behavioral Brain Research 61:65–77.

Orienting and Placing

12

TIM SCHALLERT AND MARTIN T. WOODLEE

It is not difficult to document the existence of sensory or motor function asymmetries in normal rats or rats with unilateral damage to the basal ganglia, sensorimotor cortex, and related systems throughout the central nervous system, especially when the deficit on one side is near maximal. When the unilateral deficit is subtotal, quantifying the extent of asymmetry and changes over time requires unique testing methods that directly pit one hemisphere against the other.

As an illustration, if a person is slightly hard of hearing in one ear across all tones, how would you determine which ear is better and by how much? A simple test would be to put headphones on the person, play the same level of sound simultaneously in each ear, and determine which side the sound seems to be coming from. Because sound localization is influenced by relative intensity, this method could be used to confirm the existence of a sensory asymmetry. If the sound appears to come from the left, it can be concluded that the right ear (or left hemisphere) is impaired relative to the left ear. To determine the magnitude of the asymmetry, you could then raise the intensity of the sound presented to the relatively impaired ear and/or reduce the intensity of the sound presented to the better ear until the sound seemed to come from neither the left nor right side. The ratio of the sound intensity presented to the impaired ear relative to that of the sound presented to the better ear would quantify the extent of asymmetry. This two-part method is essentially the approach one can take in as-

sessing sensorimotor asymmetries in rats with partial unilateral damage to the brain, and in evaluating treatments.

Behavioral deficits in Parkinson's disease and stroke often can be traced to both sensory and movement initiation problems or an impaired ability to make appropriate motor responses to simple sensory events. In animals, unilateral damage to the sensorimotor cortex, striatum, or nigrostriatal pathway appears to have the perceptual effect of dulling somatosensory and proprioceptive sensory input on one side and, in some cases, enhancing the input from the other side. Asymmetrical sensory deficits, motor reactivity to bilateral sensory input, or predominantly motor dysfunctions can be examined with tests using a two-part method in which an asymmetry is first identified and then the extent of the asymmetry is quantified.

In animal models, it is important to select sensorimotor tests that are sensitive to the brain damage and treatment effects. This chapter describes behavioral tests that have been useful for examining the potential clinical efficacy of interventions that might be beneficial for neurological disorders. It is important to be able to distinguish whether an intervention promotes brain repair mechanisms, saves cells, enhances motor learning and retraining, or reduces the extent of secondary degeneration of tissue. We have chosen to include a subset of sensorimotor tests that we and others have found to be reliable, sensitive, quantitative, and easy to use in rat neurological models. The tests also cover the range of cellular degeneration typical of focal ischemic injury, nigrostriatal terminal loss, and cervical spinal trauma.

ENVIRONMENTAL ENRICHMENT AND SENSORIMOTOR BEHAVIOR

Most wild rats live in a very complex environment that requires them to navigate obstacles, avoid predators, manipulate objects and circumstances to gain access to food and mates, and so forth, using a wide array of motor skills. By contrast, standard laboratory housing is severely lacking in this sort of stimulation, and laboratory "enriched" environments are still less complex than the rat's natural habitat (Greenough et al., 1976; Jones et al., 2003; Schallert et al., 2003). Even the most sedentary of people do not experience as impoverished an environment as a rat living in an isolated home cage. Therefore, to study sensorimotor behavior in the rat, it may be prudent to make some effort to house animals so that behaviors analogous to natural rat behavior are encouraged.

BILATERAL TACTILE STIMULATION TEST

Rats compulsively groom themselves and respond vigorously to any foreign substance that becomes stuck to some part of their bodies. The adaptive advantages of this behavior may include thermoregulation and maintenance against insects. Somatosensory asymmetries have been effectively determined using a test that involves reacting to, and removing, small sticky stimuli from the forelimbs. It is a two-part test; however, few investigators take advantage of both parts, which are needed to evaluate sensory function independent of the motor component. Practice effects and motor learning play a partial role in the motor aspects of this test but do not affect the sensory side, which can be investigated independently.

SENSORY ASYMMETRY

Small adhesive paper stimuli (Avery adhesive-backed labels, 113 mm^2) are attached to the relatively hairless distal-radial aspect of each of the rat's forelimbs (Schallert et al., 1982, 1983, 2000; Schallert and Whishaw, 1984; Lindner et al., 2003; Fleming et al., 2003) (Fig. 12–1). The rat is placed back into its home cage so that it is not distracted by a novel environment, and it quickly uses its teeth to remove these dots one at a time. In some animals there is a small preoperative bias; in these cases, the hemisphere selected for injury can be opposite to the bias. Also, postoperative outcome can be compared against baseline values for each rat. Rats receiving unilateral lesions to brain areas subserving sensorimotor functions, especially those of the forelimbs, develop an immediate bias for removing adhesive stimuli of similar size from the unimpaired limb first. The order of contacting the ipsilateral versus contralateral stimulus reflects that there is a bias, but the magnitude of the sensory asymmetry requires further evaluation (see later). The latency to remove the stimuli can be used as a measure of motor capacity and is sensitive to practice effects, unlike the order of contact (Schallert and Whishaw, 1984).

Each trial ends when the rat removes both stimuli, or after 2 minutes has elapsed. To avoid habituation to the stimuli, individ-

Figure 12–1. Attaching adhesive stimuli (dots) to a rat's forelimbs in preparation for the bilateral tactile stimulation test.

ual trials should occur at intervals of no less than 5 minutes. In addition, the rats used should be well handled and have received several practice trials with the test before preoperative data are collected. Experience with the test calms the rats and makes the stimuli easier to apply but does not appear to affect actual performance.

This test is generally used to examine sensorimotor integration, although, as indicated earlier, it is possible to some degree to distinguish between the sensory and motor components involved (Schallert et al., 2002). For example, a change in the latency between initial contact and subsequent removal of a dot (i.e., how much time it takes to remove the stimulus) can be an index of sensorimotor function. As in many of the tests presented here, however, it is important that such a change be represented as an asymmetry between the impaired and unimpaired limbs in unilateral lesion models to control for nonmotor and nonsensory factors (e.g., motivational state, alertness) that could have a global influence on latencies to contact and remove the dot. The contralateral (impaired limb) *motor* component of this test is best assessed by determining the time point at which the animal makes contact with the stimulus on the impaired side and scoring how much time *after* that time point it takes to remove that stimulus. This difference then would be compared with a comparable score of intact control animals (i.e., how long after a control rat contacts a given stimulus before it is removed, again controlling for practice effects by equating extent of experience).

MAGNITUDE OF SENSORY ASYMMETRY

The second part of this test is used as a means of measuring the degree of sensory asymmetry. In this part, the size of the dot placed on the impaired limb is progressively increased (by overlapping two dots), while the dot on the unimpaired limb is made smaller (by cutting down one dot). The dot sizes are in-

creased or decreased by 14 mm^2, as illustrated in Figure 12–2, allowing for area ratios ranging from 1.3:1 to 15:1 between the impaired and unimpaired limbs, respectively. A sufficient increase in this ratio leads to a neutralization, and even a *reversal* (with a slightly higher ratio), in the bias for the limb that is contacted first, and the ratio at which this occurs is used as the measure of severity of the sensory asymmetry. This measure is correlated with the amount of brain damage (Schallert et al., 1983; Schallert and Whishaw, 1984; Barth et al., 1990); indeed, a small asymmetry can be detected in rats with simple burr holes in the skull. Animals are started at the 2.2:1 ratio (level 3). If the stimulus is removed from the unimpaired limb first, animals then are tested at *two* levels higher. If the stimulus is removed from the impaired limb first, the animal is tested at *one* level lower. This process is continued until the experimenter has determined between which two levels the bias exists and assigned the rat a score that reflects this ratio (e.g., a score of 2.5 is given if the animal's bias reverses between levels 2 and 3).

Acute and chronic asymmetries on this test have been demonstrated in models of cortical injury and ischemia, parkinsonism, and spinal cord injury (Schallert et al., 2000). As recovery occurs, the ratio defining the magnitude of asymmetry becomes smaller inde-

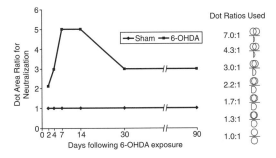

Figure 12–2. (*Left*) Data from an animal model of 6-hydroxydopamine–induced parkinsonism, in which sensory asymmetries measured with the bilateral tactile stimulation test improved over time but did not return to sham-operated control levels. (*Right*) Schematic of the different area ratios of the "dots" used in the test.

pendent of how much or how little practice occurs. Depending on the degree of striatal or nigrostriatal damage, full recovery can occur, even in hemidecorticate rats (Schallert and Whishaw, 1984). However, small changes in the testing environment (e.g., partially opening the home cage while testing) can partially reverse recovery so that the ratio becomes larger, possibly because the striatum is being taxed. This is important because it suggests that the testing environment can have a major influence on measures of functional outcome.

LIMB-USE ASYMMETRY ("CYLINDER") TEST

The limb-use asymmetry test evaluates the forelimb use of rats placed in a transparent Plexiglas cylinder. It has been used in a wide variety of motor system injury models, including middle cerebral artery occlusion, spinal cord injury, traumatic brain injury, parkinsonian models, cortical ablation, and focal cortical ischemia (Schallert et al., 2000; Schallert and Tillerson, 2000; Tillerson et al., 2001, 2002; Lindner et al., 2003). A notable feature is a high degree of sensitivity to chronic deficits not noticeably masked by postlesion compensatory behaviors, as well as to chronic sensorimotor deficits that many tests fail to detect. The test is also easy to use and score, has a high interrater reliability, is well correlated with the extent of lesions, including a wide range of dopamine depletion (even 50% or less) (Tillerson et al., 2001), and is relatively unaffected by practice effects or, it seems, the compensatory strategies often adopted by animals after motor system insults (Schallert et al., 2002).

Rats are tireless explorers, in both their natural environments and laboratory home cages. They often explore vertical surfaces by rearing up on their hindlimbs and exploring the surface with their front paws and vibrissae (Gharbawie et al., 2003). The cylinder test takes advantage of this tendency and of the common

impairment in the initiation of movement and control of static stable equilibrium, especially center of gravity (Schallert et al., 1979, 1992). A rat is placed in an upright Plexiglas cylinder, open at both ends and measuring 30 cm high by 20 cm in diameter, that rests on a tabletop. The number of independent placements observed for either the right or left forelimb, as well as the number of "both limb" (i.e., simultaneous or near-simultaneous) placements, made onto the inner wall of the cylinder during rears is recorded. These limb placements occur when the rat shifts its weight, touches the cylinder wall, or steps to regain center of gravity during lateral movements along the cylinder wall ("wall stepping").

The data can be recorded over a set period of time in the cylinder or until a certain number of placements has been made. (We prefer the latter technique because different rats, and especially different strains, can vary widely in their activity levels in the cylinder.) To film the rat's behavior for later rating, (1) a camera is placed over the cylinder (Fig. 12–3A), (2) a camera is positioned to the side of the cylinder with a mirror angled behind and to the side to enable the experimenter to see the rat from all angles during live rating so that no limb movement is missed, or (3) the cylinder is placed atop a raised, transparent surface with a mirror positioned beneath at a 45° angle, with the camera aimed at the mirror to film the limb placements from below (Fig. 12–3B). Care should be taken so that the rats do not habituate to the cylinder lest they become inactive. This can be avoided by testing during the dark cycle and by dividing long trials into shorter segments separated by several minutes, during which the rat is placed back in the home cage.

Limb use is scored as the percentage of left, right, or both-limb wall placements relative to the total number of placements observed. One can also obtain a single limb-use asymmetry score by subtracting the percent independent use of the impaired limb from the percent independent use of the

A

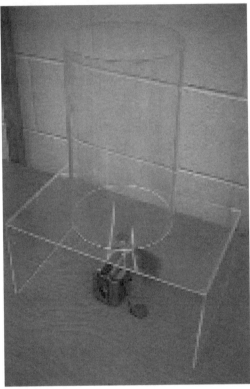

Setup 1:

Place camera directly beneath cylinder.

Setup 2:

Place a mirror at 45° angle beneath cylinder and film at 45° angle to mirror.

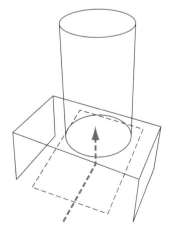

B

Figure 12–3. (A) Top-down view of a rat making placements in the cylinder, filmed from a camera placed over the cylinder. (B) Alternate setup that can be used to film the rat from below the cylinder.

unimpaired limb. Higher numbers indicate a greater bias for use of the unimpaired limb. The former scoring method is advantageous in that it provides more information about both- versus independent-limb use events. It should be noted, however, that even with the latter scoring method, a large number of both-limb use events lower the asymmetry score, albeit not as much as would an equal number of independent impaired-limb placements. An alternative formula is one that we recently adopted because it reduces variability even further and sets a nonbias at 50%:

$$[(\text{ipsi} + {}^{1}/_{2}\text{ both}) \text{ divided by}$$
$$(\text{ipsi} + \text{contra} + \text{both})] \times 100$$

An additional measure can be obtained from this test. The use of a single limb to make lateral weight-shifting movements, independent of the other limb, is reduced in the impaired limb and enhanced in the nonimpaired limb and reflects a very high degree of functional integrity. That is, when a rat rears and places one forelimb on the wall of the cylinder and then makes a lateral movement to another location on the wall during the same rear sequence, this is considered an independent lateral weight-shifting movement, as opposed to a simple limb placement on the wall (which for the contralateral limb may show some recovery). The number of independent-limb weight-shifting movements along the wall for the ipsilateral forelimb can be compared with that of the contralateral forelimb. After unilateral injury to the sensorimotor cortex, striatum, or other motor areas, such movements are rarely observed in the affected forelimb but are commonly chronic in the unaffected forelimb (where they are even more frequently observed than in either limb of control animals, suggesting a reorganization in the intact hemisphere).

Preoperative baseline values should be obtained before animals undergo surgery or other experimental manipulations. Although there is no consistent population bias in limb

preference in the cylinder, some rats do display a predilection for independent use of one limb. When this occurs, experimental lesions can be applied contralateral to this preferred limb so that experimental effects are not confounded by the preexisting limb-use bias. Animals without a preoperative bias can be randomly assigned the lesion side.

Motivation differs between strains of rats. Long-Evans hooded rats, for example, are more active and thus might be considered preferable as animal models, all other considerations being equal. Some rats, especially Sprague-Dawley rats (in our experience), may not initially engage in an adequate amount of wall exploratory behavior in the cylinder. By and large, however, the behavior can be encouraged in any rat with the use of any number of "tricks" that do not affect the limb-use asymmetry score itself, including the following:

- Momentarily turning out the lights in the testing room and testing during red light
- Blowing into or tapping the top of the cylinder
- Placing a dark cage cover (especially the rat's own) over the cylinder
- Placing shavings from the rat's home cage into the cylinder
- Scooting the cylinder (with the rat inside) gently a few cm along the tabletop
- Lightly touching a pencil eraser or cotton tipped applicator to the rat's nose
- Dangling another rat into the cylinder momentarily
- Presenting novel scents or treats at the top of the cylinder
- Picking up the rat and replacing it into the cylinder
- Placing the rat in a new cylinder
- Picking up the rat, flipping over the cylinder, and putting the rat back in

TESTS OF FORELIMB PLACING

Researchers have developed a variety of forelimb-placing tests. Limb placing is usually triggered by visual or vestibular cues or by contacting the limb being tested with a surface (Wolgin and Kehoe, 1983; Marshall, 1982). Rats use their vibrissae to gain bilateral information about the proximal environment, and this information is integrated between the hemispheres. When the bottoms of all four feet indicate that there is no stable surface for support, the rat is motivated to respond to the first object that one set of vibrissae contacts. In exploring its natural world, the rat frequently encounters surfaces that are unstable or, in the case of a cliff, unsuitable for locomotion. All four limbs must be able to respond to information from either set of vibrissae.

The test we describe next is the vibrissae-elicited forelimb placing test, which uses stimulation of the rat's vibrissae to trigger a placing response (Barth et al., 1990; Schallert et al., 2000; Lindner et al., 2003). This is a nice feature in light of the very important role that the vibrissae play in the rat's sensory environment—indeed, they are thought to be one of the primary tools rats use to explore their world. In addition, the test can be adapted to investigate neural events in the sensorimotor system that occur across the midline (as described next), a feature that is more difficult to implement using other placing triggers.

In this test, the rat's torso is supported by the investigator and suspended such that all four legs hang freely in the air. The experimenter then brings the rat toward the edge of a tabletop or another flat surface, taking care to avoid abrupt movements that might trigger placing due to a vestibular response. If such responses are noted, they should be extinguished by taking the rat through the testing motions in open space (i.e., away from the tabletop) a few times. In the traditional, same-side version of this test, the rat's vibrissae are brushed against the table edge on the same side of the body in which forelimb placing is being evaluated. The percentage of trials in which the rat successfully places its forepaw onto the tabletop is recorded for each side. In addition, the triggering stimulus can be provided by moving the rat head on toward the table edge, thus providing chin-based and/or bilateral vibrissae stimulation, or by holding the rat on its side and stimulating the whiskers *opposite* the limb being evaluated (Fig. 12–4 demonstrates these different types of vibrissae stimulation). In all of these testing scenarios, the experimenter should gently restrain the limb not being tested. Naturally, this requires a tame rat that has been well handled for some time before testing, and which ideally has had a chance to acclimate to the test and the experimenter before being introduced to the experimental manipulation. Trials should be counted only when the rat is relaxed and does not struggle, and achieving this can require a great deal of practice on the part of the experimenter. Intact rats will place with 100% success in all variants of this test.

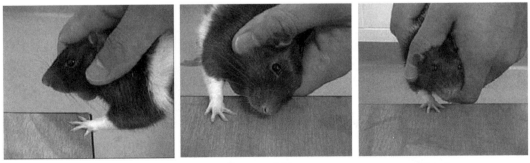

| Same-side | Cross-midline | Head-on |

Figure 12–4. Forms of vibrissae-elicited forelimb placing, demonstrating the proper grip and orientation of the rat for this test.

The same-side version of the test has been used in the evaluation of many central nervous system injury models (Schallert et al., 2000). At our laboratory, we have begun to investigate the recovery of the cross-midline type of placing response in rats receiving cortical (via middle cerebral artery occlusion or focal ischemia to the forelimb area of sensorimotor cortex) or nigrostriatal (via 6-hydroxydopamine infusions to the nigrostriatal bundle) injury. One striking feature noted here is that vibrissae stimulation applied to the "good" (i.e., ipsilesional) side of the body is able to trigger a placing reaction in the impaired forelimb long before stimulation of the contralesional vibrissae can. In contrast, lesions to the nigrostriatal system lead to a complete failure of placing in the contralesional limb in this test, consistent with parkinsonian akinesia. Also, the placing deficit recovers over a period of weeks in the cortical injury models (the rate of recovery depending on the extent of damage to the forelimb area of the sensorimotor cortex and especially to the extent of striatal damage) but persists chronically in parkinsonian models (Felt et al., 2002; Woodlee et al., 2003).

For example, after middle cerebral artery occlusion that damages the striatum, the contralateral forelimb no longer responds to information from the vibrissae about the location of stable surfaces, although the ipsilateral forelimb can respond appropriately to information from the contralateral vibrissae (suggesting that the deficit is not due to a pure sensory impairment associated with the contralateral vibrissae). Moreover, except for severe damage to nigrostriatal dopamine terminals, in which the contralateral forelimb is akinetic, the contralateral forelimb recovers placing in response to ipsilateral vibrissae stimulation. That is, sensory information sent to the intact hemisphere can eventually control motor function associated with the damaged hemisphere, which is typical of normal rats.

TESTS OF HINDLIMB FUNCTION

Rats do not normally use their hindlimbs to initiate or execute complex movement. In this regard, we like to think of rats as being "front-wheel drive," a phenomenon that is illustrated in Figure 12–5, wherein rats supported solely

Figure 12–5. Rats operate primarily via "front-wheel drive." This series depicts a rat remaining stationary for 10 seconds when supported only on its hindlimbs (rats do not walk on their hindlimbs even if given more time) but proceeding to move briskly along a beam when flipped over so that it can support its weight on its forelimbs.

on the forelimbs or hindlimbs initiate movement only when on the forelimbs (Schallert and Woodlee, 2003). In part, this is because the vibrissae are not in contact with the ground and thus there is essentially a "stop" signal. This makes testing of hindlimb function rather difficult, but some tests have been developed that provide reliable measures of hindlimb function. This is good news for researchers of spinal cord injury, because experimental animal models of these conditions often involve lesioning caudal to the thoracic cord so that animals can maintain forelimb function and thus continue to care for themselves postoperatively. Also, research into sciatic nerve injury, a widely used model of peripheral nerve damage, can benefit from well-developed tests of hindlimb function.

A relatively new hindlimb test that we developed is the ledged tapered beam–walking test (Schallert et al., 2002). In this test, rats are trained to traverse an elevated beam that is tapered along its extent and has an underhanging ledge (2 cm wide, dropped 2 cm below the upper beam surface) that the rat can use as a crutch if it slips (see Figure 12–6 for dimensions and setup). Footfaults (slips) made with the hindlimbs can be measured as an index of hindlimb function. A footfault can be rated as a half-fault if the paw slips off the upper sur-

face of the beam without falling all the way to the ledge or as a full fault if the paw is placed fully on the ledge. The difficulty of the rat's traverse increases as it moves along the narrowing beam, thus leading to more footfaults. For this reason the beam can be divided into three "bins" of difficulty along its extent, and these can be scored separately or weighted relative to each other to develop a single score. We generally run rats for five trials on a given day of testing, with several minutes between each trial (during which the rat's cagemate, for example, can be tested) to avoid habituation to the test.

An important feature of this test is that the presence of the ledge allows the rat to display a deficit that it might normally make compensatory adjustments to hide. Rats are well known for compensating to overcome lesion-induced deficits, and indeed this can make the development of good behavioral tests difficult. Some compensatory motor adjustments appear to be more automatic in that they appear immediately in response to an impairment, whereas other adjustments require new learning. If one wishes to test the direct effect of a therapeutic intervention on the system in question, it is important to have tests that will target the deficit directly and be minimally affected by these compensatory be-

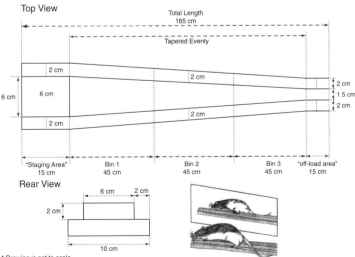

Figure 12–6. The ledged tapered beam, showing dimensions for construction. (*Lower right*) Rat running the beam, making a full left hindlimb footfault (slip) onto the ledge.

haviors. If the test is influenced by compensation, it may not be clear if the therapy is actually ameliorating the deficit per se rather than enhancing motor learning mechanisms that allow for development of the compensatory behavior. Beam-walking tasks that do not use a ledge are plagued by this problem, because rats very quickly learn to make compensatory postural adjustments to keep themselves from falling off the beam. Limb dysfunction may still exist, but the shift in body weight can hide it. With the ledged beam, there is less threat of falling and therefore compensation is less of a problem. In fact, a beam with a detachable ledge can be used to measure the ability of the rat to learn compensatory skills. For example, even several weeks after the insult, rats sustaining brain damage due to middle cerebral artery occlusion (a commonly used stroke model) continue to show a stable deficit on the ledged beam test (Schallert et al., 2002). If the ledge is removed, however, the rats learn to compensate over the course of just a few trials until they are running the beam successfully with no footfaults. This does not necessarily indicate recovery of limb function, though, because rats will begin to make foot faults again if the ledge is subsequently replaced. The speed with which rats are able to shift between displaying a deficit on the ledged beam and learning to compensate in the ledge's absence may be reflective of the level of impairment and the capacity of motor learning circuits that may or may not have been affected by the lesion.

Some hints make the use of the beam more successful. Preoperatively, rats must be trained to run the beam without fault, and preferably without stopping to explore the beam or its surroundings during the run. There is no prescribed number of training trials needed to achieve this result; each rat can simply be trained until this criterion is reached. Good training eases the testing phase, because stopping to encourage the rat to traverse the beam becomes less necessary.

When setting up the beam, the experimenter may want to place the rat's home cage at the end to serve as a reinforcer. The cage may also be covered with a dark cloth to make it more enticing. During the initial training, the experimenter can encourage the rat to run by tapping on the beam in front of the rat, picking up the rat's tail from behind to encourage it to move away, or "tucking" the rat's hindquarters with the experimenter's hands. On early trials, the rat frequently stops to sniff the beam or have a look around the testing room, but this generally ceases during the course of pretraining. Objects should not be placed to the side or below the beam because these tend to distract the animal. A comprehensive review of the setup, use, and scoring of the beam can be found in Schallert et al. (2002).

Other opportunities exist for testing hindlimb function. Although rats rely primarily on their forelimbs for most movement, the hindlimbs are used in behaviors such as jumping, swimming (in which the forelimbs usually stay stationary as the hindlimbs paddle [Whishaw et al., 1981; Kolb and Tomie, 1988; Stoltz et al., 1999]), and backing out of tunnels or other tight areas in which the rat cannot turn around. They can also use the hindlimbs to maintain balance during a rear; one can also quantify hindlimb stepping in the cylinder test (see earlier) as an index of hindlimb function, if the cylinder is set up to be filmed from below (Fleming et al., 2002). In the cylinder, rats with unilateral nigrostriatal system damage mimicking a hemiparkinsonian state tend to leave the impaired hindlimb planted in one place and pivot around this akinetic limb by stepping with the unimpaired limb.

CONCLUSION

The sensorimotor tests described earlier are certainly not the only ones that should be considered useful, but in our experience these qualify as among the best for assessing functional outcome after unilateral focal ischemic injury, nigrostriatal degeneration, traumatic head injury, damage to the intrinsic neurons of striatum, and cervical spinal hemisection. It is possible to use aspects of these tests along with others to determine the location and extent of injury and the degree of improvement over time. With practice, investigators can reliably and rapidly evaluate treatment effects. Our Web site (http://www.schallertlab.org) has downloadable videos and information that can help new researchers adopt these and related tests of sensory and motor function.

REFERENCES

Barth TM, Jones TA, Schallert T (1990) Functional subdivisions of the rat somatic sensorimotor cortex. Behavioural Brain Research 39:73–95.

Felt BT, Schallert T, Shao J, Liu Y, Li X, Barks JD (2002) Early appearance of functional deficits after neonatal excitotoxic and hypoxic-ischemic injury: Fragile recovery after development and role of the NMDA receptor. Developmental Neuroscience 24:418–425.

Fleming SM, Delville Y, Schallert T (2003) An intermittent, controlled-rate, slow progressive degeneration model of Parkinson's disease suitable for evaluating neuroprotective therapies: Effects of methylphenidate. Behavioural Brain Research, in press.

Fleming SM, Woodlee MT, Schallert T (2002) Chronic hindlimb and forelimb deficits differ between rat models of Parkinson's disease and stroke: A motion picture poster. 2002 Abstract viewer, program No. 885.4. Orlando, FL: Society for Neuroscience Conference.

Gharbawie OA, Whishaw PA, Whishaw IQ (2003) The topography of three-dimensional exploration: A new quantification of vertical and horizontal exploration, postural support, and exploratory bouts in the cylinder test. Behavioural Brain Research, in press.

Greenough WT, Fass B, DeVoogd T (1976) The influence of experience on recovery following brain damage in rodents: Hypotheses based on developmental research. In: Environments as therapy for brain dysfunction (Walsh RN and Greenough WT, eds.), pp. 10–50. New York: Plenum Press.

Jones TA, Bury SD, Adkins-Muir DL, Luke LM, Allred RP, Sakata JT (2003) Importance of behavioral manipulations and measures in rat models of brain damage and brain repair. ILAR Journal 44:144–152.

Kolb B and Tomie JA (1988) Recovery from early cortical damage in rats. IV: Effects of hemidecortication at 1, 5, or 10 days of age on cerebral anatomy and behavior. Behavioural Brain Research 28:259–274.

Lindner MD, Gribkoff VK, Donlan NA, Jones TA (2003) Long-lasting functional disabilities in middle-aged rats with small cerebral infarcts. Journal of Neuroscience 23:10913–10922.

Marshall JF (1982) Sensorimotor disturbances in the aging rodent. Journal of Gerontology 37:548–554.

Schallert T, De Ryck, M, Whishaw IQ, Ramirez VD, Teitelbaum P (1979) Excessive bracing reactions and their control by atropine and L-dopa in an animal analog of parkinsonism. Experimental Neurology 64:33–43.

Schallert T, Fleming SM, Leasure JL, Tillerson JL, Bland ST (2000) CNS plasticity and assessment of forelimb sensorimotor outcome in unilateral rat models of stroke, cortical ablation, parkinsonism, and spinal cord injury. Neuropharmacology 39:777–787.

Schallert T, Norton D, Jones TA (1992) A clinically relevant unilateral rat model of parkinsonian akinesia. Journal of Neural Transplantation and Plasticity (currently Neural Plasticity) 3, 332–333.

Schallert T and Tillerson JL (2000) Intervention strategies for degeneration of dopamine neurons in parkinsonism: Optimizing behavioral assessment of outcome. In: CNS diseases: Innovate models of CNS diseases from molecule to therapy (Emerich DF, Dean RLI, Sanberg PR, eds.), pp. 131–151. Totowa, NJ: Humana Press.

Schallert T, Upchurch M, Lobaugh N, Farrar SB, Spirduso WW, Gilliam P, Vaughn D, Wilcox RE (1982) Tactile extinction: Distinguishing between sensorimotor and motor asymmetries in rats with unilateral nigrostriatal damage. Pharmacology, Biochemistry, and Behavior 16:455–462.

Schallert T, Upchurch M, Wilcox RE, Vaughn DM (1983) Posture-independent sensorimotor analysis of inter-hemispheric receptor asymmetries in neostriatum. Pharmacology, Biochemistry, and Behavior 18:753–759.

Schallert T and Whishaw IQ (1984) Bilateral cutaneous stimulation of the somatosensory system in hemidecorticate rats. Behavioral Neuroscience 98:518–540.

Schallert T and Woodlee MT (2003) Brain-dependent movements and cerebral-spinal connections: Key targets of cellular and behavioral enrichment in CNS injury models. Journal of Rehabilitation Research and Development 40(1S):9–18.

Schallert T, Woodlee MT, Fleming SM (2002) Disentangling multiple types of recovery from brain injury. In: Pharmacology of cerebral ischemia (Krieglstein J and Klumpp S, eds.), pp. 201–216. Stuttgart: Medpharm Scientific Publishers.

Schallert T, Woodlee MT, Fleming SM (2003) Experimental focal ischemic injury: Behavior-brain interactions and issues of animal handling and housing. ILAR Journal 44:130–143.

Stoltz S, Humm JL, Schallert T (1999) Cortical injury impairs contralateral forelimb immobility during swimming: A simple test for loss of inhibitory motor control. Behavioural Brain Research 106:127–132.

Tillerson JL, Cohen AD, Caudle WM, Zigmond MJ, Schallert T, Miller GW (2002) Forced nonuse in unilateral parkinsonian rats exacerbates injury. Journal of Neuroscience 22:6790–6799.

Tillerson JL, Cohen AD, Philhower J, Miller GW, Zigmond MJ, Schallert T (2001) Forced limb-use effects on the behavioral and neurochemical effects of 6-hydroxydopamine. Journal of Neuroscience 21:4427–4435.

Whishaw IQ, Schallert T, Kolb B (1981) An analysis of feeding and sensorimotor abilities of rats after decortication. Journal of Comparative and Physiological Psychology (currently Behavioral Neuroscience) 95:85–103.

Wolgin DL and Kehoe P (1983) Cortical KCl reinstates forelimb placing following damage to the internal capsule. Physiology and Behavior 31:197–202.

Woodlee MT, Choi SH, Zhao X, Aronowski J, Grotta JC, Chang J, Hong JJ, Lin T, Redwine GG, Schallert T (2003) Distinctive behavioral profiles and stages of recovery in animal models of stroke and Parkinson's disease. 2003 Abstract viewer, program No. 947.5. New Orleans, LA: Society for Neuroscience Conference.

Grooming

13

J. WAYNE ALDRIDGE

Natural grooming in rodents is an advantageous behavioral model useful for studying the organization and neural mechanisms of movement sequences. Grooming consists of complex strings of movements to clean and maintain the fur and skin of the body; these movements include wiping, licking, and scratching. Grooming is natural and ubiquitous. It is observed readily; rats spend up to half of their time during waking hours engaged in grooming (Bolles, 1960). Most grooming bouts are initiated by paw-licking or face-washing movements that proceed to grooming of the fur around the head, neck, and body in a cephalocaudal stepwise pattern (Richmond and Sachs, 1978). Unitary grooming actions such as scratching or direct contact with the trunk are emitted on their own in some instances; however, the cephalocaudal succession of grooming actions across the body surface is most frequent and well established.

Kent Berridge, John Fentress, and their colleagues and students made critical breakthroughs in our understanding of the functional organization of grooming sequences (Berridge et al., 1987; Berridge and Fentress, 1987a, 1987b; Berridge and Whishaw, 1992; Aldridge et al., 1993; Cromwell and Berridge, 1996). They demonstrated that grooming patterns have lawful relationships among the individual components and that the basal ganglia play a key role in sequence implementation (Berridge et al., 1987). By meticulous transcriptions of the timing and serial order of individual movements from continuous videotape recordings and by evaluations of the statistical predictability of sequence elements and the probabilities of sequential patterns, they showed that the temporal structure of grooming actions has predictable organizational features. Grooming is not random; rather, it exhibits a marked serial dependence (Berridge et al., 1987; Berridge, 1990).

Most grooming sequences consist of flexibly ordered mixtures of strokes, licking, scratches, etc.; however, occasionally rats emit a fixed pattern, or "chain," of grooming actions. These occasional chain sequences are composed of the same movements as flexible grooming patterns, but the serial structure of chains is relatively fixed in order and time. The chains are consistent and repeatable in contrast to "nonchain" grooming patterns, which are more flexible in their sequential composition and serial structure.

The stereotyped grooming sequence has approximately 25 contiguous movements lasting approximately 5 seconds in total. This chain sequence has a stable serial order of four phases (Berridge et al., 1987; Berridge, 1990) (Fig. 13–1, top). Phase 1 consists of 5-9 rapid elliptical strokes over the nose and mystacial vibrissae lasting for about 1 second. Phase 2 is short (0.25 second) and consists of small asymmetrical strokes of increasing amplitude. Phase 3 consists of large bilateral strokes that take 2 to 3 seconds for the animal to complete. The chain concludes with phase 4, which consists of a postural turn followed by a period (1 to 3 seconds) of body licking directed to the flank. The last phase varies more in length than other phases and often ends by blending

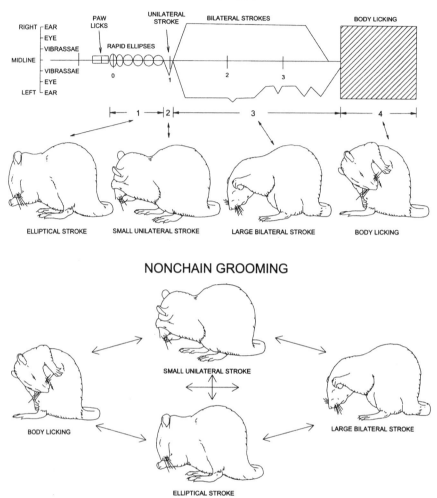

Figure 13–1. Syntactic and nonchain grooming. (*Top*) The four syntactic phases—elliptical strokes, unilateral strokes, bilateral strokes, and body licking. The choreographical timeline has forepaw movement as distance from the midline (right indicates up; left, down) as a function of time (x-axis, tics = 1 second) for a typical syntactic chain (left paw represented by line below the axis, right paw represented by line above the axis). (*Bottom*) Nonchain grooming consists of the same movements. Unlike chain grooming, the serial order of grooming actions is flexible and movements occur in varying combinations.

into subsequent nonchain grooming in which chains are embedded. For practical purposes, the signature rapid elliptical strokes of phase 1 provide a reliable marker for the stereotyped syntactic chain. In flexible nonchain grooming, elliptical strokes are usually unitary.

Thus, rodent grooming has two notable sequence patterns: (*1*) a fixed chain sequence and (*2*) nonchain sequences in more variable and flexible patterns (Fig. 13–1, bottom). Both grooming patterns are composed of the same actions, except the rigidity of the sequence structure differs (Fig. 13–1). Stereotyped chain grooming is less frequent, occurring at rates of 2 to 15 chains per hour and comprising a total of approximately 10 to 75 seconds of grooming. In contrast, nonchain grooming is more frequent with total durations up to or more than 20 times greater than chain grooming.

In a seminal paper, Karl Lashley (1951)

noted the importance of action sequences and their implications for probing underlying neural mechanisms. All behavior is sequential, but some sequences exhibit *syntax*. A syntactic sequence follows rules that determine the temporal progression of its elements. These rules impart lawful predictability to the sequence. Language has syntax. Given an arbitrary word, one can predict with some level of probability the next word in a sequence of words. Other behaviors can be described as having properties of syntax if one can demonstrate lawful sequential dependencies. Chain sequence grooming, as noted by Berridge and colleagues (1987), exhibits syntax. It follows predictable rules for its components and structure. Once the stereotyped grooming chain begins, each remaining phase can be predicted with greater than 90% accuracy. This entire syntactical chain occurs with a frequency that is more than 13,000 times greater than could be expected by chance (based on the relative probabilities of the component 25 actions obtained from grooming outside of this syntactic chain). A comparative phylogenetic analysis of rat, mouse, hamster, gerbil, guinea pig, and ground squirrel grooming patterns (Berridge, 1990) has demonstrated that grooming syntax is a basic biological trait conserved across related species.

Syntactic grooming chains were discovered through laborious moment-by-moment transcriptions of behavior from continuous videotape records. An alternative and, indeed, the more common method for cataloging behavior relies on *sampling*. In sampling, the action in which the animal is engaged at every measurement interval, typically 15 seconds or longer, is recorded to build a distribution of behavioral events in time. Sampling methods have clear advantages in many applications. More animals can be studied per session because a single observer can scan multiple cages. In contrast, continuous tabulation relies on tedious examination of all movements on a one-by-one basis. Sampling methods have been particularly productive in identify-

ing the basic elements of grooming behavior (Bolles, 1960; Spruijt et al., 1992) and the effects of drug manipulations on grooming behavior (Spruijt et al., 1986; Molloy and Waddington, 1987).

Sampling methods, however, have disadvantages for reconstructing detailed sequential organization patterns. Unless the time between samples is extremely short, the detailed temporal structure of grooming may be missed. To capture the multiple strokes of phase 1 or the phase 2 transition in syntactic grooming chains, for example, sampling intervals would need to be less than 1 second. Syntactic grooming chains typically have a duration of 5 seconds for four phases of up to 25 strokes. To even make "hits" on chain sequences, sampling must occur at less than 5 second intervals. Because chain and nonchain grooming strokes are similar, the special sequential properties of grooming chains might be missed with a sampling method approach. Without information about preceding and following actions, it would not be clear whether a grooming movement occurred within a syntactic chain or a flexible nonchain sequence. Under some circumstances, the effects of drugs or other manipulations on grooming might be inadvertently confounded across the two sequence types. Flexible nonchain grooming dominates normal grooming with syntactic chains interspersed irregularly at rates of about 10 chains per 2 hour session in normal, undisturbed animals (J. W. Aldridge, unpublished observations). Thus, observation periods and recordings of 1 to 2 hours are typical minimum durations necessary to expose the syntactic chain and its properties. Although sampling procedures are useful for some behavioral investigations, unless they are fast enough to approximate continuous recording, they may result in syntactic chains being missed altogether or under counted.

A particular advantage of rodent grooming as a model system for evaluating the functional organization of behavior sequences and underlying neural mechanisms is the fact the grooming actions of the syntactical chain sequence occur in unpredictable order and flexible combinations outside of the syntactical chain sequence. Thus, the same movements can be studied in two different sequence contexts: (1) the syntactical chain and (2) flexible combinations of nonchain grooming. The similarity of individual grooming actions and sequences of actions among individuals and even across species (Berridge, 1990) facilitates comparisons of kinematically similar movements in syntactic and nonsyntactical, flexible sequences. In addition, grooming sequences do not depend on learning and memory. Learned sensorimotor sequences, which are commonly used in behavioral neuroscience, are valuable and productive tools for complex cognitive testing, however, in some instances, it may be difficult to dissociate coding for a sequence function from a memory function. In contrast, innate movement sequences such as grooming, do not depend on explicit training and thus provide a window on behavioral organization and neuronal mechanisms independent of memory and explicit training.

SYNTACTICAL GROOMING: IMPLICATIONS FOR NERVOUS SYSTEM FUNCTION

Berridge and colleagues have shown that syntactic grooming chains had a neural substrate in the basal ganglia and brain stem (Berridge and Fentress, 1987b; Berridge and Whishaw, 1992; Aldridge et al., 1993; Cromwell and Berridge, 1996). The basal ganglia play a critical role in the organization of grooming syntax, as evidenced by the fact that striatal lesions have a potent impact on syntactic grooming. The number of grooming chains "completed" drops by more than 50% (Berridge and Fentress, 1987a) after a striatal lesion, even while the number of chains *initiated* is not reduced. Rats appeared unable to implement the syntactic rule despite "attempts" to do so. The crucial regions of the

striatum were "mapped" by small bilateral lesions (≤1 mm) (Cromwell and Berridge, 1996). Cromwell and Berridge found that small lesions in the dorsolateral quadrant of the neostriatum led to sequence deficits as significant as large striatal lesions. This dorsolateral quadrant is part of the dorsal "motor" circuit defined by Nauta and others (Nauta and Domesick, 1984; Alheid and Heimer, 1988). It is noteworthy that grooming chains are still attempted after striatal lesions even though they fail to be implemented effectively. This finding suggests that the striatum may play a role in implementing or facilitating the sequence rather than a role in sequence initiation.

The pontine hindbrain also contributes to generating sequential patterns. Decerebrated rats still occasionally generate the basic pattern of syntactic chains (Berridge, 1989a). Complete syntactic chains were never seen in *myelencephalic* decerebrates, which lacked a pons and cerebellum and had only a medulla remaining. Myelencephalic decerebrates still emit grooming actions with postural support but show grossly degraded sequential organization; the *syntax* of grooming is lost. *Mesencephalic* decerebrates (intact midbrain and hindbrain) and *metencephalic* decerebrates (hindbrain only) still produce occasional syntactic chains with structural sequence errors. Decerebrates complete less than half of the chains that they begin, suggesting that the basal ganglia may control implementation and completion of syntactic chain grooming sequences. In contrast, the brain stem may have a role in generating the kinematic details of the individual actions and contributing to the sequence in a rudimentary form. Normal execution of syntactic grooming chains requires an intact neostriatum.

The critical importance of the basal ganglia to syntactic grooming sequences was affirmed by comparisons of lesions of neostriatum, motor cortex, and cerebellum and complete decortication (Berridge and Whishaw, 1992). *Only neostriatal lesions produced perma-*

nent impairment of the serial organization of syntactic grooming chains. All other lesions (secondary motor cortex, combined primary and secondary motor cortex, complete decortication, cerebellar ablation) produced only minor temporary disruption of sequential organization, related to *nonsequential* motor deficits, such as deficits of fine coordination of forelimb trajectories, movement timing, or posture. The basic structure of the sequence remained intact if the neostriatum was spared. The failure of primary and secondary motor cortex lesions to induce sequential deficits in grooming syntax is interesting, given the "parallel loops" between cortex–striatum–thalamus–cortex. The contribution of motor cortical regions to grooming behavior remains to be determined in future work.

The implication from this work is that the striatum and basal ganglia are crucial to the *implementation* of motor sequences. The fact that decerebrate animals emit occasional syntactic chains, albeit poorly structured, suggests that the neostriatum is not the "central pattern generator" of syntactic grooming chains. Instead, the pattern generator circuitry must be contained to a considerable extent within the brainstem. The neostriatum must be intact to implement the pattern on the flow of normal behavior. This special role in implementing sequential behavior may have evolved to include learned behavioral sequences (Rapoport, 1989; Aldridge et al., 1993), and it may account for the characteristic breakdown in sequence organization in human basal ganglia disorders in humans, such as occurs in Parkinson's disease. Other recent studies in animals and model systems make it clear that sequencing of action and cognition into syntactic patterns of serial order is an important behavioral function mediated by the basal ganglia (Kermadi and Joseph, 1995; Mushiake and Strick, 1995; Beiser and Houk, 1998; Berns and Sejnowski, 1998; Matsumoto et al., 1999; Lieberman, 2000).

NEURAL SUBSTRATES FOR CODING OF GROOMING SYNTAX

The activation profiles of neurons in the neostriatum and substantia nigra pars reticulata (SNpr) during grooming further support a basal ganglia role in coding and implementing grooming syntax (Aldridge and Berridge, 1998; Meyer-Luehmann et al., 2002). Whether neuronal activation was correlated to grooming movements depended on the *sequential context* in which the movement occurred (syntactic chain versus nonchain flexible grooming sequences). Furthermore, the dorsolateral striatum site crucial to syntactic grooming had stronger activation than did ventromedial regions (Aldridge and Berridge, 1998).

These findings are based on neuronal recordings from rats implanted with permanent multisite recording electrodes in neostriatum (dorsolateral or ventromedial) or SNpr. The implant, which is connected by a flexible cable to a commutator, allows animals to move about and to groom freely with no restraint (Aldridge and Berridge, 1998; Meyer-Luehmann et al., 2002). Spontaneous behavior for 1 or more hours of normal grooming and free movement is recorded on time-synchronized videotape from under a glass floor with simultaneous recording of neuronal spike activity on a computer. A frame-by-frame analysis of the videotaped grooming sequences is done off-line (Aldridge and Berridge, 1998; Meyer-Luehmann et al., 2002) to demarcate syntactic chain grooming and flexible nonchain grooming and to determine the onset and end times of syntactic grooming phases and individual grooming movements within phases. Changes in neuronal activity in relation to grooming actions is assessed by perievent time histograms to average neural spike activity over 5 to 10 repetitions of each movement. Recording sites are verified histologically after the completion of recording.

The activity of 41% of striatal cells was preferentially related to the sequential pattern of syntactic chains; that is, neurons were acti-

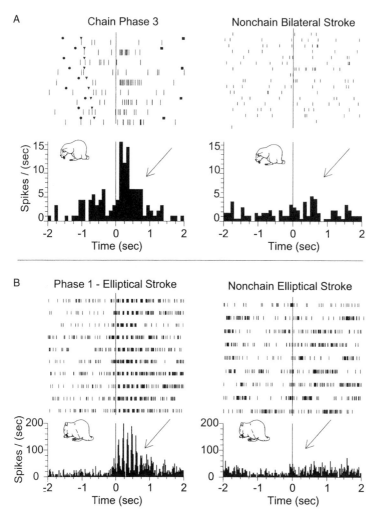

Figure 13–2. Neural activation during syntactic grooming. (*A*) Dorsolateral striatal neuron illustrating an increase of activity associated with the onset of phase 3 (bilateral strokes) during chain grooming (*left, arrow*). The same neuron was not activated during nonchain bilateral strokes (*right, arrow*). (*B*) Nigral neuron is activated during the elliptical strokes of phase 1 during chain grooming (*left, arrow*) but not during the elliptical strokes in nonchain grooming (*right, arrow*).

vated during movements in the context of syntactic chain grooming but not during equivalent movements emitted in flexible nonchain grooming (Aldridge and Berridge, 1998) (Fig. 13–2A). Only 14% of neurons had a pattern of activation suggesting they could code simple motor properties of grooming movements—that is, activated during movement in both contexts, inside and outside of sequential chains.

Regional differences in neuronal activation patterns were apparent (Aldridge and Berridge, 1998). Neurons in the dorsolateral striatal site crucial for syntactic grooming increased their firing rates by 116% during syntactic chains compared with 30% in the ventromedial region. Dorsolateral neurons also

seemed to code phase-specific aspects of the syntactic sequence of grooming. In the crucial dorsolateral region, more neurons responded during multiple phases of the sequence than in the ventromedial region (18% of dorsolateral neurons and 5% of ventromedial neurons).

To determine if activation during grooming was strictly a correlate of movement, the activation patterns of neurons responsive during syntactic chains were examined during similar movements emitted during flexible nonchain (nonsyntactic) grooming. Most neurons exhibited context sequence–dependent activation patterns (see Fig. 13–2A). Few neurons (16%) could be categorized as simply movement related, that is, as neurons active during movement without regard to the sequence in which it occurred. Neuronal firing rates affirmed the important influence of sequence context. During syntactic chain grooming, rates were significantly higher than during nonchain grooming (17%; paired t test, $p < 0.001$) or quiet resting (44%; $p < .001$).

The SNpr is an important output structure of the basal ganglia. The SNpr receives massive projections from dorsolateral striatum (Deniau et al., 1996); it is not surprising that neuronal activity profiles of SNpr, like the striatum, reflect a dependence on the context of sequential syntax (Meyer-Luehmann et al., 2002). Fifty-five percent (n = 26) of SNpr neurons were active during grooming behavior, with most activation (73%, 19 of 26 responsive neurons) occurring during the first two phases of the syntactic sequence. The onset of the syntactic grooming sequence is a dominant feature with vigorous activation during phase 1 (96%, 25 of 26 responsive neurons). The importance of sequential context is again clear. Many neurons (36%) responsive during phase 1 did not respond at all during similar nonchain grooming movements (see Fig. 13–2B). Even the neurons that were active during elliptical strokes of nonchain grooming had significantly faster firing rates when the strokes occurred in the context of syntactic grooming (50 versus 28 spikes per second).

In addition to phase 1, SNpr neurons, like the striatum, exhibited sequence-dependent activation patterns during subsequent grooming phases. In contrast to the striatum, however, SNpr had more vigorous activation during some nonchain grooming strokes than during kinematically similar syntactic grooming. For example, the proportion of activated neurons was higher during bilateral strokes of nonchain grooming (65% of neurons) and firing rates were faster during phases 2 and 3 than during kinematically similar strokes of syntactic chain grooming. Thus, SNpr neurons preferentially coded the onset of the syntactic sequence (phase 1) and then seemed to diminish or inhibit activation related to mo-

tor parameters of phases 2 and 3. Although the direction of neuronal activation differs in the nigra compared with the striatum, these results demonstrate that neurons at the output of the basal ganglia, like the striatal input region, are modulated by the sequence context during the performance of instinctive grooming movements. The specificity of SNpr activation to the *onset* of the syntactic chain pattern and relatively diminished activity to later phases of the chain may reflect a sequence-dependent balance between excitatory subthalamic nucleus and inhibitory striatal inputs.

ROLE OF DOPAMINE IN SEQUENTIAL CHAIN GROOMING

Evidence suggesting a role for dopaminergic neural mechanisms in syntactic chain grooming includes the following. Lesions of dopaminergic afferents to the striatum with the neurotoxin 6-hydroxydopamine (6-OHDA) (Berridge, 1989b) have consequences as severe as destruction of the striatum itself. In the normal ontogeny of rats, the motor components of grooming appear much earlier than syntactic chain sequence even though it uses the same movements. The appearance of syntactic grooming develops in parallel with the development of the dopaminergic markers in the maturing striatum (Colonnese et al., 1996). Finally, dopamine D1 receptor agonists enhance the quantity of grooming (Starr and Starr, 1986); this is supported by evidence that dopamine D1 agonists given either systemically or intraventricularly (Berridge and Aldridge, 2000a; Berridge and Aldridge, 2000b) enhance syntactic grooming over and above the increase in overall grooming.

Dopamine agonists have a potent effect on syntactic grooming. In general, D1 agonists enhanced and D2 agonists impaired syntactic grooming sequences (Berridge and Aldridge, 2000a). Corticotropin (ACTH), which is known to elicit excessive grooming

(Dunn, 1988; Dunn and Berridge, 1990), also reduced the amount of grooming and the likelihood that the sequential pattern would be completed. Most interesting, D1 activation specifically enhances stereotyped grooming over and beyond increases in movement itself (*superstereotypy*). Both the relative frequency of syntactic chain emission and the percentage rate of syntactic chain completion were increased relative to changes in nonchain grooming. The efficacy of different D1 agonists depends on the route of administration and whether the agent is a full or partial agonist; however, the main effects were clear. D1 activation enhances syntactic grooming over and above the increase in flexible grooming. This stands in contrast to the decrease associated with D2 activation and other peptides by any route or dose (Berridge and Aldridge, 2000a).

CONCLUSION

Rodent grooming behavior is a particularly useful model system with which to study the organization and neuronal mechanisms of behavioral sequences. Because it has syntactical and nonsyntactical modes and each contains essentially the same component actions, rodent grooming is an ideal system for dissociating motor from sequence control properties. It is possible to compare, on the same neurons, coding mechanisms related to movements in stereotyped and syntactical sequences versus flexible, less rigidly structured sequences. The clinical importance of these brain regions for normal and pathological sequence control cannot be overstated. Parkinson's disease, Huntington's disease, and Tourette syndrome all have basal ganglia disturbances associated with movement sequences. A better understanding of the neural mechanisms related to sequence control might lead to new therapeutic tactics for neurological treatments.

REFERENCES

Aldridge JW and Berridge KC (1998) Coding of serial order by neostriatal neurons: A "natural action" approach to movement sequence. Journal of Neuroscience 18:2777–2787.

Aldridge JW, Berridge KC, Herman M, Zimmer L (1993) Neuronal coding of serial order: Syntax of grooming in the neostriatum. Psychological Science 4:391–395.

Alheid GF and Heimer L (1988) New perspectives in basal forebrain organization of special relevance for neuropsychiatric disorders: The striatopallidal, amygdaloid, and corticopetal components of substantia innominata. Neuroscience 27:1–40.

Beiser DG and Houk JC (1998) Model of cortical-basal ganglionic processing: Encoding the serial order of sensory events. Journal of Neurophysiology 79:3168–3188.

Berns GS and Sejnowski TJ (1998) A computational model of how the basal ganglia produce sequences. Journal of Cognitive Neuroscience 10:108–121.

Berridge KC (1989a) Progressive degradation of serial grooming chains by descending decerebration. Behavioural Brain Research 33:241–253.

Berridge KC (1989b) Substantia nigra 6-OHDA lesions mimic striatopallidal disruption of syntactic grooming chains—A neural systems analysis of sequence control. Psychobiology 17:377–385.

Berridge KC (1990) Comparative fine structure of action: Rules of form and sequence in the grooming patterns of six rodent species. Behaviour 113:21–56.

Berridge KC and Aldridge JW (2000a) Super-stereotypy. I: Enhancement of a complex movement sequence by systemic dopamine D1 agonists. Synapse 37:194–204.

Berridge KC and Aldridge JW (2000b) Super-stereotypy. II: Enhancement of a complex movement sequence by intraventricular dopamine D1 agonists. Synapse 37:205–215.

Berridge KC and Fentress JC (1987a) Deafferentation does not disrupt natural rules of action syntax. Behavioural Brain Research 23:69–76.

Berridge KC and Fentress JC (1987b) Disruption of natural grooming chains after striatopallidal lesions. Psychobiology 15:336–342.

Berridge KC, Fentress JC, Parr H (1987) Natural syntax rules control action sequence of rats. Behavioural Brain Research 23:59–68.

Berridge KC and Whishaw IQ (1992) Cortex, striatum, and cerebellum: control of serial order in a grooming sequence. Experimental Brain Research 90:275–290.

Bolles RC (1960) Grooming behavior in the rat. Journal

of Comparative and Physiological Psychology 53: 306–310.

Colonnese MT, Stallman EL, Berridge KC (1996) Ontogeny of action syntax in altricial and precocial rodents: Grooming sequences by rat and guinea pig pups. Behaviour 113:1165–1195.

Cromwell HC and Berridge KC (1996) Implementation of action sequences by a neostriatal site: A lesion mapping study of grooming syntax. Journal of Neuroscience 16:3444–3458.

Deniau JM, Menetrey A, Charpier S (1996) The lamellar organization of the rat substantia nigra pars reticulata: Segregated patterns of striatal afferents and relationship to the topography of corticostriatal projections. Neuroscience 73:761–781.

Dunn AJ (1988) Studies on the neurochemical mechanisms and significance of ACTH-induced grooming. Annals of the New York Academy of Sciences 525:150–168.

Dunn AJ and Berridge CW (1990) Physiological and behavioral responses to corticotropin-releasing factor administration: Is CRF a mediator of anxiety or stress responses? Brain Research—Brain Research Reviews 15:71–100.

Kermadi I and Joseph JP (1995) Activity in the caudate nucleus of monkey during spatial sequencing. Journal of Neurophysiology 74:911–933.

Lashley KS (1951) The problem of serial order in behavior. In: Cerebral mechanisms in behavior (Jeffress LA, ed.), pp. 112–146. New York: Wiley.

Lieberman P (2000) Human Language and Our Reptilian Brain: The Subcortical Bases of Speech, Syntax, and Thought. Harvard University Press, Cambridge.

Matsumoto N, Hanakawa T, Maki S, Graybiel AM, Kimura M (1999) Nigrostriatal dopamine system in learning to perform sequential motor tasks in a predictive manner. Journal of Neurophysiology 82: 978–998.

Meyer-Luehmann M, Thompson JF, Berridge KC, Aldridge JW (2002) Substantia nigra pars reticulata neurons code initiation of a serial pattern: Implications for natural action sequences and sequential disorders. European Journal of Neuroscience 16:1599–1608.

Molloy AG and Waddington JL (1987) Assessment of grooming and other behavioural responses to the D-1 dopamine receptor agonist SK & F 38393 and its R- and S-enantiomers in the intact adult rat. Psychopharmacology (Berl) 92:164–168.

Mushiake H and Strick PL (1995) Pallidal neuron activity during sequential arm movements. Journal of Neurophysiology 74:2754–2758.

Nauta WJH and Domesick VB (1984) Afferent and efferent relationships of the basal ganglia. In: Functions of the basal ganglia. Ciba Foundation Symposium 107, pp. 3–29. London: Pitman.

Rapoport JL (1989) The boy who couldn't stop washing. New York: Penquin Books.

Richmond G and Sachs BD (1978) Grooming in Norway rats: The development and adult expression of a complex motor pattern. Behaviour 75:82–96.

Spruijt BM, Cools AR, Ellenbroek BA, Gispen WH (1986) Dopaminergic modulation of ACTH-induced grooming. European Journal of Pharmacology 120:249–256.

Spruijt BM, VanHooff JARA, Gispen WH (1992) Ethology and neurobiology of grooming behavior. Physiological Reviews 72:825–852.

Starr BS and Starr MS (1986) Differential effects of dopamine D1 and D2 agonists and antagonists on velocity of movement, rearing and grooming in the mouse. Implications for the roles of D1 and D2 receptors. Neuropharmacology 25:455–463.

Locomotion

GILLIAN MUIR

14

Locomotion is one of the most common behaviors in which rats engage, and thus the assessment of locomotor abilities is an essential component of many behavioral analyses. This is particularly true for rodent models of diseases such as spinal injury and stroke, where the recovery of locomotion is the ultimate goal of experimental therapies. In addition, many of the more complicated behavioral tasks that are assessed in the laboratory require that rats move over ground and be able to use their limbs well. Results from such analyses may be confounded if rats have locomotor impairments that affect their performance on behavioral tasks. The measurement of locomotion in rats is difficult, however, because of their small size and the rapidity with which they can move and maneuver. A thorough understanding of both the musculoskeletal and neural requirements to move over ground is necessary to properly assess locomotor behavior in these animals. This chapter provides an overview of the current knowledge on the mechanics and neural control of locomotion in the rat and outlines the methods available for measuring locomotor abilities in this species.

MECHANICS OF LOCOMOTION

To move over ground, rats, like all terrestrial animals, need to support their body weight against gravity, maintain their posture and equilibrium, and provide propulsion in the direction of their movement (Grillner, 1975). In addition, rats need to alter their speed and navigate uneven terrain (Grillner, 1975, 1981). This section discusses the mechanical aspects of overground locomotion, including the movements of the individual limbs and the coordination between limbs that allows rats to move over ground at different speeds.

THE STEP CYCLE

During locomotion in all limbed terrestrial animals, including rats, individual limbs move repeatedly through a *step cycle* that consists of a *stance phase* and a *swing phase*. During the stance phase, the limb is in contact with the ground surface and thus contributes to weight support and propulsion of the body relative to the ground (left forelimb in Fig. 14–1A). Limb action during the stance phase consists of an initial flexion of the limb joints as the limb yields under the weight of the body (Philipson's E_2 phase) (Grillner, 1975); then the limb joints extend, pushing the rat forward (E_3). During the swing phase, the limb is flexed and moved forward relative to the motion of the body (Philipson's F phase), and in the final stages of the swing phase, the limb joints are extended to prepare for the subsequent stance phase (E_1) (left forelimb in Fig. 14–1B). Each limb moves through the step cycle only once for each *stride*, which is defined as the complete pattern of limb movements that begins with the onset of ground contact of an individual limb and ends with the subsequent ground contact of the same limb.

Stride parameters, like almost all mea-

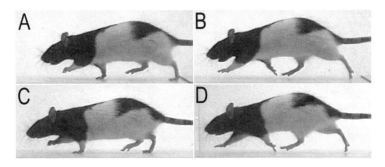

Figure 14–1. Rat trotting unrestrained for a food reward. During each stride, diagonal limb pairs make contact with the ground sequentially (*A*, right forelimb plus left hindlimb; *C*, left forelimb plus right hindlimb), separated by two brief aerial phases (*B* and *D*). For each limb, the stride is composed of a stance phase (e.g., right forelimb in *A*) and a swing phase (e.g., right forelimb in *B* through *D*). (Photograph credit: Laura Taylor.)

surements related to locomotion, vary greatly with the forward speed of movement. *Speed*, by definition, is a function of the stride length and the stride duration. Rats, like other animals, increase speed both by lengthening the stride and by shortening the stride duration. Because the durations of both the stance phase and the swing phase make up the stride duration, these parameters also change with the speed of locomotion. Importantly, they do not change similarly—the stance phase shortens dramatically as velocity increases, whereas the duration of the swing phase changes very little (Fig. 14–2). The shortening of the stance phase has important consequences for the events that occur during this phase, particularly the forces that must be exerted on the ground to support and propel the rat forward. These forces, the ground reaction forces, must be exerted over increasingly shorter time periods as the speed of locomotion increases (Fig. 14–3).

The actual movement of the limbs and limb segments throughout the step cycle is produced by action of the limb muscles, which are activated in a particular pattern so as to rotate the limb segments on each other. The characteristic movement pattern of the limb segments has been accurately measured in the rat for different speeds of locomotion (Fischer et al., 2002). In general, locomotor limb action in rats is similar to that of other

small mammals in that they maintain a crouched limb posture at all speeds, providing increased maneuverability and stability compared with larger animals with more upright limbs (Biewener, 1989, 1990, 1983; Fischer et al., 2002).

In addition to the dependence of individual limb parameters on speed, the coordination between the limbs (i.e., *gait*) also changes with speed of locomotion. There are many different gaits that quadrupedal animals use, and these are well described for species such as the horse (Adams, 1987). Nevertheless, walking, trotting, and galloping are the basic gaits used by most quadrupeds, including rats. The following discussion describes the energetics and limb dynamics of these gaits.

WALKING

During walking, the limbs are used as rigid struts. During the first half of the stance phase for each limb, the body initially rises up over each limb and then falls during the latter half of stance. At the same time, during the first half of stance, the limb produces a braking force on the body; then, as the body moves forward of the limb, propulsive force is produced which accelerates the body. In this way, the rat is rising and falling, decelerating and accelerating with each step. This oscillating pattern provides a means for reducing the en-

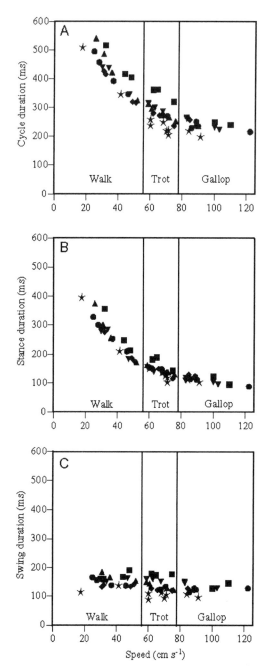

Figure 14–2. Effect of speed on stride, stance, and swing duration during rat locomotion. Stride duration decreases with increasing locomotor speed (A). This decrease is largely due to the reduction in stance duration (B) as swing duration remains relatively constant over a large speed range (C). Different symbols represent data from different individuals. (From Gillis and Biewener, 2001, with permission.)

ergetic costs of locomotion. Forward kinetic energy is initially converted to potential energy during the first half of the stance phase, as the rat slows down and rises up. In the last half of the stance phase, potential energy is transferred to forward kinetic energy as the rat falls and speeds up (Cavagna et al., 1977). This alternating transfer can reduce the energetic cost of overground locomotion at the walk by up to 75% (Heglund et al., 1982).

During walking in quadrupeds, either two or three limbs contact the ground at any one time. The walking gait used by most rats is referred to as a *lateral walk* (Gillis and Biewener, 2001). During this gait, the order of limb contact in one stride (beginning with the left hindlimb as reference) is left hindlimb, left forelimb, right hindlimb, and right forelimb. On a treadmill, rats walk to move at speeds between 0 and 55 cm/sec (Gillis and Biewener, 2001). Over ground, rats walk at the same speed range (0 to 55 cm/sec), but it is difficult to record consistent walking for several strides, because rats moving at these slow speeds are usually exploring at the same time and frequently stop and start and change directions.

TROTTING

During walking, the maximum stride length attainable, and therefore the maximum speed attainable, is limited by leg length, so animals, including rats, must incorporate an aerial phase into the gait to further increase speed. This requires a switch to trotting in quadrupeds, or to running in bipeds. During these gaits, the limbs act as springs rather than as struts. During the first half of the stance phase, the body is slowing down as the limb produces a braking force on the body, just as for walking (see Fig. 14–3, fore-aft force). Unlike during walking, however, the body is also falling during early stance and the forward kinetic energy of the body is converted not to potential energy but to elastic energy in tendons and ligaments as the limb is loaded. This elastic energy is then released as forward ki-

Figure 14–3. The effect of locomotor speed on ground reaction forces in rats during trotting at 55 and 90 cm/sec. As speed increases, peak ground reaction forces also increase because these forces are exerted for an increasingly shorter time span with each ground contact. Forces in three orthogonal directions (vertical, fore-aft, mediolateral) are expressed in Newtons per kilogram of body weight. Each force recording shows forces exerted by a forelimb (stance duration indicated by *black horizontal bar*) followed by the ipsilateral hindlimb (*gray horizontal bar*). Vertical forces demonstrate that the body weight is borne relatively equally between the forelimbs and the hindlimbs. Fore-aft forces demonstrate that the forelimbs produce most of the braking force, whereas the hindlimbs produce most of the propulsive forces. Mediolateral forces are mainly directed laterally. Force recordings are from a single individual.

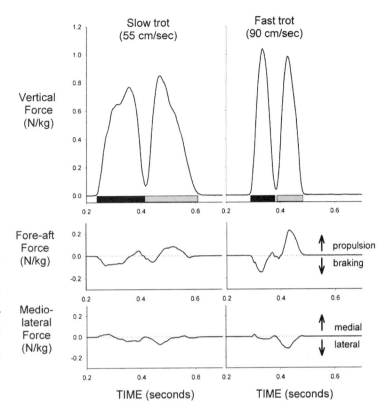

netic energy again as the rat springs off the limb in the latter half of stance.

During trotting, only two limbs contact the ground at any one time—the rat moves from one diagonal limb pair (e.g., right forelimb and left hindlimb) to the next diagonal pair (left forelimb and right hindlimb). At faster trotting speeds, the rat may actually jump from one diagonal limb pair to the other, so that there would be two phases during a trotting stride when the rat is not in contact with the ground (see Fig. 14–1). As demonstrated in Figure 14–1, the order of limb contacts during a trotting stride would be right forelimb and left hindlimb (aerial phase), left forelimb and right hindlimb (aerial phase). Although the forelimb and hindlimb diagonal pair may contact the ground simultaneously, often the forelimb or the hindlimb contacts the ground slightly earlier (i.e., about 20 milliseconds) than its diagonal partner. Which limb makes contact first

depends in part on the speed of movement as well as differences between individual rats. On a treadmill, rats trot to move at speeds between 55 and 80 cm/sec (Gillis and Biewener, 2001). Rats that have been trained to move over ground for a food reward generally use a trotting gait and move at speeds from 50 to 90 cm/sec (Muir and Whishaw, 1999b, 2000; Webb and Muir, 2002, 2003a, 2003b).

GALLOPING

Galloping or bounding gaits occur at the highest speeds in quadrupedal animals, including rats. During these gaits, stride length is further extended by incorporating movements of the torso, which extends and flexes alternately throughout the stride. The hindlimbs and forelimbs act more in synchrony during galloping than during slower gaits. At the beginning of the galloping stride, the trunk is flexed as hindlimbs are brought forward and placed on the ground. The torso then extends as the rat stretches forward onto the forelimbs and then flexes again as the hindlimbs leave the ground and are brought under the body. An aerial phase is incorporated into the gait as the rat moves off the forelimbs and before the hindlimbs contact the ground for the next stride. In bounding gaits, there is a second aerial phase during each stride, which occurs as the trunk is extending and the rat jumps from the hindlimbs onto the forelimbs. The energetics of galloping gaits combine the strategies of both walking and trotting, in that forward kinetic energy is converted to, and recovered from, gravitational potential energy and elastic strain energy at different phases in the stride (Cavagna et al., 1977).

During galloping gaits, the precise coordination of the limbs differs at different speeds, and there are either one, two, or no limbs on the ground at any one time during the stride. At slower gallops, the hindlimbs and the forelimbs do not act completely in synchrony; that is, one hindlimb makes contact with the ground earlier than the opposite hindlimb. The disparity between contact times is generally greater for the forelimb pair than for the hindlimbs. For each limb pair, the limb that makes contact with the ground slightly earlier than the other is referred to as the *trailing limb*, and the opposite limb is the *leading limb*. The order of limb contact in a slow gallop would be trailing hindlimb, leading hindlimb, trailing forelimb, leading forelimb. In some instances, the leading hindlimb and the trailing forelimb may contact the ground simultaneously (a gait referred to as a *canter* in horses). At faster gallops, the forelimbs as well as the hindlimbs move increasingly in synchrony, until the rat begins to bound from the forelimbs onto the hindlimbs and then onto the forelimbs again. On a treadmill or over ground, most rats begin to gallop at speeds greater than 80 cm/sec (Muir and Whishaw, 2000; Gillis and Biewener, 2001). For both treadmill and over-ground locomotion, there is a wide range of speeds, from approximately 70 to 100 cm/sec, at which rats either trot or gallop.

NEURAL CONTROL OF LOCOMOTION

OUTPUT FROM SPINAL CIRCUITRY CAN PRODUCE THE BASIC STEPPING PATTERN

During locomotion, limb muscles need to be activated in a particular pattern such that this activity can (1) move limbs in a cyclical manner (i.e., through the step cycle), (2) provide alternation of right and left limbs, and (3) coordinate forelimbs and hindlimbs according to the gait being used. It is well established in vertebrates that the circuitry required to produce the neural output for the first two of these tasks—movement of the limbs through the step cycle and alternation of right and left limbs—is contained completely within the spinal cord (Grillner and Wallen, 1985). Cats with a complete spinal transection are able to produce stepping movements that are identical to those in the intact animal (Grillner and Zangger, 1979; Belanger et al., 1988). Even after removal of all sensory feedback from the limb, the limb is still able to move through the step cycle, indicating that the spinal cord itself is able to generate the oscillating output required to activate muscles in an pattern appropriate for stepping.

Similar experiments have not been repeated in the adult rat, although there is much information on spinal pattern generation in the neonatal rat (Cazalets et al., 1995; Cowley and Schmidt, 1997; Kiehn and Kjaerulff, 1998; Ballion et al., 2001). The spinal locomotor circuitry for the hindlimbs in rats appears to be distributed throughout the lumbar enlargement and the lower thoracic cord, and likely consists of many rhythm generators that control different muscles or joints, although the most excitable rhythmic activity is located in the rostralmost part of the lumbar enlargement (Kiehn and Kjaerulff, 1998). For the forelimbs, the locomotor circuitry responsible for generating rhythmic activity is located in the lower cervical and uppermost thoracic spinal segments (Ballion et al., 2001).

Of course, to produce functional locomotion, spinal locomotor circuits normally require input from two important sources. First, segmental afferent feedback from the limb acts to constantly regulate and reinforce ongoing muscle activity as well as to provide information to control the transitions between the stance and swing phases of the step cycle. Second, supraspinal inputs from the brain stem and higher regions of the brain are required for the initiation and ongoing control of locomotion. The influence of both of these sources of input is discussed in more detail. Much of the research in this area has been acquired in the cat model, but specific information regarding the rat is supplied when available.

NORMAL STEPPING REQUIRES PERIPHERAL AFFERENT INPUT FROM THE LIMB

Segmental afferent feedback arises from several receptors distributed throughout the limb, including muscle spindles, Golgi tendon organs, and receptors located in the skin and joint capsules. Muscle spindles provide information regarding muscle length and velocity of length changes, whereas Golgi tendon organs provide information about tendon forces. Input from both of these receptors have been shown to influence the level of extensor muscle activity during the stance phase, such that loss of this input substantially reduces the strength of extensor muscle activity (Pearson and Collins, 1993; Guertin et al., 1995; Hiebert and Pearson, 1999). Inputs from cutaneous receptors, especially those on the footpads, can also result in an increase in extensor activity during the stance phase (Duysens and Pearson, 1976). Importantly, the influence of both muscle and cutaneous receptors on limb muscle activity is not constant but instead is modulated throughout the step cycle (Forssberg, 1979; Drew and Rossignol, 1985, 1987). During the swing phase, for example, stimulation

of cutaneous receptors of the paw results in limb flexion rather than limb extension (Drew and Rossignol, 1985, 1987).

Another major role of afferent input is in the regulation of transitions between stance and swing (for a review, see Pearson et al., 1998). During the latter half of stance, the limb begins to move caudally and bears less weight, and there is increased activity in muscle spindles within hip flexor muscles as well as a decrease in Golgi tendon organ activity within limb extensors. Both of these inputs contribute to the decreased activity in extensors and an activation of limb flexors, resulting in limb flexion and thus the onset of the swing phase (Pearson et al., 1998).

SUPRASPINAL INPUT IS REQUIRED FOR INITIATION AND ONGOING CONTROL OF LOCOMOTION

The other major source of input onto spinal locomotor circuits arises from the brain. Spinal circuits receive direct input from the cerebral cortex, red nuclei, vestibular nuclei, and numerous nuclei in the pons and medulla, including the locus coeruleus and raphe nuclei. In addition, there are many areas in the brain that are involved in locomotor control but do not project directly to the spinal cord; these include the cerebellum, the basal ganglia, and several areas collectively known as *locomotor regions*. These are areas in the brain that, when stimulated with electric current, are able to initiate locomotion in decerebrate animals, including rats (Atsuta et al., 1990, 1991). These areas are located in the mesencephalon (mesencephalic locomotor region), the hypothalamus, and the deep cerebellar nuclei. For an in-depth discussion of the possible roles of these locomotor regions during various forms of locomotion, see Jordan (1998).

In the rat, the locomotor contributions of brain regions with direct spinal input have been examined by lesioning various spinal funiculi. Axons arising from nuclei in the pons and medulla travel in the ventral half of the spinal cord. These inputs likely play an important role in locomotor control, as large ventral lesions produce severe impairment of overground locomotion in the rat (Loy et al., 2002; Schucht et al., 2002). There appears to be some functional redundancy in these ventral pathways, in that small lesions involving only portions of the ventral funiculi had mild effects on overground locomotion (Loy et al., 2002). Of course, the axons affected by ventral lesions arise from many different nuclei in the brain stem, and it is possible that more detailed information on the contributions of individual nuclei can be obtained using more sensitive techniques with which to measure locomotor abilities.

Descending inputs located in the dorsal half of the spinal cord arise largely from the cerebral cortex and the red nucleus. These inputs are thought to be primarily required for skilled locomotion, such as making adjustments for uneven terrain or avoiding obstacles. Large dorsal lesions do not greatly affect gross locomotor abilities overground in the rat but do produce large deficits in skilled locomotion, such as ladder walking (Schucht et al., 2002).

More specific lesions of dorsal pathways have distinguished the contributions of the corticospinal and rubrospinal tracts. The corticospinal tract does not appear to contribute to over ground locomotion but is required for skilled locomotion (Metz et al., 1998; Muir and Whishaw, 1999a; Metz and Whishaw, 2002; Whishaw and Metz, 2002). The rubrospinal tract also contributes to skilled locomotion but in addition has been shown to play some role during overground locomotion in rats, in that unilateral lesions of the red nucleus or of the tract itself result in a permanent locomotor asymmetry (Muir and Whishaw, 2000; Webb and Muir, 2003b).

MEASURING LOCOMOTION IN THE LABORATORY

Much of what we know regarding quantification of locomotion in rats has arisen from research focusing on central nervous system dis-

orders such as spinal cord injury or stroke. There are several reviews that have outlined appropriate methods for measuring locomotor recovery in rats after spinal cord injury (Goldberger et al., 1990; Kunkel et al., 1993; Metz et al., 2000; Muir and Webb, 2000). These methods can be applied to locomotor analysis in normal rats and in rat models of many different diseases. The following section describes the methods available for measuring locomotion in rats. For further details on specific techniques, readers are referred to the associated references.

LOCOMOTOR RATING SCALES

Locomotor rating scales, such as the BBB scale and the Tarlov scale, are ordinal rating scales devised to assess locomotor movements of the hindlimbs after thoracic spinal injury (Basso et al., 1995; Fehlings and Tator, 1995). Animals are observed in a open area and given a score based on criteria such as the movements of the hindlimbs, the tail position, and the coordination between forelimbs and hindlimbs. Such assays are quick to perform, require a minimum of equipment, and are general enough to include animals with a wide range of functional abilities. Nevertheless, it is important to note that locomotor scales are relatively specific for the type of injury for which they were designed and may not accurately measure recovery in different injury models or after different treatments. For example, the BBB scale is useful for assessing locomotion after thoracic contusion injuries or thoracic dorsal hemisections, but is less successful at describing the functional states of animals with other spinal cord lesions (Metz et al., 2000; Loy et al., 2002; Schucht et al., 2002; Webb and Muir, 2002). Finally, a limitation of any locomotor assay that relies on the observation of freely moving animals is the dependence on the motivation of individual animals; rats have been shown to display some functional "deficits" such as toe-dragging during exploratory behavior that completely disappear when the same animals are performing a motivated behavior such as trotting on a runway for a food reward (Webb and Muir, 2002). The following sections describe the measurements that are best to apply to animals that have been trained to complete a locomotor task.

KINEMATIC MEASUREMENTS

Kinematic measurements encompass a wide range of measures, including distances, angles, velocities, and accelerations of the body, of the limbs, and of the limb segments. Stride characteristics, such as stride and step lengths, and stride durations are considered to be kinematic measurements, as are limb joint angles. Kinematic measurements are essentially a quantitative and detailed description of the animal's movements. Many kinematic measures, as for other locomotor parameters, normally vary with the speed and gait of the animal, so it is important that speed of movement is recorded along with the measure(s) of interest.

Step and stride characteristics can be obtained in several ways. Animals can be videotaped as they move over ground or on a treadmill. During overground locomotion, the camera may be positioned for a lateral view (see Fig. 14–1) or a caudal view or, with the help of a transparent floor surface and a mirror placed at 45° beneath the floor, can provide a ventral view that clearly shows the placement of the footpads (Cheng et al., 1997; Webb and Muir, 2003a). Animals should be trained to run in a straight path so as to improve the accuracy of the measurements, particularly the distance measurements. Frame-by-frame analysis of the videotape can then provide step and stride lengths and durations, as well as durations of the swing and stance phases. Even more simply, use of an inkpad and paper allows the measurement of step lengths and foot placement, although temporal measurements are not available (Kunkel and Bregman, 1990).

Analysis of limb joint angles, such as shoulder, elbow, or knee angles, requires the

use of video or digital cameras and frame-by-frame analysis. Specialized computer equipment is available to assist in capturing video or digital frames, thus supplying a time series of limb positions during locomotor movements. Three-dimensional analysis, using at least two cameras simultaneously, is also possible, although normal movement of limbs in a single plane during locomotion makes two-dimensional analysis acceptable.

A significant problem with the measurement of joint angles during locomotion arises from the inability to accurately identify limb segment positions. Markers placed on the skin move as the skin moves over the limbs during normal locomotion. This movement introduces systematic errors in the location of the joint positions, particularly the proximal joints, which are compounded in small mammals such as rats that possess a crouched limb posture. The only solution is cineradiography, which provides a clear view of the limb bones during movement. Normal kinematics using this technique have been recorded for the rat (Fischer et al., 2002). Although cineradiography is impractical to use for most studies, the information obtained from such studies could be used to calculate correction factors for skin movements as has been done for other species (van den Bogert et al., 1990).

KINETIC MEASUREMENTS

Muscles move limbs by exerting forces on limb segments; the limbs then exert forces against surfaces to move the animal. *Kinetics* is the measurement of these forces. Force or strain transducers, either implanted within the limb in muscle tendons or on bone surfaces, or positioned on external surfaces, can be used to provide a sensitive measure of the ways in which animals move (Gillis and Biewener, 2002; Biewener and Blickhan, 1988; Biewener et al., 1988; Biewener and Taylor, 1986; Muir and Whishaw, 1999a, 1999b, 2000).

Ground reaction forces are the forces exerted through the limb on the ground during locomotion (see Fig. 14–3). They are measured with force platforms and provide a sensitive, quantitative, and noninvasive method for measuring locomotion. This method is especially useful for the analysis of rat locomotion because even though rats are capable of making quick maneuvers and adjustments to their gait, the slightest changes are manifest through the ground reaction forces. Recording of these forces thus provides unique and quantifiable information that is not obtainable with other methods. Measurement of ground reaction forces has demonstrated the subtle and characteristic methods that rats use to compensate for different central nervous system lesions (Muir and Whishaw, 1999a, 1999b, 2000; Webb and Muir, 2002, 2003b). Measurement of these forces, however, requires specialized equipment that must be customized or custom-built to provide the appropriate size and sensitivity for studies in the rat.

ELECTROMYOGRAPHY

Recording of muscle electrical activity during locomotion provides valuable information on how muscles are used to move the limbs through the step cycle. Simultaneous recording from many muscles during locomotion is possible with available techniques, although most studies in the rat have been limited to the measurement of one or two muscles (Cohen and Gans, 1975; Loeb and Gans, 1986; Roy et al., 1991; de Leon et al., 1994; Gorassini et al., 1999, 2000; Gramsbergen et al., 2000; Gillis and Biewener, 2001; Kaegi et al., 2002; Schumann et al., 2002). Deviations from the normal pattern after lesioning or treatments can help determine the ways in which the new movement pattern differs from normal (Gramsbergen et al., 2000; Kaegi et al., 2002).

Importantly, however, there is a wide range of normal muscle activity patterns. This is due in part to the large number of muscles that span each joint, so that the same movement can be produced by through the recruitment of several different combinations of muscles. In addition, electromyography does not provide direct information about the strength of muscle contraction because of the complex relationship between individual muscle fiber activation and force production (Basmajian and De Luca, 1985). Recording of electromyographic data entails some disadvantages because of the invasive nature of the electrodes and associated recording devices, which may alter the animal's normal locomotor behavior. Implantation of electrodes into the muscle also requires surgery as well as careful postoperative care to prevent infection and maintain the electrodes in position for the duration of the study.

LOCOMOTOR TASKS

The previous section discussed measurement techniques applicable to overground locomotion, but of course rats can be trained to locomote in a number of situations, each of which is amenable to measurement using these same methods. Rats can be trained to locomote on a treadmill and, because the animal is essentially stationary, kinematic and electromyographic measurements are somewhat easier to obtain compared with overground locomotion. Treadmill speed can be controlled precisely, as can the level of incline, so that uphill or downhill locomotion can be examined (Gillis and Biewener, 2002).

Rats can also be trained in more skilled locomotor tasks, such as ladder or beam walking. In addition to the kinematic, kinetic, and electromyographic measurements previously discussed, other measures of locomotor ability can be recorded for these tasks. These include counts of footslip errors and scoring of the paw positions on the ladder rungs or beams (Muir and Webb, 2000; Metz and Whishaw, 2002). To alter the degree of difficulty of skilled locomotor tasks, it is possible to vary or randomize the spacing between ladder rungs or to vary the width of the beam (Metz et al., 2000; Metz and Whishaw, 2002). Rats can also be trained to climb inclines or ropes, behaviors that also lend themselves to the measurement techniques previously discussed (Thallmair et al., 1998; Ramon-Cueto et al., 2000). Their natural motor abilities can be exploited for the investigation of a variety of tasks that, in combination with different measurement techniques, can provide a comprehensive analysis of locomotor skills in this species.

REFERENCES

Adams OR (1987) Natural and artificial gaits. In: Adam's lameness in horses (Stashak TS, ed.), pp. 834–839. Philadelphia: Lea & Febiger.

Atsuta Y, Garcia-Rill E, Skinner RD (1990) Characteristics of electrically induced locomotion in rat in vitro brain stem-spinal cord preparation. Journal of Neurophysiology 64:727–735.

Atsuta Y, Garcia-Rill E, Skinner RD (1991) Control of locomotion in vitro: I. Deafferentation. Somatosensory Motor Research 8:45–53.

Ballion B, Morin D, Viala D (2001) Forelimb locomotor generators and quadrupedal locomotion in the neonatal rat. European Journal of Neuroscience 14:1727–1738.

Basmajian JV and De Luca CJ (1985) EMG signal amplitude and force. In: Muscles alive: Their functions revealed by electromyography, pp. 187–200. Baltimore: Williams and Wilkins.

Basso DM, Beattie MS, Bresnahan JC (1995) A sensitive and reliable locomotor rating scale for open field testing in rats. Journal of Neurotrauma 12:1–21.

Belanger M, Drew T, Rossignol S (1988) Spinal locomotion: A comparison of the kinematics and the electromyographic activity in the same animal before and after spinalization. Acta Biologica Hungarica 39:151–154.

Biewener AA (1983) Locomotory stresses in the limb bones of two small mammals: The ground squirrel and chipmunk. Journal of Experimental Biology 103:131–154.

Biewener AA (1989) Scaling body support in mammals: Limb posture and muscle mechanics. Science 245:45–48.

Biewener AA (1990) Biomechanics of mammalian terrestrial locomotion. Science 250:1097–1103.

Biewener AA and Blickhan R (1988) Kangaroo rat locomotion: Design for elastic energy storage or acceleration? Journal of Experimental Biology 140:243–255.

Biewener AA, Blickhan R, Perry AK, Heglund NC, Taylor CR (1988) Muscle forces during locomotion in kangaroo rats: Force platform and tendon buckle measurements compared. Journal of Experimental Biology 137:191–205.

Biewener AA and Taylor CR (1986) Bone strain: A determinant of gait and speed? Journal of Experimental Biology 123:383–400.

Cavagna GA, Heglund NC, Taylor CR (1977) Mechanical work in terrestrial locomotion: Two basic mechanisms for minimizing energy expenditure. American Journal of Physiology 233:R243–R261.

Cazalets JR, Borde M, Clarac F (1995) Localization and organization of the central pattern generator for hindlimb locomotion in newborn rat. Journal of Neuroscience 15:4943–4951.

Cheng H, Almstrom S, Gimenez LL, Chang R, Ove OS, Hoffer B, Olson L (1997) Gait analysis of adult paraplegic rats after spinal cord repair. Experimental Neurology 148:544–557.

Cohen AH and Gans C (1975) Muscle activity in rat locomotion: Movement analysis and electromyography of the flexors and extensors of the elbow. Journal of Morphology 146:177–196.

Cowley KC and Schmidt BJ (1997) Regional distribution of the locomotor pattern-generating network in the neonatal rat spinal cord. Journal of Neurophysiology 77:247–259.

de Leon R, Hodgson JA, Roy RR, Edgerton VR (1994) Extensor- and flexor-like modulation within motor pools of the rat hindlimb during treadmill locomotion and swimming. Brain Research 654:241–250.

Drew T and Rossignol S (1985) Forelimb responses to cutaneous nerve stimulation during locomotion in intact cats. Brain Research 329:323–328.

Drew T and Rossignol S (1987) A kinematic and electromyographic study of cutaneous reflexes evoked from the forelimb of unrestrained walking cats. Journal of Neurophysiology 57:1160–1184.

Duysens J and Pearson KG (1976) The role of cutaneous afferents from the distal hindlimb in the regulation of the step cycle of thalamic cats. Experimental Brain Research 24:245–255.

Fehlings MG and Tator CH (1995) The relationships among the severity of spinal cord injury, residual neurological function, axon counts, and counts of retrogradely labeled neurons after experimental spinal cord injury. Experimental Neurology 132:220–228.

Fischer MS, Schilling N, Schmidt M, Haarhaus D, Witte H (2002) Basic limb kinematics of small therian mammals. Journal of Experimental Biology 205:1315–1338.

Forssberg H (1979) Stumbling corrective reaction: A phase-dependent compensatory reaction during locomotion. Journal of Neurophysiology 42:936–953.

Gillis GB and Biewener AA (2001) Hindlimb muscle function in relation to speed and gait: In vivo patterns of strain and activation in a hip and knee extensor of the rat (Rattus norvegicus). Journal of Experimental Biology 204:2717–2731.

Gillis GB and Biewener AA (2002) Effects of surface grade on proximal hindlimb muscle strain and activation during rat locomotion. Journal of Applied Physiology 93:1731–1743.

Goldberger ME, Bregman BS, Vierck-CJ J, Brown M (1990) Criteria for assessing recovery of function after spinal cord injury: Behavioral methods. Experimental Neurology 107:113–117.

Gorassini M, Bennett DJ, Kiehn O, Eken T, Hultborn H (1999) Activation patterns of hindlimb motor units in the awake rat and their relation to motoneuron intrinsic properties. Journal of Neurophysiology 82:709–717.

Gorassini M, Eken T, Bennett DJ, Kiehn O, Hultborn H (2000) Activity of hindlimb motor units during locomotion in the conscious rat. Journal of Neurophysiology 83:2002–2011.

Gramsbergen A, IJkema-Paassen J, Meek MF (2000) Sciatic nerve transection in the adult rat: Abnormal EMG patterns during locomotion by aberrant innervation of hindleg muscles. Experimental Neurology 161:183–193.

Grillner S (1975) Locomotion in vertebrates: Central mechanisms and reflex interaction. Physiology Review 55:247–304.

Grillner S (1981) Control of locomotion in bipeds, tetrapods and fish. In: Handbook of physiology, section I: The nervous system (Brooks VB, ed.), pp. 1179–1236. Bethesda: American Physiological Society.

Grillner S and Wallen P (1985) Central pattern generators for locomotion, with special reference to vertebrates. Annual Review of Neuroscience 8:233–261.

Grillner S and Zangger P (1979) On the central generation of locomotion in the low spinal cat. Experimental Brain Research 34:241–261.

Guertin P, Angel MJ, Perreault MC, McCrea DA (1995) Ankle extensor group I afferents excite extensors throughout the hindlimb during fictive locomotion in the cat. Journal of Physiology 487(Pt 1):197–209.

Heglund NC, Cavagna GA, Taylor CR (1982) Energetics and mechanics of terrestrial locomotion. III. Energy changes of the centre of mass as a function of speed and body size in birds and mammals. Journal of Experimental Biology 97:41–56.

Hiebert GW and Pearson KG (1999) Contribution of sensory feedback to the generation of extensor activity during walking in the decerebrate cat. Journal of Neurophysiology 81:758–770.

Jordan LM (1998) Initiation of locomotion in mammals.

Annals of the New York Academy of Science 860: 83–93.

Kaegi S, Schwab ME, Dietz V, Fouad K (2002) Electromyographic activity associated with spontaneous functional recovery after spinal cord injury in rats. European Journal of Neuroscience 16:249–258.

Kiehn O and Kjaerulff O (1998) Distribution of central pattern generators for rhythmic motor outputs in the spinal cord of limbed vertebrates. Annals of the New York Academy of Science 860:110–129.

Kunkel BE and Bregman BS (1990) Spinal cord transplants enhance the recovery of locomotor function after spinal cord injury at birth. Experimental Brain Research 81:25–34.

Kunkel BE, Dai HN, Bregman BS (1993) Methods to assess the development and recovery of locomotor function after spinal cord injury in rats. Experimental Neurology 119:153–164.

Loeb GE and Gans C (1986) Electromyography for experimentalists. Chicago: University of Chicago Press.

Loy DN, Magnuson DS, Zhang YP, Onifer SM, Mills MD, Cao QL, Darnall JB, Fajardo LC, Burke DA, Whittemore SR (2002) Functional redundancy of ventral spinal locomotor pathways. Journal of Neuroscience 22:315–323.

Metz GA, Dietz V, Schwab ME, van de Meent MH (1998) The effects of unilateral pyramidal tract section on hindlimb motor performance in the rat. Behavioural Brain Research 96:37–46.

Metz GA, Merkler D, Dietz V, Schwab ME, Fouad K (2000) Efficient testing of motor function in spinal cord injured rats. Brain Research 883:165–177.

Metz GA and Whishaw IQ (2002) Cortical and subcortical lesions impair skilled walking in the ladder rung walking test: A new task to evaluate fore- and hindlimb stepping, placing, and co-ordination. Journal of Neuroscience Methods 115:169–179.

Muir GD and Webb AA (2000) Mini-review: Assessment of behavioural recovery following spinal cord injury in rats. European Journal of Neuroscience 12:3079–3086.

Muir GD and Whishaw IQ (1999a) Complete locomotor recovery following corticospinal tract lesions: Measurement of ground reaction forces during overground locomotion in rats. Behavioural Brain Research 103:45–53.

Muir GD and Whishaw IQ (1999b) Ground reaction forces in locomoting hemi-parkinsonian rats: A definitive test for impairments and compensations. Experimental Brain Research 126:307–314.

Muir GD and Whishaw IQ (2000) Red nucleus lesions impair overground locomotion in rats: A kinetic analysis European Journal of Neuroscience 12:1113–1122.

Pearson KG and Collins DF (1993) Reversal of the influence of group Ib afferents from plantaris on activity in medial gastrocnemius muscle during locomotor activity. Journal of Neurophysiology 70:1009–1017.

Pearson KG, Misiaszek JE, Fouad K (1998) Enhancement and resetting of locomotor activity by muscle afferents. Annals of the New York Academy of Science 860:203–215.

Ramon-Cueto A, Cordero MI, Santos-Benito FF, Avila J (2000) Functional recovery of paraplegic rats and motor axon regeneration in their spinal cords by olfactory ensheathing glia. Neuron 25:425–435.

Roy RR, Hutchison DL, Peirotti DJ, Hodgson JA, Edgerton VR (1991) EMG patterns of rat ankle extensors and flexors during treadmill locomotion and swimming. Journal of Applied Physiology 70:2522–2529.

Schucht P, Raineteau O, Schwab ME, Fouad K (2002) Anatomical correlates of locomotor recovery following dorsal and ventral lesions of the rat spinal cord. Experimental Neurology 176:143–153.

Schumann NP, Biedermann FH, Kleine BU, Stegeman DF, Roeleveld K, Hackert R, Scholle HC (2002) Multi-channel EMG of the M. triceps brachii in rats during treadmill locomotion. Clinical Neurophysiology 113:1142–1151.

Thallmair M, Metz GA, Z'Graggen WJ, Raineteau O, Kartje GL, Schwab ME (1998) Neurite growth inhibitors restrict plasticity and functional recovery following corticospinal tract lesions. Nature and Neuroscience 1:124–131.

van den Bogert AJ, van Weeren PR, Schamhardt HC (1990) Correction for skin displacement errors in movement analysis of the horse. Journal of Biomechanics 23:97–101.

Webb AA and Muir GD (2002) Compensatory locomotor adjustments of rats with cervical or thoracic spinal cord hemisections. Journal of Neurotrauma 19:239–256.

Webb AA, Gowribai K, Muir GD (2003a) Fischer (F-344) rats have different morphology, sensorimotor and locomotor abilities compared to Lewis, Long-Evans, Sprague-Dawley and Wistar rats. Behavioural Brain Research 144:143–156.

Webb AA and Muir GD (2003b) Unilateral dorsal column and rubrospinal tract injuries affect overground locomotion in the unrestrained rat. European Journal of Neuroscience 18:412–422.

Whishaw IQ and Metz GA (2002) Absence of impairments or recovery mediated by the uncrossed pyramidal tract in the rat versus enduring deficits produced by the crossed pyramidal tract. Behavioural Brain Research 134:323–336.

Prehension

15

IAN Q. WHISHAW

Following Peterson's (1932) description of reaching in the rat, rats have been trained to reach through slots, down tubes, onto rotating tables, onto conveyor belts, off shelves, and through bars to get food. They have performed tasks of manipulating puzzle latches, pushing force transducers, and pressing bars (Whishaw and Micklyaeva, 1996). They have grasped and snapped pasta to measure reach length, force, and strength (Ballermann et al., 2000, 2001; Remple et al., 2001). Their paw movements have been observed as they pick up and manipulate food pellets, variously shaped pieces of pasta, nuts, and fruits (Whishaw and Coles, 1966). Their limb movements have even been analyzed in predatory acts of catching, manipulating, and eating crickets (Ivanco et al., 1996).

The movements made by rats when they use their digits, paws, and forelimbs for catching, manipulating, and holding objects are called *skilled movements*. Early views of the evolutionary origins of skilled paw movements proposed that they evolved in the primate lineage by modifications of movements used by the forelimbs for grasping branches. The rat's dexterity refutes this notion. Skilled movements are widely used by terrestrial vertebrates and have been lost in some mammalian orders and have been elaborated in other mammalian orders (Iwaniuk and Whishaw, 2000). The almost 2000 rodent species, which comprise approximately half of all mammalian species, are members of an order with well-developed skilled movements. The laboratory rat is a worthy representative of this order.

Skilled movements are special with respect to their neural control. For an animal to use its forelimbs to reach for and manipulate objects, the limbs must be released from their function of supporting the body against gravity. The neural structures that control forelimb use must be partially different from those used for supporting body weight and locomotion (Metz et al., 1998; Muir and Whishaw, 1999). Thus, the forebrain not only contains circuits whose function is in part to suppress movement subsystems of postural support and walking but also must contain circuits that allow the limbs to be used for manipulating objects. Because skilled movements have a long phylogenetic history, they are probably conserved across mammalian species. For this reason, the skilled movements of the laboratory rat are of interest in lines of research directed toward modeling human neurological disorders, many of which manifest themselves with compromised skilled movements.

SKILLED MOVEMENTS AND LIMB STRUCTURE AND MOVEMENT

The skeletal and muscular structures of the rat forelimb are illustrated by Green (1963). The rat forepaw has five digits. The first digit (thumb) is small, but it has a nail, whereas digits 2 through 5 have claws. Nails typically indicate that a digit is used for skilled movements. Rats can move digit 1 medially toward the palm in a precision grasping pattern for holding an object between the thumb pad and the pads of the other digits (Whishaw and Coles, 1996). For example, rats use this pincer grip to hold spaghetti (Fig. 15–1). Rats have well-developed pads on the tips of the digits and on the palm, and when they grasp and

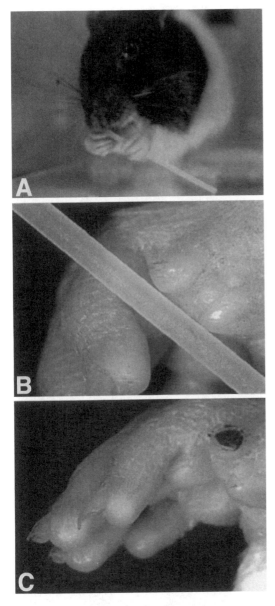

Figure 15–1. Grasp patterns during pasta eating in the rat. (A) The rat is presenting pasta to its mouth with its right paw and pushing the food with the left paw. (B) The grasp pattern displayed by the right paw involves holding the pasta between the pads of digit 1 (thumb) and digit 2. (C) Digit 1 has a nail, whereas digits 2 through 5 have claws. The presence of a nail on digit 1 suggests that it may receive considerable use in grasping objects. (Based on Whishaw and Coles, 1996.)

manipulate objects, they typically do so using the digit pads, although larger objects will be held against the palm. Rats have a few sinus hairs on the medial surface of the lower arm that project toward the palm of the paw. The sinus hairs are likely useful for detecting objects an animal is about to grasp and are especially useful for sensing movements of a live prey object that is held in the paws.

The degrees of freedom of movement of the arm of the rat are similar to that of primates, but there are differences in the way that freedom of movement is achieved. Although humans have a ball-and-socket joint at the shoulder that allows the upper arm a wide range of movement, the rat has a scapula that is tethered by muscles, thus permitting almost the same range of movement. Pronation and supination of the hand of humans are achieved by rotating the radius and ulna with respect to each other. The radius and ulna of the rat are fused. Humans cannot rotate the hand around the wrist, but the rat is able to do so. Thus, the rat limb has almost the same range of movement as the human limb, but the mechanisms underlying movement freedom are partially different (Whishaw and Micklyaeva, 1966).

Limb musculature of rats and humans is also similar, as rats have intrinsic muscles in the paw but its extensor–flexor movements and opening–closing movements are controlled by forearm muscles. As is the case for other animals, each forearm muscle is controlled by a column of spinal cord motor neurons, with more distal musculature controlled by more caudally located motor neuron columns (McKenna et al., 2000). Wise and Donoghue (1986) reviewed the neuroanatomy of brain structures that control motor neurons, and in the rat they are similar to those of primates.

FOOD HANDLING

Rats consume a wide range of foodstuffs, including grasses, nuts, leaves, roots, and almost any type of food discarded by humans. They pick up food using a characteristic sequence of five movements (Whishaw et al., 1992). (1) When a rat encounters food, it first sniffs and palpates the food with its vibrissae and perioral receptors. (2) It then grasps the food in its mouth and sits back onto its haunches and transfers the food to its paws. (3) To take the food from the mouth with the paws, the forelimbs are positioned so that the elbows are brought toward the midline of the body and the palms of the forelimbs are rotated so that they face medially. Then, by adduction of the upper arms, both paws are moved medially to grasp the food. (4) As the paws approach the food, the digits are adjusted so that their aperture is appropriate to the size of the piece of food that is being grasped. (5) Food is grasped with the tips of the digits and is manipulated with the tips of the digits for eating. The digits may adopt a large number of postures (including different postures for each paw) depending on the size and shape of the food item. The five components of spontaneous eating appear similar in different rodent species and so may comprise a "rodent-common pattern."

There is a modification to this pattern of movements for predation (Ivanko et al., 1996). When catching crickets, rats catch the cricket with a forepaw. They then sit back and hold the cricket in the forepaws to prepare it for eating. They manipulate the cricket with the digits while plucking the wings, limbs, and head from the cricket, and then they rotate the ventrum of the cricket toward their mouth for eating.

SENSORY CONTROL

Rats use olfactory and tactile information to locate food (Whishaw and Tomie, 1989). For example, they sniff food before picking it up with their mouth, they locate prey items with their vibrissae before grasping with a forepaw, and they use tactile information to shape the digits for grasping (Whishaw et al., 1992).

When transferring food from the mouth to the paws, the anticipatory shaping of the digits must be commanded by sensory information from the perioral region (Fig. 15–2). Primates use vision to locate items and to shape the digits for grasping. It is interesting in this regard that insectivores have large olfactory bulbs and a small cortex relative to prosimians, which have relatively smaller olfactory bulbs and a larger cortex. Perhaps the transfer of sensory control of skilled movements from olfaction and tactile systems to the visual system required a massive rewiring of the forebrain to the visual system in prosimians and their descendents, with the consequent increased growth of the cortex relative to the rest of the brain (Whishaw, 2003).

Olfaction is also used by rats to locate food before reaching for it with a forepaw (Whishaw and Tomie, 1989). If the eyes are patched, a rat continues to locate food as quickly and as accurately as it did when sighted. If olfaction is eliminated, however, a rat acts as if blind, systematically reaching

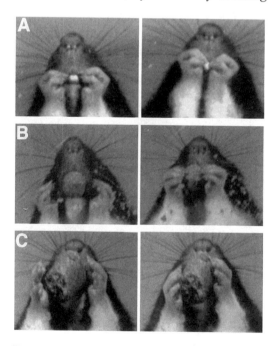

Figure 15–2. Digits are preshaped to take food from the mouse according to food size. Food: top, rice; middle, 500 mg food pellet; bottom, laboratory chow. (Based on Whishaw, Dringenberg, and Pellis, 1992.)

through each set of bars, often beginning at one end of the cage and working toward the other.

The guidance of the limb to food is not under olfactory control. To reach, a rat must lift its nose away from the target. Thus, the advance of the limb must be under central rather than olfactory control. It is interesting in this respect that humans, although aware of the target for which they are reaching, are unaware of the movements made by their forearm. Guidance of the limb of rodents and primates may be different in another way. Many neurons in the motor cortex of primates respond in relation to the direction of forearm movement, whereas other neurons respond to the force and torque of the movement. If the directionally sensitive neurons represent visual control of the limb, it is unlikely that they will be present in the rat, which always reaches to the former location of its nose.

REACHING MOVEMENTS

Most animals show asymmetries in the selection of a limb at the individual level, and some show asymmetries at the population level. Although humans display limb dominance, in that the right limb is favored in about 90% of the population, rats display only individual asymmetries (Peterson, 1932; Whishaw, 1992). About 13% of rats are ambidextrous, and the remainder is almost equally left and right limbed. Ambidexterity itself may not be so much an absence of limb preference as a lack of motor skill. Limb preference is displayed as soon as a rat begins to reach and may be an accident of learning rather than reflecting lateralization of central control. There are no sex differences.

Skilled reaching movements have the appearance of an action pattern and so are recognizable (Metz and Whishaw, 2000). Eshkol-Wachman movement notation (EWMN) can be used to describe the relations of body segments (Whishaw and Pellis, 1990), whereas

Laban movement analysis can be used to describe qualitative aspects of movement (Whishaw et al, 2003). These analyses indicate that reaching can be subdivided into a number of movements.

The stepping movements made by a rat as it approaches and reaches are central to its ability to advance its limb to the food. As a rat begins to reach, its reaching forelimb and its contralateral hindlimb move forward together. The diagonally coupled movement not only advances the forelimb but also brings the contralateral hindlimb beneath the body so that the rat can sit back onto its haunches with the food that it retrieves.

The movement of the forelimb in reaching is composed of 10 movement subcomponents (Fig. 15–3).

1. *Digits to the midline.* Using mainly the upper arm, the reaching limb is lifted from the floor so that the tips of the digits are aligned with the midline of the body.
2. *Digits flexed.* As the limb is lifted, the digits are flexed, the paw is supinated, and the wrists are partially flexed.
3. *Elbow in.* Using an upper arm movement, the elbow is adducted to the midline while the tips of the digits retain their alignment with the midline of the body.
4. *Advance.* The limb is advanced directly through the slot toward the food target.
5. *Digits extend.* During the advance, the digits extend so that the digit tips are pointing toward the target.
6. *Arpeggio.* When the paw is over the target, the paw pronates from digit 5 (the outer digit) through to digit 2, and at the same time, the digits open.
7. *Grasp.* The digits close and flex over the food, with the paw remaining in place, and the wrist is slightly extended to lift the food.
8. *Supination I.* As the paw is withdrawn, the paw supinates by almost 90°.
9. *Supination II.* Once the paw is withdrawn from the slot to the mouth, the paw further supinates by about 45° to place the food in the mouth.
10. *Release.* The mouth contacts the paw and the paw opens to release the food. In order to evaluate

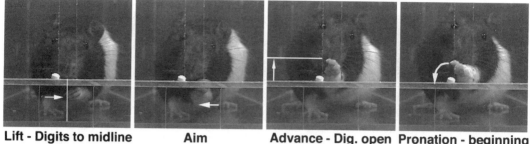

Lift - Digits to midline	**Aim**	**Advance - Dig. open**	**Pronation - beginning**

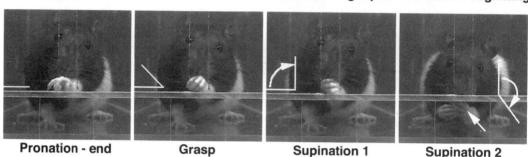

Pronation - end	**Grasp**	**Supination 1**	**Supination 2**

Figure 15-3. Movement components that comprise a reach. *Lifting* involves raising the forepaw, supinating the paw so that the palm faces inward and the digits are aligned with the midline of the body. *Aiming* involves bringing the elbow to the midline while the digits maintain their midline alignment. *Advance of the limb* is produced by a movement at the shoulder and the digits are extended. *Pronation* is achieved with an adduction of the elbow and rotation around the wrist, and the digits are opened. *Grasp supination* occurs in two stages: to withdraw the paw and to present the food to the mouth (Based on Whishaw, 2000.)

reaching performance, movements can be scored as being present/absent or present and impaired (Whishaw et al., 1993; Whishaw, 2000).

Many aspects of the reaching movement are complex in that they require fixations: holding one portion of the limb in a fixed bodywise or topographic position while other portions of the limb are moved. For example, after the digits are brought to the midline of the body, adjustment in the limb must occur to keep them at this location while the elbow moves to the midline location (movements of the elbow would otherwise displace the digits). When the head is raised to allow the limb to advance to the food, compensatory adjustments must take place in the advancing limb so that it continues toward the target and is not carried away by movement of the head and trunk. After the food is retrieved, the paw must maintain a fixed midline position so that the mouth can retrieve the food, otherwise the limb would move away as the mouth turned toward it. Electromyo-

graphic recording from forelimb muscles show that a surprising number of muscles are active during all phases of reaching (Hyland and Reynolds, 1993). It is likely that much of this concurrent activity of the musculature is related to fixations. After nervous system damage, breakdown in movement is seen in fixations.

ARPEGGIO MOVEMENT

High-speed video recording (60 frames/sec on normal replay and 120 fields/sec on replay) show that the rat grasps food with an arpeggio movement (Whishaw and Gorny, 1994). As the paw is pronated to grasp a food pellet from a shelf, the fifth digit is placed onto the shelf first, followed in succession by digits 4, 3, and 2 in an arpeggio pattern (Fig. 15–4). As each digit is placed onto the shelf, it is opened (spread away) from the preceding digit, and so covers a wide area of the shelf. The paw is

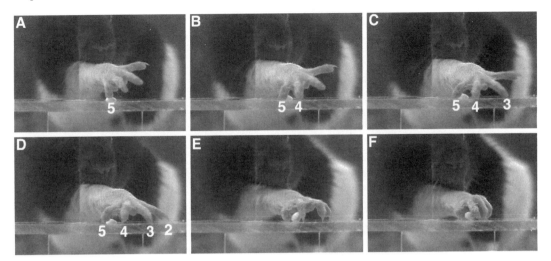

Figure 15–4. Relatively independent use of the digits in grasping a 20 mg food pellet. After the paw is pronated over the food pellet (A), the arpeggio movement brings digit 4 into contact with the food pellet (B). The food is then grasped between digit 4 and digit 5, with digit 5 moving medially in a relatively independent movement as might a primate thumb. As the paw is supinated, the other digits eventually close. (Based on Whishaw and Gorny, 1994.)

then pushed down onto the shelf in a palpating motion. If food is not present, the paw is withdrawn with closing but not flexure of the digits and then the reach is repeated. When the food is present, the pattern of grasping depends in part on the size of the food pellet. Larger pieces of food are contacted by the pad of digit 3 and grasped against the palm by digits 3 and 4. Smaller food pellets are contacted by the pad of digit 4 and grasped between digits 4 and 5.

There is some degree of independent digit movement during grasping. When large food pellets are grasped, flexure of digits 3 and 4 appears to precede flexion of the other digits, and when smaller food pellets are grasped, flexure of digits 4 and 5 appears to precede flexure of the other digits. The most striking independent digit movement is made by digit 5, which turns medially to grasp, thus playing a role like a human thumb. (There are many other situations in which independent digit movements might occur, including climbing, handling prey or food, and grasping the fur when grooming, but there are no studies of these movements.)

STRAIN DIFFERENCES

Although there are many strains of laboratory rats, there are only a couple of studies of strain differences in skilled movements. Nikkhah et al. (1998) describe strain differences in success in reaching down a staircase for food. Outbred albino Sprague-Dawley rats were among the most successful of the strains studied. In the single pellet–reaching task, impairments in reaching success have been described in albino and inbred (brother–sister matings) Fisher 344 rats relative to Long-Evans rats (VandenBerg et al., 2002). Other work shows that Sprague-Dawley and Long-Evans rats have equivalent reaching success on the single pellet–reaching task, but the movements used by the strains are different (Whishaw et al., 2003). Similarly, we have observed that albino Wistar rats display movements that are similar to those of Sprague-Dawley rats. These results suggest that there may be neural differences in the motor system of albino rats, perhaps consisting of misrouting of fibers, as has been described in the visual system.

SKILLED MOVEMENT IN NEUROINVESTIGATION AND DISEASE

There are a number of studies relating central nervous system structure and function to skilled movements. Damage to a surprisingly large number of structures impairs skilled reaching and changes the movements used; structures include the motor cortex (Whishaw et al., 2000), pyramidal tract (Whishaw et al., 1993), dorsolateral caudate nucleus (Piza and Cry, 1990), red nucleus (Whishaw and Gorny, 1992; Whishaw et al., 1998), and dorsal columns (McKenna et al., 1999). For this reason, skilled reaching is useful for the study of many disease conditions, including stroke (Whishaw, 2000), spinal cord injury (McKenna and Whishaw, 1999; Ballermann et al., 2001) and Parkinson's disease (Metz et al., 2001). Selective damage to these structures does not completely abolish skilled movements, but movement quality changes. What do these changes reveal about the organization of the motor system?

If success measures are used to evaluate nervous system injury, changes in success scores can be obtained, but for many kinds of injury, rat, with practice, will return to preoperative success. An analysis of the movements used during reaching shows that success is not regained by true recovery but through the use of compensatory behaviors (Whishaw et al., 1991, 1993, 2000). For example, the movements of pronation and supination usually produced by intrinsic limb movement can also be achieved by body rotation: pronation with contraversive body rotation and supination with ipsiversive body rotation (Fig. 15–5). Dorsal and ventral movements usually made by the limb can likewise be assisted or achieved with movements of

Control   **Motor Cortex**

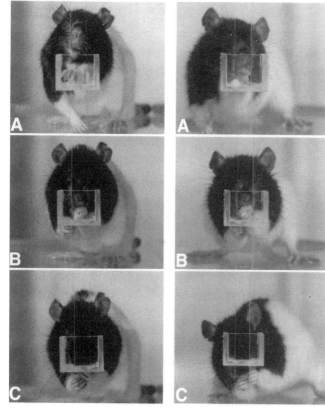

Figure 15–5. Reaching in a control rat and the use of compensatory movements in a rat reaching with the paw contralateral to a motor cortex lesion. Note the absence of body rotation in the control rat versus rotation in the motor cortex rat (ipsiversive on limb advance and contraversive to release the food to the mouth) and the wide base of support. (Based on Whishaw et al., 1991.)

the trunk. If a rat is unable to achieve the fixation of holding the paw to the mouth to retrieve the food pellet, the unimpaired paw can be used to hold the impaired paw. Descriptive movement scoring can provide a good assessment of compensatory movements. Many of the pathways of the motor system are crossed, and so studies of both paws in skilled reaching can provide insights into the contributions of crossed pathways. For example, damage to the pyramidal tract crossed pathway produces no detectable impairment in skilled reaching (Whishaw and Metz, 2002), whereas damage to crossed dopaminergic projections does produce impairments (Vergara et al., 2003).

Because skilled reaching is an acquired motor act, the behavior can be used for studies of motor learning. According to the principle of proper mass, the topographic representation of movement in motor cortex should correspond to movement ability, and studies do find that with the acquisition of motor skill there are changes in representation of the distal relative to the proximal cortical representations of the limb. Similarly, skill acquisition is accompanied by morphological changes in such structures as the motor cortex, including changes in dendritic arbor, synapse number, and synaptic function (Kleim et al., 2002; Kolb et al., 2003).

ACKNOWLEDGMENTS

This research was supported by The Canadian Stroke Network of Canada and the Natural Sciences Engineering Council of Canada.

REFERENCES

Ballermann M, Metz GA, McKenna JE, Klassen F, Whishaw IQ (2001) The pasta matrix reaching task: A simple test for measuring skilled reaching distance, direction, and dexterity in rats. Journal of Neuroscience Methods 30:39–45.

Eshkol N and Wachmann A (1958) Movement notation. London: Weidenfeld and Nicholson.

Green EC (1963) Anatomy of the rat. New York: Hafner Publishing.

Hyland BJ and Reynolds JN (1993) Pattern of activity in muscles of shoulder and elbow during forelimb reaching in the rat. Human Movement Science 12:51–70.

Ivanco TL, Pellis SM, Whishaw IQ (1996) Skilled forelimb movements in prey catching and in reaching by rats (Rattus norvegicus) and opossums (Monodelphis domestica): Relations to anatomical differences in motor systems. Behavioural Brain Research 79:163–181.

Kleim JA, Barbay S, Cooper NR, Hogg TM, Reidel CN, Remple MS, Nudo RJ (2002) Motor learning-dependent synaptogenesis is localized to functionally reorganized motor cortex. Neurobiology of Learning and Memory 77:63–77.

McKenna JE and Whishaw IQ (1999) Complete compensation in skilled reaching success with associated impairments in limb synergies, after dorsal column lesion in the rat. Journal of Neuroscience 19:1885–1894.

McKenna JE, Prusky GT, Whishaw IQ (2000). Cervical motoneuron topography reflects the proximodistal organization of muscles and movements of the rat forelimb: A retrograde carbocyanine dye analysis. Journal of Comparative Neurology 419:286–296.

Metz GA, Dietz V, Schwab ME, van de Meent H (1998). The effects of unilateral pyramidal tract section on hind limb motor performance in the rat. Behavioural Brain Research 96:37–46.

Metz GA and Whishaw IQ (2000) Skilled reaching an action pattern: Stability in rat (Rattus norvegicus) grasping movements as a function of changing food pellet size. Behavioural Brain Research 116:111–122.

Miklyaeva EI, Woodward NC, Nikiforov EG, Tompkins GJ, Klassen F, Ioffe ME, Whishaw IQ (1997) The ground reaction forces of postural adjustments during skilled reaching in unilateral dopamine-depleted hemiparkinson rats. Behavioural Brain Research 88:143–152.

Muir GD and Whishaw IQ (1999) Complete locomotor recovery following corticospinal tract lesions: Measurement of ground reaction forces during overground locomotion in rats. Behavioural Brain Research 103:45–53.

Nikkhah G, Rosenthal C, Hedrich HJ, Samii M (1988) Differences in acquisition and full performance in skilled forelimb use as measured by the 'staircase test' in five rat strains. Behavioural Brain Research 92:85–95.

Peterson GM (1932–1937 Mechanisms of handedness in the rat. Comparative Psychological Monographs 9:21–43.

Pisa M and Cyr J (1990) Regionally selective roles of the

rat's striatum in modality-specific discrimination learning and forelimb reaching. Behavioural Brain Research 37:281–292.

Remple MS, Bruneau RM, VandenBerg PM, Goertzen C, Kleim JA (2001) Sensitivity of cortical movement representations to motor experience: Evidence that skill learning but not strength training induces cortical reorganization. Behavioural Brain Research 123:133–141.

VandenBerg PM, Hogg TM, Kleim JA, Whishaw IQ (2002) Long-Evans rats have a larger cortical topographic representation of movement than Fischer-344 rats: A microstimulation study of motor cortex in naive and skilled reaching-trained rats. Brain Research Bulletin 59:197–203.

Whishaw IQ (1992) Lateralization and reaching skill related: Results and implications from a large sample of Long-Evans rats. Behavioural Brain Research 52:45–48.

Whishaw IQ (2000) Loss of the innate cortical engram for action patterns used in skilled reaching and the development of behavioural compensation following motor cortex lesions in the rat. Neuropharmacology 39:788–805.

Whishaw IQ (2003) Did a change in sensory control skilled movements stimulate the evolution of the primate frontal cortex. Behavioural Brain Research 146:31–41.

Whishaw IQ and Coles BL (1996) Varieties of paw and digit movement during spontaneous food handling in rats: Postures, bimanual coordination, preferences, and the effect of forelimb cortex lesions. Behavioural Brain Research 77:135–148.

Whishaw IQ, Dringenberg HC, Pellis SM (1992) Spontaneous forelimb grasping in free feeding by rats: Motor cortex aids limb and digit positioning. Behavioural Brain Research 48:113–125.

Whishaw IQ and Gorny B (1994) Arpeggio and fractionated digit movements used in prehension by rats. Behavioural Brain Research 60:15–24.

Whishaw IQ and Gorny B (1996) Does the red nucleus provide the tonic support against which fractionated movements occur? A study on forepaw movements used in skilled reaching by the rat. Behavioural Brain Research 74:79–90.

Whishaw IQ, Gorny B, Sarna J (1998) Paw and limb use in skilled and spontaneous reaching after pyramidal tract, red nucleus and combined lesions in the rat: Behavioral and anatomical dissociations. Behavioural Brain Research 93:167–183.

Whishaw IQ, Gorny B, Foroud A, Jeffrey A. Kleim JA (2003) Long-Evans and Sprague-Dawley rats have similar skilled reaching success and topographic limb representations in motor cortex but use different movements as assessed by EWMN and Laban movement analysis. Behavioural Brain Research 145:221–232.

Whishaw IQ and Metz GA (2002) Absence of impairments or recovery mediated by the uncrossed pyramidal tract in the rat versus enduring deficits produced by the crossed pyramidal tract. Behavioural Brain Research 134:323–336.

Whishaw IQ and Miklyaeva EI (1996) A rat's reach should exceed its grasp: Analysis of independent limb and digit use in the laboratory rat. In: Measuring movement and locomotion: From invertebrates to humans (Ossenkopp K-P, Kavaliers M, Sandberg RP, eds.), pp. 130–146. New York: RG Landes Co.

Whishaw IQ, O'Connor RB, Dunnett SB (1986). The contributions of motor cortex, nigrostriatal dopamine and caudate-putamen to skilled forelimb use in the rat. Brain 109:805–843.

Whishaw IQ and Pellis SM (1990) The structure of skilled forelimb reaching in the rat: Proximally driven movement with a single distal rotatory component. Behavioral Brain Research 41:49–59.

Whishaw IQ, Pellis SM, Gorny B, Kolb B, Tetzlaff W (1993) Proximal and distal impairments in rat forelimb use in reaching follow unilateral pyramidal tract lesions. Behavioural Brain Research 56:59–67.

Whishaw IQ, Pellis SM, Gorny BP, Pellis VC (1991) The impairments in reaching and the movements of compensation in rats with motor cortex lesions: An endpoint, videorecording, and movement notation analysis. Behavioural Brain Research 42:77–91.

Whishaw IQ, Pellis SM, Pellis VC (1992) A behavioral study of the contributions of cells and fibers of passage in the red nucleus of the rat to postural righting, skilled movements, and learning. Behavioural Brain Research 52:29–44.

Wise SP and Donoghue JP (1986) Motor cortex of rodents. In: Jones EJ, Peters A, eds. Sensory-motor areas and aspects of cortical connectivity: Cerebral Cortex, Vol 5. New York: Plenum, pp 243–265.

Locomotor and Exploratory Behavior

16

ILAN GOLANI, YOAV BENJAMINI,
ANNA DVORKIN, DINA LIPKIND,
AND NERI KAFKAFI

Rat exploratory behavior includes motor, locomotor, motivational, and cognitive aspects; it consists of a stimulating combination of stochastic and lawful elements. As technology improves, it becomes increasingly more accessible for data acquisition and analysis. With the advancement of statistical and computational data analyses, ethological knowledge, previously the exclusive property of experienced observers, becomes widely accessible by being captured by formal algorithms appropriate for automated analysis.

The study of rat exploratory behavior can commence at two levels of resolution: (1) that of the trajectory traced by the animal's whole body and (2) that in which the animal is conceived as a linkage of body parts moving in relation to each other.

Studies relating to the animal's trajectory in the environment and relating to interlimb coordination are reviewed. In each section, we start from the stage of automated data acquisition and proceed through the isolation of patterns of movement to global regularities.

TRAJECTORY ANALYSIS

SETUP

To highlight intrinsic aspects of behavior, we use a large circular empty arena. To obtain a

slow, gradual buildup of behavior, however, one might introduce a shelter into the arena, so as to induce polarity between it and the large open area. Repeated exposures, each of a long duration, further extend the buildup process.

DATA ACQUISITION

To allow the computation of velocity and acceleration and to ensure the capture of short-duration stops (which last as little as 0.2 second), rats have to be tracked at a rate of at least 25 to 30 frames/sec.

DATA PREPARATION AND ANALYSIS

Smoothing

To obtain a smooth path that would allow a meaningful derivation of moment-to-moment velocities without eliminating short, but behaviorally meaningful, arrests, we use two different statistical smoothing tools: one for the arrests and one for the progression segments (Hen et al., 2004). First, we capture the arrests by using a running medians robust smoothing method (Tukey, 1977). Only after smoothing the arrests to zero velocity, and indexing them, we use another method of Robust Lowess (Cleveland, 1977) to smooth the remainder of the location time series. In this way we end up with a record that has both the richness of the original time series and the smoothness necessary for the computation of velocities (which are computed and stored for each data point). To visualize the smoothing process see http://www.tau.ac.il/ilan99/see/help.

Segmentation into Lingering Episodes and Progression Segments

The velocity and acceleration of the animal are the outcome of all of the concurrent "forces" that act on it. Conversely, the attraction or repulsion exerted by a wall, a cliff, an edge, or an attractive place is revealed by the momentary values of these parameters. The momentary velocity of the rat can tell us whether it "thinks" it is running away or toward a familiar place. The locations and the kinematics of stops (close to zero velocity episodes), during which scanning occurs, may betray perceptual and even cognitive aspects of the behavior.

The rat's path (Fig. 16–1A) is punctuated by arrests. We classify the interarrest intervals by the maximal velocity attained in them and obtain a density plot of these velocities. This plot is typically composed of one Gaussian of low maximal speed segments (Fig. 16–1B left) and one Gaussian consisting of high maximal speed segments (Fig. 16–1B right). With the aid of a Gaussian mixture model and an appropriate algorithm (the expectation-maximization (EM) algorithm) (Everitt, 1981), we establish a cutoff value at the deep between the two Gaussians. The segments to the left of the cutoff value are defined as lingering episodes (episodes of stops or staying in place behavior), and the segments to the right of it are defined as progression segments. Treating the path as a string of discrete building blocks rather than a continuous series of coordinates (Fig. 16–1C) allows a more straightforward

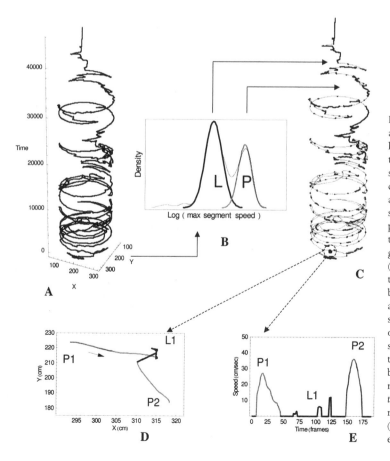

Figure 16–1. Principles of SEE analysis: The time series of the rat's location in the arena is automatically tracked in a rate of 25 to 30 Hz and smoothed. (A) The path in a three-dimensional representation of X, Y, and TIME. (B) The distribution of speed peaks (thin line) is used to parse the data into segments of two types: slow local movements ("lingering" [L], in black) and progression (P, in gray). (C) The Data can be treated as a string of these discrete behavioral units. (D,E) The path plot and speed profile of two progression segments (P1 and P2) separated by one lingering episode (L1) is demonstrated. The typical properties of these units are used to quantify the behavior. For example, the "segment acceleration" (Fig. 16–2, bottom) is estimated by dividing the segment's speed peak by its duration. (Adapted from Benjamini and Kafkafi et al., submitted)

analysis of each of the pattern types (Drai and Golani, 2001; Kafkafi et al., 2001, 2003a, 2003b). Once the relatively smooth progression segments are separated from the jittery lingering episodes, the road is open for a meaningful computation of velocity, acceleration, heading, curvature, and other measures of progression segments (Fig. 16–1D,E).

The behavior of an animal, a strain, or a species may be characterized by the cutoff value that distinguishes between lingering episodes and progression segments. After segmentation (Drai et al., 2000), both segment types can be characterized by simple quantitative measures (termed endpoints in behavior genetics) such as their length, duration, maximal speed, acceleration, and other measures derived from these (an example is shown in Fig. 16–2).

Separation between Wall (or Edge) and Center Behavior

Having at hand a sequence of progression and lingering segments defined only in terms of their kinematics, we can now examine the relationship of these patterns to the environment. The first obvious reference for location in the environment is the arena's wall or edge. Instead of using the arbitrary criterion of 15 to 25 cm from wall to distinguish wall from center behavior, we customize the cutoff point to the particular species, strain, and even individual animal, by using radial speed and distance from wall as classifying criteria. Since

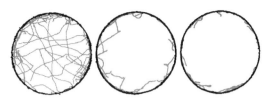

Figure 16–3. Results of the wall/center separation procedure in three ½ h sessions of three male Long-Evans hooded rats. Plots of all data points for progression segments in each of the animals. Black: Movement along the wall; Gray, movement in the center.

the radial speeds of an animal running along the wall are distributed around and close to zero cm/sec, it is possible to separate movement along the wall from movement in the center by applying the segmentation procedures described above to radial speed and distance from wall (Lipkind et al., in press).

The end product of the separation process for a 30 minute session of 3 Long-Evans hooded rats is presented in Figure 16–3. The procedure supplies a number of new behavior patterns—segments of progression along the wall, segments of progression in the center, lingering episodes along the wall, lingering episodes in the center and incursions (forays into the center). These patterns can be characterized by length, duration, maximal speed, acceleration, and other properties (see endpoints 22 to 27 in Table 16–1).

Clustering of Lingering Episodes into Operational Places

The coordinates acquired by the tracking system specify a topographical, but not necessarily an ethologically significant, location for the animal. We reserve the term *place* for a neighborhood of x,y locations with a unified operational significance for the animal. Clearly, although the animal is always located in a specified x,y location, it is not necessarily visiting a place that is meaningful to it.

Because lingering episodes are defined kinematically, we can examine the locations in which they occur. In rats, even in a bare arena devoid of objects, the x,y locations of lingering episodes often tend to be clustered

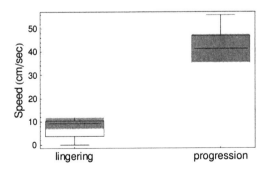

Figure 16–2. Boxplot graphs of medians of maximal speed in lingering and progression segments in a 2.5 m diameter circular arena.

Table 16–1. Locomotor and Exploratory Behavior

General Endpoints	Mean ± SE
1 Distance traveled (cm)	8312.28 ± 967.832
2 Lingering mean speed (cm/sec)	1.31668 ± 0.169496
3 Proportion of time spent more than 15 cm away from wall	0.0189796 ± 0.00586172
4 Proportion of lingering time spent away from wall	0.0084797 ± 0.0024573
5 No. of progression segments (segments)	93.9167 ± 11.9065
6 Median spatial spread of lingering episodes (cm)	6.22084 ± 0.580274
7 Median length of progression segments (cm)	45.892 ± 4.78906
8 Median duration of lingering episodes (sec)	3.08667 ± 0.293726
9 Quantile 95 of duration of progression segments (sec)	5.41 ± 0.466115
10 Quantile 95 of progression segment maximum speed (cm/sec)	68.3786 ± 3.68062
11 Median segment acceleration to maximum speed (cm/sec^2)	24.6988 ± 1.33266
12 Dart (see Kafkafi et al., 2003b)	1.30355 ± 0.0664854
13 Latency to maximum half speed (sec)	33.2467 ± 10.2184
14 Center activity proportion	0.0872066 ± 0.0280449
15 Center rest proportion	0.0263737 ± 0.00646564
16 Time proportion of lingering episodes	0.884496 ± 0.0153624
17 Activity proportion of lingering episodes	0.269275 ± 0.028709
18 Maximum spatial spread of progression segments (cm)	230.391 ± 5.36781
19 No. of stops per distance (segments/cm)	0.0115714 ± 0.000602242
20 Lingering progression threshold speed (cm/sec)	24.3093 ± 1.40721
21 No. of stops per excursion quantile 90 (stops/excursion)	12.1667 ± 1.42931

Wall-Center Endpoints	
22 Incursion maximum distance from wall (cm)	13.4907 ± 1.14988
23 Median incursion length (cm)	31.5759 ± 4.86886
24 Ratio of mean speed to and from center per incursion	0.765158 ± 0.0475105
25 Ratio of center to wall mean speed	1.11172 ± 0.0652121
26 Wall ring thickness (cm)	8.27086 ± 0.366949

Twenty-six selected behavioral measures (endpoints) characterize male Long-Evans hooded rat locomotor and exploratory behavior in a 2.5 m diameter walled circular and empty arena. The endpoints were computed with the SEE software (see for example, Kafkafi et al., 2003b; Lipkind et al., in press; Benjamini and Kafkafi et al., submitted; http://www.tau.ac.il/ilan99/see/help).

in relatively circumscribed neighborhoods. Such clusters of lingering episodes have been termed by Tchernichovski et al. (1996) as *principal*, or *preferred*, *places*.

The Home Base. There are one or two neighborhoods of locations that stand out from all the other neighborhoods in the arena in terms of the cumulative time spent in them and number of visits (stops, lingering episodes) paid to them. These neighborhoods, which were formerly demonstrated by manual scoring of the video (Eilam and Golani, 1989), are now picked up by using an algorithm that calculates the cumulative amount of time spent in different sections of the arena and the number of lingering episodes performed per section (see Fig. 16–4).

In an arena devoid of any objects or places that suggest shelter, each rat establishes its home base in a different place, early in the session (Eilam and Golani, 1989). This suggests that home base location involves the use of spatial memory. If, however, there is an object present that is clearly more conspicuous, most rats establish their home base near it. This property has been used to standardize home base location (Tchernichovski and Golani, 1995; Tchernichovski et al., 1998).

Because of its relative stability, the home base is a natural candidate for an origin of axes of a frame of reference for the examination of the rat's path. Visits to the home base are thus used to partition the path into *excursions*—sequences of stop-and-go behavior that start

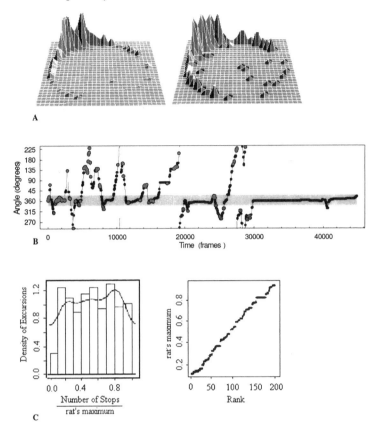

Figure 16–4. (*A*) The height of the graph at a particular location in the circular 2.5 m diameter arena represents: cumulative dwell time of a Lewis rat during lingering episodes at that location in the course of a 30 minute session (*left*) and number of lingering episodes (stops) in that location (*right*). As shown, one neighborhood in the empty arena (at 11 o'clock) stands out in the magnitude of these two measures. This neighborhood is the rat's home base. (*B*) Time series of the angular component of polar coordinates of a Long-Evans hooded rat's 30 minute session in same arena, highlighting home base location (gray band), lingering episodes (black dots near-wall, gray dots away from wall), and excursions performed from home base (line). Thick line depicts near-wall behavior, and thin line, away-from-wall behavior. Note the gradual buildup in excursion length. (*C*) (*Left*) Histogram and estimated density function of the pulled data (n = 15), of wild rats, 1/2 h sessions, on number of stops per round trip in rats that established a single home base, divided by the rat session's maximum. (*Right*) Quantile plot of the (same) pooled data on number of stops per round-trip in rats having a single home base. Straight line implies a uniform distribution. (From Golani et al., 1993.)

and end at the home base. Rats establishing more than one base have both roundtrips to the same home base and excursions that start at one home base and end at another. In single home base rat sessions, excursions have two basic features: (*1*) their outbound portion is slow and intermittent, whereas their inbound portion is fast (Eilam and Golani, 1989; Tchernichovski and Golani, 1995), and (*2*) the number of stops per excursion is constrained (Golani et al., 1993).

Number of Stops per Excursion

We have noted that after a rat leaves the home base, it never performs more than about 12 stops before returning home. This apparent upper bound could, however, be a result of very different modes of behavior: excursions could contain a typical number of, say, 8 stops. In such case, each excursion would contain *about* 8 stops, the frequency distribution of the number of stops per excursion would have a bell-shaped form with a peak frequency at

8 stops, and the number of stops would rarely exceed 12 stops. A very large sample of excursions would nevertheless yield from time to time a higher number, revealing that the upper bound is not real.

Another option is that after each stop, the rat decides whether to go back home or to proceed to the next stop. With a probability of returning home of, say, 1:2, 50% of the excursions would include one stop, 25% (half of the remaining half) would include two stops, 12.5% would include three stops, and so on; the frequency distribution would steeply trail off, again creating the false impression of an upper bound.

Still another option is that the probability of returning home would *increase* after each stop. Only in this mode would the number of excursions containing 1, 2, 3 . . . n stops be similar, thus yielding a uniform distribution. We have found that rats use the third option. The maximal number of stops per excursion does not further increase, regardless of how much we increase the number of examined rat sessions, even when the arena was increased from 4 m² to 64 m² (Golani et al., 1993). Scaling interstop distances so as to fit into increasingly larger arenas has been replicated in voles (Eilam, 2003; Eilam et al., 2003).

Other Stop-Related Findings. The "hyperactivity" attributed to rats with Fimbria-Fornix lesions is due to the abundance of stops of shorter duration (and thus more movement segments) (Whishaw et al., 1994). The distribution of stops and the number of home bases can be modified with D-amphetamine (Eilam and Golani, 1994; Cools et al., 1997, Gingras and Cools, 1997) and with quinpirole (Eilam and Szechtman, 1997). Both excursion length and number of stops per excursion are shortened and consolidated into topographically stereotyped chunks that are performed en bloc under D-amphetamine (Eilam and Golani, 1990).

Buildup of Arena Occupancy

When rats were exposed daily for a 1 hour session to a large (6.5 m diameter) empty outdoor arena, they manifested a gradual buildup of arena occupancy. Excursion length is defined in Figure 16–5A as the maximal angular distance along the wall reached by the rat within an excursion. The figure presents a summary of excursion length values for all rats as a function of their temporal order of performance, for each session separately, session by session. As shown, excursion length increases both within and across sessions. The most prominent increase occurs during the

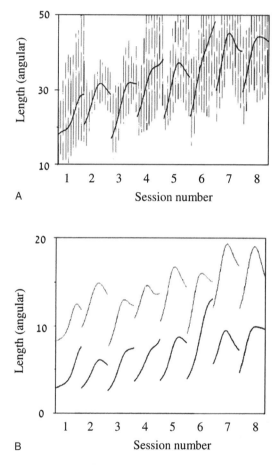

A

B

Figure 16–5. (A) The average values of the excursion lengths within the first eight sessions. Data are smoothed and are presented together with SD bars, computed at 10 parts of each session. The x-axis represents the temporal order of excursions, and the y-axis represents angular distance from the home base. The numbers below the x-axis represent the temporal sequence of sessions. (B) Smoothed average attraction distance (AD, represented by higher series of curves) and average repulsion distance (RD, represented by lower series of curves) within the first eight sessions. Data are computed at 10 parts of each session. The x-axis is as in A, and the y-axis represents angular distance from home base.

first half of the session, whereas later the length stabilizes or even decreases.

Examination of the high variability (reflected in the high standard deviation values in Figure 5A) reveals that during the first stage of the first session each rat has its own characteristic length of excursions, but all rats share a similar rate of excursion growth both within and across sessions. The exploratory process is additive from one session to the next, and the rate of increase is constant and independent of the previous experience of the rat. An additional effect is that of progression within a session. The dynamics of excursion length might reflect intrinsic constraints on the amount of novelty a rat can handle per excursion. It may also be an expression of the buildup observed in many species-specific behaviors (Lorenz, 1937).

Velocity across the Buildup Process. In the large outdoor arena the rats mostly progressed back and forth along the wall. This allowed us to compare, for each excursion at each location, the outbound velocity with the inbound velocity at that location. A negative value of the algebraic sum indicated that at that location the inbound speed was higher than the outbound speed, whereas a positive value indicated that at that location the outbound speed was higher. In this way, the excursion was divided into portions with negative values and portions with positive values. When outbound speed is slow and inbound speed is high (negative sum), it is as though the rat moved upstream on the way out, against the attraction exerted on it by the home base, whereas on the way in, it moved downstream, in cooperation with that attraction. Conversely, when outbound speed was high and inbound speed low (positive sum), it was as though the rat moved on the way out downstream, in cooperation with a repulsion exerted on it by the home base, whereas on the way in, it moved against that repulsion.

By computing dwell time at each location, we found that high home base attraction characterized low exposure (i.e., unfamiliar) portions of the arena circumference, whereas low

attraction or even repulsion characterized high exposure (familiar) portions. By computing the portions of the arena circumference from which the home base became attractive and the corresponding portions from which it became repulsive, we found that the velocity pattern of the rat changed concurrently with the increase in excursion length and in correlation with the familiarity of places. The primitive velocity pattern consisted of a slow outbound and a fast inbound pattern. During exposure, the asymmetry in velocity was inverted. The inversion spread across successive excursions from the home base out. Both excursion length and the attraction–repulsion dynamics might reflect the same intrinsic constraints on the amount of novelty a rat could handle per and across excursions (see Fig. 16–5B). An analytic model developed by Tchernichovski and Benjamini (1998) explains both the progressive increase in excursion length and the dynamics of velocity.

The progressive nonmonotonic expansion of rat movement from excursion to excursion is a general characteristic of species-specific behavior patterns (Lorenz, 1937). This growth pattern shows several similarities to the searching behavior of the desert ant (Cataglyphis spp.) (Wehner, 2003) and other invertebrates (Hoffmann, 1978, 1983). Although Tchernichovski et al. (1998) and Tchernichovski and Benjamini (1998) attribute the expansion of excursion length and the dynamics of velocity to a familiarization process, they also point out that it could well develop on an idiothetic basis.

An idiothetic basis during early phases of exploration has been suggested in several studies (e.g., McNaughton et al., 1996). It has also been indicated that increased velocity in the inbound portion of excursions has an idiothetic basis (Whishaw et al., 2001). Control and fimbria-fornix rats had similar outward velocities in segments starting from home base, but the return paths of the fimbria-fornix rats were significantly slower, more circuitous, and more variable. This was independent of light or dark conditions. The lack of dependence on allothetic cues thus suggested that rats used dead reckoning navigational strategies to initiate the homeward portion of excursions (see Chapter 38).

A Selection of Automatically Measured Parameters (Endpoints)

Having cleaved the behavior into the ethologically distinct patterns described in this review and having defined these patterns algorithmically, one can readily compute the durations, spatial properties, velocities, frequencies, etc., characterizing these patterns (a selection of endpoints is given in Table 16–1; download SEE at http://www.tau.ac.il/~ilan99/see/help).

MULTI-SEGMENTAL MOVEMENT ANALYSIS

Capturing the relations and changes in relation between the parts of the body in freely moving rats is followed by the stages of a closer examination, analysis, and integration

of the multiple measurements into few key kinematic parameters. The values taken by these parameters can subsequently be characterized in various situations and preparations related to locomotor and exploratory behavior. High-resolution video-based tracking of multiple points on the animal's body, algorithmic analysis combined with EW movement notation analysis (Eshkol and Wachmann, 1958; Eshkol, 1990) and a dynamical systems approach (Kafkafi et al., 1996; Kafkafi and Golani, 1998) are combined to yield an articulated description of the organization of rat locomotor and exploratory behavior.

DATA ACQUISITION

A bottom view of a freely moving rat walking on a glass platform provides access to the rat's trunk and all legs. A sampling rate of 50–60 Hertz provides an appropriate temporal resolution (Fig. 16–6A).

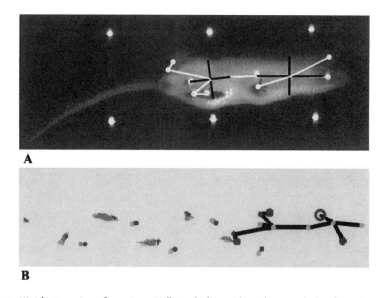

A

B

Figure 16–6. (A) A bottom view of an automatically tracked rat with markers attached to five points on its trunk and six points on its feet. Hindpaw and forepaw positions are respectively measured throughout the step cycle in reference to a frame whose origin of axes is schematically placed at the point lying between the hip (for hindpaws) and the shoulder joints (for forepaws). The frame's vertical axis coincides with the direction of progression of the midsaggital axis of the respective trunk part. (B) A snapshot taken from an animated reconstruction of the relations and changes in relation between the parts of the trunk, the feet, and their respective footprints during forward progression. Gray disks represent tracked markers; full discs on feet stand for feet in support phase, and open circles stand for feet in swing phase. The footprint trail of each foot is shown in shades of gray, with light gray representing establishment of contact, and dark gray release of foot contact with substrate. Arrows indicate heel-to-digits direction of hindfeet (see http://www.tau.ac.il/~ilan99/see/multilimb).

DATA ANALYSIS

The process of compression of the multiple measurements into fewer key parameters of whole-body unrestrained locomotor and exploratory behavior is described in several studies (Kafkafi et al., 1996; Kafkafi and Golani, 1998; Golani et al., 1999). Here we highlight three key parameters whose interaction generates a behavioral growth pattern, reminiscent of that described for excursions from the home base, which involves a progressive increase in the amplitude of movements, and a progressive increase in the animal's behavioral repertoire.

Warm-up

Infant rats respond to a novel environment by becoming immobile and then showing a process of motorial expansion called *warm-up*. Starting from immobility, horizontal movement, forward movement, and finally vertical movement are successively incorporated into the behavior. Within each of these movement parameters separately, the parts of the trunk and the legs are recruited into the movement in a cephalocaudal order, head first and hindlegs last. The head and neck never move forward unless they already moved horizontally (laterally), and they never move up unless they already moved forward. The same rule applies to the chest and to the pelvis (Figs. 16–7 and 8). Concurrent with the increase in the animal's repertoire (and hence in the unpredictability of behavior) and concurrent with a repetition of movements that were performed earlier, there is an increase in the amplitude of movements. This results in a grad-

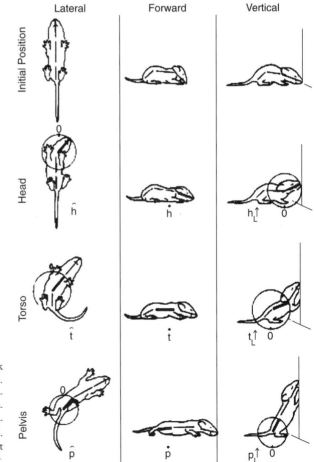

Figure 16–7. The infant rat's trunk as a linkage of articulated axes. Columns represent the three mobility gradient key parameters, and horizontal rows represent the most caudal part of the trunk that moved. The axis of the most caudal part that moved is represented by a thick bar.

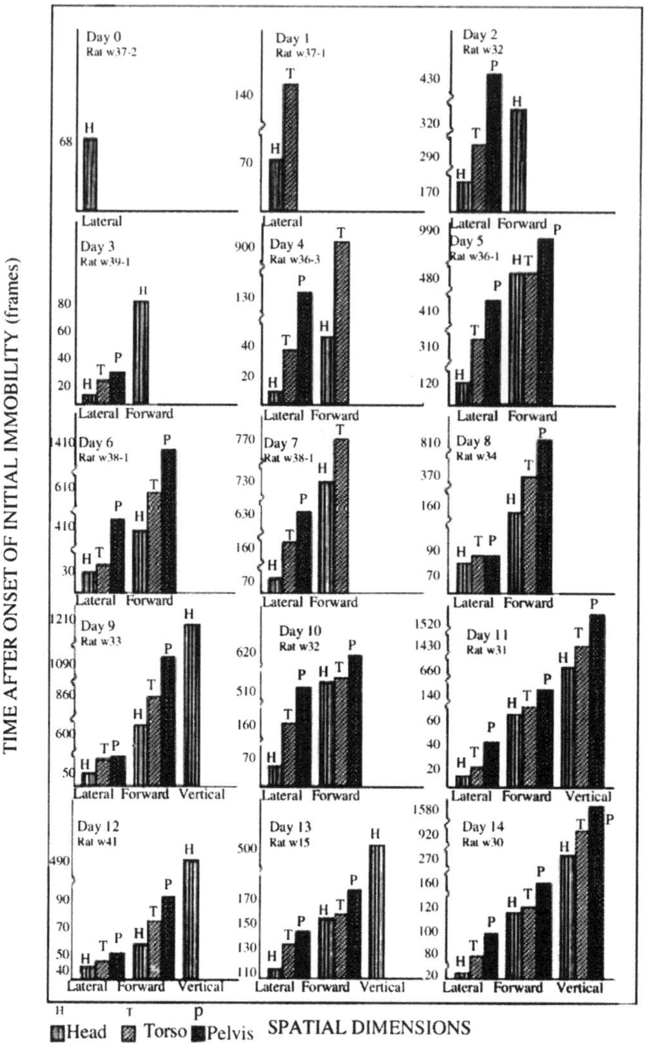

Figure 16–8. Timing of first appearance of movement of the parts of the trunk along the three spatial parameters. One example was selected at random for each developmental day. Within each moment-to-moment warm-up sequence, the parts of the body are recruited along the three spatial parameters (lateral, forward, and vertical) in a cephalocaudal order. (From Eilam and Golani, 1988.)

ual expansion of the explored portion of the environment (Golani et al., 1981; Eilam and Golani, 1988).

The warm-up process is a particular manifestation of a more general "mobility gradient" that characterizes the transition from immobility to increasing complexity and unpredictability of behavior (Golani, 1992). In particular, a progression in the opposite direction, with decreasing spatial complexity and increased stereotypy, that is termed "shutdown" occurs under the influence of the nonselective dopaminergic drugs apomorphine and amphetamine and in part the selective dopaminergic agonist quinpirole (Szechtman

et al., 1985; Eilam et al., 1989; Adani et al., 1991; Golani, 1992). The mobility gradient behavior appears to be mediated by a family of basal ganglia-thalamocortical circuits and their descending output stations (Golani, 1992).

The observation that, after pronounced immobility, the first forward movement of a part of the trunk must be preceded by a lateral movement of that part, suggested that the performance of one type of movement enables (potentiates) the performance of the next type, which enables the next, and so on, without necessarily eliciting that next type (Golani, 1992). Chevalier and Deniau (1990) attributed an enabling function to the striatum. A reciprocal relationship between horizontal scanning and forward progression has been observed in rats: as soon as forward progression stops, lateral head scans appear (during lingering episodes) (Drai et al., 2000). More recently, it has been shown that there is an inverse relationship between stepping and scanning at the level of the central nervous system. This relationship is mediated by the hypothalamus (Sinnamon et al., 1999).

CONCLUSION

A "mobility gradient" involving increasing complexity and unpredictability unfolds in rat locomotor and exploratory behavior at the levels of the path (location), trajectory (velocity), and interlimb coordination. Computerized data acquisition, appropriate preparation of these data for analysis, and cleavage of the stream of behavior into intrinsically defined parameters and patterns make this global growth pattern more accessible for study.

ACKNOWLEDGMENTS

This study was supported by a grant from the Israel Academy of Sciences, Israel Science Foundation, and by National Institutes of Health grant 5-R01-NS040234-03.

REFERENCES

Adani N, Kirtati N, Golani I (1991) The description of rat drug-induced behavior—kinematics versus response categories. Neuroscience and Biobehavioral Reviews 15:455–460.

Benjamini Y, Kafkafi N, Sakov A, Elmer G, Golani I (submitted) Genotype-environment interactions in mouse behavior: A way out of the problem.

Chevalier G and Deniau JM (1990) Disinhibition as a basic process in the expression of striatal functions. Trends in Neuroscience 13:277–280.

Cleveland WS (1977) Robust locally weighted regression and smoothing scatterplots. Journal of American Statistical Association 74:829–836.

Cools AR, Ellenbroek BA, Gingras MA, Engbersen A,

Heeren D (1997) Differences in vulnerability and susceptibility to dexamphetamine in Nijmegen high and low responders to novelty: A dose-effect analysis of spatio-temporal programming of behavior. Psychopharmacology 132:181–187.

Drai D, Benjamini Y, Golani I (2000) Statistical discrimination of natural modes of motion in rat exploratory behavior. Journal of Neuroscience Methods 96:119–131.

Drai D and Golani I (2001) SEE: A tool for the visualization and analysis of rodent exploratory behavior. Neuroscience and Biobehavioral Reviews 25:409–426.

Eilam D and Golani I (1989) Home base behavior of rats (Rattus norvegicus) exploring a novel environment. Behavioural Brain Research 34:199–211.

Eilam D and Golani I (1988) The ontogeny of exploratory-behavior in the house rat (Rattus rattus) the mobility gradient. Developmental Psychobiology 21:679–710.

Eilam D and Golani I (1990) Home base behavior in amphetamine-treated tame wild rats (Rattus norvegicus). Behavioural Brain Research 36:161–170.

Eilam D and Golani I (1994) Amphetamine-induced stereotypy in rats: its morphogenesis in locale space from normal exploration. In: Ethology and psychopharmacology. (Cooper SJ and Hendrie CA, eds.). New York: John Wiley & Sons.

Eilam D, Golani I, Szechtman H (1989) D-2-agonist quinpirole induces perseveration of routes and hyperactivity but no perseveration of movements. Brain Research 490:255–267.

Eilam D (2003) Open-field behavior withstands changes in arena size. Behavioural Brain Research 142:53–62.

Eilam D, Dank M, Maurer R (2003) Voles scale locomotion to the size of the open-field by adjusting the distance between stops: A possible link to path integration. Behavioural Brain Research 141:73–81.

Eilam D and Szechtman H (1997) A plausible rat model of obsessive-compulsive disorder: Compulsive checking behavior is induced in rats chronically injected with quinpirole. Neuroscience letters 48:S16–S16.

Eshkol N and Wachmann A (1958) Movement Notation. London: Weidenfield & Nicholson.

Eshkol N (1990) Angles and Angels. Tel-Aviv: The Movement Notation Society.

Everitt BS (1981) Finite mixture distributions. London: Chapman and Hall.

Gingras MA and Cools AR (1997) Different behavioral effects of daily or intermittent dexamphetamine administration in Nijmegen high and low responders. Psychopharmacology 132:188–194.

Golani I (1992) A mobility gradient in the organization of vertebrate movement: The perception of movement through symbolic language. Behavioral and Brain Sciences 15:249–308.

Golani I, Benjamini Y, Eilam D (1993) Stopping behavior: Constraints on exploration in rats (*Rattus norvegicus*). Behavioural Brain Research 53:21–33.

Golani I, Bronchti G, Moualem D, Teitelbaum P (1981) "Warm-up" along dimensions of movement in the ontogeny of exploration in rats and other infant mammals. Proceedings of the National Academy of Sciences USA 78:7226–7229.

Golani I, Kafkafi N, Drai D (1999) Phenotyping stereotypic behaviour: Collective variables, range of variation and predictability. Applied Animal Behavior Science 65:191–220.

Hen I, Sakov A, Kafkafi N, Golani I, Benjamini Y (2004) The dynamics of spatial behavior: How can robust smoothing techniques help? Journal of Neuroscience Methods 133:161–172.

Hoffmann G (1978) Experimentelle und theoretische analyse eines adaptiven Orientierungsverhaltens: die 'optimale' Suche der Wustenassel Hemilepistus reaumuri, Audouin und Savigny (Crustacea, Isopoda, Oniscoidea) nach ihrer Hohle. PhD thesis, Regensburg

Hoffmann G (1983) The random elements in the systematic search behavior of the desert isopod Hemilepistus reaumuri. Behavioral Ecology and Sociobiology 13:81–92.

Kafkafi N and Golani I (1998) A traveling wave of lateral movement coordinates both turning and forward walking in the ferret. Biological Cybernetics 78:441–453.

Kafkafi N, Levi-Havusha S, Golani I, Benjamini Y (1996) A stereotyped motor pattern as a stable equilibrium in a dynamical system. Biological Cybernetics 74:487–495.

Kafkafi N, Lipkind D, Benjamini Y, Mayo CL, Elmer GI, Golani I (2003a) SEE locomotor behavior test discriminates C57BL/6J and DBA/2J mouse inbred strains across laboratories and protocol conditions. Behavioral Neuroscience 117:464–477.

Kafkafi N, Mayo C, Drai D, Golani I, Elmer G (2001) Natural segmentation of the locomotor behavior of drug-induced rats in a photobeam cage. Journal of Neuroscience Methods 109:111–121.

Kafkafi N, Pagis M, Lipkind D, Mayo CL, Bemjamini Y, Golani I, Elmer GI (2003b) Darting behavior: A quantitative movement pattern designed for discrimination and replicability in mouse locomotor behavior. Behavioural Brain Research 142:193–205.

Lipkind D, Sakov A, Kafkafi N, Elmer GI, Benjamini Y, Golani I (in press) New replicable anxiety-related

measures of wall versus center behavior of mice in the open field. Journal of Applied Physiology.

Lorenz KZ (1937) Uber die Bildung des Instinktbegriffes. Naturwissenschaften 25:289–331.

McNaughton BL, Barnes CA, Gerrard JL, Gothard K, Jung MW, Knierim JJ, Kudrimoti H, Qin Y, Skaggs WE, Suster M, Weaver KL (1996) Deciphering the hippocampal polyglot: The hippocampus as a path integration system. Journal of Experimental Biology 199:173–185.

Sinnamon HM, Karvosky ME, Ilch CP (1999) Locomotion and head scanning initiated by hypothalamic stimulation are inversely related. Behavioural Brain Research 99:219–229.

Szechtiman H, Ornstein K, Teitelbaum P, Golani I (1985) The morphogenesis of stereotyped behavior induced by the dopamine receptor agonist apomorphine in the laboratory rat. Neuroscience 14: 783–798.

Tchernichovski O, Benjamini Y, Golani I (1996) Constraints and the emergence of free exploratory behavior in rat ontogeny. Behaviour 133:519–539.

Tchernichovski O, Benjamini Y, Golani I (1998) The dynamics of long-term exploration in the rat—Part I. A phase-plane analysis of the relationship between location and velocity. Biological Cybernetics 78: 423–432.

Tchernichovski O and Benjamini Y (1998) The dynamics of long-term exploration in the rat—Part II. An analytical model of the kinematic structure of rat exploratory behavior. Biological Cybernetics 78:433–440.

Tchernichovski O and Golani I (1995) A phase plane representation of rat exploratory behavior. Journal of Neuroscience Methods 62:21–27.

Tukey JW (1977) Exploratory data analysis. Boston: Addison-Wesley.

Wehner R (2003) Desert ant navigation: How miniature brains solve complex tasks. Journal of Comparative Physiology A 189:579–588.

Whishaw IQ, Cassel JC, Majchrzak M, et al. (1994) Shortstops in rats with fimbria fornix lesions—Evidence for change in the mobility gradient. Hippocampus 4:577–582.

Whishaw IQ, Hines DJ, Wallace DG (2001) Dead reckoning (path integration) requires the hippocampal formation: Evidence from spontaneous exploration and spatial learning tasks in light (allothetic) and dark (idiothetic) tests. Behavioural Brain Research 127:49–69.

Circadian Rhythms

17

MICHAEL C. ANTLE AND RALPH E. MISTLBERGER

The day–night cycle caused by the rotation of the Earth about its axis poses many survival challenges. Not surprisingly, most organisms have evolved specializations to exploit temporal niches in their environment. For example, diurnal (day-active) animals tend to have visual systems adapted for acuity at the expense of sensitivity, whereas the visual systems of nocturnal (night-active) animals, such as the rat, typically sacrifice acuity in the interests of sensitivity. For these and other specializations to be adaptive, animals must have a mechanism to appropriately coordinate their behavior and physiology with the day–night cycle. The mechanism that has evolved is a system of circadian (from the Latin *circa*, which means "about," and *dies*, which means "day") oscillators, located in the brain and in other organs, that generates daily rhythms and that actively synchronizes these rhythms with environmental cycles. Circadian rhythms are ubiquitous in rat behavior and physiology and, since the pioneering work of Richter (1922), the rat has been an important animal model for elucidating the functional properties and neural mechanisms of circadian regulation.

MEASUREMENT AND ANALYSIS

The tool of choice for measuring circadian behavioral rhythmicity in rats and other rodents has been the running wheel. Rats are avid runners, and wheels are simple to install and relatively cheap. Traditionally, running wheels

were connected to Esterline Angus pen and paper recorders (Slonaker, 1908; Richter, 1922). Today, computers and specialized sensors are available by which to measure circadian rhythms in activity and other functions. The most commonly used devices are mechanical microswitchs (for running wheels and tilt-floors), photobeams (for activity or feeding), electrical contact circuits (for eating or drinking meters), and implantable radio-telemetry transmitters (for general activity, body temperature, blood pressure, and heart rate). Sleep–wake states can be measured electrophysiologically by implanted skull electrodes connected to cables or radiofrequency transmitters.

Wheel running data are conveniently displayed in the "actogram" format (Fig. 17–1). In this format, consecutive days of activity are aligned vertically, permitting easy visual determination of two fundamental parameters of a daily rhythm: its phase (position within a cycle) and its period (average duration of a cycle, the reciprocal being frequency). As illustrated in Figure 17–1, wheel running activity in rats exhibits a very robust daily rhythm. In a standard light–dark (LD) cycle (12 hour light:12 hour dark) running is concentrated at night, and the onset of running is typically abrupt and predictable near the time of lights-off. The onset of running thus is an easily observable phase of this activity rhythm. The period of the rhythm can then be quantified by measuring the average interval between successive onsets. This can be done with linear regression (the slope of a line fit to a set of

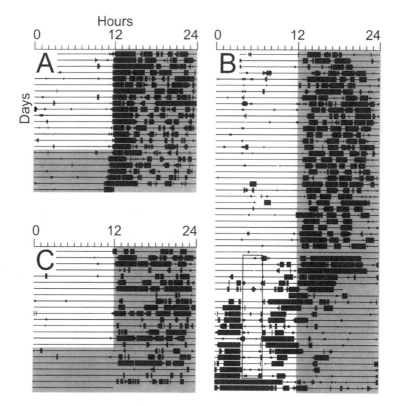

Figure 17–1. Wheel running activity in single-housed male (A and B) and female (C) rats maintained in a 24 hour light-dark cycle (dark indicated by shading) with free access to a 33 cm running wheel. Each line represents 1 day, with time plotted from left to right in 10 minute bins and consecutive days aligned vertically. Time bins in which wheel revolutions were detected are indicated by small vertical deflections (creating a heavy line when running is continuous). (A and C) Nocturnal concentration of wheel running typical of rats, and the persistence of this daily rhythm with an approximately 24 hour periodicity during 1 week in constant dark. The female rat (C) also shows a prominent increase of running every 4 to 5 days, marking the night of behavioral estrous. (B) Dramatic effect of restricting food to the daytime. This rat runs almost exclusively at night when food is available ad libitum but exhibits a prominent bout of running in anticipation of mealtime when food is restricted to 3 hours in the middle of the lights-on period (indicated by the open rectangle). In this example, the entire circadian profile appears inverted, and this is sustained during the last 2 days, when food was omitted completely. More commonly, rats express both daytime food anticipatory running and a persisting but lower level of nocturnal activity. Ablation of the hypothalamic suprachiasmatic nucleus eliminates circadian rhythms entrained to light–dark cycles or expressed in constant dark but does not eliminate the daily rhythm of food anticipatory running that emerges if meals are provided at circadian intervals.

consecutive onsets is the average deviation from 24 hours). In an LD cycle, the average period is 24 hours, and the phase is stable (onsets occur near lights-off). Thus, the LD cycle controls the phase and period, a phenomenon known as *entrainment*. The phase of entrainment may differ by individual or by age; that is, some rats may begin running just before lights-off (phase advanced relative to LD), and others may begin slightly after lights-off (phase delayed; e.g., Fig. 17–1B). Phase may

be altered by aging, reproductive status (e.g., estrous, Fig. 17–1C), and other factors.

In the absence of an LD cycle or other significant periodic environmental stimuli ("constant conditions"), the daily wheel running rhythm persists, indicating that it is generated by an internal timekeeping device (a self-sustaining oscillator). The oscillator can march to the beat of the LD cycle, but in the absence of periodic environmental signals, it "free-runs" and expresses its own intrinsic periodicity. In rats (as in humans), activity onsets in constant conditions tend to occur a bit later each "day" (i.e., relative to local time), indicating that the intrinsic period of the internal oscillator is slightly longer than 24 hours (thus the designation "circadian"). The period is not immutable and exhibits individual differences, changes with age and hormonal status, and a dependence on light intensity; the brighter the constant light, the slower the circadian oscillator runs. The free-running phase also can be shifted (advanced or delayed) by those stimuli capable of entraining circadian rhythms (see later). Phase shifts can be quantified by comparing regression lines fit to activity onsets before and after the stimulus.

Other variables, typically those that vary continuously such as core body temperature or levels of some hormones, are more easily visualized as waveforms. Simple waveforms can be described by their period and their amplitude. The amplitude is the difference between the peak (or trough) value and the mean value (the range is the difference between the peak and the trough). Period and amplitude can be quantified by curve-fitting procedures. The peak (acrophase) of a fitted sine wave is a definable phase that can be used for regression analyses. Besides linear regression, the two most common approaches to quantifying circadian period are the fast Fourier transform and the chi-squared periodogram (Refinetti, 1993).

CIRCADIAN REGULATION OF BEHAVIOR AND PHYSIOLOGY

SLEEP AND WAKE

Rats sleep predominantly during the daily light period (about 80% of a 12 hour light period) but also significantly at night (about 30% of a 12 hour dark period; Fig. 17–2). Two primary stages of sleep are recognized in all mammals: rapid eye movement (REM) sleep and non-REMS (NREM) sleep. Because eye movements are generally not measured in rat sleep studies, the REM stage is often referred to as "paradoxical" sleep, in recognition of its paradoxical defining characteristics of a waking-like electroencephalogram (EEG) combined with high arousal threshold and active inhibition of antigravity muscles. NREM sleep is often referred to as slow-wave sleep (SWS), but this is a misnomer because the cortical EEG during much of NREM sleep may have few slow waves (i.e., those in the 1 to 4 Hz range). SWS should refer only to that portion of NREM with a high concentration of slow waves. SWS typically is maximal early in the sleep period and declines exponentially. It is also greatly enhanced by prior sleep loss and is considered to be a correlate of sleep intensity or important sleep recovery processes. REM sleep, by contrast, gradually increases in bout duration and frequency as the sleep period progresses. NREM sleep and REM sleep tend to alternate at about 20 minute intervals in rats (90 minutes in humans), but this "ultradian" sleep cycle is weak and variable in the rat.

The strength of the circadian modulation of sleep can be further appreciated in conflict experiments, in which rats are sleep deprived and then allowed to recover during their usual wake period (Mistlberger et al., 1983; Fig. 17–2). Sleep (particularly SWS and REM) is significantly elevated relative to the normal amount of sleep at this phase (taken as evidence for homeostatic regulation of sleep), but total sleep in 2 hour blocks throughout the night remains much lower by comparison with sleep during the usual lights-on period. If recovery begins during the light period, there is little elevation of total sleep time, indicating that the circadian clock maintains sleep at a physiological ceiling during the usual sleep phase.

LOCOMOTION

In rats, spontaneous wheel running is more strictly nocturnal than is waking. Thus, al-

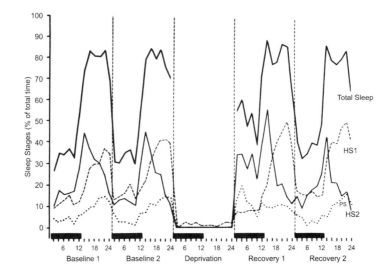

Figure 17–2. Group mean wave forms of polygraphically recorded sleep–wake states in rats maintained in a 12 hour light–dark cycle for 2 days before and 2 days after 24 hours of total sleep deprivation on a slowly rotating drum over water. Some sleep (*heavy line*) is evident at all times of day, but the total amount is greatest during the day. Total sleep time is elevated by prior sleep deprivation but also is constrained by circadian regulation. Non–rapid eye movement sleep is divided into high (HS2) and low (HS1) electroencephalographic amplitude substages. These vary inversely with time of day and with time after sleep deprivation; HS2 is considered to be functionally high-intensity sleep. Paradoxical sleep (PS also known as rapid eye movement sleep) is maximal at the end of the usual sleep period but significantly elevated immediately after total sleep loss. (Reprinted from Sleep, Vol 6, Mistlberger et al., pp. 217–233, Copyright 1983, with permission from Associated Professional Sleep Societies.)

though rats do engage in some waking activities during the day, wheel running is minimal. At night, the pattern of running activity varies by individual and by strain. Members of some rat strains tend to run primarily early in the night and others late in the night, and some exhibit bimodal or trimodal patterns, with two or three peaks at 12 or 6 hour intervals (Wollnik, 1991). These patterns are evident in sleep–wake states and general locomotion and are not secondary to the bioenergetic consequences of wheel running.

The amount, latency, and spatial distribution of locomotor activity in certain situations are often used as metrics for inferred psychological states such as anxiety or depression. There is some evidence that the response of rats in these situations (e.g., an open field or elevated plus maze) may vary with time of day and that this circadian influence may vary by metric (e.g., Jones and King, 2001; Andrade et al., 2003). It is thus recommended that in all such tests, time of day be considered as a potentially significant independent variable.

FEEDING AND DRINKING

Rats with free access to food eat in discrete bouts. These bouts are both longer and more frequent at night, when about 75% of total daily food intake occurs (Rosenwasser et al., 1981). Often the daily feeding rhythm is bimodal, with food intake concentrated at the start and the end of the dark period, but this may vary with strain of rat (Glendinning and Smith, 1994). Circadian regulation of feeding is also evident in the compensatory response to food deprivation. Thus, if a rat is food deprived for 42 hours, the size of the first "recovery" meal is significantly smaller if it occurs during the day rather than during the night (Bellinger and Mendel, 1975). However, rats are opportunistic feeders, and if food is restricted to a particular time of day, the feeding rhythm adjusts. If food is available only for a few hours in the daytime, the amount of food eaten during this time gradually increases over about 1 week, so that initial weight loss is reversed. Rats also exhibit activity in anticipation of the mealtime (Mistlberger, 1994; e.g., Fig. 17–1B). This anticipatory activity is circadian clock driven; once it is established, it reappears at the same time each day even if the animal is food deprived. Adaptive plasticity in the circadian programming of food seeking and ingestive behavior is undoubtedly one of the reasons that rats are such a widely distributed and successful species.

Drinking is also controlled by the circadian clock and tends to be more strictly nocturnal, with 85% of fluid intake occurring at night. Compensatory drinking in response to an osmotic challenge (e.g., hypertonic saline injection) is significantly greater at night (Johnson and Johnson, 1991). Rats also show circadian anticipatory activity to a daily opportunity to drink, but this is much less prominent than is food-anticipatory activity (Mistlberger, 1994).

THERMOREGULATION

In an LD cycle, body temperature (Tb) in rats varies from a daytime mean of about $37.3°$ C to a nighttime mean of about $38.1°$ C. Although Tb is influenced by behavioral state, the rhythm is not secondary to daily variations in behavior. First, Tb begins to rise about 2 hours before the daily active period, with the steepest increase observed from 30 minutes before to 60 minutes after waking (Refinetti and Menaker, 1992). Second, in constant light (LL), circadian rhythms in rats damp out over a period of weeks to months, but the Tb rhythm typically persists longer than does the activity rhythm (Eastman and Rechtschaffen, 1983).

Thermoregulation in the rat is accomplished by both autonomic and behavioral means. Notably, rats show a circadian rhythm in self-selected ambient temperature, preferring about $28°$ C during their usual rest period, when the Tb is low, and about $22°$ to $24°$ C during their usual active phase, when Tb is high. Thus, the circadian rhythm of preferred ambient temperature is in antiphase with the circadian rhythm of Tb. This indicates that the circadian rhythm of Tb is not due to a circadian modulation in "setpoint," as is the case with fever (Refinetti and Menaker, 1992).

REPRODUCTIVE BEHAVIOR

Many aspects of reproductive behavior exhibit a circadian rhythm. In the presence of a receptive female, male rats copulate more frequently and with shorter latency at night than during the day (Beach and Levinson, 1949). When tested at various times throughout the night, the greatest number of intromissions (Harlan et al., 1980) and the shortest latency to mounting, intromission, and ejaculation (Dewsbury, 1968) are observed during the last half of the dark phase. A circadian rhythm in spontaneous seminal emissions (producing seminal plugs) has also been reported (Kihlstrom, 1966; Stefanick, 1983).

Female rats also exhibit circadian rhythms in reproductive behavior and physiology. Females become sexually receptive and exhibit their highest level of mating behavior (behavioral estrus) on the night of ovulation, which occurs every 4 to 5 days in rats entrained to an LD cycle (Ball, 1937). The estrous cycle remains coupled to circadian rhythms in constant conditions, with behavioral estrous recurring every four or five circadian cycles.

Although rats are not a photoperiodic species (i.e., breeding is not dependent on day length), the length of the estrous cycle is affected by photoperiod. In an LD cycle with a 12 hour light period, about 70% of rats have a 4-day estrous cycle, whereas only 10% have a 5-day cycle. If lights are on for 16 hours per day, 21% of rats have a 4-day cycle, whereas 46% have a 5-day cycle (Hoffmann, 1968).

Timing of birth is also regulated by the circadian clock and tends to occur during the daytime on gestational day 22 or 23, whether in an LD cycle or in constant dark (DD) (Lincoln and Porter, 1976). This rhythm of parturition is eliminated by either ablation of the maternal circadian clock in the suprachiasmatic nucleus (SCN) or removal of the brains of the fetuses during late gestation (Reppert et al., 1987).

Daily rhythms may be problematic for scientists engaged in animal husbandry, particularly if access to pups immediately after birth is desired. The time of peak breeding efficiency is in antiphase with the time of parturition. If the animals are housed in a reverse LD cycle, breeding can be done during regular 9 AM–to–5 PM work hours, but births will occur during the rats' "day," which will be the researcher's night. Using a normal LD cycle, births will occur conveniently for the investigator, but the time of peak mating efficiency will occur at night. Mayer and Rosenblatt (1997) have devised a husbandry protocol that overcomes these issues. Animals are bred in a reverse LD cycle. After 1 week, the pregnant dams are transferred at mid-dark phase to a room with a normal LD cycle. Not only are all pups born during the daytime on gestational day 22 but also 75% of births occur during the middle third of the day. Although this protocol may simplify the work schedule of a researcher, it should be noted that inverting the LD cycle of a pregnant dam may have some effect on the pups' physiology, such that they may differ from pups produced under a more conventional protocol.

LEARNING AND MEMORY

In rats, circadian rhythms are relevant to learning and memory in at least three ways. First, there are circadian variations in acquisition or recall on some tasks. Second, phase shifts of circadian rhythms can disrupt learning and

memory. Third, circadian oscillators can provide time of day cues that can be used by rats to enable time–place learning; that is, circadian oscillators can be used as continuously consulted clocks to recognize time of day in the absence of environmental time cues.

The effect of time of day on learning, recall, and extinction are varied and may be task dependent. Performance of rats in a passive avoidance task was better during the light period, although testing and training occurred at the same phase, making it impossible to distinguish acquisition from recall (Davies et al., 1973). Rats appear to be better at acquiring an active avoidance task during the night under some situations. Extinction also appears to be more rapid at night under some situations (Novakova et al., 1983). Another study found no rhythm in acquisition of a free-operant avoidance task but did find that rats were more efficient responders during the night (Ghiselli and Patton, 1976). A more recent study noted that old rats, but not young rats, had performance deficits when tested during the late night, compared with the early night, on both inhibitory avoidance and delayed alternation tasks (Winocur and Hasher, 1999).

Disruption of circadian rhythms has varying effects on learning in rats. LD phase shifts or exposure to LL impairs performance on a passive avoidance task learned before the light cycle change (Tapp and Holloway, 1981; Fekete et al., 1985). LD shifts also impair recall, but not acquisition, on a water-maze task (Devan et al., 2001). By contrast, LD phase shifts may facilitate extinction on an active avoidance task (Fekete et al., 1985) and have no affect on social memory (Reijmers et al., 2001).

Rats appear to perform best on passive and active avoidance tasks when tested at the same circadian phase at which they were trained (Holloway and Wansley, 1973a, 1973b; Wansley and Holloway, 1975). This may indicate that time of day (circadian phase) is encoded in the memory for these tasks. Encoding of circadian phase is also suggested by evidence that rats can discriminate time of day. Rats can learn to press one lever at one time of the day and another lever at another time of the day for food access (Boulos and Logothetis, 1990; Mistlberger et al., 1996). Time-place learning may be more difficult to demonstrate on other types of tasks such as place preferences, water-maze escape routes, and radial arm mazes (Thorpe et al., 2003).

PHARMACOLOGY

Circadian rhythms in the response to drugs are to be expected, due to rhythms in the absorption/clearance ratio and in the susceptibility of the target tissue. For orally administered drugs, absorption is influenced by stomach content, which changes over the course of the day according to the circadian rhythm of feeding behavior. Absorption is also influenced by intestinal enzyme activity, gastric activity, and rate of glucose uptake into the circulatory system, all of which exhibit circadian rhythms. Clearance is a product of excretion and metabolism/inactivation, all of which again have a circadian rhythm (for a review, see Moore-Ede et al., 1982). There is a rhythm in the activity of liver enzymes. The pH of urine is also rhythmic, which influences the movement of drugs from the blood into urine.

There are many anecdotal reports of researchers inadvertently discovering a circadian rhythm in the action of a drug (Moore-Ede et al., 1982). In these cases, large variability in response over trials could be explained only when time of administration was taken into account. For those testing new pharmaceuticals, it is imperative that time of day be taken into account. For some drugs, peak efficacy occurs at one phase while peak toxicity occurs at a different phase. Appropriate timing of administration may therefore maximize clinical effect while minimizing adverse effects.

Two classes of drugs of particular interest to those working with rats are anesthetics and analgesics, both of which have circadian rhythms in efficacy. The concentration of halothane required to maintain anesthesia in rats is lower during the day than it is during the early night (Munson et al., 1970). However, the lethal concentration of halothane is lower during the early night (Matthews et al., 1964). The efficacy/toxicity phase–response curves for the anesthetic pentobarbitol are similarly inverted (Moore-Ede et al., 1982). These two examples illustrate how the therapeutic index of a drug can have a circadian rhythm. For drugs such as pentobarbital, it may be advisable to restrict use to the phases when the therapeutic index is at its broadest (i.e., the last half of the day in this case).

Surgical protocols frequently require the inclusion of an analgesic in addition to the anesthetic. Unfortunately, the trough in the circadian rhythm of analgesia provided by a given dose of morphine coincides with the phase when the therapeutic index for pentobarbital is at its broadest. This means that the time of day when pentobarbital is the safest to use is the same time of day when morphine provides the least pain relief.

ENVIRONMENTAL AND BEHAVIORAL INFLUENCES

Circadian rhythms influence many physiological systems relevant to phenomena of interest to behavioral neuroscientists. Thus, time of day is usually an important methodological consideration. Given that circadian rhythms are entrained by environmental time dues, attention to environmental factors that may alter circadian timing is vital.

LIGHT

Light affects behavioral rhythmicity in rats in two ways. First, acute light exposure inhibits activity and promotes sleep. Constant light tonically suppresses activity, but sleep duration does not remain enhanced (sleep is self-limiting). The acute affect of light is often re-

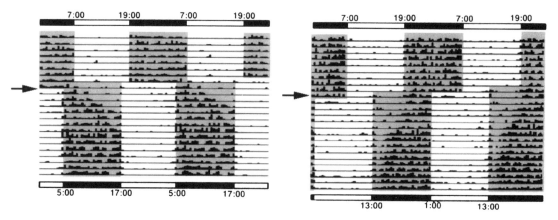

Figure 17–3. Wheel running activity of rats subjected to (A) an 8 hour delay in the light/dark cycle, simulating travel west, or (B) an 8 hour advance of the light/dark cycle, simulating travel east. Reentrainment to the delay shift is accomplished more quickly. In addition to direction of shift, many other factors can affect the rate of reentrainment (see text). (Modified from Journal of Neuroscience, Vol 23, Nagano et al., pp. 6141–6151, Copyright 2003, with permission from Society for Neuroscience.)

ferred to as *masking*, because it may mask the true phase of the rat's circadian rest–activity cycle. The second major effect of light is its role as a dominant entraining stimulus (a *zeitgeber*, which is a German word for "time-giver") for the circadian clock. Light entrains the clock because of its ability to induce phase shifts, which compensate each day for the discrepancy between the circadian period of the clock and the 24 hour period of the LD cycle. Light exposure in the evening or early night normally phase delays the clock, whereas light exposure in the morning or late night phase advances the clock. Light during the day has little effect. Thus, the circadian clock has a circadian rhythm of sensitivity to light. The net result is that any LD cycle with a period in the circadian range will "capture" the free-running clock and prevent it from drifting by daily phase adjustments opposing the direction of natural drift. For more detail on the formal mechanism of entrainment, see Mistlberger and Rusak (2000). Suffice it to say, turning on the light during the usual night can alter the timing of circadian rhythms in rats. Moreover, the light need not be bright or prolonged; even a few seconds each day can entrain free-running rhythms in nocturnal animals otherwise maintained in DD.

In LL, rats initially free-run with a period that is proportional to light intensity. Over weeks to months, bright LL can attenuate and may eliminate circadian rhythms. Circadian rhythmicity can be restored by one cycle of LD (Eastman and Rechtschaffen, 1983).

The time required to re-entrain to a shifted LD cycle, such as may occur when animals are shipped from a breeding facility to a research facility, depends on a number of factors, including the direction of shift (less time if the LD cycle is delayed rather than advanced; e.g., see Fig. 17–3), the magnitude of shift, and the housing conditions (e.g., less time if light is brighter or if rats have a running wheel in their home cage). A rule of thumb is to allow 1 day for every hour of LD cycle shift, but complete reentrainment can take 3 to 4 weeks.

FEEDING

Along with light, the timing of food intake is the most important determinant of circadian timing in rats. As noted earlier, rats can adapt to a scheduled daily meal by increasing activity in anticipation of mealtime and by a gradual increase in meal size. Mealtime is the dominant *zeitgeber* controlling the phase of daily rhythms of digestive processes and metabolism. Feeding schedules are widely used in behavioral studies of rats. For example, to motivate appetitive learning, rats are typically maintained at some reduced percentage of free-feeding weight. Typically, rats are weighed and fed at a similar time each day, and this can be expected to affect circadian timing of behavior and physiology (see Mistlberger, 1990, 1994).

LOCOMOTOR ACTIVITY

In DD or LL, the period of free-running rhythms in rats is shortened if they are provided ad libitum access to a running wheel (Yamada et al., 1986). Some physiological correlate of spontaneous behavioral activity can feed back to alter the rate at which the circadian clock cycles. Rats in DD can also be entrained by a daily bout of forced exercise on a treadmill (Mistlberger, 1991). However, induction of locomotor activity in the middle of the usual sleep phase does not induce phase shifts in rats, as has been shown for hamsters.

SOCIAL INFLUENCES

Rats that cohabitate in a seminatural environment have been reported to organize locomotor and feeding activity according to social status, with subordinates being forced to feed more during the light period. It is not known whether this represents a change in phase of the subordinate's circadian clock (for a review, see Mistlberger and Skene, 2004).

STRESS AND AROUSAL

Social defeat is considered a highly stressful stimulus for rats. A single social defeat can decrease the amplitude of locomotor, temperature, eating, drinking, and heart rate rhythms for a number of weeks (Meerlo et al., 2002). Other types of stress, such as surgical stress, chronic mild stress, forced swimming, restraint, and foot shocks, have similar effects.

However, these stimuli do not phase shift the clock. Furthermore, attenuation of circadian amplitude in overt rhythms results from processes downstream from the circadian clock and is not caused by attenuation of the amplitude of the circadian clock itself (Meerlo et al., 2002).

NEURAL MECHANISMS OF CIRCADIAN RHYTHMS

A MASTER CIRCADIAN PACEMAKER IS LOCATED IN THE HYPOTHALAMUS

Curt Richter, a pioneer of circadian rhythms research, spent many decades searching for the organ responsible for generating circadian rhythms in rats. He used a wide range of interventions, including brain lesions and removal of glands (e.g., adrenals, gonads, pituitary, thyroid, pineal, and pancreas). Only ventral hypothalamic lesions eliminated free-running rhythms in activity, drinking and feeding (Richter, 1967). The site of a master circadian pacemaker in rats was later localized to the SCN of the ventral anterior hypothalamus (Moore and Eichler, 1972; Stephan and Zucker, 1972). Much of the convergent evidence that the SCN is the site of the circadian clock comes from studies of rats. The SCN expresses rhythms in metabolic and single-unit neural activity, and individual rat SCN cells in dissociated cell cultures can maintain circadian rhythmicity for weeks at a time (Welsh et al., 1995). Transplantation of fetal SCN into an SCN-lesioned rat can restore behavioral rhythmicity (e.g., Lehman et al., 1987). The SCN receives direct and indirect projections from the retina, and many of its neurons respond to retinally mediated light. The SCN is thus considered to be the site of a master circadian clock that mediates light-entrainable circadian rhythms. It receives a variety of other inputs, notably from the median raphe and thalamic intergeniculate leaflet, that may mediate feedback effects of behavioral activity (for a review, see Mistlberger et al., 2000). However, it is not necessary for entrainment of circadian rhythms to feeding schedules. SCN-ablated rats are arrhythmic when fed ad libitum, but they develop robust food anticipatory rhythms if feed once or twice per day (Mistlberger, 1994).

CIRCADIAN CLOCK GENES HAVE BEEN IDENTIFIED AND ARE EXPRESSED IN MANY TISSUES

Circadian rhythms in single SCN neurons are driven by autoregulatory transcription–translation feedback loops, involving a set of so-called circadian clock genes and their protein products (Reppert and Weaver, 2001). These

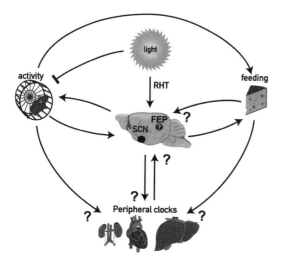

Figure 17–4. A simplified conceptual model of the mammalian circadian system. At the heart of the system is the suprachiasmatic nucleus (SCN), which consists of a heterogeneous population of circadian clock cells that generate circadian rhythms and that are entrained to environmental light/dark cycles via a retinohypothalamic tract (RHT) originating in photoreceptive retinal ganglion cells. The SCN drives daily activity rhythms but also is subject to feedback from activity, which can alter its circadian period or control its phase in the absence of light/dark cycles. A food-entrainable circadian pacemaker (FEP) is located at an unknown site outside of the SCN. Other circadian oscillators exist in peripheral organs and tissues. These are also entrainable by food, via unknown pathways, and possibly also by the SCN. In rats, light directly inhibits activity and promotes sleep. (Adapted from Cell, Vol 111, Schibler and Sassone-Corsi, pp. 919–922 Copyright 2002, with permission from Elsevier.)

genes are also expressed in a variety of other tissues, including elsewhere in the brain and in tissues in skeletal muscle, heart, lungs, and liver (Yamazaki et al., 2000). In vitro, circadian oscillations in these tissues damp out in about 4 days but can be reinitiated by changing the culture media. In vivo, the phase of peripheral oscillators (but not the SCN) is controlled by feeding time. The rat (and mammalian) circadian system can thus be characterized as multioscillatory, anatomically distributed, and sensitive to photic and nonphotic entraining stimuli (Fig. 17–4).

REFERENCES

Andrade MM, Tome MF, Santiago ES, Lucia Santos A, de Andrade TG (2003) Longitudinal study of daily variation of rats' behavior in the elevated plus-maze. Physiology and Behavior 78:125–133.

Ball J (1937) A test for measuring sexual excitability in the female rat. Comparative Psychology Monographs 14:1–37.

Beach FA and Levinson G (1949) Diurnal variation in the mating behavior of male rats. Proceedings of the Society for Experimental Biology and Medicine 72:78–80.

Bellinger LL and Mendel VE (1975) Effect of deprivation and time of refeeding on food intake. Physiology and Behavior 14:43–46.

Boulos Z and Logothetis DE (1990) Rats anticipate and discriminate between two daily feeding times. Physiology and Behavior 48:523–529.

Davies JA, Navaratnam V, Redfern PH (1973) A 24-hour rhythm in passive-avoidance behaviour in rats. Psychopharmacologia 32:211–214.

Devan BD, Goad EH, Petri HL, Antoniadis EA, Hong NS, Ko CH, Leblanc L, Lebovic SS, Lo Q, Ralph MR, McDonald RJ (2001) Circadian phase-shifted rats show normal acquisition but impaired long-term retention of place information in the water task. Neurobiology of Learning and Memory 75:51–62.

Dewsbury D (1968) Copulatory behavior of rats—Variations within the dark phase of the diurnal cycle. Communications in Behavioral Biology 1:373–377.

Eastman C and Rechtschaffen A (1983) Circadian temperature and wake rhythms of rats exposed to prolonged continuous illumination. Physiology and Behavior 31:417–427.

Fekete M, van Ree JM, Niesink RJ, de Wied D (1985)

Disrupting circadian rhythms in rats induces retrograde amnesia. Physiology and Behavior 34:883–887.

Ghiselli WB and Patton RA (1976) Diurnal variation in performance of free-operant avoidance behavior of rats. Psychology Reports 38:83–90.

Glendinning JI and Smith JC (1994) Consistency of meal patterns in laboratory rats. Physiology and Behavior 56:7–16.

Harlan RE, Shivers BD, Moss RL, Shryne JE, Gorski RA (1980) Sexual performance as a function of time of day in male and female rats. Biology Reproduction 23:64–71.

Hoffmann JC (1968) Effect of photoperiod on estrous cycle length in the rat. Endocrinology 83:1355–1357.

Holloway FA and Wansley RA (1973a) Multiple retention deficits at periodic intervals after active and passive avoidance learning. Behavioral Biology 9:1–14.

Holloway FA and Wansley RA (1973b) Multiphasic retention deficits at periodic intervals after passive-avoidance learning. Science 180:208–210.

Johnson RF and Johnson AK (1991) Drinking after osmotic challenge depends on circadian phase in rats with free-running rhythms. American Journal of Physiology 261:R334–R338.

Jones N and King SM (2001) Influence of circadian phase and test illumination on pre-clinical models of anxiety. Physiology and Behavior 72:99–106.

Kihlstrom JE (1966) Diurnal variation in the spontaneous ejaculations of the male albino rat. Nature 209:513–514.

Lehman MN, Silver R, Gladstone WR, Kahn RM, Gibson M, Bittman EL (1987) Circadian rhythmicity restored by neural transplant. Immunocytochemical characterization of the graft and its integration with the host brain. Journal of Neuroscience 7:1626–1638.

Lincoln DW and Porter DG (1976) Timing of the photoperiod and the hour of birth in rats. Nature 260:780–781.

Matthews JH, Marte E, Halberg F (1964) A circadian susceptibility-resistance cycle to fluothane in male B1 mice. Canadian Anaesthetists' Society Journal 11:280–290.

Mayer AD and Rosenblatt JS (1997) A method for regulating the duration of pregnancy and the time of parturition in Sprague-Dawley rats (Charles River CD strain). Developmental Psychobiology 32:131–136.

Meerlo P, Sgoifo A, Turek FW (2002) The effects of social defeat and other stressors on the expression of circadian rhythms. Stress 5:15–22.

Mistlberger RE (1990) Circadian pitfalls in experimental paradigms employing food restriction. Psychobiology 18:23–29.

Mistlberger RE (1991) Effects of daily schedules of forced

activity on free-running rhythms in the rat. Journal of Biological Rhythms 6:71–80.

Mistlberger RE (1994) Circadian food-anticipatory activity: Formal models and physiological mechanisms. Neuroscience and Biobehavioral Reviews 18:171–195.

Mistlberger RE and Rusak B (2000) Circadian rhythms in mammals: Formal properties and environmental influences. In: Principles and practice of sleep medicine, 3rd Edition (Kryger MH, Roth T, Dement WC, eds.), pp. 321–333. Philadelphia: WB Saunders.

Mistlberger RE and Skene DJ (2004) Social influences on circadian rhythms in man and animal. Biological Reviews, in press.

Mistlberger RE, de Groot MH, Bossert JM, Marchant EG (1996) Discrimination of circadian phase in intact and suprachiasmatic nuclei-ablated rats. Brain Research 739:12–18.

Mistlberger RE, Antle MC, Glass JD, Miller JD (2000) Behavioral and serotonergic regulation of circadian rhythms. Biological Rhythm Research 31:240–283.

Mistlberger RE, Bergmann BM, Waldenar W, Rechtschaffen A. (1983) Recovery sleep following sleep deprivation in intact and suprachiasmatic nuclei-lesioned rats. Sleep 6:217–233.

Moore RY and Eichler VB (1972) Loss of a circadian adrenal corticosterone rhythm following suprachiasmatic lesions in the rat. Brain Research 42:201–206.

Moore-Ede MC, Sulzman FM, Fuller CA (1982) The clocks that time us: Physiology of the circadian timing system. Cambridge, Mass: Harvard University Press.

Munson ES, Martucci RW, Smith RE (1970) Circadian variations in anesthetic requirement and toxicity in rats. Anesthesiology 32:507–514.

Nagano M, Adachi A, Nakahama K, Nakamura T, Tamada M, Meyer-Bernstein E, Sehgal A, Shigeyoshi Y. (2003) An abrupt shift in the day/night cycle causes desynchrony in the mammalian circadian center. Journal of Neuroscience 23:6141–6151.

Novakova V, Sterc J, Knez R (1983) The active avoidance reaction of laboratory rats: Differences between experiments carried out in the phase of motor activity and inactivity. Physiologica Bohemoslovaca 32:38–44.

Refinetti R (1993) Laboratory instrumentation and computing: Comparison of six methods for the determination of the period of circadian rhythms. Physiology and Behavior 54:869–875.

Refinetti R and Menaker M (1992) The circadian rhythm of body temperature. Physiology and Behavior 51:613–637.

Reijmers LG, Leus IE, Burbach JP, Spruijt BM, van_Ree JM (2001) Social memory in the rat: Circadian variation and effect of circadian rhythm disruption. Physiology and Behavior 72:305–309.

Reppert SM and Weaver DR (2001) Molecular analysis of mammalian circadian rhythms. Annual Review of Physiology 63:647–676.

Reppert SM, Henshaw D, Schwartz WJ, Weaver DR (1987) The circadian-gated timing of birth in rats: Disruption by maternal SCN lesions or by removal of the fetal brain. Brain Research 403:398–402.

Richter CP (1922) A behavioristic study of the activity of the rat. Comparative Psychology Monographs 1:1–55.

Richter CP (1967) Sleep and activity: Their relation to the 24-hour clock. Research Publications—Association for Research in Nervous and Mental Disease 45:8–29.

Rosenwasser AM, Boulos Z, Terman M (1981) Circadian organization of food intake and meal patterns in the rat. Physiology and Behavior 27:33–39.

Schibler U and Sassone-Corsi P (2002) A web of circadian pacemakers. Cell 111:919–922.

Slonaker JR (1908) Description of an apparatus for recording the activity of small mammals. The Anatomical Record 2:116–122.

Stefanick ML (1983) The circadian patterns of spontaneous seminal emission, sexual activity and penile reflexes in the rat. Physiology and Behavior 31:737–743.

Stephan FK and Zucker I (1972) Circadian rhythms in drinking behavior and locomotor activity of rats are eliminated by hypothalamic lesions. Proceedings of the National Academy of Science U S A 69:1583–1586.

Tapp WN and Holloway FA (1981) Phase shifting circadian rhythms produces retrograde amnesia. Science 211:1056–1058.

Thorpe CM, Bates ME, Wilkie DM (2003) Rats have trouble associating all three parts of the time-place-event memory code. Behavioral Processes 63:95–110.

Wansley RA and Holloway FA (1975) Multiple retention deficits following one-trial appetitive training. Behavioral Biology 14:135–149.

Welsh DK, Logothetis DE, Meister M, Reppert SM (1995) Individual neurons dissociated from rat suprachiasmatic nucleus express independently phased circadian firing rhythms. Neuron 14:697–706.

Winocur G and Hasher L (1999) Aging and time-of-day effects on cognition in rats. Behavioral Neuroscience 113:991–997.

Wollnik F (1991) Strain differences in the pattern and intensity of wheel running activity in laboratory rats. Experientia 47:593–598.

Yamada N, Shimoda K, Takahashi K, Takahashi S (1986) Change in period of free-running rhythms determined by two different tools in blinded rats. Physiology and Behavior 36:357–362.

Yamazaki S, Numano R, Abe M, Hida A, Takahashi R, Ueda M, Block GD, Sakaki Y, Menaker M, Tei H (2000) Resetting central and peripheral circadian oscillators in transgenic rats. Science 288:682–685.

Regulatory Systems

IV

Eating

18

PETER G. CLIFTON

For many, perhaps most, of those who study the rat in a laboratory context, feeding is a means to an end. A hungry rat rapidly learns to discriminate two tones, acquire spatial information in an eight-arm radial maze, or run down an alley for food reward. The behavior that allows the rat to ingest that food may only be "observed" in proxy form as the delivery of a food pellet into an operant chamber. Yet, feeding in the rat has a richness and a complexity that is worth study for its own sake, with relevance to those whose interest in the rat has a quite different focus. In addition, the rat is widely used in applied studies of feeding that extend our understanding of obesity in humans. My intention in this chapter is to give a brief description of the salient features of rat feeding behavior relevant to laboratory-based studies of this species.

LIFETIME PATTERNS OF INTAKE

Young rats, like all mammals, receive their initial nutrition from their mother in the form of milk. In a laboratory context, they typically begin to take solid food from day 16 of life (Thiels et al., 1990) and are usually weaned from the mother by day 21. However, if weaning is not enforced, the young continue to suckle, although progressively less frequently, until day 34 (Thiels et al., 1990). Food intake relative to body mass is high during early life, gradually decreasing as the rate of increase in body weight declines after sexual maturity. Asymptotic body weight and food consumption show a clear sex difference, with males being 1.2 to 1.5 times the weight of females as adults. In addition, there are considerable differences in asymptotic body weight between commonly used laboratory strains. Among pigmented animals, which are used more often in behavioral studies, typical adult weights vary from 300 grams in the dark agouti strain to 550 grams in the Lister hooded strain. Daily food intake is sensitive to environmental variables and especially to reductions in environmental temperature (Leung and Horwitz, 1976).

Food intake in sexually mature female rats varies significantly over the estrus cycle. On the night of estrus, when estradiol levels are high, the female reduces food intake and becomes more active. In mice, the intake of regular chow follows the same pattern, but they ingest more highly palatable food at estrus if it is made available (Petersen, 1976). The evidence in rats is less clear. For example, estradiol reduces the intake of palatable sucrose solutions as well as chow (Geary et al., 1995), but taste reactivity studies (see later) suggest a slightly different picture. Food intake increases markedly in response to the physiological demands of pregnancy and lactation. Both have been intensively studied, particularly in relation to their physiology (Hansen and Ferreira, 1986; Linden, 1989).

Although ad libitum feeding of rats is very common in laboratory settings, it may not be the ideal for longer-term experimental studies. Moderate food restriction reduces obesity, increases longevity, and decreases the incidence of neoplasms (Koolhaas, 1999) and may be justified on welfare grounds for a wide range of studies. Standard rodent diets may contain a higher proportion of fat and protein than is optimum, with additional variation in formulation, even from the same supplier, depending on geographical area.

DAILY PATTERNS OF INTAKE

Rats are crepuscular feeders, although their intake patterns are highly maleable. Thus, when food and water and freely available, lighting is on a 12 hour schedule, and disturbance is minimal, food intake is typically high around the period of lights-off and for 2 to 3 hours thereafter and then again before lights-on. Regular scheduling of food availability, or the provision of palatable food at particular times, leads to rapid adaptation of diurnal patterns of intake. For example, rats provided with a palatable mash made by mixing standard powdered chow with water will eat up to 10 grams (dry weight) over a 40 minute period during a time of the day when intake is usually minimal; this represents about 50% of total daily intake, and no food deprivation is required to produce this response. Total daily intake is also typically enhanced by this manipulation. In a similar way, rats can be provided with ad libitum access to chow in their living environment and the opportunity to visit an extremely cold area ($-15°C$), away from their cage, to forage for palatable food items at particular times. They will take a substantial proportion of their total daily calorie intake when the palatable food is made available in this way (Cabanac and Johnson, 1983). Substantial changes in the diurnal patterning of food intake are associated with physiological adaptation in hormonal rhythms within the hypothalamo-pituitary-adrenal axis. For example, corticosterone levels are usually maximal in the early dark phase of the photoperiod but will shift to anticipate feeding time when rats are fed on a daily schedule (Gallo and Wein-

berg, 1981). It is therefore important to allow sufficient time during the habituation phase of an experiment for such physiological changes to stabilize.

Intakes of water and food are well correlated over the diurnal cycle. This is likely to be promoted by both behavioral and physiological mechanisms. Rats, especially when housed alone for detailed feeding studies, may be relatively inactive for much of the time. Thus, all active behavior patterns are likely to be mutually correlated over time. Rats may also learn to drink while feeding on dry food to avoid the unpleasant consequences of dry food within the stomach (Lucas et al., 1989). In addition, there are specific physiological mechanisms that may stimulate drinking as a consequence of the presence of food in the stomach (Kraly, 1983).

MEAL PATTERNS

A tendency to feed at high intensity over relatively short periods is characteristic of many mammalian species from groups as diverse as primates, ruminants, carnivores, and rodents. These periods of intense feeding behavior are usually termed "meals." It might be hypothesized that meals are simply epiphenomena resulting from patchy food availability in the environment. However, rats in a laboratory setting with food and water continuously available nevertheless structure their intake into a sequence of meals.

Studies of meal patterning first require some method of continuously monitoring the intake of food, and preferably water intake as well. This can be achieved in a variety of ways. One technique involves the use of an automatic dispenser delivering single 45 mg pellets to a feeding niche where their presence is monitored until they are removed by the rat (Kissileff, 1970). Alternatively, the availability of cheap, but accurate, strain gauges allows online measures of the weight of a food hopper. The pellet dispenser has the advantage of greater temporal precision and the delivery of nonvarying food items with identical handling characteristics but makes it more difficult to vary the diet constituents. By contrast, direct weighing tends to reduce temporal resolution because of the requirement for the rat to move away from the food hopper but does allow for easier variation in diet constituents. The choice of technique depends on the particular questions of interest to the experimenter.

Once a series of feeding observations over time is available, a number of decisions have to be made before meals can be defined within the record. For example, is there a relatively clear drop in the probability of feeding again as the elapsed time since consumption of the last food item increases? The log-survivor technique represents an easy graphical approach to this problem and provides a simple, if implausible, associated null model (Clifton, 1987). A point close to the maximum inflection provides one estimate of a criterion that can be used to separate within-meal from between-meal intervals (Lester and Slater, 1986). However, if the modeling of these distributions is of primary interest, then the experimenter will wish to explore alternatives to this approach, fitting Weibull, log-normal, or a variety of other distributions (Sibly et al., 1990; Yeates et al., 2001). For those who simply wish to extract meal structure, it may be sufficient to choose a criterion and then repeat the analyses with other meal criteria that vary around that chosen to indicate the robustness of the conclusions (Castonguay et al., 1986). The criteria chosen in recent studies vary from 2 minutes to 10 minutes. Some earlier studies used a criterion as long as 30 minutes (Le Magnen and Tallon, 1966), which is likely to conflate individual meals.

After a minimum interval has been chosen, a data set can be processed as series of meals and the intervals between them. One further decision may be critical. Do meals defined in this way consist of two qualitatively different types of feeding behavior in which there are "snacks" (very short meals) and true

meals? Some authors exclude meals below some criterion size (e.g., 0.1 gram), and this decision may be especially important when the raw data are obtained from weighing systems that respond to both actual feeding and to exploratory behavior that may include climbing over the food container.

A number of parameters can be used to describe particular features of the meal structure. Within-meal feeding rate slows slightly during a meal (Clifton, 2000) and may be one useful index of satiety. However, it is important to note that the changes in feeding rate during a freely initiated meal are minimal by comparison with those seen in deprived rats working for food in an operant session or drinking a palatable solution (see later). In studies involving drug manipulations, feeding rate can be a very sensitive measure of motor impairment. Meal size is often taken a as a measure of within-meal satiation, whereas intermeal intervals may reflect the between-meal increase in hunger. The ratio of meal size to intermeal interval is termed the *satiety ratio* and may provide a more sensitive index of the satiating capacity of food (Clifton, 2000).

Although stable meal patterns are found in rats with free access to food, these patterns are also greatly affected by food availability. An extensive series of studies by George Collier provides some of the most striking examples of this type of effect (Collier et al., 1972; Collier, 1987). Rats were required to work for access to food by operating a conventional operant lever and lived continuously in this situation, rather than simply being exposed to it for a short period each day. The results depended on whether the rats either had to work for each food item or simply had to initiate a meal. In the latter case, once a single food item had been earned, subsequent items were free provided that feeding did not cease for 10 minutes or longer. This led to a substantial change in meal patterns in which, at extreme ratios (e.g., fixed ratio [FR] 5120), the rats took a single meal each day. Collier provided a strong functional interpretation of these data, suggesting that the rats responded to food availability by minimizing the work necessary for survival. However, such schedules may also lock an animal into repeated cycles of food deprivation, involving prolonged periods of operant responding, substantial consumption when food becomes available, and a long succeeding period of nonresponding. Although the effects of imposing a work requirement for each item of food are less dramatic than those produced by having to work for the initiation of a meal, they are clear. Meal size decreases as the work requirement for each food item within the meal is increased (Clifton et al., 1984; Timberlake et al., 1988). It seems most likely that this effect arises from interference with positive feedback processes that serve to reinforce feeding as a meal begins.

BEHAVIOR THAT ANTICIPATES FEEDING

Ethologists have long made the distinction between an initial flexible appetitive phase in motivated behavior and the subsequent more stereotyped consummatory phase (Craig, 1918). Casual observation of a rat that expects food to arrive suggests general behavioral activation coupled with response to cues that identify the location of food or the exact time at which it will arrive. Such behavior can be measured in a variety of test situations. For example, Blackburn et al. (1987) describe a simple Pavlovian conditioning paradigm in which rats come to expect a liquid diet reward after presentation of a compound light/buzzer stimulus with a duration of 150 seconds. While the conditioned stimulus is present, the rat makes anticipatory nose-pokes into the aperture where the diet will be delivered at the offset of the stimulus. A low dose of the dopamine D2 receptor antagonist pimozide (0.4 mg/kg) substantially attenuated this anticipatory nose-poke response but had no effect on intake of the liquid diet when this was provided ad libitum during a 20 minute test

session (Blackburn et al., 1987). In a similar vein, Gallagher et al. (1990) described a simple food conditioning paradigm in which the orientation response of a rat to a light cue that predicts the arrival of a food pellet is measured. In addition, they measured the rats' investigatory responses at the location where food is actually delivered. Although both responses develop as the rat is exposed to increasing numbers of conditioning trials, it turns out that the underlying neural mechanisms that subserve the two responses are rather different. Lesion studies suggest that the central nucleus of the amygdala makes an important contribution to the development of responding to the light cue but not to the investigatory responses at the food cup.

Runways provide another way of measuring appetitive responses to food. Speed of running from a start to a goal box can easily be measured and declines with repeated trials during a single test session. Several drugs that, at least under some circumstances, enhance food intake also enhance running speed in the early trials of session. They include the atypical antipsychotic olanzapine, whose clinical use is associated with the development of obesity (Thornton-Jones et al., 2002). Running speed can also be enhanced by cues that predict a particular trial is to be rewarded. Although dopamine antagonists reduced instrumental responding for food, the increased running speed in the presence of appropriate stimuli is unaffected by pretreatment with the dopamine antagonist haloperidol (McFarland and Ettenberg, 1998).

Conditioned place preference is another technique that is often used to study appetitive aspects of feeding. Although the paradigm has been generally used to assess drug reward, it is also appropriate for natural reinforcers such as food (Perks and Clifton, 1997) or sex (Everitt, 1990). Typically the task involves pairing particular contextual cues with the presence or absence of food. In test sessions, the rat is allowed to choose between the two environments, and the duration of time spent in each, or an appropriate ratio of these, provides a convenient measure of preference for food-associated cues. The task has a considerable advantage over T-maze studies that attempt to demonstrate the same phenomenon, in that spontaneous alternation between trials is not a problem. Preference measured in this way is sensitive to devaluation of the food reinforcer by sickness and to motivational state (Perks and Clifton, 1997). The paradigm has also been used to separate the role of different brain circuits in the approach to food-related contextual cues. For example, disconnection of the basolateral amygdala and ventral striatum impairs the expression of a food-reinforced place preference (Everitt et al., 1991).

More conventional operant studies have also distinguished the neurochemical systems that underlie the appetitive and consummatory phases of response to food. The selective dopamine D2 receptor antagonist raclopride, at approximately the same doses (approximately 0.5 mg/kg), strongly reduces lever pressing for food (Nakajima and Baker, 1989) but stimulates feeding on such pellets when they are available ad libitum (Clifton et al., 1991). In a more recent set of studies using similar drug manipulations, Salamone and colleagues (Cousins et al., 1994) provided rats with a concurrent choice of either working for 45 mg pellets or eating larger pieces of regular chow scattered on the floor of the operant cage. Untreated animals mostly work for pellets, whereas drug-treated animals switch to eating the less preferred chow on the cage floor.

FEEDING AND THE BEHAVIOR THAT FOLLOWS FEEDING

The details of food handling and drinking are covered elsewhere in this volume (Chapters 15 and 19). However, studies of taste reactivity and the microstructure of ingestion of liquid diets have been very influential in the study of feeding behavior.

The first group of taste reactivity studies, recently reviewed by Berridge (2000), emphasize the detailed analysis of facial expressions that rats exhibit while ingesting solutions with different sensory or nutritive characteristics. Although rats show such behavior patterns in many of the paradigms described elsewhere in this review, they are often difficult to observe and score accurately. In a typical taste reactivity study, the test solution is infused into the mouth through a previously implanted oral catheter and the behavior is recorded from below through a clear-bottom cage. Ingestion of a palatable solution is associated with a cluster of behavior patterns, including rhythmic tongue protrusion, paw licking, and lateral tongue movements. Infusion of unpalatable solutions evokes very different behavior patterns, including gapes, head shakes, face wipes, and chin rubs. This paradigm has been especially valuable in separating hedonic and incentive components of feeding. For example, pretreatment of rats with a dopamine antagonist, such as pimozide, has no effect on the proportion of ingestive and aversive responses elicited by a sucrose solution (Pecina et al., 1997). Interestingly, taste reactivity measures also suggest that female rats, at a point in the estrus cycle when estradiol levels are high, are more responsive to both attractive and unattractive taste cues (Clarke and Ossenkopp, 1998).

A second group of studies was pioneered by Jack Davis (1998). A rat presented with a nutritive or palatable solution during a 30 minute test session may ingest the same volume of solution in quite different ways. Thus, a highly palatable solution that is cleared slowly from the gut will be ingested rapidly at the beginning of the session, but then intake will quickly decline. A less-palatable solution that clears more quickly will be ingested more uniformly throughout the session. Davis discusses the use of negative exponential fitting techniques to characterize these effects. The method has been widely to study both enhancement of and reduction in intake. For example, the selective serotonin reuptake inhibitor fluoxetine produces reductions in intake that are similar to those observed after slowed gut emptying rather than reduced palatability, a finding that is consistent with the hypothesis that the drug enhances satiation processes (Lee and Clifton, 1992). Davis and Smith (1992) described a second method for analyzing this type of drinking record that relies on the way in which the rat performs short bursts of licking while ingesting a palatable solution. For example, an increase in the concentration of a sucrose solution increased the size of bouts of licking, whereas sham feeding increased the number of bouts rather than the size of bouts. In fact, Davis and Smith distinguished between two levels of organization, which they termed *bursts* and *clusters*. Subsequent authors have often used a single level of description (Spector et al., 1998).

The duration and intensity of feeding behavior may be strongly influenced by extrinsic as well as interoceptive cues. For example, presentation of cues previously associated with food intake may enhance feeding by apparently sated rats (Weingarten, 1984). This procedure has also been used to investigate the role of amygdala nuclei in the processing of conditioned stimuli that may facilitate feeding (Petrovich et al., 2002). Hungry rats were initially exposed to a situation in which they learned that one light conditioned stimulus predicted food, whereas the presentation of a second, different conditioned stimulus was not correlated with food presentations. Subsequently, the same rats were given consumption tests while sated. The conditioned stimulus previously associated with food consumption strongly enhanced eating in this situation. However, the effect was not seen in rats in which the basolateral amygdala and lateral hypothalamus had been disconnected.

Cues associated with specific features of the diet may also influence the quantity of food consumed in a meal or test session. The phenomenon of conditioned satiety provides a clear example (Booth, 1972). In this study, rats were given repeated daily training sessions in which a low-calorie diet was associated with one specific flavor, alternating with sessions in which a high-calorie diet was associated with a different flavor. In one condition of the subsequent test sessions, the rats were tested on an isocaloric diet of intermediate value to those used in training. The diet was flavored with either the flavor associated with low- or high-caloric content during training. The rats consumed more of the diet that had been flavored with the low-calorie flavor, despite their current identical calorific value, illustrating that consumption had come under control of the conditioned flavor cue.

Casual observation of rats that have just been fed suggests that a relatively stereotyped sequence behavior follows the cessation of eating. A rat may first appear quite active moving around the cage. It then settles down to a prolonged bout of grooming (see Chapter 13), beginning with the whiskers and face region and moving on down the body. Within a few minutes, the rat is likely to be quiet and inactive in one corner of the cage. These regularities in behavior were noted in early studies (Bolles, 1960), and it was then suggested that this sequence of behavior was characteristic of satiety in the rat and might be used to distinguish between experimental treatments that enhanced satiety from those that reduced feeding for other reasons (Antin et al., 1975). Since then, the so-called behavioral satiety sequence has been widely used to characterize changes in feeding behavior after pharmacological or neural manipulations. For example, drugs that primarily stimulate 5-hydroxytryptamine (serotonin) $(5\text{-HT})_{2C}$ receptors appear to advance the satiety sequence in the rat but preserve its overall form. The evidence supports this contention for some drugs that have nonspecific effects on serotonin systems, including the 5-HT releaser fenfluramine (Halford et al., 1998) and the selective serotonin reuptake inhibitor fluoxetine (Clifton et al., 1989), as well as more selective 5-HT_{2C} receptor agonists such as *m*CPP (Halford et al., 1998). By contrast, drugs such as DOI, which have substantially greater action at 5-HT_{2A} receptors in addition to their effects at 5-HT_{2C} receptors, produce increases in locomotor activity and a disruption of the normal satiety sequence (Simansky and Vaidya, 1990).

There are several comprehensive reviews of studies and methodology in this area (Clifton, 1994; Halford et al., 1998). The latter authors correctly place an emphasis on the advantages that may be gained from a complete video transcription of behavior from individual rats. However, depending on the particular purpose of the study, investigators may wish to consider using a time-sampling procedure that eliminates video records and allows a number of animals to be scored simultaneously (Clifton et al., 1989). In many circumstances, the decreased time required for such sampling permits an increase in statistical power by adding extra subjects, drug doses, or nondrug control conditions to the study.

DIET CHOICE

Although it is often assumed that rats will self-select a nutritionally appropriate diet in cafeteria-style experiments, the experimental evidence is less than compelling (Galef, 1991) and is not reviewed here. There has been considerable interest in the way in which rats select different proportions of protein, fat, and carbohydrate, the three macronutrient constituents of the diet. A variety of paradigms have been used to study such dietary selection. In one version, the rat is allowed to choose from three almost pure macronutrient sources. Each is supplemented with an appropriate mix of minerals and vitamins so that a range of diet choices will not impair general health. In one widely cited group of studies of this type, Liebowitz and her colleagues reported that treatment with drugs, such as fenfluramine, that enhance serotonergic neuro-

transmission reduced overall food intake but spared consumption of protein relative to that of carbohydrate (Shor-Posner et al., 1986; Weiss et al., 1990). However, there are substantial methodological issues in conducting such studies. If fat is provided as lard or a vegetable fat, protein as casein powder, and carbohydrate as dextrin or starch, then macronutrient content is confounded with many other factors, including taste, smell, texture, and water content. As a consequence, preference may vary substantially across the different diets. In addition, individual rats may show stable but idiosyncratic differences in the proportions of each diet that they consume. Other studies, which appear superficially similar to these, have generated very different patterns of results. For example, it has been reported that fenfluramine selectively *suppresses* fat consumption and *spares* carbohydrate consumption (Smith et al., 1998). A similar result has been obtained using the selective serotonin reuptake inhibitor fluoxetine (Heisler et al., 1999). It seems likely that these and similar inconsistencies arise from the complex nature of the diet selection paradigm in which a number of factors vary with macronutrient content.

In an attempt to disentangle these factors, a number of alternative diet selection techniques have been used. For example, rats may be provided with a carbohydrate supplement of polycose in addition to a standard chow diet (Lawton and Blundell, 1992). Under these conditions, the effects of fenfluramine varied markedly with the water content of the two diet components. Fenfluramine suppressed the consumption of dry polycose relative to chow presented as a wet mash, yet spared consumption of either sucrose or polycose solutions relative to dry chow. Data such as these reinforce the critical point made in a recent review of studies in this area (Thibault and Booth, 1999). Results for any single diet selection paradigm do not allow conclusions that can be discussed in terms of general effects on macronutrient selection. Instead, they must be interpreted solely as effects on the particular test diets that have been chosen for the study.

NEOPHOBIA AND DIET VARIETY

As an opportunistic omnivore, it might be expected that rats would sample potentially valuable, but novel, food items. Although this does occur, rats also show a strong neophobia to novel foods, even to those that are closely related to familiar foods (Barnett, 1963). Thus a first presentation of "palatable" mash made by soaking standard laboratory chow with water will be sampled but may also show signs of rejection (pushed to the back of the cage, covered with bedding material). After several days, intake increases rapidly, especially in situations that allow social facilitation between individuals (Galef et al., 1997). Once the items of a diet have become familiar, provision of several foods that vary in macronutrient content or sensory characteristics may enhance intake and, with chronic exposure, increase body weight. Such effects can be produced by presenting rats with a sequence of "courses" within a meal that vary in a single sensory characteristic (Treit et al., 1983) or even by simple alternation of two differently flavored courses (Clifton et al., 1987).

IS THERE A ROLE FOR SIMPLE INTAKE TESTS?

Simple intake tests over short periods of time remain the most common form of data presented in studies that explore feeding behavior in the rat. When used with care, they provide valuable preliminary data. However, it is always important to consider possible floor and ceiling effects when designing such an experiment. Tests using regular chow in nondeprived rats are not likely to reveal suppression of intake. Moderate food deprivation, or provision of more palatable food, is likely to

be more successful. Equally, an increase in food intake in nondeprived animals may be harder to demonstrate when either a palatable food or a fresh supply of regular chow is provided as the test meal. Both procedures are likely to take food intake to ceiling levels. Within-subject designs using scheduled food availability may promote the development of tolerance with repeated drug administration and reduce statistical power even when appropriate balancing is used. More generally, after completing initial studies using simple intake measures, consider the experimental hypotheses that are under consideration. Do they predict differential effects on the appetitive and consummatory phases of feeding behavior? Do they suggest that basic intake might remain similar but that the behavioral trajectory that leads to similar intake might be quite different? If the answer to these or similar questions is "yes," then consider adopting one or more of the paradigms described here for further investigation.

CONCLUSION

Feeding behavior in the rat has a complexity that should be mirrored in the experimental paradigms that are used for its observation and measurement. A combination of approaches, combining the detailed observational techniques of ethology with traditional behavioral analysis derived from experimental psychology, provides a good starting point. Appropriate choice of test situation, in conjunction with techniques that measure or manipulate aspects of the rat's physiology or neural function, has the capacity to reveal a good deal about the relationships between brain, physiology, and behavior in this species.

ACKNOWLEDGMENT

The author is grateful to Dr. Liz Somerville for her insightful comments on an earlier draft of this chapter.

REFERENCES

Antin J, Gibbs J, Holt J, Young RC, Smith GP (1975) Cholecystokinin elicits the complete behavioral sequence of satiety in rats. Journal of Comparative and Physiological Psychology 89:784–790.

Barnett S (1963) A study in behaviour. London: Methuen.

Berridge KC (2000) Measuring hedonic impact in animals and infants: microstructure of affective taste reactivity patterns. Neuroscience and Biobehavioral Reviews 24:173–198.

Blackburn JR, Phillips AG, Fibiger HC (1987) Dopamine and preparatory behavior: I. Effects of pimozide. Behavioral Neuroscience 101:352–360.

Bolles R (1960) Grooming behaviour in the rat. Journal of Comparative and Physiological Psychology 53:306–310.

Booth DA (1972) Conditioned satiety in the rat. Journal of Comparative and Physiological Psychology 81: 457–471.

Cabanac M and Johnson KG (1983) Analysis of a conflict between palatability and cold exposure in rats. Physiology and Behavior 31:249–253.

Castonguay TW, Kaiser LL, Stern JS (1986) Meal pattern analysis: Artifacts, assumptions and implications. Brain Research Bulletin 17:439–443.

Clarke SN and Ossenkopp KP (1998) Taste reactivity responses in rats: Influence of sex and the estrous cycle. American Journal of Physiology 274:R718–R724.

Clifton P (1987) Analysis of feeding and drinking patterns. In: Feeding and drinking (Toates F and Rowland N, eds.). Oxford: Elsevier.

Clifton PG (1994) The neuropharmacology of meal patterning. In: Ethology and psychopharmacology (Cooper SJ and Hendrie CA, eds.). Chichester: Wiley.

Clifton PG (2000) Meal patterning in rodents: psychopharmacological and neuroanatomical studies. Neuroscience and Biobehavioral Reviews 24:213–222.

Clifton PG, Popplewell DA, Burton MJ (1984) Feeding rate and meal patterns in the laboratory rat. Physiology and Behavior 32:369–374.

Clifton PG, Burton MJ, Sharp C (1987) Rapid loss of stimulus-specific satiety after consumption of a second food. Appetite 9:149–156.

Clifton PG, Barnfield AM, Philcox L (1989) A behavioural profile of fluoxetine-induced anorexia. Psychopharmacology (Berlin) 97:89–95.

Clifton PG, Rusk IN, Cooper SJ (1991) Effects of dopamine D1 and dopamine D2 antagonists on the free feeding and drinking patterns of rats. Behavioral Neuroscience 105:272–281.

Collier G (1987) Operant methodologies for studying feeding and drinking. In: Feeding and drinking (Toates F and Rowland N, eds.). Oxford: Elsevier.

Collier G, Hirsch E, Hamlin PH (1972) The ecological determinants of reinforcement in the rat. Physiology and Behavior 9:705–716.

Cousins MS, Wei W, Salamone JD (1994) Pharmacological characterization of performance on a concurrent lever pressing/feeding choice procedure: Effects of dopamine antagonist, cholinomimetic, sedative and stimulant drugs. Psychopharmacology (Berlin) 116:529–537.

Craig W (1918) Appetites and aversions as constituents of instincts. Biological Bulletin of Woods Hole 34:91–107.

Davis J (1998) A model for the control of ingestion—20 Years on. Progress in Psychobiology and Physiological Psychology 17:127–173.

Davis JD and Smith GP (1992) Analysis of the microstructure of the rhythmic tongue movements of rats ingesting maltose and sucrose solutions. Behavioral Neuroscience 106:217–228.

Everitt BJ (1990) Sexual motivation: A neural and behavioural analysis of the mechanisms underlying appetitive and copulatory responses of male rats. Neuroscience and Biobehavioral Reviews 14:217–232.

Everitt BJ, Morris KA, O'Brien A, Robbins TW (1991) The basolateral amygdala-ventral striatal system and conditioned place preference: Further evidence of limbic-striatal interactions underlying reward-related processes. Neuroscience 42:1–18.

Galef BG Jr (1991) A contrarian view of the wisdom of the body as it relates to dietary self-selection. Psychological Reviews 98:218–223.

Galef BG Jr, Whiskin EE, Bielavska E (1997) Interaction with demonstrator rats changes observer rats' affective responses to flavors. Journal of Comparative Psychology 111:393–398.

Gallagher M, Graham PW, Holland PC (1990) The amygdala central nucleus and appetitive Pavlovian conditioning: Lesions impair one class of conditioned behavior. Journal of Neuroscience 10:1906–1911.

Gallo PV and Weinberg J (1981) Corticosterone rhythmicity in the rat: interactive effects of dietary restriction and schedule of feeding. Journal of Nutrition 111:208–218.

Geary N, Trace D, Smith GP (1995) Estradiol interacts with gastric or postgastric food stimuli to decrease sucrose ingestion in ovariectomized rats. Physiology and Behavior 57:155–158.

Halford JC, Wanninayake SC, Blundell JE (1998) Behavioral satiety sequence (BSS) for the diagnosis of drug action on food intake. Pharmacology, Biochemistry, and Behavior 61:159–168.

Hansen S and Ferreira A (1986) Food intake, aggression, and fear behavior in the mother rat: Control by neural systems concerned with milk ejection and maternal behavior. Behavioral Neuroscience 100:64–70.

Heisler LK, Kanarek RB, Homoleski B (1999) Reduction of fat and protein intakes but not carbohydrate intake following acute and chronic fluoxetine in female rats. Pharmacology, Biochemistry, and Behavior 63:377–385.

Kissileff HR (1970) Free feeding in normal and "recovered lateral" rats monitored by a pellet-detecting eatometer. Physiology and Behavior 5:163–173.

Koolhaas J (1999) The laboratory rat. In: The care and management of laboratory animals (Poole T, ed.), pp. 313–330. London: Blackwell.

Kraly FS (1983) Histamine plays a part in induction of drinking by food intake. Nature 302:65–66.

Lawton CL and Blundell JE (1992) The effect of d-fenfluramine on intake of carbohydrate supplements is influenced by the hydration of the test diets. Behavioural Pharmacology 3:517–523.

Le Magnen J and Tallon S (1966) La periodicite spontanee de la prise d'aliments ad-libitum du rat blanc. Journal of Physiology (Paris) 58:323–349.

Lee MD and Clifton PG (1992) Partial reversal of fluoxetine anorexia by the 5-HT antagonist metergoline. Psychopharmacology (Berlin) 107:359–364.

Lester NP and Slater PJB (1986) Minimising errors in splitting behaviour into bouts. Behaviour 79:153–161.

Leung PM and Horwitz BA (1976) Free-feeding patterns of rats in response to changes in environmental temperature. American Journal of Physiology 231:1220–1224.

Linden A (1989) Role of cholecystokinin in feeding and lactation. Acta Physiologica Scandinavica Supplementum 585:i–vii, 1–49.

Lucas GA, Timberlake W, Gawley DJ (1989) Learning and meal-associated drinking: meal-related deficits produce adjustments in postprandial drinking. Physiology and Behavior 46:361–367.

McFarland K and Ettenberg A (1998) Haloperidol does not affect motivational processes in an operant runway model of food-seeking behavior. Behavioral Neuroscience 112:630–635.

Nakajima S and Baker JD (1989) Effects of D2 dopamine receptor blockade with raclopride on intracranial self-stimulation and food-reinforced operant behaviour. Psychopharmacology (Berlin) 98:330–333.

Pecina S, Berridge KC, Parker LA (1997) Pimozide does not shift palatability: Separation of anhedonia from sensorimotor suppression by taste reactivity. Pharmacology, Biochemistry, and Behavior 58:801–811.

Perks SM and Clifton PG (1997) Reinforcer revaluation and conditioned place preference. Physiology and Behavior 61:1–5.

Petersen S (1976) The temporal pattern of feeding over the oestrous cycle of the mouse. Animal Behaviour 24:939–955.

Petrovich GD, Setlow B, Holland PC, Gallagher M (2002) Amygdalo-hypothalamic circuit allows learned cues to override satiety and promote eating. Journal of Neuroscience 22:8748–8753.

Shor-Posner G, Grinker JA, Marinescu C, Brown O, Leibowitz SF (1986) Hypothalamic serotonin in the control of meal patterns and macronutrient selection. Brain Research Bulletin 17:663–671.

Sibly R, Nott HMR, Fletcher DJ (1990) Splitting behaviour into bouts. Animal Behaviour 39:63–69.

Simansky KJ and Vaidya AH (1990) Behavioral mechanisms for the anorectic action of the serotonin (5-HT) uptake inhibitor sertraline in rats: Comparison with directly acting 5-HT agonists. Brain Research Bulletin 25:953–960.

Smith BK, York DA, Bray GA (1998) Chronic d-fenfluramine treatment reduces fat intake independent of macronutrient preference. Pharmacology, Biochemistry, and Behavior 60:105–114.

Spector AC, Klumpp PA, Kaplan JM (1998) Analytical issues in the evaluation of food deprivation and sucrose concentration effects on the microstructure of licking behavior in the rat. Behavioral Neuroscience 112:678–694.

Thibault L and Booth DA (1999) Macronutrient-specific dietary selection in rodents and its neural bases. Neuroscience and Biobehavioral Reviews 23:457–528.

Thiels E, Alberts JR, Cramer CP (1990) Weaning in rats: II. Pup behavior patterns. Developmental Psychobiology 23:495–510.

Thornton-Jones Z, Neill JC, Reynolds GP (2002) The atypical antipsychotic olanzapine enhances ingestive behaviour in the rat: A preliminary study. Journal of Psychopharmacology (Oxford, England) 16:35–37.

Timberlake W, Gawley DJ, Lucas GA (1988) Time horizons in rats: The effect of operant control of access to future food. Journal of the Experimental Analysis of Behavior 50:405–417.

Treit D, Spetch ML, Deutsch JA (1983) Variety in the flavor of food enhances eating in the rat: A controlled demonstration. Physiology and Behavior 30:207–211.

Weingarten HP (1984) Meal initiation controlled by learned cues: Basic behavioral properties. Appetite 5:147–158.

Weiss GF, Rogacki N, Fueg A, Buchen D, Leibowitz SF (1990) Impact of hypothalamic d-norfenfluramine and peripheral d-fenfluramine injection on macronutrient intake in the rat. Brain Research Bulletin 25:849–859.

Yeates MP, Tolkamp BJ, Allcroft DJ, Kyriazakis I (2001) The use of mixed distribution models to determine bout criteria for analysis of animal behaviour. Journal of Theoretical Biology 213:413–425.

Drinking

19

NEIL E. ROWLAND

Food and fluid intakes often are considered as examples of behaviors with a homeostatic foundation. This chapter is organized around a construct of *hydromineral homeostasis* and discusses only water and mineral consumption. Other liquids that are often used in rat studies include liquid diets, sugars, and alcoholic drinks; these are not mentioned explicitly, but the same general principles and procedures may be applied to them. Water, sometimes with dissolved trace minerals, is the only naturally occurring fluid source. Sodium appetite is also much studied in the laboratory, as an innate and specific appetite. Sodium is inextricably related to water in the body and is the primary mineral in "hydromineral." Most laboratory studies use solutions of sodium salts, as is discussed, but it is likely that natural sources of sodium are not in fluid form.

Mammals have no mechanism for fluid storage, so states of physiological fluid need must engage powerful behavioral mechanisms (e.g., motivation, thirst) that drive the animal to fluid. Most laboratory studies deliberately minimize this motivational component and instead offer the fluids without effort in a safe environment; in this case the drinking is more reflexive than motivated. There is a relative lack of fluid studies in more naturalistic or effortful environments (Marwine and Collier, 1979; Quartermain et al., 1967). Under noneffortful conditions, need-related drinking is termed *primary* or *homeostatic* and is in contrast to *secondary* or *nonhomeostatic* drinking that occurs without identified need (Fitzsimons, 1979). The latter could in principle fall within a category of predictive homeostasis (Rowland, 1990), but there is no direct behavioral evidence that rats can predict future fluid needs (Stricker et al., 2003).

PHYSIOLOGY OF FLUID BALANCE

This section provides a brief overview of the main body fluid compartments, of how fluids are gained and lost by those compartments, and of relevant neural and hormonal signals. Design of experiments in fluid intake requires a working knowledge of these homeostatic principles. Thirst is caused by multiple factors, so the choice of specific stimulus to use in an experiment is of great theoretical importance.

INTRACELLULAR AND EXTRACELLULAR FLUID COMPARTMENTS

Fluids compose about 69% of a rat's total body weight. Approximately two-thirds of these body fluids are contained inside cells (*intracellular*), and one-third, outside cells (*extracellular*). Extracellular fluid is distributed between vascular (blood plasma) and interstitial (tissue) subcompartments in about a 1:3 ratio (Fig. 19–1). Solutes dissolved in these fluids give rise to osmotic pressure; net flow of water across cell walls is driven by associated differences in osmotic pressure. Under conditions of perfect balance (*euhydration*), intracellular and extracellular compartments have the

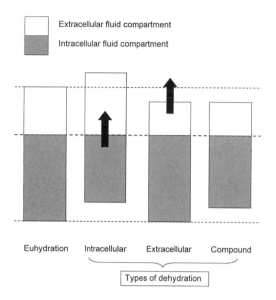

Figure 19–1. Schematic representation of the relative sizes of intracellular and extracellular body fluid compartments in states of hydromineral balance (euhydration) and three types of dehydration. For simplicity, vascular and interstitial extracellular fluids ($\approx$1:3 ratio) are not distinguished. *Arrows* indicate the initial net movement or loss of fluid. Deprivation of fluid produces compound dehydration, which involves loss of extracellular fluid and movement of cellular water to the extracellular space.

same osmotic pressure (about 290 milliosmoles per liter, or isotonic) and there is no net water movement between compartments. However, the solutes producing osmolar load (osmolytes) differ substantially between the intra- and extracellular compartments, as shown in Table 19–1. For this discussion, the main extracellular solute is sodium chloride. An isotonic solution of NaCl is thus approximately 0.15 molar (M).

Table 19–1. Body Fluid Compartments and Constituents

Property	Intracellular	Extracellular
Volume (% body wt)	46	23*
Na^+ (mEq/L)	12	145
K^+ (mEq/L)	150	4
Ca^{2+} (mEq/L)	0.001	5
Cl^- (mEq/L)	5	105
HCO_3^- (mEq/L)	12	25
Phosphates (P_i, mEq/L)	100	2

*ECF is $\approx$75% interstitial fluid and $\approx$25% plasma.

INTRACELLULAR DEHYDRATION THIRST

Intracellular dehydration occurs when the concentration of extracellular solute is increased above isotonic (a hypertonic condition) and water is then pulled from inside cells until the osmotic pressure is again equalized on the two sides of the cell wall (Fig. 19–1). This causes physical shrinkage of cells; cells called *osmoreceptors* have stretch receptors that transduce stretch into biological signals. Both peripheral (e.g., gut, liver) and central (e.g., forebrain) osmoreceptors play roles in hydromineral balance in rats. The principal way to produce intracellular dehydration is through the administration of hypertonic NaCl, because the extra sodium ions are largely trapped outside the cells. The administration of impermeable hypertonic solutions, like NaCl, produces water intake in proportion to the cell shrinkage; comparable hypertonic solutions of permeable solutes (e.g., glucose, urea) do not cause drinking. Signals from cell shrinkage are integrated in the brain to produce the state of thirst that in turn motivates water-seeking behavior. The term *osmotic thirst* is commonly used, but *intracellular dehydration thirst* is accurate.

EXTRACELLULAR DEHYDRATION THIRST

Extracellular dehydration occurs when isotonic extracellular fluid is lost without change in osmotic pressure (*hypovolemia*); there is no net fluid movement across cell membranes. The vascular and interstitial (tissue) extracellular compartments are in rapid exchange equilibrium; hypovolemia thus reduces blood volume. Serious losses compromise the delivery of adequate blood to tissues and can rapidly become life threatening. Low-pressure (venous) vessels of the circulatory system have elastic walls that allow for these vessels to change diameter in response to increased or decreased blood volume. Stretch receptors or mechanoreceptors in the walls of these vessels transduce this volume status into a neu-

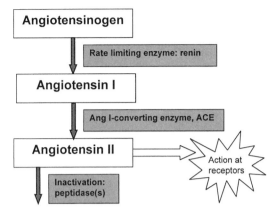

Figure 19–2. Major components of renin-angiotensin systems. In the circulation, renin from the kidney is the rate-limiting step in synthesis of the decapeptide angiotensin I, which is very rapidly cleaved to the principal biologically active form, angiotensin II (an octapeptide). Angiotensin I–converting enzyme (ACE) inhibitors (e.g., captopril) slow this cleavage. Angiotensin II activates specific receptors in many locations, including in select brain regions such as the subfornical organ.

ronal signal that generates local (reflexive) and central (e.g., thirst) responses. Reduced blood pressure also causes release of renin from the kidney into the circulation (Fig. 19–2); renin then catalyzes the synthesis of angiotensin II (Ang II), a peptide whose circulating concentration is related to hypovolemia (Fitzsimons, 1998). Rats exhibit extracellular dehydration thirst and drinking (Stricker 1968); this is also known as *volumetric thirst*.

COMPOUND DEHYDRATION

Many naturally occurring situations that cause physiological dehydration are not purely intracellular or purely extracellular but are instead a mixture (see Fig. 19–1). Experimentally imposed water deprivation, whether acute or on a daily schedule, is such a compound stimulus. The principal stimulus of thirst during deprivation of water is the amount and type of food consumed during that time. Rats that are deprived of both food and water have minimal fluid needs. The usual laboratory food is commercial rat chow, which has a relatively high (approximately 0.5%) content of NaCl. When

food is consumed, there are transient changes as fluid is secreted into the gastrointestinal tract, intermediate changes as the solutes are absorbed and cause intracellular dehydration, and late changes as the NaCl and other waste products from the food are excreted in urine causing hypovolemia. Additionally, as duration of deprivation increases, physiological anorexia occurs (Watts, 2000), and so the further intake of solutes is slowed. Under many conditions, intracellular and extracellular signals combine to produce an integrated thirst signal (see Fig. 19–3), observations that form the basis of a dual-depletion model of thirst (Rowland, 2002).

HORMONAL SIGNALS

Both intracellular and extracellular dehydration stimulate the release of vasopressin from nerve terminals in the posterior pituitary gland. These are the terminals of neurosecretory magnocellular neurons in the supraoptic and paraventricular nuclei of the hypothalamus. These cells have osmoreceptor properties and have afferents from peripheral osmoreceptor and baroreceptor elements. The relevant receptors for circulating vasopressin are of the V1 subtype in the kidney, and their activation causes the retention of water. This is a primary reason that hypovolemic rats excrete little or no urine (anuria). However, in the case of osmotic sodium loads, this is counteracted by atrial natriuretic factor (secreted from the atrium of the heart), which causes the excess NaCl load to be excreted in urine (natriuresis) and coincidental loss of some water because there is an upper limit on the concentrating capacity of rat kidney. Thus, after the injection of NaCl, osmolality first increases and then, even if water intake is not allowed, decreases as natriuresis occurs. This loss of fluid leaves a hypovolemic condition, so if a delay is imposed between injection of NaCl and access to water, the dehydration is complex. In fact, the amount of drinking that occurs is less than theoretically needed to dilute the salt load to isotonicity, except if uri-

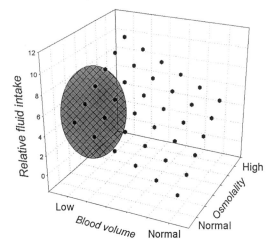

Figure 19–3. Schematic of fluid intake as a function of increasing plasma osmolality and decreasing blood volume. The expected intake is approximately the algebraic sum of the intake produced by each component alone. The shaded area indicates the region in which sodium appetite is observed.

nary excretion is prevented (e.g., by nephrectomy, surgical removal of the kidneys).

The peptide hormone Ang II also is made from renin that is released during hypovolemia (Fig. 19–2). Ang II has many effects, including increasing blood pressure through direct actions on receptors in blood vessels, but also gives rise to a central signal via Ang II type 1 (AT1) receptors in the subfornical organ, a brain circumventricular organ that is accessible to circulating Ang II. Although Ang II alone can stimulate thirst, it also causes release of the sodium-retaining hormone aldosterone from the adrenal cortex, which, together with Ang II, produces sodium appetite.

DOCUMENTATION OF PHYSIOLOGICAL EFFECT

Simple blood assays are often advisable to document the efficacy of treatments. Small volumes can be taken from a tail nick (a local anesthetic agent may be necessary) into a capillary tube. These samples may then be centrifuged for measurement of hematocrit ratio (packed cell volume), and plasma can be removed for measurement of protein concentration (a hand refractometer is simple to use), sodium concentration (flame photometer or ion electrode), and/or osmolality (freezing point depression). Each of these requires only small volumes of plasma. Plasma sodium concentration and osmolality are indices of osmotic imbalance. Fractional increases in protein and hematocit are approximations of volume depletion. Plasma hormones such as aldosterone and vasopressin can be measured using radioimmunoassay, as can renin activity, which, because renin is the rate-limiting step in Ang II synthesis, is correlated with Ang II concentration. Urinary volume and sodium and potassium concentrations in urine are useful measures; for these, metabolic cages or stands are needed that allow separation of urine from feces and/or spilled food.

SPECIFIC PROCEDURES AND STIMULI TO INDUCE WATER INTAKE IN RATS

INTRACELLULAR DEHYDRATION

Administration of hypertonic NaCl reliably causes drinking, often within a few minutes. The amount of water that would need to be consumed to dilute the salt load to isotonicity is equal to the solute load (in milliosmoles) divided by the initial plasma osmolality (in milliosmoles per liter). Thus, an injection of 1 ml of 1 mol/L NaCl would require an intake of 5 to 6 ml of water to dilute to isotonicity. However, as noted earlier, observed intake is typically less than 50% of this amount because of concurrent sodium excretion. Intake is related to the dose of NaCl, and the threshold rise in plasma osmolality for drinking to occur is approximately 2%. To study thresholds, the best method for the administration of hypertonic NaCl is through an indwelling catheter in the general circulation, such as the jugular vein (Fitzsimons, 1963). Although this method requires surgery, it allows remote and painless infusions into freely moving rats. A catheter may also be placed into the hepatic portal vein (which drains the intestinal field into the liver and so is the natural route of entry of ingested solutes), and this method has been used to study the contribution of hepatic osmoreceptors to thirst. An elegant dual-catheter method, in which NaCl is infused into the portal vein at the same time that water is infused into the general circulation, allows the experimenter to effectively restrict the stimulus to the liver (Morita et al., 1997). Infusion procedures usually are performed in short (e.g., 1 to 2 hours long) sessions.

More conveniently, acute intraperitoneal or subcutaneous injections of hypertonic NaCl induce reliable drinking in rats, although such injections appear to be temporarily painful. Two procedures can be done to minimize the potential distress, which can interfere with the expression of drinking behaviors. First, the rats should be handled and accustomed to injection procedures. Second, unless it will interfere with the experimental goal, a small amount of local anesthetic (e.g., bupivacaine, lidocaine) may be added to the injection. Under ideal conditions, rats start to drink within 10 to 20 minutes of injection and drinking is complete within 60 to 90 minutes.

A chronic version, producing a sustained osmotic load, can be simply realized by adding NaCl to the food: for example, adding 3% NaCl to powdered food produces an approximately 50% increase in daily water intake without significant anorexia. Higher concentrations of NaCl are tolerated but may be associated with reduced food intake. This mode of administering the stimulus will preferentially stimulate visceral osmoreceptors (Stricker et al., 2003; see also "Meal-Associated Drinking").

EXTRACELLULAR DEHYDRATION DRINKING

The most direct way of reducing blood volume is hemorrhage; indeed, wounded people with high blood loss often experience intense thirst. In the laboratory, this should be facilitated using an indwelling vascular catheter. However, this method removes critical blood constituents, leaving a weakened animal, so it is rarely used to study drinking.

The method of choice to produce extracellular dehydration is injection of the colloid polyethylene glycol (PEG). This sequesters isotonic filtrate of plasma at the injection site, visible as an edema, for several hours. PEG is best if administered subcutaneously, in the loose skin of the scapular region, as a solution in water or isotonic saline. High formula weight PEG ($>20,000$) should be used, typically in 20% or 30% solution (weight/volume) and at a dose of 1% to 2% body weight (Stricker, 1968). Solutions should be warmed to body temperature to dissolve and for injection. These solutions are very viscous, so a large-diameter needle is needed for injection. Unlike hypertonic saline, injection of PEG is not painful, but to obtain the best edema the bolus should be palpated gently to spread from the injection site. Brief gas anesthesia may be used for this, but it is not necessary in well-handled rats.

The edema, and consequent hypovolemia, takes 1 to 2 hours to develop fully, and the onset of thirst is correspondingly slow. The hypovolemia is accompanied by anuria, so any water consumed dilutes the extracellular fluid (hyponatremia), which is an inhibitory signal for drinking (Stricker, 1969). Thus, water intake in this model is self-limiting because of dilutional hyponatremia. Typically, water intake starts 2 to 3 hours after injection; salt intake is delayed. The best control for this procedure is a sham injection rather than saline (which constitutes a volume expansion).

Hypovolemia also may be produced by natriuretic agents such as furosemide. Functionally, these cause the kidney to lose a relatively large amount of near-isotonic urine, and the attendant loss of sodium causes hypovolemia. Restoration of extracellular volume requires the intake of both water and NaCl, and so the relationship of water to salt intake is of theoretical importance. Although diuretics can cause thirst, they are more often used to stimulate sodium appetite (see "Appetite").

As discussed earlier, Ang II is a dipsogen (Fitzsimons, 1998). Because it has a short biological half-life, the best route by which to administer Ang II is intravenous infusion. The threshold dose for drinking under normal laboratory conditions is approximately 100 ng/kg body weight per minute and a drinking latency of about 10 minutes. Procedurally simpler, but not useful for questions about thresholds, subcutaneous bolus administration at doses of greater than 50 μg/kg also causes a robust drinking response. Other putative dipsogenic substances can be screened using these procedures.

Ang II is also dipsogenic when administered either acutely or chronically into the brain. For such injections, a cannula must be surgically implanted into either into the cerebral ventricles or specific brain regions. Drinking often occurs within a few seconds of injection.

WATER DEPRIVATION

As noted earlier, the primary cause of physiological dehydration during water deprivation is the concurrent intake of food. Thus, for studies involving water deprivation, it is advisable to record concurrent food intake because differences (say, between experimental groups) could produce a change in drinking that is only secondary to changes in food intake. Some studies may call for more rigorous measurement of intakes and urinary excretion using metabolic cages. Water deprivation in excess of 24 hours is not normally approved by Institutional Animal Care and Use Committees (IACUCs).

MEAL-ASSOCIATED DRINKING

Under conditions of continuous access to food and water, rats take discrete meals (usually about 10 per day) that are either interrupted or followed immediately by episodes of drinking. In fact, about 80% of spontaneous water intake occurs in this prandial manner. Direct measurement of prandial drinking requires continuous recording of food and water intake, either qualitatively using sensors of when the commodities are accessed (e.g., lickometers, photobeams) or quantitatively using weighing devices. A different aspect of meal-associated drinking can be measured more simply by forcing a single meal, such as by prior food restriction, and then measuring volumetrically the water intake that occurs with that meal. In this case, the water-to-food ratio (given in, for example, milliliters per gram) is a useful derivative measure.

THE TEST ENVIRONMENT

Not only are the choice and mode of administration of the dipsogenic stimulus important, but so is the test environment. Rats are naturally apprehensive of novel environments and commodities. They should be adapted to the cage in which drinking will occur and to the fluid(s) to be presented.

THE ENVIRONMENT

The choice of environment is usually between either home cage or a test arena. If using the latter, the rats should be allowed at least one (preferably more) previous drinking episode in that environment. If a behavior more complex than reflexive licking (e.g., a lever press operant) is required, more training is needed. A second important variable is time of day: rats naturally do most of their drinking at night, when certain thresholds may be functionally lower. However, the intake of food (also mostly nocturnal) may have an unwanted or uncontrolled effect on drinking. For this reason, most short-term drinking studies are performed either during the light period when food intake is infrequent or by removing food 1 to 2 hours before the drinking test. A third variable is the temperature of both fluids and environment. Most animals are housed and tested at temperatures that are comfortable to humans, but if rats are moved to a separate room for testing, it should be at the same temperature, and the drinking fluid equilibrated at that temperature. A fourth variable is social factors. Most studies examine intake of individual rats; either direct or indirect interaction with conspecifics could influence drinking and under normal conditions should be avoided. Interaction with humans is an unavoidable aspect of most drinking studies, ranging from direct handling such as giving injections or placing in a test cage to indirect influences such as other human activity in the room. Rats are extremely adaptable to a range of conditions, and their drinking behavior is usually robust, but a consistent routine is essential.

The approved standard for rat housing has changed over the past decade from stainless steel mesh to plastic tubs with soft bedding. These latter may better reproduce natural burrow material, although they do not typically allow these nocturnal animals any type of shade. Drinking studies in these cages normally require sipper spouts that protrude through the metal grill top of the cage (this avoids the spout from making contact with the bedding and leaking fluid). Steel mesh cages allow the spout to protrude through the wall of the cage. Fluid intake studies also are linked to measurement of urinary output, and these have to be performed in metabolic cages with mesh floors. As noted earlier, rats should be adapted to whichever caging is chosen; if that is considered "nonstandard," then approval for the exception should be requested from the IACUC.

THE SOLUTION AND ITS PRESENTATION

In the study of water intake, one decision that should be made is between using either deionized or tap water. This decision depends in part on the experimental question and the quality of tap water in a particular laboratory. In general, if the tap water is known to have high or variable mineral content and/or has an odor to humans, it is advisable not to use it. The rats should be adapted to the water chosen for several days before the experiment. If solutes (e.g., NaCl) are to be added, the same choice of solvent applies. If in doubt, use either deionized/distilled water or, if that is not readily available, commercially available bottled water. Added solutes should be of the highest purity available.

If rats are purchased from a commercial vendor, they are most likely accustomed to drinking from the nipple of an automatic water system, and this may also be used in your vivarium. However, such systems are unsuitable for measuring water intake, so most laboratory experiments use water bottles with metal sipper spouts to which rats must be accustomed. These can be purchased commercially. Spouts differ considerably in their characteristics, and one of the major and often overlooked sources of variance within an experiment comes from spout topography. Rats lick in bouts separated by pauses; the licking rate within a bout is about seven licks per second. Thus, the amount of fluid consumed per unit time depends both on the volume per lick and on the pause duration(s). The former depends on the diameter at the orifice of the spout and on whether a ball bearing is present. It is therefore best to use only one type of spout in your laboratory: in that way, every animal has the same type of spout and there is no possibility of day-to-day variance. Also, if more than one fluid is offered, both spouts will be identical. If the spouts are used with rubber stoppers or washers, as is usually the case, ensure that the shaft is pushed all the way through the stopper to avoid the formation of airlocks.

In the real world, rats lick water from puddles or other open surfaces. Richter tubes are drinking cylinders that have horizontal surface drinking troughs. Glass models can be purchased commercially, but they are generally more expensive and harder to clean than tubes with spouts and are used relatively little. One exception is the use of fluid dippers, which are small (e.g., 0.1 ml) troughs or cups, to deliver standard volume fluid reinforcements in operant procedures; rats readily adapt to drinking from them.

TEST DURATION

Most adult rats with standard food available drink 30 to 50 ml/day. Thus, the measurement of 24-hour intakes to the nearest ± 1 ml usually is adequate. I use 50 or 100 ml plastic graduated cylinders (with the top lip cut off by using a hacksaw) with rubber one-hole stoppers and metal sipper spouts. The start and finish volume graduations are read directly. An alternative, for which graduated tubes are not necessary, is to weigh the tube or bottle before and after. This has the advantage that most electronic scales allow the weights to be sent directly to a computer spreadsheet.

Intakes of about 10 ml or less, such as stimulated by acute dipsogens, need to be recorded with greater accuracy (± 0.1 ml). Although gravimetric measurement (as earlier) is viable, the likelihood of losing a few drops of fluid when the bottle is placed on or removed from the cage is quite high. For this reason, I recommend direct volumetric measurement while the tube is on the cage. For such measurements, I use either 25 ml graduated plastic or glass pipets with each end cut off so that a sipper spout (in a collar of plastic tubing) can be firmly wedged in one end and a small rubber stopper in the other.

Volume consumed is the principal dependent variable in many studies of fluid intake, but the pattern of intake is also of importance in some applications. Several computer-linked lick sensors are available commercially for this purpose. One design has a small infrared beam across the tip of the spout, which itself is slightly recessed so that

normally the protruded tongue will break it. Another design is a contact sensor in which the rat completes an electric circuit when it licks from the spout. The currents involved are too small to be appreciable to the rat. These may be used in acute drinking studies, including examination of the effect of tastants on behavior, but in conjunction with sensors of food intake, they can be used to study temporal relationships between food and fluid intakes over long periods.

SODIUM PREFERENCE AND APPETITE

PREFERENCE

It is conceptually important to separate preference from appetite. Preference, in this case for sodium solutions, is exhibited under conditions of sodium balance (need free) and is determined by comparing intake of sodium salt with a reference solution (e.g., water). This may be conducted either in separate but otherwise identical sessions called one-bottle tests or in sessions with both fluids available simultaneously, called two-bottle tests. Long-duration (e.g., 24 hours) preference tests may be affected by the rat learning about postingestional consequences of a particular flavor. Short-duration preference tests can overcome this pitfall but usually require fluid deprivation to induce the intake, so the possible interaction of need state with preference must be considered in the experimental design. More complex designs use more than two bottles or choices of fluid; however, the number of available options may influence choice behavior. Many rat strains show a spontaneous preference for NaCl over water in the range of approximately 0.05 to 0.2 mol/L.

APPETITE

Appetite is defined as intake in excess of that during normal or based conditions, and that has motivational characteristics. The most common way of studying sodium appetite in rats is to offer a hypertonic NaCl solution (0.3 to 0.5 mol/L) that is above the spontaneously preferred range. Studies on the taste specificity have shown that sodium appetite produced by the methods to be discussed is specific for the cation sodium but is insensitive to changes in the associated anion. Mineral-deficient mammals most likely do not encounter salt solutions in their natural habitat: rather, they obtain their minerals from food or mineral-enriched soil deposits. Curiously, very few studies have successfully found sodium appetite for salty foods in rats. However, in recent work in our laboratory, we found that salt gels work remarkably well. These are made by mixing concentrated salt solutions (we have used 0.5 to 1.5 mol/L) with gelatin powder (5% w/v) and then allow them to solidify in glass jars for presentation. Rats ingest minimal amounts of 1 mol/L or above under need-free conditions but exhibit a robust intake during sodium depletion.

PRODUCING SODIUM DEPLETION

The acute injection of a rapid-acting loop diuretic such as furosemide (also called frusemide) causes a dose-related loss of sodium and water in urine and hypovolemia. Subcutaneous single injection of 2 mg/kg or more causes a near-maximal loss of about 2 mEq sodium in an adult rat within 1 to 2 hours. This hypovolemia is associated with a sodium appetite (see Fig. 19–2), but this develops relatively slowly over the next 12 to 24 hours. Thus, one of the most used protocols involves injection of furosemide followed by a 24 hour period without available sodium. To accomplish this reliably, a fresh cage or bedding should be provided, along with distilled water and a low- or no-sodium diet. At the end of this time, rats will ingest several milliliters of hypertonic NaCl, often in substantial excess of their 2 mEq deficit.

A chronic version of this protocol may also be used. Either daily injections of furosemide or the addition of the diuretic hydrochlorothiazide to the low-sodium food produces a robust, sustained sodium appetite (Rowland and Colbert, 2003).

OTHER STIMULI OF SODIUM APPETITE

Several other procedures will produce sodium appetite; most involve the use of Ang II and/or aldosterone (Fregly and Rowland, 1985). Under natural conditions, these hormones probably work in synergy to produce sodium appetite. However, activation of either hormonal system alone is sufficient. Adrenalectomy, which removes the endogenous source of aldosterone, produces a high-Ang sodium appetite. Conversely, the administration of high doses of deoxycorticosterone, a mineralocorticoid hormone, produces sodium appetite with concurrent suppression of Ang II formation. Thus, just as the study of thirst involves a choice between stimuli of discrete component systems, a similar situation applies to sodium appetite.

MOTIVATION AND THE STRUCTURE OF SODIUM APPETITE

Several authors have made the point that sodium appetite is both innate and motivated. In regard to the latter, it has been shown that rats given some of the above stimuli will perform operant tasks in discrete sessions to obtain sodium solutions (Quartermain et al., 1967). Under free access conditions, rats showing an appetite for concentrated NaCl solution consume it in discrete bouts in close temporal proximity to spontaneous meals and prandial water (Stricker et al., 1992). Recently, using a combination of operant and free access protocols and standard rat operant cages, we have found that rats temporally structure sodium "meals" in accordance with relative cost and need. Thus, although rats will take water and salt together, this is not physiologically obligatory.

REFERENCES

Fitzsimons JT (1963) The effects of slow infusions of hypertonic solutions on drinking and drinking thresholds in rats. Journal of Physiology 167:344–354.

Fitzsimons JT (1979) The physiology of thirst and sodium appetite. Monographs of the Physiological Society #35, Cambridge University Press.

Fitzsimons JT (1998) Angiotensin, thirst, and sodium appetite. Physiology Review 78:583–686.

Fregly MJ and Rowland NE (1985) Role of renin-angiotensin-aldosterone system in NaCl appetite of rats. American Journal of Physiology Regulatory, Integrative, and Comparative Physiology 248:R1–R11.

Marwine A and Collier G (1979). The rat at the waterhole. Journal of Comparative Physiology and Psychology 93:391–402.

Morita H, Yamashita Y, Nishida Y, Tokuda M, Hatase O, Hosomi H (1997). Fos induction in rat brain neurons after stimulation of the hepatoportal Na-sensitive mechanism. American Journal of Physiology Regulatory, Integrative, and Comparative Physiology 272:R913–R923.

Quartermain D, Miller NE, Wolf G (1967) Role of experience in relationship between sodium deficiency and rate of bar pressing for salt. Journal of Comparative Physiology and Psychology 63:417–420.

Rowland NE (1990) On the waterfront: Predictive and reactive regulatory descriptions of thirst and sodium appetite. Physiology and Behavior 48:899–903.

Rowland NE (2002) Thirst and sodium appetite. In: Stevens' handbook of experimental psychology, 3rd edition, vol. 3: Learning, motivation and emotion (Pashler H and Gallistel CR, eds.), pp. 669–707. New York: Wiley.

Rowland NE and Colbert CL (2003). Sodium appetite induced in rats by chronic administration of a thiazide diuretic. Physiology and Behavior 79:613–619.

Stricker EM (1968) Some physiological and motivational properties of the hypovolemic stimulus for thirst. Physiology and Behavior 3:379–385.

Stricker EM (1969) Osmoregulation and volume regulation in rats: Inhibition of hypovolemic thirst by water. American Journal of Physiology 217:98–105.

Stricker EM, Gannon KS, Smith JC (1992) Salt appetite induced by DOCA treatment or adrenalectomy in rats: Analysis of ingestive behavior. Physiology and Behavior 52:793–802.

Stricker EM, Hoffmann ML, Riccardi CJ, Smith JC (2003) Increased water intake by rats maintained on high NaCl diet: Analysis of ingestive behavior. Physiology and Behavior 79:621–631.

Watts AG (2000) Understanding the neural control of ingestive behaviors: Helping to separate cause from effect with dehydration-associated anorexia. Hormones and Behavior 37:261–283.

Foraging

20

IAN Q. WHISHAW

Rats forage on a wide variety of foodstuffs. Some food is eaten where it is found and some is carried to havens, where it is eaten or left for later (Lore and Flannelly, 1978; Takahashi and Lore, 1980; Whishaw and Whishaw, 1996). With the exception of lactating females, rats do not cache (hoard) food. Food carrying may help rats avoid food theft, avoid predation, and redistribute food stores for the benefit of colony members. The susceptibility of a foraging rat to theft by conspecifics is documented by observations made of wild and semi-wild colonies (Barnett and Spencer, 1951). Chitty (1954) observes large rats catch and overturn smaller rats to steal grain from their mouths, and Whishaw and Whishaw (1996) observe that large "dominant" rats are less likely to carry food or be attacked by conspecifics than are smaller rats.

Optimal foraging theory proposes that foraging behavior represents a tradeoff between strategies for obtaining food and strategies for avoiding attack and predation. Rules governing eating, food theft and protection, and food carrying illustrate innovative ways that the behavior of the rat has been sculptured by optimizing principles. The following sections describe eating behavior, food protection and theft, food carrying, and some aspects of their neural control.

EATING TIME

A rat can optimize food acquisition by eating quickly. Rats vary their eating speed in response to exposure, time of day, food deprivation, and previous previous deprivation history. They display individual differences in eating speed (Whishaw et al., 1992). Increasing eating speed, however, can have digestive costs through reduced chewing and saliva wetting (Morse, 1985).

The largest influence on eating behavior is in the time required to eat. Obviously a rat will eat a small piece of food more quickly than a large piece of food, but many other factors influence eating speed. The time taken to eat a piece of food of a given size is influenced by location. In the open, such as on an open table or in a cage without a cover, eating is more rapid than it is in a shelter. In the open, rats also make many head scans, during which they continue chewing, whereas head scans are rare in sheltered environments. Eating is faster in a novel location than in a familiar location. Eating speed also varies as a function of the amount of food eaten. As the number of food pellets eaten increases, so does the average eating time per pellet. Thus, eating speeds and head scanning behavior suggest that eating rats are vigilant and sensitive to the possibility of predation or attack as well as to nutrient need.

Time of day and personal history affect eating speed (Fig. 20–1). Rats eat more quickly during the dark phase of their 24 hour day–night cycle and eat more quickly if they have been, or are, food deprived. Lighting during the day–night cycle also influences eating speed, as rats eat more slowly when the lights are off, especially in the light portion of the cycle. Finally, some rats eat very quickly and others eat more slowly independent of

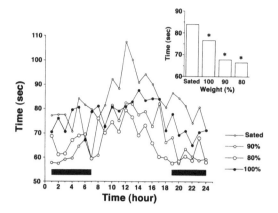

Figure 20–1. Mean time to eat a 1 g food pellet as a function of time of day and food deprivation schedule. *Inset:* Mean eating time as a function of deprivation schedule. *Black bars,* lights off. (Based on Whishaw et al., 1992.)

their deprivation level and previous feeding history. The individual differences could be related to nursing success during infancy or to prenatal and genetic influences.

Rats retain a retrospective knowledge of their feeding time (Whishaw and Gorny, 1991). After eating a food pellet, a rat scans its surroundings by sniffing and running its vibrissae over the ground. The size of the area that it scans is proportional to the size of the food pellet that it has just eaten, with larger scans following consumption of larger food pellets. If food hardness is varied independent of food size, the area that is scanned is best predicted by the time required to eat a food item. Although it is likely that scans are directed toward finding dropped crumbs, rats that are tested on mesh, through which crumbs fall, still scan.

FOOD WRENCHING AND DODGING

The immobile feeding posture of the rat, in which it sits on its haunches with the food held in its paws, leaves it vulnerable to attack from other rats. This vulnerability is amplified when it is considered that other rats are excited by a feeding rat, investigate it, pick up bits of food that it has dropped, sniff its snout, and lick crumbs from its lips (Barnett and Spencer, 1951;

Galef, 1983; Galef and Wigmore, 1983; Posadas-Andrews and Roper, 1983).

Rats are artists in food theft and theft avoidance (Whishaw, 1988; Whishaw and Tomie, 1987). A rat attempting to steal food approaches a feeding rat from the rear, walks along the side of the victim, and reaches under its snout to wrest the food from its paws (Fig. 20–2). It may grasp the victim's paw so that it can expose the food and/or knock it

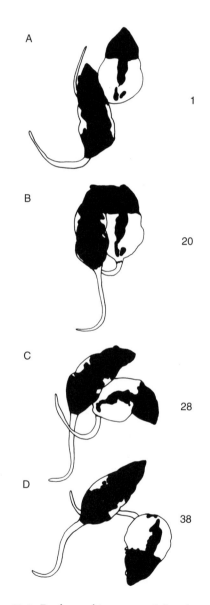

Figure 20–2. Food wrenching attempt (*left*) and a successful dodge (*right*). (Based on Whishaw and Tomie, 1987.)

free. A victim evades the robber by dodging (Whishaw and Tomie, 1987; Whishaw, 1988). A dodge consists of a turn of the head followed by steps with the hindlimbs that turn the rat away from the robber. The maneuver leaves the victim time to continue eating.

This description of an average dodge should not belie the variations in the movement. A rat might run forward, dodge backward over an approaching rat, or simply twist its head and pivot. In some dodges, the food is held in both front paws so that the rat can easily continue eating, but the food may be held in one front paw while the other assists turning, or the food may be transferred to the mouth so that the rat can use both front paws to increase the size of the dodge. The food item is transferred back to the paws for eating at the completion of the dodge. Dodges may be assisted with a hop, end with a hop, or be attached to short runs.

There is seldom overt aggression between the robber and victim. The robbing–dodging interaction can occur repeatedly, as long as one member of the pair has food to eat. If a robber wrenches the food away from a victim, it will in turn dodge, and the former victim will replace it as the robber. Experience in robber–victim interactions contributes both to the effectiveness of the victim's dodges and to the skill and aggression of the robber.

EATING TIME INFLUENCES DODGE SIZE

The victim of a theft is conservative in the movements made to avoid the robber (Whishaw and Gorny, 1994). If a rat has a small food pellet (20 to 94 mg), it picks it up by mouth and quickly chews and swallows it, and so there is no possibility of theft. If a food pellet is larger (>190 g), it is held in the paws as the rat adopts a sitting posture to eat, thus allowing the robber an opportunity to steal. Dodge distance and dodge angle in response to an attempted theft increase with increases

in the size of food. A rat with a small food pellet may avert its head; with a larger food pellet, it may make a partial turn, and with a very large pellet, it may turn completely (about 180 degrees) and sometimes run a short distance. The victim also tracks the size of the food that it is eating and adjusts its dodges to the amount food it has yet to eat. When beginning to eat a large piece of food (e.g., 1 g pellet), the victim makes a maximum sized dodge and as size of the food diminishes, the size of dodge diminishes, and eventually as the food becomes smaller, small dodges are replaced by a head turn or no movement at all.

If dodging were simply a function of the size of the food being eaten, a rat might have difficulty gauging how to protect a large, easy-to-eat piece of food relative to a small, hard-to-eat piece of food. Rats solve this problem by calibrating food size in terms of anticipated eating time (Whishaw and Gorny, 1994). When rats are given food pellets baked to different hardness, victims make larger dodges with the harder food. To further determine the role of food size and eating time in determining dodge size, we gave rats a comparison series of foods and a test series of foods. The comparison series consisted of 10 different-sized, round, commercial food pellets weighing between 20 mg and 1000 mg. The test series of foods included barley, wheat, Mung beans, and Azuki beans. In terms of size, the test series was comparable to the low end of comparison series, and in terms of time required to eat, the test series fell in the upper half of the comparison series. When dodge probability and dodge size are measured as a function of the different foods that a victim was eating, these measures were closely related to the time required to eat the food item and not to the size of the food item.

Eating time is arguably the easiest way to gauge susceptibility to food loss given the wide range of food items that a rat might consume. Of course, a rat must learn how long it will take to eat various kinds of food. When we gave a grain of rice, which is about the size of a 20 mg food pellet, to rats, they did not dodge on the first approach of a robber but did on the second approach. Thus, one eating experience is sufficient for the rat to learn that a small piece of hard food takes much longer to eat than an equal-sized soft piece of food.

SEX DIFFERENCES IN DODGING

The dodging movements made by male and female rats are different (Field et al., 1996, 1997a). Female rats move their snout through a greater spatial curvature, and the snout achieves a greater velocity, relative to the pelvis, than occurs for males. Stepping movements of females are also simpler. They step away first with the contraversive, to the robber, hind limb and then with the ipsiversive hind limb. Males first step with ipsiversive, to the robber hind limb, a movement that brings the rear of the victim toward the rear of the robber. The victim then steps away from the robber with the contraversive hind limb. Thus, the male gives the robber "the hip" before pivoting its forequarters. There is some variation in the stepping movements of both sexes, but in general the female pivots around a point more posterior on the body than a male (Fig. 20–3).

Sex differences in dodging, in which the female's dodge propels it away from the robber while the male's dodge initially propels it into the robber, may reflect the more general differences in behavior of the sexes. Dodging

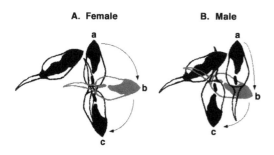

Figure 20–3. Difference in dodges by female and male rats: The female pivots around its pelvis, and the male pivots around the mid body. (Based on Field and Pellis, 1998.)

is exhibited by female rats during mating to correct ineffective approaches by a male rat (Whishaw and Kolb, 1985). The "give him the hip" behavior of the male rat is similar to the male rat's threat display (Pellis and Pellis, 1987). Thus, a "dodge" is akin to a word in a behavioral lexicon in that it can be used in different behavioral contexts while retaining the accent of the performer.

That dodging movements are sex related suggests that they are determined by hormonal influences. The hormonal influence must occur very early in development because castration of juvenile rats does not affect the male-typical patterns of movement. Neonatal castration does feminize the male's pattern of movement, whereas the early administration of testicular hormones to females masculinizes their movement (Field et al., 1997a, 1997b). Thus, it is likely that prenatal and perinatal hormonal influences affect not only the later reproductive roles of rats but also the movements that they make in the nonreproductive behavior of dodging.

FOOD CARRYING

Food carrying is an elaboration of dodging (Fig. 20–4). When food pellets of varying size are presented through a small aperture to a hungry rat in an alley, the rat produces a food size-related escalation in movement (Whishaw et al., 1990; Whishaw and Tomie, 1989). Small food pellets are swallowed as the rat grasps them with its mouth (*eat*). Intermediate-sized food pellets are transferred from the mouth to the paws and are eaten as the rat adopts a sitting posture (*sit*). As food size increases further, rats dodge farther away from the aperture (*dodge*) until, with large food pellets, they run to the far end of the alley with the food before eating (*carry*).

When given a choice of food sizes, rats are selective with respect to the size of food items that they choose. If a pile of variously sized food pellets is available, a rat first scans the pellets

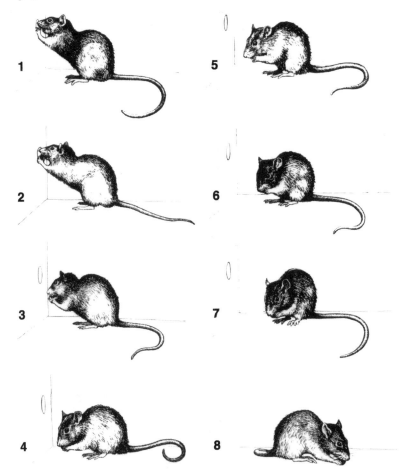

Figure 20–4. Relation between food size and movement. In the series 1 through 8, the rat receives successively larger food pellets; as it does so, it makes larger movements that lead it to eventually dodge away from the food source. (Based on Whishaw and Tomie, 1989.)

and then chooses the largest food pellets and carries them. In addition, it will attempt to carry more than one pellet of a cartable size. We have observed a rat stuffing three 1 g food pellets into its mouth while attempting to carry an additional food pellet in a forepaw, as it ran on three legs. Surprisingly, if only small food pellets are available, a rat will only eat and will not attempt to carry a number of them.

THE HOME BASE

If a refuge or covered area is available, it becomes the home base for a rat's foraging behavior, although the rat's behavior is still cal-ibrated by food size (Fig. 20–5). A rat will only leave the refuge after a "stop-sniff-look" investigation of its surroundings, and it leaves the refuge with a cautious walk, with the body held low. The eat, sit, and dodge behaviors still occur in the open if it finds smaller food pellets. If it finds a large piece of food, it gallops back to the refuge. The latency to initiate a carry response and the travel speed home increase with increases in the size of a food pellet. In addition, the latency to return after eating increases as a function of the size of a food pellet just eaten.

A refuge can be a source of potential problems (Whishaw, 1991). If another rat also uses the home base as a refuge, it might be

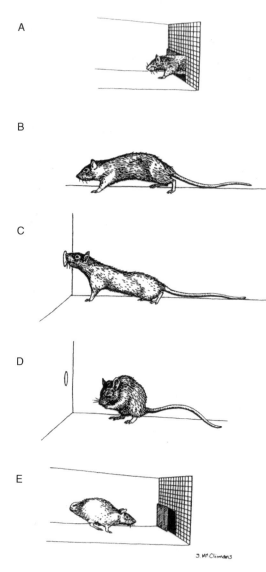

Figure 20–5. Behaviors displayed by rats in a food-carrying task. (A) Stop-sniff-look before leaving the refuge. (B) Cautious approach to food source. (C) Eat response in which a small food pellet is swallowed. (D) Sit response in which a rat sits on its haunches with an intermediate-sized piece of food held in the paws. (E) Carry response in which a rat turns back to the refuge with a large piece of food. (Based on Whishaw and Oddie, 1989.)

thought that a food-carrying rat will avoid the home base to reduce the possibility of food theft. They do not. Rats given three different refuges (covered boxes with an entrance) quickly adopt a preferred refuge. Placing a hungry rat in the preferred refuge, however, results in little change in preference. If the

preferred refuge is modified by removing its covers (black paper taped onto the Plexiglas boxes), the rats will quickly adopt a new refuge. Thus, the refuge appears to be a refuge from predation and not from conspecific competition. This finding seems consistent with the observations of wild rats (Whishaw and Whishaw, 1996). When provided with a large supply of peanuts, some of the rats carried peanuts back to their home territory only to have them repeatedly stolen by other rats. This confirms that the carrying rats fail to anticipate and avoid theft.

Rats may avoid carrying food to a home base if it is their nest area. Rats given access to a number of refuges quickly adopt one refuge as a "home" in which they sleep while they carry the food to a different refuge for eating.

EATING TIME INSTRUCTS FOOD CARRYING

A central prediction of optimal foraging theory is that a foraging animal should attempt to maximize its food intake while minimizing its exposure to attack/predation. An effective way of making optimal judgments is to use time as a measure of exposure and nutrient intake. Thus, rats should follow the rule: "Carry if eating time exceeds return trip time" (Whishaw, 1990). Using an alley with food access at one end and a refuge at the other end, we provided rats with 10 different-sized food pellets and a number of natural foods, including wheat, pearl barley, Mung beans, and Azuki beans. Three experiments were performed. In the first experiment, the 10 different-sized food pellets were either soft or hard (produced by increased baking). In the second experiment, the rats received soft food pellets and the natural foods. In both of these experiments, food carrying was predicted by eating time and not food size or hardness (Fig. 20–6).

In a third study, rats were given an adaptation experience, modeled on human psy-

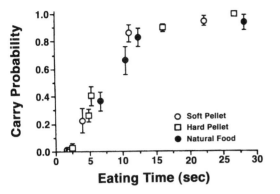

Figure 20–6. Food-carrying probability (mean and standard error) with different kinds of food plotted against eating time. Anticipated eating time rather than food size or hardness predicts the probability of food carrying. (Based on Whishaw, 1990.)

chophysical experiments. One group of rats received experience only with small pellets (sizes 1 to 7), and another group received experience only with large food pellets (sizes 4 to 10). After training, all rats were given a test with the full range of pellets. If the rats were influenced by context, intermediate-sized food pellets should be more likely to be carried by the group receiving the small comparison series and less likely to be carried by the group receiving the large comparison series. If the rats were responding to internal cues, such as, time, the contextual training should not affect food-carrying behavior. The results indicated that the rats were not influenced by context, which suggests that they were using eating time. Thus, these experiments demonstrated that rats did obey the rule: "Carry if eating time exceeds return trip time."

Optimal foraging theory also predicts that the distance to the refuge influences food carrying (Whishaw, 1993). If the distance is short, it will be worthwhile carrying small food pellets to the refuge, whereas if the distance is long, a larger piece of food should be required to initiate a return. When rats were allowed to forage over distances varying from a few centimeters to over 600 cm, variations in distance produced significant changes in food-carrying probability that were related to travel time. The probability that food was carried decreased linearly with travel distance.

Other manipulations of travel time also affected the probability of carrying food. If rats are required to walk across a short narrow beam, which increases travel time and introduces the risk of falling, they display a reduced tendency to carry food of a size that they would otherwise carry. If a frank risk is introduced, such as, the smell of a cat, the animals arrest all foraging behavior.

IMPLICATIONS FOR BRAIN FUNCTION

Because of its role in optimizing behavior in its use of time, space, and different motor acts, food handling provides a rich behavior for examining the nervous system function. Beginning with Wolfe's (1939) description of "hoarding" in laboratory rats, there has been interest in the neural control of food carrying (Mark, 1950; Munn, 1955; Ross et al., 1955). Food-carrying behavior has also been of interest in assays for neural injury (Whishaw and Tomie, 1988; Whishaw and Oddie, 1989; McNamara and Whishaw, 1990) and for applied problems such as modeling anxiety (Dringenberg et al, 1998; 2000).

Damage to limbic structures, including the medial frontal cortex, hippocampus, and nucleus accumbens, reduces hoarding. Interest in control exerted by these structures is heightened by two observations. First, context is important (Whishaw, 1993). Animals with hippocampal damage that have stopped food carrying may begin again if activated behaviorally (e.g., being startled while eating). They may also stop carrying food if their refuge is moved, but, again, carrying can be restored with training. If the distance between food and the refuge is increased, however, rats with hippocampal lesions will stop carrying at shorter distances than that needed to arrest carrying in control rats.

Early studies on food carrying focused on the influence of food deprivation on food-carrying behavior. Bindra's (1978) demonstration that both food deprived and non–food-deprived rats "hoard" suggests that not only will rats hoard food for nutrition, they also will hoard because of the food's incentive value. Whishaw and Kornelesen (1993) confirm this finding by dissociating food carrying and food hoarding with nucleus accumbens lesions. Both food-deprived control rats and rats with neurotoxic damage to the cells of the nucleus accumbens carried food from an open area to a refuge. As the rats consumed the food and became sated, the nucleus accumbens group stopped carrying food, whereas the control rats continued to carry. Thus, control rats responded to both the nutrient and incentive value of the food, whereas the nucleus accumbens group rats responded only to its nutrient value.

Taken together, the rich array of behaviors related to food handling in the rat provides a challenge to investigators of the neural control of behavior. In addition, eating behavior can be a useful measure in a wide range of studies, including those related to addiction, eating disorders, individual differences, and spatial and temporal behavior.

REFERENCES

Barnett SA and Spencer MM (1951) Feeding, social behaviour and interspecific competition in wild rats. Behavior 3:229–242.

Bindra D (1978) How adaptive behaviour is produced: a perceptual-motivational alternative to response reinforcement. Behavioural Brain Sciences 1:41–91.

Chitty D (1954) The control of rats and mice, Vols 1 and 2: Rats. Oxford: Clarendon Press.

Dringenberg HC, Wightman M, Beninger RJ. (2000) The effects of amphetamine and raclopride on food transport: Possible relation to defensive behavior in rats. Behavioral Pharmacology 11:447–454.

Dringenberg HC, Kornelsen RA, Pacelli R, Petersen K, Vanderwolf CH (1998) Effects of amygdaloid lesions, hippocampal lesions, and buspirone on black-white exploration and food carrying in rats. Behavioural Brain Research 96:161–172.

Field EF and Pellis SM (1998) Sex differences in the organization of behavior patterns: Endpoint measures do not tell the whole story. In: (Ellis L and Ebertz L, eds.). West Point, Conn: Praeger.

Field EF, Whishaw IQ, Pellis SM (1996) A kinematic analysis of evasive dodging movements used during food protection in the rat (Rattus norvegicus): Evidence for sex differences in movement. Journal of Comparative Psychology 119:298–306.

Field EF, Whishaw IQ, Pellis SM (1997a) Organization of sex-typical patterns of defense during food protection in the rat: The role of the opponent's sex. Aggressive Behavior 23:197–214.

Field EF, Whishaw IQ, Pellis SM (1997b) A kinematic analysis of sex-typical movement patterns used during evasive dodging to protect a food item: The role of testicular hormones. Behavioral Neuroscience 111:808–815.

Galef BG Jr (1983) Utilization by Norway rats (R. norvegicus) of multiple messages concerning distant foods. Journal of Comparative Psychology 97:364–371.

Galef BG Jr and Wigmore SW (1983) Transfer of information concerning distant foods: A laboratory investigation of the "information-center" hypothesis. Animal Behaviour 31:748–758.

Lore RK and Klannelly K (1978) Habit selection and burrow construction by wild Rattus norvegicus in a landfill. Journal of Comparative and Physiological Psychology 92:888–896.

Marx MH (1950) Stimulus-response analysis of hoarding habit in the rat. Psychological Review 57:80–94.

McNamara RK and Whishaw IQ (1990) Blockade of hoarding in rats by diazepam: an analysis of the anxiety and object value hypotheses of hoarding. Psychopharmacology 101:214–221.

Munn ML (1933) Handbook of psychological research on the rat. Boston: Houghton Mifflin.

Pellis SM and Pellis VC (1987) Play-fighting differs from serious attack in both target of attack and tactics of fighting in the laboratory rat Rattus norvegicus. Aggressive Behavior 13:227–242.

Posadas-Andrews A and Roper TJ (1983) Social transmission of food preferences in adult rats. Animal behavior 31:265–271.

Ross S, Smith WI, Wossner BL (1955) Hoarding: An analysis of experiments and trends. Journal of General Psychology 52:307–326.

Takahashi LK and Lore RK (1980) Foraging and food hoarding of wild Rattus norvegicus in an urban environment. Behavioral and Neural Biology 29:527–531.

Whishaw IQ (1988) Food wrenching and dodging: Use of action patterns for the analysis of sensorimotor and social behavior in the rat. Journal of Neuroscience Methods 24:169–178.

Whishaw IQ (1990) Time estimates contribute to food handling decisions by rats: Implications for neural control of hoarding. Psychobiology 18:460–466.

Whishaw IQ (1991) The defensive strategies of foraging rats: A review and synthesis. The Psychological Record 41:185–205.

Whishaw IQ (1993) Activation, travel distance, and environmental change influence food carrying in rats with hippocampal, medial thalamic and septal lesions: Implications for studies on hoarding and theories of hippocampal function. Hippocampus 3:373–385.

Whishaw IQ and Oddie SD (1989) Qualitative and quantitative analyses of hoarding in medial frontal cortex rats using a new behavioral paradigm. Behavioural Brain Research 33:255–256.

Whishaw, IQ, Oddie SD, McNamara RK, Harris TL, Perry BS (1990) Psychophysical methods for the study of sensory-motor behavior using a food-carrying (hoarding) task in rodents. Journal of Neuroscience Methods 32:123–133.

Whishaw IQ, Dringenberg HC, Comery TA (1992) Rats (Rattus norvegicus) modulate eating speed and vigilance to optimize food consumption: Effects of cover, circadian rhythm, food deprivation, and individual differences. Journal of Comparative Psychology 4:411–419.

Whishaw IQ and Gorny BP (1991) Postprandial scanning by the rat (Rattus norvegicus): The importance of eating time and an application of "warm-up" movements. Journal of Comparative Psychology 10:39–44.

Whishaw IQ and Gorny BP (1994) Food wrenching and dodging: Eating time estimates influence dodge probability and amplitude, Aggressive Behavior 20:35–47.

Whishaw IQ and Kornelsen RA (1993) Two types of motivation revealed by ibotenic acid nucleus accumbens lesions: Dissociation of food carrying and hoarding and the role of primary and incentive motivation. Behavioural Brain Research 55:283–295.

Whishaw IQ and Kolb B (1985) the mating movements of male decorticate rats: Evidence for subcortically generated movements by the male but regulation of approaches by the female. Behavioural Brain Research 17:171–191.

Whishaw IQ and Tome J (1987) Food wresting and dodging: Strategies used by rats (Rattus norvegicus) for obtaining and protecting food from conspecifics. Journal of Comparative Psychology 101:110–123.

Whishaw IQ and Tomie J (1988) Food wrenching and dodging: A neuroethological tests of cortical and dopaminergic contributions to sensorimotor behavior in the rat, Behavioral Neuroscience 102:110–123.

Whishaw IQ and Tomie J (1989) Food-pellet size modifies the hoarding behavior of foraging rats. Psychobiology 17:83–101.

Whishaw IQ and Whishaw GE (1996) Conspecific aggression influences food carrying: Studies on a wild population of Rattus norvegicus. Aggressive Behavior 22:47–66.

Wolfe JB (1939) An exploratory study of food-storing in rats. Journal of Comparative Psychology 28:97–108.

Thermoregulation

21

EVELYN SATINOFF

To avoid becoming too hot or too cold, rats build dens, nests, and burrows; huddle together; bask in the sun; lie in the shade; swim; sprawl; groom various parts of their body; move from one location to another; sleep; or become active. All of these activities come under the rubric of *behavioral thermoregulation*. Of course, rats also mobilize a host of more reflexive behaviors such as shivering, piloerection, peripheral vasoconstriction, and brown adipose tissue activation to generate and conserve heat and peripheral vasodilation to lose it. It is apparent that the range of behaviors that can be mobilized far outnumbers the available reflexive responses. This is what makes the study of thermoregulation one of the most challenging and interesting areas of investigation in the behavioral sciences.

Thermoregulatory behavior is also relevant to a wide range of other research areas, both those concerned with normal regulatory functions and those related to pathological conditions. For instance, (1) all metabolic functions of the body are affected by body temperature, and regulatory mechanisms that control body fat and its metabolism affect body temperature (Collins et al., 2001). (2) In the process of investigating neural events that supposedly related to learning and memory, on a number of occasions scientists have thought that they had discovered a central correlate only to find later that the change was due to normal changes in body temperature (Anderson and Moser, 1995). (3) While investigating compounds that might minimize the extent to which brain trauma produces brain

injury, scientists have believed that they discovered a therapeutic compound only to later find that the therapy resulted secondarily from changes in body temperature (Corbett and Thornhill, 2000). Indeed, most doses of most drugs used to investigate the neuropharmacology of any behavior also affect body temperature, and the effect on behavior may be a secondary effect of the action of the drug on body temperature (Satinoff, 1979). (4) Postures of thermoregulatory behavior may be related to symptomology associated with some pathological conditions (Schallert et al., 1978). (5) Some behaviors, such as lordosis in infant rat pups, depend on body temperature and may not be displayed when body temperature is too high or too low (Leonard, 1987; Satinoff, 1991).

An infant rat is an *ectotherm*: it does not generate heat internally and is largely dependent for homeostasis on the environmental temperature in the nest. An adult rat is an *endoderm*: it regulates its body temperature internally over a broad range of ambient conditions. This transition occurs over the first 2 months of life as the infant rat gains mobility, body hair, and body mass. Nevertheless, because of its small surface-to-mass ratio, rats are always threatened by temperature extremes, which make them highly motivated to escape challenges to normal body temperature.

Over all ages, rats can survive wide fluctuations in body temperature, from as low as 18° C, a temperature at which bodily functions almost cease and at which surgery can be performed without anesthesia (Arokina et al., 2002), to as high as 41° C, which is just short of body temperature at which heat stroke and associated physical damage occur (Lin, 1999) The purposes of this chapter are to sketch behavioral and reflexive thermoregulatory behaviors in adult and immature rats, to describe some of the methods of studying thermoregulatory behavior, and to outline some of the neural mechanisms that mediate temperature regulation.

THERMONEUTRALITY AND THE CONCEPT OF SET POINT

For rats, as for all animals, there is a range of ambient temperatures at which the basal rate of the animal's own heat production equals the rate of heat lost to the environment and at which a minimum amount of thermoregulatory effort is required to maintain a constant body temperature. The most accurate definitions of *ambient thermoneutrality* are based on both heat loss and heat production responses. Thus, a "zone of least thermoregulatory effort" can be bounded on the low end by activity that will raise body temperature and on the high end by activity that will lower body temperature (Satinoff and Hendersen, 1977).

There is some debate over the range of thermoneutrality for adult rats. It has been described as being as wide as 18° to 28° C (Poole and Stephenson, 1977) and as narrow as 29.5° to 30.5° C (Szymusiak and Satinoff, 1981; Romanovsky et al., 2002). (Of course, as with anything in science, everything depends on the measurements used to derive the results.) That is, within this temperature range, rats continue with ongoing behavior while not initiating activities to be used specifically to regulate their body temperature. This tolerance is in part related to just what they are doing at the time. While resting in a home cage, their body temperature may fall as low as 35° C, whereas when engaged in a strenuous physical activity, such as solving a problem in a maze or voluntarily running in an activity

wheel, their body temperature may rise as high as 41° C. That rats tolerate temperatures within this range before actively defending their temperature suggests that body temperature within this range defines the rat's comfort zone. When given a choice, however, rats may prefer a much narrower range of body temperatures, a range that can be referred to as the *thermoneutral zone*, or zone of thermal comfort.

Because rats and other animals regulate their body temperature around a relatively constant value, this system has been usefully described by control theory. This regulated body temperature is referred to as the "set point." In engineering terms, *set point* is the value of the input at which the output is zero (Fig. 21–1). Behaviorally defined, set point is the value of body temperature that an animal will defend—the reference, or desired, or optimal body temperature—and it can only be inferred. It is sometimes erroneously assumed to be encoded in the neural structure of thermoregulatory nuclei in the hypothalamus. But any discussion of the brain structures involved in thermoregulation is an anatomical model and set point has no anatomical correlates: it is strictly a useful descriptive device.

There are a number of conditions in which set point changes, that is, conditions in which animals will defend a different body temperature.

1. *Circadian rhythms.* Rats are active during the night portion of the day–night cycle and sleep mainly during the light portion of the cycle. They allow their body temperature to reach high levels before they defend it; at night they are primarily interested in eating, drinking, and being active. Rats mainly sleep in the light period of a light–dark cycle; their body temperature falls in the light and they allow this to happen (i.e., they do not defend the drop). This may be adaptive in that it allows them to increase metabolic rate more easily at a time when they are searching for and consuming food and to decrease metabolic rate at a time when they are resting.

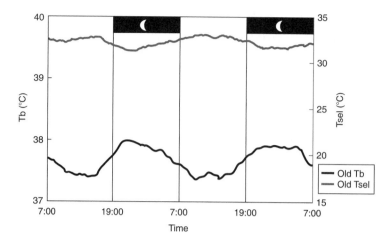

Figure 21–1. Control diagram for behavioral thermoregulation outlining what is involved when a rat wants to raise its body temperature. The learned response, such as pressing a lever to turn on a heat lamp, or moving to a warmer environment, occurs when there is a discrepancy (the error signal) between the ideal body temperature (the setpoint) and the actual body temperature. The error arises from a disturbance. (This could be a cold environment, cooling the brain, or a change in immunological compounds that give rise to fever.) The actual body temperature is monitored and the feedback loops maintain the response at a level appropriate to the existing discrepancy. The lower feedback path carries information about actual body temperature that is compared with the set temperature at the comparator, or signal mixer. The upper pathway adjusts the parameters of the response mechanisms in terms of response cost and response effectiveness and optimizes the effectiveness of the system. Low effectiveness or high response cost might be expected to lower the slope of the function relating error to response.

2. *Age.* When allowed to choose their thermal environment in a thermally graded alleyway, old rats prefer a warmer ambient temperature than do young ones (Florez-Duquet et al., 2001). Often, the circadian variation in body temperature is lower in old rats than in young ones (Fig. 21–2).

3. *Hypoxia.* In rats, resting oxygen uptake changes about 11% for every 1° C change in Tb. Therefore, when oxygen is in short supply, such as during hypoxia, a high body temperature could be injurious. Under hypoxic conditions, rats defend a lower body temperature. This appears to

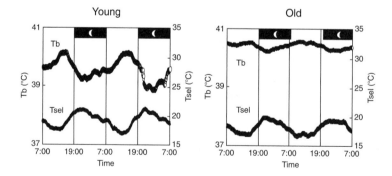

Figure 21–2. The diurnal fluctuations of body temperature and the temperature the rats selected over 48 hours in a thermal gradient. The mean body temperature (measured by telemetry) of the young rats varied from about 37.7° to 38.3° C. At the same time they selected ambient temperatures between about 24° and 31° C. Note that at night, when the rats were active, and generating more internal heat, they preferred ambient temperatures lower than in the daytime, when they were mostly sleeping. Body temperatures of old rats (average 24 months of age) was a little lower than those of the young rats, and they preferred significantly higher temperatures.

be adaptive because it lowers metabolic rate, and thus the need for oxygen (Wood and Gonzalez, 1996).

4. *Hormonal state.* Regulated temperature varies with reproductive condition. Female rats in the luteal stage of their estrous cycle appear to have an upward shift in set point and a lowered shift during the follicular phase. This shift may be related to changes in levels of progesterone or estrogen, and possibly other nongonadal hormone. Because female rats are extremely active during the luteal stage, this change may facilitate a higher level of metabolism (Kittrell and Satinoff, 1988). Because female rats cycle every 4 to 5 days, thermoregulatory changes can pose a challenge to experiments in which thermoregulatory behavior is not a primary concern.

5. *Fever.* Fever is a condition in which all behavioral and reflexive thermoregulatory responses work in parallel to raise body temperature to a new, higher set level (see Fig. 21–1). It is normally triggered by components of pathogenic organisms like bacteria, viruses, and fungi, which initiate a cascade of events that ultimately result in the release of brain prostaglandins (Ranels and Griffin, 2003). Correlational studies suggest that the febrile response has survival value in that it aids in the destruction of the pathological organism (Dantzer, 2001).

6. *Other.* The set point is apparently changed with many other agents and conditions; these include the effects of many drugs such as salicylates (Satinoff, 1972), stress (Peloso et al., 2002), and brain injury (Satinoff and Prosser, 1988).

THE NEURAL CONTROL OF TEMPERATURE REGULATION

In the late 1800s, investigators discovered that thermoregulation could be deranged after brain damage, and by the 1930s Ranson (1935) used stereotaxic surgery to localize a "heat-loss center" to the preoptic/anterior hypothalamus and a "heat-production center" to the posterior hypothalamus. Subsequent work indicated that thermoregulatory behaviors are initiated when the preoptic area is heated or cooled or is locally injected with neurotransmitter-related compounds. In addition, hypo-

thalamic thermosensitive cells alter their firing rats with changes in their own temperature or changes in the temperature in other parts of the body. Thus, substantial early work pointed to the preoptic/anterior hypothalamus as the brain "thermostat." However, the concepts of centers and thermostats in the brain are much too simplistic. The hypothalamus is importantly involved in a thermal control system that involves many parts of the brain, from the cortex to the spinal cord. Furthermore, although large lesions in this area disrupt thermoregulation, they do not prevent animals from maintaining their body temperature at near-normal levels if they have the behavioral responses to do so. Temperature regulation is too important to the life of an animal to be controlled by a simple thermostat (Satinoff, 1978, 1983).

THE DEVELOPMENT OF THERMOREGULATORY BEHAVIOR

Newborn rat pups are blind and hairless and have reduced immobility at birth. The key problems they face are a lack of thermal insulation, in terms of both fur and subcutaneous fat stores, and an unfavorable surface-to-mass ratio. They have a narrow thermoneutral zone ranging from 34° to 35° C at birth, which gradually increases to the adult range over the next 4 to 6 weeks. These ambient temperatures are much higher than the usual laboratory temperatures of 21° to 24° C. Thus, for infant rats, thermoneutrality consists of temperatures that would produce heat stress in adult rats. Although brown fat can be activated as early as the first postnatal day (Kortner et al., 1993) and infant rats can increase their metabolic rate from 10 days of age, or sooner (Nuesslein-Hildesheim and Schmidt, 1993), this capability is practically useless because they cannot conserve heat. Although it might be thought that infant rats would be better off drifting into hypothermia in the cold, hyperthermic rats develop much more slowly that do euthermic rats

(Stone et al., 1976). Therefore, it is desirable for infant rats to maintain a relatively high body temperature. The implications for behaviorists is that any work directed toward studying the behavior of infant rats requires an ambient temperature above that required for the study of adult rats.

As mentioned earlier, infant rats increase metabolic rate in response to cold stress, mostly or completely by generating heat in their brown adipose tissue. However, because of their small mass, they quickly become hypothermic at ambient temperatures below their thermoneutral zone, and metabolic responses fail. Due largely to difficulties in measurement, the extent to which infant rats can control temperature by peripheral vasoconstriction and vasodilation is uncertain. It is likely that the major improvement in temperature regulation over the first three weeks of life is due to the development of thermal

insulation (Conklin and Heggeness, 1971). Thus, rat pups are ectothermic (controlled by external temperature) and become endothermic (contribute to the control of their own body temperature) only after they develop fur. Interestingly, exposing rat pups to a cool environment hinders their pelage development compared with exposing them to a warm environment (Gerrish et al., 1998).

Pups can behaviorally thermoregulate if they do not become incapacitated by the cold. Rat pups will seek heat as early as 1 day after birth. In thermally graded alleys, rat pups orient and move along the thermal gradient from cool to warm if they are not immobilized by the cold (Kleitman and Satinoff, 1982) (Fig. 2–3). This result is not surprising because rat pups are born in litters, and litters regulate their body temperatures by active group huddling. Rat pups in a huddle move from the top of the pile down into it when too much of

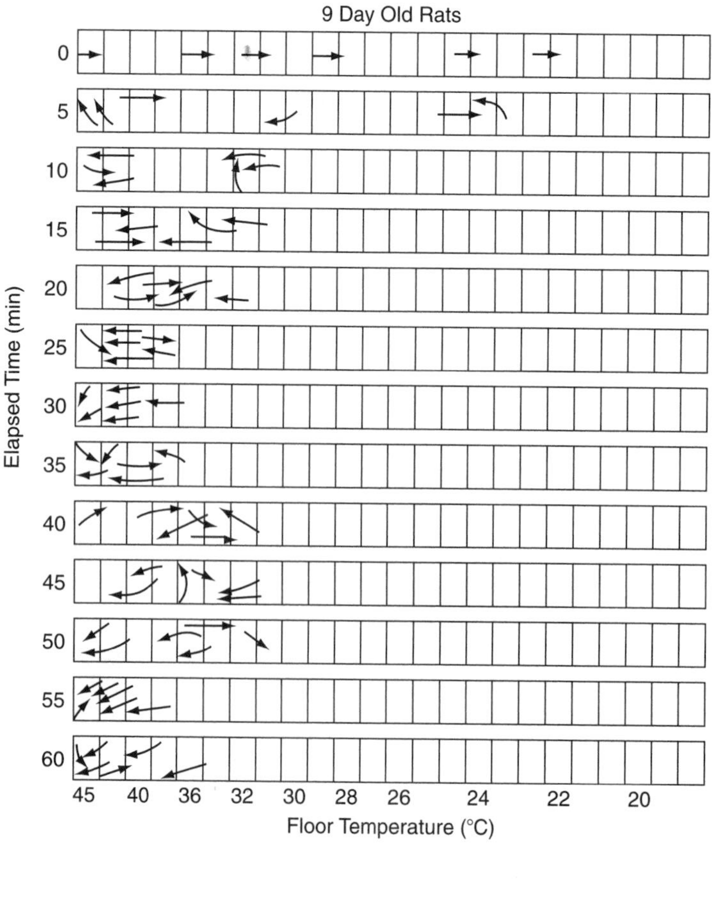

Figure 21–3. Positions of 1-day-old rat pups in a thermal gradient at 10 minute intervals for 120 minutes.

their surface area is exposed and they become cool and they move out to the edges when they become too warm (Alberts, 1978). In fact, huddling in rat pups is almost exclusively thermally directed during the first week of life. Huddling leads to lower oxygen consumption in the members of the huddle than if they were alone, largely because huddled animals expose less surface area to the cool surround and thereby lose less heat.

Thus, newborn rats can only produce and conserve heat behaviorally—by changing their position in the huddle or orienting and going toward other littermates when they are scattered around the nest side. Huddling or social aggregation continues to provide heat comfort into adulthood, as rats housed in groups may spend a large portion of their resting time in a pile.

When rat pups are tested at warm ambient temperatures, one can see the earlier appearance of behaviors that seemingly have nothing to do with thermoregulation. For instance, there had been reports that in both females and males primed with estradiol benzoate, female sexual behaviors—in particular, lordosis and ear wiggling—occurred no earlier than 2 to 3 weeks of age. However, when Williams (1987) tested pups at 33° to 35° C, both responses were seen in pups of either sex at 4 to 6 days of age. Rat pups also feed in response to food deprivation and tail pinch, drink in response to angiotensin injections, and respond to odor conditioning, but only if they are tested within their thermoneutral zone (Satinoff, 1991). Finally, neural structures have also been found to function much earlier in rat pups kept within their thermoneutral zone than in pups that are slightly cool (Horwitz et al., 1982).

BEHAVIORAL METHODS

The measurement of body temperature is not simple because temperature can vary widely within the body. Which temperature to measure depends on what question is being asked. No one really knows what is the "regulated temperature." In rats, it could be core or skin temperature or some core-to-skin gradient.

Temperature can be measured on points on the periphery of the body, such as the tail, which is highly vascularized and therefore optimal for losing or conserving body heat (Owens et al., 2002). Temperature can be measured within the body cavity, and this temperature is referred to as *core temperature.* Core temperature provides a relatively stable value that is least affected by environmental temperature fluctuations or local metabolic activity. Temperature can also be measured within the brain. Within the brain, it reflects both the core temperature of the body and the temperature generated by the activity of the neural structure within which the temperature-measuring device is inserted (DeBow and Colbourne, 2003).

Temperature can be measured with a wide range of devices. One commonly used method involves inserting a temperature-measuring probe at least 6 cm into the rectum to measure core temperature. This can be done acutely, while the rat is restrained in the hand, or chronically, if the animal is placed in a restraining box so that it cannot chew the probe. Manual insertion of temperature probes has the advantage of being simple, but the process of handling and restraining the rat will quickly and reliably produce a stress-related increase in body temperature (Eikelboom and Stewart, 1982).

There are several ways of sensing temperature, but the most common is to use a device that contains a thermistor, which is a semiconductor device made of materials whose resistance varies as a function of temperature. A recorder senses current changes as a function of changes in resistance.

Temperature probes have a number of advantages. The size of probes can vary from millimeters to microns, which make them useful for recoding skin, core, or brain temperature. They can also be used acutely or chronically implanted. They can be quickly and painlessly inserted into the rectum or placed on the skin to obtain a temperature reading, or they can be chronically fixed in place. They are also relatively inexpensive. The drawbacks of temperature probes are several: (1) a recording lead must be attached to the rat, which may limit the animal's freedom; (2) the rat may be able to reach the lead and chew it off; and finally, (3) even though it is a mild stressor, it is a stressor, and that may alter normal body temperature.

For long-term recordings, a transmitter can be implanted in an animal's body cavity and an AM radio signal from the transmitter can be recorded via computer. A signal can be recorded for periods of weeks or months, as long as the batteries on the transmitter function. These transmitters come in various sizes, and some can be inserted into the peritoneal cavity of rats that are at least 10 days old.

Measures of changes in skin temperature, which can indicate vasoconstriction and vasodilatation, can be obtained with thermocouples placed on the skin, such as on the tail, and shielded from air temperature fluctuations with insulating tape (Romanovsky et al., 2002). Tail skin temperature can also be measured noninvasively with radiotelemetry (Gordon et al., 2002). Temperature can also be measured by painting a rat, usually on the tail, with thermosensitive liquid crystal paint, although this method requires video recording and calibrating color changes (Romanovsy at al., 2002).

Shivering can be visually observed, recorded with a movement sensor (Harrod et al., 2002) or recorded using electrical recording of muscle activity (Whishaw and Vanderwolf, 1971). Finally, metabolic processes related to thermogensis can be inferred by measuring evaporative water loss and/or oxygen use in animals placed in calorimeters (Buchanan et al., 2003).

Several behavioral measures have been developed to study temperature regulation in rats. Operant methods present rats with a lever that they can press to control ambient temperature. The lever may trigger warm or cold airflow, warm or cold water spray, the onset of radiant heat, and so on. Operant methods tells an investigator how motivated a rat is to work to control its temperature. If all that one wishes to know is the ambient temperature a rat prefers, it is much easier, on both the investigator and the rat, to place the rat in a thermally graded alleyway, where it can choose to move to the place it finds thermally comfortable. If bedding material is available, rats will build nests in a cold environment and they will modulate food carrying and food consumption in response to ambient temperature changes.

As a cautionary note, some investigators have attempted to decrease core temperature by placing rats in a cold environment, but for adult rats, heat-generating mechanisms are so effective that they actually raise body temperature in response to acute cold stress. Core temperature can be reduced more effectively by lightly spraying a rat's fur with cool water or placing a rat in a cool bath. Because rats have an unfavorable body surface to mass ration, their heat generating mechanisms are unequal to the challenge of cold-water stress.

STRATEGIES FOR TEMPERATURE REGULATION

HEAT STRESS

Rats use four different behaviors when challenged with heat, such as radiant heat or warm airflow in which environmental temperature increases from about 26° C to 41° C. These responses are mediated both by activity in peripheral thermoreceptors and by central cooling of hypothalamic thermoregulatory regions via blood flow.

1. Rats have a number of reflexive responses, including vasodilatation, especially of the hairless tail, to induce body cooling.

2. Their initial behavioral response is to become more active by walking around and rearing.

3. If escape is not possible, rats engage in intense grooming during which they spread saliva over their body to assist evaporation-related cooling. Saliva spreading can be so intense that prolonged grooming can result in dehydration.

4. Faced with continued heat stress, animals give up all activity and postural support and lie prone, thus reducing heat production from metabolic and muscle activity. The extension posture consists of relaxation and elongation of the body in a prone position, with the normally curved spine straightened horizontally, forelegs placed under the neck, and hindlimbs folded outward and toward the rear. At first the head remains erect but later it rests on the floor, with the eyes partly closed and retracted (Roberts et al., 1974).

COLD STRESS

Rats have at least five different coping strategies when challenged with cold stress, such as cold airflow or body wetting.

1. They display reflexive responses, including peripheral vasoconstriction, especially of the tail, and piloerection of the fur, to increase its insulation value.

2. Their initial behavioral response is to become more active by walking around and rearing.

3. If escape is not possible, animals increase heat production using nonshivering thermogenesis. This consists of energy expenditure and heat production obtained by metabolizing brown adipose tissue (brown fat) in response to stimulation from the sympathetic nervous system. This response is very effective: total body energy expenditure doubles within minutes of a dose of a beta-adrenergic agonist.

4. Under continued cold stress that results in a lowering of core temperature, shivering thermogenesis is initiated. This consists of the adoption of a hunched body posture with postural support in which agonist and antagonist muscles contract concurrently, thus generating heat in the absence of locomotion. Shivering can be mild or so violent that an animal can be thrown off balance. Energy expenditure from brown fat and shivering can be measured indirectly as

oxygen consumption, which requires that animals be tested in a sealed environment in which airflow is controlled.

5. Fur condition contributes to thermoregulation, and rats have a gland called the Harderian gland that is adjacent to the eye and secretes Harderian liquid (Buzzell, 1996). The liquid is released when the paws apply pressure around the eye socket during grooming. The role of Harderian material in control of temperature in rats has not been investigated, but it may be similar to that described for the Mongolian gerbil (*Meriones unguiculatus*). For the gerbil, Harderian release decreases at high body temperature (when saliva production increases) and increases at low body temperature. This relationship suggests that Harderian material improves insulation of the pelage (Thiessen, 1989).

ACKNOWLEDGMENTS

Supported by National Institute of Mental Health grant R01-MH41138.

REFERENCES

Alberts JR (1978) Huddling by rat pups: Group behavioral mechanisms of temperature regulation and energy consumption. Journal of Comparative and Physiological Psychology 92:231–235.

Andersen P and Moser EI (1995) Brain temperature and hippocampal function. Hippocampus 95:491–498.

Arokina NK, Potekhina IL, Volkova MF (2002) Development of deep hypothermia in rats with limited motor activity. Rossiiskii Fiziologicheskii Zhurnal Imeni I.M. Sechenova/Rossiiskaia Akademiia Nauk 88:1477–1484.

Buchanan JB, Peloso E, Satinoff E (2003) Thermoregulatory and metabolic changes during fever in young and old rats. American Journal of Physiology (Regulatory and Integrative Comparative Physiology) 285:R1165–R1169.

Buzzell GR (1996) The Harderian gland: Perspectives. Microscope Research Techniques 34:2–5.

Collins S, Cao W, Daniel KW, Dixon TM, Medvedev AV, Onuma H, Surwit R (2001) Adrenoceptors, uncoupling proteins, and energy expenditure. Experimental Biology and Medicine 226:982–990.

Conklin P and Heggeness FW (1971) Maturation of temperature homeostasis in the rat. American Journal of Physiology 220:333–336.

Corbett D and Thornhill J (2002) Temperature modulation (hypothermic and hyperthermic conditions) and its influence on histological and behavioral outcomes following cerebral ischemia. Brain Pathology 10:145–152.

Dantzer R (2001) Cytokine-induced sickness behavior: Mechanisms and implications. Annals of the New York Academy of Sciences 933:222–234.

DeBow S and Colbourne F (2003) Brain temperature measurement and regulation in awake and freely moving rodents. Methods 2:167–171.

Eikelboom R and Stewart, J (1982) Conditioning of drug-induced physiological responses. Psychological Reviews 89:507–528.

Florez-Duquet M, Peloso E, Satinoff E (2001) Fever and behavioral thermoregulation in young and old rats. American Journal of Physiology (Regulatory and Integrative Comparative Physiology) 280:R1457–R1461.

Gordon CJ, Puckett E, Padnos B (2002) Rat tail skin temperature monitored noninvasively by radiotelemetry: Characterization by examination of vasomotor responses to thermomodulatory agents. Journal of Pharmacological and Toxicological Methods 47: 107–114.

Harrod S, Metzger M, Stempowski N, Riccio D (2002) Cold tolerance: Behavioral differences following single or multiple cold exposures. Physiology and Behavior 76:27–39.

Horwitz J, Heller A, Hoffmann PC (1982) The effect of development of thermoregulatory function on the biochemical assessment of the ontogeny of neonatal dopaminergic neuronal activity. Brain Research 235:245–252.

Kittrell EM and Satinoff E (1988) Diurnal rhythms of body temperature, drinking and activity over reproductive cycles. Physiology and Behavior 42:477–484.

Kleitman N and Satinoff E (1981) Thermoregulatory behavior in rat pups from birth to weaning. Physiology and Behavior 29:537–541.

Kortner G, Schildhauer K, Petrova O, Schmidt I. (1993) Rapid changes in metabolic cold defense and GDP binding to brown adipose tissue mitochondria of rat pups. American Journal of Physiology (Regulatory and Integrative Comparative Physiology) 264:R1017–R1023.

Lin MT (1999) Pathogenesis of an experimental heat-stroke model. Clinical Experimental and Pharmacological Physiology 26:826–837.

Nuesslein-Hildesheim B and Schmidt I (1994) Is the circadian core temperature rhythm of juvenile rats due to a periodic blockade of thermoregulatory thermogenesis? Pflugers Archives 427450–4.

Owens NC, Oootsuka Y, Kanosue K, McAllen RM (2002) Thermoregulatory control of sympathetic fibres supplying the rat's tail. Journal of Physiology (London) 543:849–858.

Peloso E, Wachulec M, Satinoff E (2002) Stress-induced hyperthermia depends on both time of day and light condition. Journal of Biological Rhythms 17:164–170.

Poole S and Stephenson JD (1977) Body temperature regulation and thermoneutrality in rats. Quarterly Journal of Experimental and Cognitive Medical Sciences 62:143–149.

Ranels HJ and Griffin JD (2003) The effects of prostaglandin E2 on the firing rate activity of thermosensitive and temperature insensitive neurons in the ventromedial preoptic area of the rat hypothalamus. Brain Research 64:42–50.

Ranson SW (1935) The anatomy of the nervous system from the standpoint of development and function. Philadelphia: W.B. Saunders Co.

Roberts WW, Mooney RD, Martin JR (1974) Thermoregulatory behaviors of laboratory rodents. Journal of Comparative and Physiological Psychology 86:693–699.

Romanovsky AA, Ivanov AI, Shimansky YP (2002) Ambient temperature for experiments in rats: A new method for determining the zone of thermal neutrality. Journal of Applied Physiology 92:1–21.

Satinoff E (1972) Salicylate: Action on normal body temperature in rats. Science 176:532–533.

Satinoff E (1978) Neural organization and evolution of thermal regulation in mammals. Science 201:16–22.

Satinoff E (1979) Drugs and thermoregulatory behavior. In: Body temperature, drug effects and therapeutic implications (Lomax P and Schonbaum E, eds.), pp. 151–181. New York: Marcel Dekker.

Satinoff E (1983) A reevaluation of the concept of the homeostatic organizatin of temperature regulation. In: Handbood of behavioral neurobiology (Satinoff E and Teitelbaum P, eds.), pp. 443–467. New York: Plenum Press.

Satinoff E (1991) Developmental aspects of behavioral and reflexive thermoregulation. In: Developmental psychobiology: New methods and changing concepts. (Shair HN, Barr GA, Hofer MA, eds.), pp. 169–188. New York: Oxford.

Satinoff E and Hendersen R (1977) Thermoregulatory behavior. In Handbook of operant behavior (Honig WK and Staddon JER, eds.), pp. 153–173. Englewood Cliffs, Colo.: Prentice-Hall.

Satinoff E and Prosser RA (1988) Suprachiasmatic nuclear lesions eliminate circadian rhythms of drinking and activity, but not of body temperature, in male rats. Journal of Biological Rhythms 3:1–22.

Schallert T, Whishaw IQ, DeRyck, Teitelbaum P (1978). The postures of catecholamine-depletion catalepsy: Their possible adaptive value in thermoregulation. Physiology and Behavior 21:817–820.

Stone EA, Bonnet KA, Hofer MA (1976) Survival and development of maternally deprived rats: Role of body temperature. Psychosomatic Medicine 38:242–249.

Szymusiak R and Satinoff E (1981) Maximal oxygen consumption defines a narrower thermoneutral zone than does minimal metabolic rate. Physiology and Behavior 26:689–690.

Thiessen GM (1989) The possible interaction of Harderian material and saliva for thermoregulation in the Mongolian gerbil, Meriones unguiculatus. Perception Motor Skills 68:3–10.

Whishaw IQ and Vanderwolf CH (1971) Hippocampal EEG and behavior: Effects of variation in body temperature and relation of EEG to vibrissae movement, swimming and shivering. Physiology and Behavior 6:391–397.

Williams CL (1987) Estradiol benzoate facilitates lordosis and ear wiggling of 4- to 6-day-old rats. Behavioral Neuroscience 101:718–723.

Wood SC and Gonzales R (1996) Hypothermia in hypoxic animals: Mechanisms, mediators, and functional significance. Comparative Biochemistry and Physiology B Biochemistry and Molecular Biology 113:37–43.

Stress

22

JAAP M. KOOLHAAS, SIETSE F. DE BOER, AND BAUKE BUWALDA

A considerable part of our current knowledge of stress and the pathophysiology of stress-related disorders is based on experimental studies in rats. The scientific rationale of these studies is that the mechanisms underlying stress and adaptation have a common biological basis in animals and humans. Despite the wealth of data and publications on stress research in the rat and the wide variety of stress paradigms, one may criticize the validity of some of these studies. Studies often seem to fail in particular with respect to their face validity, which means that the model fails to sufficiently mimic both the etiology and the symptomatology of human stress-related disorders. For example, many animal studies use stressors that bear little or no relationship to the biology of the species, that is, to the situations an animal may encounter in its everyday life in a natural habitat. If we want to improve and refine our understanding of the causal mechanisms of stress-related disorders, we need behavioral similarity models that experimentally exploit the common biological basis of animals and humans.

In this chapter, the biology of the rat and its natural defense mechanisms are used as a starting point in the evaluation and description of stress models in the rat. In other words, we will focus at tests that explore the capacity of rats to cope with ecologically relevant problems. The individual capacity to deal with everyday problems in the natural environment is considered to be one of the driving forces of evolution and speciation. Organisms have become adapted to a dynamic and complex natural environment in which they have to find their food, deal with conspecifics, or react to changes in climate. The capacity to cope with environmental challenges largely determines the individual survival in the natural habitat. In the course of evolution, animals have developed a wide variety of defense mechanisms to deal with such environmental challenges. Central in the biology of the rat is its social nature. Several studies in free ranging social groups of animals indicate that the stability of social environment is an important factor in health and disease. Unstable social groups and failure of social adaptive capacities may lead to serious forms of stress pathology. This is reflected in the relationship between position in the social hierarchy and the incidence of certain stress pathologies.

The first studies on the relationship between social position and stress pathology mainly concentrated on cardiovascular disease. For example in mice (Ely, 1981; Lockwood and Turney, 1981; Henry and Stephens-Larson, 1985), rats (Henry et al., 1993; Fokkema et al., 1995), and monkeys (Manuck et al., 1983), it has been demonstrated that hypertension and cardiovascular abnormalities are more frequent in socially unstable groups and occur predominantly in the dominant and subdominant males of the social group. Stomach ulcers are mainly found in social outcasts of colonies (Calhoun, 1962; Barnett, 1987). Similar observations on the relation between social position and pathology were reported with respect to immune system–mediated diseases (Spencer et al., 1996; Stefanski et al., 2001).

Before embarking on a description of the various stress paradigms used in rats, we discuss some issues that are fundamental to all stress models.

STRESS

Central to modern stress research are the terms *controllability* and *predictability*. These terms date back to a series of experiments by Weiss in the late 1960s (Weiss, 1972). He demonstrated that it is not so much the physical nature of an aversive stimulus that induces stress pathology but rather the degree in which the stimulus can be predicted and controlled. Rats that cannot predict and have no control over a stressor, such as an electric shock, appeared to have severe damage to the stomach wall, show signs of immunosuppression (Keller et al., 1981; Weiss et al., 1989), and have the largest rise in plasma corticosterone. Using this type of evidence, the following definitions of stress can be formulated:

> *Acute stress* is the state of an organism after a sudden decrease in the predictability and/or the controllability of relevant environmental factors.

> *Chronic stress* is the state of an organism that occurs when relevant environmental aspects have a low predictability and are not, or not very well, controllable over a long period of time.

Although the concept of controllability and predictability has strongly contributed to the present knowledge and insights into the development of stress pathology, we would like to make some remarks on the way in which these concepts are generally used in an experimental setting.

In most experiments, controllability is operationally defined as a binary factor, that is, as full control or complete loss of control. However, in everyday life situations, controllability is generally graded from absolute control via threat to control in various degrees to loss of control. Few studies consider the im-

portance of a different degree of control in the development of stress pathology. The importance of such a distinction is demonstrated in experiments aimed at understanding the development of hypertension. These experiments show that the crucial factor is *threat to control* rather than *loss of control* (Koolhaas and Bohus, 1989). Apart from the graded nature of controllability and predictability, the *frequency* and *duration* of stressors are also matters of concern. Usually a distinction is made between acute and chronic stress. It is well accepted that the chronic character of stressors may indeed result in various forms of stress pathology. However, chronic stress is not a very well defined concept. Many animal models specifically aimed at chronic stress use a series of intermittent acute stressors that may change daily rather than using the continuous presence of a stressor. Moreover, chronic stress studies rarely control for the factor time after the start of the stress procedure. The human literature indicates that acute stressors or major life events may have long-term consequences, ultimately leading to a higher incidence of disease. Recent studies show that in rats, the experience of a single uncontrollable event is sufficient to induce changes in behavior and stress physiological parameters that range from hours to days, weeks, or months (Koolhaas et al., 1997b).

INDIVIDUAL VARIATION IN COPING STRATEGIES

A wide variety of medical, psychological, and biological studies in both humans and animals demonstrate that individuals can differ considerably in their vulnerability for stress-related disorders. Apparently, they differ in their capacity to cope with environmental demands. Factors that have been shown to affect the individual coping capacity include genotype, ontogeny, adult experience, age, social support, and so on. Individual differentiation in behavior and physiology is a well-known phenomenon in many animal species. Several attempts have been made to classify individual variation into personalities, temperaments, or coping styles that may predict the response to environmental stressors. The scarce literature in a number of feral animal species suggests that in nature, the dimension of proactive and reactive coping strategies can be distinguished (Koolhaas et al., 1999). Authors may use different terms to characterize phenotypes, such as shyness and boldness or proactive and reactive or active and passive, but they all seem to share the same basic characteristics. Proactive coping is characterized by an active control of the environment (i.e., active avoidance, offensive aggression, nest building, etc.). Reactive coping, however, is characterized by a more readily acceptance of the environment as it is. Detailed analysis of coping strategies in rats and mice indicates that the most fundamental difference is the degree of behavioral flexibility. Reactive coping males are flexible, whereas the proactive coping is characterized by rigidity and routine formation. Recent studies in feral populations indicate that this differential flexibility may have its origin in a differential survival value in nature. The challenge for the future is to understand the functional significance of individual variation in nature by integrating ethological, physiological, and ecological approaches in the study of coping strategies.

From a biomedical point of view, the concept of coping strategies implies that different animals have a differential vulnerability to stress-related disorders. Negative health consequences might arise if an animal cannot cope with the stressor or requires very demanding coping efforts. In view of the differential neuroendocrine reactivity and neurobiological make-up of proactive and reactive coping animals, one may expect different types of stress-pathology to develop under conditions in which a particular coping strategy fails. Indeed, there are indications that coping strategies differ in susceptibility to develop cardiovascular pathology, ulcer forma-

tion, stereotypies, depression, and infectious disease (Koolhaas et al., 1999).

STRESS MODELS

We discuss some stress models used in adult rats that challenge the defense mechanisms and hence call on the natural adaptive capacity of the animal. Most of these models involve stressors of the social and physical environment. Few studies use food availability as stressor. This is surprising in view of the fact that the controllability and predictability of food are crucial to survival in nature and may strongly challenge the adaptive capacities. Indeed, both food restriction and increased caloric intake are reported to affect stress physiology (Rupp, 1999; Seres et al., 2002).

ACUTE STRESS MODELS

Acute Social Stress

Acute social stress can be studied in the resident–intruder paradigm. This model is based on the fact that a male rat will defend its territory against an unfamiliar male intruder. Territorial behavior develops within 1 week when an adult male rat (i.e., >3 months of age) is housed with a female rat in a large cage of about 0.5 m². When, after this period, an unfamiliar male conspecific of the same strain and weight is introduced into the home cage, the resident male will attack the intruder and a fight develops. Usually, the resident is the victor of this social interaction. This paradigm allows analysis of both the winner and the loser of the conflict.

Defeat. When the intruder animal is used as the experimental animal, one can study the consequences of social defeat or loss of social control as stressor. Social defeat by a male conspecific induces an acute increase in heart rate, blood pressure, and body temperature; strong neuroendocrine responses in plasma catecholamines, corticosterone, prolactin, and testosterone; and changes in central nervous serotonergic neurotransmission (Koolhaas et al., 1997b; Berton et al., 1999; Sgoifo et al., 1999a). These responses, including the behavioral reaction (flight, immobility), can be considered as part of the classic response to an acute stressor. However, a comparison of a range of different stressors reveals that social defeat may be one of the most severe stressors measured in terms of the magnitude of the corticosterone and catecholamine response (Koolhaas et al., 1997a). More important, however, is the time course of these stress responses.

Recent studies using more chronic recordings indicate that the various stress parameters have a different time course. The cardiovascular and catecholaminergic response to a 1 hour social defeat diminishes within 1 or 2 hours after the defeat, but the corticosterone response lasts for longer than 4 hours. After an initial rise, plasma testosterone drops below baseline levels and remains at extremely low levels for at least 2 days. A single social defeat appears to induce a reduction in the circadian variation in body temperature, growth, sexual interest, and open field exploration that may last from 2 to 10 days after the social stress (Koolhaas et al., 1997b). Miczek et al. (1990) found changes in opiate analgesia that last for at least 1 month after the defeat. Although many of these changes can be considered as part of the symptomatology of human depression, we emphasize that social defeat induces changes in a variety of physiological and behavioral parameters, each of which may have different temporal dynamics. Hence, an acute stressor may have chronic consequences. These lasting changes may be adaptive but may just as well be considered as the early signs of stress pathology. The bottom line is that the social defeat model allows further analysis of the changes in time of factors known to be involved in stress and adaptation. By manipulating the frequency, intensity, and type of previous social experiences, one can obtain insight into the (mal)adaptive nature of these changes.

Victory. By using the resident male in the resident–intruder paradigm as experimental animal, one can study the consequences of threat to control. Although the resident ultimately controls its social environment, this is preceded by a certain degree of unpredictability and threat to control. This is clearly indicated by the fact that the stress response—in terms of plasma corticosterone and catecholamines, heart rate, and blood pressure—is initially almost as high as in the defeated intruder, but these stress parameters rapidly return to baseline levels as soon as the dominance relationship becomes clear. Typical for the winner of the social interaction are the cardiovascular abnormalities observed in the electrocardiogram immediately after the interaction. These abnormalities indicate a strong shift in the autonomic balance toward high sympathetic dominance (Sgoifo et al., 1999b).

Defensive Burying
Defensive burying refers to the natural behavior of rats of displacing bedding material with typical alternating forward-pushing movements of their forepaws (threading or thrusting) and shoveling movements of their heads directed at localized sources of unfamiliar, aversive stimulation/threat. Harmful and noxious objects so buried include electrified prods (Treit et al., 1981), rat chow pellets coated with quinine, spouts of bottles containing unpleasant tasting liquids such as pepper sauce and liquids to which the rats have developed taste aversion, flash cubes that discharge near them, tubes that direct airbursts into their faces or deliver noxious smells, dead conspecifics, and predators (De Boer and Koolhaas, 2003). Although different in form, function, and intensity, rats may also bury seemingly harmless objects such as nonelectrified prods, flash cubes that do not flash, and marbles (Treit, 1985). By burying unfamiliar and/or harmful objects, individuals can successfully avoid or remove

aversive and possibly life-threatening dangers from their habitat. Together with flight, freezing, and certain forms of agonistic behavior, defensive burying constitutes the unconditioned (innate) species-specific defensive behavioral repertoire in rats.

The procedure of the shock-prod defensive burying test is quite simple and basically unchanged since its original description by Pinel and Treit (1978). In a test chamber (either the home cage or a familiarized test cage after several habituation trials) with sufficient suitable bedding material on the floor, subjects are confronted with a wire-wrapped prod/probe ($\phi = 1$ cm; 6 to 7 cm long) that is inserted through a small hole 2 cm above the bedding in one of the test chamber walls. The noninsulated wires of the prod are connected to a shock source, and whenever the subject touches the prod with its forepaws or snout, it receives an electric shock (manually operated or automatically delivered). The prod either remains electrified during the entire test period or is deactivated after the first contact. After the first contact with the electrified prod, the animal's behavior is observed and/or recorded on video for a 10 to 15 minute test session. During this observation period, a variety of behaviors can be quantified. The repertoire of behavioral reactions is well delineated and catalogued in rats and mice (Tsuda et al., 1988; De Boer et al., 1991), and methods for its reliable measurement have become standard equipment in behavioral-physiological and pharmacological laboratories. It is important to notice that there is a large individual variation in the amount of time spent burying. Because burying behavior is generally negatively correlated with immobility behavior, it seems that this differentiation is based on a differential expression of anxiety (De Boer and Koolhaas, 2003).

Predator
Rats readily display various forms of defensive behavior when confronted with a predator such as a cat. Because even rats that never previously saw a cat show the response and because the response does not habituate, it is considered to be an evolutionary ancient and innate reaction. Although a live cat produces a much stronger response, in a laboratory experimental setting, usually the odor of a cat is used. Cat odor can be presented to the rat by first rubbing a cat with cotton wool for a standard period of time and then putting the wool on top of the rat's cage. Dielenberg and McGregor (2001) developed another way of presenting cat odor to a rat. They used a fabric collar that had been worn by a cat for 3 weeks and presented this in a specific "cat odor avoidance" apparatus. This test cage allows the measurement of various avoidance behaviors and risk assessment behavior. It is important to notice that several studies now indicate that the response to cat odor differs from the response to fox odor. Fox odor, which can be obtained as the synthetic compound trimethylthiazoline (TMT), seems to act more as a generally aversive stimulus (McGregor et al., 2002; Blanchard et al., 2003).

Novelty
Novelty is often used as a minor stressor. It usually does not induce a maximal stress response and for that reason is suitable to determine stress reactivity. There is a considerable individual difference in the tendency of rats to explore a novel environment or novel objects that is more generally related to coping style and hypothalamus-pituitary-adrenal axis reactivity (Steimer and Driscoll, 2003). This individual differentiation is considered to be an animal model for the sensation-seeking trait in humans (Dellu et al., 1996). There are various ways to measure the response to novelty. One can introduce a novel object into the home cage of the experimental animal and measure the behavioral and neuroendocrine response. However, the response to clean bedding material can also be used as a standard novelty stress. In this way one takes the advantage of the fact that cages have to be regularly cleaned anyway.

Forced Swimming

Feral rats, in contrast to mice, often live close to water and like to swim. They often swim voluntarily, and the activation of the neuroendocrine stress response is mainly related to the increase in physical activity during swimming. Porsolt et al. (1977) was the first to use forced swimming in a tank that had no possibilities for the animal to escape, as a stressor to induce signs of depression. Indeed, the forced swim test strongly activates the neuroendocrine stress response; when the animal is replaced in the test apparatus 24 hours later, they more readily give up attempts to escape. Without the opportunity to escape, the physiological stress response appears to be unrelated to the physical activity (Abel, 1994a, 1994c). The forced swim test was subsequently somewhat modified and is now widely used as a test to measure the efficacy of potential antidepressant drugs. Briefly, the test uses a cylindrical tank with a diameter of about 25 cm and a height of about 50 cm filled with 30 cm of water at 25° C (Abel, 1994b). Three main types of behavior can be observed when the rat is put into the water: swimming, attempts to climb the wall, and floating. This floating behavior increases over the test time and is generally considered as a form of despair in which the animal has given up its attempts to escape. However, one may also argue that the animal has two alternative ways to cope with the situation by either actively trying to escape or quietly floating on the water surface and saving the energy of useless escape attempts. A recent study using proactive and reactive coping mice confirmed the idea that the test seems to measure differential ways of coping with an inescapable stressor rather than signs of despair and depression (Veenema et al., 2003). Indeed, treatment with an antidepressant drug reduced floating in the reactive coping male as predicted by a wealth of pharmacological literature, but the same treatment also reduced climbing and swimming in the proactive coping male (unpublished observation).

CHRONIC STRESS MODELS

Social Stress

Because of its permanent nature, the social environment is frequently used as a way to induce naturalistic chronic stressors. The variety of models used to study chronic social stress may differ in complexity and degree of experimental control.

Social Groups. Social stress can be studied in its most complex form using groups or colonies of rats. This allows an analysis of the relationship between the position of individuals in the social hierarchy and stress parameters. The design of cages used for colonies may range from large outdoor enclosures to cages of several square meters with nest boxes or much smaller cages without any further facilities. One of the most extensively studies paradigms is the visible burrow system developed by Blanchard et al. (1995). The visible burrow system (VSB) consists of a large central square and some nest chambers that are connected to the central square by means of Plexiglas tunnels. Colonies generally consist of four adult males and two adult females to facilitate territoriality and aggression. Blanchard et al. (1995) observed clear signs of chronic stress in the subordinate males in particular, whereas dominant males developed symptoms of stress only when the group consisted of a relatively large number of highly aggressive males. The advantage of such a system is that the model allows the animals to express their natural behaviors and defense mechanisms. The disadvantage is the limited degree of experimental control and the limited possibilities for continuous monitoring of physiological parameters.

Social Instability. The degree of social stress in groups of rats depends mainly on the stability of the social structure. In stable social groups, hardly any signs of stress pathology can be observed. Therefore, researchers tend to increase stress by reducing the social stability. This can be achieved by forming groups consisting of only aggressive males. This leads to serious and regular dominance fights. However, this procedure might give a considerable bias to the experimental results due to the selection procedure. Another way to manipulate social stability is to mix groups on a regular basis. Lemaire and Mormede (1995) successfully applied this method in rats. They could demonstrate that regular mixing of unfamiliar groups leads to the development of hypertension, in particular, in a more socially active strain of rats.

Chronic Subordination. Living as a subordinate male in the presence of a dominant male is generally considered to be a chronic stressor. Usually, subordination is studied in larger groups of males such as the visible burrow system mentioned earlier. Occasionally, groups of three or even two males are used, living together for a prolonged period of time. Subordination is determined by direct observation of the social interactions between the group members.

In species like mice and tree shrews, a *sensory contact model* is frequently used. In this model, the subordinate male is housed together with a dominant neighbor, but the two animals are separated by a wire mesh screen, allowing visual, auditory, and olfactory contact. Dominance relationships are established or reconfirmed by removing the wire mesh screen every day or week for a short period of time (Fuchs et al., 1993; Veenema et al., 2003). Although this is a useful model in these species, we are not aware of its application in rats.

CONCLUDING REMARKS

The stress models described here have in common that they somehow challenge the natural defense mechanisms and hence call on the adaptive capacity of the animal. We would like to emphasize the word *natural*, because its means that, on the basis of the specific evolutionary biology of the species, one might expect the animal to have an adequate answer to a given challenge. This distinguishes the selected stress models from models using foot shocks, centrifuges, tail suspension, loud noise, and so on. Although these latter models certainly activate stress physiological systems, one may question the adaptive nature of these responses. By definition, there is a mismatch in these models between the challenge or the stressor and the available repertoire of behavioral, neuroendocrine, and neurobiological adaptive mechanisms. Because such a mismatch may be fundamental to the development of stress pathology, this may be the reason that these models are so popular in stress research. However, we believe that the field of stress research has to move toward a more subtle understanding of the factors and processes underlying the development of stress pathology.

Rather than pushing the animal toward a stress physiological ceiling, it might be far more informative to explore the natural factors that determine and modulate the individual adaptive capacity. These factors include not only perinatal and adult (social) experience but also factors that affect the speed of recovery after a stressor. One of these latter factors is social support. It appears that the long-term consequences of social defeat are strongly influenced by the social housing conditions after the defeat; that is, defeat combined with social isolation has a much stronger impact in rats than either of the two alone (Ruis et al., 1999). A second important question that needs to be addressed and that requires more naturalistic stress models concerns the adaptive nature of the stress response. Implicit in the interpretation of many stress experiments is that the observed changes somehow contribute to the development of pathology. We would argue that one needs a more naturalistic setting to obtain experimental evidence for the (mal)adaptive consequences of a given stress response. Quite likely, the stress response of an organism will be a trade-off between the short-term benefits and the long-term costs in terms of chances to pathology and evolutionary fitness. Unfortunately, such questions have rarely been addressed in the rat.

Finally, we would like to address a more ethical issue. As in many scientific experiments, stress research aims at certain effects as predicted by specific hypotheses. However, when the predicted effect of a stressor is not found, there is a strong tendency to blame the stressor rather than the hypothesis. This may lead to the use of extreme stressors that go far beyond the biological range and hence fall short in face validity. These include stressors such as prolonged restraint (days), high-intensity shocks, prolonged conditions of extreme crowding, and so on. Again, the biology and ecology of the species should be the guidelines for contemporary stress research.

REFERENCES

Abel EL (1994a) A further analysis of physiological changes in rats in the forced swim test. Physiology and Behavior 56:795–800.

Abel EL (1994b) Behavioral and physiological effects of different water depths in the forced swim test. Physiology and Behavior 56:411–414.

Abel EL (1994c) Physical activity does not account for the physiological response to forced swim testing. Physiology and Behavior 56:677–681.

Barnett SA (1987) The rat: A study in behavior. Chicago: University of Chicago Press.

Berton O, Durand M, Aguerre S, Mormede P, Chaouloff F (1999) Behavioral, neuroendocrine and serotonergic consequences of single social defeat and repeated fluoxetine pretreatment in the Lewis rat strain. Neuroscience 92:327–341.

Blanchard DC, Markham C, Yang M, Hubbard D, Madarang E, Blanchard RJ (2003) Failure to produce conditioning with low-dose trimethylthiazoline or cat feces as unconditioned stimuli. Behavioral Neuroscience 117:360–368.

Blanchard DC, Spencer RL, Weiss SM, Blanchard RJ, McEwen B, Sakai RR (1995) Visible burrow system as a model of chronic social stress: Behavioral and neuroendocrine correlates. Psychoneuroendocrinology 20:117–134.

Calhoun JB (1962) The ecology and sociology of the Norway rat. Washington, D.C.: Governement Printing Office.

De Boer SF and Koolhaas JM (2003) Defensive burying in rodents: Ethology, neurobiology and psychopharmacology. European Journal of Pharmacology 463:145–161.

De Boer SF, van der Gugten J, Slangen JL (1991) Behavioral and hormonal indices of anxiolytic and anxiogenic drug action in the shock-prod defensive burying paradigm. In: Animal models in psychopharmacolgy, pp. 81–96. Basel: Birkhauser Verlag.

Dellu F, Piazza PV, Mayo W, Le Moal M, Simon H (1996) Novelty-seeking in rats—biobehavioral characteristics and possible relationship with the sensation-seeking trait in man. Neuropsychobiology 34:136–145.

Dielenberg RA and McGregor IS (2001) Defensive behavior in rats towards predatory odors: A review. Neuroscience and Biobehavioral Reviews 25:597–609.

Ely DL (1981) Hypertension, social rank, and aortic arteriosclerosis in CBA/J mice. Physiology and Behavior 26:655–661.

Fokkema DS, Koolhaas JM, van der Gugten J (1995) Individual characteristics of behavior, blood pressure, and adrenal hormones in colony rats. Physiology and Behavior 57:857–862.

Friedman MJ and Schnurr PP (1995) The relationship between trauma, post-traumatic stress disorder, and physical health. In: Neurobiological and clinical consequences of stress: From normal adaptation to PTSD (Friedman MJ, Charney DS, Deutch AY, eds.), pp. 507–524. Philadelphia: Lippincott-Raven Publishers.

Fuchs E, Jöhren O, Flügge G (1993) Psychosocial conflict in the tree shrew: Effects on sympathoadrenal activity and blood pressure. Psychoneuroendocrinology 18:557–565.

Henry JP, Liu YY, Nadra WE, Qian CG, Mormede P, Lemaire V, Ely D, Hendley ED (1993) Psychosocial stress can induce chronic hypertension in normotensive strains of rats. Hypertension 21:714–723.

Henry JP and Stephens-Larson P (1985) Specific effects of stress on disease processes. In: Animal stress (Moberg GP, ed.), pp. 161–173. Bethesda: American Physiological Society.

Keller SE, Weiss JM, Schleifer SJ, Miller NE, Stein M (1981) Suppression of immunity by stress: Effect of a graded series of stressors on lymphocyte proliferation. Science 213:1397–1400.

Koolhaas JM, De Boer SF, De Ruiter AJ, Meerlo P, Sgoifo A (1997a) Social stress in rats and mice. Acta Physiologica Scandinavica 161:69–72.

Koolhaas JM and Bohus B (1989) Social control in relation to neuroendocrine and immunological responses. In: Stress, personal control and health (Steptoe A and Appels A, eds.), pp. 295–304. Brussels: John Wiley & Sons Ltd.

Koolhaas JM, Korte SM, De Boer SF, Van Der Vegt BJ, Van Reenen CG, Hopster H, De Jong IC, Ruis MA, Blokhuis HJ (1999) Coping styles in animals: Current status in behavior and stress-physiology. Neuroscience and Biobehavioral Reviews 23:925–935.

Koolhaas JM, Meerlo P, Boer SFd, Strubbe JH, Bohus B (1997b) The temporal dynamics of the stress response. Neuroscience and Biobehavioral Reviews 21:775–782.

Layton B and Krikorian R (2002) Memory mechanisms in posttraumatic stress disorder. Journal of Neuropsychiatry and Clinical Neuroscience 14:254–261.

Lemaire V and Mormede P (1995) Telemetered recording of blood pressure and heart rate in different strains of rats during chronic social stress. Physiology and Behavior 58:1181–1188.

Lockwood JA and Turney T (1981) Social dominance and stress induced hypertension: Strain differences in inbred mice. Physiology and Behavior 26:547–549.

Manuck SB, Kaplan JR, Clarkson TB (1983) Behaviorally induced heart rate reactivity and atherosclerosis in cynomolgous monkeys. Psychosomatic Medicine 45:95–108.

McEwen BS (2002) The neurobiology and neuroendocrinology of stress. Implications for post-traumatic stress disorder from a basic science perspective. The Psychiatric Clinics North America 25:469–494, ix.

McGregor IS, Schrama L, Ambermoon P, Dielenberg RA (2002) Not all 'predator odours' are equal: Cat odour but not 2,4,5 trimethylthiazoline (TMT; fox odour) elicits specific defensive behaviours in rats. Behavioural Brain Research 129:1–16.

Miczek KA, Thompson ML, Tornatzky W (1990) Short and long term physiological and neurochemical adaptations to social conflict. In: NATO ASI Series D: Behavioural and social sciences (Puglisi-Allegra S and Oliverio A, eds.), pp. 15–30. Dordrecht: Kluwer.

Porsolt RD, Le Pichon M, Jalfre M (1977) Depression: A new animal model sensitive to antidepressant treatment. Nature 266:730–732.

Ruis MA, te Brake JH, Buwalda B, De Boer SF, Meerlo P, Korte SM, Blokhuis HJ, Koolhaas JM (1999) Housing familiar male wildtype rats together reduces the long-term adverse behavioural and physiological effects of social defeat. Psychoneuroendocrinology 24:285–300.

Rupp H (1999) Excess catecholamine syndrome. Pathophysiology and therapy. Annals of the New York Academy of Science 881:430–444.

Seres J, Stancikova M, Svik K, Krsova D, Jurcovicova J (2002) Effects of chronic food restriction stress and chronic psychological stress on the development of adjuvant arthritis in male Long Evans rats. Annals of the New York Academy of Science 966:315–319.

Sgoifo A, Koolhaas J, De Boer S, Musso E, Stilli D, Buwalda B, Meerlo P (1999a) Social stress, autonomic neural activation, and cardiac activity in rats. Neuroscience and Biobehavioral Reviews 23:915–923.

Sgoifo A, Koolhaas JM, Musso E, De Boer SF (1999b) Different sympathovagal modulation of heart rate during social and nonsocial stress episodes in wild-type rats. Physiology and Behavior 67:733–738.

Spencer RL, Miller AH, Moday H, McEwen BS, Blanchard RJ, Blanchard DC, Sakai RR (1996) Chronic social stress produces reductions in available splenic type II corticosteroid receptor binding and plasma corticosteroid binding globulin levels. Psychoneuroendocrinology 21:95–109.

Stefanski V, Knopf G, Schulz S (2001) Long-term colony housing in Long Evans rats: Immunological, hormonal, and behavioral consequences. Journal of Neuroimmunology 114:122–130.

Steimer T and Driscoll P (2003) Divergent stress responses and coping styles in psychogenetically selected Roman high-(RHA) and low-(RLA) avoidance rats: Behavioural, neuroendocrine and developmental aspects. Stress 6:87–100.

Treit D (1985) Animal models for the study of anti-anxiety agents: A review. Neuroscience and Biobehavioral Reviews 9:203–222.

Treit D, Pinel JPJ, Fibiger HC (1981) Conditioned defensive burying: A new paradigm for the study of anxiolytic agents. Pharmacology, Biochemistry, and Behavior 15:619–626.

Tsuda A, Yoshishige I, Tanaka M (1988) Behavioral field analysis in two strains of rats in a conditioned defensive burying paradigm. Animal Learning Behavior 16:354–358.

Veenema AH, Meijer OC, de Kloet ER, Koolhaas JM (2003) Genetic selection for coping style predicts stressor susceptibility. Journal of Neuroendocrinology 15:256–267.

Weiss JM (1972) Influence of psychological variables on stress-induced pathology. In: Physiology, emotion and psychosomatic illness. CIBA Foundation Symposium (Porter R and Knight J, eds.). Amsterdam: Elsevier.

Weiss JM, Sundar SK, Becker KJ, Cierpal MA (1989) Behavioral and neural influences on cellular immune responses: Effects of stress and interleukin-1. Journal of Clinical Psychiatry 50:43–55.

Yehuda R, McFarlane AC, Shalev AY (1998) Predicting the development of posttraumatic stress disorder from the acute response to a traumatic event. Biological Psychiatry 44:1305–1313.

Immune System

23

HYMIE ANISMAN AND
ALEXANDER W. KUSNECOV

The major function of the immune system is to monitor the organism's internal environment for signs of tissue damage and microbial infiltration (e.g., bacteria and viruses). In considering the factors that govern immune functioning, it seems that in the main, immune processes are similar across species (or strains within a species), although characteristic differences occur within certain immune components. In this report we provide a brief overview of the functioning of the immune system, followed by a description of how immune alterations can affect central nervous system (CNS) processes and behavior in the rat. We also describe how factors that have an impact on psychological processes, most notably stressors, may come to affect immune functioning. In so doing, we introduce numerous caveats concerning the conditions and limitations that determine the nature of the effects observed, making it clear that disentangling the impact of various manipulations on immune activity is complex.

In addition to experimental stressor manipulations that affect immune functioning, laboratory conditions or experimental paradigms may act as stressors that can affect immune competence (or activate products of the immune system, such as cytokines, that could affect CNS activity). In general, stressors can be subdivided into those of a processive nature (i.e., those that involve appraisal of a situation or higher order sensory cortical processing, including events such as exposure to a novel environment, change of social conditions, predator exposure, restraint, cold swim,

foot or tail shock, or conditioned fear cues), which are either psychogenic or neurogenic (being purely psychological in nature or involving physical or painful stimuli, respectively). In addition, stressors may include systemic (e.g. metabolic) insults, such as bacterial or viral infection (Herman and Cullinan, 1997), which in rats induce many (although not all) of the central neurochemical alterations provoked by processive stressors, and thus might also be expected to provoke behavioral changes that often are elicited by aversive events (Anisman et al., 2002). The nature of the immune and cytokine changes induced is dependent on several stressor and organismic characteristics. Of particular note in this regard is that even mild stressors may have an impact on immune functioning of the rat, including standard laboratory manipulations, such as handling and injection procedures, and transport from the breeder. Of course, intense insults such as surgery, especially that involving direct brain manipulations (e.g., lesions, stimulation, cannulation), may have particularly potent actions on central cytokine functioning (Fassbender et al., 2001) and hence may influence neuronal activity and behavioral output.

A BRIEF PRIMER ON
IMMUNITY IN THE RAT

Stimulation of immune cells, as depicted in Figure 23–1, results in an exquisitely orchestrated immune response that exhibits antigen

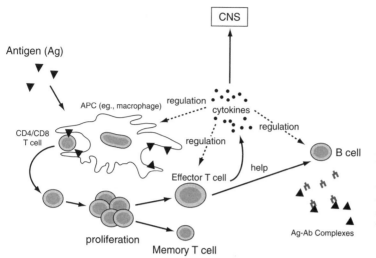

Figure 23–1. Simplified diagram of cellular interactions in the immune system. An immune response is initiated on the presentation of antigen (Ag) by antigen presenting cells (APC) to T and B lymphocytes. On stimulation with Ag, lymphocytes enter a proliferative phase, out of which they emerge as effector cells capable of producing Ab, if B cells, or aiding the latter, if T cells. Regulation of lymphocyte proliferative and effector functions is mediated by cytokines that are produced by most cells of the immune system and can affect central nervous system function.

specificity, learning, and a high degree of autocrine and paracrine regulation. Immunological memory in the immune system, demonstrated by the rapid and robust response to an antigen (i.e., foreign molecule) that the immune system has previously encountered is mediated by T and B lymphocytes (henceforth called T and B cells), the main cells of the immune system responsible for its adaptive or acquired characteristics.[1]

Other cells of the immune system serve accessory functions in that they can present antigen to lymphocytes, regulate the activity of lymphocytes, and enzymatically digest foreign cells (e.g., bacteria) and necrotic tissue through a process called *phagocytosis*. Accessory cells include dendritic cells, which specialize as antigen-presenting cells (APCs), and polymorphonuclear phagocytes (monocytes and macrophages). It should be noted that macrophages are actually monocytes in a mature, more differentiated state that includes phagocytic function—for example, in the CNS, microglial cells assume the role of resident macrophages.

In any given immune response, there are three identifiable processes: induction, activation, and effector function. The first two processes represent stages in which antigen is presented to and stimulates lymphocytes, culminating in their activation. Once activated,

lymphocytes can differentiate into blast cells that mediate effector functions, namely antibody and cytokine production, and cytotoxicity. Antibody production is the primary function of B cells, which can also serve as APCs, whereas T cells perform regulatory and cytotoxic functions.[2] Specifically, there are two subtypes of T cells: T helper (aka CD4-positive cells) and T cytotoxic cells (aka CD8-positive cells). The helper CD4$^+$ T cells (Th cells) were originally defined by their ability to help B cells produce antibody against antigenic stimuli, although it is now known that "help" has a broader definition in that immune responses can be suppressed or downregulated by Th cells, an important self-regulatory function of the immune system that can prevent excessive inflammation and reduce the risk of autoimmunity. This inhibitory or anti-inflammatory function is mediated by the Th2 subset of Th cells, with the Th1 subset principally identified with proinflammatory immune responses.

Activation of B cells leads to production of soluble immunoglobulin (Ig) molecules measurable in bodily fluids (e.g., blood plasma/serum). There are five classes of Ig molecules (IgA, IgD, IgE, IgM, and IgG), which represent the *antibodies* induced by antigen. Most classes of antibody are produced during the primary immune response, although IgM predominates. Reactivation of B cells by antigen results in a secondary humoral response of greater affinity and avidity for the antigen and is mostly IgG.

All T cells express a T-cell receptor (TCR), which recognizes specific antigenic peptides and initiates intracellular signals that result in proliferation and differentiation. Additional surface molecules, CD4 and CD8, also mediate helper and cytotoxic T-cell functions. Thus, $CD4^+$ Th cells are regulatory, whereas $CD8^+$ cytotoxic T cells directly lyse infected and neoplastic cells. Lytic functions are also carried out by large granular lymphocytes called natural killer (NK) cells.

Induction of a T-cell response requires a recognition step in which antigen specifically binds to the TCR. The recognition step between antigen and TCR involves a physical interaction between the T cell and an APC. For example, macrophages, after digesting a large foreign substance by phagocytosis, load the digested peptide fragments onto intracellular class II molecules of the major histocompatibility complex (MHC) that can then present the antigenic peptide to a T cell that expresses a TCR with specific complementarity to the peptide sequence. This results in activation and differentiation of the T cell.

Much of the effector function of the immune response involves the production of soluble substances that promote the elimination of antigen (e.g., antibody molecules). Cytokines, which play a fundamental role in this respect, are protein molecules synthesized and secreted by cells to serve as autocrine/paracrine signaling and growth and differentiation factors. Fibroblasts and endothelial cells secrete cytokines, but so do cells of the immune system (and as is now well known, the CNS), including T cells, monocytes, and macrophages. The major characteristics of cytokines are that (1) they are rarely constitutively expressed, (2) cellular effects are receptor mediated, and (3) they are generally pleiotropic (Thompson, 1998). The latter characteristic is particularly germane to neu-

roscience, because cytokines both affect and are produced in the CNS. Indeed, one of the fastest growing areas in neuroscience research is the role of cytokines in CNS pathology, as well as in normal behavior.

Disturbances of various arms of the immune system are associated with different disease states. Diminished activity of T and B cells has been associated with acquired immunodeficiency syndrome, whereas as reduced NK-cell activity may be related to viral illness and neoplastic disease. Furthermore, it is possible that the recognition abilities of the immune cell may become compromised, or suppressor cells (or inhibitory cytokines) may not operate adequately, leading to the immune system attacking the self, hence favoring the development of autoimmune disorders, such as lupus erythematosus and rheumatoid arthritis (Abbas et al., 2000). Furthermore, dysfunction within any given system can be effected through different processes. Lymphocytes in circulation may be altered, proliferation of these lymphocytes may be compromised, or the killing potential of the lymphocytes may be impaired. Such changes may develop owing to dysfunctions within the immune system itself, or they may reflect the influence of processes (e.g., neuroendocrine) that ordinarily regulate immune activity (Besedovsky and Del Rey, 2001).

In considering analyses of immune functioning, a variety of in vitro and in vivo procedures are available in the rat, but the ultimate test of whether a treatment has positive or negative effects concerns its effects on the organism's well-being in the face of challenge; that is, has vulnerability to pathology been altered? This can be examined in terms of altered vulnerability to viral or bacterial insults, recovery from illnesses or rate of wound healing, and development or exacerbation of pathological symptoms, both in randomly bred animals or in those with specific vulnerabilities (e.g., Fisher 344 versus Lewis rats). In addition, one can examine the effects of the treatment on specific attributes or arms of the

immune system; this includes simply counting the number of specific types of immune
cells (per fixed volume of blood). Although
this does not characterize functional changes
that may have occurred, it is certainly important to know whether a given treatment, such
as a stressor, alters the cell trafficking patterns
of immune cells. Functional attributes, in contrast, include the level of T- or B-cell proliferation in response to specific stimulants (mitogens) that are present in secondary immune
organs (e.g., spleen) or in circulation, antibody
responses to antigenic stimuli, or the killing
ability of specific immune cells on being appropriately challenged (e.g., NK cell toxicity).

INTERACTIONS BETWEEN THE IMMUNE SYSTEM AND THE BRAIN

Multidirectional communication, involving
concurrent and sequential interactions, is
known to occur among the immune, endocrine, autonomic, and central nervous systems, likely involving several processes. For instance, cytokines released from activated
immune cells may influence CNS activity without actually entering the brain parenchyma but
by stimulating afferent vagal fibers (Dantzer,
2001) or receptors at circumventricular and
other vascular regions (Nadeau and Rivest,
1999). Cytokines may also gain access to the
brain at circumventricular organs (Nadeau and
Rivest, 1999) that lack an efficient blood–brain
barrier or through saturable carrier-mediated
transport mechanisms (Banks, 2001), eventually reaching distal sites through volume diffusion (Konsman et al., 2000). Finally, cytokines
may nonselectively stimulate cells of large
blood vessels and small capillaries, thereby permitting greater passage of cytokines into the
brain (Rivest et al., 2000).

Once in the brain, cytokines can bind to
specific cytokine receptors (Cunningham and
De Souza, 1993), thus influencing neuronal
functioning (Anisman and Merali, 1999). The
routes by which these cytokines come to af-

fect limbic processes (e.g., stimulation of the
central amygdala) remain to be determined,
although stimulation of the parabrachial nucleus and paraventricular hypothalamus may
be particularly important in this regard (Buller
and Day, 2002). Ultimately, cytokines affect
neuroendocrine and central neurotransmitter
processes, and vice versa (Dunn, 1995; Anisman and Merali, 1999), and thus it is to be expected that dysfunctions of one system may
have repercussions on other systems.

Signaling of the CNS by peripheral cytokines is complemented by cytokine synthesis within the brain, most likely within astro-
cytes and microglia but possibly within
neuronal tissues (Rivest and Laflamme, 1995).
Indeed, in rats, central cytokine bioactivity increases appreciably in response to a variety of
physical and chemical insults (e.g., brain injury, cerebral ischemia, kainic acid or 6-hy-
droxydopamine lesions, seizure) (Rothwell
and Luheshi, 2000), as well as systemic or central challenge with bacterial endotoxin or
viruses (e.g., Nadeau and Rivest, 1999). Moreover, stressors may influence central cytokine
expression or protein levels (Nguyen et al.,
1998). These various challenges may affect the
regulation of ligand–receptor interactions and
may affect acute-phase reactions (Black, 2002;
Nguyen et al., 2002), thus influencing a wide
range of neurological disease states (Nguyen
et al., 2002). However, it is still unclear under
what conditions cytokines act in a reparatory
capacity or when they promote neuronal
damage.

Aside from contributions to neurological
disorders, inflammatory immune processes
may contribute to psychological disturbances,
such as depression (Maes, 1999). To be sure,
depressive illness has been thought to stem
from neurochemical alterations, such as variations of norepinephrine and serotonin, as
well as neuropeptides such as corticotropin-
releasing hormone (CRH), which may be provoked by stressful encounters. Interestingly,
when administered systemically, cytokines influence hypothalamus-pituitary-adrenal (HPA)

activity and promote hypothalamic and extrahypothalamic neurotransmitter changes reminiscent of those elicited by stressors (Anisman and Merali, 1999, 2002). These neurochemical changes may in turn promote behavioral changes characteristic of depressive states (e.g., induction of anhedonia [Anisman et al., 2002]), which may be antagonized by antidepressant treatments (Merali et al., 2003). Essentially, the view was adopted that the immune system may be part of a regulatory loop that, by virtue of its effects on neurochemical processes, might contribute to the symptoms of mood and anxiety-related disorders (Anisman and Merali, 2002). Indeed, as in rats, in humans receiving cytokine immunotherapy (e.g., interleukin-2 or interferon-α administration in the treatment of certain cancers and hepatitis C), profound depressive symptoms may evolve that may be attenuated by treatment with the selective serotonin reuptake inhibitor paroxetine (Musselman et al., 2001). Of course, it is possible that the cytokines themselves are insufficient to engender depression but rather reflect an interaction between the cytokine treatment and the distress being experienced by the patient with cancer or hepatitis C, especially as the effects of an immunological challenge may be appreciably increased if administered on the backdrop of a sustained stressor regimen (Tannenbaum et al., 2002).

STRESS, CENTRAL PROCESSES, AND IMMUNOLOGICAL ALTERATIONS

Just as immune activation can affect CNS functioning, it appears that stressors, via their impact on neuroendocrine, autonomic, and central nervous system processes, may affect immune activity across a range of different species, including laboratory (e.g., mice, rats, monkeys) and agricultural animals (e.g., fowl, pigs, cattle). It is generally thought that acute stressors provoke several biological defensive responses of adaptive significance. Acute

stressors of modest severity increase HPA activity, neuropeptide changes at hypothalamic and extrahypothalamic sites, and monoamine (norepinephrine, dopamine, and serotonin) synthesis and utilization (Anisman and Merali, 1999; Anisman et al., 2002). With more chronic stressors, further compensatory neurochemical changes evolve, which may also serve to maintain the organism's well-being (Lopez et al., 1999). Yet, if the stressor persists, the wear and tear on biological systems may become excessive (allostatic overload), and the availability of resources sufficiently diminished, culminating in increased vulnerability to pathology (McEwen, 2000).

Individual differences in response to adverse events have long been appreciated, and some rats appear to be relatively vulnerable to stressor-provoked neurotransmitter and neuroendocrine changes, and one can imagine that vulnerability to allostatic load would vary in a similar fashion. Furthermore, stressor encounters, as well as cytokine challenges, may result in the sensitization of neuronal functioning, so that later stressor experiences may elicit more profound neurochemical changes (Tilders and Schmidt, 1999). It has similarly been observed that rat pups subjected to immunological challenges within a few days postpartum may, as adults, display increased stressor reactivity (Shanks et al., 2000), just as early life traumatic events may promote such outcomes in humans (Heim and Nemeroff, 2001). These early life sensitization effects are not limited to neuroendocrine changes but are also evident with respect to immune responses mounted after adult stressor experiences (Shanks et al., 2000).

Just as stressors affect neuroendocrine and central neurotransmitter functioning, both psychogenic and neurogenic insults affect various aspects of immune functioning in the rat, such as suppression of splenic NK cell activity, mitogen-stimulated cell proliferation, the plaque-forming cell response following antigenic stimulation, and macrophage activity (Kusnecov et al., 2001). The specific changes observed are dependent on factors such as stressor severity and history, experiential, organismic, genetic and ontogenetic factors, and the specific immune parameters and compartments (e.g., spleen versus blood) being examined. Moreover, many of the immune alterations are independent of tissue damage associated with stressors, because they are provoked by psychogenic stressors (e.g., stressor-related odors, cues associated with stressors, and psychosocial stressors) (Kusnecov and Rabin, 1994; Kusnecov et al., 2001; Moynihan and Stevens, 2001).

It is important to distinguish between the immune changes exerted as a function of stressor severity and chronicity. For instance, in rats, mild stressors may augment immunity, whereas more protracted stressors may have the opposite effect. From an adaptive perspective, this seems to make sense. Stressful events, as clear threats to the organism's well-being, should encourage increased cell trafficking, cell proliferation, and cytotoxicity. However, as the stressor continues, and the allostatic load increases, the reduced resource availability may render the organism more vulnerable to immunological disturbances (Dhabbar and McEwen, 1999). For instance, acute stressors enhanced the delayed-type hypersensitivity response (DTH), an in vivo measure of T-cell–mediated immunity (Dhabbar and McEwen, 1999; Dhabhar, 2000). This measure requires initial sensitization with antigen, followed days to weeks later by challenge with the sensitizing antigen. This promotes an inflammatory response characterized by increased redness and swelling of the challenged part of the body (typically the footpad or pinnae of the ear). When rats received a single session of restraint immediately before challenge with the sensitizing chemical DNFB, the DTH response of challenged rats was increased, and this effect was sensitive to the severity of the stressor (Dhabhar and McEwen, 1999; Dhabhar, 2000).

It should be underscored that in these stud-ies, the immune-enhancing effects of stressors were examined in relation to elicitation of the DTH response in rats that had previously been exposed to the sensitizing antigen. Thus, the findings may be related to how acute stressors affect the memory T-cell response, rather than how stressors influence the immune response in otherwise naïve T cells. In this regard, acute stressors impaired the DTH response following introduction of sheep erythrocytes into the lung (Blecha et al., 1982), raising the possibility that the nature of the outcomes observed may be related to either differences in the specific antigens used or the localization of antigens to different immunological compartments (dermal tissue versus upper respiratory mucosal surfaces). Furthermore, stressor effects on the DTH response in rats (Flint and Tinkle, 2001) may differ from those seen in mice (Wood et al., 1993), although it is difficult to equate stressor severities across species.

In considering the factors that influence stressor-elicited immune changes, both Kusnecov et al. (2001) and Moynihan and Stevens (2001) made the point that the effects of chronic stressors on immune functioning may be related to the specific immune response being examined. Using in vitro assessment of cell-mediated immunity, such as lymphocyte proliferation to T-cell mitogens, concanavalin A (ConA) or phytohemagglutinin (PHA,) it has commonly been reported that acute stressors suppress rat splenocyte proliferation. Yet, the antigen-specific proliferative response to cholera toxin–sensitized spleen cells can be enhanced by the same stressor (Kusnecov and Rabin, 1993).

As mentioned, acute stressors cause reductions in splenic and blood lymphocyte cellular proliferation to the nonspecific T-cell mitogens (Kusnecov and Rabin, 1994). Such an outcome, however, is not consistently observed after more prolonged stressors (e.g., foot shock, immobilization, isolation). Short periods of social isolation or swim stress inhibited lymphocyte proliferation in blood and spleen lymphocytes, whereas more prolonged stressor exposure reversed this effect (Kusnecov et al., 2001). However, studies using immobilization/restraint or electric shock indicated that the immunosuppressive effects of acute stressors were retained after a chronic stressor regimen (Lysle et al., 1987; Batuman et al., 1990). In explaining these divergent findings, the nature and severity of the stressors used may prove to be relevant. Once again, however, important species differences may also exist in this respects, as acute stressors enhanced the activity of splenic B cells or accessory cells in mice (Lu et al., 1998; Shanks and Kusnecov, 1998).

Of the studies that have examined the effects of stressors on immune functioning, the majority involved neurogenic stressors, whereas some used psychogenic insults. Because these stressors may activate different neural circuits (Lopez et al., 2001; Anisman et al., 2002) and may have different effects on peripheral processes, the fact that both types of stressors generally elicited similar effects is consistent with the notion that the psychological ramifications of stressors were responsible for the observed outcomes. However, it is important to note that not all psychogenic stressors exert similar effects, because neural circuitry activated by predatory stressors may not be the same as that associated with other psychogenic stressors. Indeed, one of the most potent stressors for rats (and humans) is social disruption, a perturbation that may have particularly marked effects on immune functioning and susceptibility to infectious disease (Sheridan, 1998). Interestingly, social defeat in rats followed by lipopolysaccharide administration 1 week later resulted in a diminished corticosterone response, increased circulating interleukin-1 levels, and increased mortality in response to the lipopolysaccharide (Carborez et al., 2002). Similarly, 1 week of social disruption provoked resistance to the immunosuppressive effects of glucocorticoids (Stark et al.,

2001), a situation that can result in excessive production of cytokines in response to lipopolysaccharide and ultimately increased mortality (Quan et al., 2001). Thus, protracted stress does not necessarily vanquish host defense mechanisms against environmental pathogens but may actually give rise to changes that may result in unexpectedly serious health complications.

CONCLUSION

The original conception of the immune system as an autonomously functioning system has been invalidated by research demonstrating neurohormonal regulation of immune cells. This is supported further by the "hardwired" infrastructure of noradrenergic and peptidergic innervation of lymphoid organs. Together with the hormones of the neuroendcrine system, these sympathetic and parasympathetic connections to the immune system are subject to influences from upstream neural processes that respond to environmental stressors, such that any given stressor is capable of communicating to and altering the functional state of the immune system.

As discussed, a number of in vitro and in vivo parameters of immune function can be altered by stressor exposure, although the specific immunological readout depends on a range of stressor and organismic variables that await full characterization. It should be noted that regardless of the immunological impact of the CNS response to a psychogenic or neurogenic stressor, there may be important implications for CNS function in terms of compensatory changes in the immune system. As we discussed, cytokines are the chief regulatory products of the immune system, and it is well established that a number of cytokines influence various motivational and cognitive behaviors, which includes the recruitment of distinct stress-related pathways in the brain, including the locus coeruleus, amygdala, hippocampus, and hypothalamus. Although acute stressors can increase immune responding, this has largely been demonstrated using benign immunological stimuli. However, in principle, among stressed organisms the immune response to an infectious agent could result in excess production of proinflammatory cytokines that serve to differentially affect the CNS relative to nonstressed organisms. Moreover, this would have an impact on a CNS already engaged in adjusting to the psychogenic stressor that initially altered the immune response and thus might affect CNS pathology.

ACKNOWLEDGMENTS

This work was supported by the Canadian Institutes of Health Research (H.A.), U.S. Public Health Service grants DA14186 (A.W.K.) and MH60706 (A.W.K.), and National Institute for Environmental and Health Science Rutgers University and the University of Medicine and Dentistry of New Jersey Center Grant P30 ES05022. The authors are indebted to Zul Merali and Shawn Hayley for their comments. H.A. holds a Canada Research Chair in Neuroscience.

NOTES

1. More detailed information on the immune system can be found in Abbas et al. (2000) and Janeway and Travers (2001).

2. All immune cells, including lymphocytes, originate in the bone marrow, although T cells require further maturation in the thymus gland; B cells mature largely in the bone marrow and fetal liver.

REFERENCES

Abbas AK, Lichtman AH, Pober JS (2000) Cellular and molecular immunology, 4th ed. Philadelphia: WB Saunders Company.

Anisman H, Kokkinidis L, Merali Z (2002) Further evidence for the depressive effects of cytokines: Anhedonia and neurochemical changes. Behavior and Immunity 16:544–556.

Anisman H and Merali Z (2002) Cytokines, stress, and depressive illness. Brain, Behavior and Immunity 16:513–524.

Anisman H and Merali Z (1999) Anhedonic and anxiogenic effects of cytokine exposure. Advances in Experimental and Medica Biology 461:199–233.

Banks WA (2001) Cytokines, CVSs, and the blood-brain-barrier. In: Psychoneuroimmunology, 3rd ed. (Ader R, Felten DL, Cohen N, eds.), pp. 483–498. New York: Academic Press.

Batuman OA, Sajewski D, Ottenweller JE, Pitman DL, Natelson BH (1990) Effects of repeated stress on T cell numbers and function in rats. Behavior and Immunity 4:105–117.

Besedovsky HO and Del Rey A (2001) Cytokines as mediators of central and peripheral immune-neuroendocrine interactions. In Psychoneuroimmunology, 3rd ed. (Ader R, Felten DL, Cohen N, eds.), pp. 483–498. New York: Academic Press.

Black PH (2002) Stress and the inflammatory response: A review of neurogenic inflammation. Behavior and Immunity 16:622–653.

Blecha F, Barry RA, Kelley KW (1982) Stress-induced alterations in delayed-type hypersensitivity to SRBC and contact sensitivity to DNFB in mice. Proceedings of the Society of Experimental and Biological Medicine 169:239–246.

Buller KM and Day TA (2002) Systemic administration of interleukin-1beta activates select populations of central amygdala afferents. Journal of Comparative Neurology 452:288–296.

Carborez SG, Gasparotto OC, Buwalda B. Bohus B (2002) Long-term consequences of social stress on corticosterone and IL-1beta levels in endotoxin-challenged rats. Physiology and Behavior 76:99–105.

Cunningham ET and De Souza EB (1993) Interleukin 1 receptors in the brain and endocrine tissues. Immunology Today 14:171–176.

Dantzer R (2001) Cytokine-induced sickness behavior: Mechanisms and implications. Annals of the New York Academy of Science 933:222–234.

de Groot J, Ruis MA, Scholten JW, Koolhaas JM, Boersma WJ (2001) Long-term effects of social stress on antiviral immunity in pigs. Physiology and Behavior 73:145–158.

Dhabhar FS (2000) Acute stress enhances while chronic stress suppresses skin immunity. The role of stress hormones and leukocyte trafficking. Annals of the New York Academy of Science 917:876–893

Dhabhar FS and McEwen BS (1999) Enhancing versus suppressive effects of stress hormones on skin immune function. Proceedings of the National Academy of Sciences 96:1059–1064.

Fassbender K, Schneider S, Bertsch T, Schlueter D, Fatar M, Ragoschke A, Kuhl S, Kischka U, Hennerici M (2001) Temporal profile of release of interleukin-1β in neurotrauma. Neuroscience Letters 284:135–138.

Flint MS and Tinkle SS (2001) C57BL/6 mice are resistant to acute restraint modulation of cutaneous hypersensitivity. Toxicological Science 62:250–256.

Heim C and Nemeroff CB (2001) The role of childhood trauma in the neurobiology of mood and anxiety disorders: Preclinical and clinical studies. Biological Psychiatry 49:1023–1039.

Herman JP and Cullinan WE (1997) Neurocircuitry of stress: Central control of hypothalamo-pituitary-adrenocortical axis. Trends in Neuroscience 20:78–84.

Janeway C and Travers P (2001) Immunobiology, 5th ed. New York: Garland Publishing.

Konsman JP, Parnet P, Dantzer R (2002) Cytokine-induced sickness behaviour: mechanisms and implications. Trends in Neuroscience 25:154–159.

Kusnecov AW and Rabin BS (1994) Stressor-induced alterations of immune function: mechanisms and issues. International Archives of Allergy and Immunology 105:107–121.

Kusnecov AW and Rabin BS (1993) Inescapable footshock exposure differentially alters antigen- and mitogen-stimulated spleen cell proliferation in rats. Journal of Neuroimmunology 44:33–42.

Kusnecov AW, Sved A, Rabin B (2001) Immunologic effects of acute versus chronic stress in animals. In: Psychoneuroimmunology, 3rd ed. (Ader R, Felten DL, Cohen N, eds.), pp. 265–278. New York: Academic Press.

Lopez JF, Akil H, Watson SJ (1999) Neural circuits mediating stress. Biological Psychiatry 146:461–471.

Lu ZW, Song C, Ravindran AV, Merali Z, Anisman H (1998) Influence of a psychogenic and a neurogenic stressor on several indices of immune functioning in different strains of mice. Brain, Behavior and Immunity 12:7–22.

Lysle DT, Lyte M, Fowler H, Rabin BS (1987) Shock-induced modulation of lymphocyte reactivity: Suppression, habituation, and recovery. Life Science 41:1805–1814.

Maes M (1999) Major depression and activation of the inflammatory response system. Advances in Experimental and Medical Biology 461:25–45.

McEwen BS (2000) Allostasis and allostatic load: Implications for neuropsychopharmacology. Neuropsychopharmacology 22:108–124.

Merali Z, Brennan K, Brau P, Anisman H (2003) Dissociating anorexia and anhedonia elicited by interleukin-1β: Antidepressant and gender effects on responding for "free chow" and "earned" sucrose intake. Psychopharmacology 165:413–418.

Moynihan JA and Stevens SY (2001) Mechanisms of stress-induced modulation of immunity in animals. In: Psychoneuroimmunology, 3rd ed. (Ader R, Felten DL, Cohen N, eds.), pp. 227–249. New York: Academic Press.

Musselman DL, Lawson DH, Gumnick JF, Manatunga A, Penna S, Goodkin R, Greiner K, Nemeroff C, Miller AH (2001) Paroxetine for the prevention of the depression and neurotoxicity induced by high dose interferon alpha. The New England Journal of Medicine 344:961–966.

Nadeau S and Rivest S (1999) Regulation of the gene encoding tumor necrosis factor alpha (TNF-alpha) in the rat brain and pituitary in response in different models of systemic immune challenge. Journal of Neuropathology and Experimental Neurology 58:61–77.

Nguyen MD, Julien J-P, Rivest S (2002) Innate immunity: The missing link in neuroprotection and neurodegeneration. Nature Reviews 3:216–227.

Nguyen KT, Deak T, Owens SM, Kohno T, Fleshner M, Watkins LR, Maier S (1998) Exposure to acute stress induces brain interleukin-1β protein in the rat. Journal of Neuroscience 19:2799–2805.

Quan N, Avitsur R, Stark JL, He L, Shah M, Caligiuri M, Padgett DA, Marucha PT, Sheridan JF (2001) Social stress increases the susceptibility to endotoxic shock. Journal of Neuroimmunology 115:36–45.

Rivest S and Laflamme N (1995) Neuronal activity and neuropeptide gene transcription in the brains of immune-challenged rats. Journal of Neuroendocrinology 7:501–525.

Rivest S, Lacroix S, Vallieres L, Nadeau S, Zhang J, Laflamme N (2000) How the blood talks to the brain parenchyma and the paraventricular nucleus of the hypothalamus during systemic inflammatory and infectious stimuli. Proceedings of the Society for Experimental Biology and Medicine 223:22–38.

Rothwell NJ and Luheshi G (2000) Interleukin 1 in the brain: biology, pathology and therapeutic target. Trends in Neuroscience 23:618–625.

Shanks N and Kusnecov AW (1998) Differential immune reactivity to stress in BALB/cByJ and C57BL/6J mice: In vivo dependence on macrophages. Physiology and Behavior 65:95–103.

Shanks N, Windle RJ, Perks PA, Harbuz MS, Jessop DS, Ingram CD, Lightman SL (2000) Early-life exposure to endotoxin alters hypothalamic-pituitary-adrenal function and predisposition to inflammation. Proceedings of the National Academy of Sciences 97:5645–5650.

Sheridan JF (1998) Norman Cousins Memorial Lecture 1997. Stress-induced modulation of anti-viral immunity. Brain Behavior and Immunity 12:1–6.

Song C, Merali Z, Anisman H (1999) Variations of nucleus accumbens dopamine and serotonin following systemic interleukin-1, interleukin-2 or interleukin-6 treatment. Neuroscience 88:823–836.

Stark JL, Avitsur R, Hunzeker J, Padgett DA Sheridan JF (2002) Interleukin-6 and the development of social disruption-induced glucocorticoids resistance. Journal of Neuroimmunology 124:9–15.

Tannenbaum B, Tannebaum G, Anisman H (2002) Neurochemical and behavioral alterations elicited by a chronic intermittent stressor regimen: Implications for allostatic load. Brain Research 953:82–92.

Thomson A (1998) The cytokine handbook, 3rd ed. San Diego: Academic Press.

Tilders FJH and Schmidt ED (1999) Cross-sensitization between immune and non-immune stressors. A role in the etiology of depression? Advances in Experimental Medicine and Biology 461:179–197.

Wood PG, Karol MH, Kusnecov AW, Rabin BS (1993) Enhancement of antigen-specific humoral and cell-mediated immunity by electric footshock stress in rats. Brain Behavior and Immunity 7:121–134.

Development

Prenatal Behavior

SCOTT R. ROBINSON AND
MICHELE R. BRUMLEY

For much of the history of comparative, physiological, and developmental psychology, the experimental demonstration that a particular behavioral capacity could be expressed soon after birth was sufficient to conclude that the behavior developed without benefit of experience—in other words, that the behavior was "innate." Development before birth was mostly viewed as a process of maturation, involving cellular and tissue-level interactions that were guided by regulatory genes. A new perspective on fetal development has emerged, however, from research that has demonstrated complex behavioral organization, sensory responsiveness, and capacity for learning during the prenatal period. In large part, this research has been advanced by the use of animal models of fetal behavioral development, prominently including studies of the rat fetus (Robinson and Smotherman, 1992a, 1995; Smotherman and Robinson, 1997).

Like other placental mammals, rats are born after a period of physiological dependency on the mother. Because the fetus's life support derives from the placental connection to the mother's uterus, researchers face a significant challenge to gain experimental access to fetal subjects for behavioral study. This challenge has been overcome by methods that involve blockade of the spinal cord of the pregnant rat, permitting surgical exteriorization of the uterus and fetuses while avoiding the activity-suppressing effects of general anesthesia (Smotherman and Robinson, 1991). When immersed in a bath containing physiological saline

maintained at maternal body temperature (37.5° C), this preparation provides visual and experimental access to individual fetal subjects. Fetal rats can be observed without manipulation through the semitransparent wall of the uterus (in utero). Clearer visualization and more direct experimental access are provided by delivering individual fetuses from the uterus into the saline bath, either with the amniotic sac intact (in amnion) or after the embryonic membranes have been removed (ex utero). These methods permit experiments involving video recording of fetal motor behavior, presentation of chemical and tactile stimuli, administration of drugs, or surgical manipulation of the central nervous system (CNS) in test sessions lasting up to 2 hours. Developmental changes in fetal behavior are measured by cross-sectional experimental designs, with fetal subjects provided by pregnancies at different gestational ages from the inception of fetal movement (E16, or 16 days postconception) through term (E21–22). The advent of these methods for studying the behavior of the rat fetus has provided a window on prenatal development through which the fetus may be studied in naturalistic environments.

ECOLOGY OF FETAL DEVELOPMENT

Adult behavior represents a continuous and changing interaction between the animal and its environment. Ethologists have long maintained that it is imperative to recognize the relevant features of an animal's environment if its behavior is to be fully understood. Although some have argued that the same perspective must be adopted in studying the behavior of the fetus (Smotherman and Robinson, 1988; Ronca et al., 1993), too often prenatal behavior is treated as though the developing embryo existed in an environmental vacuum. The prenatal environment is buffered from some perturbations arising in the world outside of the mother. But the fetal milieu also constitutes a rich and dynamically changing environment, in which the fetus develops in relation to the mother and to siblings that share the same needs for space and life support.

The key maternal element of the fetal environment is the uterus, comprising the highly vascularized endometrium that provides a site of attachment for the placenta and a muscular myometrium that elastically constrains the fetus and provides periodic physical stimulation during nonlabor contractions (Jenkin and Nathanielsz, 1994). Within the uterus, each fetus is surrounded by its own embryonic membranes—the chorion and amnion—which maintain a volume of amniotic fluid that bathes the fetus throughout gestation. The concentric envelopes of the maternal abdomen, uterus, chorion, amnion, and amniotic fluid create a series of barriers that block or attenuate sensory stimuli originating in the outside world (particularly visual stimuli, but to a lesser extent, acoustic, mechanical, and chemical stimuli). At the same time, the inner world provides a rich source of chemical stimuli in amniotic fluid, which contains hundreds of chemical compounds (Wirtschafter and Williams, 1957), and mechanical stimuli arising from uterine contractions and locomotion, changes of posture, grooming, and other active movements of the mother (Ronca et al., 1993). Siblings in utero provide another source of both chemical and mechanical stimulation that is likely to influence fetal behavior and development. Androgens produced by male fetuses masculinize the behavior and morphology of female siblings in utero (Meisel and Ward, 1981), and movements of

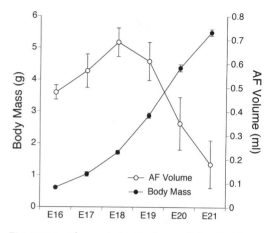

Figure 24–1. Changes in fetal body mass (left axis) and amniotic fluid volume (right axis) during the last 6 days of gestation in the rat (E16-E21). Points show means; bars depict SEM.

adjacent fetuses can affect the amount and pattern of fetal activity (Brumley and Robinson, 2002).

Sensory stimuli arising from the internal environment are likely to exert differential effects at different times during gestation, owing to pronounced changes in fetal growth in relation to the physical environment in utero. For example, in late gestation the fetal rat doubles in mass every 1.4 days, and this growth rate occurs at the same time that amniotic fluid peaks and declines in volume, producing a marked reduction in free space available for fetal movement (Fig. 24–1). Fetuses are sensitive to differences in the physical environment, as evidenced by different rates of spontaneous movement observed in utero, in amnion, and ex utero (Smotherman and Robinson, 1986). For these reasons, studies involving fetal subjects should carefully select and control environmental conditions at the time of behavioral testing.

MOVEMENT AND SENSATION DURING PRENATAL DEVELOPMENT

Perhaps the most prominent aspect of fetal behavior is the fact that all fetuses move spontaneously before birth. Early researchers were in agreement that fetal motility appears to involve a random and purposeless collection of kicks, jerks, and twitches (Hamburger, 1973). Application of quantitative methods for characterizing fetal movements and video technology for analyzing movement sequences in detail have forced a revision in the conclusion that fetal activity lacks organization. Rat fetuses begin to express movement on E16, when small-amplitude movements of the forelimbs, head, and body trunk can be seen. Over the next few days, the rate of movement increases markedly and then remains stable from about E18 through term (Smotherman and Robinson, 1986). Not only does the amount of activity increase, but also movements become more organized, showing quantitative patterning in the form of cyclicity and synchrony. Cyclic motor activity, which involves waxing and waning periods of activity of about one cycle per minute, is characteristic of the spontaneous movements of a variety of species, including fetal and infant rats (Smotherman et al., 1988). Because it continues to be expressed in both forelimbs and hindlimbs after midthoracic spinal cord transection, cyclicity appears to be generated from numerous sources in the CNS (Robertson and Smotherman, 1990).

Synchrony is another form of temporal patterning that involves the nearly simultaneous movement of two or more body parts (Robinson and Smotherman, 1988). Interlimb synchrony is expressed among all pairwise limb combinations, with movements becoming more tightly coupled (intermovement intervals <0.2 second) as gestation proceeds (Kleven et al., in press). High levels of interlimb synchrony are evident between forelimbs as early as E18 of gestation, with synchrony becoming prominent in hindlimbs and between girdles 1 day later (E19). Like cyclicity, interlimb synchrony does not depend on neural control from rostral sources in the CNS; fetuses continue to exhibit synchronous bouts of limb activity after cervical transection of the spinal cord (Robinson et al., 2000).

At the same time that rat fetuses are showing pronounced developmental changes in the organization of spontaneous movement, they also are capable of expressing responsiveness to sensory stimulation in tactile, chemical, and proprioceptive modalities. Fetal rats exhibit simple motor responses to cutaneous tactile stimulation on E16. Tactile stimuli can be presented in a controlled manner through the use of von Frey filaments, which are calibrated to bend when a specified force is applied at a point. Fetuses show simple withdrawal reflexes to punctate stimuli at younger ages but express coordinated wiping or scratching responses directed at the site of stimulation by E20 (Smotherman and Robinson, 1992). General tactile stimulation can be provided by stroking an area of the skin with a soft paintbrush. Fetal responses to such stimulation vary by age and locus of application; gentle stroking has been shown to facilitate coordinated body and hindlimb movements in fetuses near term (Robinson and Smotherman, 1994), including expression of a stereotypic leg extension response (Robinson and Smotherman, 1992a). This response consists of elevation of hindquarters and extension of the hindlegs after brush strokes to the anogenital region, which mimics the licking behavior of the mother directed toward newborn rats (Moore and Chadwick-Dias, 1986).

Responsiveness to chemical stimuli has been extensively studied by infusing small volumes (20 μl) of chemosensory fluids into the mouth of the fetus through a cannula inserted through the lower jaw, with a flanged tip resting on the midline of the tongue (Smotherman and Robinson, 1991). Fetuses show motor responses to chemosensory infusion as early as E18, and by E20 they express coordinated action patterns (Fig. 24–2). Intraoral infusion of lemon extract, or other fluids with strong olfactory components, reliably evokes facial wiping responses in the E20 and E21 rat fetus (Robinson and Smotherman, 1991a). Similarly, infusion of milk can elicit components of postnatal suckling behavior

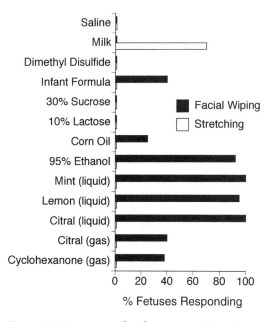

Figure 24–2. Percentage of rat fetuses responding to intra-oral infusions of various chemosensory stimuli (in liquid or gas phase) with facial wiping or stretch responses.

(Robinson and Smotherman, 1992b). It is likely that fetal responses to such compound chemical stimuli are mediated by multiple chemosensory systems. Stimuli that are predominantly perceived in one modality, such as sucrose or citral (an artificial lemon odor), can promote changes in motor activity in the E20 fetus, suggesting that gustatory and olfactory systems are functional by this time. Further, fetuses that are prepared by surgical transection of the olfactory bulbs show a reduction, but not elimination, of responsiveness to complex chemosensory stimuli such as lemon, suggesting that other orosensory systems, including the trigeminal system, play a role in modulating fetal chemosensory responsiveness (Smotherman and Robinson, 1992).

FETAL ACTION PATTERNS

Highly sequenced and coordinated forms of behavior in the rat fetus, commonly referred to as *fetal action patterns*, can be evoked by ex-

perimental presentation of appropriate stimuli. A prominent example of a fetal action pattern is facial wiping. Facial wiping involves placement of the forepaw onto the face followed by a downward stroke of the paw across the face, generally from ear to nose (Robinson and Smotherman, 1991a). This response, which resembles grooming and aversion responses in adult rats (Richmond and Sachs, 1980), can be reliably evoked in rat fetuses by intraoral infusion of novel chemosensory fluids on the last 2 days of gestation (E20 and E21). On E19, however, fetuses show facial wiping when they are tested inside the amniotic sac but not after the membranes are stripped away (Robinson and Smotherman, 1991a). Thus, E19 fetuses appear to require additional biomechanical support to stabilize the head for paw–face contact to be established (Fig. 24–3, left).

Punctate tactile stimulation applied to the perioral region also can evoke facial wiping in the fetus. Facial wiping in response to chemosensory or tactile stimulation has been used as a bioassay to assess neurochemical- and stimulus-induced changes in fetal sensory thresholds (Smotherman and Robinson, 1992). For example, administration of opioidergic drugs, such as the opioid agonist morphine, reduces fetal responsiveness to sensory stimuli. Endogenous opioid activity can be evoked by presenting biologically relevant fluids, such as milk or amniotic fluid, into the mouth of the fetus. When challenged with an intraoral lemon infusion 60 seconds after exposure to milk or amniotic fluid, rat fetuses show diminished facial wiping responses, but wiping is reinstated if subjects are pretreated with an opioid receptor antagonist, such as naloxone (Smotherman and Robinson, 1992; Korthank and Robinson, 1998).

Facial wiping is just one of a collection of organized action patterns expressed by the rat fetus. These behaviors have proved to be useful in the study of the ontogeny and neural substrates of behavior in the prenatal rat (Robinson and Smotherman, 1992a). Experi-

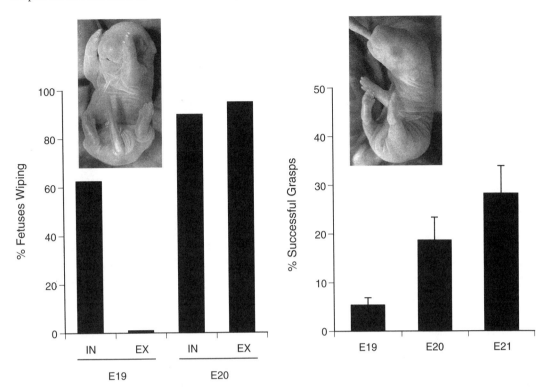

Figure 24–3. (*Left*) Percentage of E19 and E20 rat fetuses exhibiting facial wiping to lemon when tested in amnion (IN) or ex utero (EX). (*Right*) Percentage of mouthing responses that resulted in oral grasping of an artificial nipple by E19, E20, and E21 rat fetuses.

mental manipulations that mimic patterns of postnatal stimulation have revealed that many action patterns have their neurobehavioral origins in the fetus.

ORAL GRASP RESPONSE

Fetuses that are presented with an artificial nipple near the mouth respond with lateral head movements, oral behavior, and active seizing of the nipple (Robinson et al., 1992). Continuous improvement in the ability to grasp the nipple occurs between E19 and E21 (Fig. 24–3, right), just days before functional suckling at the lactating dam's nipple is required of the newborn pup. Oral grasping of an artificial nipple provides an alternative method for infusing fluids into the mouth of the fetus or neonate and has been used as a conditioned stimulus in classical conditioning experiments with fetal rats (see later).

STRETCH RESPONSE

Shortly after milk is delivered into the mouth of the newborn rat during suckling, the pup exhibits dorsiflexion of the back, elevation of the head, and coordinated hindleg extension. Stretch responses can be experimentally elicited in E20 and E21 rat fetuses by the intraoral infusion of bovine light cream, which is similar in fat, water, and other constitutents to mature rat milk. In contrast to the prompt stretch of neonates, the fetal stretch response occurs with an average latency of 180 seconds after infusion of milk (Robinson and Smotherman, 1992b).

ORAL ACTIVITY

Various forms of oral activity in the fetus are evoked after chemosensory and perioral tactile stimulation. These include mouthing in response to oral presentation of milk and mouthing, licking, sucking, and biting in response to perioral presentation of an artificial nipple (Robinson et al., 1992). Oral activity also occurs during spontaneous activity when fetuses remain within the amniotic sac, surrounded by their own amniotic fluid.

ALTERNATED STEPPING

Administration of various neurochemicals to the fetus has been shown to produce bouts of alternated stepping behavior. L-DOPA (L-beta-3,4-dihydroxyphenylalanine) elicits forelimb stepping (Robinson and Kleven, in press) that is similar to the air-stepping of neonatal rats (van Hartesveldt et al., 1990), and serotonin agonists elicit hindlimb stepping (Brumley et al., 2003). It is presumed that these neuroactive drugs engage neural substrates that give rise to functional locomotion in the postnatal animal.

RESPONSE TO UMBILICAL CORD COMPRESSION

Rat fetuses show a stereotyped behavioral and physiological response to acute hypoxia produced by compression of the umbilical cord (Robinson and Smotherman, 1992a). After occlusion of the umbilical cord with a microvascular clamp, fetuses show an initial decrease and then pronounced but transient increase in spontaneous activity, accompanied by heart rate deceleration. Neonatal rats do not show a comparable response to hypoxia, suggesting that the fetal response may be an adaptation unique to the prenatal period.

BRADYCARDIA

Heart rate is a physiological measure that is used as an index of sensory responsiveness in many animal models. Rat fetuses exhibit a transient bradycardia after oral infusion of a novel chemosensory stimulus such as lemon, immediately before expression of the milk-

evoked stretch response, and during hypoxia induced by umbilical cord occlusion. Episodes of tachycardia in response to stimulation have not been reported in the rat fetus but can be expressed by rat pups within 5 to 7 days after birth (Smotherman et al., 1991).

PROPRIOCEPTIVE STIMULI AND MOTOR LEARNING

The expression of species-typical action patterns demonstrates that fetal rats are responsive to exteroceptive stimuli, such as cutaneous, olfactory, and gustatory cues. However, fetuses also are responsive to stimuli arising from their own movements, as demonstrated in a paradigm developed to probe the sensorimotor dynamics of limb movement in the rat fetus (Robinson and Kleven, in press). This motor learning paradigm is similar to experiments conducted with human infants (Thelen, 1994), in which subjects are trained to learn a new pattern of interlimb coordination by restricting the movement of two limbs with an interlimb yoke. To effect yoke motor learning, two legs are fitted with a length of thread that provides a physical connection between the limbs for a 30 minute training period. During training, movement of one leg causes the yoked leg also to move. Thus, proprioceptive feedback is generated by both active and passive movement. Conjugate movements occur when the two legs initiate movement at the same time and follow parallel spatial trajectories. Accordingly, E20 rat fetuses show a significant, gradual increase in conjugate movements between yoked limbs during the training period, suggesting that they can use kinesthetic feedback (in the absence of explicit reinforcement) to alter patterns of interlimb coordination during spontaneous motor activity (Robinson and Kleven, in press). After the yoke is removed and the legs are no longer physically coupled, fetuses continue to show conjugate leg movements for the next 15 to 25 minutes. Such persistence makes it unlikely that the yoke is eliciting only reflexive movements. The elasticity and degree to which the yoke effectively couples the two legs influence the number of conjugate movements during the training period (Fig. 24–4), suggesting that even minor changes in the quality of movement-related feedback can alter the expression of coordinated motor behavior in the rat fetus.

EXPOSURE LEARNING

Considerable evidence indicates that fetuses can modify their behavior as a consequence of prenatal sensory experience. For instance, the rat fetus can habituate to repeated exposures of a sensory stimulus (Robinson and Smotherman, 1995). On E20 and E21, fetuses initially exhibit an increase in motor activity and bradycardia in response to oral infusion of lemon, but responsiveness wanes after 5 to 10 infusions. Dishabituation by infusion of a novel mint solution reinstates fetal responses to lemon, confirming that the response decrement is due to central responsiveness and not

to peripheral effects such as receptor adaptation or effector fatigue.

Evidence also suggests that exposure learning in the fetus provides a scaffold for later behavior. Familiar tastants experienced in utero can be distinguished from novel tastants as measured by overall fetal motor activity a few days later (Robinson and Smotherman, 1991b) and by mouthing and licking behavior in later-term fetuses and neonates (Mickley et al., 2000). Cues present in amniotic fluid (Hepper, 1987) or maternal diet (Hepper, 1988) are preferred during taste or odor tests conducted at postnatal ages. In addition, prenatal exposure to a particular odor cue can influence subsequent associative learning with that odor in newborn pups (Chotro et al., 1991). Because many constituents of maternal diet can cross the placenta to enter fetal circulation and amniotic fluid, simple exposure learning may play an important role in the establishment of dietary preferences during prenatal development (Robinson and Smotherman, 1991b).

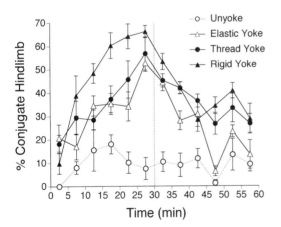

Figure 24–4. Conjugate movements expressed as a percentage of hindlimb activity during yoke motor learning in the E20 rat fetus. During training, the hindlimbs are fitted with an interlimb yoke made of elastic thread (elastic), flexible suture (thread), or thread stiffened with cyanoacrylate (rigid). Following the 30 minute training period, the yoke is removed (*dotted vertical line*). Control subjects are fitted with a yoke that is cut at the beginning of training (Unyoke). Points show mean percent conjugate movements in successive 5 minute intervals; bars depict SEM.

ASSOCIATIVE LEARNING

Associative learning also can be acquired and expressed before birth. In the classical conditioning of activity in the fetal rat (Robinson and Smotherman, 1991b), a neutral conditioned stimulus (sucrose CS) is paired with an unconditioned stimulus that activates fetal behavior (lemon US). After four CS–US pairings, E20 fetuses respond to the CS alone with a conditioned increase in motor activity. Physiological responses also can be conditioned in the fetal rat. For example, intraoral infusions of milk reduce fetal sensory responsiveness by promoting activity in the endogenous opioid system (Smotherman and Robinson, 1992). After paired presentations of an artificial nipple (CS) and milk (US), fetuses exhibit a reduction in cutaneous responsiveness when reexposed to the nipple CS alone, an effect that is mediated by a conditioned increase in opioid activity (Robinson and Smotherman, 1995).

The effects of associative learning in the fetus can have lasting effects on fetal behavior and response to stimulation. In a taste/odor aversion paradigm of perinatal learning, intraperitoneal (IP) injection of LiCl is used as an aversive US (Robinson and Smotherman, 1991b). On E17, injection of mint odor (a neutral CS) into the amniotic sac followed by IP injection of LiCl has an immediate effect of suppressing fetal activity and evoking body curls. When reexposed to the mint CS 2 days later, E19 fetuses exhibit a conditioned motor response that resembles the unconditioned response of E17 fetuses. Likewise, newborns that have been exposed as fetuses to apple juice paired with LiCl injection are less likely to suckle at nipples painted with apple juice and to exhibit longer latencies to move across a short runway suffused with apple odor to suckle from a lactating dam. Therefore, associative learning, like exposure learning, can be expressed by rat fetuses before birth and is capable of influencing postnatal behavioral development.

CONCLUSIONS

Study of the rat fetus in vivo provides a simple mammalian system that permits investigation of early neurobehavioral development. Unlike other model systems that are prominently used in basic neuroscience research, such as surgically reduced preparations or phylogenetically simpler organisms, study of the rat fetus provides an alternative animal model that offers the unique advantage of developmental continuity with behavioral functions in the adult mammal. Research has confirmed that behavioral organization, coordinated movement, sensory responsiveness, and learning all can be expressed in the prenatal rat. Study of the rat fetus thus is well situated to complement existing research approaches with adult rats as well as simpler

systems to understand basic problems in behavioral neuroscience and neurobehavioral development.

ACKNOWLEDGMENT

S.R.R. is supported by National Institute of Child Health and Human Development grant HD-33862.

REFERENCES

Brumley MR, Fleenor RA, Simmons LL, Robinson SR (2003) Serotonin agonists alter motor activity and promote hindlimb stepping in the intact and mid-thoracic transected E20 rat fetus (abstract). Developmental Psychobiology 43:249.

Brumley MR and Robinson SR (2002) Responsiveness of rat fetuses to sibling motor activity: Communication in utero? (abstract). Developmental Psychobiology 41:73.

Chotro MG, Cordoba NE, Molina JC (1991) Acute prenatal experience with alcohol in the amniotic fluid: Interactions with aversive and appetitive alcohol orosensory learning in the rat pup. Developmental Psychobiology 24:431–451.

Hamburger V (1973) Anatomical and physiological basis of embryonic motility in birds and mammals. In: Behavioral embryology (Gottlieb G, ed.), pp. 51–76. New York: Academic Press.

Hepper PG (1987) The amniotic fluid: An important priming role in kin recognition. Animal Behaviour 35:1343–1346.

Hepper PG (1988) Adaptive fetal learning: prenatal exposure to garlic affects postnatal preferences. Animal Behaviour 36:935–936.

Jenkin G and Nathanielsz PW (1994) Myometrial activity during pregnancy and parturition. In: Textbook of fetal physiology (Thorburn GD and Harding R, eds.), pp. 405–414. Oxford: Oxford University Press.

Kleven GA, Lane MS, Robinson SR (in press) Development of interlimb movement synchrony in the rat fetus. Behavioral Neuroscience.

Korthank AJ and Robinson SR (1998) Effects of amniotic fluid on opioid activity and fetal responses to chemosensory stimuli. Developmental Psychobiology 33:235–248.

Meisel RL and Ward IL (1981) Fetal female rats are masculinized by male littermates located caudally in the uterus. Science 220:437–438.

Mickley GA, Remmers-Roeber DR, Crouse C, Walker C, Dengler C (2000) Detection of novelty by perinatal rats. Physiology and Behavior 70:217–225.

Moore CL and Chadwick-Dias AM (1986) Behavioral responses of infant rats to maternal licking: Variations with age and sex. Developmental Psychobiology 19:427–438.

Richmond G and Sachs BD (1980) Grooming in norway rats: the development and adult expression of a complex motor pattern. Behaviour 75:82–96.

Robertson SS and Smotherman WP (1990) The neural control of cyclic activity in the fetal rat. Physiology and Behavior 47:121–126.

Robinson SR and Kleven GA (in press) Learning to move before birth. In: Prenatal development of postnatal functions (Hopkins B and Johnson S, eds.). Westport, CT: Greenwood Publishing Group.

Robinson SR and Smotherman WP (1988) Chance and chunks in the ontogeny of fetal behavior. In: Behavior of the Fetus (Smotherman WP and Robinson SR, eds.), pp. 95–115. Caldwell, NJ: Telford Press.

Robinson SR and Smotherman WP (1991a) The amniotic sac as scaffolding: Prenatal ontogeny of an action pattern. Developmental Psychobiology 24: 463–485.

Robinson SR and Smotherman WP (1991b) Fetal learning: Implications for the development of kin recognition. In: Kin recognition (Hepper PG, ed.), pp. 308–334. Cambridge: Cambridge University Press.

Robinson SR and Smotherman WP (1992a) Fundamental motor patterns of the mammalian fetus. Journal of Neurobiology 23:1574–1600.

Robinson SR and Smotherman WP (1992b) Organization of the stretch response to milk in the rat fetus. Developmental Psychobiology 25:33–49.

Robinson SR and Smotherman WP (1994) Behavioral effects of milk in the rat fetus. Behavioral Neuroscience 108:1139–1149.

Robinson SR and Smotherman WP (1995) Habituation and classical conditioning in the rat fetus: Opioid involvements. In: Fetal development: A psychobiological perspective (Lecanuet JP, Krasnegor NA, Fifer WP, Smotherman WP, eds.), pp. 295–314. New York: Lawrence Erlbaum & Associates.

Robinson SR, Blumberg MS, Lane MS, Kreber L (2000) Spontaneous motor activity in fetal and infant rats is organized into discrete multilimb bouts. Behavioral Neuroscience 114:328–336.

Robinson SR, Hoeltzel TCM, Cooke KM, Umphress SM, Murrish DE, Smotherman WP (1992) Oral capture and grasping of an artificial nipple by rat fetuses. Developmental Psychobiology 25:543–555.

Ronca AE, Lamkin CA, Alberts JR (1993) Maternal contributions to sensory experience in the fetal and newborn rat (*Rattus norvegicus*). Journal of Comparative Psychology 107:61–74.

Smotherman WP and Robinson SR (1986) Environmental determinants of behaviour in the rat fetus. Animal Behaviour 34:1859–1873.

Smotherman WP and Robinson SR (1988a) The uterus as environment: The ecology of fetal experience. In: Handbook of behavioral neurobiology, vol 9, Developmental psychobiology and behavioral ecology (Blass EM, ed.), pp. 149–196. New York: Plenum Press.

Smotherman WP and Robinson SR (1991) Accessibility of the rat fetus for psychobiological investigation. In: Developmental psychobiology: New methods and changing concepts (Shair HN, Hofer MA, Barr G, eds.), pp. 148–164. New York: Oxford University Press.

Smotherman WP and Robinson SR (1992) Prenatal experience with milk: Fetal behavior and endogenous opioid systems. Neuroscience and Biobehavioral Reviews 16:351–364.

Smotherman WP and Robinson SR (1997) Prenatal ontogeny of sensory responsiveness and learning. In: Comparative psychology: A handbook (Greenberg G and Haraway MM, eds.), pp. 586–601. New York: Garland Press.

Smotherman WP, Robinson SR, Hepper PG, Ronca AE, Alberts JR (1991) Heart rate response of the rat fetus and neonate to a chemosensory stimulus. Physiology and Behavior 50:47–52.

Smotherman WP, Robinson SR, Robertson SS (1988) Cyclic motor activity in the fetal rat (*Rattus norvegicus*). Journal of Comparative Psychology 102:78–82.

Thelen E (1994) Three-month-old infants can learn task-specific patterns of interlimb coordination. Psychological Science 5:280–285.

van Hartesveldt C, Sickles AE, Porter JD, Stehouwer DJ (1990) L-DOPA-induced air-stepping in developing rats. Developmental Brain Research 58:251–255.

Wirtschafter ZT and Williams DW (1957) The dynamics of protein changes in the amniotic fluid of normal and abnormal rat embryos. American Journal of Obstetrics and Gynecology 74:1022–1028.

Infancy

25

JEFFREY R. ALBERTS

Born after a mere 22 days of gestation, the Norway rat pup is understandably immature. Over the next 3 to 4 weeks, development is rapid and dramatic. The newborn (day 0 to 4), infant (day 5 to 10), juvenile (days 11 to 17), and weanling (days 17 to 28 and beyond) stages are distinct and dynamic. The present chapter offers brief overviews of pup growth and differentiation of behavioral systems. There follows a kind of "ethogram" of early postnatal behavior in *Rattus norvegicus*, intended to put into a more natural context some of the processes of sensory and motor development, with an emphasis on information that can be used for designing and interpreting tests with immature rats.

SOME DEVELOPMENTAL SEQUENCES

There are a variety of descriptive studies of the young Norway rat's behavior. Willard Small's diary-like daily observations is replete with rich descriptions and some simple but revealing tests (Small, 1899). The account by Bolles and Woods (1964) is more objectively observational, based mostly on daily time-samples of a mother and litter in a laboratory cage. Altman and Sudarshan (1971) described postnatal rat development by reporting the results of a comprehensive battery of motor tests. There are, in addition, accounts and analyses of different features of development—sensory and perceptual ontogenesis (Alberts, 1984), physiological and regulatory development (Adolph, 1971), and different

topics in behavioral maturation. Integrative studies of maternal behavior and pup development are also available (e.g., Rosenblatt, 1965; see also Chapter 27).

With such resources at hand, it is only practical here to recognize some of the striking sequences that comprise the Norway rat's rapid and dramatic behavioral ontogeny from birth to postnatal week 3 or 4, after which it weans to independence.

GROWTH AND SEQUENTIAL CHANGES IN APPEARANCE

Domesticated rat pups weigh about 7 grams at birth (day 0). Once they are cleaned by the dam and begin stable breathing, they develop a rich, red coloration, visible over the furless body. The glabrous pups' skin is so thin that the mother's milk can be seen in their stomach through the abdominal wall. The newborn has mere bulges for eyes and folds of skin where the external ears (pinnae) will form.

Close inspection reveals arrays of fine whiskers on the mystacial pads. The nares (nostrils) are open, because breathing occurs through the nose. There are milk teeth, but the incisors have yet to appear. Paws and claws are nicely formed but the pups can not stand, grip, or ambulate.

Postnatal growth is rapid (20 grams by day 10, 30 grams on day 15). Each day, the pup's general appearance changes. The steady increase in body mass includes skeletal growth and differentiation accompanied by muscle growth by which antigravitational support and movements are sustained. Subcutaneous fat adds some insulation. Beginning on about day 5, there are signs of fur growth and coat coloration is visible. By day 10, the pup is covered with fur, although a coat containing both guard hairs and underhairs is attained only after at least 3 weeks of age, if not later (Gerrish and Alberts, 1995). The pinna of the external ear separates from the head by about day 10, and the external auditory meatus opens on day 12. Eyelids unseal by day 15.

SEQUENCES OF SENSORY DEVELOPMENT

At birth, some, but not all, sensory systems are functional. Moreover, those that are functional in the neonate are not functionally complete (i.e., within each modality, the pups continue to develop responsiveness to a greater range of stimuli, increased sensitivity to lower levels of stimulation, and improved acuity, discrimination, and recognition). Thus, it is important to distinguish between onset of function and the subsequent development of function (Alberts, 1984).

Onset of sensory function is thought to occur via an immutable sequence, possibly universal for all vertebrates (Gottlieb, 1971; Alberts, 1984), with tactile, vestibular, auditory, and visual function beginning in this order. Other modalities have not been sequenced this way, but at birth, a Norway rat pup also displays rudimentary function in its tactile, vestibular, thermal, and chemosensitive systems.

Tactile function is present only on some body regions. Pups respond to punctate probing with a von Frey hair in the perioral area, on the forepaws, and around the anogenital region. There is a general, rostral-to-caudal topographic spread of tactile sensitivity. The vibrissae (whiskers) may be functional as "tactile hairs," but this has not been studied behaviorally. Rather, it has been found that on the day of birth, after 30 minutes of stroking of a vibrissal pad, there was an increase in 2-deoxyglucose uptake in the trigeminal sensory nucleus (Wu and Gonzalez, 1997) ipsilateral to the stimulation.

Prenatal *vestibular* responses to tilting stimulation (angular acceleration) has been demonstrated by movement reactions and by tachycardia reflexes (Ronca and Alberts, 1994, 2000). Newborns demonstrate righting responses (although these become more reliable and robust with age). Indeed, the strength of most vestibular reflexes appears to increase ontogenetically (see Altman and Sudarshan, 1974). Geotaxis has long been considered one of the infant rats' characteristic responses (e.g., Crozier and Pincus, 1929), but this behavior has been reconsidered (Krieder and Blumberg, 1999) and reinterpreted (Alberts et al., in press).

The *chemical* senses are also operative at birth but undergo extensive development. Because the rat is an obligate nose breather, the development of nasal sampling (e.g., sniffing) is significant (Welker, 1964; Alberts and May, 1980a). In one of the only olfactometric assays of chemosensitivity in rat pups, it was found that sensitivity to natural and chemical odors increases gradually, until at least 17 days of age (Alberts and May, 1980b). Trigeminal, vomeronasal, and main (cranial nerve I) chemosensitive receptors transmit olfactory information, and each of these has an ontogenetic timetable. Experience and level of stimulation are components to the timing of these developments (Alberts, 1981; Brunjes, 1994).

Taste function has been demonstrated by introducing fluid stimuli into the mouth and measuring behavioral responses (e.g., Hall and Bryan, 1981). There also are electrophysiological data that support the presence of gustatory function in the rat pup (e.g., Hill, 1987). Again, increased range and levels of sensitivity follow. The pups' impressive abilities to integrate developmentally into their behavior complex information from the chemical senses is exemplified by the findings that cues in amniotic fluid and mother's milk can capture control of olfactory and taste-guided behaviors, such as suckling (Pedersen and Blass, 1982) and food recognition (Galef and Sherry, 1973).

Temperature sensitivity is seen in a newborn's thermogenic response to cooling. Although there are no quantitative assessments of the temperature sense, simply moving a pup from one incubator to another, within a 23° C room, is sufficient to trigger an episode of nonshivering thermogenesis (Efimova, et al., 1992). Temperature sensitivity is also seen in pups' movements on a thermal gradient (Kleitman and Satinoff, 1982). A pup's sensitivity to temperature gradients, indicated by the ability to spatially discriminate temperature differences, probably refelects discrimination of temperature differences across body areas. Related to this, perhaps, is the finding that conductive warmth to the ventrum is a potent reinforcer for newborn rats in an operant task (Flory et al., 1997).

Pups can respond to *acoustic* stimulation before their ears open, but there is a dramatic increase in sensitivity when, around day 12, the auditory meatus opens and drains. Auditory sensitivity increases earlier for low frequencies, with sensitivity to high frequencies developing later (Brunjes and Alberts, 1981).

Rat pups are sensitive to *light* before the eyelids unseal: Pups tend to move away from sources of light (negative geotaxis). When the eyes open, the pups can resolve displays of at least 1°21' visual angle, as evidenced by an optokinetic head nystagmus to a moving array (Brunjes and Alberts, 1981). Depth perception also develops and is experience dependent (Tees, 1976).

Eye opening does not necessarily signal the onset of *pattern sensitivity*. For example, hyperthyroidism accelerates eye opening in rats, but the precocially unsealed lids expose a visual system that is not similarly advanced in its development. The milestone of eye opening cannot be used as a marker for visual development (Brunjes and Alberts, 1981).

SEQUENCES OF MOVEMENT AND POSTURAL DEVELOPMENT

Respiratory movements are prominent in the newborn's behavioral repertoire. The onset and establishment of regular breathing move-

ments occur within the first hour after parturition. Respiration rate in the calm, 1-day-old resting pup is about 2 cycles per second (cps). This basal rate increases to about 4 cps by day 7 to 9 (Alberts and May, 1980a).

The breathing pup also behaves by curling and extending its body along the sagittal plane. When actively curling, it forms a C shape. Conversely, the extension can be so complete that the pup's spine forms the shape of lordosis. Stretching and extending by a newborn within the first couple of hours of postnatal existence create the squirming appearance of newborn rats.

On day 1, the ventrum-down orientation is a pup's primary orientation (c.f., Fraenkel and Gunn, 1940). This is seen when a pup is placed on its back. They actively reorient and right themselves and resume their primary orientation. Pellis et al. (1991) traced ontogenetic changes in the pups' righting (turning from supine to prone). There are developmental sequences of righting strategies, which they analyzed in terms of a sequence of forms, each triggered by tactile and vestibular stimuli. Their detailed account of ontogenetic sequence is striking and provides recognizable behavioral modules that appear in other settings.

When a newborn or infant is prone, its legs are often splayed outward. It may move its head from side to side. Such scanning movements may be a form of olfactory, tactile, or thermal sampling. By day 4 to 5, scanning movements recruit additional spinal segments and the forelimbs. Pups frequently turn their head to one side and extend the contralateral forelimb, effectively pushing, or punting, their body into a partial rotation. Because the hindlimbs tend to be inert at these ages, the pups' "punting behavior" moves them in rough circles.

Pups begin using a quadraped stance around day 10 and then begin to crawl (day 10 to 11) and then walk (day 12 to 13) and run (day 15). The developmental kinematics of limb and interlimb movments involved in locomotion have been described (e.g., Bekoff and Trainer, 1979; Stehouwer and Van Hartesveldt, 2000).

Limb movements during face- and body-washing sequences have been analyzed developmentally (Richmond and Sachs, 1980). Ventral grooming can occur when the pup assumes a stereotyped vertical pose, balanced on hindlimbs and tail. In this position, it can lean forward and groom its ventral surface, even reaching the genital area. Self-grooming increases and then differentiates sexually, with males engaging in more self-stimulation of the anogenital area during the prepubertal period (Moore and Rogers, 1984).

Locomotor activity has been measured developmentally with a variety of "stabilimeter" devices. Under several different, well-controlled conditions, isolated pups gradually increase level of activity from day 1 to at least day 12. Then there is a 10-fold increase in activity, followed by a dramatic diminution to lower, more stable levels (Campbell et al., 1969). Interestingly, when the ontogeny of general activity is measured in the presence of an anesthetized rat dam, the 15-day activity peak is not seen; under such conditions, there is a generally linear increase in activity from days 5 to 30 (Randall and Campbell, 1976). With this observation of different results when developmental measures are made under more natural conditions, we turn to a review of the developing pups' "species-typical" behavior.

AN ETHOGRAM OF A NORWAY RAT'S EARLY LIFE

The newborn rat is a stunningly immature and incomplete creature, when its appearance, morphology, sensory capacities, motor abilities, coordination, and behavioral repertoire are compared with those of the adult. But the same features that are "incomplete" in relation to the adult phenotype can simultaneously be seen as "complete," well articulated, and serving as adaptation when evaluated within the context of newborn's world.

The fetal rat is expelled from the mother's body during a 6 hour labor, with uterine and abdominal contractions that squeeze and push the fetus through the birth canal to the outside world (Ronca et al., 1993). Each newborn is encased within an amniotic sac that the mother orally removes and consumes, along with amniotic fluid and placenta. After the umbilical connection is severed, the newborn rat begins postnatal behavior. Interestingly, the ontogeny of postnatal breathing is facilitated *not by the physiological stimulus* of hypoxia, *but rather by the sensory stimulus* of the birth process, which is composed of specific tactile, proprioceptive, vestibular, and thermal events (Ronca and Alberts, 1995a, 1995b). The operational status of the perinates' sensory-perceptual systems, however immature it may be, is sufficiently sensitive across modalities and adequately "tuned" within each modality that the newborn can respond adaptively to its new world (Alberts and Ronca, 1993).

When the rat dam finishes parturition and has consumed the placentas and cleaned the pups, she gathers the pups and assembles them beneath her body, usually within a constructed nest. This form of contact behavior, or brooding, allows for conductive heat exchange from the dam's body to those of the pups with which she is in contact. Under these conditions, augmented by an insulative nest, the pups' body temperatures rise to about 35° C.

SUCKLING

If breathing is a pup's initial adaptive postnatal behavior, suckling is probably the next. The first nipple attachment, like those that follow throughout the 3 to 4 weeks of suckling, is under olfactory control. Anosmic pups do not suckle (see Alberts, 1976). A maternal cue for suckling was demonstrated by Teicher and Blass (1976; see also Hofer et al., 1976). Washing the dam's ventrum eliminates suckling and painting the nipples with a distillate of the wash reinstates the pups' behavior (Teicher and Blass, 1978). Nevertheless, the olfactory cue or cues that stimulate suckling do *not* attract pups to the nipples. Rather, the odor activates the pups, and their increased activity (extending, scanning, squirming, probing) brings them into contact with the dam and a nipple. Oral reflexes, triggered by perioral tactile cues, are responsible for nipple apprehension.

The specific odors that can activate the pup to suckle appear to be learned. Normally, amniotic fluid and saliva are potent stimuli (Teicher and Blass, 1978). Pederson and Blass (1976) showed that a pup's initial nipple attachment can be activated by an odor introduced into the amniotic fluid on E17, about 5 days before term. The underlying mechanisms of this experience-dependent process and the extent to which this represents a naturalistic form of classical conditioning remain topics of study, but much has been learned about the initiation and expression of suckling (Blass and Teicher, 1980). One implication of these findings is that the pups' normal responses to amniotic fluid are similarly learned.

The typical sequence for mother-initiated nursing bouts begins when the dam investigates the pups, sniffing and licking them and handling them with her forepaws (Rosenblatt, 1965). This stimulation activates them. The pups' stretching and probing induce the kyphosis posture in the dam (Stern, 1988), and the pups attach to nipples. The litter's suckling behavior creates a series of neuroendocrine events in the dam that culminate in simultaneous release of milk to all 12 nipples. The pups in turn display such a dramatic, reflexive, whole-body response "stretch response" to the receipt of milk that the pups' stretch response can be used as a bioassay of intramammary pressure and milk letdown (Lincoln et al., 1973).

After a milk letdown and stretch response, pups often release the nipple and begin to squirm and root into the dam's ventrum. Because a common event in the dam simultaneously triggers the behavior of all of the pups, after a letdown the litter often becomes a seathing mass that settles as the pups reattach to a nipple. This sequence of letdowns in the dam and the explosive behavior of the pups occurs every 6 to 9 minutes during a nursing bout. Viewed in the broader context of developmental time, the greatest frequency and duration of nursing bouts are early in the postnatal period. Letdowns begin to decrease around day 18 (Cramer et al., 1990).

Pups live in close association with the mother's body for much of each day from day 0 to day 15, after which the mother begins to withdraw behaviorally. Leon et al. (1978) suggested that thermal factors limit the mother's ability to remain in prolonged contact with the litter. Stern and Azzara (2000) dispute this interpretation. The issue may involve the thermal parameters that are normally experienced by the dam and the litter in the nest; body heat and metabolic demands of lactation and development are, under any circumstances, central factors in ontogenetic regulations.

Until about day 15, the major limiting factor in the pups' milk intake is the availability of mother's milk (Friedman, 1975; Hall and Rosenblatt, 1977). Satiety mechanisms in the pups do not determine when the pups cease ingesting milk. At early ages, under natural conditions with the dam, the pups are essentially suckling machines that ingest all of the available milk. Nevertheless, regulatory mechanisms for ingestion can be revealed under experimental conditions (Hall and Williams, 1983).

HUDDLING

Contact behavior, or huddling, is also prevalent in rat repertoire. Huddling begins soon after birth and, under most circumstances, is maintained throughout adult life. Figure 25–1 illustrates some typical huddles of pups at two ages. Few behaviors are so fundamental and enduring. As the mother makes longer and more frequent excursions from the nest (Cramer et al., 1990), the huddle becomes the pups' immediate environment (Alberts and Cramer, 1988).

Figure 25–1. Huddles of rat pups at 5 and 20 days of age. Despite the dramatic sensory and motor developments, huddling is a continuous feature of the pups' behavioral repertoire.

Much is known about huddles of pups. A huddle is far more than a pile of bodies. The behavior of individual pups actively forms and maintains the group. Through the behavior of individuals, a group behavior emerges (Alberts, 1978a; Schank and Alberts, 1997a, 1997b). This group displays a form of *group-regulatory behavior* in which its total exposed surface area varies with ambient temperature. By huddling this way, body heat can be retained and metabolic energy conserved (Alberts, 1978a).

The pups' huddling behavior is under multisensory control (Alberts, 1978b) at all ages, with age-related hierarchies of sensory dominance. For instance, thermal stimuli are more salient than olfactory cues for eliciting huddling by 5- and 10-day-old rats, whereas olfactory cues are dominant in the 15- and 20-day-olds (Alberts and Brunjes, 1978). More recent analyses indicate that 7-day-old pups are indifferent to the activity state of adjacent pups, but by day 10 their huddling is affected by the movements and activity state of their littermates (Schank and Alberts, 1997b).

The cutaneous contact derived from huddling is a form of thermotactile stimulation that serves as a reinforcer for the associative learning of odors. Although suckling per se

and the ingestion of milk can both function as reinforcers of behavior, the formation of odor preferences for huddling is not affected by suckling or milk but is apparently induced by thermotactile stimulation (Alberts and May, 1984). Under different conditions, tactile stimulation alone (stroking) can also have reinforcing properties (Sullivan and Hall, 1988).

VOCAL EMISSIONS AND SOME REFLEX-LIKE REACTIONS

When infant and juvenile rat pups are isolated and cool, they emit vocalizations in the 40 to 50 kHz range that are anthropomorphically termed "ultrasounds," although the emissions are easily detectable by adult rats. A burst of high-frequency or ultrasonic vocalization (USV) appears to attract the attention of a mother and can lead to retrieval behaviors (Allin and Banks, 1972).

Two types of analysis prevail in the controls of USV production by rat pups. One is thermal and regulatory; the other emphasizes isolation and inferred internal states, such as distress. Although these views are not mutually exclusive, the literature appears divided and contradictory.

Without doubt, the loss of body heat and decreased body temperature provide a potent stimulus for the onset of USV production. Heating the pup reduces such USVs. A regulatory explanation of infant rat USV posits that cooling creates a homeostatic perturbation in the pup, leading to responses that involve changes in respiration. The pup's altered respiratory maneuvers can produce as a byproduct laryngeal noises in the high-frequency range (due simply to the small size of the pup's anatomy). The oxygen demands of increased metabolic heat production have been hypothesized as a basis of USV; more recently, abdominal maneuvers that enhance blood flow to the heart under cool conditions that alter blood viscosity have also been identified (see Chapter 35).

The alternate view emphasizes the removal and replacement of social or maternal cues (e.g., Hofer and Shair, 1978, 1987). Many of the studies that were intended to demonstrate USV responses to isolation failed to control for the rapid heat loss sustained by the small, thermally fragile rat pups, and it has become clear that these pups are behaviorally and physiologically sensitive to small and brief episodes of cooling (Blumberg et al., 1992). There do appear, however, to be instances in which nonthermal factors can stimulate or augment USV production, requiring more

systematic and integrative studies to attain a solid understanding of this aspect of the pup's behavior.

USVs appear to play a role in attracting the dam to an isolated pup and eliciting retrieval or transport. Brewster and Leon (1980) described the appearance and ontogenetic disappearance of the transport reflex in rat pups, whereby stimulation to the skin on the back of young pups elicits a reflexive vetroflexion of the tail, relative immobility, and raising of the hindlimbs. Such responses augment the mother's efficiency in carrying the pup. Figure 25–2 illustrates some of their findings, including how the transport reflex wanes developmentally and diminishes greatly after day 24, when pups are independently mobile and presumably no longer require maternal transport.

Other "disappearing reflexes" present compelling cases of developmentally coordinated timing and function in relation to the cycle of maternal behavior. At birth, pups are generally incapable of urinating or defecating spontaneously. Nevertheless, in response to anogenital stimulation, they display a micturition reflex. The mother's licking releases urination from the pup, and this behavior not only serves the immediate needs of the pups but it maintains an interactive bond between mother and pups and in fact serves the hy-

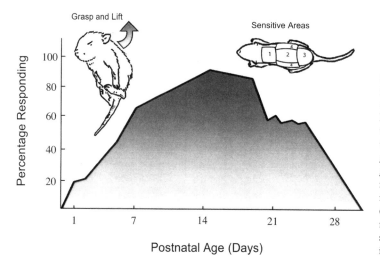

Figure 25–2. The transport reflex (*upper left*) facilitates the mother's ability to carry pups because their limbs are retracted and held close to the body. The reflex is triggered by the tactile stimulation involved in grasping and lifting them; dorsal regions 1 and 2 (*upper right*) were the most sensitive areas for evoking the reflex. The graph shows the strength (probability) of the pups' transport reflex, emerging as they increase in size and waning as they approach independence.

drational needs of the dam (Friedman et al., 1981; Alberts and Gubernick, 1983). Lactating rats have a robust sodium appetite (Richter and Barelare, 1938; Alberts and Gubernick, 1983; Gubernick and Alberts, 1983) and avidly seek and consume the pups' hypotonic urine. As the pups' kidneys develop, solute concentration increases, and as urine becomes hypertonic, spontaneous urination develops and maternal licking disappears.

Moore and Chadwick-Diaz (1986) describe another reflex that supports mother–offspring interactions during licking and that disappears with development. The dam requires access to a pup's anogenital area to efficiently stimulate the micturition reflex. The pup's anogenital region is most accessible when the pup is supine, but, as we have noted, the righting reflex is strong and reliable. A dramatic inhibition of the righting response can be produced by tactile pressure to the pup's ventrum. This is the sort of stimulation that the mother's snout and paws provide during anogenital licking of a supine pup. This inhibition of righting wanes developmentally, with the kind of coordinate timing that suggests ontogenetic adaptation.

WEANING

Weaning is a developmental process unique to and universally represented in mammalian young. Formally, weaning is the shift from ingestion of mother's milk to independent ingestion of solid food and water. In the Norway rat, weaning occurs naturally across postnatal days 14 to 34, or so. It is common laboratory practice "to wean" a litter (i.e., to separate offspring from the dam) around day 21. This can be done, but here we discuss a more gradual, naturally occurring process. In general, weaning represents the achievement of independence from reliance on maternal or parental resources.

In the domain of ingestive behavior, weaning is composed of two processes that proceed simultaneously but largely independently. One is the abandonment of suckling, and the other is the onset of independent feeding and drinking. Hall and Williams (1983) synthesized a large body of data and argue that indeed suckling and feeding are separable behaviors, controlled by different stimuli and mediated by different neural pathways. The developmental sequences during weaning seem consistent with such an analysis.

The mother's milk production peaks around day 15. A litter of eight pups may take 60 ml of milk per day around that time (Friedman et al., 1981). Time spent nursing is about 10 hours per day (Cramer et al., 1990). Contrary to some hypotheses, weaning in rats is *not* caused by a deficit in the dam's ability to produce sufficient milk relative to the nutritive requirements of the developing litter. At 15 days of age, litters consume all of the milk available from a dam, whereas 20-day-old litters do not (Thiels et al., 1988). The diminution in suckling derives from the behavioral interactions with the dam. As the dam's pattern of maternal behavior creates more frequent and longer departures from the nest and litter, pups suckle less. The less they suck, the less likely they are to resume suckling (Thiels et al., 1988). Conversely, the act of suckling maintains itself. Thus, it appears that the decrease in a pup's suckling response to a dam diminishes as a consequence of decreased access. In addition, there is a separate but coordinate set of experiences that lead to the onset of feeding, food site selection, and taste preferences. Homeostatically regulated aspects of intake also develop, in part on the basis of experience.

Onset of feeding begins with sampling of solid foods. Depending on the configuration and characteristics of the pups' environment, such sampling begins between days 14 and 16. Galef and associates provided a rich account of the cues and behavioral processes that contribute to the sampling and selection of diet by weanling rats. After 2 weeks of postnatal life, a rat pup has experienced a variety of chemical cues in its environment, including flavors in mother's milk; odors associated with maternal and sibling urine, feces, saliva, and breath; and secretions of the sebaceous and preputial glands. Thus, when the young rat ventures from the nest, there are numerous cues that it can and does recognize as familiar. For example, pups readily detect "residual cues" left by conspecifics at a feeding site (Galef and Heiber, 1976).

Wild and domesticated rat pups reared in seminatural environments in a laboratory make their first departures from the natal nest between days 16 and 19 (Galef and Clark, 1971). The mother is usually outside the nest entrance when pups egress, but experimental analyses indicate that the pups are not seeking the dam. Rather, they respond to visual, acoustic, and/or olfactory cues (Leon and Moltz, 1972; Alberts and Leimbach, 1980; Galef, 1983). The cues that trigger egression from the nest and approach to feeding sites are remarkably general. There are no preferences for the mother in relation to another adult (Galef and Clark, 1971), and it seems that mere stimulus strength determines probability of approach (Gerrish and Alberts, 1995).

Adult rats tend to congregate at feeding sites. When pups approach conspecifics, it is likely that they approach an area where adults are feeding. It appears that the onset of solid food intake may involve the power of a general approach mechanism bringing the pup into the vicinity of safe, ingesta that has been or is being eaten by adult conspecifics. Once the pups are in the vicinity of food that adults are eating or an area where rats have eaten and where there remain some residual cues, they may make recognize flavors that have been present in mother's milk (Galef and Henderson, 1972) and sample such foods if the rat they have suckled from has eaten there. They also learn associations between food odors borne by rats, based on chemicals in the breath of the conspecifics (Galef et al., 1988). By the time pups egress from the nest, they are capable of rapidly recognizing and learning which substances are associated with

postingestional signals of nutritive content. In addition, they are equipped with basic homeostatic controls for regulated intake of calories, electrolytes, and water. With such capabilities, the pups are capable of sustaining themselves after only a few weeks of postnatal development.

DESIGNING AND INTERPRETING TESTS WITH RAT PUPS

A full discussion of the strategies for testing young rats is unfortunately beyond the scope of the present chapter. It can be inferred from the material presented briefly herein that the special features of the immature rat and their rapid developmental transformations present both challenges and opportunities for researchers who want to probe them for systematic data and insights into developmental status.

Context, both global and proximal, can be a powerful factor in testing the immature rat, more so than for the adult. Small changes can make big differences in outcome. Air and surface temperatures are vitally important and may require careful control and monitoring for reliable results. The behavioral and physiological effects and the hedonic value of temperatures vary with age, so valid cross-age comparisons may require testing at different temperatures if behavioral state or metabolic rate is to be equated. Newborns display unique behavioral capabilities when testing is conducted at warm air temperatures (Hall, 1979; Johanson and Hall, 1980).

Olfactory context also can be important. Testing in the presence of familiar odors can enhance learning and memory performance.

Knowledge of the pups' behavioral repertoire during development can be used to craft useful and robust measures. Similarly, an appreciation of the components and timing of maternal activities can be very important in calibrating schedules of deprivation or reward. The pups' physiological characteristics can require careful calibration for the effects of privation, including dehydration, stomach clearing and stomach fill, across ages.

Knowledge of the developmental sequences in relation to developmental niches (Alberts and Cramer, 1988) can be used to craft useful and robust stimuli for eliciting and rewarding behavior. It is always important to be aware of the task demands with pups at each point in their development so that performance deficits are not misattributed.

This information, although detailed and multileveled, is not cause for despair. Awareness of the pups' natural environment and their species- and age-typical repertoires can usually explain, if not predict, relations and patterns that are coherent and manageable and that can be made tractable for testing and understanding.

References

Alberts JR (1976) Olfactory contributions to behavioral development in rodents. In: Mammalian olfaction, reproductive processes, and behavior, pp. 67–93. New York: Academic Press.

Alberts JR (1978a) Huddling by rat pups: Group behavioral mechanisms of temperature regulation and energy conservation. Journal of Comparative and Psysiological Psychology 92:231–240.

Alberts JR (1978b) Huddling by rat pups: Multisensory control of contact behavior. Journal of Comparative and Physiological Psychology 92:220–230.

Alberts JR (1981) Ontogeny of olfaction: Reciprocal roles of sensation and behavior in the development of perception. In: Development of perception: Psychobiological perspectives, Vol 1 (Aslin RN, Alberts JR, Petersen MR, eds.), pp. 321–357. New York: Academic Press.

Alberts JR (1984) Sensory perceptual development in the Norway rat: A view toward comparative studies. In: Comparative perspectives on memory development (R Kail and N Spear, eds.), pp. 65–101. New York: Plenum.

Alberts JR and Brunjes PC (1978) Ontogeny of thermal and olfactory determinants of huddling in the rat. Journal of Comparative and Physiological Psychology 92:897–906.

Alberts JR and Cramer CP (1988) Ecology and experience: Sources of means and meaning of developmental change. In: Handbook of behavioral neurobiology, Vol 9 (Blass EM, ed.), pp. 1–39. New York: Plenum Publishing Corp.

Alberts JR and Gubernick DJ (1983) Reciprocity and resource exchange: A model of parent-offspring relations. In: Symbiosis in parent-offspring interactions (Rosenblum LA and Moltz H, eds.). New York: Plenum Press.

Alberts JR and Leimbach MP (1980) The first foray: Maternal influences in nest egression in the weanling rat. Developmental Psychobiology 13:417–429.

Alberts JR and May B (1980a) Development of nasal respiration and sniffing in the rat. Physiology and Behavior 24:957–963.

Alberts JR and May B (1980b) Ontogeny of olfaction: Development of the rat's sensitivity to urine and amyl acetate. Physiology and Behavior 24:965–970.

Alberts JR and May B (1984) Nonnutritive, thermotactile induction of filial huddling in rat pups. Developmental Psychobiology 17:161–181.

Alberts JR, Motz B, and Schank JC (2004) Positive geotaxis in rats: A natural behavior and an historical correction. Journal of Comparative Psychology 118:123–132.

Alberts JR and Ronca AE (1993) Fetal experience revealed by rats: Psychobiological insights. Early Human Development 35:153–166.

Allin JT and Banks EM (1972) Functional aspects of ultrasound production by infant albino rats (Rattus norvegicus). Animal Behavior 20:175–185.

Altman J and Sudarshan K (1975) Postnatal development of locomotion in the laboratory rat. Animal Behaviour 23:896–920.

Bekoff A and Trainer W (1979) The development of interlimb co-ordination during swimming in postnatal rats. Journal of Experimental Biology 83:1–11.

Blass EM and Teicher MH (1980) Suckling. Science 210:15–22.

Blumberg MS, Efimova IV, Alberts JR (1992) Thermogenesis during ultrasonic vocalization by rat pups isolated in a warm environment: A thermographic analysis. Developmental Psychobiology 25:497–510.

Bolles RC and Woods PJ (1964) The ontogeny of behaviour in the albino rat. Animal Behaviour 12:427–441.

Brewster J and Leon M (1980) Facilitation of maternal transport by Norway rat pups. Journal of Comparative and Physiological Psychology 94:80–88.

Brunjes PC (1994) Unilateral naris closure and olfactory system development. Brain Research Reviews 19:146–160.

Brunjes PD and Alberts JR (1981) Early auditory and visual function in normal and hyperthyroid rats. Behavioral and Neural Biology 31:393–412.

Campbell BA, Lytle LD, Fibiger HC (1969) Ontogeny of adrenergic arousal and cholinergic inhibitory mechanisms in the rat. Science 166:637–638.

Cramer CP, Thiels E, Alberts J (1990) Weaning in rats: I. Maternal behavior. Developmental Psychobiology 23:479–493.

Crozier WJ and Pincus G (1929a) Analysis of the geotropic orientation of young rats. I. Journal of General Physiology 13:59–80.

Fleming A (1965) Maternal behavior. In: Determinants of infant behavior, Vol 3 (Foss BM, ed.). London: Methuen & Co Ltd.

Flory GS, Langley CM, Pfister JF, Alberts JR (1997) Instrumental learning for a thermal reinforcer in 1-day-old rats. Developmental Psychobiology 30:41–47.

Fraenkel GS and Gunn DL (1940) The orientation of animals. London: Oxford University Press.

Friedman MI (1975) Some determinants of milk ingestion in suckling rats. Journal of Comparative and Physiological Psychology 89:636–647.

Friedman MI, Bruno JP, Alberts JR (1981) Physiological and behavioral consequences in rats of water recycling during lactation. Journal of Comparative and Physiological Psychology 95:26–35.

Galef BG (1983) Utilization by Norway rats (R. norvegicus) of multiple messages concerning distant foods. Journal of Comparative Psychology 97:364–371.

Galef BG and Clark MM (1971) Social factors in the poison avoidance and feeding behavior of wild and domesticated rat pups. Journal of Comparative and Physiological Psychololgy 75:341–357.

Galef BG and Heiber L (1976) Role of residual olfactory cues in the determination of feeding site selection and exploration patterns of domestic rats. Journal of Comparative and Physiological Psychology 90: 727–739.

Galef BG and Henderson PW (1972) Mother's milk: A determinant of the feeding preferences of weaning rat pups. Journal of Comparative and Physiological Psychology 78:213–219.

Galef BG, Mason JR, Preti G, Bean NJ (1988) Carbon disulfide: A semiochemical mediating socially-induced diet choice in rats. Physiology and Behavior 42:119–124.

Galef BG and Sherry DF (1973) Mother's milk: A medium for transmission of cues reflecting the flavor of the mother's diet. Journal of Comparative and Physiological Psychology 83:374–378.

Gerrish CJ and Alberts JR (1995) Differential influence of adult and juvenile conspecifics on feeding by weanling rats (Rattus norvegicus): A size-related explanation. Journal of Comparative Psychology 109:61–67.

Gottlieb G (1971) Ontogenesis of sensory function in birds and mammals. In: The biopsychology of development (Tobach E, Aronson LR, Shaw E, eds.). New York: Academic Press.

Gubernick DJ and Alberts JR (1983) Maternal licking of young: Resource exchange and proximate controls. Physiology and Behavior 31:593–601.

Hall WG (1979a) Feeding and behavioral activation in infant rats. Science 205:206–209.

Hall WG (1979b) The ontogeny of feeding in rats: I. Ingestive and behavioral responses to oral infusions. Journal of Comparative and Physiological Psychology 93:977–1000.

Hall WG and Bryan TE (1980) The ontogeny of feeding in rats: II. Independent ingestive behavior. Journal of Comparative and Psychological Psychology 94:746–756.

Hall WG and Rosenblatt JS (1977) Suckling behavior and intake control in the developing rat pup. Journal of Comparative and Physiological Psychology 91:1232–1247.

Hall WG and Williams CL (1983) Suckling isn't feeding, or is it? A search for developmental continuities. Advances in the Study of Behavior 13:219–254.

Hill DL (1987) Development of taste responses in the rat prabrachial nucleus. Journal of Neurophysiology 57:481–495.

Hofer MA, Brunelli SA, and Shair H (1993) Ultrasonic vocalization responses of rat pups to acute separation and contact comfort do not depend on maternal cues. Developmental Psychobiology 26:81–95.

Hofer MA and Shair H (1978) Ultrasonic vocalization during social interaction and isolation in 2-week-old rats. Developmental Psychobiology 11:495–504.

Johanson IB and Hall WG (1980) The ontogeny of feeding in rats: III. Thermal determinants of early ingestive responding. Journal of Comparative and Physiological Psychology 94:977–992.

Kleitman N and Satinoff E (1982) Thermoregulatory behavior in rat pups from birth to weaning. Physiology and Behavior 29:537–541.

Kreider JC and Blumberg MS (1999) Geotaxis in 2-week-old Norway rats (Rattus norvegicus): A reevaluation. Developmental Psychobiology 35:35–42.

Leon M, Croskerry PG, Smith GK (1978) Thermal control of mother-young contact in rats. Physiology & Behavior 21:793–811.

Leon M and Moltz H (1972) Maternal pheromone: Discrimination by pre-weaning albino rats. Physiology and Behavior 7:265–267.

Lincoln DW, Hill A, Wakerly JB (1973) The milk-injection reflex of the rat; An intermittent function not abolished by surgical levels of anesthesia. Journal of Endocrinology 57:459–476.

Moore CL and Chadwick-Diaz AM (1986) Behavioral responses of infant rats to maternal licking: Variations with age and sex. Developmental Psychobiology 19:427–438.

Moore CL and Rogers SA (1984) Contribution of self-grooming to onset of puberty in male rats. Developmental Psychobiology 17:243–253.

Pedersen PE and Blass EM (1982) Prenatal and postnatal determinants of the 1st suckling episode in albino rats. Developmental Psychobiology 15:349–355.

Pellis VC, Pellis SM, Teitelbaum P (1991) A descriptive analysis of the postnatal development of contact-righting in rats (Rattus norvegicus). Developmental Psychobiology 24:237–267.

Randall PK and Campbell BA (1976) Ontogeny of behavioral arousal: Effect of maternal and sibling presence. Journal of Comparative and Physiological Psychology 90:453–459.

Richmond G and Sachs BD (1980) Grooming in Norway rats: The development and adult expression of a complex motor pattern. Behaviour 75:82–96.

Richter CP and Barelare B (1938) Nutritional requirements of pregnant and lactating rats studies by the self-selection method. Endocrinology 23:15–24.

Ronca AE and Alberts JR (1994) Sensory stimuli associated with gestation and parturition evoke cardiac and behavioral responses in fetal rats. Psychobiology 22:270–282.

Ronca AE and Alberts JR (2000) Effects of prenatal spaceflight on vestibular responses in neonatal rats. Journal of Applied Physiology 89:2318–2324.

Ronca AE and Alberts JR (1995) Simulated uterine contractions facilitate fetal and newborn respiratory behavior in rats. Physiology and Behavior 58:1035–1041.

Ronca AE and Alberts JR (1995) Maternal contributions to fetal experience and the transition from prenatal to postnatal life. In: Fetal behavior: A psychobiological perspective (Lecanuet J-P, Krasnegor N, Smotherman WP, eds.), pp. 331–351. Hillsdale, NJ: Lawrence Erlbaum Associates.

Ronca AE, Lamkin CA, Alberts JR (1993) Maternal contributions to sensory experience in the fetal and newborn rat. Journal of Comparative Psychology 107:61–74.

Rosenblatt JS (1965) The basis of synchrony in the behavioral interaction between the mother and her offspring in the laboratory rat. In: Determinants of infant behavior, Vol 3 (Foss BM, ed.), pp. 3–44. London: Methuen & Co Ltd.

Schank J and Alberts J (1997) Self-organized huddles of rat pups modeled by simple rules of individual behavior. Journal of Theoretical Biology 189:11–25.

Schank JC and Alberts JR (1997) Aggregation and the emergence of social behavior in rat pups modeled by simple rules of individual behavior. In: International Conference on Complex Systems (Bar-Yam Y, ed.), pp. 1–8. Nashua, NH: New England Complex Systems Institute.

Small WS (1899) Notes on the psychic development of the young white rat. American Journal of Psychology 11:80–100.

Stehouwer DJ and Van Hartesveldt C (2000) Kinematic analyses of air-stepping in normal and decerebrate preweanling rats. Developmental Psychobiology 36:1–8.

Stern JM (1988) A revised view of the multisensory control of maternal behaviour in rats: Critical role of tactile inputs. In: Ethoexperimental analysis of behaviour (Blanchard RJ, Brain PF, Blanchard DC, Parmigiani S, eds.). Il Ciocco: Martinus Nijhoff.

Stern JM and Azzara AV (2000) Thermal control of mother-young contact revisited: Hyperthermic rats nurse normally. Physiology and Behavior 77:11–18.

Stone EA, Bonnet KA, Hofer MA (1976) Survival and development of maternally deprived rats: Role of body temperature. Psychosomatic Medicine 38:242–249.

Sullivan RM and Hall WG (1988) Reinforcers in infancy: Classical conditioning using stroking or intraoral infusions of milk as USC. Developmental Psychobiology 21:215–224.

Tees R (1976) Perceptual development in mammals. In: Studies on the development of behavior and the nervous system, Vol 3 (Gottlieb G, ed.), pp. 282–326. New York: Academic Press.

Teicher MH and Blass EM (1976) Suckling in newborn rats: Eliminated by nipple lavage, reinstated by pup saliva. Science 193:422–425.

Thiels E, Cramer CP, Alberts JR (1988) Behavioral interactions rather than milk availability determine decline in milk intake of weanling rats. Physiology and Behavior 42:507–515.

Welker WI (1964) Analysis of sniffing of the albino rat. Behaviour 22:223–244.

Wu CC and Gonzalez MF (1997) Functional development of the vibrissae somatosensory system of the rat: (14c)2-Deoxyglucose metabolic mapping study. Journal of Comparative Neurology 384:323–336.

Adolescence

RUSSELL W. BROWN

The rapid development of cognitive and motor behavior in the laboratory rat is the essence of survival for the animal. Rats are weaned from the mother at approximately 21 days of age. Therefore, within 21 days, the animal must have the ability to avoid predators, locate food caches, and consume food and is preparing to establish a home territrory. Based on this rapid development, the rat provides an excellent model for the development of behavior and the brain structures that mediate these behaviors. The following sections describe the appearance of motor abilities, social and play behavior, sexual maturity, and sensory function, followed by a section on cognitive function, including a brief discussion of the development of simple stimulus associations made in conditioned taste aversion followed by complex associations necessitated in spatial memory. Finally, a brief discussion is included on the development of brain structures and their possible role in these behaviors.

APPEARANCE OF BEHAVIORS

AMBULATORY ABILITIES

The newborn rat on the first day of life (postnatal day 1 [P1]) is essentially motionless but very quickly develops behaviors that can help it to secure and locate food later in life. By P8, the animal is able to crawl but does not have much use of its hindlimbs. By P12 and P13, the raised posture necessary for walking develops, although the animal does not yet move swiftly and the hindlimbs often slip and are dragged

behind the body. The important act of rearing on the hindlimbs presupposes functional maturation of the hindlimbs and does not emerge with appreciable frequency in the open field until P18. Rearing on the hindlimbs is typically observed in aroused adult rats and may represent an acute form of investigatory response, which may reflect a preparatory move for climbing. By P21, most animals are able to traverse a 0.5 cm wide path, ascend and descend a wire mesh or ladder surface, and ascend or descend a rope with few errors (Altman and Sudarshan, 1975). Therefore, the rat develops many ambulatory skills that are at least approaching adult-like abilities by the time it is weaned from the mother at P21.

TESTING

One of the simplest tasks used to test gross ambulatory ability in the rat is the locomotor box, or locomotor arena. The arena used to test locomotor behavior is typically covered with a network grid of infrared beams. Each time an infrared beam is broken by the rat, an activity count is scored, either manually by the experimenter or a computer program attached to the arena. This task has been used prevalently in behavioral pharmacology research to model addiction behavior (for a review, see Kalivas and Pierce, 1997), and automated computer programs have been developed to provide a variety of measurements for locomotor behavior, including horizontal and vertical activity, bouts of movement, and total distance traveled to name a few.

ABILITY TO EAT AND OBTAIN FOOD

Handling and Food-Eating Posture

As adults, rats are able to use skilled movements to reach, manipulate, and eat food. Handling food with the forepaws requires bimanual coordination in the rat, which is the ability of an organism to use both forelimbs while carrying out two different behaviors to solve a task. A recent systematic study of the development of food-obtaining behavior was performed by Brenda Coles and Ian Whishaw (Brenda Coles, unpublished master's thesis). The typical eating posture is characterized by a rat sitting back on its haunches using the hindlimbs as the base of support for the body in a slightly splayed manner. The forelimbs are held underneath the snout and are used to hold and manipulate food (Fig. 26–1). Therefore, for a rat to be able to eat food in a typical adult manner, forelimbs must be sufficiently developed to manipulate the food to eat, and the hindlimbs must be sufficiently mature to support the animal's body weight. It is not until about P18 that pups begin to sit up on their haunches when eating small food pellets; they are still close to the ground with the hindlegs splayed outward in an exaggerated manner, providing a wide base of support. By P21, eating posture becomes sexually dimorphic, in that male rats are able to sit unaided in an adult-like posture to eat all sizes

Figure 26–1. Adult-like eating posture in the rat. When a rat is eating a piece of food, the hindlimbs provide support for the body, whereas the forelimbs are free to manipulate the piece of food. (Adapted from Coles BLK and Whishaw IQ, unpublished data.)

of food pellets, but females are still using littermates or objects for support to attain the same goal. By P24, this sexual dimorphism has disappeared, and both female and male rats are able to eat with an adult-like posture.

Reaching

Another skilled motor behavior that develops in the rat is reaching for food. A reach consists of lifting the forelimbs from the ground, positioning the elbows inward, so that the paws are adjacent to the mouth and clasping the food with the digits. These movements are executed mainly with the upper forelimb. As the limb is positioned for grasping, the aperture of the digits is adjusted to anticipate the size of the food and the food is grasped and manipulated with the tips of the digits. The food is then supinated approximately 90° as the paw is retracted in toward the body and then supinated 90° again for the rat to retrieve the food with its mouth from the paw (Whishaw and Tomie, 1989). Results have shown that rats do not attempt to reach for food until P19, but at this point reaches are not successful. This may be due to the fact that rats are unable to aim their body or demonstrate pronation. Reaching attempts become more frequent by P21. By P23, rats attempt to use different types of strategies to become somewhat succesful at reaching for the food, but the precise reaching does not develop until about P26.

Reaching can be studied in at least two different ways. In the tray-reaching test, rats can be trained to reach for chicken feed placed in a tray mounted approximately 1 cm in front of a cage equipped on one side with metal bars spaced approximately 0.5 cm apart. A more complex version of the skilled reaching task is the *single-pellet precision reaching task*. In this task, the rat must reach through a narrow slot to obtain a single pellet of food placed on a shelf. Both of these tasks have proved useful for testing of skilled motor behavior in rats, with animals learning the tray-reaching task more rapidly than the single-pellet precision reaching task, indicating that precision skilled reaching may be the more complex motor behavior to perform for the rat.

Securing of the Food from Conspecifics

There are two behaviors that have been studied in laboratory rats to secure food from conspecifics: robbing and dodging. *Robbing* behavior is characterized by a rat walking up alongside the other rat's body from the rear and attempting to grasp the food either with the mouth or with a reach of one of the forelimbs. *Dodging* is a defensive tactic in which the feeding rat uses forequarter rotation and hindlimb stepping movements to escape from a conspecific attempting to steal the food (Whishaw and Tomie, 1989). Successful dodging requires a full 180° turn initiated by a contraversive rotation of the front half of the body away from the robber, with steps being taken ipsilateral to the turn direction by both the forelimb and hindlimb. Robbing behavior appears prevalent beginning at P17, and rats are successful at stealing food from conspecifics. One reason that robbers are so successful at this age is that pups do not seem to be able to dodge the perpetrator, likely due to the fact that the hindlimbs have not yet reached maturity to aid in dodging the robber. Animals begin to attempt to dodge the robber at P19 but are not very successful, and dodging does not become effective or near maturity until approximately P25 (B. L. K. Coles and I. Q. Whishaw, unpublished data).

Social/Play Behavior

Rats engage in various forms of play behavior, with social play in the form of playing fighting being among the most commonly reported. The frequency of play fighting reaches its peak during adolescence and declines after puberty. In rats, play fighting involves attack and defense of the nape of the neck, which, if contacted, is gently nuzzled. As juveniles (beginning around P30), the most frequently used defensive tactic is to rotate to the supine position when the nape

is contacted. This results in an on-top/on-bottom orientation referred to as *pinning*. Different play behaviors develop at different rates, and play fighting becomes rougher, especially in males, as the animal reaches adult maturity (approximately P60). One of the more frequent play behaviors observed immediately after weaning is *wrestling*, in which two animals roll and tumble with one another. Another frequent play behavior that often initiates play is *pouncing*; animals exhibiting this behavior invariably engage in a play bout that is long enough for the recipient animal responds. This typically precedes any other behavior in the play sequence. As animals reach late adolescence (around P50), *boxing* becomes prevalent, in which two animals stand upright facing one another and make pawing movements toward each other (Meaney and Stewart, 1981).

Sensory Function

When the rat is born, both its eyes and ears are closed. The only sensory functions that are functional on the first day of life are taste, odor, and touch. This is known because the rat can learn associations involving taste, odor, or touch stimuli on the first day of life and even has the abilities before birth (Smotherman, 1982). Obviously, the animal is not yet able to forage for its own food, and essentially the animal is completely dependent on the mother for food and protection. The ears do not open until approximately P8 to P9, and the eyes do not open until approximately P15 to P16. Therefore, gradually through the first 3 weeks of life, all sensory abilities appear to allow the animal to obtain and forage food for itself.

Mating Behavior

In male rats, the onset of mounting typically occurs between P41 and P45. Frequency of male mounting of the female increases between P46 and P50 and decreases between P51 and P55. The onset of mounting is associated with anogenital sniffing and chase behaviors. *Lordosis* is the behavior in which a female indicates to the male that she is viable for mating. This behavior is defined as the female rat elevating its anogenital region accompanied by a downward arching of its back and twitching of the ears. Lordosis begins to appear in female rats at about P42 and is demonstrated with increasing frequency through P55.

In conclusion, the development and appearance of behaviors to obtain and secure food seem to follow the rostrocaudal gradient of maturation, with the forelimbs leading hindlimb maturation by a few days. A timeline of behavioral development is presented in Figure 26–2. A discussion of the correlation of the development of the central nervous system and behavior is included at the end of the chapter.

COGNITION

One of the most important and complex aspects of a rat's behavioral repertoire is the ability to locate food in the natural environment and avoid predators. For the rat to perform this ability, it must learn associations between stimuli. Cognitive abilities develop rapidly

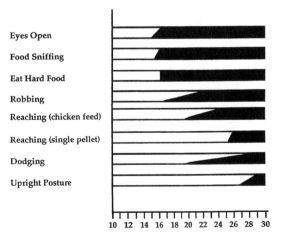

Figure 26–2. Timeline of behavioral development. (Adapted from Coles BLK and Whishaw IQ, unpublished master's thesis.)

during the first 3 weeks of life, and for many cognitive associations, animals are able to perform at near-adult levels by the time they are weaned. Based on the fact that cognitive abilities in rats covers a vast area of research, this section focuses primarily on the development of the ability to avoid an aversive stimulus (a taste) and the ability to find a goal location, defined as spatial memory. Avoiding an aversive stimulus involves relatively simple stimuli associations, whereas spatial memory involves more complex stimulus associations. This section concludes with a discussion of the infantile amnesia phenomenon, which is rapid forgetting that appears to occur during early development.

CONDITIONED TASTE AVERSION

A unique type of learning that has been demonstrated in the laboratory rat as well as humans is *conditioned taste aversion* (CTA). One of the basic abilities that the rat must acquire early in development is the ability to avoid ingesting poisonous substances. In the CTA paradigm, rats are trained with a taste (such as saccharin or sucrose) that is temporally paired with an injection of lithium choloride (LiCl) that produces illness. As in humans, CTA appears to be a unique form of learning is that adult rats have shown the ability to acquire an aversion to a particular taste in one trial (a single pairing of the taste and LiCl injection), and the memory for this association is persistent for weeks.

Interestingly, rats have demonstrated the ability to acquire taste aversion as early as before birth in utero (Smotherman, 1982) and has demonstrated the ability to avoid a particular taste as early as P1 (Schweitzer and Green, 1984). Training adult animals on this task is relatively simple. Based on the fact that rats prefer saccharin and sucrose solutions, these substances can be mixed in with the drinking water. The animal is then presented with this solution, and verification of imbibing the solution can be done with a specialized drinking tube that allows for the measuring of volume. Immediately after presentation of the sweetened water, the animal is given an intraperitoneal injection of LiCl (20 mg/kg). On a retention test the next day, the animal should demonstrate a drastic reduction in consumption of the sweetened water.

In preweanling animals, CTA is a bit more difficult to perform. At very early ages, before about 10 days of age, a gavage is made through insertion of a tube through the cheek of the young animal and injection of the solution through the gavage into the animal's mouth. Another way to administer the solution is through an injection directly into the animal's mouth. As in adults, immediately after presentation of the substance, an intraperitoneal injection of LiCl is given. Therefore, CTA can be acquired at a very early age and is a fairly simple learning paradigm to use in the laboratory rat.

INFANTILE AMNESIA

Although rats can learn many different types of complex associations, early in development these learned associations appear to be quite susceptible to rapid forgetting in infant rats, referred to in the literature as *infantile amnesia* (Spear and Riccio, 1994). Infantile amnesia occurs for all altricial mammals in which it has been tested.

There is an interesting contradiction in the literature concerning infantile amnesia. Smotherman (1982) demonstrated in an elegant series of studies that rats can learn to avoid a taste associated with illness as early as embryonic day 19 and that this association appears to be well retained in memory, as rats tested 2 to 3 weeks after birth still demonstrate learning of the aversion. On the other hand, studies have shown that infant rats demonstrate rapid forgetting of the CTA association, as animals trained on the CTA paradigm at P1, P10, or P18 demonstrate rapid forgetting of this association tested just a few days later.

However, animals slightly older at P20 demonstrate retention of the taste aversion when tested 25 days after the conditioned stimulus–unconditioned stimulus pairing (Schweitzer and Green, 1982). What this literature seems to suggest is that brain structures that mediate CTA are still developing up to 18 days of age but reach maturity at about weanling age. This seems to make sense evolutionarily, as rats must begin searching for their own food around this time point in development.

How is this contradiction in the literature resolved? Is it possible that an animal can learn and remember an association better in utero than it can after birth? In both studies, the taste was presented through injection by the experimenter, and the animal was tested through imbibing the water themselves. It certainly seems that acquiring such an aversion is hard-wired, as animals can learn this association before birth. However, retaining it in memory seems to be the larger problem. The resolution to the contradiction likely lies in experimental methodology. In the one case, the taste is injected into the amniotic fluid, whereas in the other case, the taste is presented through the mouth. Therefore, it may be that rats can learn certain associations to avoid particular stimuli in utero, however, the action of drinking of the fluid and its association with illness may not develop until much later. This suggests that brain mechanisms that underlie CTA do not develop until around weaning age, but other biologically important stimuli that the animal can learn to avoid may actually be acquired in utero. This also demonstrates that the ability to learn associations and retain them in memory depends on the nature of the stimulus presentation and the type of stimuli that must be associated. Depending on the association to be learned and the age of the animal, brain mechanisms may not yet have developed to mediate remembering of certain types of associations.

SPATIAL MEMORY

Memory for a spatial location in the rat is an extremely important survival skill, as rats must remember the location of food caches and the availability of these resources. There are two primary skills that have been tested in the young rat. One skill is the use of distal cues to locate a hidden spatial location, otherwise known as "place" navigation, and the other is the use of a visible cue to locate a visible spatial location, otherwise known as "cued" navigation. Clearly, finding a hidden location through the use of distal cues appears to be a much more difficult task than simply approaching a visible cue to find a spatial location. Indeed, adult animals demonstrate a more rapid learning curve in cued navigation than in place navigation, and it appears the cued task is simpler than the place task.

A very effective spatial memory task is the Morris water maze (MWM). The MWM is named for its creator, Richard G. M. Morris, who published a now seminal paper on methodologies used to train animals on this task (1981). This task involves training rats to locate a "hidden" or "cued" platform in a pool of water. In both versions of the MWM, the water is colored with powdered paint or powdered milk so that the animal must use extramaze cues to locate the platform, which is located approximately 1 to 2 cm below the water surface. In the cued version, the platform is cued through placement of a wooden block on top of the platform so that its location can easily be visualized by the swimming rat.

There are several advantages of this task over other spatial tasks, such as the radial arm maze, including the fact that animals do not have to be placed on food restriction to be motivated to locate the platform and training can be completed within 1 day. One disadvantage of this task, according to some investigators, is that rats are being removed from their "natural" environment (land) to locate a safe location. The argument against this notion is

that rats are amphibious animals and are able to swim quite well even on the first day that the eyes are open on P15 to P16. Another disadvantage is that rats can experience hypothermia while swimming, especially when they are young and not able to thermoregulate as well as do adults. This problem can typically be handled by drying the animals between trials and keeping the water at an ambient temperature for this age of rat (around 23° to 25° C for young rats instead of 19° to 20° C for adult rats) (Brown and Whishaw, 2000).

Finally, proper training methodology must be used with young rats. Very short intertrial intervals that are 1 minute or less can often cause fatigue in the young animal, producing poor performance not necessarily related to cognitive ability. One training technique that has been shown to be effective and relatively efficient is to begin training on P17, and animals are trained for 3 consecutive days. Two trial blocks of four trials are given on each day, however, these trial blocks are spaced apart by approximately 2 to 3 hours, and individual trials are spaced apart by at least 5 minutes. This allows the animal proper time to rest between trials, and the animal should demonstrate an asymptote of learning equivalent to adult levels by the last day of training (P19) (Kraemer and Randall, 1996). One final note is to make certain that there are proper extramaze cues around the pool, including posters, desks, the experimenter, and other visual cues that are relatively close to the edge of the pool (within 1 meter). Using proper extramaze cues helps to facilitate the animal's performance, because rats use these cues to locate the platform.

There has been a considerable amount of debate as to the spatial abilities in the preweanling and early postweanling rat. Rudy and colleagues (1987) demonstrated that rats cannot begin to learn to rely on distal cue navigation until approximately 20 days of age and are not adept at using distal cues to locate a spatial location until about 23 days of age. However, several other studies have reported that rats can learn a spatial discrimination based on distal cues at an earlier age. Brown and Whishaw (2001) reported that both proximal and distal cue navigation appear to develop at the same rate, and both are functional by 19 days of age but are not mature at 18 days of age. This suggests that similar mechanisms may mediate both place and cue navigation, contradicting the idea that these two types of navigation may develop at different rates.

Although the 19-day-old rat can find a spatial location, it appears the ability to retain a memory for that location does not develop until much later. In fact, animals trained on a spatial task at 19 to 21 days of age demonstrate complete forgetting of that location just 3 days after training (Brown and Kraemer, 1997). In contrast, adult rats have demonstrated memory for a spatial location as long as 3 months after testing (Sutherland and Dyck, 1984). Therefore, it appears that that the ability to learn a spatial location develops at an early age, even before weaning, but the ability to remember a spatial location may not develop until adulthood.

WHAT DOES THIS TELL US ABOUT DEVELOPMENT OF THE BRAIN AND BRAIN FUNCTION?

Throughout the rat's first 3 weeks of life, the nervous system of the rat undergoes a rapid process of development. Surprisingly, there have been few attempts to correlate motor behavior with the details of anatomical changes in the brain. In general, the appearance of motor behaviors seems to follow the rostrocaudal gradient of maturation of the cerebral cortex; therefore, the mouth and forelimbs mature before the more caudally located hindlimbs receive their spinal cord connections.

MOTOR DEVELOPMENT

Regarding the motor cortex in the rat, this brain area has a representation of the body

that is composed of a large forelimb area located in the anterior portion of the motor cortex and the hindlimb area that is located more posterior. Following the rostrocaudal gradient of development, the more anterior forelimb areas are formed first, and thus mature first before the hindlimbs. The corticospinal tract is the projection from the motor cortex to the spinal cord; it also is thought to contribute to skilled limb movements. Growth of the corticospinal tract follows the same neurogenetic principle of growth: the axons from cortex terminate in an anterior (older) to posterior (younger) pattern. Thus, the anatomical connections for the forelimbs mature before those for the hindlimbs (see Fig. 26–3). This pattern of development is in accord with the behavioral literature, which has demonstrated that forelimb use matures before the hindlimbs (Bayer and Altman, 1991).

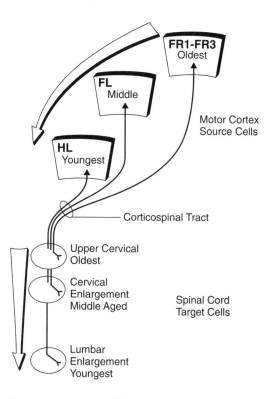

Figure 26–3. Diagram of the formation of the development of anatomical connections between layer V output neurons of the corticospinal tract and terminations in the spinal cord. (Adapted from Bayer SA and Altman J, 1991.)

CONDITIONED TASTE AVERSION

There have been three primary brain structures implicated in CTA: the parabrachial nucleus (PBN), amygdala, and insular cortex. All three of these areas are known to be important in modulating gustatory responses. The PBN is located in the brain stem. It sends direct major projections to the amygdala and to the insular cortex via the ventromedial nuclei of the thalamus and is known to play a major role in taste. The amygdala is the central brain structure known to be involved in emotion, is part of the limbic system, and is located in the temporal lobe. Ablations of specific nuclei within the amygdala have been shown to produce deficits in aversive conditioning, demonstrating the importance of this brain area in this type of learning. The insular cortex is the area of the cerebral cortex that surrounds the rhinal sulcus. The insular cortex has been referred to as the visceral cortex because it receives taste and visceral information from the thalamus. It has been postulated that the insular cortex receives convergences of primary sensory inputs not seen within any other sensory areas of the cortex. It appears that these three brain areas—the PBN, amygdala, and insular cortex—form a circuit in the brain that is responsible for the acquisition of conditioned taste aversions (Bermudez-Rattoni, 1995). There have not been any attempts to map out the development of the brain circuitry between these three areas and to correlate it with CTA. However, a connection between the amygdala and medial prefrontal cortex develops at P19, suggesting that the connectivity of the amygdala to the cortex is maturing at this age (Cunningham et al., 2003).

SPATIAL MEMORY

It has been shown that an important brain structure to mediate cognition, especially spatial memory, is the hippocampus. Although all areas of the hippocampus proper and hippocampal formaton have been implicated in spatial memory function, the maturity of the mossy fiber system at P18 may be especially important. The mossy fibers are axons of the dentate granule cells projecting onto the pyramidal cells of the dentate hilus (CA4) and CA3 pyramidal cells, and this mossy fiber system appears to be mature by approximately P18 (Zimmer, 1978). Based on past results that have shown animals develop spatial abilties near adult levels by P19, it appears that maturity of the mossy fiber connections of the dentate gyrus–CA4 and dentate gyrus–CA3 areas at this age may be especially important in mediating spatial memory performance at this age.

In summary, this chapter just touches on the vast literature describing the development of cognitive and motor abilities in the laboratory rat. There are several important issues to keep in mind when training preweaning or early postweaning laboratory rats. The first issue is training methodology. Rats at an early age have an ever-expanding behavioral repertoire, but to comprehend their abilities, the animal must be trained properly. This means that longer intertrial intervals may need to be used, and special provisions in terms of environment may be needed to be taken into consideration. Another important issue involves the complexity and rapid development of cognitive and motor abilities in the rat. A rat that does not have an ability on a particular day of development may have that ability at near-adult levels on the next day. With this rapid development in abilities and brain structures, the developing rat provides an ideal but complex subject to study the ontogeny of learning, memory, and motor abilities. Finally, it is always important to keep in mind the stress incurred on the young animal. Taking a preweanling away from the mother may produce undue stress, especially if the animal is away from the mother for an extended period of time. Attempting to understand a young animal's abilities is a daunting task, and one in which every aspect of the training environment and methodology must be kept in mind.

REFERENCES

Altman J and Sudarshan K (1975) Postnatal development of locomotion in the laboratory rat. Animal Behavior 23:8096–8920.

Bayer SA and Altman J (1991) Neocortical development. New York: Raven Press.

Bermudez-Rattoni F (1995) The role of the insular cortex in the acquisition and long lasting memory for aversively motivated behavior. In: Plasticity in the central nervous system (McGaugh JL, Bemudez-Rattoni F, Prado-Alcala RA, eds.). Mahwah, NJ: Erlbaum.

Brown RW and Kraemer PJ (1997) Ontogenetic differences in retention of spatial learning tested with the Morris water maze. Developmental Psychobiology 30:329–341.

Brown RW and Whishaw IQ (2000) Similarities in the development of place and cue navigation by rats in a swimming pool. Developmental Psychobiology 37:238–245.

Coles BLK and Whishaw IQ (1996) Neural changes in forelimb cortex and behavioural development. Unpublished master's thesis, Lethbridge, Alberta, Canada: University of Lethbridge.

Cunningham MG, Bhattacharya S, Benes FM (2002) Amygdalo-cortical sprouting continues into early adulthood: Implications for the development of normal and abnormal function during adolescence. Journal of Comparative Neurology 453:116–130.

Kraemer PJ and Randall CK (1995) Spatial learning in preweanling rats trained in a Morris water maze. Psychobiology 23:144–152.

Pierce RC and Kalivas PW (1997) A circuitry model of the expression of behavioral sensitization to amphetamine-like psychostimulants. Brain Research Brain Research Reviews 25:192–216.

Rudy JW, Stadler-Morris S, Albert P (1987) Ontogeny of spatial navigation behaviors in the rat: Dissociation of "proximal"- and "distal"-cue-based behaviors. Behavioural Neuroscience 101:62–73.

Schweitzer L and Green L (1982) Acquisition and extended retention of a conditioned taste aversion in preweanling rats. Journal of Comparative and Physiological Psychology 96:791–806.

Smotherman WP (1982) Odor aversion learning by the rat fetus. Physiology and Behavior 29:769–771.

Spear NE and Riccio DC (1994) Memory: Phenomena and principles. Needham Heights, Mass.: Allyn & Bacon.

Whishaw IQ and Tomie J-A (1989) Food-pellet size modifies the hoarding behavior of foraging rats. Psychobiology 17:93–101.

Zimmer J (1978) Development of the hippocampus and fascia dentata: Morphological and histochemical aspects. Maturation of the Nerv Sys, Progress in Brain Research, Vol. 48. MA Corner, Ed. Elsevier/North Holland Press: Amsterdam.

Maternal Behavior

27

STEPHANIE L. REES, VEDRAN LOVIC,
AND ALISON S. FLEMING

Parental or maternal behavior is an ultimate test of individual fitness, as the ability of individuals to raise healthy offspring is necessary both for the individuals' capacity to pass on their genes and for species survival. Although the ultimate function of raising healthy offspring is common among different species, there is considerable variability in the expression of maternal behavior across mammalian species, with the most marked distinctions occurring between altricial and precocial species. In the altricial species, such as rodent species, the young are born in an immature state, often in litters and usually into a stable nest or home environment where the young remain for a considerable period before weaning (Weisner and Sheard, 1933; Fleming and Li, 2002). In contrast, in some precocial species, the young have fully developed sensory and motor abilities within hours of birth (Gonzalez-Mariscal and Poindron, 2003). In this chapter, only the behavior of laboratory rats, *Rattus norvegicus*, is discussed, a familiar, well-studied example of an altricial species.

Laboratory rats have proved to be a good model for the study of hormonal (Rosenblatt, 2002), sensory (Stern, 1996), neural (Numan and Sheehan, 1997), experiential (Li and Fleming, 2003), and developmental (Fleming et al., 2002) factors that control maternal behavior. The study of maternal behavior itself has use because it is a highly organized behavior that can be used as a model of social behavior. The study of maternal behavior can be used in the analysis of the effects of exposure to drugs of abuse (cocaine: Mattson et al., 2003), to ther-

apeutic agents such as antipsychotics (Li et al., in press), or to prenatal and/or postnatal stress, alcohol, maternal separation, or "enrichment" (Kuhn and Schanberg, 1998). Maternal behavior in and of itself is now known to have marked effects on the neurological and behavioral development of offspring, with the quantity of licking by the mother altering the development of the pups stress and endocrine systems (Liu et al., 1997), brain development, and cognitive, affective, and social behaviors (Hofer, 1994; Fleming et al., 2002).

This chapter describes maternal behavior of the laboratory rat and outlines various methods of observing and quantifying this behavior. Although in some rodent biparental species males also show parental behavior, this is not the case for most rodents, including *R. norvegicus*. However, under certain experimental conditions (see later), males also show many of the components of behavior normally shown by the mother rat (Rosenblatt et al., 1996; Rosenblatt and Ceus, 1998). We describe the general and specific methods for the testing of maternal behavior. Also, several environmental and situational factors that affect the expression of maternal behavior must be considered.

DESCRIPTION OF MATERNAL BEHAVIOR

Based on a long history of research in the area (Weisner and Sheard, 1933), there is a relatively complete picture of the phenomenology of rat maternal behavior. Under the influence of maternal hormones, the new mother rat is maternally responsive to newborn pups as soon as they emerge from the birth canal (Rosenblatt and Lehrman, 1963). Typically rats give birth to litters of 8 to 16 pups, and these litters tend to consist of approximately an equal number of males and females. At parturition (birth), the mother pulls off the amniotic sac, eats the placentas, and cleans the pups (Hudson et al., 1999). Within the first 30 minutes after parturition, she retrieves all of the pups to a nest site, mouthes and licks them, and adopts a nursing posture over them. No prior experience with pups is needed for this immediate maternal responsiveness (Fleming and Rosenblatt, 1974). Pups remain with the mother rat until weaning which occurs normally between postnatal days 22 and 30. As the pups develop, they move toward independence, first eating crumbs from around the mother's mouth and fur and eventually eating rat chow at a distance from the nest and drinking from the provided water bottle. Over time, pups spend less time in the nest and with the mother rat and mothers often actively move away from pups when they approach.

Once maternal behavior has been exhibited at parturition under hormonal influences, the behaviour is sustained through experiences with the pups acquired during the postpartum period. This effect of experience is also influenced by input from the pups themselves (Li and Fleming, 2003).

Described here are some of the more important maternal behaviors displayed by rat dams, especially during the first 10 days postpartum.

RETRIEVAL

Retrieval of pups consists of the mother rat carrying a pup to the present nest or to a new location where a new nest will be constructed. This behavior is almost always observed after parturition, when pups tend to be scattered around the nest. Pup retrieval, however, is also evident during the retrieval test, and this behavior is thought to be a measure of a rat's "motivation" to be maternal. A very maternally responsive rat retrieves the pups quickly and efficiently to the nest site, whereas a nonmaternal rat does not retrieve the pups. A mother rat will frequently pick up pups while she and the pups are in the nest and reposition the pups within the nest. This type of behavior is not considered retrieval; rather, it is referred to as *pup pick-ups* or *mouthing*.

PUP LICKING

Pup licking is an important source of stimulation for newborn pups. In the past several decades, a number of studies have demonstrated the important effect of pup licking on offspring's emotionality (Francis and Meaney, 1999), cognition (Liu et al., 2000; Lovic and Fleming, 2003), and physiology (Kuhn and Schanberg, 1998), as well as propagation of maternal behavior (Fleming et al., 2002). There are two types of licking: pup body licking and pup anogenital licking. Body licking can be observed during various circumstances in the maternal cage (e.g., before retrieval, between retrievals, during nursing), whereas anogenital licking tends to be observed while pups are nursing and are on their backs. During anogenital licking, pups show reflexive hind extremities extensions in response to anogenital stimulation. This type of stimulation is important to ensure pup urination and defecation and plays an important role in sexual development of the male pup (Moore, 1984). The ability to observe pup licking, and to differentiate between two types of licking, is initially difficult but can be acquired within a few hours of observing maternal behavior.

NURSING POSTURES

The purpose of crouching postures is to allow the pups access to teats and milk, to regulate their temperature, and to protect them from environmental elements. The definition of *crouching* differs greatly across studies and laboratories. There are two general types of nursing postures: hovering and crouching. *Hovering* is a posture in which a rat is positioned over some or all of the pups in the nest, but the female is not quiescent. She is actively licking pups, moving the nest material, self-grooming, or moving pups within the litter while hovering. Despite the mother rat being active, at least some pups have access to her teats (Fig. 27–1A). *Crouching* is considered to be a quiescent posture, and it usually occurs in response to sufficient stimulation by pups. A mother rat tends to stop other activities (although anogenital licking is sometimes observed) and develops a characteristic posture with her extremities spread out and back arched. Crouching is sometimes divided into low crouching and high crouching postures, depending on the degree of the arch of the spinal column (Figs. 27–1B and 27–1C). A third nursing posture that is rarely observed during testing, especially with younger pups, is a supine posture. Here the mother rat lies on her side, giving the pups access to her nipples. This posture is observed during longer periods of undisturbed maternal behavior in the nest with older pups (>10 days old; see Fig. 27–1D).

NEST BUILDING AND OTHER MATERNAL BEHAVIORS

Other maternal behaviors that are typically observed and recorded during a maternal test are nest building (in which the mother rat gathers nesting material to a nest site) and

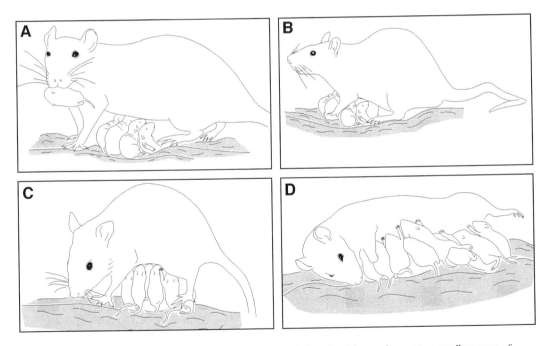

Figure 27–1. Illustrations of different nursing postures. (*A*) *Hovering*. The mother rat is over all or most of the pups while engaging in other behaviors such as mouthing, repositioning, licking pups, fixing the nest, or self-grooming. (*B*) *Low crouch*. Soon after the mother rat settles in the nest, the frequency of behaviors other than low crouching decreases. She extends her limbs with her back slightly arched allowing the pups access to her ventrum. (*C*) *High crouch*. This posture is characterized by high arching of the back and significant limb extensions by the mother rat that allow the rooting pups access to her ventrum. Typically, mother rats cease all other activity during high crouch nursing. (*D*) *Supine*. This posture is characterized by the mother rat lying/sleeping on her side with pups attached to her teats. This posture is observed with older pups and after longer periods of undisturbed nest environment.

sniffing pups. For nest building, often a nest rating can be used (in which 1 indicates no nest; 2, some nest [some nest material arranged in a single location]; 3, moderate nest [low walls of a nest are evident]; 4, good nest [walls of nest are distinct and surround a cavity]; and 5, excellent nest [walls of nest are high and can conceal pups]). The time that the mother rat spends in the nest can also be recorded.

NONMATERNAL BEHAVIORS IN MATERNAL CONTEXT

Mother rats engage in a number of nonmaternal behaviors in a maternal testing context, and it is a good practice to record these behaviors while testing for maternal behavior, to assess general activity level and what the mother is doing when she is not responding to the pups. Mother rats often groom themselves, usually after retrieving the pups. Mother rats also engage in a number of explorative behaviors such as sniffing air and digging of the bedding. They eat, occasionally settle in a cage corner, and sleep away from pups.

CONDITIONS OF OBSERVING AND QUANTIFYING MATERNAL BEHAVIOR

To assess maternal behavior in rats, a number of standardized procedures have been developed that vary somewhat from one laboratory to another, usually in the duration of the observations. The following described observation strategies are derived from work in several laboratories.

TEST CAGE AND ENVIRONMENT

Testing should be conducted in a large, clear Plexiglas cage. Rats should be transferred to the test cage either during late pregnancy, or at least 1 day before testing begins, for habit-

uation and to reduce effects of novelty. Typical dimensions of test cages are approximately 45 × 40 × 20 cm, although cages of different dimensions can be used. It is important that the test cage is transparent, to allow a clear view of behaviors, and that the type of cage is consistent across testing, because the size and condition of the cage can have a significant impact on maternal behaviors. Wood shavings bedding should be spread over the bottom of the cage at an approximate depth of 1.5 cm, and nesting material should also be provided (e.g., two paper towels shredded into 2 to 3 cm pieces). In addition to the cage setting, laboratory setting should be standardized, with ambient temperature maintained at about 22° C, 40% to 50% humidity, and 12 hour light–dark photo period. Testing is usually done during the light phase or inactive phase of the cycle. Observations can be done during the dark phase as well, but rats tend to be more active during this phase. For all tests, litters should be culled to an equal number of pups; eight pups is standard (four males and four females).

METHODS OF RECORDING BEHAVIOR

There are two general methods of quantifying maternal behavior in the laboratory setting. One method involves periodic sampling of ongoing maternal behavior, sometimes referred to as *spot checks* (e.g., recording what the mother does at 100 different 5 second points during the day) (Francis et al., 2002). This type of testing is almost always done with rats whose nests have not been disturbed. The other type of testing involves a continuous observation period (10 to 120 minutes) with either a nondisrupted or disrupted maternal nest. Several continuous observations can be made over several days to get a good idea of changes in maternal behavior over time. The method of choice should be based on the nature and needs of the experiment, as well as the resources and constraints at hand.

Paper and Pencil Technique
(One-Zero Time Sampling)

A table can be created, with columns representing time and rows representing behaviors. This creates boxes that represent each behavior at every time point (typically a 5 second time point). By checking each behavior that is seen during a time point, the frequency of each behavior over the test can be determined when the number of checks over each row are added. This method of recording is convenient but not as precise as the use of an event recorder.

Event Recording of Ongoing Behavior

This method of recording consists of using a computer with an event-recording program (e.g., Behavioral Evaluation Strategy and Taxonomies [BEST] Program) that can be tailored for each user in terms of length of the test and number of behaviors recorded. Each behavior to be recorded is represented by a key on the keyboard. When the mother rat performs a behavior, the designated key is pressed until the mother rat ceases to perform this behavior. This allows the tester to record both frequency (each time a key is pressed) and duration (how long a key is pressed) of each behavior over the test.

MATERNAL TEST PROTOCOLS

The following are not the only types of maternal tests used, but they are the simplest and most typical. The test of choice should be based on the nature and needs of the experiment.

Continuous Observation of
Nondisturbed Maternal Behavior

A nondisturbed maternal test is a basic observation of natural, nondisturbed maternal behaviors. At no point during this test are the mother rat and her litter disturbed and the tester can choose to start observations at any point during the day by simply beginning to record ongoing behaviors. These tests can range in length from 10 minutes to several hours. Rats can be tested on one particular day or over several continuous or alternate days. Unless processes of weaning are of interest, testing is usually undertaken over the first 2 weeks postpartum.

This test protocol allows the tester to determine whether experimental manipulations affect the natural display of maternal behavior—that is, how much time the mother rat spends with the pups and if her pattern of behaviors is similar to that shown by a nonmanipulated mother rat.

Maternal Behavior Test with Retrieval

Maternal behavior test with retrieval is similar to an undisturbed maternal test, except that the mother and pups are separated for a brief period before the beginning of the test. The tester removes the pups from the nest, leaving the nest relatively intact. Pups are removed for about 5 minutes and must be maintained at room temperature. After this brief separation period, pups are returned to the test cage in the corner diagonally opposite the nest. The recordings should start immediately as maternally responsive rats will start retrieving the pups without a delay.

This test is often performed in combination with subsequent spot checks across the day. These spot checks are done at approximately 2 hour intervals after the retrieval test. The retrieval test mainly assesses retrieval behavior that is an indication of maternal motivation. Retrieval behavior varies greatly between postpartum rats, sensitized virgin rats, and juvenile rats, so the degree of maternal motivation can easily be assessed with this test. Less motivated rats, such as virgin rats, show little, if any, retrieval behavior, whereas more motivated rats, such as newly parturient rats that are under the influence of maternal hormones, show efficient retrieval of all pups in the cage.

Maternal Memory Test

The role of maternal hormones subsides over the first few days after parturition. Then maternal behavior is maintained through experience, sometimes referred to as *maternal memory* or *maternal experience effect* (Li and Fleming, 2003). The latency of maternal behavior after a period of separation from pups is shorter if rats have had an earlier maternal interaction with pups than if they have not.

The maternal memory test is a variation of the retrieval test except that it measures how well rats "remember" their experience with pups during the postpartum period or earlier induction period. In general, much like other tests of memory, maternal memory test has two phases: the experience or exposure phase and the test phase.

After parturition, pups are taken away from the mother rat at 15 minute intervals, not allowing her to have the experience of being maternal. After a certain period, from several hours later to the next day, the mother rat is exposed to foster pups obtained from a donor mother rat. Most likely, this mother rat will readily display maternal behavior. The mother rat is left with the pups for at least 1 hour and then separated from any pup-related cues for several days (e.g., 10). After this period of isolation, the mother rat is tested again with new foster pups.

Mother rats with intact "maternal memory" display maternal behavior within 1 or 2 days, whereas mother rats treated with some kind of manipulation (e.g., treatment with protein synthesis inhibitors or lesions of the nucleus accumbens [Li and Fleming, 2003]) will not show maternal behavior. Alternatively, mother rats with poor maternal memories have long latencies to show maternal behavior, similar to virgin rats.

Maternal Motivation and Preference for Pup-Related Stimuli Tests

Although rat maternal behavior is organized, the mechanisms of its control are quite numerous and dependent on each other, as motor, motivational, experiential, attentional, and other factors all play a role in the expression of maternal behavior (e.g., Mattson et al.,

2003). Maternal behavior tests in the maternal context are useful for assessing actual maternal responsiveness and performance (i.e., appetitive component of maternal behavior), however, they are not necessarily the best tools for the assessment of individual mechanisms (e.g., motivation) that contribute to maternal behavior. A description follows of a testing procedure that can be used to assess pup hedonic values and avoidance of pup-associated stimuli.

One test that can assess motivation is the conditioned place preference test, and it has been used to assess how rewarding pups are. The test is useful because it can discriminate between rats that show similar maternal behavior during a typical maternal test, as outlined earlier, yet do not find pups equally rewarding (Morgan et al., 1999; Mattson et al., 2003). Testing is done in an apparatus that consists of two white Plexiglas boxes (22 × 40 × 30 cm), with each box differing in wall patterns (horizontal black bars versus vertical black bars) and texture of the floor (smooth versus perforated). The test consists of two phases: the exposure phase and the test phase. During the *exposure phase*, rats are randomly assigned to have one specific box associated with pups. On two alternate days, rats are exposed to pups in the assigned box, and on the other two alternate days, rats are exposed to the other box (Mattson et al., 2003).

During the *test phase*, on test day, the rats are placed in the test apparatus without barriers between the boxes, allowing them to freely spend time in either of the boxes. Entries and time spent in each box is recorded during a 10 minute testing period. Rats that find pups rewarding will spend significantly more time in the box that was previously associated with pups than in the other box. For example, multiparous and primiparous rats find pups more rewarding than virgin rats and thus spend more time in the box associated with pup-related cues.

Another way to determine how maternally "motivated" rats are, without actually testing maternal behavior, is to test rats in a *preference* or *choice* test. Two stimuli (pup urine odors versus diestrous female urine odors or cool vocalizing pups versus warm nonvocalizing pups) are placed simultaneously at two ends of a Y maze, or single stimuli are presented consecutively on successive trials, and the relative times spent in proximity to each stimulus during a 5 minute test is computed (Fleming et al., 1989; see also Farrell and Alberts, 2002).

Testing of Maternal Behavior in Nonpostpartum Rats

Although initially "fearful" and avoidant of pups, virgin female rats can be induced to become maternal, without showing lactation, through continuous exposure to pup stimulation. We describe two procedures that are used to induce maternal behavior in virgin rats.

Sensitization. Not having the advantage of hormonal priming associated with pregnancy and parturition, virgin females and males are not maternally responsive when first presented with newborn foster pups (Rosenblatt, 1967; Fleming and Luebke, 1981; Rosenblatt and Ceus, 1998). After 5 to 10 days of continuous exposure to pups, virgin rats eventually begin to respond maternally (Rosenblatt, 1967), showing a pattern of behavior that closely resembles, but does not perfectly parallel, the behavior of a new mother rat (Lonstein et al., 1999). To be considered maternal during the sensitization procedure, virgin rats are continuously exposed to foster pups until the maternal criterion is reached—that is, the virgin rat retrieves all foster pups to the nest site over 2 consecutive days during retrieval tests. Maternal tests of sensitized virgin rats should follow the protocols outlined earlier. Foster pups need to be replaced daily because virgin rats do not lactate and pups need milk to survive (donor mothers should not be left with fewer than six pups because lactation must be maintained in these mothers). One

problem that may occur during the sensitization procedure is cannibalization. Because virgin rats find pups aversive, virgin rats occasionally kill the pups. If this occurs during the test, the test should be stopped, injured pups should be euthanized, and healthy pups should be returned to their mothers. A virgin that cannibalizes 2 days in succession should be removed from the experiment.

Hormonal Priming. Virgin rats can also be induced to become maternal through the administration of maternal hormones, namely estrogen and progesterone (Bridges, 1984). Virgin rats are ovariectomized and subcutaneously implanted with a 2 mm Silastic capsule of estrogen in the dorsal region of the neck. Three days after these initial procedures, three 30 mm Silastic capsules of progesterone are subcutaneously implanted in the same area. Ten days after this procedure, the progesterone capsules are removed and the estrogen capsule is left. One day after the removal of the progesterone capsules, the virgin rat can be exposed to pups for maternal testing. As in the sensitization procedure, pups must be replaced daily because hormonally primed virgin rats do not lactate. Although closer to paralleling maternal behavior of the postpartum rat than the behavior of the sensitized virgin, hormone priming of virgin rats does not account for the actions of other hormones, such as prolactin, oxytocin, and corticosterone, which are also altered in the lactational/postpartum state.

To assess relative sensitivities to hormones in rats under different conditions, the hormone regimen can be either lengthened or shortened (Bridges, 1984).

Tests of Juvenile Maternal Behavior

Immediately after weaning, male and female juvenile rats spontaneously show maternal behavior (Rees and Fleming, 2001), which subsides around the onset of puberty. In natural environments, mother rats are typically giving birth to a second litter when their first litter is being weaned (Gilbert et al., 1983), and hence, juvenile rats are exposed to pups around weaning.

Juvenile maternal testing is similar to the virgin sensitization procedure, in that over several days, juvenile rats are exposed to pups and juvenile behavior is recorded during a maternal test. There are, however, some differences between adult and juvenile maternal behavior because juvenile rats are much smaller than adult rats and tend to drag rather than carry pups to the nest.

Also due to their size, juvenile rats cannot adopt a rigid arched back posture over pups and instead tend to lie on top of the pups. Juvenile rats may show all components of maternal behavior before they show retrieving, which is the opposite pattern observed in adult maternal rats. Juvenile maternal behavior can be of interest because it allows the tester to investigate the development of maternal behavior and, in a more general sense, of early social behavior.

ISSUES TO BE CONSIDERED WHEN TESTING

There are some factors to consider when testing maternal behavior. If possible, these factors should remain constant (or should be controlled for) across experimental groups because confounds may arise.

Separation of Mother Rat from Pups
One of the most well-documented experimental manipulations of maternal care is early maternal separation (Lehmann and Feldon, 2000). It is well known that any type of separation can alter pup development (Hofer, 1994) and the mother's behavior toward pups (Pryce et al., 2001).

Very short periods of separation (<15 minutes), termed "early handling," increases the mother's behavior toward pups over several time points after the separation period. On the other hand, longer periods of separation (>1 hour) do not increase maternal re-

sponsiveness (Pryce et al., 2001). If some type of manipulation that separates the mother rat from pups, for example drug administration, is being investigated in conjunction with maternal behavior, the disruptive effects of separation should be considered in the behavioral analysis.

Cage Size
When testing adult males or virgin females or juveniles of either gender, in the induction paradigm, the size of the cage affects the speed with which rats begin to show retrieval behavior. Very small cages, where the rat and the pups are forced into close proximity, results in faster latencies to initiate retrieving (across days), than in larger cages, where rats can maintain a distance from the pups.

Circadian Rhythms
Maternal behavior, like other behaviors, has a circadian rhythm. Typically, mother rats interact more with their pups during the day or light part of the cycle (Leon et al., 1984).

Size of the Litter
One important factor that should be and can easily be controlled in studies investigating maternal behavior is the size of the litter. It has been demonstrated that the larger the litter, the less licking is received by the litter, per pup and per litter (Fleming et al., 2002). Mother rats, however, spend more time with larger litters than with smaller litters (Deviterne et al., 1990). To be able to compare results across studies, the litter size must remain constant, because changes in the litter size also change pup growth rate and development (Agnish and Keller, 1997).

Age of Litter
Maternal behavior changes as the pups age (Stern and MacKinnon, 1978). There is more maternal behavior, especially retrieval, received by 1-week-old pups than by older pups, perhaps because pups are more active and independent as they age. Hence, if donor pups

are used during testing, it is recommended that they be young (1 to 5 days old).

Gender Ratio of the Litter

Another issue to consider is the gender ratio of the litter. Before testing, a litter should be culled so that there is an equal number of female and male pups. In a litter with an equal gender ratio, there are gender differences in licking received by pups. Male pups receive more anogenital licking from the mother than do female pups (Moore, 1985). This preference is based on many factors, one being testosterone and another being pheromones (Moore, 1986).

After the first postpartum week, mother rats will also begin to retrieve male pups before female pups during a retrieval test (Deviterne and Desor, 1990).

Also, one study shows that female rats raised in unisexual litters have lower frequencies of pregnancy, but when pregnancy and birth occur, litter sizes are much larger than those of rats raised in bisexual litters (Sharpe, 1975). If gender is an important issue being investigated, the difference in maternal behavior received between genders can easily be assessed by marking female and male pups with different colors and tracking the behavior of the mother rat toward each gender and color.

Strain Differences

There are differences in maternal behavior across strains of rats. For example, Long-Evans mother rats show more licking, especially anogential licking, toward their pups than do Fisher 344 mother rats, although gender differences in licking are retained (Moore et al., 1997). Long-Evans mother rats also show a more ordered litter, more physical contact with their pups, and more frequent nursing postures than both Wistar and Sprague-Dawley mother rats (MacIver and Jeffrey, 1967). These differences in maternal behavior across strains remain constant regardless of the strain of foster pup (MacIver and Jeffrey, 1967). Finally, in different laboratories, there are marked differences in the latency with which rats begin to show maternal behavior during the induction procedure (Terkel and Rosenblatt, 1971). These differences may well be related to differences between the strains in "emotionality" or "neophobia" (Fleming and Luebke, 1981). Hence, when comparing studies, strain of rat should be taken into account due to the fact that there are strain differences in maternal behavior.

Use of Foster Pups

Although maternal rats accept any pups presented to them, evidence suggests that they can differentiate between their own and foster pups and that they treat them differently. For example, foster pups from split litters, in which half of the pups are foster pups and half are pups from the mother being tested, weigh much less by postnatal day 30 and show a reduced capacity to survive food deprivation than biological pups (Ackerman et al., 1977), suggesting that foster pups are treated differently than the mother's own offspring.

Parity of the Mother

Maternal behavior changes with parity, or number of births, because primiparous (gave birth to one litter) and multiparous (gave birth to more than one litter) rats will react differently to pups (Wright et al., 1977). There are also differences in the neural circuit of maternal rats as a function of parity with increases in GFAP (Featherstone et al., 2000) and opiate receptors (Bridges and Hammer, 1992) in the medial preoptic area. This demonstrates that unless parity is being compared, all rats must be of the same reproductive condition (nulliparous, primiparous, or multiparous) to control for the potential confound of parity.

CONCLUSION

Maternal behavior can easily be recorded through careful observations and well-planned procedures. Many factors should be considered while planning a maternal behavior study because maternal behavior is a complex behavior, influenced by many factors and regulated by multiple mechanisms. Being a reliable and robust behavior, it can also be used in laboratory demonstrations to illustrate, for example, the role of hormones or of olfactory cues in the regulation of behavior, in the expression of learning, or in the principles of reinforcement within a species-specific characteristic context.

ACKNOWLEDGMENTS

We give many thanks to Alison Diaz for production of the illustrations used in this chapter.

REFERENCES

Ackerman SH, Hofer MA, Weiner H (1977) Some effects of a split litter cross foster design applied to 15 day old rat pups. Physiology and Behavior 19:433–436.

Agnish ND and Keller KA (1997) The rationale for culling of rodent litters. Fundamental and Applied Toxicology 38:2–6.

Bridges RS (1984) A quantitative analysis of the roles of dosage, sequence, and duration of estradiol and progesterone exposure in the regulation of maternal behavior in the rat. Endocrinology 114:930–940.

Bridges RS and Hammer RP Jr (1992) Parity-associated alterations of medial preoptic opiate receptors in female rats. Brain Research 578:269–274.

Ceger P and Kuhn CM (1998) Responses to maternal separation: Mechanisms and mediators. International Journal of Developmental Neuroscience 16:261–270.

Deviterne D and Desor D (1990) Selective pup retrieving by mother rats: Sex and early development characteristics as discrimination factors. Developmental Psychobiology 23:361–368.

Deviterne D, Desor D, Krafft B (1990) Maternal behavior variations and adaptations, and pup development within litters of various sizes in Wistar rats. Developmental Psychobiology 23:349–360.

Farrell WJ and Alberts JR (2002) Maternal responsiveness to infant Norway rat (Rattus norvegicus) ultrasonic vocalizations during the maternal behavior cycle and after steroid and experiential induction regimens. Journal of Comparative Psychology 116:286–296.

Featherstone RE, Fleming AS, Ivy GO (2000) Platicity in the maternal circuit: Effects of experience and partum condition on brain astrocyte number in female rats. Behavioral Neuroscience 114:158–172.

Fleming AS, Kraemer GW, Gonzalez A, Lovic V, Rees S, Melo A (2002) Mothering begets mothering: The transmission of behavior and its neurobiology across generations. Pharmacology, Biochemistry, and Behavior 73:61–75.

Fleming AS and Li M (2002) Psychobiology of maternal behavior and its early determinants in nonhuman mammals. In Handbook of parenting: Biology and ecology, Vol 2 (Borenstein MH, ed.). Mahwah, NJ: Lawrence Erlbaum Associates.

Fleming AS and Luebke C (1981) Timidity prevents the virgin female rat from being a good-mother: Emotionality differences between nulliparous and parturient females. Physiology and Behavior 27:863–868.

Fleming AS and Rosenblatt JS (1974) Maternal behavior in the virgin and lactating rat. Journal of Comparative and Physiological Psychology 86:957–972.

Francis DD and Meaney MJ (1999) Maternal care and the development of stress responses. Current Opinion in Neurobiology 9:128–134.

Francis DD, Young LJ, Meaney MJ, Insel TR (2002) Naturally occurring differences in maternal care are associated with the expression of oxytocin and vasopressin (V1a) receptors: Gender differences. J Neuroendocrinol 14:349–353.

Gilbert AN, Burgoon DA, Sullivan KA, Adler NT (1983) Mother-weanling interactions in Norway rats in the presence of a successive litter produced by postpartum mating. Physiology and Behavior 30:267–271.

Gonzalez-Mariscal G and Poindron P (2002) Parental care in mammals: immediate internal and sensory factors of control. In: Hormones, brain and behavior, Vol. 1 (Eds, Pfaff DW, Arnold AP, Etgen AM, Fahrbach SE, Rubin RT). Elsevier Science: San Diego.

Hofer MA (1994) Early relationships as regulators of infant physiology and behavior. Acta Pediatrica 397:S9–S18.

Hudson R, Cruz Y, Lucio A, Ninomiya J, Martinez-Gomez M (1999) Temporal and behavioral patterning of parturition in rabbits and rats. Physiology and Behavior 66:599–604.

Lehmann J and Feldon J (2000) Long-term biobehavioral effects of maternal separation in the rat: Consistent or confusing? Reviews in Neuroscience 11:383–408.

Leon M, Adels L, Coopersmith R, Woodside B (1984) Diurnal cycle of mother-young contact in Norway rats. Physiology and Behavior 32:999–1003.

Li M and Fleming AS (2003) Differential involvement of nucleus accumbens shell and core subregions in maternal memory in postpartum female rats. Behavioral Neuroscience 117:426–445.

Liu D, Diorio J, Day JC, Francis DD, Meaney MJ (2000) Maternal care, hippocampal synaptogenesis and cognitive development in rats. Nature Neuroscience 3:799–806.

Liu D, Diorio J, Tannenbaum B, Caldji C, Francis D, Freedman A, Sharma S, Pearson D, Plotsky PM, Meaney MJ (1997) Maternal care, hippocampal glucocorticoids receptors, and hypothalamic-pituitary-adrenal responses to stress. Science 277:1659–1662.

Lonstein JS, Wagner CK, De Vries GJ (1999) Comparison of the "nursing" and other parental behaviors of nulliparous and lactating female rats. Hormones and Behavior 36:242–251.

Lovic V and Fleming AS (2004) Artificially reared female rats show reduced prepulse inhibition and deficits in the attentional set-shifting task—reversal of effects with maternal-like licking stimulation. Behavioural Brain Research, 148(1–2):209–219.

Mattson BJ, Williams SE, Rosenblatt JS, Morrell JL (2003) Preferences for cocaine- or pup-associated chambers differentiates otherwise behaviorally identical postpartum maternal rats. Psychopharmacology (Berlin) 167:1–8.

McIver AH and Jeffrey WE (1967) Strain differences in maternal behavior in rats. Behaviour 28:210–216.

Morgan HD, Watchus JA, Milgram MW, Fleming AS (1999) The long lasting effects of electrical simulation of the medial preoptic area and medial amygdala on maternal behavior in female rats. Behavioral Brain Research 99:61–73.

Moore CL (1984) Maternal contributions to the development of masculine sexual behavior in laboratory rats. Developmental Psychobiology 17:347–356.

Moore CL (1985) Sex differences in urinary odors produced by young laboratory rats (Rattus norvegicus). Journal of Comparative Psychology 99:336–341.

Moore CL (1986) A hormonal basis for sex differences in the self-grooming of rats. Hormones and Behavior 20:155–165.

Moore CL, Wong L, Daum MC, Leclair OU (1997) Mother-infant interactions in two strains of rats: Implications for dissociating mechanism and function of a maternal pattern. Developmental Psychobiology 30:301–312.

Numan M and Sheehan TP (1997) Neuroanatomical circuitry of mammalian maternal behavior. Annals of New York Academy of Sciences 807:101–125.

Pryce CR, Bettschen D, Feldon J (2001) Comparison of the effects of early handling and early deprivation on maternal care in the rat. Developmental Psychobiology 38:239–251.

Rees SL and Fleming AS (2001) How early maternal separation and juvenile experience with pups affect maternal behavior and emotionality in adult postpartum rats. Animal Learning and Behavior 29:221–233.

Rosenblatt JS (1967) Nonhormonal basis of maternal behavior in the rat. Science 156:1512–1514.

Rosenblatt JS (2002) Hormonal basis of parenting in mammals. In: Handbook of parenting, Vol 2 (Bornenstein MH, ed.). Mahwah, NJ: Lawrence Erlbaum Associates.

Rosenblatt JS and Ceus K (1998) Estrogen implants in the medial preoptic area stimulate maternal behavior in male rats. Hormones and Behavior 33:23–30.

Rosenblatt JS, Hazelwood S, Poole J (1996) Maternal behavior in male rats: Effects of medial preoptic area lesions and presence of maternal aggression. Hormones and Behavior 30:201–215.

Rosenblatt JS and Lehrman DS (1963) Maternal behavior in the laboratory rat. In: Maternal behavior in mammals (Rheingold HL, ed.). New York: Wiley.

Sharpe RM (1975) The influence of the sex of litter-mates on subsequent maternal behavior in Rattus norvegicus. Animal Behavior 23:551–559.

Smotherman WP, Wiener SG, Mendoza SP, Levine S (1976) Pituitary-adrenal responsiveness of rat mothers to noxious stimuli and stimuli produced by pups. CIBA Foundation Symposium 45:5–25.

Stern JM (1996) Somatosensation and maternal care in Norway rats. Advances in the Study of Behavior 25:243–293.

Stern JM and MacKinnon DA (1978) Sensory regulation of maternal behavior in rats: Effects of pup age. Developmental Psychobiology 11:579–586.

Trekel J and Rosenblatt JS (1971) Aspects of nonhormonal maternal behavior in the rat. Hormones and Behavior 2:161–171.

Wiesner BP and Sheard NM (1933) Maternal behavior in the rat. Edinburgh: Oliver.

Play and Fighting

28

SERGIO M. PELLIS AND VIVIEN C. PELLIS

Fighting involves attack by one animal and defense by another. After a successful deflection of an opponent's attack, the defender may then launch its own attack; this leads to a defense by the other animal. The relative skills in attack and defense determines whether an attacker succeeds (Geist, 1971). Rats are no exception to these general principles and have been a favorite subject for the study of aggression (Blanchard and Blanchard, 1994). A conceptual and descriptive difficulty has arisen because rats engage in both agonistic and playful forms of fighting. The problem has been in identifying whether they differ and how and by what criteria an observer can determine when one grades into the other.

Serious fighting is most often seen in adults, and play fighting is most often seen in juveniles (Pellis and Pellis, 1987). To establish the basic parameters of these behaviors, the discussion begins with the serious fighting of adults; this is then contrasted with the playful fighting of juveniles. We show that serious and playful fighting are distinct behaviors in rats, with play derived from sex, not aggression. Further, we show that rats have modified the sexual content of play fighting and have co-opted its use in adulthood for quasi-aggressive purposes. To understand the subtle, but important, differences between serious and playful fighting, an understanding of the targets that rats compete for during these contests is essential.

TARGETS OF AGONISTIC ATTACK

In launching an attack, an attacker seeks to gain some advantage over their opponent; this may involve throwing the opponent off balance or striking or biting the opponent (Geist, 1978). Often, blows with feet or specialized structures—such as horns and antlers—or bites are directed to particular body targets. If body targets have been subjected to a long history of such attacks, thickening of the skin (to limit penetration of teeth or horns) in that region or of the underlying skeleton (which absorbs the impact of a blow) occurs (Pellis, 1997). Rats direct their bites to specific areas of the opponent's body.

To facilitate the analysis of serious fighting, and so identify the targets being attacked and the tactics being used for attack and defense of those targets, a resident–intruder paradigm is useful. In this paradigm, an unfamiliar male rat is placed into the home enclosure of a resident male; in this situation, the resident does most of the attacking and the intruder does most of the defending (Blanchard and Blanchard, 1990; Kemble, 1993).

Using the resident–intruder paradigm, it has been shown repeatedly that the resident directs most of its bites to the lower dorsum and flanks and that the intruder directs retaliatory bites to the resident's face. This pattern of bite delivery for attackers and defenders has been shown to be true for a variety of domestic strains as well as for wild-caught rats (Blanchard and Blanchard, 1990). Furthermore, a pattern of scarring consistent with these targets has been found in free-living rats (Blanchard et al., 1985).

Analyses of the fighting movements adopted by contestants of several species of murid rodents (the family to which rats belong; see Chapter 1) show that the face is a target of both attack and defense (Pellis, 1997). The defender launches a bite at the attacker's face either as retaliation for a bite to their lower flanks or dorsum, or when approached and hemmed in by an attacker (Pellis and Pellis, 1987; Blanchard and Blanchard, 1994). That is, a defensive bite to the face occurs in response to an attack by an opponent.

In contrast, it is the attacker that maneuvers itself into a position from which to launch an attack at the defender's face (Pellis and Pellis, 1992). Although for most encounters attackers are more likely to launch bites to the flanks and lower dorsum, the possibility of a shift to an attack to the face poses functional challenges to the defender. Similarly, the possible retaliation by the defender in delivering bites to its opponent's face poses a problem to the attacker.

EFFECTS OF TARGET LOCATION ON TACTICS OF ATTACK AND DEFENSE

In the typical encounter between resident and intruder male rats, the resident approaches in a lateral orientation and moves toward its opponent's rump. In response, the intruder rears onto its hind feet, fans out its vibrissae, and faces the opponent. From this position, the intruder can continually turn to face the resident as it attempts to move to the defender's rear.

A biting lunge to the opponent's rump can be countered with a biting lunge to the attacker's face by the defender. The attacker may then stand on all fours and orient laterally in front of the upright defender for a protracted period of time, seemingly motionless. The lack of movement, however, is only apparent; frame-by-frame analyses of videorecordings reveal that small movements by one animal are

countered by small movements by the other (Blanchard and Blanchard, 1994). To break this stalemate, the attacker has to risk retaliation.

The most typical maneuver used by the attacker to break this stalemate is for it to press forward against the intruder while maintaining the lateral orientation (Fig. 28–1). In

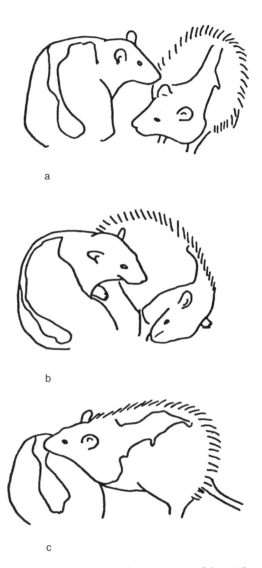

Figure 28–1. The lateral attack tactic in serious fighting. The attacker, usually standing on all fours, orients laterally in front of a defender (*a*). Then, as the defender orients to the attacker's face, the attacker presses, with its lower flank, against the defender's exposed ventrum (*b*). If the attacker manages to unbalance the defender, the attacker may then seize this opportunity to lunge and bite at the defender's exposed lower dorsum (*c*). (Adapted from Pellis and Pellis [1987]).

this way, the attacker can press its lower flank against the defender's ventrum, while keeping its own head out of reach, as the defender would have to relinquish its postural stability by lunging obliquely toward the attacker's head. To do so would then put the defender in a vulnerable position for a counterattack by the attacker. From this position, the attacker can lunge at the defender's lower flank. Even if unsuccessful, this tactic of attack can lead to the defender losing its footing. The attacker can then lunge and bite the defender's lower flank or rump, with a reduced chance of being bitten on the face by the defender. The lateral attack maneuver can be seen as incorporating a defensive component that protects the attacker's head from a retaliatory strike by the defender (Pellis, 1997).

From the defender's perspective, there is a potential tradeoff between standing its ground versus turning to flee as soon as the attacker's last attack is checked. Although standing its ground may lead the attacker to switch to a head attack, fleeing exposes the target on the defender most likely to be bitten, the rump. The same applies to another commonly used defensive tactic, that of turning to supine (Adams, 1980). As is the case for the upright tactic, lying supine protects the dorsum from attack, and facing the attacker with vibrissae fanned out provides a barrier against the attacker's continued attempts to gain access. Unfortunately for the defender, a highly motivated attacker may forgo further attempts to gain access to the dorsum and flanks and switch its attack to the face. From the supine position, the defender is even less well positioned than when facing its opponent from a standing position to block such an attack. Again, there is a tradeoff by the defender in either remaining supine or fleeing the moment that turning to supine has deflected an attack to its rump.

The tradeoff between different tactics is influenced by several factors, including differences in the defender's prior experience with a particular attacker as well as the physical constraints of the test enclosure (Pellis et al., 1989; Pellis and Pellis, 1992). For example, rats are more likely to use the supine defense in smaller enclosures (Boice and Adams, 1983). Presumably, in confined conditions, the option to turn and flee is more dangerous.

This brief overview of the tactics of attack and defense shows that such tactics are used to compete for access to particular body targets. It is clear that the movements performed during aggressive encounters need to be analyzed functionally in relationship to those body targets (Pellis, 1997). Unfortunately, much of the comparative and experimental literature continues to provide mostly numerical frequencies of the occurrence of prespecified tactics, such as the lateral, upright, and supine positions (Alleva, 1993). The problem is that a change in the frequency of any particular tactic cannot be interpreted given that the change could be due to the action of the attacker, defender, or both (Cools, 1985).

Several approaches can be used to analyze interactions so as to discern how the movements of the interactions are related to the attack and defense of particular targets. One such approach is to use choreographic notation techniques to monitor the spatiotemporal pattern of movements by the body parts of both animals simultaneously (Pellis, 1989; Foroud and Pellis, 2003).

Another approach is to identify the moments in an encounter—referred to as "decision points"—where one animal, in committing itself to an action, can no longer influence the subsequent action made by his opponent (Pellis, 1989). For example, consider an attacker approaching from the rear and lunging at the opponent's rump. If the attacker leaps into the air as it lunges, then what the opponent chooses to do to defend its rump can no longer be influenced by the attacker, who is now in the air and has limited its own maneuverability. In this decision point, the typical murid rodent has three tactics of defense from which to choose: (1) to leap or run away,

(2) to rotate to supine and face the attacker, or (3) to rotate while standing to an upright position and face the attacker (Pellis et al., 1992).

These techniques for evaluating the tactics of attack and defense are contingent on knowing the body targets over which the animals compete. Naturally, the simplest way to identify such targets is to score the opponent's wounds (Blanchard and Blanchard, 1990). Such a direct approach can be helpful, but it has a limitation. If both opponents are skilled at combat, an attack by one may be successfully countered by the other; this may result in a situation where neither animal succeeds in delivering a bite (Geist, 1971). Also, successful parrying of an attack by a defender may lead the attacker to deliver a bite opportunistically on some other body location (Pellis and Pellis, 1988). Therefore, wound marks may not necessarily reveal the body targets around which tactics of attack and defense are organized (Pellis, 1997).

Choreographic descriptive methods can be used to determine the targets by identifying those body parts that are approached by one combatant and withdrawn by the other (Pellis, 1989). In turn, the targets so revealed can be substantiated by techniques that directly assess the targeting of the attacker, such as by placing an anesthetized intruder into the home cage of a resident, and then recording the body targets that are bitten. Given that the intruder cannot move, the choice of target by the resident is not constrained by the intruder's defensive actions (e.g., Blanchard et al., 1977; Pellis and Pellis, 1992). Similarly, models of intruders, such as euthanized rats that are frozen into a specific posture, can be placed into the resident's cage (Kruk et al., 1979).

Regardless of the combination of techniques used, knowing the body parts targeted by combatants facilitates the analysis of how the tactics that are used are organized to attack and defend those targets. The framework provided by such an analysis of targets and tactics provides a basis from which to evaluate the causal mechanisms accounting for species, sex, and age differences and experimental effects on overt fighting behavior (Pellis, 1997).

TARGETS IN THE PLAY FIGHTING OF RATS

When the content of play fighting is scored in terms of the behavior patterns used, rats are found to rotate to supine, stand upright, face the partner in a lateral orientation, and lunge and flee (Poole and Fish, 1975). That is, the content of play involves the same behavior patterns that are seen in serious fighting (Grant and MacIntosh, 1966). Some researchers have argued that on the basis of the relative frequency and probabilistic relationship between these behavior patterns in play fighting compared with serious fighting, that the two types of fighting are causally generated by different motivational systems (Poole and Fish, 1975; Panksepp, 1981). Others have claimed that the small differences between the play fighting of juveniles and the serious fighting of adult rats reveal that play fighting is an immature form of serious fighting (Silverman, 1978; Taylor, 1980). Both playful and serious fighting involve competitive interactions that use the same species-typical behavior patterns of aggression. Scoring the relative frequency of these behavior patterns and their sequential organization leads to a debate of "Just how different is different?"

A descriptive approach using choreographic methods revealed that during play fighting, rats compete for access to the partner's nape. This nape contact rarely involves biting; rather, the attacker nuzzles its partner's nape with its snout (Pellis, 1988). That is, the attacking partner uses tactics of attack to gain access to the nape; the defending partner then uses tactics of defense to avoid such contact and to launch its own attacks on the original attacker's nape, which then defends and so on

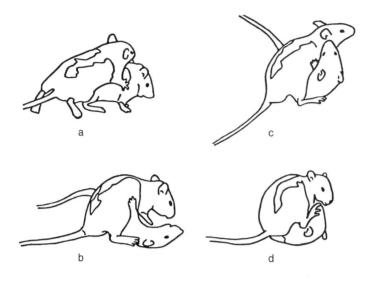

Figure 28–2. In play fighting, the attacker uses tactics of attack, such as lunging, to gain access to its partner's nape (a). In turn, the defender uses tactics of defense, such as rotating to supine, to avoid this contact (b). From this supine position, the defender may then launch its own attack on the original attacker's nape (c). The original attacker may then use defensive tactics to block such an attack (d). (Adapted from Pellis and Pellis [1987]).

(Fig. 28–2). Thus, although the targets in serious fighting are the lower dorsum and flanks and, to a lesser extent, the face, during play fighting the target is the nape area, which is nuzzled rather than bitten (Pellis and Pellis, 1987).

The difference between the two forms of fighting in rats therefore is not a quantitative difference, as was debated earlier, but rather is a qualitative one. Indeed, although a play fight may occasionally escalate to serious fighting, when it does, the partners stop competing for nuzzling contact of the nape and switch to biting attacks to their partner's rump (Takahashi and Lore, 1983; Pellis and Pellis, 1991). The similarities in postures between play fighting and serious fighting reflect that there is a standard species-typical set of tactics for attack and defense, which may be used in competitive interactions not only in conspecific aggression and play but also in sexual and predatory encounters (Pellis, 1988).

Analysis of those species-typical postures or behavior patterns reveals that they are modified in play fighting compared with serious fighting to accommodate the differences in targets (Pellis and Pellis, 1987). These findings support the view that play fighting is a distinct behavioral system, and not merely immature aggression. However, the findings for

rats need to be reconciled with those from a broader comparative literature that show that for a wide range of mammals and for some birds, play fighting in juveniles involves the same targets of attack and defense as in adult serious fighting (Aldis, 1975). Why are rats seemingly aberrant in this regard?

THE ORIGINS OF PLAY FIGHTING IN RATS

In rats, nuzzling of the nape can occur during adult sexual encounters, and partners may use a variety of defensive tactics to avoid such contact (Pellis, 1988). Therefore, one possibility is that the proponents of play fighting as immature serious fighting may have been right, except that they have used the wrong adult behavioral system for their comparison. That is, play fighting in rats may simply be immature sexual behavior. Indeed, comparing the play behavior of rats with that of other murid rodents confirms this possibility (Pellis, 1993). During play fighting, murid rodents do not compete for access to agonistic targets but rather compete for those typically contacted during the precopulatory phase of sexual encounters. The targets competed for during play and sexual behavior can be very different from those attacked during serious fighting. For example, in Djungarian hamsters, serious fighting involves bites directed at the rump and sides of the face, whereas during courtship and play fighting, it is the front of the mouth that is nuzzled (Pellis and Pellis, 1989).

Earlier researchers likely missed the connection of play fighting to sexual behavior in rats and other murid rodents because they focused on scoring behavior patterns, rather than on the targets around which those behavior patterns are organized (Fagen, 1981; Hole and Einon, 1984). Furthermore, students of animal play have tended to label play as sexual only when the behavior patterns used by the participants are those obviously derived from sexual contexts, such as mounting (Mitchell, 1979; Fagen, 1981). Because juvenile play fighting in murid rodents mimics the adult precopulatory phase of sexual behavior (Pellis, 1993), mounting is rare, and so the commonality in targets would have had to be identified to make the connection (Pellis, 1988).

TACTICS USED IN PLAY FIGHTING AND THEIR DEVELOPMENT

Play fighting in rats, from its earliest onset— in the days before weaning—involves nape contact (Pellis and Pellis, 1997). It is only in the latter stages of the juvenile phase, as rats approach puberty, that they occasionally mount one another after nape contact (Pellis and Pellis, 1990). This gradual onset, from more appetitive components to more consummatory ones, supports the possibility that juvenile play fighting is an immature stage of the sexual motivational system. Comparative data support this hypothesis.

Both montane and prairie voles compete for access to the nape during play fighting and during precopulatory behavior (Pellis et al., 1989; Pierce et al., 1991). During sexual encounters, female montane voles are more

likely to use upright defensive tactics to block their partner's access to their own napes than are female prairie voles; instead, females of this species are more likely to rotate to supine to avoid nape contact than are montane voles (Pierce et al., 1991). Correspondingly, during juvenile play fighting, prairie voles use the supine defense more often than do montane voles, whereas montane voles use the upright defense more often than do prairie voles (Pellis et al., 1989). That is, in the organization of defense in both of these vole species, play fighting resembles the species-specific pattern of their adult sexual behavior (Pellis and Pellis, 1998). Rats, however, differ.

During adult precopulatory encounters, the most likely defensive response of a female rat to a male's contact with her nape is for her to evade by leaping or running away or by laterally dodging away. Rotating around the longitudinal axis of the body, thus moving the nape away while turning to face the male, is the least likely response (<10% of cases) (Pellis and Iwaniuk, 2004). In marked contrast, in juvenile play fighting, only around 20% to 30% of defense involves evasion, whereas 60% to 80% of cases involve rotating around the longitudinal axis to supine or nearly supine (Pellis and Pellis, 1990). That is, in rats, play fighting inverts the pattern typical of adult sexual behavior (Pellis and Pellis, 1998). Therefore, unlike what may be the case for some other species of murid rodents, in rats, play fighting has been modified in that it is not simply an immature version of adult sexual behavior. That play fighting in juvenile rats represents a novel behavioral system is supported by two lines of evidence.

Rats, but not other species of murid rodents that do not have the rat-like modified patterns of play fighting, experience emotional and cognitive deficits if deprived of play fighting in the juvenile phase (Einon et al., 1978). Furthermore, rats, but not other murid rodents, use play fighting in adulthood in nonsexual contexts (Pellis, 1993; 2002). The novel functions of play fighting in adult rats show that a pattern of play derived from sexual behavior can be co-opted for use in quasi-aggressive situations. In turn, this shows that the link between play fighting and aggression in rats that was identified by some earlier researchers may not have been entirely illusory (Silverman, 1978; Taylor, 1978).

PLAY FIGHTING AS QUASI-AGGRESSION IN RATS

Play fighting in rats, as defined by competition for snout-to-nape contact, persists well beyond puberty, into adulthood (Pellis and Pellis, 1990). In the juvenile phase, the most common response when contacted on the nape is for the rat to rotate around its longitudinal axis to a fully supine position. For female rats, even as the frequency of play fighting wanes after puberty, the rotation to supine remains the most likely defensive response. The situation for male rats differs.

With the onset of puberty, males are more likely to rotate only partially around the longitudinal axis of the body—thus keeping at least one hindpaw in contact with the ground—than they are to rotate fully to supine. From this partially rotated position, the defending male can use its flank to push laterally against the attacker, or it can rear upright and face the attacker. All male rats undergo this change in their preferred defensive strategy, but after puberty, they also modify their pattern of defense depending on the identity of their play partner (Pellis and Pellis, 1992).

When attacked playfully by a female or a subordinate male, a male rat most likely uses the partial rotation tactic. However, when attacked by a dominant male, the subordinate male most likely uses the complete rotation tactic. Indeed, two subordinate males behave symmetrically toward one another, with both likely to use the partial rotation tactic when attacked by the other. However, if one becomes dominant after the removal of the resident dominant male, they then behave asym-

metrically toward one another, in that the subordinate male now rotates fully to supine (Pellis and Pellis, 1990, 1991; Pellis et al., 1993).

From an analysis of this modulation of play fighting in rats, and with the absence of such nonsexual uses of play in adulthood by other species of murid rodents, it appears that among adult male rats, play fighting is used for social assessment and manipulation (Pellis, 2002). Among familiars, subordinate males initiate more play with the dominant male and, when contacted playfully by him, respond in a more juvenile fashion. That is, subordinate males appear to use play for maintaining affiliative bonds with the dominant male. In contrast, a subordinate male may play more roughly so as to challenge the status of the dominant male. Similarly, both dominants and subordinates can use play fighting to challenge the status of an unfamiliar male encountered in a neutral arena (Pellis and Pellis, 1991, 1992; Pellis et al., 1993; Smith et al., 1999).

CONCLUSION

Play fighting and serious fighting superficially resemble each other because both use many of the same tactics with which to compete for access to bodily targets. In rats, serious and playful fighting differ qualitatively as they involve competition over different targets, with the playful targets having been derived from sexual behavior. These findings suggest that serious and playful fighting have different causal and functional properties. Surprisingly, the two may overlap in ways that do not seem to be the case for other murid rodents. Rats have evolved novel control mechanisms over play fighting, and this emancipation from its regulatory origins has allowed play fighting to be co-opted for nonsexual uses, one of which is aggressive competition for enhanced social status (Pellis, 2002; Pellis and Iwaniuk, 2004).

Interestingly, the physiological profile of rats, in terms of their endocrinological responses when engaged in play fighting, is similar to that which is present when they are engaged in serious aggression, with the physiological distinction between the two becoming increasingly blurred with age (Hurst et al., 1996). Nonetheless, neonatal treatment with testosterone propionate leads to an elevation of play fighting, but the changes in play do not affect which rats become dominant, suggesting that different developmental processes regulate playful and serious fighting (Pellis et al., 1992). Therefore, although at some levels of analysis playful and serious fighting are distinct, at other levels, the two are similar. It appears that in rats, then, there has been a complex evolutionary relationship between serious and playful fighting. These two forms of fighting may have had separate origins, but some of their current functions overlap, as do some of the neurobehavioral mechanisms that regulate them.

REFERENCES

Adams DB (1980) Motivational systems of agonistic behavior in muroid rodents: A comparative review and neural model. Aggressive Behavior 6:295–346.

Aldis O (1975) Play fighting. New York: Academic Press.

Alleva E (1993) Assessment of aggressive behavior in rodents. In: Methods in neurosciences. Paradigms for the study of behavior, Vol 14 (Conn PM, ed.), pp. 111–137. New York: Academic Press.

Blanchard DC and Blanchard RJ (1990) The colony model of aggression and defense. In: Contemporary issues in comparative psychology (Dewsbury DA, ed.), pp. 410–430. Sunderland, Mass.: Sinauer Associates, Inc.

Blanchard RJ and Blanchard DC (1994) Environmental targets and sensorimotor systems in aggression and defence. In: Ethology and psychopharmacology (Cooper SJ and Hendrie CA, eds.), pp. 133–157. New York: John Wiley & Sons.

Blanchard RJ, Blanchard DC, Pank L, Fellows D (1985) Conspecific wounding in free ranging *Rattus norvegicus*. The Psychological Record 35:329–335.

Blanchard RJ, Blanchard DC, Takahashi T, Kelly MJ (1977) Attack and defensive behaviour in the albino rat. Animal Behaviour 5:622–634.

Boice R and Adams N (1983). Degrees of captivity and aggressive behavior in domestic Norway rats. Bulletin of the Psychonomic Society 21:149–152.

Cools AR (1985) Brain and behavior: hierarchy of feedback systems and control of input. In: Perspectives in ethology. Mechanisms, Vol 6 (Bateson PPG and Klopfer PH, eds.), pp. 109–168. New York: Plenum Press.

Einon DF, Morgan MJ, Kibbler CC (1978) Brief periods of socialization and later behavior in the rat. Developmental Psychobiology 11:213–225.

Fagen R (1981) Animal play behavior. New York: Oxford University Press.

Foroud A and Pellis SM (2003) The development of "roughness" in the play fighting of rats: A Laban movement analysis perspective. Developmental Psychobiology 42:35–43.

Geist V (1971) Mountain sheep. Chicago: University of Chicago Press.

Geist V (1978) On weapons, combat and ecology. In: Advances in the study of communication and affect, Vol 4 (Krames LP, Pliner P, Aloway T, eds.), pp. 1–30. New York: Plenum Press.

Grant EC and MacIntosh JM (1966) A comparison of some of the social postures of some common laboratory rodents. Behaviour 21:246–259.

Hole GT and Einon DF (1984) Play in rodents. In: Play in animals and man (Smith PK, ed.), pp. 95–117. Oxford: Basil Blackwell.

Hurst JL, Barnard CJ, Hare R, Wheeldon EB, West CD (1996) Housing and welfare in laboratory rats: Time-budgeting and pathophysiology in single sex groups. Animal Behaviour 52:335–360.

Kemble ED (1993) Resident-intruder paradigms for the study of rodent aggression. In: Methods in neurosciences. Paradigms for the study of behavior, Vol 14 (Conn PM, ed.), pp. 138–150. New York: Academic Press.

Kruk MR, van der Poel AM, de Vos-Frerichs TP (1979) The induction of aggressive behavior by electrical stimulation in the hypothalamus of male rats. Behaviour 70:292–322.

Mitchell G (1979) Behavioral sex differences in nonhuman primates. New York: Van Nostrand Reinhold Co.

Panksepp J (1981) The ontogeny of play in rats. Developmental Psychobiology 14:327–332.

Pellis SM (1988) Agonistic versus amicable targets of attack and defense: Consequences for the origin, function and descriptive classification of play-fighting. Aggressive Behavior 14:85–104.

Pellis SM (1989) Fighting: The problem of selecting appropriate behavior patterns. In: Ethoexperimental approaches to the study of behavior (Blanchard RJ, Brain PF, Blanchard DC, Parmigiani S, eds.), pp. 361–374. Dordrecht, the Netherlands: Kluwer Academic Publishers.

Pellis SM (1993) Sex and the evolution of play fighting: A review and model based on the behavior of muroid rodents. Play Theory and Research 1:55–75.

Pellis SM (1997) Targets and tactics: The analysis of moment-to-moment decision making in animal combat. Aggressive Behavior 23:107–129.

Pellis SM (2002) Keeping in touch: Play fighting and social knowledge. In: The cognitive animal: empirical and theoretical perspectives on animal cognition (Bekoff M, Allen C, Burghardt GM, eds.), pp. 421–427. Cambridge, Mass.: MIT Press.

Pellis SM and Iwaniuk AN (2004) Evolving a playful brain: A levels of control approach. International Journal of Comparative Psychology 17:90–116.

Pellis SM and Pellis VC (1987) Play fighting differs from serious fighting in both the target of attack and tactics of fighting in the laboratory rat *Rattus norvegicus*. Aggressive Behavior 13:227–242.

Pellis SM and Pellis VC (1988) Play-fighting in the Syrian golden hamster *Mesocricetus auratus* Waterhouse and its relationship to serious fighting during post-weaning development. Developmental Psychobiology 21:323–337.

Pellis SM and Pellis VC (1989) Targets of attack and defense in the play fighting by the Djungarian hamster *Phodopus campbelli*: Links to fighting and sex. Aggressive Behavior 15:217–234.

Pellis SM and Pellis VC (1990) Differential rates of attack, defense and counterattack during the developmental decrease in play fighting by male and female rats. Developmental Psychobiology 23:215–231.

Pellis SM and Pellis VC (1991) Role reversal changes during the ontogeny of play fighting in male rats: Attack versus defense. Aggressive Behavior 17:179–189.

Pellis SM and Pellis VC (1992) Juvenilized play fighting in subordinate male rats. Aggressive Behavior 18:449–457.

Pellis SM and Pellis VC (1997) The pre-juvenile onset of play fighting in rats (*Rattus norvegicus*). Developmental Psychobiology 31:193–205.

Pellis SM and Pellis VC (1998) The play fighting of rats in comparative perspective: A schema for neurobehavioral analyses. Neuroscience and Biobehavioral Reviews 23:87–101.

Pellis SM, Pellis VC, Dewsbury DA (1989) Different levels of complexity in the play fighting by muroid rodents appear to result from different levels of intensity of attack and defense. Aggressive Behavior 15:297–310.

Pellis SM, Pellis VC, Kolb B (1992) Neonatal testosterone augmentation increases juvenile play fighting but does not influence the adult dominance relationships of male rats. Aggressive Behavior 18:437–447.

Pellis SM, Pellis VC, McKenna MM (1993) Some subordinates are more equal than others: Play fighting amongst adult subordinate male rats. Aggressive Behavior 19:385–393.

Pellis SM, Pellis VC, Pierce JD Jr, Dewsbury DA (1992) Disentangling the contribution of the attacker from that of the defender in the differences in the intraspecific fighting of two species of voles. Aggressive Behavior 18:425–435.

Pierce JD Jr, Pellis VC, Dewsbury DA, Pellis SM (1991) Targets and tactics of agonistic and precopulatory behavior in montane and prairie voles: Their relationship to juvenile play fighting. Aggressive Behavior 17:337–349.

Poole TB and Fish J (1975) An investigation of playful behaviour in *Rattus norvegicus* and *Mus musculus* (Mammalia). Journal of the Zoological Society, London 175:61–71.

Silverman P (1978) Animal behaviour in the laboratory. New York: Pica Press.

Smith LK, Fantella S-L, Pellis SM (1999) Playful defensive responses in adult male rats depend on the status of the unfamiliar opponent. Aggressive Behavior 25:141–152.

Takahashi LK and Lore RK (1983) Play fighting and the development of agonistic behavior in male and female rats. Aggressive Behavior 9:217–227.

Taylor GT (1980) Fighting in juvenile rats and the ontogeny of agonistic behavior. Journal of Comparative and Physiological Psychology 94:953–961.

Sex

WILLIAM J. JENKINS AND JILL B. BECKER

The genes that an animal inherits set off a cascade of events that result in the sexual differentiation of the external genitalia and central nervous system. Simply, genetic sex determines gonadal sex and gonadal sex determines phenotypic sex.[1] During the late prenatal and early postnatal development of the rat, exposure to testosterone produced by the maturing testes produces "organizational effects" on neuronal differentiation, growth, and survival that result in masculinization of the brain. In the absence of testosterone, the brain is feminized[2] (Breedlove et al., 2002).

As an adult, the animal responds to its own gonadal hormones (testosterone for the male, estradiol and progesterone for the female) with the initiation of sex-typical behaviors, including reproductive behaviors. Sexually differentiated brain structures generate sexually dimorphic copulatory behaviors as a function of "activational effects" of circulating hormone levels of the adult. For example, male rats that are castrated neonatally display limited masculine copulatory behaviors and exhibit lordosis if treated with estradiol and progesterone in adulthood. Furthermore, female rats that are treated with testosterone or estradiol during early development show increased male-typical and decreased female-typical sexual responses (Breedlove et al., 2002).

To study sexual behavior in the rat, it is necessary to understand the endocrinology of the reproductive system of both females and males. Hormones produced by the gonads feed back to the brain to stimulate reproductive behavior. In this chapter, we provide an overview of the reproductive systems and the neural systems that mediate sexual behavior and discuss how to study sexual behavior in rats. We address these issues in females first and in males next.

SEXUAL BEHAVIOR IN THE FEMALE RAT

Sexual behavior in the female rat is a function of a complex interplay of hormones and environmental circumstances. Female rats are nonseasonal, spontaneous ovulators that display a 4 to 5 day estrous cycle (Fig. 29–1) composed of four basic stages: diestrus 1 (or metestrus), diestrus 2 (or diestrus), proestrus, and estrus (when the female rat is sexually receptive). This is one of the most rapid ovarian cycles among mammals, and it is made possible by truncating the cycle immediately after ovulation.

THE ESTROUS CYCLE

Follicular Phase

The ovarian cycle begins with the development of follicles from oocytes in the ovary. During the follicular phase, the hypothalamus produces a pulsatile release of gonadotropin-releasing hormone (GnRH) that stimulates the release of two primary gonadotropins from the anterior pituitary: follicle-stimulating hormone (FSH) and luteinizing hormone (LH). These gonadotropins stimulate the growth of

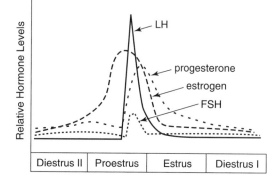

Figure 29–1. Plasma concentrations of hormones across the female rat estrous cycle. LH, luteinizing hormone; FSH, follicle-stimulating hormone. (Adapted from McCarthy and Becker, 2002.)

follicles and estradiol synthesis in the ovarian granulosa cells. There also is increased steroidogenesis in the ovary, and estradiol secretion increases gradually during this phase. In the rat, this stage is 2 days long. The first day is called *diestrus 1* or *metestrus*, and the second day is called *diestrus 2* or just *diestrus*.

Periovulatory Period

The time just before and after ovulation is dynamic. Estradiol reaches its peak concentrations of 100 to 150 pg/ml around noon on the day of proestrus approximately 6 to 12 hours before ovulation. The peak of estradiol triggers a surge of GnRH from the hypothalamus and the release of prolactin from the anterior pituitary that induces LH and FSH release to reach their surge levels. Eight to 10 hours later, the mature follicle releases the ovum and transforms into the *corpus luteum*. Progesterone also increases a few hours before ovulation about 4 to 6 hours after the estradiol surge, during the afternoon of *proestrus* (Freeman, 1994).

Estrus is the period of sexual receptivity and the day of ovulation. Onset of sexual receptivity occurs shortly after the start of the dark phase of the light–dark cycle and precedes ovulation by a few hours in most rats. Ovulation, induced by the LH surge on proestrus, occurs 4 to 6 hours after nightfall,

and sexual receptivity persists for 12 to 20 hours (depending on whether the female mates). Note that behavioral receptivity occurs 36 to 48 hours after estradiol begins to increase and 4 to 6 hours after the increase in progesterone. Baseline serum concentrations of estradiol at "vaginal estrus" or behavioral estrus are approximately 3 to 12 pg/ml (McCarthy et al., 2002).

In many female mammals, ovulation is followed by an additional phase called the *luteal phase*, which is maintained by hormones produced by the corpus luteum. In rats, the corpus luteum becomes functional only if the female engages in sexual behaviors that activate a progestational reflex (twice-daily surges of prolactin). These surges of prolactin maintain the corpus luteum so that progesterone is released and implantation can occur if the eggs are fertilized. The corpus luteum is maintained for approximately 12 days, if the rat is pregnant, the placenta then assumes responsibility for the maintenance of progesterone secretion. In the event of an unfertile mating, the female rat may exhibit "pseudopregnancy," a 12-day period of anestrus induced by the secretion of progesterone from the corpus luteum (Smith et al., 1975; Gunnet et al., 1983).

Stimulation of the vaginocervical area by the male or by the experimenter can induce pseudopregnancy. This is mentioned, because to determine where a female rat is in her ovarian cycle the investigator must obtain cells from the vaginal epithelium. If one is too forceful during this procedure, pseudopregnancy can be the result.

Determining Estrous Cycle Phase by Vaginal Lavage

The stages of the estrous cycle can be determined by examining morphological changes in vaginal epithelial cells under light microscopy (Figs. 29–2 and 29–3). This is usually done by flushing the vagina with saline using an eyedropper and 0.9% saline. The tip of the eyedropper is filled with a small amount (1 to

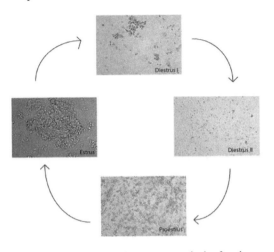

Figure 29–2. Vaginal cytology varies with the female rat estrous cycle. See Table 29–1 for description of cell types.

2 drops) of saline and then inserted into the vagina of the female rat. The vagina is flushed two or three times with the saline, or until the saline becomes cloudy, then the fluid is placed onto a slide. The eyedropper is thoroughly rinsed with distilled water, and the next rat is sampled. For obtaining data from a large number of rats at a time, we use a piece of Plexiglas about 4 × 8 cm onto which Silastic adhesive has been affixed to make barriers about 1 × 1 cm. With this plate, samples from 32 rats can be obtained at one time. The resulting samples (vaginal smear) are examined

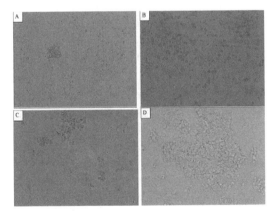

Figure 29–3. Vaginal cytology as assessed by saline lavage under light microscopy. *A,* Diestrus I; *B,* diestrus II; *C,* proestrus; *D,* estrus. See Table 29–1 for description of cell types.

under a light microscope while the sample is still wet (see later).

How to Hold the Rat for Vaginal Smears

There are two ways to hold the rat to obtain the sample. The ventral approach is used most commonly in rats that have been handled. The female is picked up and held in one hand with the belly exposed, using the little finger to hold out one back leg. The experimenter's other hand is used to insert the eyedropper into the vagina (it is a good idea to hold the tail with the lower fingers of the hand, holding the eyedropper to steady the hand and keep the tail from flailing). The other approach is a dorsal approach in which the female is standing facing away from the experimenter and then her hindquarters are lifted slightly by grasping the base of the tail; the eyedropper is then inserted into the vagina under the tail.

The greatest concern about collecting vaginal lavage samples is to not be overly aggressive so that the animal does not become pseudopregnant. It may seem that the dorsal approach is easier for those not experienced at handling rats, but the rats tend to struggle quite a lot with this approach, and because they have two feet on a surface, they can achieve quite a lot of force if so motivated. It is also not as easy to see how far the eyedropper has been inserted. We generally have better success in obtaining good samples and at not inducing pseudopregnancy with the ventral approach, even if the animal handlers are inexperienced.

How to Read the Slides

Staining the cells is not necessary. Examine the liquid obtained under a ×10 to ×20 objective with a blue or green filter. Interpreting the stage of estrous cycle from the cell morphology requires practice and experience with a particular rat (Table 29–1; see Figs. 29–2 and 29–3). Exactly what the cells will look like at a particular stage of the cycle varies with the time of day (especially on proestrus) and the

Table 29–1. Description of Estrous Cycle Changes in Vaginal Cell Morphology

Day of Cycle	Cell Morphology in Smear
Diestrus 1 (metaestrus)	Small, round cells called leukocytes (occasionally during early metaestrus, cells can be seen that look like the nucleated large cells of proestrus. The only way to tell the difference is in retrospect—the next day the animal has leukocytes and is not in estrus).
Diestrus 2 (diestrus)	Small, round cells called leukocytes plus some large, round cells without nuclei are seen.
Proestrus	Proestrus is characterized by a predominance of *nucleated epithelial cells*, which are large and round and have an easily visible nucleus. These appear individually and in clusters and can be interspersed with cornified squamous epithelial cells. Leukocytes are not present in the smear. To see the nuclei, the examiner must focus up and down; the cells look like donuts. Some spindly cells may be seen as well.
Estrus	Large, irregularly shaped cells are referred to as cornified cells because occasionally they clump together and resemble an ear of corn in the husk. *Cornified squamous epithelial cells* are large and irregular, have no visible nucleus, and contain a granular cytoplasm. These appear individually and in very large clusters. Nucleated epithelial cells and leukocytes are not present on the smear.
Pseudopregnant	Usually a combination of all of the above cell types is seen, or a smear looks like one from diestrus with irregularly shaped cells. This persists for about 12 days.

rat. Rather than trying to guess whether a rat is in a particular stage of the cycle, it is better to describe the cell types that appear in the smear. After 7 to 8 days, it is usually possible to determine the stage of the cycle of a female and what the cytology looks like for each stage. We collect data for at least two complete cycles before using a female rat in an estrous cycle–timed experiment, because it can take this long to be confident, based on the cytology, of the stage of the cycle of a rat. In addition, when we begin collecting estrous cycle data, we group together house rats that are in the same phase of the estrous cycle to take advantage of the effects of pheromones of estrous cycle synchrony (Schank et al., 1992).

SEXUAL BEHAVIOR OF THE FEMALE RAT

Mating behavior occurs in the female rat during the period of sexual receptivity that begins shortly after lights-out (for rats on a 12 hour light–dark cycle) on the evening of proestrus through the early morning of estrus (Barnett, 1975). Copulation in rats consists of three primary interactions between the male and female. These are defined by the behavior of the male, which consists of *mounting, intromission,*

and *ejaculation*. These events can be reliably distinguished under experimental conditions as a function of various characteristics and qualities associated with each (see later). The male mounts the female from the rear, clasping and palpating her flanks with his forepaws. When the female is sexually receptive, this mounting results in the female assuming a reflexive posture known as *lordosis*. When in lordosis, the female arches her back and dorsiflexes her tail. If the mount is accompanied by insertion of the penis into the vagina, it is referred to as an *intromission*. The mating sequence is characterized by repeated intromissions before the male ejaculates.

Lordosis and Proceptive Behaviors

Lordosis is usually quantified by determining the lordosis quotient (LQ), which is the percentage of time that the female exhibits lordosis when the male mounts. The LQ is calculated by dividing the number of times the female displays lordosis by the total number of male contacts, multiplied by 100%. The LQ is the most common measure of sexual receptivity used for female rats. Some investigators give a qualitative rating to the intensity of lordosis, but it is more common to use the LQ.

To determine the LQ, it is necessary to carefully observe the female rat's posture. This is most easily accomplished by video-recording the behavioral interaction, as this allows for a given sequence to be observed multiple times and at slow motion if necessary. When a female rat assumes the lordotic posture, she arches her back and often elevates her head. Lordosis is accompanied by temporary immobility that sometimes persists even after the mount, intromission, or ejaculation is over. Lordosis can also be elicited by an experimenter who grasps the female lightly with the index finger and thumb along the lateral region of the hindquarters. Sexually receptive females will have an LQ that approaches 100%; however, researchers often operationally define sexual receptivity as an LQ that is greater than 50%.

Lordosis may be reflexive in nature, but the female is not a passive participant in the copulatory bout. Female rats engage in a series of behaviors that include approaching the male, presenting, hopping, darting, and ear wiggling to solicit sexual behavior from the male. These proceptive, or solicitation, behaviors may represent an index of female sexual initiative or motivation, and during these behaviors the female emits ultrasonic vocalizations that are thought to attract the male (Beach, 1976; White et al., 1993). These behaviors are observed under laboratory testing conditions, although their display depends on the context in which copulation occurs. Specifically, the behaviors described earlier occur when the rats mate in pairs in relatively small test cages (Erskine, 1989).

Under more natural circumstances, in which rats mate in groups in larger, more complex arenas, the female displays a different set of solicitational behaviors, including approach to, orientation to, and run away from the male. Because of the nature of these behaviors, they are observed under conditions in which there is sufficient physical space for them to occur. To quantify these behaviors, the incidence of each behavior (ear wiggling, hopping, and darting) is recorded from visual observation or videorecording of a sexual interaction (Erskine, 1989).

Pacing Behavior

Under seminatural conditions, the female is able to use solicitational behaviors sequentially; under these circumstances, the female controls or "paces" the rate of copulation. This is important for the female rat because the interval between intromissions is critical for triggering the neuroendocrine reflex that results in the progestational state necessary for pregnancy. Males and females differ in their optimal latencies between intromissions during copulation. Males prefer regular, rapid intromissions that lead to a fairly quick ejaculation. Females, on the other hand, require longer intervals between intromissions to optimize vaginocervical stimulation received during the copulatory bout (Erskine, 1989).

In the wild, both males and females achieve their preferred rate of copulation by engaging in group mating. In the laboratory, females can also pace the rate of copulation if sex occurs in an environment where the female rat both approaches and avoids the male. In our laboratory, the female rat paces in a testing arena that is divided into two chambers with an opaque barrier (Fig. 29–4). The male is physically tethered to the larger portion of the arena so that he has complete mobility up to the barrier but he cannot cross the barrier. The female, on the other hand, has full access to both sides of the chamber. It should be noted, however, that if a tethered male is used during a test of sexual behavior, it is important that the male is trained with the tether before experimental data are collected. Often, when a male is placed in a tether for the first time, he is more interested in trying to get out of the tether than he is in the estrous female!

Observations of paced mating behavior have led researchers to focus on two components of paced copulatory behavior that may reflect two different aspects of the interaction:

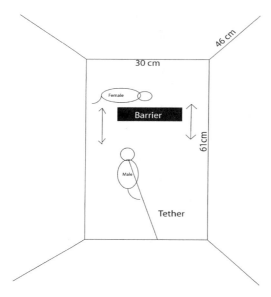

Figure 29–4. Schematic of pacing apparatus. The apparatus is made of clear Plexiglas that measures 61 cm long × 30 cm wide × 46 cm high. The barrier is placed approximately one third of the length of the chamber away from the wall. It is made of an opaque Plexiglas and measures 20 cm long × 1 cm wide × 25 cm high. The female rat has free access to either side of the chamber, while the male is resticted through the use of a flexible tether and harness to the larger side of the chamber. For additional details see Xiao and Becker (1997).

(1) percent exits (number of times a female leaves the male after a given coital stimulus/total number of coital stimuli × 100%) and (2) return latency (amount of time female avoids male after a given coital stimuli). Both of these measures increase as sexual stimuli become more intense (mount < intromission < ejaculation). Therefore, pacing depends on the ability of the female to discriminate between sexual stimuli and to perform appropriate motoric behavioral responses as a result.

Two measures of pacing are determined from behavioral observations. *Percent exits* are calculated from the number of times that the female rat exits the male side of the chamber after receiving a mount, intromission, or ejaculation. If using an apparatus like that shown in Figure 29–4, an exit is determined by the female completely crossing the opaque barrier to avoid the male rat. The number of exits after a mount (for example) is divided by the number of mounts (×100%) to obtain percent exits after mounts. After a female exits the male side following a given coital stimulus, we also monitor the number of seconds that the female stays away from the male before crossing over to the male. Add the total time that the female avoids the male and divide by the total number of exits for a given coital stimulus to obtain the mean *return latency* for mounts, intromissions, and ejaculations. Again, when a female rat is pacing, both percent exits and return latencies increase with the intensity of the coital stimuli.

IMPORTANCE OF PACING BEHAVIOR

As mentioned, the female rat requires vaginocervical stimulation to initiate the neuroendocrine reflex that maintains the immature corpora lutea. It is not simply a matter of a minimum amount of vaginocervical stimulation to trigger this event, however; rather, the pattern in which the vaginocervical stimulation is received is critical for the female rat. Female rats exhibit high rates of pseudopregnancy with as few as five intromissions when they were allowed to pace the rate of copulation compared with females that receive five nonpaced intromissions. This discrepancy decreases as the number of intromissions received increases, nevertheless, females pacing the rate of intromissions always show higher rates of pseudopregnancy (Gilman et al., 1979). Furthermore, when female rats pace the rate of intromissions, behavioral estrus is truncated compared with those rats that have nonpaced sex (Erskine et al., 1982). These results indicate that although pacing the rate of copulation is not necessary for the induction of pregnancy or pseudopregnancy, the female can optimize the vaginocervical stimulation received under circumstances in which she actively controls the rate of copulation and that doing so results in physiologically relevant consequences.

NEURAL SUBSTRATES OF RECEPTIVE, PROCEPTIVE, AND PACED SEXUAL BEHAVIOR

Female sexual behavior in the rat is contingent on the presence of the ovarian hormones estradiol and progesterone. Mating behavior in the female is disrupted by ovariectomy and is reestablished by estradiol and progesterone acting sequentially, These hormones activate the neural systems necessary for these behaviors by binding to intracellular receptors that activate genomic processes that result in the induction of specific proteins (Pfaff et al., 1994).

The importance of estradiol and progesterone for sexual receptivity has been demonstrated in experiments with ovariectomized rats to remove the endogenous source of ovarian hormones. If no hormones are given, an ovariectomized female will not be receptive and will reject attempts by a male rat to mate with her. If an ovariectomized female is treated with estradiol alone, after about 6 to 10 days she will exhibit sexual receptivity, but the behavior does not appear normal. If progesterone alone is given, sexual behavior is never induced even after weeks of progesterone treatment.

On the other hand, when estradiol is administered about 48 hours before progesterone, as occurs during the estrous cycle, normal-appearing sexual receptivity results. This is because estradiol primes the brain to be sensitive to progesterone by inducing the synthesis of progesterone receptors in the hypothalamus. Progesterone then acts on these receptors to induce sexual behavior (McCarthy et al., 2002).

The hypothalamus is where neuroendocrine information is integrated to initiate female receptive behavior. The ventromedial hypothalamus (VMH) and the medial preoptic area of the hypothalamus (MPOA) are two sites believed to be critical to the display of sexual behavior in the female rat. Copulatory behavior induces consistent and reliable in-

creases in the immediate-early gene c-fos in these two structures (Erskine, 1993; Pfaus et al., 1993; Coolen et al., 1996).

The VMH is crucial to the display of lordosis in the female rat. Lesions to the VMH disrupt sexual receptivity. Electrical stimulation of the VMH facilitates the lordosis response as a function of hormone treatment. Direct estradiol implants into the VMH induce sexual receptivity and facilitate sexual responses in ovariectomized female rats in conjunction with systemic administration of progesterone. Sequential implants of estradiol and progesterone into VMH induce sexual behavior in female rats that is comparable to levels observed by systemic treatment with estradiol and progesterone (Pfaff et al., 1994; McCarthy et al., 2002).

Implants of an estradiol antagonist into the VMH block the induction of sexual receptivity by systemically administered hormones, and damage to the VMH that blocks sexual receptivity also reduces the amount of time that a female spends near a tethered male. Similarly, progesterone implants into the VMH elicit hopping, darting, and ear wiggling from female rats. The efferents projecting from the VMH to various midbrain sites appear to be especially important for the display of lordosis, however, although lesions to these areas have no consistent effect on solicitational behaviors displayed by the female (Erskine, 1989; Pfaff et al., 1994; McCarthy et al., 2002).

If the VMH is involved in the facilitation of lordosis, it appears that the MPOA normally inhibits this reflex. Lesions to the MPOA reduce the amount of estradiol required to induce lordosis and lead to higher lordosis quotients as a function of hormonal priming. In contrast, electrical stimulation of the MPOA disrupts lordosis. This role of the MPOA is not without controversy, however, as different effects of MPOA lesions have been observed as a function of testing conditions (Erskine, 1989; Pfaff et al., 1994; McCarthy et al., 2002).

There is also evidence that the MPOA is involved in the display of proceptive behaviors. Both nonspecific and axon-sparing lesions of the MPOA diminish proceptive behaviors (Whitney, 1986; Hoshina et al., 1994). When the female rat is tested under conditions that allow the female to escape the male, lesions to the MPOA result in the female spending less time with the male and receiving fewer intromissions. Lesions to the MPOA are also associated with an increase in the rate at which females leave a male after a given sexual interaction, as well as the amount of time they stay away from the male after this interaction. These behavioral effects are observed without apparent effect on the LQs of these females (Yang et al., 2000). The MPOA may also be involved in encoding vaginocervical stimulation, and lesions of the MPOA cause pseudopregnancy in the rat. Together these studies suggest that lordosis is under the control of both the VMH and MPOA, which act in opposition to one another, whereas proceptive behaviors are associated with both (Gunnet et al., 1983; Haskins et al., 1983).

The neural substrates of pacing behavior have received less attention; however, the display of this behavior depends on the ability of the female rat to interpret the vaginocervical stimulation received during the copulatory bout. Vaginocervical stimulation is relayed through the pelvic and pudenal nerves, and transsection of these nerves disrupts the female's ability to pace copulation (Erskine, 1992) without affecting the display of sexual receptivity or proceptivity (Rowe et al., 1993). Vaginocervical stimulation is associated with increased metabolic activity and Fos-immunoreactivity in the MPOA, mesencephalic reticular formation, bed nucleus of the stria terminalis dorsal raphe, and medial amygdala (Erskine, 1993; Pfaus et al., 1993; Rowe et al., 1993; Wersinger et al., 1993; Polston et al., 1995). The medial amygdala appears to be critical for processing vaginocervical stimulation and relaying information to hypothalamic centers that mediate neuroendocrine function (Polston et al., 2001a, 2001b).

Pacing behavior relies on the female's ability to interpret coital stimuli, engage in an appropriate motoric response, and temporally sequences contacts with the male. Given this complex interplay among sensorimotor function and motivation, the striatum and nucleus accumbens (NAcc) have been investigated for their role in pacing behavior (Becker, 1999).

Extracellular dopamine concentrations increase in both striatum and NAcc in female rats that are pacing the rate of copulation compared with those having nonpaced sex (Mermelstein et al., 1995). In females that are induced into sexual receptivity with sequential implants of estradiol and progesterone into the VMH, estradiol in the striatum enhances percent exits and estradiol in the NAcc increases return latencies (Xiao et al., 1997). Lesions of the striatum affect the percent exit measure after ejaculations (Jenkins et al., 2001). Lesions of the NAcc increase the female rat's rate of rejection of a male attempting to mount, without affecting LQs exhibited by these females and in some instances result in a complete avoidance of the male (Rivas et al., 1990, 1991; Jenkins et al., 2001). Previous sexual experience results in an enhanced increase in dopamine concentrations in the NAcc of female hamsters (Kohlert et al., 1999). These studies provide evidence that the striatum and NAcc serve important roles as neuroanatomical substrates of paced copulatory behavior in the female rat.

SEXUAL MOTIVATION IN THE FEMALE RAT

Female sexual behavior has traditionally been thought of in terms of receptive versus proceptive behaviors. We have stressed that the female is not a passive participant in copulatory behavior. It is only recently, however, that researchers have begun to look at sexual motivation in females (despite the fact that the word *proceptive* implies "female initiative" to engage in sexual behavior [Beach, 1976]). This discrepancy is probably due in large part to the fact that female rats who have received hormone priming require stimulation from a male to exhibit lordosis. As such, they have often been viewed as victims of their hormonal status and environmental circumstance, rather than as initiators of the sexual interaction.

Nevertheless, female rats in estrus will cross an electric grid to gain access to an intact male (Meyerson et al., 1973). They also show a preferences for odors of sexually active males to those of estrous females (Bakker et al., 1996), and they prefer to spend time with intact, sexually active males (Pfeifle et al., 1983). Female rats even perform operant responses to gain access to sexually active males, develop conditioned place preferences to chambers where they have had sex versus chambers where they were alone, and develop conditioned place preferences for paced sex versus nonpaced sex (French et al., 1972; Oldenburger et al., 1992; Matthews et al., 1997; Paredes et al., 1999; Jenkins et al., 2003). Under traditional testing conditions, however, the female's full display of proceptive behavior is not observed. In fact, the view of the female as a passive participant in the copulatory bout may be more a function of testing context than reflective of female sexual behavior. Obviously, experiments that allow the female to pace the rate of copulation have provided new insight into female sexual behavior and motivation.

SEXUAL BEHAVIOR IN THE MALE RAT

Although the female shows sexual receptivity only during behavioral estrus, the male will mate at any time that he comes into contact with a receptive female. Typically, a male will perform 8 to 10 intromissions before achieving an ejaculation, which involves the deposition of a sperm plug into the female's vagina. During the mating sequence, the male emits ultrasonic vocalizations. There are two calls: (1) with the initial contacts, the male emits the

mating call, and (2) just before ejaculation, he emits the *preejaculatory call*. These calls promote immobility of the female during lordosis and in the wild may serve to attract females to the male (McClintock et al., 1982; White et al., 1990). After ejaculation, there is a *postejaculatory call* that is thought to serve to keep the female away from the male for a while (Anisko et al., 1978). This is important because after an ejaculation, the male enters a quiescent period of inactivity known as the *postejaculatory refractory period*. This can last from 2 to 5 minutes, at which point the male usually reinitiates sexual behavior with the female. The pattern repeats itself until the male becomes sexually sated (Bermant, 1967; Adler, 1969). As a side note, sexually sated males often reinitiate copulatory behavior if presented with a novel receptive female, a phenomenon humorously referred to as the "Coolidge effect" (Bermant et al., 1968).

HOW TO MEASURE MALE SEXUAL BEHAVIOR

Having to differentiate among mounts, intromissions, and ejaculations can seem like a guessing game. After closely observing the sexual interactions between the male and female rat, however, researchers can quickly and reliably distinguish each. Again, it is a good idea to videorecord sessions to score sexual behavior. The key to determining what the male has just done is to observe his behavior during the dismount. After a mount, male rats typically dismount the female gently, simply moving off of the female. They often groom themselves and remount the female quickly. Intromissions, on the other hand, are a little flashier. After intromitting, the male generally springs off of the female. Just before the dismount, there is a sharp pelvic thrust, and the male often flails his forepaws out to the side during the thrust and just before the dismount. Grooming of the genital region often follows intromissions, and sometimes the erect penis can still be seen outside the body cavity immediately after the intromission. To the novice, differentiating between mounts and intromissions is often the most difficult part of scoring sexual behavior in the male rat.

The ejaculation is often the easiest sexual interaction to determine early on. The ejaculation involves several pronounced pelvic thrusts. Instead of dismounting right away, males typically continue to grasp the female after achieving an ejaculation and the female often has to kick him away. There is marked decrease in muscle tension in the male rat after an ejaculation, which can be described as a "melting away" of tension to people new to observing sexual behavior in the rat. Ejaculations are generally followed by prolonged grooming bouts; the male rat often avoids the female and sometimes takes a little nap. Once the male recovers from this "refractory period," he often reinitiates sexual behavior; however, the number of intromissions that precede an ejaculation can be decreased dramatically as the duration of the copulatory bout increases.

In addition to the type of interaction (i.e., mount, intromission, ejaculation) in which the male engages, researchers often collect information on the latency to first mount, intromission, and ejaculation. Information is collected on the number of intromissions preceding an ejaculation, and noncontact erections are often used as an index of the male's motivation to engage in sexual behavior.

NEURAL SUBSTRATES OF SEXUAL BEHAVIOR IN THE MALE: MOTIVATION VERSUS ABILITY

Many of the same structures implicated in female sexual behavior are also important for the sexual behavior in males. For example, the MPOA is critical for male sexual behavior. Lesions to the MPOA disrupt the male's ability to perform copulatory behaviors (Heimer et al., 1966, 1982), although there is some evidence that such disruption is most likely in those rats that are sexually naïve (DeJonge et al., 1989).

Furthermore, the neuroanatomical substrates of the ability to engage in sexual behavior are dissociable from those involved in the motivation to engage in sexual behavior. This dissociation was demonstrated by an ingenious experiment conducted by Everitt and colleagues (1990). In this experiment, male rats were trained to bar press for a sexually receptive female. Once the behavior was firmly established, males received lesions to either the MPOA or basolateral amygdala, and the males were placed into the operant chamber. Males receiving lesions to the MPOA readily bar pressed for receptive female rats. Once the female was presented, however, the male failed to copulate with her. On the other hand, males receiving lesions to the amygdala did not bar press as much for the female, but if the female was presented to them, they would copulate. Clearly, the male's motivation to engage in sexual behavior with the female is dissociable from his ability to engage in sexual behavior with the female (Everitt, 1990).

As in the female, it appears that the MPOA serves more than a consummatory role in sexual behavior. Work from Elaine Hull's laboratory suggests that the dopamine systems of the MPOA are also associated with anticipatory aspects of sexual behavior (Hull et al., 1999). A dopamine agonist infused into the MPOA facilitates male sexual behavior (Hull et al., 1986), and a dopamine antagonist infused into MPOA reduces the number of ejaculations achieved by the male (Pehek et al., 1988). Treatment with dopamine antagonists also reduces the number of intromissions that precede ejaculation and delays the initiation of copulation (Pehek et al., 1988; Pfaus et al., 1989). MPOA lesions in the male rat can also alter partner preferences, certainly suggesting a role for MPOA in the motivation to engage in sexual behavior (Paredes et al., 1998).

SEXUAL MOTIVATION IN THE MALE RAT

The striatum and NAcc are also involved in male sexual behavior. As in the female rat, the

NAcc is implicated in the motivation to engage in sexual behavior. In experiments from the Everitt laboratory, animals with basolateral amygdala lesions that had ceased bar pressing for access to females reinitiated bar pressing when amphetamine was infused into the NAcc, indicating that increased dopamine in the NAcc is important for sexual motivation in the male as well as the female (Everitt, 1990). Extracellular concentrations of dopamine increase in the NAcc when male rats are presented with a receptive female rats behind a wire screen and then again during copulation itself. Increases in extracellular concentrations of dopamine in the striatum, however, are observed only during copulatory behavior (Pfaus et al., 1990), and the reinitiation of sexual behavior when a new female is presented (i.e., the Coolidge effect) is associated with increases in NAcc dopamine (Fiorino et al., 1997). Certainly, other neural structures are involved in the sexual behavior of the male rat, but the MPOA, amygdala, and NAcc have received an enormous amount of attention from researchers interested in understanding the neural substrates of male sexual behavior, and motivation (for a discussion of other neural structures and neurotransmitters involved in male sexual behavior, see Bitran et al., 1987; Coolen et al., 1998; Hull et al., 1999; Pfaus, 1999).

The male rat's motivation to engage in sexual behavior has never been doubted. Various measures have been used to demonstrate the male rat's motivation to engage in sexual behavior. Males will perform operant behaviors to gain access to sexually receptive females (Everitt, 1990), exhibit noncontact erections as a function of estrous female rat cues (e.g., Liu et al., 1998), and develop conditioned place preferences for areas in which they have engaged in sexual behavior (e.g., Mehara et al., 1990). Researchers also use level searching (Mendelson et al., 1989) and latency to initiate sexual contact as an index of the male rat's motivation to engage in sexual behavior (Fiorino et al., 1999), although correlational and factor analyses indicate that an-

ticipation of and initiation of sexual behavior may represent different conceptual mechanisms (Pfaus et al., 1990).

CONCLUDING REMARKS

Because the female assumes a reflexive posture during copulation, researchers have assumed that the male rat not only initiates sexual contact but also controls the rate at which the copulation occurs (see Bermant, 1967). We now know that that the male is not the only active participant in the copulatory bout provided the testing arena allows for the female to pace, but there remain some long-held conceptions of male sexual behavior and motivation that have been challenged. For instance, although it is clear that female rats find only sexual behavior that occurs at their preferred interval rewarding (Jenkins et al., 2003), it has traditionally been thought that males find sexual behavior rewarding under any circumstance provided they achieve an ejaculation. Recently, however, Martinez and Paredes (2001) demonstrated that males, like females, develop conditioned place preferences only for sex that occurs at their preferred rate (Martinez et al., 2001). Therefore, the idea that males are motivated to engage in sex under any circumstances may have to be modified. Further research and investigation should clarify this issue.

We have seen that the male and female rat share many commonalities in terms of the anatomical substrates of sexual behavior. Another common feature in the sexual behavior of males and females is the role that the striatum and NAcc play in mediating sexual behavior and motivation.

NOTES

1. This concept has been the dominant theory in neuroendocrinology for more than 40 years. Recent evidence, however, indicates that there may also be con-

tributions from the Y and/or X chromosomes that determine if there are sex differences in the brain (independent of the ~*Sry* gene that determines if an animal has testes; Carruth LL, Reisert I, et al. (2002) Sex chromosome genes directly affect brain sexual differentiation. Nature Neuroscience 5:933–934; and Xu J, Burgoyne P, et al. (2002) Sex differences in sex chromosome gene expression in mouse brain. Human Molecular Genetics 11:1409–1419.

2. There is also evidence that complete feminization requires exposure to low levels of estradiol during development, so the story becomes more complicated all the time (Fitch RH and Denenberg VH [1998] A role for ovarian hormones in sexual differentiation of the brain. Behavioral and Brain Sciences 21:311–352).

REFERENCES

Adler NT (1969) Effects of the male's copulatory behavior on successful pregnancy of the female rat. Journal of Comparative Physiology and Psychology 69:613–622.

Anisko JJ, Suer SF, et al (1978) Relation between 22-kHz ultrasonic signals and sociosexual behavior in rats. Journal of Comparative Physiology and Psychology 92:821–829.

Bakker J, Van Ophemert J, et al (1996) Sexual differentiation of odor and partner preference in the rat. Physiology and Behavior 60:489–494.

Barnett SA (1975) Reproductive behavior. In: The rat: A study in behavior, p. 138. Chicago: The University of Chicago Press.

Beach FA (1976) Sexual attractivity, proceptivity, and receptivity in female mammals. Hormones and Behavior 7:105–138.

Becker JB (1999) Gender differences in and influences of reproductive hormones on dopaminergic function in striatum and nucleus accumbens. Pharmacology, Biochemistry, and Behavior 64:803–812.

Bermant G (1967) Copulation in rats. Psychology Today. 1:52–60.

Bermant G, Lott D, et al (1968) Temporal characteristics of the Coolidge effect in male rat copulatory behavior. Journal of Comparative and Physiological Psychology 65:447–452.

Bitran D and Hull EM (1987) Pharmacological analysis of male rat sexual behavior. Neuroscience and Biobehavior Reviews 11:365–389.

Breedlove SM and Hampson E (2002) Sexual differentiation of the brain and behavior. In: Behavioral endocrinology (Becker JB, Breedlove SM, Crews D, McCarthy MM, eds.), pp. 75–115. Cambridge, Mass.: MIT Press.

Carruth LL, Reisert I, et al (2002) Sex chromosome genes directly affect brain sexual differentiation. Nature Neuroscience 5:933–934.

Coolen LM, Peters HJ, et al (1998) Anatomical interrelationships of the medial preoptic area and other brain regions activated following male sexual behavior: A combined for and tract-tracing study. Journal of Comparative Neurology 397:421–435.

Coolen LM, Peters HJPW, et al (1996) Fos immunoreactivity in the rat brain following consummatory elements of sexual behavior: A sex comparison. Brain Research 738:67–82.

DeJonge FH, Louwerse AL, et al (1989) Lesions of the SDN-POA inhibit sexual behavior of male Wistar rats. Brain Research Bulletin 23:483–492.

Erskine MS (1989) Solicitation behavior in the estrous female rat: A review. Hormones and Behavior 23:473–502.

Erskine MS (1992) Pelvic and pudendal nerves influence the display of paced mating behavior in response to estrogen and progesterone in the female rat. Behavioral Neuroscience 106:690–697.

Erskine MS (1993) Mating-induced increases in FOS protein in preoptic area and medial amygdala of cycling female rats. Brain Research Bulletin 32:447–451.

Erskine MS and Baum MJ (1982) Effects of paced coital stimulation on termination of estrus and brain indoleamine levels in female rats. Pharmacology, Biochemistry, and Behavior 17:857–861.

Everitt BJ (1990) Sexual motivation: A neural and behavioural analysis of the mechanisms underlying appetitive and copulatory responses of male rats. Neuroscience and Biobehavioral Reviews 14:217–232.

Fiorino DF, Coury A, et al (1997) Dynamic changes in the nucleus accumbens dopamine efflux during the coolidge effect in male rats. Journal of Neuroscience 17:4849–4855.

Fiorino DF and Phillips AG (1999) Facilitation of sexual behavior and enhanced dopamine efflux in the nucleus accumbens of male rats after D-amphetamine-induced behavioral sensitization. Journal of Neuroscience 19:456–463.

Fitch RH and Denenberg VH (1998) A role for ovarian hormones in sexual differentiation of the brain. Behavioral and Brain Sciences 21:311.

Freeman ME (1994) The neuroendocrine control of the ovarian cycle of the rat. In: The physiology of reproduction, 2nd ed (Knobil E and Neill JD, eds.). New York: Raven Press, Ltd.

French D, Fitzpatrick D, et al (1972) Operant investigations of mating preference in female rats. Journal of Comparative Physiology and Psychology 81:226–232.

Gilman DP, Mercer LF, et al (1979) Influence of female copulatory behavior on the induction of pseudopregnancy in the female rat. Physiology and Behavior 22:675–678.

Gunnet JW and Freeman ME (1983) The mating-induced release of prolactin: A unique neuroendocrine response. Endocrinology Reviews 4:44–61.

Hansen S, Kohler C, et al (1982) Effects of ibotenic acid-induced neuronal degeneration in the medial preoptic area and the lateral hypothalamic area on sexual behavior in the male rat. Brain Research 239:213–232.

Haskins JT and Moss RL (1983) Action of estrogen and mechanical vaginocervical stimulation on the membrane excitability of hypothalamic and midbrain neurons. Brain Research Bulletin 10:489–496.

Heimer L and Larsson K (1966) Impairment of mating behavior in male rats foloowing lesions in the pre-optic-anterior hypothalamic continuum. Brain Research 3:248–263.

Hoshina Y, Takeo U, et al (1994) Axon-spring lesion of the preoptic area enhances receptivity and diminishes proceptivity among components of female rat sexual behavior. Behavioural Brain Research 61: 197–204.

Hull EM, Bitran D, et al (1986) Dopaminergic control of male sex behavior in rats: Effects of an intra-cerebrally-infused agonist. Brain Research 370:73–81.

Hull EM, Lorrain DS, et al (1999) Hormone-neurotransmitter interactions in the control of sexual behavior. Behavioural Brain Research 105:105–116.

Jenkins WJ and Becker JB (2001) Role of the striatum and nucleus accumbens in paced copulatory behavior in the female rat. Behavioural Brain Research 121:119–128.

Jenkins WJ and Becker JB (2003) Females devlop conditioned place preference for sex at their preferred interval. Hormones and Behavior 43:503–507.

Kohlert JG and Meisel RL (1999) Sexual experience sensitizes mating-related nucleus accumbens dopamine responses of female Syrian hamsters. Behavioural Brain Research 99:45–52.

Liu Y, Sachs BD, et al (1998) Sexual behavior in male rats after radiofrequency or dopamine-depleting lesions in the nucleus accumbens. Pharmacology, Biochemistry, and Behavior 60:585–592.

Martinez I and Paredes RG (2001) Only self-paced mating is rewarding in rats of both sexes. Hormones and Behavior 40:510–517.

Matthews TJ, Grigore M, et al (1997) Sexual reinforcement in the female rat. Journal of Experimental Analysis of Behavior 68:399–410.

McCarthy MM and Becker JB (2002) Neuroendocrinology of sexual behavior in the female. In: Behavioral endocrinology (Becker JB, Breedlove SM, Crews D, McCarthy MM, eds.), pp. 117–151. Cambridge, Mass.: MIT Press/Bradford Books.

McClintock MK, Anisko JJ, et al (1982) Group mating among Norway rats. II. The social dynamics of copulation: Competition, cooperation, and mate choice. Animal Behavior 30:410–425.

Mehara BJ and Baum MJ (1990) Nalozone disrupts the expression but not the acquisition by male rats of a conditioned place preference response for an oestrous female. Psychopharmacology 101:118–125.

Mendelson SD and Pfaus JG (1989) Level searching: A new assay of sexual motivation in the male rat. Physiology and Behavior 45:337–341.

Mermelstein PG and Becker JB (1995) Increased extra-cellular dopamine in the nucleus accumbens and striatum of the female rat during paced copulatory behavior. Behavioral Neuroscience 109:354–365.

Meyerson BJ and Lindstrom L (1973) Sexual motivation of in the female rat. Acta Physiologica Scandinavica (Supplement) 389:1–80.

Oldenburger WP, Everitt BJ, et al (1992) Conditioned place preference induced by sexual interaction in female rats. Hormones and Behavior 26:214–228.

Paredes RG, Tzschentke T, et al (1998) Lesions of the medial preoptic area anterior hypothalamus (MPOA/AH) modify partner preference in male rats. Brain Research 813:1–8.

Paredes RG and Vazquez B (1999) What do female rats like about sex? Paced mating. Behavioural Brain Research 105:117–127.

Pehek EA, Warner RK, et al (1988) Microinjection of cis-flupenthixol, a dopamine antagonist, into the medial preoptic area impairs sexual behavior of male rats. Brain Research 443:70–76.

Pfaff DW, Schwartz-Giblin S, et al (1994) Cellular and molecular mechanisms of female reproductive behaviors. In: The physiology of reproduction, 2nd ed (E Knobil and JD Neill, eds.), pp. 107–220. New York: Raven Press.

Pfaus JG (1999) Neurobiology of sexual behavior. Current Opinions in Neurobiology 9:751–758.

Pfaus JG, Damsma G, et al (1990) Sexual behavior enhances central dopamine transmission in the male rat. Brain Research 530:345–348.

Pfaus JG, Kleopoulos SP, et al (1993) Sexual stimulation activates c-fos within estrogen concentrating regions of the female rat forebrain. Brain Research 624:253–267.

Pfaus JG, Mendelson SD, et al (1990) A correlational and factor analysis of anticipatory and consummatory measures of sexual behavior in the male rat. Psychoneuroendocrinology 15:329–340.

Pfaus JG and Phillips AG (1989) Differential effects of dopamine receptor antagonists on the sexual behavior of male rats. Psychopharmacology 98:363–368.

Pfeifle JK and Edwards DA (1983) Midbrain lesions eliminate sexual receptivity but spare sexual motivation in female rats. Physiology and Behavior 31:385–389.

Polston EK and Erskine MS (1995) Patterns of induction of the immediate-early genes c-fos and egr-1 in the female rat brain following differential amounts of mating stimulation. Neuroendocrinology 62:370–384.

Polston EK and Erskine MS (2001a) Excitotoxic lesions of the medial amygdala differentially disrupt prolactin secretory responses in cycling and mated female rats. Journal of Neuroendocrinology 13:13–21.

Polston EK, Heitz M, et al (2001b) NMDA-mediated activation of the medial amygdala initiates a downstream neuroendocrine memory responsible for pseudopregnancy in the female rat. Journal of Neuroscience 21:4104–4110.

Rivas FJ and Mir D (1990) Effects of nucleus accumbens lesion on female rat sexual receptivity and proceptivity in a partner preference paradigm. Behavioural Brain Research 41:239–249.

Rivas FJ and Mir D (1991) Accumbens lesion in female rats increases mount rejection without modifying lordosis. Revista Espanola de Fisiologia 47:1–6.

Rowe DW and Erskine MS (1993) c-Fos proto-oncogene activity induced by mating in the preoptic area, hypothalamus and amygdala in the female rat: Role of afferent input via the pelvic nerve. Brain Research 621:25–34.

Schank JC and McClintock MK (1992) A coupled-oscillator model of ovarian-cycle synchrony among female rats. Journal of Theoretical Biology 157:317–362.

Smith MS, Freeman ME, et al (1975) The control of progesterone secretion during the estrous cycle and early pseudopregnancy in the rat: Prolactin, gonatropin and steroid levels associated with rescue of the corpus luteum of pseudopregnancy. Endocrinology 96:219–226.

Wersinger SR, Baum MJ, et al (1993) Mating-induced FOS-like immunoreactivity in the rat forebrain: A sex comparison and a dimorphic effect of pelvic nerve transection. Journal of Neuroendocrinology 5:557–568.

White NR, Cagiano R, et al (1990) Changes in mating vocalizations over the ejaculatory series in rats (Rattus norvegicus). Journal of Comparative Psychology 104:255–262.

White NR, Gonzales RN, et al (1993) Do vocalizations of the male rat elicit calling from the female? Behavior and Neural Biology 59:76–78.

Whitney JF (1986) Effect of medial preoptic lesions on sexual behavior of female rats is determined by test situation. Behavioral Neuroscience 100:230–235.

Xiao L and Becker JB (1997) Hormonal activation of the striatum and the nucleus accumbens modulates paced mating behavior in the female rat. Hormones and Behavior 32:114–124.

Xu J, Burgoyne P, et al (2002) Sex differences in sex chromosome gene expression in mouse brain. Human Molecular Genetics 11:1409–1419.

Yang LY and Clemens LG (2000) MPOA lesions affect female pacing of copulation in rats. Behavioral Neuroscience 114:1191–1202.

Environment

30

ROBBIN L. GIBB

Desirable qualities in rats have been selected for through captive breeding. These qualities include tameness, curiosity, and reduced fear and aggression in response to handling (Barnett, 1975). Genetic selection may have made laboratory rats well suited to the laboratory, but the question arises of what exactly are *laboratory conditions*. For example, within a colony, standard housing may range from single housing in Plexiglas shoeboxes to group housing. Across laboratories, variation may be even greater. I discuss how environmental parameters influence the physiology and behavior of rats and thus the experimental outcome. This review considers lighting, humidity, airflow, noise, cage construction, diet, and social opportunity. Maternal influences, age at weaning, animal care, exercise regimens, and enrichment are also discussed. These factors can induce changes in brain weight, cerebral vascularization, adrenal size, and body weight.

HOUSING CONSIDERATIONS

The environment of the rat can be considered at the macroenvironment and microenvironment levels. *Macroenvironment* refers to the ambient conditions in the animal colony; these include lighting, temperature, humidity, airflow, and noise. *Microenvironment* refers to the conditions within the cage. The material used to construct the cage, the design and size, the type of lid and bedding, access to food, and social opportunity are variables that affect microenvironment. Cage design also influences the ventilation, lighting, temperature, and noise level that its occupants perceive. For example, hanging wire cages are well ventilated and reduce animal contact with their excreta but do not allow the animal to modify its microenvironment. Macroenvironmental conditions have a greater impact on animals housed in this type of cage. Shoebox cages with bedding are less well ventilated but allow the animal to make alterations in its environment. In addition, bedding affords the animal opportunity to dig, a natural behavior that laboratory rats share with their wild predecessors (Canadian Council on Animal Care [CCAC], 1984). A study by Krohn et al. (2003) used telemetry to monitor heart rate, body temperature, and blood pressure of rats kept on three types of flooring: grid floor, plastic floor, or bedding. Grid flooring caused elevations in blood pressure and heart rate and thus was rated most undesirable. Bedding was most acceptable. Anzaldo and colleagues (1994) allowed rats to inhabit cages that were equipped with L-shaped partitions (high-perimeter housing), cages that allowed increased floor space with a three-dimensional design, or standard cages. The three-dimensional cage was least preferred; the high-perimeter cage was most preferred. The preference of the rats for the "high-perimeter" caging may have reflected their thigmotactic (edge-using) tendencies. As the population density within the cages increased, the preference for the high-perimeter caging over standard caging was reduced. This study showed that rats preferred social interaction

321

and security over increased floor space and that population density within a cage can alter the choice of spatial design.

LIGHTING

Three characteristics of light that affect housing conditions are intensity, quality (wavelength), and photoperiod (CCAC, 1993). Monitoring the intensity of light in the animal colony ensures adequate illumination for providing animal care without causing blindness in the rats. Lighting that is considered normal for humans can cause retinal damage in rats, especially albinos (Bellhorn, 1980). A study by Wasowicz et al. (2002) showed that retinas of pigmented animals such as Long-Evans rats are affected by prolonged exposure to a moderate light source. Light intensity varies between cages near the light source and those closer to the floor. In some animal rooms, as much as an 80-fold difference in illumination can occur in the vertical dimension (Schofield and Brown, 2003). Animals housed in the uppermost levels of racks are more susceptible to blindness than are those housed lower down. Eye problems could interfere with experiments that depend on the animal detecting visual cues.

Few studies have been conducted on the effects of light quality on rats. Spalding et al. (1969) showed the wavelength of ambient light influenced wheel running in mice, and the degree of the effect was dependent on the strain of mouse. A study of the effects of different types of fluorescent lighting (full spectrum, cool white, black, etc.) on organ and body weights in mice showed that the type of lighting affected both organ and body weight in males but not in females (Saltarelli and Coppola, 1979). The CCAC (1993) recommends that light used in animal housing be as close to natural sunlight as possible.

Light synchronizes circadian rhythm with environmental time through photo-transduction by retinal ganglion cells (Berson et al., 2002). Circadian rhythms control the sleep–wake cycle of an animal and can influence its performance in an experiment, especially in older animals (Winocur and Hasher, 1999; Poulos and Borlongan, 2000). Most animal housing revolves on a 12 hour light–dark cycle. If lights are switched on during the dark cycle or left on for a 24 hour period, retinal ganglion cells respond by altering the circadian cycle. Breeding cycles of rats can be affected by circadian timing. Hoffman (1973) reported that a 12 hour light–dark cycle produces a 4 day estrous period in Sprague-Dawley rats, whereas a 16–8 hour light–dark cycle increases the estrous period to 5 or more days. Circadian rhythms influence the physiology of rats. Changes in body temperature, corticosterone levels, neurotransmitter receptor binding, drug sensitivity, size of experimentally induced cortical infarct, and motor activity have been associated with the circadian cycle (Ixart et al., 1977; Vinall et al., 2000; Benstaali et al., 2001; Rebuelto et al., 2002). As such, schedules for behavioral testing and surgical procedures should be consistent to reduce variation in experimental outcome.

TEMPERATURE

Although rats have fur coats, they are sensitive to fluctuations in ambient temperature. The normal temperature for a rat room is in the range of 20° to 24° C. This range allows optimal growth of rats and seems most compatible with their behavioral preferences (Allmann-Iselin, 2000). Temperatures outside this range induce activity and metabolic changes that can affect experimental design. Dose–response curves for drugs can be shifted by changes in ambient temperature. A 4° C variation in temperature can cause a 10-fold variation in drug toxicity (Clough, 1987). Shifts in ambient temperature also affect the amount of food and water that an animal consumes. Changes in ingestive behaviors can alter the effective dose of an administered drug.

The number of inhabitants in the cage influences the temperature of the microenvironment. Body temperature can be influenced by changes in animal care personnel, stormy weather, and handling (Clough, 1987).

HUMIDITY

Relative humidity in an animal facility is recommended to be approximately 50%, although a range from 40% to 70% can be tolerated (CCAC, 1993). Airborne microorganisms are less viable at a relative humidity of 50%. Low humidity can cause health problems such as dry skin and ringtail, whereas high humidity can increase ammonia production from the cages (Clough, 1987), thereby increasing the incidence of respiratory distress.

VENTILATION

Ventilation within the animal room influences temperature, humidity, and air quality. Cage design and placement also affect the airflow at the microenvironment level. Draft-free ventilation that allows 15 to 20 air exchanges per hour is recommended. Rats housed in shoebox cages with filter tops require monitoring to ensure that ammonia from soiled bedding does not reach toxic levels. High levels of ammonia in the environment are associated with respiratory distress or disease (Broderson et al., 1976). The human threshold level for detection of ammonia (8 parts per million [ppm]) is above the concentration capable of inducing pathology (Schofield and Brown, 2003; CCAC, 1993). Bedding should also be free from aromatic carcinogens and pesticides. Both contaminants are associated with sawdust bedding (Clough, 1987).

NOISE

Whether a sound has damaging effects depends on its loudness, frequency, and duration. Sounds of 160 decibels will cause damage to hearing in rats and in humans. It is recommended that animal room noise does not exceed 85 decibels, although auditory damage in rats has been found after intermittent exposure to sounds at 83 decibels (CCAC, 1993). Rats hear sounds that range in frequency from 1000 Hz to 100,000 Hz, depending on the strain (Gamble, 1982). Thus, they are insensitive to lower-frequency tones that fall within the human auditory range but their upper threshold is well beyond the human range. This makes monitoring noise in animal facilities more difficult. Sounds that we are incapable of detecting in addition to those that we hear may cause changes in plasma corticosterone levels, immune system function, reproductive fitness, and body weight in rats (Clough, 1987). Certain sudden, loud sounds can induce a startle response or audiogenic seizure in rats and mice. Nursing dams have been known to cannibalize their young after exposure to sudden, loud noise. Sounds can also induce aggression or changes in tolerance to electric shock (Gamble, 1982).

Rodents produce ultrasonic vocalizations to communicate during mating, aggressive behaviors, and maternal care (Harding and McGinnis, 2003; CCAC, 1984; Von Fritag et al., 2002; Smotherman et al., 1974). Excessive noise can reduce the effectiveness of this means of communication.

Noise in the environment can influence the development of audition in young rodents. Chang and Merzenich (2003) demonstrated that rats reared in the presence of continuous moderate noise showed delayed auditory cortical maturation. This effect was reversed by returning the animals to a normal acoustic environment. Thus, animal holding facilities should be well away from sources of mechanical noise, because constant exposure to sounds can alter the timing of normal auditory development in young rats.

DIETARY CONSIDERATIONS

The nutritional requirements of an animal can be influenced by many factors. Genetic strain, sex, age, physiological status, and environment contribute to the nutrient requirements of the rat. Rats of different genetic strains grow at variable rates and thus have specific nutritive requirements. Because male rats grow faster and have a higher proportion of body protein than do females, they require a higher proportion of protein in their diet. Similarly, growing, lactating, and postoperative animals require a higher percentage of dietary protein than do adult animals that are simply maintaining their body weight. Rats living in cooler conditions increase food intake to maintain a constant body temperature, whereas rats living in warmer conditions reduce food intake and may require higher nutritional density in their food.

DIETARY RESTRICTION OR OPTIMIZATION

The negative consequences of ad libitum feeding has now been established in every outbred, inbred, and hybrid cross strain of rat examined (Keenan et al., 2000). Rat diet formulations are based on the nutritional requirements of weanling rats and lactating dams and contain between 18% and 23% protein. Animals in the growing or nursing phases of life require approximately 15% protein in their diet, whereas adult animals in a maintenance phase require 5% to 12% protein (Keenan et al., 2000). Animals recovering from surgical procedures also require more protein in their diet, to ensure rapid healing. Because it is simpler to give all rats in a colony the same food, most research facilities overnourish their adult inhabitants. Unrestricted access to food is "unnatural" and compromises the health of the animal. Among laboratory animals, only rodents are commonly given ad libitum access to food. Other species have their food intake restricted in accordance with good scientific and veterinary practices (Keenan et al., 2002).

Dietary restriction has a positive impact on the health of rats. Feeding rats ad libitum highly nutritious rat chow causes obesity, diabetes, and tumors; shortens the life span; and tends to reduce cognitive performance particularly as the animal ages (Means et al., 1993). Formation of free radicals and/or glycation reactions of sugars with proteins may be responsible for the aging effects associated with ad libitum feeding. Dietary restriction is associated with increased production of proteins known to enhance neuroplasticity and confer resistance to metabolic insult such as brain-derived neurotrophic factor (Mattson et al., 2002; Mattson et al., 2003). Anson and colleagues (2003) have shown that the pattern of feeding dietary restricted animals affects the degree of benefit derived from the procedure. Mice fed every other day ate the same amount as unrestricted animals and maintained their body weight but showed an increased resistance of neurons in the brain to the effects of excitotoxic stress. The number of dendritic spines found on neurons in the rat neocortex decline with aging, but 24-month-old rats that were restricted to every other day feeding had the same number of dendritic spines as did 6-month-old rats fed ad libitum (Moroi-Fetters et al., 1989).

A study by Markowska and Savonenko (2002) showed that the effectiveness of dietary restriction varies with the genetic strain of the rat. For example, Fisher 344 rats failed to show significant benefit from dietary restriction but offspring of a Fisher 344 and Brown Norway cross showed improvement on tests of both cognitive and sensorimotor behaviors.

CONTROL OF ENVIRONMENT

Joffe et al. (1973) raised rats in an environment in which the animals were able to control food, light, and water availability by bar pressing. These rats, compared with animals raised in standard housing under identical food, light, and water availability conditions, were more exploratory, less emotional, and more confident in open-field testing. This study supports the notion that animals prefer to exert control on their environment and that having such control reduces their affective response to stress.

SOCIAL OPPORTUNITY

Rats are social creatures, and they benefit from opportunities for social interaction. Social isolation is known to cause changes in behavior (i.e., alcohol consumption) and temperament, as well as physiology, including changes in size of the adrenal and thyroid glands (Baker et al., 1979) in adult rats. A study by Hurst and colleagues (1997) examined the effects of housing male rats as singles or in groups of three in two joined but divided cages. The type of barrier that was used to divide the cages was used to vary the degree of social contact between the two cages. Rats housed in isolation engaged more frequently in behaviors related to escape or seeking social information. Singly housed male rats were more aggressive if not exposed to neighbors or other cagemates, yet they showed reduced corticosterone concentration and organ pathology compared with group-housed rats. Although single housing may reduce social stress, animals thus housed are motivated to seek social interaction. Sharp and colleagues (2003) examined the effect of group housing on stress responses to witnessing common experimental procedures and husbandry. They found that group housing reduced the stress response to witnessing tail injections, restraint, cage changes, and decapitation. Group housing thus is preferable to solitary housing; if housing rats singly is part of the experimental design, it should be justified to and approved by the local animal care committee (CCAC, 1993). Short-term social isolation of rats (4 to 7 days) has been shown to increase the frequency of social interaction when the opportunity arises (Niesink and van Ree, 1982), suggesting that rats find social interaction rewarding. However, overcrowding leads to stress and increased aggression.

MATERNAL INFLUENCES

Natural variations in maternal care can influence the cognitive development of offspring. Meaney and his group have shown that mothers that spend more time in an arched-back nursing posture and licking and grooming their pups have offspring that show enhanced spatial learning in adulthood (Liu et al., 2000). These animals have elevated levels of glutamate receptors and growth factors in the hippocampus. Levine (1967) demonstrated that removing the pups from the nest for brief time intervals during the early postnatal period resulted in reduced response to stress in adulthood. This procedure was called "handling." It is now known that "handling" reduces basal levels of corticosterone (Levine, 1967; Beane et al., 2002) and alters expression of glucocorticoid receptors in the hippocampus and frontal cortex (Diorio et al., 1993; Bodnoff et al., 1995; Liu et al., 2000). It has been shown that mothers who experienced "handling" in infancy had offspring that showed reduced plasma steroids in response to novel stimuli (Denenberg and Whimbey, 1963; Levine, 1967). This finding indicates that early experiences of a mother can affect stress responses in subsequent generations.

We have shown that complex housing of a pregnant dam throughout the duration of her pregnancy ameliorates the behavioral devastation normally associated with perinatal cortical lesion. Both normal and frontal lesion offspring showed enhanced spatial cognition after prenatal condominium experience (Gibb et al., 2001). Similar results were found with prenatal tactile stimulation. Pregnant dams were "petted" with a soft hairbrush three times a day for 15 minutes throughout the duration of their pregnancy. Offspring that were given postnatal day 3 lesions of frontal cortex showed marked improvement on behavioral tests. Both prenatal complex housing and tactile stimulation altered neuronal morphology in sham-operated and lesioned animals. Thus, experiences of the mother rat can have an impact on the behavior and physiology of her offspring. Social isolation of rats during the preweaning period of life (3 to 6 hours per day for 5 days) alters the behavior of both the pups and the mother rat by increasing their activity and the number of mother–pup interactions (Zimmerberg et al., 2003). Taken together, these studies show that although there is some natural variation in maternal care that can influence the behavior of offspring, it is prudent to attempt to control the early experiences of rat pups to reduce the confounding effects of variations in experience.

AGE AT WEANING

Preweanling rats are sensitive to a maternal pheromone present in the mother's feces and respond by eating maternal feces to promote development of myelin in the brain. Rats respond to this pheromone until 27 days of age, but rats deficient in myelin continue to respond to the pheromone beyond this age (Schumacher and Moltz, 1985). Weaning rat pups too early has been shown to have serious physiological and behavioral consequences. Increased susceptibility to gastric pathology and delayed maturation of responses to restraint stress were noted in pups weaned at 15 days rather than at 22 days of age (LaBarba and White, 1971; Ackerman et al., 1975; Milkovic et al., 1975).

HUSBANDRY VARIABLES

Many environmental factors can act as uncontrolled variables in an experiment: music in the colony, strong smells, different care conditions, animal transportation, and even frequency of cage cleaning.

Using music in the colony as a controlled source of sound is thought to be helpful in reducing the disruptiveness of uncontrolled noise. Sounds associated with normal husbandry procedures have a greater impact on the magnitude of the subsequent stress response in animals accustomed to silence than in those used to constant sounds. Music can also provide a form of enriching experience for rats. Rauscher, Robinson, and Jens (1998) showed that rats exposed in utero and 60 days postpartum to Mozart compositions were able to complete a maze more quickly and with fewer errors than were rats exposed to white noise or silence.

Rats possess a highly specialized sense of smell. Just as much of our behavior is guided by sight, rats use their keen sense of smell to familiarize themselves with their environment. Pheromones are smells that help rats identify the presence of neighbors and can give signals that affect development, reproductive fitness, and some behaviors of nearby rats. Strong cleaning odors, ammonia buildup, and the use of perfumes by laboratory personnel can all interfere with the acquisition of odor information by rats.

Another important variable to consider is the handling of the animal provided by both experimenter and animal care personnel. Some animal health care technicians treat the animals they care for like pets and handle them a great deal, whereas others treat them as though they are wild and handle them with reticence. Likewise, some technicians talk to and handle animals in the cage they are cleaning, whereas others avoid contact. These disparate methods of handling can cause the animals to mount varying degrees of a stress response, which could affect their performance when subjected to testing procedures.

Transportation of animals via airplane or truck affects corticosterone levels and immune function (CCAC, 1993). It is recommended that a minimum period of adjustment of 2 days be allowed to ensure stabilization of physiological parameters. Timed-pregnant females that are subjected to transportation stress may have offspring that are very different behaviorally and neuroanatomically from offspring from mothers that do not experience this stress during pregnancy (Stewart and Kolb, 1988).

Frequency of cage cleaning affects the fitness and number of useable rats at weaning. Litters that had their cages changed twice per week had more healthy survivors than did litters exposed to once-a-week bedding changes (Cisar and Jayson, 1967). This may have resulted from increased exposure to ammonia in the once-a-week litters or from increased handling in the twice-a-week litters, but the frequency of cage cleaning may affect the experimental outcome.

EXERCISE

A rat's natural inclination is to explore its environment for food and mating opportunities. Access to a running wheel provides laboratory animals with a means for exploration beyond the limits of their caged environment. Although exercise is not normal behavior of animals in the wild, there is mounting evidence that exercise is beneficial to the health of rats. Exercise has been shown to increase the production of neurotrophic factors in the central nervous system (Gomez-Pinilla et al., 2001) and neurons in the hippocampus and motor cortex (van Praag et al., 1999; Galvez et al., 2002). There also is evidence that exercise reduces an animal's response to stress (Greenwood et al., 2002). In addition, exercise has been shown to be therapeutic but not prophylactic for rats that have sustained cortical injury (Gentile et al., 1987).

ENRICHMENT

In the 1940s, Donald Hebb raised a group of laboratory rats in his home. When he tested these animals as adults in a maze, he found that they made fewer errors than did animals raised under standard laboratory conditions. This was the first demonstration that an enriching environment can influence the behavioral performance of rats. Rosenzweig and his colleagues (1971) extended this finding by

showing that the brains of "enriched" animals were heavier and showed an increase in cortical thickness, acetylcholinesterase activity, synaptic contacts, and dendritic arborization. Rats housed in a complex environment also undergo brain changes that include increases in glial density and vasculature (Black et al., 1987). It is interesting to note that environmental enrichment has a greater effect on open-field behavior and body weight of wild rats than on their domesticated counterparts (Huck and Price, 1975). It appears that genetic changes accompanying domestication have made laboratory rats more resistant to the influences of experience.

Greenough and Black (1992) proposed that environment can influence brain morphology in one of two ways: experience-dependent and experience-expectant changes in the brain. Experience-expectant changes occur during development and require proper input for a system like the visual system to develop normally. This involves stabilizing useful synapses and deleting redundant ones. Experience-dependent changes are those that allow experiences to alter the animal throughout its life span. Such experiences include maze learning and motor learning.

We have determined that the impact of complex housing on neuroanatomical changes in the cerebral cortex varies with age and sex in rats (Kolb et al., 2003). Animals placed in enriched environments as adults showed an increase in spine density, whereas animals enriched at weaning showed a decrease in spine density. Male rats at all ages showed increases in dendritic length, yet only adult females showed similar increases.

Enriching experience is not only derived from complex housing. Sensory stimulation and behavioral testing can also be considered forms of enrichment. Brief periods of tactile or olfactory stimulation after brain injury in rats can improve their behavioral outcome (Gibb and Kolb, 2000; C. L. R. Gonzalez, unpublished observations) not only during development but also in adulthood. Participation in an experiment exposes an animal to a variety of experiences to which animals in the colony are not subjected. These experiences have the potential to alter the subsequent behavior and neuroanatomy of test subjects. Kolb and colleagues (2003) conducted an experiment to determine the relative effects of solving the place version of the Morris water task (learning condition), swimming in the pool without a platform (yoked condition: animals were allowed to swim for the same length of time that it took the animals in the learning condition to find the platform), and no behavioral testing on spine density and dendritic arbor in the occipital cortex. Rats that solved the problem had the greatest dendritic arbor and spine density, yet rats that swam for an equal length of time showed a significant elevation of these measures above the baseline found for nontested animals. This result shows that simple participation in the experiment was sufficient to change cortical circuitry.

CONCLUSION

Enrichment can take many forms: access to running wheels, handling, sensory stimulation, group housing, or complex housing (Fig. 30–1). There is a debate as to whether standard housing for rats should include some form of enrichment. Proponents of this view believe that standard housing produces "impoverished" animals that have underdeveloped brains and a limited behavioral repertoire. But despite the minimal stimulation provided by standard housing, very little pathological behavior has been ascribed to rats raised in these conditions. Although environmental stimulation produces a smarter rat, the value of "intelligence" to animals that do not need to compete for food, housing, or mating opportunities is difficult to assess. The use of laboratory rats that have been reared in standard housing as a baseline group to study the effects of the environment on brain

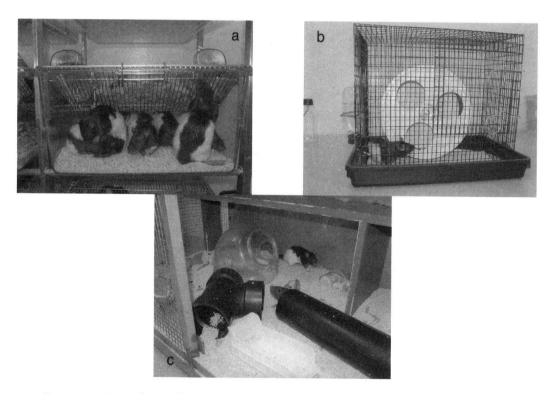

Figure 30–1. Various forms of housing for laboratory rats. *a*, Group housing. *b*, Running wheel access. *c*, Complex housing.

development and function have yielded many valuable insights over the past 60 years. Although it is important to consider optimal housing conditions for the experimental subjects, one should be mindful of the impact of adding enriching devices or protocols to standard laboratory rat care. Minimal changes in environmental conditions can have huge effects on the behavior and physiology of lab animals.

REFERENCES

Ackerman SH, Hofer MA, Weiner H (1975) Age at maternal separation and gastric erosion susceptibility in the rat. Psychosomatic Medicine 37:180–184.

Allmann-Iselin I (2000) Husbandry. In: The Laboratory Rat (Krinke GJ, ed.), pp. 45–72. London: Academic Press.

Anson RM, Guo Z, De Cabo R, Iyun T, Rios M, Hagepanos A, Ingram DK, Lane MA, Mattson MP (2003) Intermittent fasting dissociates beneficial effects of dietary restriction on glucose metabolism and neuronal resistance to injury from calorie intake. Proceedings of the National Academy of Science U S A 100:6216–6220.

Anzaldo AJ, Harrison PC, Maghirang R-G, Gonyou HW (1994) Increasing welfare of laboratory rats with the help of spatially enhanced cages. Animal Welfare Information Center Newsletter 5(3):1–2.

Baker HJ, Lindsey JR, Wiesbroth SH (1979) The laboratory rat. London: Academic Press.

Barnett SA (1975) The rat: A study in behavior. Chicago: The University of Chicago Press.

Beane ML, Cole MA, Spencer RL, Rudy JW (2002) Neonatal handling enhances contextual fear conditioning and alters corticosterone stress responses in young rats. Hormones and Behavior 41:33–40.

Bellhorn RW (1980) Lighting in the animal environment. Laboratory Animal Science 20:440–450.

Benstaali C, Mailloux A, Bogdan A, Auzeby A, Touitou Y (2001) Circadian rhythms of body temperature and motor activity in rodents their relationships with the light-dark cycle. Life Sciences 68:2645–2656.

Berson DM, Dunn FA, Takao M (2002) Phototransduction by retinal ganglion cells that set the circadian clock. Science 295:1070–1073.

Black JE, Sirevaag AM, Greenough WT (1987) Complex experience promotes capilliary formation in young rat visual cortex. Neuroscience Letters 83:351–355.

Bodnoff SR, Humphreys AG, Lehman JC, Diamond DM, Rose GM, Meaney MJ (1995) Enduring effects of chronic corticosterone treatment on spatial learning, synaptic plasticity, and hippocampal neuropathology in young and mid-aged rats. Journal of Neuroscience 15:61–69.

Broderson JR, Lindsey JR, Crawford JE (1976) The role of environmental ammonia in respiratory mycoplamosis of rats. American Journal of Pathology 85:115–130.

Canadian Council on Animal Care (1984) CCAC guide. Ottawa: CCAC.

Canadian Council on Animal Care (1993) CCAC guide. Ottawa: CCAC.

Chang EF and Merzenich MM (2003) Environmental noise retards auditory cortical development. Science 300:498–502.

Cisar CF and Jayson G (1967) Effects of frequency of cage cleaning on rat litters prior to weaning. Laboratory Animal Care 17:215–218.

Clough C (1987) Quality in laboratory animals. In: Laboratory animals: An introduction for new experimenters (Tuffery AA, ed.), pp. 79–97. Chichester: Wiley-Interscience.

Denenberg VH and Whimbey AE (1963) Behavior of adult rats ils modified by the experiences their mothers had as infants. Science 142:1192–1193.

Diorio D, Viau V, Meaney MJ (1993) The role of the medial prefrontal cortex (cingulate gyrus) in the regulation of hypothalamic-pituitary-adrenal responses to stress. Journal of Neuroscience 13:3839–3847.

Galvez R, Soskin PN, Cho JH, Grossman AW, Greee (2002) Voluntary exercise increases the number of new neurons in the adult rat motor cortex in a time dependent fashion. In: Society for Neuroscience, program No. 662.610. Orlando, FL: Society for Neuroscience.

Gamble MR (1982) Sound and its significance for laboratory animals. Biological Reviews 57:395–421.

Gentile AM, Beheshti Z, Held JM (1987) Enrichment versus exercise effects on motor impairments following cortical removals in rats. Behavioral and Neural Biology 47:321–332.

Gibb R and Kolb B (2000) Comparison of the effects of pre- and postnatal tactile stimulation on functional recovery following early frontal cortex lesions. Society for Neurosciences Abstracts. 26 Program number 366.9.

Gibb R, Gonzalez CLR, Kolb B (2001) Prenatal enrichment leads to improved functionaol recovery following perinatal frontal cortex injury: Effects of maternal complex housing. In: Society for Neurosciences. Abstracts. Vol 27 Program number 476.4.

Gomez-Pinilla F, Ying Z, Opazo P, Roy RR, Edgerton VR (2001) Differential regulation by exercise of BDNF and NT-3 in rat spinal cord and skeletal muscle. European Journal of Neuroscience 13:1078–1084.

Greenough WT and Black JE (1992) Induction of brain structure by experience. Substrates for cognitive development. In: Minnesota Symposium on Child Development (Nelson CA, ed.), pp. 155–200. Hillsdale: Lawrence Erlbaum.

Greenwood BN, Hinde JL, Nickerson M, Thompson K, Fleshner M (2002) Freewheel running changes the brain's response to acute, uncontrollable stress. Society for Neurosciences Abstracts. Vol. 28 Program number 843.10.

Harding SM and McGinnis MY (2003) Effects of testosterone in the VMN on copulation, partner preference, and vocalizations in male rats. Hormones and Behavior 43:327–335.

Hoffman JC (1973) The influence of photoperiods on reproductive functions in female mammals. In: Endocrinology. II. Handbook of Physiology. Section 7, pp. 57–77. Washington, D.C.: American Psychological Society.

Huck UW and Price EO (1975) Differential effects of environmental enrichment of the open-field behavior of wild and domestic Norway rats. Journal of Comparative Physiology and Psychology 89:892–898.

Hurst JL, Barnard CJ, Nevison CM, West CD (1997) Housing and welfare in laboratory rats: Welfare implications of isolation and social contact among caged males. Animal Welfare 6:329–347.

Ixart G, Szafarczyk A, Belugou JL, Assenmacher I (1977) Temporal relationships between the diurnal rhythm of hypothalamic corticotrophin releasing factor, pituitary corticotrophin and plasma corticosterone in the rat. Journal of Endocrinology 72:113–120.

Joffe JM, Rawson RA, Muclik JA (1973) Control of their environment reduces emotionality in rats. Science 180:1383–1384.

Keenan KP, Ballam GC, Haught DG, Laroque P (2000) Nutrition. In: The laboratory rat (Krinke GJ, ed.), pp. 57–75. London: Academic Press.

Kolb B, Gibb R, Gorny G (2003) Experience-dependent changes in dendritic arbor and spine density in neocortex vary with age and sex. Neurobiology of Learning and Memory 79:1–10.

Krohn TC, Hansen AK, Dragsted N (2003) Telemetry as a method for measuring the impact of housing conditions on rats' welfare. Animal Welfare 12:53–62.

LaBarba RC and White JL (1971) Litter size variations and emotional reactivity in BALB/c mice. Journal of Comparative and Physiological Psychology 75:254–257.

Levine S (1967) Maternal and environmental influences on the adrenocortical response to stress in weanling rats. Science 156:258–260.

Liu D, Caldji C, Sharma S, Plotsky PM, Meaney MJ (2000) Influence of neonatal rearing conditions on stress-induced adrenocorticotropin responses and norepinepherine release in the hypothalamic paraventricular nucleus. Journal of Neuroendocrinology 12:5–12.

Markowska AL and Savonenko A (2002) Retardation of cognitive aging by life-long diet restriction: Implication for genetic variance. Neurobiology of Aging 23:75–86.

Mattson MP, Duan W, Guo Z (2003) Meal size and frequency affect neuronal plasticity and vulnerability to disease: Cellular and molecular mechanisms. Journal of Neurochemistry 84:417–431.

Mattson MP, Duan W, Chan SL, Cheng A, Haughey N, Gary DS, Guo Z, Lee J, Furukawa K (2002) Neuroprotective and neurorestorative signal transduction mechanisms in brain aging: modification by genes, diet and behavior. Neurobiology and Aging 23:695–705.

Means LW, Higgins JL, Fernandez TJ (1993) Mid-life onset of dietary restriction extends life and prolongs cognitive functioning. Physiology and Behavior 54:503–508.

Milkovic K, Paunovic JA, Joffe JM (1975) Effects of pre- and postnatal litter size on development and behavior of rat offspring. Developmental Psychobiology 9:365–375.

Moroi-Fetters SE, Mervis RF, London ED, Ingram DK (1989) Dietary restriction suppresses age-related changes in dendritic spines. Neurobiology and Aging 10:317–322.

Niesink RJ and van Ree JM (1982) Short-term isolation increases social interactions of male rats: A parametric analysis. Physiology and Behavior 29:819–825.

Poulos S and Borlongan C (2000) Artificial lighting conditions and melatonin alter motor performance in adult rats. Neuroscience Letters 280:33–36.

Rauscher FH, Robinson D, Jens JJ (1998) Improved maze learning through early music exposure in rats. Neurological Research 20:427–432.

Rebuelto M, Ambros L, Montoya L, Bonafine R (2002) Treatment-time-dependent difference of ketamine pharmacological response and toxicity in rats. Chronobiology International 19:937–945.

Rosenzweig MR (1971) Effects of environment on development of brain and behavior. In: The biopsychology of development (Tobach E and Shaw E, eds.), pp. 303–342. New York: Academic Press.

Saltarelli CG and Coppola CP (1979) Influence of visible light on organ weights of mice. Laboratory Animal Science 29:319–322.

Schofield JC and Brown MJ (2003) Animal care and use: A nonexperimental variable. In: Essentials for animal research: A primer for research personnel (Bennett B, Brown M, and Schofield J, eds) Second Edition. [Book available online]. Retrieved February 10, 2004 from the

World Wide Web: http://www.nal.usda.gov/awic/pubs/noawicpubs/essentia-htm.

Schumacher SK and Moltz H (1985) Prolonged responsiveness to the maternal pheromone in the post-weanling rat. Physiology and Behavior 34:471–473.

Sharp J, Zammit T, Azar T, Lawson D (2003) Stress-like responses to common procedures in male rats housed alone or with other rats. Contemporary Topics In Laboratory Animal Science 41:8–14.

Smotherman WP, Bell RW, Starzec J, Elias J, Zachman TA (1974) Maternal responses to infant vocalizations and olfactory cues in rats and mice. Behavior and Biology 12:55–66.

Spalding JF, Archuleta RF, Holland LM (1969) Influence of the visible colour spectrum on activity in mice. Laboratory Animal Care 19:50–54.

Stewart J and Kolb B (1988) The effects of neonatal gonadectomy and prenatal stress on cortical thickness and asymmetry in rats. Behavioral and Neural Biology 49:344–360.

van Praag H, Kempermann G, Gage FH (1999) Running increases cell proliferation and neurogenesis in the adult mouse dentate gyrus. Nature Neuroscience 2:266–270.

Vinall PE, Kramer MS, Heinel LA, Rosenwasser RH (2000) Temporal changes in sensitivity of rats to cerebral ischemic insult. Journal of Neurosurgery 93:82–89.

Von Fritag JC, Schot M, van den Bos R, Spruijt BM (2002) Individual housing during the play period results in changed responses to and consequences of a psychosocial stress situation in rats. Developmental Psychobiology 41:58–69.

Wasowicz M, Morice C, Ferrari P, Callebert J, Versaux-Botteri C (2002) Long-term effects of light damage on the retina of albino and pigmented rats. Investigations in Ophthalmology and Visual Science 43:813–820.

Winocur G and Hasher L (1999) Aging and time-of-day effects on cognition in rats. Behavioural Neuroscience 113:991–997.

Zimmerberg B, Rosenthal AJ, Stark AC (2003) Neonatal social isolation alters both maternal and pup behaviours in rats. Developmental Psychobiology 42:52–63.

Defense and Social Behavior

VI

Antipredator Defense

31

D. CAROLINE BLANCHARD AND ROBERT J. BLANCHARD

NATURAL BEHAVIORS

A focal aspect of the class of defensive behaviors of the rat is that they are "natural." This implies, first, that they occur in nature, that is, in the real world, outside a laboratory setting. In such real-world settings, behaviors are subject to their natural consequences; some have good outcomes for the animals that have performed them, whereas others may result in a range of woes from inconvenience to disaster. In line with the tenets of Darwinian evolution, the conditions under which such behaviors occur have an important effect on the success of their outcome, such that particular behaviors are adaptive if and only if they occur in situations that are conducive to their success. Thus, the concept of "natural behavior" suggests a strong relationship between both environmental (physical and social) stimuli and stimuli arising from the animal itself, and particular behaviors. A third aspect of "natural" to many biologists and psychologists is that evolution and some degree of genetic determination are involved in shaping such a behavior; that is, it is not solely the product of specific learning.

DEFENSE IN RATS: SPECIFIC BEHAVIORS

In both wild and laboratory rats (*Rattus norvegicus*), the specific defensive behaviors that have been described (e.g., Blanchard and Blanchard, 1969, 1989; Pinel and Treit, 1978; Blanchard RJ et al., 1980, 1989, 1990, 1991; Blanchard DC et al., 1981, 1991; Dielenberg et al., 1999, 2001; McGregor et al., 2002) include the following.

Flight: rapid movement (typically running) away from a threat source

Hiding or sheltering: entering and remaining in a place where the animal is less visible or that the threat cannot enter

Freezing: immobility, also called *crouching* when a specific posture is maintained

Alarm cries: cries of about 22 kHz that are emitted when familiar conspecifics are present, and to which they may respond defensively

Defensive threat: defensive upright or standing posture facing the oncoming threat, typically accompanied by tooth exposure and screams

Defensive attack: biting at the oncoming threat, often after a jump toward it

Risk assessment: a pattern of investigation of the threat source, including scanning it from a distance while *freezing*; and a "stretch attend" or "stretch approach" behavior in which the animal adopts a stretched-out, low-back posture while oriented toward the threat source and may show short bursts of movement interspersed with periods of immobility. Closer approaches and even contact may occur, again typically involving the low-back stretched posture.

Defensive burying: discrete threat stimuli are often covered with substrate or other materials. It is not certain if this is specifically a defensive response or a more general response to dis-agreeable objects, because rats also bury dead conspecifics, unpalatable food, and novel ob-jects (Wilkie et al., 1979; Blanchard RJ et al., 1989). "Burying" may also involve an element of risk assessment, in that tossing objects onto animals or other stimuli with ambiguous threat characteristics may cause these stimuli to respond in such a fashion as to make their threat potential clearer.

This list is likely to be incomplete, with new categories of defense awaiting descrip-tion and analysis. In addition, subtle variations and combinations of the above may occur un-der particular circumstances. For example "tunnel guarding" behavior of colony subor-dinate males appears to combine sheltering (in the tunnels of a *visible burrow system*) with defensive threat and attack toward the dominant, should the latter attempt to en-ter the tunnel in which the subordinate is lodged (Blanchard RJ et al., 2001A). These behaviors make obvious a point that we elaborate later, that features of the situation have a very strong influence on the form of the defense seen: No tunnel, no tunnel guarding.

STIMULI THAT ELICIT DEFENSE

TYPES OF THREAT STIMULI

Defensive behaviors occur in response to threats to an animal's life or body. These threats can be divided into four categories: (1) predators, (2) attacking conspecifics, (3) threatening features of the environment (lightning, fire, high places, and water), and (4) nonconspecific but dangerous resource competitors. The list suggests several impor-tant points. First, it is active danger to life or limb that elicits defensive behavior, not a pas-sive threat such as hunger, nor a threat or chal-lenge to some resource that does not involve immediate danger to the animal. Second, the distinction between animate threats and inan-

imate threats will turn out to be important for both the type of defensive behavior offered and for laboratory models attempting to measure these defenses.

An additional point concerning types of threat stimuli is that some of these are clear, immediate, salient, and embodied in a specific object, whereas others are ambiguous. Cer-tain sounds or odors may serve as distinct cues that danger is or has been near but provide lit-tle information on the exact location or iden-tity of the danger or whether it is still present. As Table 31–1 indicates, these features are as-sociated with very different types of defense. A localizable threat is important for some types of defensive behavior, such as flight, to be effective. Simply running around without reference to the location of the threat source may attract attention without removing the subject from the presence of the threat. Sim-ilarly, an embodied threat is necessary for de-fensive attack to be effective; biting at sounds or odors does not reduce the threat they may signal.

Threat Ambiguity and Risk Assessment

The major type of defense to ambiguous threat is risk assessment, an active investi-gation of the threat stimulus or situation that typically involves a stretched, low-back posture that appears to minimize the chance of the animal being detected as it goes about the business of investigating the threat. The stretched posture is very similar to the pos-ture adopted by a stalking animal; another situation in which the joint goals of ap-proach to another animal while escaping de-tection by that animal might be inferred. In addition, the movement parameters of the stretch approach are such as to minimize de-tection: the animal's forward movement is brief and interspersed with periods of im-mobility, during which detection by the threat is less likely. What is clear is that risk assessment is associated with threat-related information gathering. A very clever set of studies by Pinel et al. (1989) indicated that

Table 31–1. Stimuli, Behaviors, Outcomes, and Affect Associated With Threat Detection or with Defense to Specific Threats

Aspect	Threat detection	Defense
Stimuli	Animate movement	Predator
	Predator odors	Attacking conspecific
	Alarm cries	
	Conspecific odors	
Behaviors	Risk assessment	Flight
	Stretch attend	Freezing
	Stretch approach	Defensive threat and attack
	Hiding-avoidance	Hiding-avoidance
		Conspecific back defense
		Alarm vocalizations
Outcome	Danger detected	Escape
	or	Avoidance of detection
	Safety detected	Frightening off threat
		Protection of self
		Protection of body parts
		Warning conspecifics (respectively)
Affect	Anxiety	Fear

proximity to a stimulus while in this stretch attend/approach mode was associated with learning of defensiveness to that stimulus, whereas equal or greater proximity without risk assessment was not.

Ambiguity of threat does not necessarily imply that the stimulus is itself amorphous or disembodied. The attack potential of a conspecific or even a predator may be ambiguous, even if the animal itself is clearly present. In such cases, the threatened animal's first response may be risk assessment. When a dominant rat is tethered such that it cannot approach a subordinate, the subordinate reacts by stretch-attending or stretch-approaching the dominant, presumably to check out the threat features of the latter (unpublished observations). Similarly, a predator at a distance and not approaching the subject may elicit risk assessment rather than flight. When approached from a distance by the human experimenter, a wild rat first freezes while oriented to the predator, with an abrupt transition to flight when the latter is about 3 meters away, presumably when its threat potential becomes clear (Blanchard DC et al., 1981).

Risk assessment behavior is adaptive because defensive behaviors are costly in terms of time and energy. Displaying them or continuing to display them when there is no threat may be very wasteful. Risk assessment is part of the process of deciding when to stop being defensive. Not being defensive when there is a real threat can be disasterous, however, and risk assessment may help the animal to determine that a danger is genuine, leading to the expression of more specific defenses. Finally, displaying the wrong defensive behavior for a particular threat and situation may be as dangerous as displaying no defense. Risk assessment helps the animal to determine which defensive behavior to express. Thus, risk assessment is crucial to all of the cognitive and decision-making aspects of defense.

THE ROLE OF "EXPEDITING" STIMULI IN CONTROLLING SPECIFIC DEFENSES

As noted earlier, features of the social and physical environment in which the threat is encountered may affect the type of defense that is offered. The presence of an exit route facilitates flight (Blanchard RJ et al., 1989), manipulable substrate facilitates defensive burying (Pinel and Treit, 1979), shelters promote hiding and tunnel guarding (Dielenberg et al., 1999; Blanchard RJ et al., 2001b), and both a shelter and the presence of conspecifics enhances antipredator alarm vocalizations (Blanchard RJ et al., 1991). While there is nothing to prevent an animal from performing the actions associated with a particular defense in the absence of the relevant support stimulus, these defensive behaviors tend not to occur without their particular support stimuli.

Most laboratory studies of the responsivity of rats to threat stimuli include none of the preceding features: not an exit from the test situation nor a shelter, no conspecifics, not even a specific, embodied threat stimulus. These omissions sharply reduce levels of flight, hiding, defensive threat/attack, burying, and alarm vocalizations, which leaves freezing and risk assessment. The latter is typically not measured, which leaves freezing. This situation is partly responsible for the widespread treatment of freezing as the primary, or even sole, measure of defensiveness in the rat.

THE ROLE OF THREAT INTENSITY IN CONTROLLING SPECIFIC DEFENSES

An alternative explanation for how defensive behaviors are determined was offered by Fanselow and his coworkers. Fanselow and Lester (1988) conceptualized a threat intensity or stimulus imminence-based differentiation among three sets of defensive behaviors: (1) preencounter defenses (represented by compromises in normal activities), (2) postencounter defenses (freezing), and (3) circastrike defenses (flight, vocalization, biting, and high activity responses such as those to shock), further suggesting (Fanselow, 1994) that this intensity/imminence parameter determines the form of defense offered in particular situations.

Threat stimulus intensity does affect the magnitude of defensive behavior. Large pieces of a cloth rubbed on a cat elicit more defensive behavior in rats than do small pieces of this cloth, even when both are hidden in containers such that only cat odor remains as the effective stimulus (L. Takahashi, personal communication). While Zangrossi and File (1992) found that exposure to cat odor produces anxiety-like responses on the elevated plus-maze for an hour, but not a day, afterwards, Adamec & Swallow (1993) reported a much longer-lasting anxiety-like plus-maze response after cat exposure, a difference that likely reflects greater threat intensity for a cat, compared to cat odor alone. However, it is also clear that specific changs in environmental stimuli can rapidly and dramatically alter the expression of specific defenses, even when the threat stimulus itself is unchanged. For example, closing a door to block an escape route instantaneously changes flight to freezing in wild rats (Blanchard DC et al., 1981). While this change might possibly be interpreted as *increasing* threat intensity, in the stimulus intensity/imminence schema an *increase* in intensity changes freezing to flight, not the opposite. Such considerations suggest that the determinants of defensive behaviors are more complex than a single intensity dimension. They agree with data indicating that that the patterning of defense is responsive to "expediting" features of both the threat stimulus and the local environment.

OUTCOMES OF DEFENSIVE BEHAVIORS

The outcomes of defense provide the mechanism by which these behaviors evolve. Rats are fossorial, nocturnal, and, to some degree, colonial. The *fossorial (burrow-dwelling)* feature facilitates the adaptive value of running away or hiding from danger, particularly if a tunnel of some sort is available. The *nocturnal* feature promotes the use of nonvisual senses, and olfaction in the rat is especially useful in the context of defense. *Colonial* rodents are particularly noted for their alarm cries, and these are prominent in rats.

A intriguing aspect of evolutionary analysis of behavior relates to extended responsivity to olfactory threat stimuli in the rat. Blanchard DC et al. (2003b) suggest that aversive odorants may elicit different patterns of defensive behavior, depending on the ability of the odor to predict the presence of a predator. Cat fur and skin odors (obtained by rubbing a cat with a cloth or by using a collar worn by a cat) elicit a range of defenses, including avoidance, freezing, and risk assessment. When exposed later to the stimulus or situation where the odor was initially encountered but is now absent, rats show defensive behaviors to these conditioned stimuli (e.g., Blanchard RJ et al., 2001b; Dielenberg et al., 2001; McGregor et al., 2002). When fresh fecal material from the same cat that donated the fur or skin odor is presented, the immediate defensive behaviors are virtually identical to those seen to the fur or skin odor. However, no conditioned defensiveness occurs when the rat is again confronted by the situational and specific stimuli with which the odors of feces had originally been associated. This is consonant with findings that synthetic feces/anal gland odors also fail to serve as effective unconditioned stimuli for Pavlovian defensive conditioning (Blanchard DC et al., 2003b). These differences are interpreted as reflecting that fur and skin odors dissipate quickly and thus have much greater ability to predict the presence of a predator than do feces, which, along with its odors, dissipates very slowly. Thus, the ability of odors to serve as unconditioned stimuli for defensive conditioning may reflect whether these odors accurately predict danger.

LABORATORY MODELS OF RAT DEFENSIVE BEHAVIORS

There are many laboratory tasks designed to measure some aspect of behavior potentially related to defense. Most "anxiety models" or tasks measuring "anxiety-like behaviors" (see Rodgers, 1997, for a review) tap some aspect of defensiveness, as do many of the tasks created to measure depression (Willner, 1991). Although these may or may not be adequate for the purposes for which they were devised, in preclinical testing of drugs, such tasks do not provide a precise analysis of how their measures relate to the range of natural behaviors shown by rats. Attempts to elicit and measure defensive behaviors are relatively recent. Nevertheless, early reports (Yerkes, 1913; Stone, 1932) detailing the response of wild and laboratory rats to human touch and handling documented that wild rats flee or bite if picked up, whereas laboratory rats are less reactive. Curti (1935) reported that exposure to a cat is capable of eliciting some fear-related responses in the rat, although her conclusion that the odor of a cat was not an adequate specific stimulus for these responses suggests that her test situation was not a sensitive one. Defensive behaviors of rats to conspecific attack (which involve some specific elements not found in antipredator defense) were well described by Grant and his colleagues (Grant and Chance, 1958; Grant, 1963; Grant and MacKintosh, 1963). Although freezing to a cat or to shock stimuli had earlier been described in some detail (e.g. Curti, 1935; Blanchard and Blanchard, 1969; Fanselow, 1980), the first specific attempt to outline the array of defensive responses to an oncoming predator involved wild rats, in the Fear/Defense Test Battery (FDTB).

FEAR/DEFENSE TEST BATTERY

The FDTB was run in an oval runway with a high barrier down the center of its long axis that did not extend to either end wall. This created an endless runway of 5 or 6 meters on a side that enabled the animal to run around the end of the barrier and into the next straight side, indefinitely, while being chased by a human experimenter. The closing of a door to block the runway at one end trapped the rat, enabling its response to the oncoming experimenter to be evaluated at varying distances of 5, 4, 3, 2, 1, 0.5, and 0 (contact) meters. Response to attempted pick-up was also measured (see Blanchard DC, 1997, for a review of procedures and findings).

In this test, the responses of wild rats were extremely consistent. In several studies, more than 97% of wild rats fled from the oncoming experimenter when the runway was open, permitting flight. The average distance between the experimenter and the subject when the subject turned to move away was about 2.7 meters. When the door was closed, precluding flight, the rat quickly froze, typically in an upright posture and always facing the experimenter. Freezing was seen during about 80% of observations from 5 meters to about 2 meters, but the rats showed a startle response to sudden sounds (hand clap or firing of starter's pistol) that increased in amplitude as the experimenter approached, suggesting that this freezing response also involves preparation for violent action that increases as contact becomes imminent. At about 1 meter, defensive threat screams suddenly began to occur, and at about 0.5 meter, some animals jumped at or past the experimenter, with biting should contact occur. When the experimenter attempted to pick up the subjects, 100% of wild rats bit. These relationships are schematized in Figure 31–1.

Laboratory rats in the FDTB showed less of every response than did wild rats, except for freezing, differences that were paralleled precisely in rats bred over 35 generations for "tameness" or "wildness" from wild-trapped stock in Novosibirsk, Siberia (Blanchard DC et al., 1994). This wild–laboratory rat difference appears to reflect selective breeding of rats for failure to show active defenses such as defensive threat/attack and flight, during the process of domestication. The same general findings were noted to human handling (Takahashi and Blanchard, 1982) as those in response to an attacking conspecific. Moreover, the responses of wild rats reared in a laboratory setting were gener-

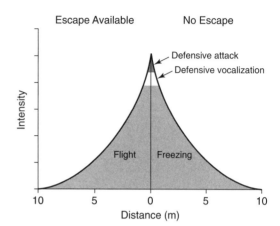

Figure 31–1. Schematic of the intensity of defensive behaviors (flight, freezing, defensive threat vocalizations, and defensive attack) as a function of distance between an animal and a threat stimulus, in situations where escape is available or not available. Based on responses of wild *R. norvegicus* to an oncoming experimenter in the FDTB

ally more similar to those of wild-trapped wild rats than to those of laboratory rats, suggesting that the wild–laboratory rat difference is largely genetic (Blanchard DC, 1997).

ANXIETY/DEFENSE TEST BATTERY

What the F/DTB does not do very well is to elicit risk assessment. It uses a very embodied threat stimulus, the human experimenter, and focuses largely on the specific defenses of flight, freezing, and defensive threat/attack. In contrast, the Anxiety/Defense Test Battery (A/DTB) was devised specifically to elicit and permit measurement of risk assessment, with focuses on stretch/attend and stretch/approach behaviors and on reductions of normal activities in response to the presence of threat. The term *anxiety* in the name of the battery reflects that risk assessment activities are extremely close to specific behaviors associated with generalized anxiety disorder ("vigilance and scanning" [American Psychiatric Association, 1987]) and that anxiety has traditionally been regarded as associated with ambiguity of threat and danger cues rather than with the clear presence of danger (Freud, 1930; Estes and Skinner, 1941). Various tests were devised to enable measurement of risk assessment and also were used to evaluate the effects of various pharmacological agents on this measure. The most often used test from the original A/DTB is a test of responsivity to cat odors, with a parallel test involving a live cat as the stimulus.

Cat Odor and Cat Exposure Tests

Because of the difficulties of keeping a live cat in a laboratory setting, cat odor tests have come to be more commonly used than cat exposure. The cat odor test involves a 1 meter long alley with a cat odor stimulus (obtained by rubbing a cloth on a cat) at one end, in which avoidance of the odor stimulus, and risk assessment are the major measures. If a hide box is added, cat odor (from a cat collar) elicits hiding as well as avoidance and risk assessment. A single 10 minute exposure to odor is sufficient to produce conditioned defensiveness in the situation, 24 hours later (McGregor et al., 2002). Newton Canteras (personal communication) has further modified this test to elicit strong and prolonged risk assessment in a situation in which a cat was previously encountered, and he is using this prolonged risk assessment to investigate the neural systems involved in this specific behavior.

RELEVANCE TO UNDERSTANDING EMOTION

Tests of defensive behavior are relevant to an understanding of emotional responses to danger on a behavioral and neural level. Behaviorally, they outline a system of innate, unconditioned responses to various types of threat that may also be conditioned to appropriate stimuli in a single, brief exposure (Blanchard RJ et al., 2001b; Blanchard DC et al., 2003b; Dielenberg et al., 2001; McGregor et al., 2002). Although the status of these responses as innate has not been researched in humans, recent scenario studies (Blanchard DC et al., 2001b) suggest that human re-

sponsivity to threat includes all of the defenses outlined in rats, and more, and that these conserved defensive behaviors occur in much the same situations as those analyzed for rats.

In neural investigations, Canteras and his colleagues used combinations of c-fos, tract-tracing, and lesioning techniques to outline a set of potential neural circuits involved in responsivity to a cat (Canteras, 2002). They suggested the importance of particular hypothalamic structures in responsivity to a cat. Lesions of one such area, the dorsal premammillary nucleus, dramatically reduces reactions to a cat or cat odor (Canteras et al., 1997; Blanchard DC et al., 2003c) but not to foot shock (Blanchard DC et al., 2003c). The dorsal premammillary nucleus, in turn, projects both directly and indirectly to the periaqueductal gray, an area in which a number of defensive behaviors are differentially elicitable by electrical or excitatory amino acid stimulation (Depaulis and Bandler, 1991). Thus, the circuitry underlying many defensive behaviors provides justification for their conceptualization as differentiable entities rather than as behaviorally equivalent components of an single motivational system.

Beginning with rat defensive behaviors, and later with similar defensive behaviors in mice (the Mouse Defense Test Battery, which incorporates features of both the F/FDTB and the A/FDTB), a body of evidence has been gathered that suggests particular defensive behaviors may show greater or lesser responsivity to drugs effective against particular anxiety disorders. Risk assessment and defensive threat/attack respond selectively to drugs effective against generalized anxiety disorder, whereas antipanic drugs reduce flight, propanic drugs enhance flight, and drugs with no effect on panic also have no effect on flight (Blanchard DC et al., 2001A, 2003A). These findings, in combination with systems analyses of defense, raise the possibility of much more specific information on the brain substrate of defensive behaviors, suggesting they may facilitate precision and selectivity in the design of physiologically acting treatments for psychopathologies.

Defensive behaviors also respond to, and change with, experience. Their strong and rapid conditioning to some types of stimuli associated with threat may be a factor in the etiology of a number of different anxiety syndromes. Because conditioning of defensive behaviors, with the exception of freezing, is only beginning to be investigated, it is not clear how these phenomena might be related to the development of threat-related learning (either "normal" or "abnormal") or how experience-based therapies might best be used to modulate these phenomena. Information on conditioning of defensive behaviors using both full and partial predator stimuli and conditioned stimuli of different types, may be useful in understanding behavioral differences in human anxiety disorders and the possibilities of experiential as well as drug treatments for these conditions.

REFERENCES

Adamec RE and Shallow T (1993) Lasting effects on rodent. Anxiety of a single exposure to a cat. Physiology and Behavior 54:101–109.

American Psychiatric Association (1987) DSM-IIIR: Diagnostic and statistical manual of mental disorders, 3rd edition (revised). Washington, D.C.: The Association.

Blanchard DC (1997) Stimulus and environmental control of defensive behaviors. In: The functional behaviorism of Robert C. Bolles: Learning, motivation and cognition. (Bouton M and Fanselow M, eds.), pp. 283–305. Washington, D.C.: American Psychological Association.

Blanchard DC, Blanchard RJ, Rodgers RJ (1991) Risk assessment and animal models of anxiety. In: Animal models in psychopharmacology (Olivier B, Mos J, Slangen JL, eds.), pp. 117–134. Basel: Birkhauser Verlag AG.

Blanchard DC, Griebel G, Blanchard RJ (2001a) Mouse defensive behaviors: Pharmacological and behavioral assays for anxiety and panic. Neuroscience and Biobehavioral Reviews 25:205–218.

Blanchard DC, Griebel G, Blanchard RJ (2003a) The

mouse defense test battery: Pharmacological and behavioral assays for anxiety and panic. European Journal of Psychology 463:97–116.

Blanchard DC, Hynd AL, Minke KA, Blanchard RJ (2001b) Human defensive behaviors to threat scenarios show parallels to fear- and anxiety-related defense patterns of nonhuman mammals. Neuroscience and Biobehavioral Reviews 25:761–770.

Blanchard DC, Lee EMC, Williams G, Blanchard RJ (1981) Taming of Rattus norvegicus by lesions of the mesencephalic central gray. Physiological Psychology 9:157–163.

Blanchard DC, Li C-I, Hubbard D, Markham C, Yang M, Takahashi LK, Blanchard RJ (2003c) Dorsal premammillary nucleus differentially modulates defensive behaviors induced by different threat stimuli. Neuroscience Letters 345:145–148.

Blanchard DC, Markham C, Yang M, Hubbard D, Madarang E, Blanchard RJ (2003b) Failure to produce conditioning with low-dose TMT, or, cat feces, as unconditioned stimuli. Behavioral Neuroscience 117:360–368.

Blanchard DC, Popova NK, Plyusnina IZ, Velichko IV, Campbell D, Blanchard RJ, Nikulina J, Nikulina EM (1994) Defensive behaviors of "wild-type" and "domesticated" wild rats in a fear/defense test battery. Aggressive Behavior 20:387–398.

Blanchard DC, Rodgers RJ, Hori K, Hendrie CA, Blanchard RJ (1989) Attenuation of defensive threat and attack in wild rats (Rattus rattus) by benzodiazepines. Psychopharmacology 97:392–401.

Blanchard RJ and Blanchard DC (1969) Crouching as an index of fear. Journal of Comparative Physiological Psychology 67:370–375.

Blanchard RJ and Blanchard DC (1989) Anti-predator defensive behaviors in a visible burrow system. Journal of Comparative Psychology 103:70–82.

Blanchard RJ, Blanchard DC, Agullana R, Weiss SM (1991) Twenty-two kHz alarm cries to presentation of a predator, by laboratory rats living in visible burrow systems. Physiology and Behavior 50:967–972.

Blanchard RJ, Blanchard DC, Hori K (1989) Ethoexperimental approaches to the study of defensive behavior. In: Ethoexperimental approaches to the study of behavior (Blanchard RJ, Brain PF, Blanchard DC, Parmigiani S, eds.), pp. 114–136. Dordrecht: Kluwer Academic Publishers.

Blanchard RJ, Blanchard DC, Weiss SM, Mayer S (1990) Effects of ethanol and diazepam on reactivity to predatory odors. Pharmacology Biochemistry and Behavior 35:775–780.

Blanchard RJ, Dulloog L, Markham C, Nishimura O, Compton JN, Jun A, Han C, Blanchard DC (2001a) Sexual and aggressive interactions in a visible burrow system with provisioned burrows. Physiology and Behavior 72:245–254.

Blanchard RJ, Kleinschmidt CF, Fukunaga-Stinson C, Blanchard DC (1980) Defensive attack behavior in male and female rats. Animal Learning and Behavior 8:177–183.

Blanchard RJ, Yang M, Li C-I, Garvacio A, Blanchard DC (2001b) Cue and context conditioning of defensive behaviors to cat odor stimuli. Neuroscience and Biobehavioral Reviews 26:587–595.

Canteras NS (2002) The medial hypothalamic defensive system: Hodological organization and functional implications. Pharmacology Biochemistry and Behavior 71:481–491.

Canteras NS, Chiavegatto S, Valle LE, Swanson LW (1997) Severe reduction of rat defensive behavior to a predator by discrete hypothalamic chemical lesions. Brain Research Bulletin 44:297–305.

Curti MW (1935) Native responses of white rats in the presence of cats. Psychological Monographs 46:76–98.

Depaulis A and Bandler R (1991) The midbrain periaqueductal grey matter: Functional, anatomical and immunohistochemical organization. NATO ASI Series A Vol 213. New York: Plenum.

Dielenberg RA, Arnold JC, McGregor IS (1999) Low-dose midazolam attenuates predatory odor avoidance in rats. Pharmacology Biochemistry and Behavior 62:197–201.

Dielenberg RA, Carrive P, McGregor IS (2001) The cardiovascular and behavioral response to cat odor in rats: Unconditioned and conditioned effects. Brain Research 897:228–237.

Estes WK, Skinner BF (l94l) Some quantitative properties of anxiety. Journal of Experimental Psychology 29:390–400.

Freud S (1930) Inhibitions, symptoms, and anxiety. London: Hogarth Press.

Grant EC (1963) An analysis of the social behaviour of the male laboratory rat. Behaviour 21:260–281.

Grant EC and Chance MRA (1958) Rank order in caged rats. Animal Behavior 6:183–194.

Grant EC and MacKintosh JH (1963) A comparison of the social postures of some common laboratory rodents. Behaviour 21:246–259.

Fanselow MS (1980) Conditioned and unconditional components of post-shock freezing. Pavlovian Journal of Biological Science 15:177–182.

Fanselow MS (1994) Neural organization of the defensive behavior system responsible for fear. Psychonomic Bulletin and Review 1:429–438.

Fanselow MS and Lester LS (1988) A functional behavioristic approach to aversively motivated behavior: Predatory imminence as a determinant of the topography of defensive behavior. In: Evolution and learning (Bolles RC and Beecher MD, eds.), pp. 185–211. Hillsdale, NJ: Erlbaum.

Griebel G, Blanchard DC, Jung A, Blanchard RJ (1995) A model of a antipredator defense in Swiss-Webster mice: Effects of benzodiazepine receptor ligands with different intrinsic activities. Behavioural Pharmacology 6:732–745.

McGregor IS, Schrama L, Ambermoon P, Dielenberg RA (2002) Not all 'predator odours' are equal: Cat odour but not 2,4,5 trimethylthiazoline (TMT; fox odour) elicits specific defensive behaviours in rats. Behavioral Brain Research 129:1–16.

Pinel JPJ, Mana M, Ward J'AA (1989) Stretched-approach sequences directed at a localized shock source by Rattus norvegicus. Journal of Comparative Psychology 103:140–148.

Pinel JPJ and Treit D (1978) Burying as a defensive response in rats. Journal of Comparative and Physiological Psychology 92:708–712.

Pinel JPJ and Treit D (1979) Conditioned defensive burying in rats: Availability of burying materials. Animal Learning and Behavior 7:392–396.

Rodgers RJ (1997) Animal models of 'anxiety': Where next? Behavioral Pharmacology 8:477–496.

Stone CP (1932) Wildness and savageness in rats of different strains. In: Studies in the dynamics of behavior (Lashley KS, ed.), pp. 3–55. Chicago: University of Chicago Press.

Takahashi LK and Blanchard RJ (1982) Attack and defense in laboratory and wild Norway and black rats. Behavioral Processes 7:49–62.

Wilkie DM, MacLennan AJ, Pinel JP (1979) Rat defensive behavior: Burying noxious food. Journal of the Experimental Analysis of Behavior 31:299–306.

Willner P (1991) Behavioural models in psychopharmacology: Theoretical, industrial and clinical perspectives. Cambridge: Cambridge University Press.

Yerkes RM (1913) The heredity of savageness and wildness in rats. Journal of Animal Behavior 3:286–296.

Zangrossi H and File SE (1992) Behavioral consequences in animal tests of anxiety and exploration of exposure to cat odor. Brain Research Bulletin 29:381–388.

Aggressive, Defensive, and Submissive Behavior

32

KLAUS A. MICZEK AND SIETSE F. DE BOER

The statements that "rats have very little social life" and "are not particularly influenced by each other's action" in the previous *Handbook of Psychological Research on the Rat* (Munn 1950) has been corrected. Some 40 years ago, Barnett (1963) gave a succinct account of the social interactions of wild rats (*Rattus norvegicus*) in the classic work *The Rat: A Study in Behavior*. When food supplies, nesting opportunities, and infectious and predatory pressures are favorable, wild rats crowd together in colonies that may number many hundreds.

Nevertheless, life in the colony is not simple. Colonies crash periodically due to intense social conflict despite conducive environmental conditions. Usually, one adult male rat dominates a small group of females and young rats by defending the region around their feeding and nest sites from intruders, and, in this sense, Barnett defined the region that is defended as *territory*. Outside the territories, neutral areas exist where fighting is minimal and avoidance occurs. In general, a colony comprises a number of territories and neutral areas. Wild resident rats patrol their territory and mark these landmarks by urine deposition (Eibl-Eibesfeldt, 1950; Telle, 1966). Although it is mainly the dominant male who drives away male intruders, lactating female rats defend their nest, against both males and females. The forces of cohesion and dispersal are apparent in various developmental and reproductive phases of the rat: It is instructive to see rats huddle and sleep together in a group and to engage in chases, threats, attack leaps, and bites, prompting escapes and even-

tual emigration of the loser from the group, particularly when there are definite boundaries to their territory such as under captive conditions. Barnett (1963) contrasted the social and aggressive interactions among feral rats (*R. norvegicus*) with those of laboratory rats, characterizing the latter as being "unlikely to attack conspecifics with the vigor displayed by wild rats" and to fail "to behave in a normal way."

Subsequent analyses of the salient social signals, aggressive acts, and postures during situations of conflict in small breeding colonies of laboratory rats have found the differences in aggressive behavior between laboratory and feral rats to be mostly in degree rather than in kind (Grant and Mackintosh, 1963; Luciano and Lore, 1975; Zook and Adams, 1975; Blanchard et al., 1977; Miczek, 1979; Boice, 1981; de Boer and Koolhaas, 2003) (Fig. 32–1). As among feral rats (Steiniger, 1950), within breeding colonies of laboratory rats, a dominant, or *alpha*, rat typically is defined by prevailing most often over rival, or *beta*, males and over subordinate, or *omega*, animals in aggressive confrontations. Even in small breeding colonies of laboratory rats that are provisioned with clumped sources for food and water, subordinate members need to be rescued periodically to ensure their survival (Blanchard et al., 1985). Repeated conflict in unstable social groups of laboratory rats, as for feral rats (Calhoun, 1948), increases the risk of injuries, compromises the immune system, diverts energy from reproductive activities and foraging, disrupts circa-

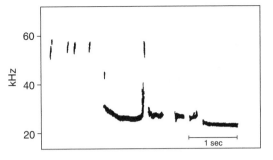

Figure 32–1. Sonogram of 50 to 60 and 20 to 25 kHz vocalizations during a confrontation between a resident and an intruder rat.

dian and physiological rhythms, places prolonged demands on endocrine functions that result in gonadal atrophy and adrenal hypertrophy, and ultimately shortens the life span (Fleshner et al., 1989; Stefanski, 2001). Not surprisingly, social instability and/or chronic subordination in rats and other animals is used as a chronic social stress model in fundamental research of the mechanisms underlying stress pathology in humans (see Chapter 22. Koolhaas et al. in this book). Intermittent exposure to aggressive behavior that results in brief episodes of social defeat stress sensitizes intruder rats, resulting in their increased drug self-administration (Miczek and Mutschler, 1996; Covington and Miczek, 2001).

Aggressive behavior, although infrequent, is part of life in a colony, particularly during its formation, and these interactions within the group are referred to as *dominance* or *within-group aggressive behavior*. As in their feral counterparts, the most potent trigger for aggressive behavior in resident male laboratory rats is the intrusion by an unfamiliar adult male, and these interactions are called *resident–intruder aggression*. In some sense, the aggression directed toward an intruder can be referred to as *territorial*, because it is more likely in marked surroundings than in unfamiliar locales. The probability of aggressive behavior increases when the resident male cohabitates with a female, although the female does not have to be present during the actual con-

frontation (Barnett et al., 1968; Flannelly and Lore, 1977). Resident–intruder confrontations typically occur between the male resident and a male intruder or, less frequently, between a female resident and a female intruder. Aggressive behavior by a female resident is more likely in the initial postpartum period, at which time the dam attacks both male and female intruders (*maternal aggression*). Prior positive (winning) or negative (losing) aggressive experiences promote the subsequent display of offensive or defensive/submissive aggressive behavior, respectively.

PROVOCATIVE SIGNALS

When the resident has been instigated by a male intruder that is protected from attacks, the intensity and frequency of subsequent aggressive behavior may escalate to high levels. This instigation results from exposure to provocative olfactory, visual, auditory, and tactile signals originating from the intruding opponent and has been attributed to increased aggressive arousal (Potegal, 1992).

The initial contact usually occurs via olfactory signals (pheromones) that provide information about the rat's sex, age, reproductive status, and recent nutritional history and other relevant events. Both the resident and the intruder visibly move their whiskers, indicating *sniffing,* and elongate their neck in a *stretched attend posture.* The resident and intruder explore each other with *nasonasal contact (Schnauzenkontrolle).* This special type of sniffing and *anogenital contact* or inspection serve to identify the olfactory signature of the opponent. Pheromones that convey information about the breeding status of the opponent guide subsequent interactions with the intruder. Intruders that are mature breeding males are attacked most readily, whereas attacks toward juvenile males or weanlings are less likely. The duration of actively exploring the intruder can serve as a quantitative index of social memory of previous encounters.

After olfactory identification, the resident male may *crawl under* the intruder, who in turn freezes. Occasionally, the intruder can be seen to crawl under the resident. This type of tactile stimulation may inhibit further escalation to fighting, although there is evidence for facilitative effects of tactile contact on initiating attack behavior. *Walking over* represents a second type of tactile signal, sometimes accompanied by the deposition of urine.

In the minutes leading up to intense attack bites, the resident rat emits brief pulses of ultrasonic vocalizations in the 50 kHz range that may reflect high excitement. During this preattack phase, tooth-chattering frequently occurs as well. Different types of ultrasonic vocalizations are emitted by the intruder, although these calls are more prominent in later phases of the aggressive encounters. The intruder's vocalizations are more prolonged and monotonous and have most of their energy in the 20 to 25 kHz range (Fig. 32–1).

Figure 32–2. Allogrooming (aggressive neck grooming) of the intruder rat by an aggressive resident.

INITIATION OF AGGRESSIVE BOUT

The attack bite is frequently preceded by *allogrooming*, especially of the neck region of the intruder; this is also referred to as *aggressive neck grooming* (Fig. 32–2). Slow motion video analysis reveals that the resident seizes folds of neck skin, while the intruder remains immobile in a *crouch* posture. Any sudden movement by the intruder triggers a bite by the resident, often accompanied by kicking movements of the rear legs. The probability that aggressive neck grooming leads to subsequent attack exceeds chance significantly, but it is not certain.

A potent trigger for the resident's attack is rapid locomotion by the intruder, and this locomotion may occur in the form of an *escape*. Even when the experimenter has arranged a large area for the resident rats to establish as a colony and for the intruder to explore, the captive nature of the environment is bound to alter the behavior of the intruder. Genuine flight will be less likely and more passive coping strategies will be evident in the form of *crouch* postures (Fig. 32–3). All four feet are planted on the substrate and the intruder crouches while completely immobile, occasionally moving the head and whiskers slightly.

The resident displays the *sideways threat* (Fig. 32–3), sometimes referred to as *lateral attack,* by arching its back, extending the rear legs, with clear signs of piloerection (*Imponiergehabe*). The orientation of the resident is mostly in a right angle or in parallel to the intruder, accompanied by kicking of the rear leg that is closest to the intruder. The resident may move toward and away from the intruder while maintaining the sideways threat posture, and these movements may represent behavioral ambivalence. When the resident engages in the sideways threat more persistently, it may actually encircle the intruder, which assumes either the crouch posture or the *defensive upright* posture. This defensive upright posture by the intruder is also displayed in reaction to the *aggressive upright* posture of the resident, in which the rat stands on its hind legs in a half-erect posture. Frequently, both the resident and the intruder face each other in this upright position with vertical movements of the forepaws (*mutual*

Figure 32–3. Resident rat (*right*) displaying the sideways threat posture toward an intruder (*left*) in the defensive upright posture.

upright posture) (Fig. 32–4). Piloerection and teeth-chattering are visible signs of intense sympathetic motor activity during the sideways and upright threat postures and may be interpreted to signify high arousal.

When the intruder moves away, it is followed or *pursued* by the resident who keeps on sniffing the intruder and engages in neck grooming, once the intruder arrests in the for-

Figure 32–4. Mutual upright posture.

ward motion. Rapid pursuits or chases are characteristically seen at the very beginning of a fight or its termination.

AGGRESSIVE BOUT

The *sine qua non* criterion element of an aggressive encounter is the *attack bite* (Fig. 32–5). As evidenced by analysis of wounding patterns, the bites are most often directed toward the neck and back region of the opponent and include quick closures of the jaws puncturing the skin. It is possible that multiple bites are delivered in rapid succession. In the case of intense biting and concurrent evasive action by the opponent, the shearing action of the bite can result in lacerations. This type of injury is possible due to the kicking movements by the resident's rear legs while rapid jaw closures occur.

In its most intense form, the attack bite is preceded by an *attack jump*. The resident leaps at the intruder who typically attempts to escape. During the attack jump, the resident's legs are completely off the ground, and the resident lands on the intruder's back, which in turn evades and rapidly assumes a supine posture.

In the sequence of aggressive acts, the bite is most often preceded by a sideways threat posture and followed by an *aggressive posture*, sometimes referred to as "pinning" or

Figure 32–5. Attack bite by a resident male rat and evasive action by the intruder rat.

"on top" (Fig. 32–6). When the resident assumes the aggressive posture, it bends over the intruder, which concurrently displays the submissive supine posture. The angle between the resident's aggressive posture and the intruder's supine posture may be either orthogonal or in parallel. These postures may be maintained for a few seconds or extend for minutes. By assuming the supine posture, the intruder prevents access to the neck and back, which are the primary targets for the resident's bites. During the display of the full aggressive posture, the resident rat continues to show piloerection, tooth-chattering, and stiff forelimbs, often attempting to reach the back of the neck as a target for bites. When successfully applying bites, the jaws are seen to be closed repeatedly, while the intruder is completely limp.

Aggressive bouts are typically composed of 5 to 10 acts and postures, which implies that the resident loops through several cycles of two to four behavioral elements, comprising pursuit, sideways threat, attack bite and aggressive postures. In order to be considered as part of an aggressive bout, these acts and postures follow each other within ca. 6 seconds. The duration of a bout is quite variable, but usually does not extend for more than 30 seconds.

Figure 32–6. A resident rat displaying the aggressive posture above an intruder in the submissive-supine posture.

TERMINATION OF AGGRESSIVE BOUT

Termination of the aggressive bout comes about by the intruder's escape, although this is less successful under captive laboratory conditions. More often, the resident stops displaying the aggressive posture and moves away, while the intruder maintains the submissive supine posture, sometimes for minutes after the resident rat has departed. The continued display of the supine posture by the intruder and the display of the crouch posture, with minimal movements, are the most frequent behavioral elements for terminating attacks and aggressive postures by the resident under laboratory conditions. Michael Chance (1962) also described an upright posture, termed "sensory cut-off," in which the intruder orients away from the resident rat, and this posture renders attacks by the resident less likely.

Whether the increasingly passive mode of behavior by the intruder is the primary determinant for the termination of an aggressive bout or the inhibition of aggressive behavior originates within the aggressive resident remains to be distinguished. It is evident that inactive intruder rats that emit 22 kHz ultrasonic vocalizations are less likely to be attacked by the resident than are active intruders. At the same time, it is evident that the duration and composition of aggressive bursts are predictable parameters that may be based on endogenous control mechanisms.

SEQUENTIAL STRUCTURE

The salient acts and postures displayed by a rat when confronting an opponent are organized in terms of time and sequence. The temporal and sequential structure of aggressive behavior becomes apparent when the probability of transition from one act to the next one is analyzed with such tools as the lag sequential analysis, cluster analysis, and log survivor analysis (van der Poel et al., 1989; Miczek et al., 2002). The probability of a pursuit leading to a sideways threat is more than twice as likely as chance, and the probability of a sideways threat being followed by an attack bite is even higher (Miczek et al., 1989, 1992) (Fig. 32–7).

This high-probability sequence is characteristic of aggressive behavior for a particular strain of laboratory rats, and it can serve as a template for comparison to detect potentially deviant and excessive types of aggressive behavior. Like the operational definitions of acts and postures, the temporal and sequential patterns of these behaviors are species-normative characteristics amenable to quantitative analysis.

LABORATORY RESIDENT–INTRUDER TEST

The display of offensive and defensive aggressive behaviors during a social conflict between a resident and an intruder rat can be evoked within the laboratory using the resident–intruder aggression test (i.e., Olivier, 1977; Olivier et al., 1994; Miczek, 1979; Koolhaas et al., 1980). In this test, male and female rats are typically housed in pairs with mates of the same strain for 21 days, usually in larger

cages with unrestricted supply of food and water and objects for marking. Males or females are selected for consistent attack behavior through the following procedure. After removal of the cagemate and pups, a stimulus animal of the same strain and sex, usually of lesser weight and without a history of fighting, is introduced for a confrontation between the resident and the intruder. The characteristics of the intruder animal are closely defined in terms of its age, size, and behavioral history to provide a standard stimulus. During the initial encounters, the resident–intruder confrontation is terminated when (1) the resident rat has delivered 10 attack bites, (2) the intruder displays the supine posture for 5 consecutive seconds and emits ultrasonic vocalizations, or (3) 5 minutes has elapsed. Typically, intruder rats display submission within 90 seconds after being attacked. Resident rats engage in distinctive species-typical agonistic behaviors as described earlier, including pursuit, threats, and attacks, whereas intruders show escape and defensive behavior. In addition, intruder rats emit ultrasonic distress vocalizations (Olivier, 1981; Thomas et al., 1983; van der Poel and Miczek, 1991). With increasing experience, the latency to the first attack by resident rats confronting an intruder becomes very short and is less informative as an index of readiness to display this

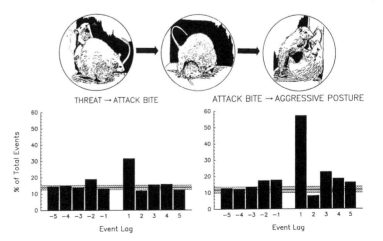

Figure 32–7. Lag sequential analysis of sideways threat, attack bite, and aggressive posture. The probability of each specific behavior following (lag +1, +2, +3, . . .) or preceding (lag −1, −2, −3, . . .) as the first, second, third, fourth, or fifth element to another specific behavioral element is shown by the vertical bars. The expected level of a random sequence is shown by the stippled horizontal bands. (Adapted from Miczek et al., 1992.)

behavior. During the encounter with the intruder, the full range of behavioral elements displayed by both animals is videotaped and subsequently analyzed in detail using computerized behavioral recording and analysis systems. By recording the frequencies, durations, latencies, and temporal and sequential patterns of all of the observed behavioral acts and postures, a detailed quantitative picture (ethogram and/or sequential structure) of agonistic behavior is obtained.

DEVELOPMENT OF PATHOLOGICAL OR DEVIANT FORMS OF RESIDENT–INTRUDER AGGRESSION

A considerable part of our current knowledge of the ethology, pharmacology, and neurobiology of normal and functional forms of human aggression is based on experimental resident–intruder aggression in rats and other animals. Despite this wealth of data and publications on aggression research in the rat, the social and neural determinants of pathological or deviant forms of human aggression (i.e., impulsive violence) remain poorly understood. One important reason for this gap in our knowledge is the lack of good and relevant animal models of pathological aggression. Ideally, such models should demonstrate excessive, injurious, and impulsive aggressive behavior that exceeds and/or deviates from normal species-typical levels or patterns (Miczek et al., 2002; de Boer and Koolhaas, 2003). To date, in virtually all laboratory inbred or outbred rat strains, the intensity and variation of aggressive behavioral traits have been dramatically compromised as a result of selection and inbreeding during the course of the domestication process (de Boer et al., 2003). Consequently, to obtain appreciable levels of aggression in these placid and docile laboratory strains, several procedural manipulations have been used such as prolonged social isolation, brief social provocations, application of aversive stimuli, electrical brain stimulation, administration of pharmacological agents, and, more recently, deletions of specific genes. Although these experimentally heightened forms of aggressive behavior may to some extent resemble more intense forms compared with their species-typical rates of aggression, they may still fall into the normative range compared with the patterns and levels of their wild ancestors. Indeed, higher levels and wider ranges of spontaneous intraspecific aggression are encountered in feral or seminatural populations of rats compared with their laboratory-bred conspecifics (de Boer et al., 2003). Therefore, an increase in the intensity of aggression is just one component of pathological behavior. Productive and relevant animal models of pathological forms of aggression should demonstrate intense and injurious aggression that exceeds normal species-typical levels and patterns—that is, forms of aggressive behavior that are no longer subject to inhibitory control and have lost their function in social communication.

Loss of the social communicative nature of the aggressive interaction may be expressed in (1) the disappearance of the normal investigatory and threatening sequence of acts and postures, (2) persistence in the attack-biting mode even though the intruder displays the submissive supine posture, (3) severe wounding and eventually death of the intruder if not stopped by the experimenter, and (4) loss of the ability to distinguish male from female intruders, resulting in attacking the latter and/or even anesthetized opponents. The development of rat aggression models with these behavioral abnormalities may expand knowledge about the neurobiology of pathological and violent forms of human aggression.

IMPLICATIONS FOR NEUROBIOLOGICAL RESEARCH

A prerequisite for productive neurobiological inquiries into the mechanisms mediating and modulating aggressive behavior is an adequate methodology for inducing, measuring, and analyzing the pattern of aggressive acts and postures, in their species-typical as well as pathological forms. Much of neurobiological aggression research is conducted in resident rats confronting an intruder, primarily because the neuroanatomy, neurochemistry, and neuropharmacology of rats continue to be elucidated with all molecular and cellular tools of neuroscience. Among the most promising lines of inquiry are those that focused on targeting subtypes of serotonergic, GABAergic, glutamatergic, dopaminergic, and several neuropeptidergic receptors and their genes as potential targets for pharmacotherapeutic interventions (Miczek et al., 2002). Rats have afforded the opportunity to measure aminergic activity during the initiation, execution, and termination of aggressive bouts, as well as in anticipation of such confrontations via in vivo microdialysis in real time (van Erp and Miczek, 2000; Ferrari et al., 2003). This methodology promises to be relevant to the fundamental issue of integrating serotonin deficiency that characterizes certain violence-prone individuals with the phasic changes in serotonin that are triggered by aggressive behavior itself. The enduring neuroadaptive changes that result from brief aggressive episodes may encompass rapid release of amines and peptides, lead to receptor upregulation and downregulation for several days, induce functional cellular activation as seen via immediate early gene expression, and engender neurogenesis (Miczek et al., 2004).

ACKNOWLEDGMENTS

The author would like to thank J. Thomas Sopko for his exceptional technical assistance. Preparation of this review and the original research from our own laboratory were supported by U.S. Public Health Service research grants AA13983 and DA02632 and grants from the Alcoholic Beverage Medical Research Foundation (K.A.M., P.I.).

REFERENCES

Barnett SA (1963) The rat. A study in behavior. Chicago: Aldine.

Barnett SA, Evans CS, Stoddart RC (1968) Influence of females on conflict among wild rats. Journal of Zoology 154:391–396.

Blanchard RJ, Blanchard DC, Flannelly KJ (1985) Social stress, mortality and aggression in colonies and burrowing habitats. Behavioural Processes 11:209–213.

Blanchard RJ, Blanchard DC, Takahashi T, Kelley MJ (1977) Attack and defensive behaviour in the albino rat. Animal Behaviour 25:622–634.

Boice R (1981) Behavioral comparability of wild and domesticated rats. Behavioral Genetics 11:545–553.

Calhoun JB (1948) Mortality and movement of brown rats (Rattus norvegicus) in artificially supersaturated populations. Journal of Wildlife Management 12:167–172.

Chance MRA (1962) An interpretation of some agonistic postures: The role of "cut-off" acts and postures. Symposium of the Zoological Society of London 8:71–89.

Covington HE III and Miczek KA (2001) Repeated social-defeat stress, cocaine or morphine: effects on behavioral sensitization and intravenous cocaine self-administration "binges." Psychopharmacology 158:388–398.

de Boer SF, van der Vegt BJ, Koolhaas JM (2003) Individual variation in aggression of feral rodent strains: A standard for the genetics of aggression and violence. Behavior Genetics 33:481–497.

Eibl-Eibesfeldt I (1950) Beiträge zur Biologie der Haus- und der Ährenmaus nebst einigen Beobachtungen an anderen Nagern. Zeitschrift für Tierpsychologie 7:558–587.

Ferrari PF, van Erp AMM, Tornatzky W, Miczek KA (2003) Accumbal dopamine and serotonin in anticipation of the next aggressive episode in rats. European Journal of Neuroscience 17:371–378.

Flannelly K and Lore R (1977) Observations of the subterranean activity of domesticated and wild rats (Rattus norvegicus): A descriptive study. Psychological Record 2:315–329.

Fleshner M, Laudenslager ML, Simons L, Maier SF (1989) Reduced serum antibodies associated with social defeat in rats. Physiology and Behavior 45:1183–1187.

Grant EC and Mackintosh JH (1963) A comparison of the social postures of some common laboratory rodents. Behaviour 21:246–295.

Koolhaas JM, Schuurman T, Wiepkema PR (1980) The organization of intraspecific agonistic behaviour in the rat. Progress in Neurobiology 15:247–268.

Luciano D and Lore R (1975) Aggression and social experience in domesticated rats. Journal of Comparative and Physiological Psychology 88:917–923.

Miczek KA (1979) A new test for aggression in rats without aversive stimulation: Differential effects of d-amphetamine and cocaine. Psychopharmacology 60:253–259.

Miczek KA, Covington HE III, Nikulina EM, Hammer RP Jr (2004) Aggression and defeat: persistent effects on cocaine self-administration and gene expression in peptidergic and aminergic mesocorticolimbic circuits. Neuroscience and Biobehavioral Review 27:787–802.

Miczek KA, Fish EW, De Bold JF, de Almeida RMM (2002) Social and neural determinants of aggressive behavior: Pharmacotherapeutic targets at serotonin, dopamine and γ-aminobutyric acid systems. Psychopharmacology 163:434–458.

Miczek KA, Haney M, Tidey J, Vatne T, Weerts E, De-Bold JF (1989) Temporal and sequential patterns of agonistic behavior: Effects of alcohol, anxiolytics and psychomotor stimulants. Psychopharmacology 97:149–151.

Miczek KA and Mutschler NH (1996) Activational effects of social stress on IV cocaine self-administration in rats. Psychopharmacology 128:256–264.

Miczek KA, Weerts EM, Tornatzky W, DeBold JF, Vatne TM (1992) Alcohol and "bursts" of aggressive behavior: Ethological analysis of individual differences in rats. Psychopharmacology 107:551–563.

Munn NL (1950) Handbook of psychological research on the rat. An introduction to animal psychology. Boston: Houghton Mifflin.

Olivier B (1977) The ventromedial hypothalamus and aggressive behaviour in rats. Aggressive Behavior 3:47–56.

Olivier B (1981) Selective anti-aggressive properties of DU 27725: Ethological analyses of intermale and territorial aggression in the male rat. Pharmacology, Biochemistry and Behavior 14–S1:61–77.

Olivier B, Molewijk E, van Oorschot R, van der Poel G,

Zethof T, van der Heyden J, Mos J (1994) New animal models of anxiety. European Neuropsychopharmacology 4:93–102.

Potegal M (1992) Time course of aggressive arousal in female hamsters and male rats. Behavioral and Neural Biology 58:120–124.

Stefanski V (2001) Social stress in laboratory rats: Behavior, immune function, and tumor metastasis. Physiology and Behavior 73:385–391.

Steiniger F (1950) Beitrag zur Soziologie und sonstigen Biologie der Wanderratte. Zeitschrift für Tierpsychologie 7:356–379.

Telle HJ (1966) Beitrag zur Kenntnis der Verhaltensweise von Ratten, vergleichend dargestellt bei Rattus norvegicus und Rattus rattus. Zeitschrift für angewandte Zoologie 53:129–196.

Thomas DA, Takahashi LK, Barfield RJ (1983) Analysis of ultrasonic vocalizations emitted by intruders during aggressive encounters among rats (Rattus norvegicus). Journal of Comparative Biology 97:201–206.

van der Poel AM and Miczek KA (1991) Long ultrasonic calls in male rats following mating, defeat and aversive stimulation: Frequency modulation and bout structure. Behaviour 119:127–142.

van der Poel AM, Noach EJK, Miczek KA (1989) Temporal patterning of ultrasonic distress calls in the adult rat: Effects of morphine and benzodiazepines. Psychopharmacology 97:147–148.

van Erp AMM and Miczek KA (2000) Aggressive behavior, increased accumbal dopamine and decreased cortical serotonin in rats. Journal of Neuroscience 20:9320–9325.

Zook JM and Adams DB (1975) Competitive fighting in the rat. Journal of Comparative and Physiological Psychology 88:418–423.

Defensive Burying

33

DALLAS TREIT AND JOHN J. P. PINEL

In 1970, one of the authors was developing an animal model of petit mal epilepsy (Pinel and Chorover, 1972). The rats that served as subjects in this study were individually housed in Plexiglas cages, which were aligned against the wall on a long table in the laboratory. As it turned out, this alignment and the archaic design of the cages played important roles in our first observation of the conditioned defensive burying paradigm. The cages were basically closed cubes of Plexiglas with a single, small hinged door on the front wall providing the only access.

Given the alignment of the cages against the wall, the consistency of the rats' initial behavior was visually striking: during the habituation period, every rat constructed a nest from the bedding material on the floor of its cage, and every rat located its nest at the back of the cage, as far as possible from the activities of the laboratory.

The induction of the petit mal state began a few days later with the first of a series of toxic intraperitoneal injections of chlorambucil. Each rat was removed with difficulty from its cage through the small door, injected with chlorambucil, and then returned to its cage. A few hours later, the rats seemed in reasonable health, but the topography of the bedding on the floor of the cages had changed in a way that was made blatantly obvious by the alignment of the cages. Every rat had moved the accumulation of bedding at the back of its cage to the front, where it had used it to bury the entrance. It seemed that every rat had tried to block access to its cage of the "evil injecting hand."

Defensive burying had been described in only one previous report of aversive conditioning in rats, but, unfortunately, no measures of the behavior were provided (Hudson, 1950). Accordingly, in the mid 1970s, motivated by our serendipitous observation of burying behavior, we commenced a program of research that focused first on the behavior itself and then on its applications to neuroscientific research.

DEVELOPMENT OF DEFENSIVE BURYING PARADIGMS

The paradigm that we developed for the study of the defensive burying response was semi-natural in two key respects (Pinel and Treit, 1978). First, the floor of the test environment was covered with a particulate material (usually bedding material), and second, the "threatening" stimulus was always an object—rather than an ethereal stimulus such as a light or a tone. In the wild, painful stimuli typically emanate from dangerous objects, and this spatial contiguity is likely a critical factor in making it easy for animals to learn to recognize sources of danger.

In our typical study of defensive burying, rats are confronted on the test day with an unfamiliar wire-wrapped dowel, often referred to as a shock prod, mounted on the wall of a familiar test chamber. When a rat contacts the *shock prod*, it receives a single brief shock and reflexively withdraws. After a period of immobility, the rat moves forward toward the prod, pushing and spraying the ground material with its snout and rapid, alternating movements of its forepaws. Figure 33–1 provides an illustration of defensive burying behavior.

In our first series of experiments (Pinel and Treit, 1978), we found that defensive burying behavior conditioned to a shock prod was well retained after only a single shock. Rats in the experimental group were shocked once by the shock prod and were immediately removed from the chamber for 10 seconds, 5 minutes, 5 hours, 3 days, or 20 days, at which time they were returned to the chamber with the shock prod still in place but disconnected from the shock source. At all intervals, the experimental rats buried the shock source significantly more often than did nonshocked control rats. Indeed, few shocked rats moved material in any direction other than toward the shock prod. When defensive behavior is directed at a previously neutral object that has been the source of aversive stimulation, it has been termed *conditioned defensive burying* (Pinel and Treit, 1978).

In another experiment in the same series, two identical shock prods were mounted on opposite walls of the test chamber, and the rats received a single shock from one of them. Virtually all of the defensive burying was directed at the shock source rather than the identical control object. When defensive burying is selectively directed at a source of aversive stimulation in the presence of a similar control object, it has been termed *discriminated defensive burying* (Pinel and Treit, 1983).

UNCONDITIONED DEFENSIVE BURYING

Although defensive burying is most commonly studied as a conditioned response, it also occurs as an unconditioned response. For example, in one experiment, one of four different sources of aversive stimulation was mounted on the wall of the test chamber: a shock source, a length of polyethylene tubing, a flashbulb, or a mousetrap. When each experimental subject touched one of these sources with a forepaw, it was to receive a shock, an air blast, a flash, or a physical blow, respectively. The results in the shock and air blast conditions were as expected: control rats engaged in little or no burying, whereas almost every experimental rat buried the prod or the air tube. However, in the mousetrap and flashbulb conditions, rats began burying the test objects before the aversive stimulus could be administered. Habituation to the

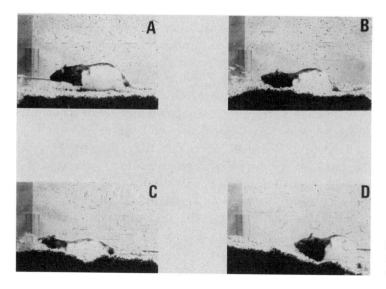

Figure 33–1. A rat burying a wall-mounted prod from which it has just received a single, brief electric shock.

mousetrap or the flashbulb eliminated this unconditioned burying (Pinel and Treit, 1983).

In another study of unconditioned burying, rats sprayed bedding material over the bodies of decaying dead conspecifics but not anethetized rats or fresh corpses (Pinel et al., 1981). The authors hypothesized that this unconditioned burying behavior was elicited by the odor of putrescine or cadaverine, two chemical stimuli associated with decaying tissue. Indeed, rats buried anesthetized conspecifics or wooden dowels that had been pretreated with putrescine or cadaverine, whereas rats rendered anosmic by intranasal injections of zinc sulfate did not.

The first few laboratory studies of defensive burying established two things: that rats enter the experimental environment with an established tendency to bury some objects but not others and that they readily learn to bury any object that has been the source of aversive stimulation.

CHARACTERISTICS AND GENERALITY OF DEFENSIVE BURYING

Subsequent studies of defensive burying established its generality (Pinel and Treit, 1983). For example, burying behavior has been observed using a variety of bedding materials, including sand, sawdust, wood shavings, ground corncob, and even wooden blocks (Pinel and Treit, 1979). The wooden block conditions were particularly informative because the blocks were large enough to preclude the burying of a wall-mounted shock prod with the typical burying response (i.e., forelimb spraying). Instead, the rats picked up blocks in their teeth and placed or threw them over the prod. In one wooden block condition, the wooden blocks were all placed in a pile at the opposite end of the chamber from the shock prod. The rats in this condition first carried, threw, or pushed the blocks to the vicinity of the shock prod before starting to bury it. Clearly, burying behavior is not re-

stricted to the stereotypical response seen with homogeneous particulate materials such as wood shavings or ground corncob.

Conditioned defensive burying also occurs in a variety of test environments. The duration of burying in rats decreases with increases in the size of the test chamber; however, it still occurs in very large chambers, where rats are not required to stay in the vicinity of the shock prod.

Conditioned defensive burying occurs reliably even in two-compartment boxes, which provide rats with the ability to escape the compartment that contains the shock prod (Pinel et al., 1980). The two-compartment apparatus is particularly useful for observing or manipulating the active (e.g., burying) and passive (e.g., spatial avoidance) components of the rats' defensive responses toward the shock prod, by opening or closing a partition separating the two sides of the chamber (Treit et al., 1986).

BURYING BEHAVIOR: ORGANISMIC VARIABLES

SPECIES

Burying behavior has been observed in a variety of different rodent species; however, there has been little systematic research in species other than rats. In general, rats tend to bury a shock prod for longer periods of time than do either mice or ground squirrels, and burying is only rarely observed in gerbils and hamsters. In one study, conditioned defensive burying of a shock prod was directly compared in Richardson's ground squirrels, thirteen-lined ground squirrels, and Long-Evans rats. Defensive burying was observed in all three rodent species, but the topography of the response was different and the duration was less in the two species of ground squirrels than in the rats (Heynen et al., 1989). Comparisons of conditioned defensive burying in various rodent species are complicated by the fact that they tend to be based on paradigms initially developed to produce robust burying in rats.

STRAINS

Several studies have compared conditioned defensive burying in different strains—in both rats and mice. In one study of burying in different strains of rats (Treit et al., 1980), Fisher rats were found to bury a shock prod more than Wistar rats, which buried more often than Long-Evans rats. In another study (Pare et al., 1992), Fisher and Wistar rats were found to bury more than were Wistar-Kyoto rats.

In one comparison of burying in three strains of mice, CF-1 mice were found to bury more often than CD-1 and BALB/c mice (Treit et al., 1980).

SEX AND AGE

In rats, burying has been observed in both sexes and a wide range of ages. There do not appear to be differences between male rats and nulliparous female rats in the degree to which they bury shock sources. In contrast, age has a substantial effect on burying. Treit et al. (1980) compared male rats that were 30, 60, or 90 days old and found the 60-day old rats to engage in significantly more defensive burying.

BURYING AS A DEFENSIVE RESPONSE IN THE WILD

Defensive burying has not been frequently studied in the wild, and burying has been studied in only a handful of the hundreds of known wild species of rodents. Indeed, the only systematic ethological studies of defensive burying were conducted in ground squirrels by Owings and Coss (1977). They found that ground squirrels drive off predatory snakes by using the defensive burying response to spray sand at them. Owings and

Coss also reported that ground squirrels use the burying response to construct walls in their burrows to block the movement of snakes. Similarly, Calhoun (1962) noted that lower-status wild Norway rats exposed to conspecific threat plugged the entrance holes to their underground nests, and Johnston (1975) observed a male golden hamster blocking the entrance to its chamber with wood shavings after it had been defeated by a higher-ranking male in a seminatural environment.

CONDUCTING CONDITIONED DEFENSIVE BURYING EXPERIMENTS

SUBJECTS

Although conditioned defensive burying can be readily observed in most rats and many other rodents, young adults tend to display the most burying (Treit et al., 1980). Females with litters also display particularly high levels (Pinel et al., 1990).

HOUSING

It is important that the subjects be reared and housed on bedding material. Pinel et al. (1989) reared rats with no opportunity to interact with particulate matter. When exposed to the conditioned defensive burying paradigm as adults, they attempted to bury the shock source but their burying behavior was sporadic, uncoordinated, and poorly directed.

HANDLING AND HABITUATION

In most aversive conditioning paradigms, it is important that the test environment does not induce confounding defensive responses. Thus, in studies of conditioned defensive burying, the subjects are typically handled for 3 days and then habituated to the test box (without the shock source) for each of the next 4 days (Treit and Fundytus, 1988).

TEST CHAMBER

Any test chamber will suffice, but burying is most robust in small chambers, where subjects are forced to stay in the vicinity of the shock source. The typical test chamber is a $40 \times 30 \times 40$ cm Plexiglas chamber with a 5 cm layer of bedding material, but smaller chambers result in even more burying (Pinel et al., 1980).

SHOCK SOURCE

Any aversive stimulus and source can be used in conditioned defensive burying experiments, but electric shock delivered from a shock prod (two wires wrapped around a wooden dowel) has been used most commonly. The dowel is typically mounted on the wall, 2 cm above the level of the bedding (see Fig. 33–1). More complex stimuli may elicit unconditioned burying.

SHOCK PARAMETERS

All shocks in conventional conditioned defensive burying experiments are very brief (about 0.1 second), as determined by the latency of the withdrawal reflex. Because there is considerable variability in the contacts made by various animals, constant current shockers should be used, and both current flow and the animals' reactions should be monitored. Criteria should be established for excluding animals that do not experience a reasonable shock. Current intensities should be selected with caution. The duration of burying by rats was shown to increase monotonically from 0.5 to 10 mA (Treit and Pinel, 1983). However, in some studies, the objective is not merely to produce a lot of burying behavior but rather to assess the effects of various treatments on burying behavior. In such cases, extremely robust burying can prove to be insensitive (see the later discussion of anxiolytics). In most conditioned defensive burying experiments, only one shock is delivered, but

in others, the shock source remains activated during the entire test period.

BEHAVIORAL MEASURES

All experimental sessions should be videotaped, to facilitate the detailed assessment of the rat's behavior. Specific measures have included the following:

- Duration of burying (i.e., total time each rat sprays bedding material toward test object)
- Frequency of burying bouts
- Latency to burying
- Number of cautious approach sequences
- Number of contacts with the test object
- Behavioral reaction to the aversive stimulus (e.g., reaction to shock has been measured on 4-point scale [Degroot and Treit, 2003])
- Duration of freezing behavior
- Height of bedding material over the prod at the end of the test session

DURATION OF THE TEST

Longer tests are associated with more burying; however, in most studies, the tests last 10 or 15 minutes.

MAINTAINING THE APPARATUS

After each animal has been removed from the chamber, the bedding material should be cleaned of feces and smoothed to a uniform height, and the shock prod should be cleaned of any moisture, dust, or debris. The entire shock circuit should also be periodically tested with a multimeter to ensure no change in conductivity has occurred.

USE OF DEFENSIVE BURYING PARADIGMS IN NEUROSCIENCE RESEARCH

Defensive burying paradigms have been used in neuroscience for a variety of purposes; however, they have most commonly been used to screen anxiolytic drugs and to study the roles of the septum, amygdala, and hippocampus in fear and anxiety. These four lines of research are discussed in the following sections.

SCREENING ANXIOLYTIC DRUGS

Several studies have shown that anxiolytic drugs (e.g., diazepam) produce a dose-dependent suppression of prod burying, with a relative potency that is consistent with their clinical effectiveness in the treatment of human anxiety. For reviews, see Treit and Menard (1998) and Treit et al. (2003).

The drug class specificity of the test is sensitive to procedural variations (Treit and Menard, 1998). For example, anxiolytics reduce shock prod burying at intermediate, but not at high, shock intensities. Suppression of defensive burying by benzodiazepine-type anxiolytics is not secondary to a general motor impairment, associative learning deficits, or analgesia; it can be blocked by benzodiazepine receptor antagonists such as flumazenil (Treit et al., 2003). Conversely, drugs that increase anxiety in humans (i.e., anxiogenic agents such as yohimbine) increase the amount of time that rats bury a shock source. These effects of anxiolytic and anxiogenic agents on defensive burying support the view that burying shock sources is a "fear" reaction in rats (Treit and Menard, 1998).

One of the strengths of shock prod burying as a screening test for anxiolytics is that several different "fear" responses can be measured within the same setting. This has proved to be particularly valuable for the detection of serotonergic-type anxiolytics, such as buspirone, which have often been difficult to detect in other screening tests (Treit et al., 2003). In one study, for example, Treit and Fundytus (1988) compared the effects of chlordiazepoxide and buspirone in a modified burying test in which the shock source remained continuously electrified. Both buspirone and chlordiazepoxide decreased the amount of

time rats buried the shock prod and concomitantly increased the number of contact-induced shocks that rats received from the prod. These bidirectional effects on prod burying and on the number of prod shocks provide convergent evidence of anxiolytic treatment effects.

THE NEURAL MECHANISMS OF FEAR AND ANXIETY

The Septum

Numerous studies have shown that ablation or pharmacological inhibition of the septum produces anxiolytic effects in the shock-prod burying test (for reviews, see Menard and Treit, 1999; Treit and Menard, 2000). Briefly, electrolytic or excitotoxic lesions of the septum decreased burying of a constantly electrified prod, without concomitant effects on general activity, reactivity to handling, reactivity to shock, or shock source avoidance. A similar pattern of effects was produced when septal activity was inhibited using intraseptal microinfusions of the benzodiazepine-type anxiolytic midazolam (Menard and Treit, 1999), the direct acting $GABA_A$ agonist muscimol (Degroot and Treit, 2003), the 5-hydroxytryptamine$_{1a}$ (serotonin$_{1a}$ agonist $(R)(+)$-8-hydroxy-2-(di-n-propylamino) tetralin, or by both N-methyl-D-aspartate (NMDA) D($-$)-2-amino-5-phosphonopentanoic acid (AP-5) and non-NMDA receptor antagonists (6-cyano-7-nitroquinoxaline; Menard and Treit, 1999, 2000).

In most of these studies, the anxiolytic effects of septal suppression were replicated in the elevated plus-maze. In the elevated plus-maze, untrained rats avoid the open arms of the elevated maze and remain in the enclosed arms (Pellow et al., 1985). The combined use of the plus-maze and the defensive burying test for studying the neural mechanisms of fear and anxiety is important for three reasons. First, the fear-inducing stimuli in the two tests are distinctly different (i.e., painful electric shock versus open elevated spaces). Second, fear reduction is primarily indicated by an *in-*

crease in a specific activity in the plus-maze (i.e., open-arm exploration) and by a *decrease* in a specific activity in the defensive burying test (i.e., shock prod burying). Thus, reductions in "anxiety" seen in *both* tests are difficult to explain in terms of nonspecific effects on general activity, arousal, pain sensitivity, or behavioral inhibition. Third, neither test involves an explicit memory requirement, a factor that can complicate the interpretation of drug or lesion effects in other paradigms (Treit, 1985).

The Amygdala

The amygdala has long been implicated in fear and anxiety (e.g., Davis, 1992; LeDoux, 1996). What is the relative contribution of the amygdala in anxiety compared with other limbic structures, such as the septum? To address this question, Treit and colleagues compared the effects of amygdala lesions with those of septal lesions in the defensive burying and plus-maze tests (for a review, see Treit and Menard, 2000). As in previous studies, septal lesions decreased shock prod burying and increased open-arm exploration, without producing effects on general activity, reactivity to handling, reactivity to shock, or shock prod avoidance. Interestingly, however, amygdalar lesions had no effects on burying behavior or plus-maze behavior but dramatically increased shock prod contacts. This selective effect of amygdalar lesions was found in all experiments, across a variety of different lesion parameters, and in the absence of any effects on general activity or shock reactivity. In addition, this selective effect of amygdalar lesions did not appear to reflect a general deficit in response inhibition or passive avoidance because the lesioned rats avoided the open arms of the plus-maze to the same extent as sham-lesion controls (Treit and Menard, 2000) (Table 33–1).

Table 33–1. Summary of Drug and Lesion Effects (see Text)

Manipulation	Site	Plus-Maze Open Arm Exploration	Shock Prod Contacts	Shock Prod Burying
Injection of anxiolytic drugs (e.g., midazolam)	Systemic (e.g., intraperitoneally)	Increased	Increased at high doses	Decreased
Injection of anxiogenic drugs (e.g., yohimbine)	Systemic (e.g., intraperitoneally)	Decreased	No effect	Increased
Lesion	Septum	Increased	No effect	Decreased
Microinfusion of midazolam		Increased	No effect	Decreased
Lesion	Amygdala	No effect	Increased	No effect
Microinfusion of midazolam		No effect	Increased	No effect
Microinfusion of physostigmine (20 μg)	Hippocampus	Not tested	No effect	Decreased
Microinfusion of muscimol (10 ng)	Septum	Not tested	No effect	Decreased
Microinfusion of combined, subeffective doses of muscimol (2.5 ng) in septum and physostigmine (5 μg) in hippocampus	Septum and hippocampus	Not tested	No effect	Decreased

A possible interpretation of the effects of amygdala lesions on defensive burying is that the rats could not learn or remember the association between the electric shock and the shock source. However, this interpretation is inconsistent with the prod-burying behavior of amygdala-lesion rats, which was well directed and indistinguishable from sham-lesion controls, a finding that has been replicated in other laboratories (Treit and Menard, 2000). In addition, neurotoxic or reversible tetrodotoxin lesions of the amygdala did not impair the ability of the rats to subsequently avoid the shock prod after a delay of several days was imposed between the initial shock and a retention test (Lehmann et al., 2000, 2003).

Taken together, the effects of septal and amygdalar lesions suggest that the amygdala and septum independently control the expression of different fear reactions. Subsequent studies have reinforced this general conclusion. Microinfusions of midazolam into the septum increased open arm exploration in the plus maze and decreased defensive burying in the shock prod test, whereas amygdalar infusions produced neither of these effects. Amygdalar infusions did, however, dramatically impair the shock prod avoidance of rats, an anxiolytic effect not found after septal infusions. Coinfusions of the benzodiazepine receptor antagonist flumazenil blocked each of these specific anxiolytic effects without producing any intrinsic activity by itself. These results suggest that benzodiazepine receptor systems within the amygdala and septum differentially mediate specific fear reactions (Treit and Menard, 2000) (Table 33–1).

The Hippocampus

Anatomically, the septum is extensively connected to the hippocampus (e.g., Risold and Swanson, 1997); together they form a substantial part of the limbic system. Functionally, according to Gray's (1982) theory, the septum and hippocampus act in concert to control fear and anxiety, as evidenced in part by the correspondence between the effects of septal or hippocampal lesions in traditional aversive learning paradigms and the effects of anxiolytic drugs in these same paradigms (Gray, 1982).

There is evidence that hippocampal cholinergic systems may be particularly important in the modulation of anxiety. For example, increases in the fear reactions of rats have been observed in a variety of tests after intrahippocampal infusions of cholinergic antagonists (e.g., File et al., 1998). One expectation, based on these antagonist studies, is that producing upregulation of cholinergic systems, for example, with the acetylcholinesterase inhibitor physostigmine, might reduce anxiety. Furthermore, given the connections between the septum and hippocampus, it seemed that septal GABAergic systems might interact with hippocampal cholinergic systems in the control of anxiety.

To test these hypotheses, Degroot and Treit (2003) examined the independent and combined effects of stimulating septal GABAergic systems and hippocampal cholingeric systems using the defensive burying test. They found the following: (1) that a 10 ng infusion of muscimol into the septum produced a significant suppression of shock prod burying, whereas lower doses (2.5 and 5.0 ng) did not, (2) that burying was significantly reduced after a 20 μg infusion of physostigmine into the hippocampus but not after a lower dose (5 and 10 μg), and (3) that the combination of subthreshold doses of physostigmine (5 μg) and muscimol (2.5 ng) significantly reduced burying. These results generally support Gray's theory and suggest that hippocampal cholinergic and septal GABAergic systems can act synergistically in the modulation of fear reactions.

CONCLUSION

Various forms of the defensive burying paradigm have proved useful in neuroscientific research. Because the defensive burying response is reliable, highly directed, and specific to a variety of aversive situations, large differences in burying between experimental and control subjects are typical. In addition, defensive burying requires no pretraining, it can be studied as an unconditioned response, it can be conditioned in a single trial, and the conditioned form is well retained. Finally, defensive burying is not restricted to a particular type of aversive stimulus, test environment, burying material, or species, strain, sex, or age of rodent. Most important, it changes in predictable ways to anxiolytic and anxiogenic drugs and brain lesions.

REFERENCES

Calhoun J (1962) The ecology and sociology of the Norway rat. Bethesda: U.S. Department of Health, Education and Welfare.

Davis M (1992) The role of the amygdala in fear and anxiety. Annual Review of Neuroscience 15:353–375.

Degroot A and Treit D (2003) Septal GABAergic and hippocampal cholinergic systems interact in the modulation of anxiety. Neuroscience 117:493–501.

File SE, Gonzalez LE, Andrews N (1998) Endogenous acetylcholine in the dorsal hippocampus reduces anxiety through actions on nicotinic and muscarinic receptors. Behavioral Neuroscience 112: 352–359.

Gray JA (1982) The neuropsychology of anxiety: An enquiry into the function of the septo-hippocampal system. Oxford: Oxford University Press.

Heynen AJ, Sainsbury RS, Montoya CP (1989) Cross-species responses in the defensive burying paradigm: A comparison between Long-Evans rats (Rattus norvegicus), Richardson's ground squirrels (Spermophilus richardsonii), and Thirteen-Lined ground squirrels (Catellus tridecemlineatus). Journal of Comparative Psychology 103:184–190.

Hudson BB (1950) One-trial learning in the domestic rat. Genetic Psychology Monographs 41:99–145.

Johnston RE (1975) Scent marking by male golden hamsters (Mesocricetus auratus), III: Behavior in a seminatural environment. Z Tierpsychol 37: 213–221.

LeDoux J (1996) Emotional networks and motor control: A fearful view. In: Progress in brain research (Holstege G, Bandler R, Saper CB, eds.), pp. 437–446. Amsterdam: Elsevier Press.

Lehmann H, Treit D, Parent MB (2000) Amygdala lesions do not impair shock-probe avoidance retention performance. Behavioral Neuroscience 114: 107–116.

Lehmann H, Treit D, Parent MB (2003) Spared anterograde memory for shock-probe fear conditioning after inactivation of the amygdala. Learning and Memory 10:261–269.

Menard J and Treit D (1999) Effects of centrally administered anxiolytic compounds in animal models of anxiety. Neuroscience and Biobehavioral Reviews 23:591–613.

Menard J and Treit D (2000) Intra-septal infusions of excitatory amino acid receptor antagonists have different effects in two animal models of anxiety. Behavioural Pharmacology 11:99–108.

Owings DH and Coss RG (1977) Snake mobbing by California ground squirrels: Adaptive variation and ontogeny. Behavior 62:50–69.

Pare WP (1992) The performance of WKY rats on three tests of emotional behavior. Physiology and Behavior 51:1051–1056.

Pellow S, Chopin P, File SE, Briley M (1985) Validation of open: Closed arm entries in an elevated plus-maze as a measure of anxiety in the rat. Journal of Neuroscience Methods 14:149–167.

Pinel JPJ and Chorover SL (1972) Inhibition of arousal of epilepsy induced by chlorambucil in rats. Nature 236:232–234.

Pinel JPJ and Treit D (1978) Burying as a defensive response in rats. Journal of Comparative and Physiological Psychology 92:708–712.

Pinel JPJ and Treit D (1979) Conditioned defensive burying in rats: Availability of burying materials. Animal Learning and Behavior 7:392–396.

Pinel JPJ and Treit D (1983) The conditioned defensive burying paradigm and behavioral neuroscience. In: Behavioral approaches to brain research (Robinson T, ed.), pp. 212–234. New York: Oxford University Press.

Pinel JPJ, Gorzalka BB, Ladak F (1981) Cadaverine and putrescine initiate the burial of dead conspecifics by rats. Physiology and Behavior 27:819–824.

Pinel JPJ, Petrovic DM, Jones CH (1990) Defensive burying, nest relocation, and pup transport in lactating female rats. The Quarterly Journal of Experimental Psychology 42B:401–411.

Pinel JPJ, Symons LA, Christensen BK, Tees RC (1989) Development of defensive burying in Rattus norvegicus: Experience and defensive responses. Journal of Comparative Psychology 103:359–365.

Pinel JPJ, Treit D, Ladak F, Maclennan AJ (1980) Conditioned defensive burying in rats free to escape. Animal Learning and Behavior 8:477–451.

Resold PY and Swanson LW (1997) Connections of the

rat lateral septal complex. Brain Research Reviews 24:115–195.

Treit D (1985) Animal models for the study of anti-anxiety agents: A review. Neuroscience and Biobehavioral Reviews 9:203–222.

Treit D and Fundytus M (1988) A comparison of buspirone and chlordiazepoxide in the shock-probe/burying test for anxiolytics. Pharmacology Biochemistry and Behavior 30:1071–1075.

Treit D and Menard J (1998). Animal models of anxiety and depression. In: Neuromethods. Vol 32, In vivo neuromethods (Boulton A, Baker G, Bateson A, eds.), pp. 89–148. Totowa, NJ: Humana Press.

Treit D and Menard J (2000) The septum and anxiety. In: The behavioral neuroscience of the septal region (Numan R, ed.), pp. 210–223. New York: Springer-Verlag Inc.

Treit D, Degroot A, Shah A (2003) Animal models of anxiety and anxiolytic drug action. In: Handbook of depression and anxiety, 2nd edition (Kasper S, den Boer JA, Sitsen JMA, eds.), pp. 681–702. New York: Marcel Dekker.

Treit D, Lolordo VM, Armstrong DE (1986) The effects of diazepam on "fear" reactions in rats are modulated by environmental constraints on the rat's defensive repertoire. Pharmacology Biochemistry and Behavior 25:561–565.

Treit D, Terlecki LJ, Pinel JPJ (1980) Conditioned defensive burying in rodents: Organismic variables. Bulletin of the Psychonomic Society 16:451–454.

Social Learning

BENNETT G. GALEF, JR.

34

Systematic observation of free-living mammals and birds often reveals differences in the behavior of species members that live in different areas. Such geographic variation in the behavior of chimpanzees and orangutans is particularly well documented (Whiten et al., 1999; van Schaik et al., 2003) and is widely known because of the attention it has received in the popular press. However, before the recent dramatic increase in field studies of the great apes, it was not unreasonable to propose, as did Steiniger (1950, p. 369), that "the [Norway] rat appears especially able to develop local traditions, more so perhaps than other more-closely examined mammals, possibly including the anthropoids."

NORWAY RATS

Norway rats are arguably the most successful, and surely the most widely distributed, nonhuman mammals on Earth. Breeding populations have been reported from Nome, Alaska, at 60 degrees North latitude, where rats feed on human garbage, to South Georgia Island, at 55 degrees South latitude, where tussock grass, beetles, and ground-nesting birds provide sustenance for colonies of Norway rats.

As the preceding two examples suggest, much of the success that rats enjoy results from the extraordinary range of foods that they are able to exploit, and as in the great apes, much of the known variation in behavior in free-living Norway rats involves foraging behavior. Rats in West Virginia catch and

eat fingerling fish in trout hatcheries, whereas those living on Norderoog island in the North Sea stalk and kill ducks and sparrows. Yet other *R. norvegicus* living along the banks of the Po River in Italy dive for and feed on mollusks living on the bottom of the river, while their fellow rats in Japan scavenge dead fish that wash up on the seashore. Such naturally occurring variability in feeding behavior has been the focus of most experimental studies of social learning in the species.

PREVIEW

I begin the present brief review of the literature on social influences on food choices of Norway rats with a description of fieldwork strongly suggesting that interactions between adult free-living rats and their young can determine which foods the young come to eat. I then describe very briefly several behavioral processes that have been shown in the laboratory to be sufficient to influence food choice in young rats. Last, I describe in somewhat greater detail a type of social influence on rats' food preferences that has already proved to be useful in studies of the physical substrates of learning and memory.

FIELD OBSERVATIONS OF NORWAY RATS

Fritz Steiniger, an applied ecologist whose professional interest lay in enhancing the efficiency with which rodent pests could be exterminated, was the first to report difficulties in controlling pest populations of Norway rats using the economically desirable method of placing permanent stations containing poisoned bait in rat-infested areas (Steiniger, 1950). Steiniger found that although rats ate ample amounts of poison bait and died in large numbers when a permanent bait station was first introduced into their colony's territory, later acceptance of the bait by colony members was very poor, and colonies targeted for extermination soon returned to their initial sizes.

Steiniger reported that permanent bait stations failed because young rats, born to colony members that had survived their initial contact with the poisoned bait and had learned to avoid eating it, refused to even taste the bait that the adults of their colony were avoiding.

A LABORATORY ANALOGUE

Avoidance by young wild rats of a food that adults of their colony have learned to avoid eating is a robust phenomenon that is easily observed in rats transferred from their natural habitats to laboratory enclosures. We captured adult wild rats (R. norvegicus) on garbage dumps in southern Ontario, transferred them to our laboratory, and placed them in groups of five or six in 2 m² enclosures that each contained nesting boxes and nesting materials and provided ad libitum access to water. For 3 hours each day, we offered each colony two foods that differed in taste, smell, texture, and color (Galef and Clark, 1971b).

To begin a typical experiment, we introduced a sublethal concentration of toxin into one of the two foods that we gave our captives to eat daily. The rats soon learned to avoid eating the poisoned food, and for weeks thereafter, they avoided eating the food that had been noxious, even when we gave them uncontaminated samples of it (Garcia et al., 1966).

After we had trained our colonies to avoid one of the two foods that we placed in their enclosure each day, we waited for female colony members to give birth and for their young to grow to weaning age. As the young approached independence, we started to observe their colony on closed-circuit television throughout daily feeding periods. When the young started to eat solid food, we recorded the frequency with which they ate each of the two foods in their cage: one that adult colony members were eating and the other that the adults had learned to avoid.

We found, without exception, that weaning rats ate only the food that the adults of their colony were eating and totally avoided the alternative food that the adults had learned to avoid. Even after we removed pups from their natal enclosures, housed them individually, and offered them the same two foods that had been available when they were in their colony cages, pups continued to eat only the food that the adults of their colony had eaten (Galef and Clark, 1971b) (Fig. 34–1).

ANALYSIS OF THE PHENOMENON

My students and I have spent much of the past 30 years determining how the food choices of adult rats might influence those of the young they rear (see reviews in Galef, 1977, 1988, 1996a, 1996b). Over those years, those working in my laboratory and in other laboratories as well have discovered many different ways in which the food choices of young rats are affected by social interactions with conspecific adults.

Prenatal Effects
Fetal rats exposed to a flavor while still in their mother's womb (through injection of a flavored solution into the dam's amniotic fluid) will, when grown, drink more of a solution containing that flavor than will control rats that lack prenatal exposure to it (Smotherman, 1982). Even feeding a food with a strong odor to a female rat while she gestates a litter

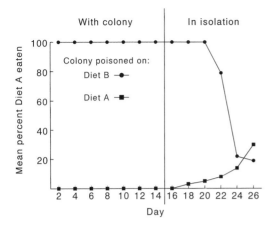

Figure 34–1. Rat pups born into colonies trained to avoid eating either diet A or diet B are offered a choice for 3 hours per day between diet A and diet B. Abscissa shows the days since pups started to eat solid food; ordinate, relative frequency with which pups from the two types of colony ate diet *A*, Pup diet choice while still in their natal colonies (*left*) and the amount of diet A eaten, as a percentage of total amount eaten, by pups after transfer to individual cages and offer of diet A and diet B for 9 hours per day (*right*). (Data from Galef and Clark [1971].)

suffices to enhance her postnatal preferences of her young for the odor of that food (Hepper, 1988).

Effects during Suckling

Flavors of foods that a rat dam eats while lactating affect the flavor of her milk, and exposure to milk flavored by the foods that a lactating dam eats while rearing her young affects the food preferences of her pups at weaning (e.g., Galef and Sherry, 1973).

Effects during Weaning

Galef and Clark (1971a) used time-lapse video recordings to observe each of nine wild rat pups that had ad libitum access to solid food take its first meal. All nine pups ate solid food for the first time under the same circumstances. Each ate at the same time that an adult member of its colony was feeding, which was highly unlikely given the temporal distribution of adult meals, and each ate at the same place an adult was feeding, not at an alternative feeding site a short distance away.

Even an anesthetized adult rat placed near one of two otherwise identical feeding sites made that site far more attractive to pups than one without an adult present (Galef, 1981).

By comparing the circumstances in which intact and visually deprived rats weaned, we found that intact pups use visual cues to approach adults from a distance when selecting a place to eat solid food for the first time.

Effects of Snatching Food from Adults

Young rats, like the young of many other mammalian species, seem to be especially interested in the particular piece of food that someone else is eating. Juvenile rats will walk across a cage floor carpeted with food pellets and steal an identical pellet from the mouth or paws of an adult or a peer that is eating it. Young rats that have stolen a pellet of unfamiliar food from the mouth of a conspecific subsequently show a greater preference for that food than do young rats that have eaten an identical food pellet taken from the floor of their cage (Galef et al., 2001).

Effects of Scent Marks and Scent Trails

While feeding, adult rats deposit olfactory cues both on and around a food they are eating (Galef and Beck, 1985). Such residual odors attract pups and, like the physical presence of an adult rat at a feeding site, cause young rats to prefer marked sites. Further, when an adult has finished eating and travels back to its burrow, it deposits a scent trail that directs young rats seeking food to the location at which the adult ate (Galef and Buckley, 1996).

IMPLICATIONS OF REDUNDANCY

Redundancy in the behavioral processes that support social influences on food choice in rats is in itself important. Such redundancy suggests that for rats, as for the honeybees studied by Karl von Frisch (1967), socially acquired information substantially increases foraging efficiency. Indeed, it is easy to demonstrate that, for naïve rats residing in an environment where foods containing needed nutrients are difficult to identify, the presence of conspecifics that have already learned to select an appropriate diet can make the difference between life and death. Young rats that would have died because of an inability to learn independently to focus their intake on the sole protein-rich food available among a cafeteria of foods available to them learned rapidly to eat that food when caged with adult conspecifics trained to do so (Beck and Galef, 1989).

IS THERE ANYTHING SPECIAL ABOUT SOCIAL LEARNING?

Our analyses have indicated that in most instances of social influence on the food choices of young rats, interaction between adult and young rats has resulted in introduction of the young to one food rather than another. Adults bias young either to initiate feeding on foods that the adults are eating, rather than on alternative foods, or to start to feed at feeding sites that the adults are visiting, rather than at alternative sites. Differences in the responses of young rats to familiar and unfamiliar foods and locations are then responsible for much of the influence of adults on the choices of juveniles with which they interact (Galef, 1971b).

Such effects of socially induced familiarity on food choice are particularly pronounced in genetically wild Norway rats that are extremely hesitant to eat unfamiliar foods (Barnett, 1958). The extreme neophobia of wild rats makes introduction of juveniles to one food rather than another a critical event in the development of their feeding repertoires (Galef and Clark, 1971b).

However, not all of the social influences on the food choices of rats reflect a simple social biasing of naïve young rats to eat one food rather than another together with neophobia. In the case discussed in the next section, so-

cially induced food preference seems to result from a behavioral process that directly alters the affective response of young rats to foods experienced in a social context (Galef et al., 1997).

FLAVOR CUES ON THE BREATH OF RATS

In the early 1980s, scientists in several laboratories demonstrated that after a naïve "observer" rat interacts with a recently fed conspecific "demonstrator," the observer exhibits a substantial enhancement of its preference for whatever food its demonstrator ate (Galef and Wigmore, 1983; Strupp and Levitsky, 1984). For example, after naïve observer rats interacted briefly with conspecific demonstrators fed either a cinnamon- or a cocoa-flavored diet, the former group of observers preferred cinnamon-flavored food, whereas the latter preferred cocoa-flavored food, if offered a choice between the two (Fig. 34–2).

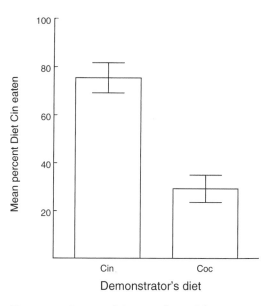

Figure 34–2. Amount of cinnamon-flavored diet (Diet Cin) eaten, as a percentage of total intake over 22 hours, by observer rats that interacted with demonstrators fed either Diet Cin or cocoa-flavored diet (Diet Coc). Error bars show 1 SEM. (Data from experiments like those described in Galef and Wigmore [1983].)

The effects of a single brief exposure to recently fed demonstrator rats on the food choices of their observers are both surprisingly powerful and surprisingly long lasting. Many observer rats taught to totally avoid ingesting a food by following its ingestion with an injection of toxin, and then placed with demonstrator rats that have eaten the food that their observers had learned to avoid, totally abandoned their aversions. Similarly, most observer rats that interacted with a demonstrator fed a diet adulterated with cayenne pepper (an inherently unpalatable taste to rats) subsequently preferred peppered to unadulterated diet (Galef, 1986b). Such effects of demonstrator rats on the food choices of their observers can be seen a month or more after a demonstrator and observer interact (Galef and Whiskin, 2003).

ANALYSIS

The behavioral process that produces such social influence on the food choices of observer rats is now quite well understood. Olfactory cues passing to observer rats from demonstrators cause observers to increase their preferences for the foods that their respective demonstrators ate (Galef and Wigmore, 1983). Observers sniff at the mouths of demonstrators, and this sampling of a demonstrator's breath is both necessary and sufficient for demonstrators to influence the later food choices of observers (Galef and Stein, 1985).

Both food-related odors escaping from the digestive tract of a demonstrator and the scent of bits of food clinging to a demonstrator's fur and vibrissae allow observers to identify the food that a demonstrator has recently eaten. And after an observer rat experiences simultaneously the scent of a food and rat breath, the observer shows an enhanced preference for the food the scent of which it experienced together with rat breath (Galef and Stein, 1985).

Gas chromatography performed on samples of rat breath has shown that it contains two sulfur compounds: carbon disulfide and carbonyl sulfide. Rats exposed to a food dusted onto either the head of an anesthetized conspecific or a piece of cloth moistened with a dilute solution of carbon disulfide subsequently show an enhanced preference for that food. To the contrary, rats exposed to a food, that had been dusted onto the head of a dead conspecific, onto the rear of a live conspecific or onto a piece of cloth moistened with distilled water do not develop a similar preference (Galef et al., 1988) (Fig. 34–3). Thus, experience of carbon disulfide, a natural constituent of rat breath, in conjunction with a food odor, like experience of rat breath in conjunction with a food odor, is sufficient to enhance preference for the food.

SYNTHESIS

The breath of humans, like the breath of rats, contains trace quantities of carbon disulfide. As would be expected on the hypothesis that experience of food odors together with car-

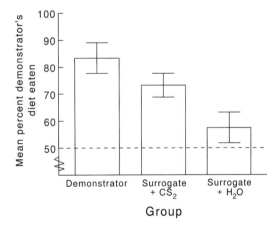

Figure 34–3. Observer rats interacted with either an anesthetized demonstrator rat or a cloth "surrogate" demonstrator. The demonstrator with which each observer interacted had been powdered with either cinnamon- or cocoa-flavored diet. Surrogates were moistened with either a dilute aqueous solution of CS_2 or an equal amount of distilled water. The figure indicates the mean percent of each observer's total intake that was the diet with which its demonstrator or surrogate had been powdered. Error bars show 1 SEM. (Data from Galef et al. [1988].)

bon disulfide induces food preferences in rats, when a human "demonstrator" eats a flavored food and breathes on a rat, the rat's preference for the food that its human demonstrator ate is markedly enhanced (Galef, 2001).

LIMITATIONS

Surprisingly, rats do not learn to avoid a food by interacting with a sick or an unconscious demonstrator that has eaten it. To the contrary, rats show an increased preference for a food that was eaten by an ill conspecific with which they interacted (Galef et al., 1990).

Further, exposure to an odor in conjunction with a conspecific does not enhance the general affinity of a rat for that odor; exposure to an odor in a social context that profoundly affects food preference has no effect on the odor preferences of rats in other contexts. For example, rats that have interacted with a conspecific that has eaten a cinnamon-flavored diet prefer cinnamon-flavored food but show no enhancement of their preference for cinnamon-scented nest materials or cinnamon-scented nest sites (Galef and Iliffe, 1994). Such findings suggest that social induction of food preference is a learning process evolved specifically to facilitate foraging rather than other activities of rats.

EXTENTIONS

Rats can use information concerning foods that other rats have eaten in some interesting ways. For example, after "observer" rats had an opportunity to learn where in a three-arm maze each of three distinctively flavored foods were to be found, we let each observer rat interact briefly with a demonstrator rat that had eaten one of those three foods. Without any specific training, the observers went directly to the arm of the maze where they had learned that the food that their demonstrator had eaten was usually located (Galef and Wigmore, 1983). Obviously, rats can integrate their cognitive map of food distribution with

socially acquired information about the current availability of foods to increase the efficiency with which they forage.

APPLICATION TO STUDIES OF NERVOUS SYSTEM FUNCTION

Socially induced enhanced diet preference provides an efficient and reliable way to induce a learned appetitive behavior in rats (or mice, gerbils, hamsters, voles or bats) that, like other types of learned behavior, can serve as a dependent variable in studies of brain function. Neuroscientists have used the socially induced change in food preference described here to study the effects of manipulations of the neural substrate on learning and memory (Burton et al., 2000; Winocur et al., 2001; see Galef, 2002, for further references). As one might expect, both direct and genetic manipulations of the nervous system affect social learning of food preferences.

There are several advantages in using socially learned food preference as a dependent measure in studies of brain function: (1) learning occurs in a single trial, (2) little or no skill is needed to train subjects, (3) no special equipment is needed to train subjects, and (4) subjects need never be deprived or stressed. The procedure for inducing social enhancement of food preference consists of three straightforward steps. First, a demonstrator rat is placed on a feeding schedule and given one of two distinctively flavored foods to eat. Second, each demonstrator is placed together with an observer, and demonstrator and observer rats are allowed to interact for 15 minutes or longer. During this period of interaction, observers have the opportunity to smell the scented food on the breath of their respective demonstrators. Last, each observer is given a choice between the two distinctively flavored foods that were offered to demonstrators in the first step (Galef, 2002) (Fig. 34–4). In the third step, observers invariably show an enhanced preference for whichever

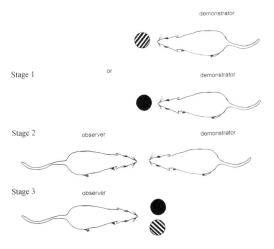

Stage 1

Stage 2

Stage 3

Figure 34–4. Schematic of the three stages of an experiment demonstrating social influence on the diet preferences of observer rats'. In stage 1, each demonstrator rat ate one of two distinctively flavored foods. (From Galef [2002]. Reprinted with permission of John Wiley & Sons, Inc.)

flavored food was eaten by their respective demonstrators.

The effect is robust. Demonstrator and observer can be male or female, young or old, previously familiar or unfamiliar with one another, and genetically related or unrelated to one another (Galef et al., 1984). Demonstrators can ingest almost any scented liquid or solid before interacting with their observers. There can be a delay of several hours between when a demonstrator is fed and when it interacts with its observer. Demonstrators can be separated from their observers by a hardware-cloth screen while they interact, and interaction can take place in the home cage of demonstrator or observer or in a neutral arena. There can be a delay of weeks between when demonstrator and observer interact and when the observer is tested. Invariably, if demonstrators that have recently ingested a distinctively flavored substance are placed for a few minutes together with observers that are otherwise unfamiliar with the flavor of the food that was eaten by their respective demonstrators, the observers subsequently show significant enhancement of their relative intake of that food.

REFERENCES

Barnett SA (1958) Experiments on "neophobia" in wild and laboratory rats. British Journal of Psychology 49:195–201.

Beck M and Galef BG Jr (1989) Social influences on the selection of protein-sufficient diet by Norway rats. Journal of Comparative Psychology 103:132–139.

Burton s, Murphy D, Qureshi U, Suton P, O'Keefe J (2000) Combined lesions of hippocampus and subiculum do not produce deficits in nonspatial sociallearning. Journal of Neuroscience 20:5468–5475.

Galef BG Jr (1977) Mechanisms for the social transmission of food preferences from adult to weanling rats. In: Learning mechanisms in food selection (Barker LM, Best M, Domjan M, eds.), pp. 123–150. Waco, TX: Baylor University Press.

Galef BG Jr (1981) The development of olfactory control of feeding site selection in rat pups. Journal of Comparative and Physiological Psychology 95:615–622.

Galef BG Jr (1986) Social interaction modifies learned aversions, sodium appetite, and both palatability and handling-time induced dietary preference in rats (Rattus norvegicus). Journal of Comparative Psychology 100:432–439.

Galef BG Jr (1988) Communication of information concerning distant diets in a social, central-place foraging species (Rattus norvegicus). In: Social learning: psychological and biological perspectives (Zentall TR and Galef BG Jr, eds.) pp. 119–140. Hillsdale, NJ: Erlbaum.

Galef BG Jr (1992) The question of animal culture. Human Nature 3:157–178.

Galef BG Jr (1996a) Social enhancement of food preferences in Norway rats. In: Social learning and imitation: the roots of culture (Heyes CM and Galef BG Jr, eds.) pp. 49–64. New York: Academic Press.

Galef BG Jr (1996b) Social influences on food preferences and feeding behaviors of vertebrates. In: Why we eat what we eat (Capaldi E, ed.) pp. 207–232. Washington, D.C.: American Psychological Association.

Galef BG Jr (2001) Analyses of social learning processes affecting animals' choices of foods and mates. Mexican Journal of Behavior Analysis 27:145–164.

Galef BG Jr and Whiskin EE (2003) Socially transmitted food preferences can be used to study long-term memory in rats. Learning and Behavior 31:160–164.

Galef BG Jr and Allen C (1995) A new model system for studying animal traditions. Animal Behaviour 50:705–717.

Galef BG Jr and Beck M (1985) Aversive and attractive marking of toxic and safe foods by Norway rats. Behavioral and Neural Biology 43:298–310.

Galef BG Jr and Buckley LL (1996) Use of foraging trails by Norway rats. Animal Behaviour 51:765–771.

Galef BG Jr and Clark MM (1971a) Parent-offspring interactions determine time and place of first ingestion of solid food by wild rat pups. Psychonomic Science 25:15–16.

Galef BG Jr and Clark MM (1971b) Social factors in the poison avoidance and feeding behavior of wild and domesticated rat pups. Journal of Comparative and Physiological Psychology 25:341–357.

Galef BG Jr and Iliffe CP (1994) Social enhancement of odor preference in rats: is there something special about odors associated with foods? Journal of Comparative Psychology 108:266–273.

Galef BG Jr (2002) Social learning of food preferences in rodents: rapid appetitive learning. Current Protocols in Neuroscience. 8.5D1–8.5D8.

Galef BG Jr, Kennett DJ, Wigmore SW (1984) Transfer of information concerning distant foods in rats: a robust phenomenon. Animal Learning and Behavior 12:292–296.

Galef BG Jr, Marczinski CA, Murray KA, Whiskin EE (2001) Studies of food stealing by young Norway rats. Journal of Comparative Psychology 115:16–21.

Galef BG Jr, Mason JR, Pretti G, Bean, NJ (1988) Carbon disulfide: a semiochemical mediating socially-induced diet choice in rats. Physiology and Behaviour 42:119–124.

Galef BG Jr, McQuoid LM, Whiskin EE (1990) Further evidence that Norway rats do not socially transmit learned aversions to toxic baits. Animal Learning and Behavior 18:199–205.

Galef BG Jr and Sherry DF (1973) Mother's milk: a medium for transmission of information about mother's diet. Journal of Comparative and Physiological Psychology 83:374–378.

Galef BG Jr and Stein M (1985) Demonstrator influence on observer diet preference: analyses of critical social interactions and olfactory signals. Animal Learning and Behavior 13:131–138.

Galef BG Jr, Whiskin EE, Bielavska E (1997) Interaction with demonstrator rats changes their observers' affective responses to flavors. Journal of Comparative Psychology 111:393–398.

Galef BG Jr and Wigmore SW (1983) Transfer of information concerning distant foods: a laboratory investigation of the information-centre" hypothesis. Animal Behaviour 31:748–758.

Garcia J, Ervin FR, Koelling RA (1966) Learning with prolonged delay of reinforcement. Psychonomic Science 5:121–122.

Hepper PG (1988) Adaptive fetal learning: prenatal exposure to garlic affects postnatal preference. Animal Behaviour 36:935–936.

Smotherman WP (1982) Odor aversion learning by the rat fetus. Physiology and Behavior 29:769–771.

Steiniger von F (1950) Beitrage zur Soziologie und sonstigen Biologie der Wanderratte. Zeitschrift fur Tierpsychologie 7:356–379.

Strupp BJ and Levitsky DA (1984) Social transmission of food preferences in adult hooded rats (*Rattus norvegicus*). Journal of Comparative Psychology 98:257–266.

Van Schaik CP, Ancrenaz M, Borgen G, Galdikas B, Singleton I, Suzuki A, Utami SS, Merill M (2003) Orangutan cultures and the evolution of material culture implications. Science 299:102–105.

Von Frisch K (1967) The dance language and orientation of bees. Cambridge, Mass.: Belknap Press.

Whiten A, Goodall J, McGrew WC, Nishida T, Reynolds V, Sugiyama Y, Tutin CEG, Wrangham RW, Boesch C (1999) Culture in chimpanzees. Nature 399:682–685.

Winocur G, McDonald RM, Moscovitch M (2001). Anterograde and retrograde amnesia in rats with large hippocampal lesions. Hippocampus 11:18–26.

Vocalization

35

GRETA SOKOLOFF
AND MARK S. BLUMBERG

Throughout their life span, rats emit vocalizations in a variety of environmental and social contexts, including nest separation in infants and sexual behavior in adults. Rats, being small and therefore having a small vocal apparatus, produce vocalizations predominantly at ultrasonic frequencies, that is, at frequencies above the range detectable by the human ear (>20 kHz). This chapter provides an overview of the contexts, mechanisms, and suggested functions of rat vocalizations. The proposed use of these vocalizations as a model for investigations of anxiety and depression is also discussed.

FREQUENCY AND TEMPORAL CHARACTERISTICS

Rats emit ultrasonic vocalizations across a frequency range of 20 to 70 kHz (Table 35–1). During early postnatal development, infants produce vocalizations with a dominant frequency of approximately 40 kHz when isolated from the nest. As the pup, and its vocal apparatus, increases in size, the dominant frequency of this vocalization decreases progressively to 25 kHz by 20 days of age (Blumberg et al., 2000a). Infant vocalizations are relatively pure tones produced by expiring air under high pressure against constricted laryngeal folds, as is also the case for the adult's 22-kHz vocalization (Roberts, 1975b, 1975c; Sanders et al., 2001). The 22-kHz vocalization, associated with a variety of contexts, including sexual behavior and aggression (Sales, 1972a, 1972b), has been referred to as a *long call* because it is produced dur-

ing prolonged expirations (>1 second). In contrast, a third category of rat ultrasonic vocalization has been referred to as a *short call* because of its relatively brief duration (<65 milliseconds); these chirp-like vocalizations, emitted by weanlings, juveniles, and adults at high frequencies (35 to 70 kHz), are associated with contexts that include vigorous activity, high levels of arousal, and social contact. Although it has been suggested that some instances of these vocalizations may be produced as a biomechanical byproduct of activity (Blumberg, 1992), many instances of this vocalization appear to be independent of activity (Knutson et al., 2002). Very little is known regarding the mechanisms that produce these vocalizations.

Just as small animals produce higher-frequency sounds, so are they able to hear higher-frequency sounds. The auditory system of adult rats exhibits a peak sensitivity range of 10 to 50 kHz (Crowley et al., 1965; Gourevitch and Hack, 1966; Brown, 1973). Similarly, pre-weanling rats detect and respond to auditory stimuli in the range of 1 to 70 kHz, with a peak sensitivity of approximately 40 kHz (Crowley et al., 1965; Crowley and Hepp-Reymond, 1966).

ENVIRONMENTAL CONTEXTS ASSOCIATED WITH ULTRASONIC VOCALIZATIONS

INFANTS

Infant rats emit broadband vocalizations comprising audible and ultrasonic frequencies dur-

Table 35–1. Four General Categories of Rat Vocalizations

Frequency	Age	Examples of Contexts and Stimuli
Broadband	Infancy to adulthood	Painful stimuli, such as tail pinch and foot shock
25 to 45 kHz (depending on age and size)	Infancy	Isolation from the nest; cooling
22 kHz	Adulthood	After ejaculation, after defeat in aggressive encounter, and after termination of foot shock
35 to 70 kHz	Juvenile period to adulthood	Arousal, such as during play, copulation, and aggression

ing maternal handling (e.g., grooming) and intense tactile stimulation (e.g., tail pinching). In contrast, ultrasonic vocalizations are evoked by isolation from the nest (Noirot, 1972). Cold is the primary feature of isolation that elicits the vocalization, as isolated pups that are kept warm do not vocalize (Allin and Banks, 1971; Okon, 1971; Blumberg et al., 1992b). Furthermore, as pups mature and are better able to thermoregulate, isolation results in fewer ultrasonic emissions (Okon, 1971; Blumberg et al., 1992a).

Infant ultrasound production increases during the first 2 postnatal weeks and declines thereafter (Noirot, 1968; Sewell, 1970; Okon, 1971; Noirot, 1972). During development, it appears that factors capable of evoking and modulating ultrasound production diversify and become more complex. For example, olfactory stimuli begin to play a role in the attenuation of ultrasound production during cold exposure. Specifically, odors associated with the nest (e.g., dam, siblings, home cage bedding) effectively reduce isolation-induced vocalizations (Oswalt and Meier, 1975; Hofer and Shair, 1987), as does exposure to an unfamiliar adult male rat (Takahashi, 1992).

One phenomenon that illustrates the emergence of complex controls of ultrasound production during the second postnatal week is maternal potentiation (Hofer et al., 1998; Kraebel et al., 2002; Shair et al., 2003). The paradigm for studying maternal potentiation begins with an initial period of isolation. The pup is then briefly reunited with the dam. Finally, the pup is again isolated, resulting in

elevated levels of ultrasound production. As with other contexts in which infant vocalizations are studied, air temperature plays a modulatory role in the expression of maternal potentiation (Kraebel et al., 2002; Shair et al., 2003).

ADULTS

Ultrasonic vocalizations can be detected in colonies of wild rats as well as in group-housed or isolated laboratory rats (Calhoun, 1962; Francis, 1977). Although these adult vocalizations are primarily associated with social interactions, isolated rats vocalize spontaneously in a circadian fashion, with peak vocalization rates occurring during the middle of the dark period (Francis, 1977).

High-frequency ultrasonic "chirps" predominate during social interactions associated with high levels of arousal. In adult rats, chirps are emitted when rats are placed together as well as when an individual rat is placed in an area previously visited by other rats (Brudzynski and Pniak, 2002). Furthermore, juveniles vocalize in social situations that have previously been associated with play (Knutson et al., 1998), and adult male rats vocalize when placed in empty cages where social contact has previously occurred (Bialy et al., 2000; Brudzynski and Pniak, 2002).

Ultrasonic vocalizations also occur during reproductive behavior. Males and females emit 50- to 70-kHz vocalizations primarily during genital investigation, chasing, and mounting (Sales, 1972a, 1972b; Barfield et al.,

1979; Thomas and Barfield, 1985). The 22 kHz vocalization of males reliably occurs during the postejaculatory refractory period (Barfield et al., 1979).

When exposed to a natural predator, like a cat, rats housed in groups emit 22-kHz vocalizations. Experiments using artificial burrows consisting of multiple rats have indicated that 22-kHz vocalizations are produced when the cat is introduced and continue to be emitted after the cat is removed (Blanchard et al., 1991). This is in contrast to individual rats that, when tested in the presence of a cat, do not vocalize regardless of whether they are able to escape (Blanchard et al., 1991).

Aggressive encounters are another context in which ultrasonic vocalizations are reliably evoked. Initially, when two male rats are introduced and during the initial phases of aggressive behavior, 50- to 70-kHz vocalizations are emitted by both animals. After the aggressive encounter, however, 22-kHz vocalizations are emitted exclusively by the submissive rat while exhibiting the belly-up submissive posture (Sewell, 1967; Sales, 1972b).

ANATOMICAL CONSIDERATIONS

Audible rat vocalizations are produced by vibration of the laryngeal folds, as with human voiced speech (Roberts, 1975a). They are elicited by stimulation of Aδ- and C-fibres (Ardid et al., 1993) as well as by stimulation of the trigeminal spinal tract nucleus (Yajima et al., 1981), suggesting that audible vocalizations are produced in response to noxious or painful stimulation. Although both audible and ultrasonic vocalizations are produced by the larynx and occur during the expiratory phase of respiration, the 40-kHz infant vocalization and the 22-kHz adult vocalization appear unique in that they are produced using a whistle-like mechanism that entails forced expiration through a constricted and nonvibrating larynx (Roberts, 1975b, 1975c). The 50- to 70-kHz vocalization is acoustically more complex and is likely produced by vibration of the laryngeal folds.

Audible vocalizations become weaker and ultrasonic vocalizations are virtually abolished by nerve cuts that denervate the laryngeal musculature (Roberts, 1975b). Specifically, unilateral or bilateral transection of the inferior laryngeal nerve, a branch of the vagus nerve, abolishes ultrasound in infant rats. In addition, transection of the superior laryngeal nerve, also a branch of the vagus nerve, changes the sound pressure and frequency of these vocalizations as well as vocalization rate (Wetzel et al., 1980).

The neural circuit that controls the larynx in rats and other species includes the midbrain periaqueductal gray (PAG). In fact, stimulation of the PAG evokes species-specific vocalizations in many animals (Zhang et al., 1994), including rats (Yajima et al., 1980). In rats, stimulation of a pathway emanating from the dorsal region of the thalamus and terminating in the dorsomedial region of the PAG elicits 22-kHz vocalizations (Yajima et al., 1980). From the PAG, projections to the dorsomedial aspect of the medullary reticular formation are also involved in ultrasound production, including a number of cranial nerve nuclei (e.g., facial, hypoglossal, and vagal) (Yajima et al., 1981). The inferior and superior laryngeal nerves, necessary for the production of ultrasound, arise from the dorsal and ventral regions of the medullary nucleus ambiguus, respectively (Wetzel et al., 1980). Audible vocalizations are evoked by stimulation of numerous regions of the hypothalamus as well as the ventromedial PAG (Yajima et al., 1980).

FUNCTIONAL SIGNIFICANCE OF ULTRASONIC VOCALIZATIONS

Many theories exist concerning the functional significance of rat ultrasonic vocalizations. The earliest theories focused on communicatory functions. Extensive work on the effects of thermal stimuli on ultrasound production in infant rats led to physiological theories of ultrasound production. More recently, motivational theories of ultrasound production have emerged that focus on the emotional contexts in which the vocalizations are produced (Table 35–2).

COMMUNICATORY HYPOTHESES

The fact that infant rats vocalize when isolated from the nest and the fact that dams retrieve isolated pups suggests an important communicatory role for these vocalizations. Today, infant rat vocalizations are commonly referred to as *distress* or *separation calls* (Oswalt and Meier, 1975; Hofer et al., 1994). In fact, even the absence of vocalizations in isolated infant rats has been ascribed a communicatory function, namely, to prevent the alerting of predators to an infant's location (Takahashi, 1992; Hofer et al., 1994).

During reproductive behavior, 50- to 70-kHz vocalizations emitted by males have been proposed to serve the communicatory function of increasing female solicitation behav-iors such as hopping and darting (Sales, 1972a; Barfield et al., 1979). In contrast, the postejaculatory 22-kHz vocalization has been hypothesized to communicate to the female a reduction in the male's sexual motivation (Barfield et al., 1979). Modulation of this vocalization by the female's presence during the postejaculatory interval has been interpreted as evidence in favor of a communicatory function (Sachs and Bialy, 2000).

Still other communication hypotheses have been posited for 22-kHz vocalizations. With respect to a male defeated in an aggressive encounter, it has been proposed that the vocalization serves to prevent further aggression by the dominant male (Sales, 1972b). With respect to group-housed rats that vocalize when exposed to a cat, it has been hypothesized that these vocalizations serve to alert conspecifics to the predator's presence (Blanchard et al., 1991). Thus, the 22-kHz vocalizations of adults, like those of infants, have been deemed *distress* or *alarm calls*.

As stated, infant ultrasonic vocalizations elicit maternal retrieval and the ultrasonic vocalizations of male rats during reproductive behavior alter female behavior (Allin and

Table 35–2. Theories of Ultrasound Production in Rats

Theory	Category	Example	Citations
Communicatory	Infant	Eliciting maternal retrieval after separation from the nest	Allin and Banks, 1972; Farrell and Alberts, 2002a
	50 to 70 kHz	Facilitation of female sexual behavior	Barfield et al., 1979
	22 kHz	Appeasement after aggressive encounter	Sales, 1972a
Physiological	Infant	Relationship to cardiopulmonary function	Blumberg and Alberts, 1990; Blumberg and Sokoloff, 2001
	22 kHz	Relationship to brain temperature	Blumberg and Moltz, 1987
Motivational/emotional	Infant	Anxiety/distress	Shair et al., 2003; Winslow and Insel, 1991
	~50 to 70 kHz	Expectation of play or social contact	Knutson et al., 1998; Brudzynski and Pniak, 2002
	22 kHz	Withdrawal from substance abuse	Vivian and Mizcek, 1993

Banks, 1972; Noirot, 1972; Sachs and Bialy, 2000; Farrell and Alberts, 2002a). Such demonstrations of communicatory effect, however, are not equivalent to demonstrations of communicatory function (Blumberg and Alberts, 1992, 1997). Furthermore, although evidence exists that seems to fit easily within a communicatory framework, other evidence does not fit so easily. For example, the apparent suppressive effect of littermates on the vocalization can be overcome by decreasing temperature until the huddle's thermoregulatory capabilities are exceeded (Blumberg et al., 1992a; Sokoloff and Blumberg, 2001). Furthermore, studies in which pup odor and acoustic cues are manipulated demonstrate that the vocalization alone is not sufficient to evoke maternal retrieval (Smotherman et al., 1974; Farrell and Alberts, 2002b). As another example, during agonistic encounters, the aggressive behavior of dominant males is not necessarily reduced by the vocalizations of subordinates, and the freezing behavior and vocalizations of the subordinate males are not reduced by the vocalizations of the dominant males (Takahashi et al., 1983). Taken together, these examples highlight the need for caution when assessing the communicatory significance of rat vocalizations.

PHYSIOLOGICAL HYPOTHESES

Many investigators have appreciated the significance of cold exposure for eliciting infant vocalizations during isolation from the nest (Allin and Banks, 1971; Okon, 1971; Oswalt and Meier, 1975). In fact, some level of cold exposure remains as an integral component of most methodological approaches in the field, even when the primary focus of a study is the ability of pharmacological agents or nest-related cues to attenuate the vocalization (Kraebel et al., 2002). The question is whether temperature is merely a cue to the pup that it is isolated from the nest or a physical stimulus that evokes a significant change in the pup's physiological functioning.

Infant rats are capable of endogenous heat production, but their small size results in rapid heat loss when exposed to standard room temperatures (i.e., 22° C). When observed at air temperatures in which brown adipose tissue thermogenesis is sufficient to maintain elevated body temperatures, however, infant rats do not vocalize (Blumberg and Stolba, 1996). Of importance is that pharmacological manipulations that augment or attenuate brown adipose tissue thermogenesis result in decreases or increases in infant vocalization rates, respectively (Blumberg et al., 1999; Farrell and Alberts, 2000).

The effects of cold exposure on the cardiopulmonary system of infant rats have led to specific hypotheses concerning the physiological mechanisms underlying ultrasound production (Blumberg and Alberts, 1990; Blumberg and Sokoloff, 2001). Infants do not vocalize during cold exposure unless the cooling is severe enough to cause a decrease in cardiac rate (Blumberg et al., 1999). In addition, at these same temperatures, blood viscosity increases significantly, further compromising cardiopulmonary function. Therefore, it has been hypothesized that infant vocalizations are acoustic byproducts of a physiological maneuver that serves to maintain cardiopulmonary function (Blumberg and Sokoloff, 2001).

In addition to cold exposure, administration of the α_2-adrenoceptor agonist clonidine results in prolonged and robust vocalization responses. The effect of clonidine is so profound that returning the pup to the nest is not sufficient to attenuate the vocalization (Kehoe and Harris, 1989). Interestingly, clonidine also produces a profound bradycardia and when the β_1-adrenoceptor agonist prenalterol was used to inoculate pups against clonidine-induced bradycardia, ultrasound production was significantly attenuated (Blumberg et al., 2000b).

Physiological correlates of adult vocalizations have also been reported. For example, the 22-kHz vocalization occurs during the chill phase of fever evoked by central administration of prostaglandin E_2; in addition, the postejaculatory vocalization is virtually abolished by prior administration of sodium salicylate, a drug that decreases body temperature (Blumberg and Moltz, 1987). The significance of this relationship between temperature and emission of vocalization is not yet clear.

MOTIVATIONAL AND EMOTIONAL THEORIES

Rat vocalizations are often suggested to be expressions of emotion. According to one theory, 50- to 70-kHz vocalizations are an index of positive affect, whereas 22-kHz vocalizations are an index of negative affect (Knutson et al., 2002). The view that infant vocalizations model human distress and anxiety has developed in lock-step with the use of infant vocalizations to examine the efficacy of pharmacological agents developed for the treatment of human psychological disorders (Miczek et al., 1995).

The contexts and manipulations that elicit ultrasound production provide the foundation for the affective classification of the different categories of vocalizations. As stated, 50- to 70-kHz vocalizations occur during social contact, such as reproductive behavior and play (Sales, 1972a; Barfield et al., 1979; Knutson et al., 1998). Electrical brain stimulation, which is known to be reinforcing, also elicits these vocalizations (Burgdorf et al., 2000), a finding that is consistent with the notion that these vocalizations are an index of positive affect (Burgdorf et al., 2000; Knutson et al., 2002). In juvenile rats, "tickling" by human handlers results in 50-kHz chirps that have been described as laughter (Panksepp and Burgdorf, 2000).

In contrast to 50- to 70-kHz chirps, 22-kHz vocalizations are typically associated with physiological and psychological stressors, such as fever (Blumberg and Moltz, 1987) and aggressive encounters (Sales, 1972b). Aversive conditioning paradigms (i.e., fear conditioning; Lee et al., 2001) and withdrawal

from morphine dependence (Vivian and Miczek, 1991) also elicit these vocalizations. These findings are consistent with the notion that the 22-kHz vocalization is an index of negative affect (Miczek et al., 1995; Knutson et al., 2002).

There are, however, pieces that do not fit so easily into the emotional framework just presented. For example, as already noted, 50- to 70-kHz vocalizations accompany agonistic encounters just as readily as they do sexual and playful ones. Are we then to suppose that rats experience positive affect while fighting? Similarly, are we to assume that the postejaculatory male, emitting the 22-kHz vocalization, experiences a negative affective state akin to that experienced by a male that has just been defeated in an aggressive encounter with a conspecific? These and other incongruities pose difficulties for any unidimensional affective theory of rat vocalizations.

THE VOCALIZING RAT AS A MODEL OF HUMAN PSYCHOLOGICAL DISORDERS

Although there remains much uncertainty concerning the mechanisms and functional significance of rat vocalizations, it is widely believed that they can be effectively used to investigate anxiety, distress, fear, and drug abuse. For this reason, psychopharmacological approaches to studying rat vocalizations have become very popular. The hope is that increased knowledge concerning the effects of psychoactive drugs on infant and adult vocalizations will increase understanding of human psychiatric disorders and their treatment (Miczek et al., 1995).

Because 22-kHz vocalizations are sometimes associated with aversive events and behaviors associated with fear (e.g., freezing), it is suggested that rats producing this vocalization can be used as a model for depression and anxiety in humans (Miczek et al., 1995; Schreiber et al., 1998). Similarly, using the isolation paradigm, ultrasonic vocalizations in infant rats have been proposed as a model of separation anxiety (Winslow and Insel, 1991). Support for this view comes from pharmacological studies in which anxiolytic agents are shown to attenuate ultrasound production under some conditions. For example, selective serotonin reuptake inhibitors (SSRIs) and benzodiazepines have been shown to reduce vocalizations in adult and infant rats (Insel et al., 1986; Olivier et al., 1998; Schreiber et al., 1998). Opioids also reduce the occurrence of 22-kHz vocalizations in adults in response to tail shock (van der Poel et al., 1989). Finally, anxiolytic agents attenuate ultrasonic vocalizations during morphine withdrawal (Vivian and Miczek, 1991).

Although the vocalizing rat, infant and adult, has been proposed as a useful model for testing anxiolytic drugs (Olivier et al., 1998), some research indicates that it is not a robust model. First, at least one anxiolytic agent, clonidine, increases ultrasound in infants (Kehoe and Harris, 1989; Blumberg et al., 1999). Second, benzodiazepines do not reduce prestimulus ultrasonic vocalizations after fear conditioning (van der Poel et al., 1989). Third, only antidepressants working via the serotonergic system attenuate ultrasound production, whereas other antidepressant compounds acting on the noradrenergic system result in anxiogenesis, as measured by increased ultrasound production, and still other compounds (e.g., amitriptyline) have no effect (Borsini et al., 2002). Regardless, we should not be surprised if the diversity of rat vocalizations do not fit neatly into conceptual categories developed for the diagnosis of human clinical disorders (Blumberg and Sokoloff, 2001).

MEASURING ULTRASONIC VOCALIZATIONS IN THE LABORATORY

The ease of making ultrasonic vocalizations audible using a bat detector, originally in-vented to study echolocation in bats, has stimulated a growth industry in the area of rat vocalizations. Today, in numerous laboratories across the world, infant and adult vocalizations are used for studies of basic physiology, separation responses, and psychopharmacology. Reflecting this surge in interest, there are a number of companies that produce ultrasound-sensitive detectors with an array of features and capabilities.

The analysis of ultrasonic vocalizations can be as simple as manually counting the total number of vocalizations, either during data collection or afterward from an audio recording. Skilled listeners typically exhibit high interrater reliabilities even when counting vocalizations in infant rats that can occur many times each second. Alternatively, automatic scoring of vocalizations can be accomplished by digitally recording the vocalizations using a data acquisition system and counting bursts that exceed a threshold value; some companies now offer systems that are designed specifically for this task. For more complex acoustic analyses (e.g., frequency modulation, peak frequency, amplitude, duration of individual calls), investigators can use any of a number of analysis programs available.

CONCLUSIONS

Ultrasonic vocalizations accompany a wide array of social behavior in the rat, from the isolation-induced vocalization of the infants to the postejaculatory vocalization of the adult males. Interestingly, there has of yet been no serious attempt to synthesize the various theories and perspectives in the field. For example, communicatory and motivational theories of ultrasound production have focused primarily on the function of this behavior (i.e., a signal for maternal retrieval or as an index of emotional state). In contrast, physiological theories have focused primarily on mechanisms underlying the behavior (i.e., reflexive cardiopulmonary compensations). Therefore, there exists a large gap in our understanding of these vocal behaviors throughout development and across the different environmental contexts that they are expressed.

The 40-kHz vocalization of the isolated infant rat and the 22-kHz vocalization of the adult rat are produced by a similar laryngeal mechanism, thus suggesting that the two vocalizations are homologous (Blumberg and Alberts, 1991; Blumberg et al., 2000a). As discussed, these vocalizations are not elicited in similar contexts or by similar stimuli, nor are they similarly modulated by pharmacological agents. For example, although clonidine evokes ultrasound production in infant rats (Kehoe and Harris, 1989; Blumberg et al., 2000a), it attenuates conditioned 22-kHz vocalizations in adults (Molewijk et al.). In contrast, cholinergic agonists increase 22-kHz vocalizations in adults rats (Brudzynski, 2001) but not 40-kHz vocalizations in infants (Kehoe et al., 2001). These two developmental differences alone point to the gaps in our understanding of the origins and mechanisms of rat ultrasound. Focusing on developmental changes in these vocalizations may be an informative approach for elucidating their underlying mechanisms.

Perhaps a synthesis of the various theoretical viewpoints will be developed that describes and explains the causes and functions of these vocalizations in all their diversity. In the meantime, attempts to promote these vocalizations as models for human psychological conditions and psychiatric disorders should be met with healthy skepticism.

REFERENCES

Allin JT and Banks EM (1971) Effects of temperature on ultrasound production by infant albino rats. Developmental Psychobiology 4:149–156.

Allin JT and Banks EM (1972) Functional aspects of ultrasound production by infant albino rats *(Rattus norvegicus)*. Animal Behaviour 20:175–185.

Ardid D, Jourdan D, Eschalier A, Arabia C, Bars DL (1993) Vocalization elicited by activation of Aδ- and C-fibres in the rat. Neuroreport 5:105–108.

Barfield RJ, Auerbach P, Geyer LA, McIntosh TK (1979) Ultrasonic vocalizations in rat sexual behavior. American Zoologist 19:469–480.

Bialy M, Rydz M, Kaczmarek L (2000) Precontact 50-kHz vocalizations in male rats during acquisition of sexual experience. Behavioral Neuroscience 114:983–990.

Blanchard RJ, Blanchard DC, Agullana R, Weiss SM (1991) Twenty-two kHz alarm cries to presentation of a predator, by laboratory rats living in a visible burrow system. Physiology and Behavior 50:967–972.

Blumberg MS (1992) Rodent ultrasonic short calls: locomotion, biomechanics, and communication. Journal of Comparative Psychology 106:360–365.

Blumberg MS, Alberts JR (1990) Ultrasonic vocalizations by rat pups in the cold: an acoustic by-product of laryngeal braking? Behavioral Neuroscience 104:808–817.

Blumberg MS and Alberts JR (1991) On the significance of similarities between ultrasonic vocalizations of infant and adult rats. Neuroscience and Biobehavioral Reviews 50:95–99.

Blumberg MS and Alberts JR (1992) Functions and effects in animal communication: reactions to Guilford & Dawkins. Animal Behaviour 44:382–383.

Blumberg MS and Alberts JR (1997) Incidental emissions, fortuitous effects, and the origins of communication. In: Perspectives in ethology (Thompson NS, ed.), pp. 225–249. New York: Plenum Press.

Blumberg MS, Efimova IV, Alberts JR (1992a) Ultrasonic vocalizations by rats pups: the primary importance of ambient temperature and the thermal significance of contact comfort. Developmental Psychobiology 25:229–250.

Blumberg MS, Efimova IV, Alberts JR (1992b) Thermogenesis during ultrasonic vocalization by rat pups isolated in a warm environment: a thermographic analysis. Developmental Psychobiology 25:497–510.

Blumberg MS, Kreber LA, Sokoloff G, Kent KJ (2000b) Cardiovascular mediation of clonidine-induced ultrasound production in infant rats. Behavioral Neuroscience 114:602–608.

Blumberg MS and Moltz H (1987) Hypothalamic temperature and the 22 kHz vocalization of the male rat. Physiology and Behavior 40:637–640.

Blumberg MS, Sokoloff G, Kent KJ (1999) Cardiovascular concomitants of ultrasound production during cold exposure in infant rats. Behavioral Neuroscience 113:1274–1282.

Blumberg MS, Sokoloff G, Kent KJ (2000a) A developmental analysis of clonidine's effects on cardiac rate and ultrasound production in infant rats. Developmental Psychobiology 36:186–193.

Blumberg MS and Sokoloff G (2001) Do infant rats cry? Psychological Review 108:83–95.

Blumberg MS and Stolba MA (1996) Thermogenesis, myoclonic twitching, and ultrasonic vocalization in neonatal rats during moderate and extreme cold exposure. Behavioral Neuroscience 110:305–314.

Borsini F, Podhorna J, Marazziti D (2002) Do animal models of anxiety predict anxiolytic-like effects of antidepressants? Psychopharmacology 163:121–141.

Brown AM (1973) High frequency peaks in the cochlear microphonic response of rodents. Journal of Comparative Physiology 83:377–392.

Brudzynski SM (2001) Pharmacological and behavioral characteristics fo 22 kHz alarm calls in rats. Neuroscience and Biobehavioral Reviews 25:611–617.

Brudzynski SM and Pniak A (2002) Social contacts and production of 50-kHz short ultrasonic calls in adult rats. Journal of Comparative Psychology 116:73–82.

Burgdorf J, Knutson B, Panksepp J (2000) Anticipation of rewarding electrical brain stimulation evokes ultrasonic vocalizations in rats. Behavioral Neuroscience 114:320–327.

Calhoun JB (1962) The ecology and sociology of the Norway rat. Bethesda: U.S. Department of Health, Education, and Welfare.

Crowley DE, Hepp-Reymond M-C (1966) Development of cochlear function in the ear of the infant rat. Journal of Comparative and Physiological Psychology 63:427–432.

Crowley DE, Hepp-Raymond M-C, Tabonite D, Palin J (1965) Cochlear potentials in the albino rat. Journal of Auditory Research 5:307–316.

Farrell WJ and Alberts JR (2000) Ultrasonic vocalizations by rat pups after adrenergic manipulations of brown fat metabolism. Behavioral Neuroscience 114:805–813.

Farrell WJ and Alberts JR (2002a) Maternal responsiveness to infant Norway rat (*Rattus norvegicus*) ultrasonic vocalizations during the maternal behavior cycle and after steroid and experiential induction regimens. Journal of Comparative Psychology 116:286–296.

Farrell WJ and Alberts JR (2002b) Stimulus control of maternal responsiveness to Norway rat (*Rattus norvegicus*) pup ultrasonic vocalizations. Journal of Comparative Psychology 116:297–307.

Francis RL (1977) 22-kHz calls by isolated rats. Nature 265:236–238.

Gourevitch G and Hack M (1966) Audibility in the rat. Journal of Comparative and Physiological Psychology 62:289–291.

Hofer MA and Shair HN (1987) Isolation distress in 2-week-old rats: influence of home cage, social companions, and prior experience with littermates. Developmental Psychobiology 20:465–476.

Hofer MA, Brunelli SA, Shair HN (1994) Potentiation of isolation-induced vocalization by brief exposure of rat pups to maternal cues. Developmental Psychobiology 26:81–95.

Hofer MA, Masmela JR, Brunelli SA, Shair HN (1998) The ontogeny of maternal potentiation of the infant rats' isolation call. Developmental Psychobiology 33:189–201.

Insel TR, Hill JL, Mayor RB (1986) Rat pup ultrasonic isolation calls: possible mediation by the benzodiazepine receptor complex. Pharmacology Biochemistry and Behavior 24:1263–1267.

Kehoe P and Harris JC (1989) Ontogeny of noradrenergic effects of ultrasonic vocalizations in rat pups. Behavioral Neuroscience 103:1099–1107.

Kehoe P, Callahan M, Daigle A, Malinson K, Brudzynski S (2001) The effect of cholinergic stimulation on rat pup ultrasonic vocalizations. Developmental Psychobiology 38:92–100.

Knutson B, Burgdorf J, Panksepp J (1998) Anticipation of play elicits high-frequency ultrasonic vocalizations in young rats. Journal of Comparative Psychology 1:65–73.

Knutson B, Burgdorf J, Panksepp J (2002) Ultrasonic vocalizations as indices of affective states in rats. Psychological Bulletin 128:961–977.

Kraebel KS, Brasser SM, Campbell JO, Spear LP, Spear NE (2002) Developmental differences in temporal patterns and potentiation of isolation-induced ultrasonic vocalizations: influence of temperature variables. Developmental Psychobiology 40:147–159.

Lee HJ, Choi J-S, Brown TH, Kim JJ (2001) Amygdalar NMDA receptors are critical for the expression of multiple conditioned fear responses. Journal of Neuroscience 21:4116–4124.

Miczek KA, Weerts EM, Vivian JA, Barros HM (1995) Aggression, anxiety and vocalizations in animals: GABA$_A$ and 5-HT anxiolytics. Psychopharmacology 121:38–56.

Molewijk HE, van der Poel AM, Mos J, van der Heyden JAM, Olivier B (1995) Conditioned ultrasonic distress vocalizations in adult male rats as a behavioural paradigm for screening anti-panic drugs. Psychopharmacology 117:32–40.

Noirot E (1968) Ultrasounds in young rodents. II. Changes with age in albino rats. Animal Behaviour 16:129–134.

Noirot E (1972) Ultrasounds and maternal behavior in small rodents. Developmental Psychobiology 5:371–387.

Okon EE (1971) The temperature relations of vocalization in infant Golden hamsters and Wistar rats. Journal of Zoology London 164:227–237.

Olivier B, Molewijk HE, van der Heyden JAM, van Oorschot R, Ronken E, Mos J, Miczek KA (1998) Ultrasonic vocalizations in rat pups: effects of serotonergic ligands. Neuroscience and Biobehavioral Review 23:215–227.

Oswalt GL and Meier GW (1975) Olfactory, thermal,

and tactual influences on infantile ultrasonic vocalization in rats. Developmental Psychobiology 8: 129–135.

Panksepp J and Burgdorf J (2000) 50-kHz chirping (laughter?) in response to conditioned and unconditioned tickle-induced reward in rats: effects of social housing and genetic variables. Behavioural Brain Research 115:25–38.

Roberts LH (1975a) The functional anatomy of the rodent larynx in relation to audible and ultrasonic cry production. Zoological Journal of the Linnaean Society 56:255–264.

Roberts LH (1975b) Evidence for the laryngeal source of ultrasonic and audible cries of rodents. Journal of Zoology, London 175:243–257.

Roberts LH (1975c) The rodent ultrasound production mechanism. Ultrasonics 13:83–85.

Sachs BD and Bialy M (2000) Female presence during postejaculatory interval facilitates penile erection and 22-kHz vocalization in male rats. Behavioral Neuroscience 114:1203–1208.

Sales GD (1972a) Ultrasound and mating behaviour in rodents with some observations on other behavioural situations. Journal of Zoology, London 168: 149–164.

Sales GD (1972b) Ultrasound and aggressive behaviour in rats and other small mammals. Animal behaviour 20:88–100.

Sanders I, Weisz DJ, Yang BY, Fung K, Amirali A (2001) The mechanism of ultrasonic vocalization in the rat. In: Society for Neuroscience. San Diego, CA. November 10–15, 2001.

Schreiber R, Melon C, Vry JD (1998) The role of 5-HT receptor subtypes in the anxiolytic effects of elective serotonin ruptake inhibitors in the rat ultrasonic vocalization test. Psychopharmacology 135:383–391.

Sewell GD (1967) Ultrasound in adult rodents. Nature 215:512.

Sewell GD (1970) Ultrasonic signals from rodents. Ultrasonics 8:26–30.

Shair HN, Brunelli SA, Masmela JR, Boone E, Hofer MA (2003) Social, thermal, and temporal influences on isolation-induced and maternally potentiated ultrasonic vocalizations of rat pups. Developmental Psychobiology 42:206–222.

Smotherman WP, Bell RW, Starzec J, Elias J, Zachman TA (1974) Maternal responses to infant vocalizations and olfactory cues in rats and mice. Behavioural Biology 12:55–66.

Sokoloff G and Blumberg MS (2001) Competition and cooperation among huddling infant rats. Developmental Psychobiology 39:1–9.

Takahashi LK (1992) Developmental expression of defensive responses during exposure to conspecific adults in preweanling rats (Rattus norvegicus). Journal of Comparative Psychology 106:66–77.

Takahashi LK, Thomas DA, Barfield RJ (1983) Analysis of ultrasonic vocalizations emitted by residents during aggressive encounters among rats (Rattus norvegicus). Journal of Comparative and Physiological Psychology 97:207–212.

Thomas DA and Barfield RJ (1985) Ultrasonic vocalization of the female rat (Rattus norvegicus) during mating. Animal Behavior 33:720–725.

van der Poel AM, Noach EJK, Miczek KA (1989) Temporal patterning pf ultrasonic distress calls in the adult rat: effects of morphine and benzodiazepines. Psychopharmacology 97:147–148.

Vivian JA and Miczek KA (1991) Ultrasounds during morphine withdrawal. Pyshcopharmacology 104:187–193.

Wetzel DM, Kelley DB, Campbell BA (1980) Central control of ultrasonic vocalizations in neonatal rats: I. Brain stem motor nuclei. Journal of Comparative and Physiological Psychology 94:596–605.

Winslow JT and Insel TR (1991) Endogenous opioids: Do they modulate the rat pup's response to social isolation. Behavioral Neuroscience 105:253–263.

Yajima Y, Hayashi Y, Yoshii N (1980) The midbrain gray substance as a highly sensitive neural structure for the production of ultrasonic vocalizations in the rat. Brain Research 198:446–452.

Yajima Y, Hayashi Y, Yoshii N (1981) Identification of ultrasonic vocalization substrates determined by electrical stimulation applied to the medulla oblongata in the rat. Brain Research 229:353–362.

Zhang SP, Davis PJ, Bandler R, Carrive P (1994) Brain stem integration of vocalization: role of the midbrain periaqueductal gray. Journal of Neurophysiology 72:1337–1356.

Cognition

VII

Object Recognition

36

DAVE G. MUMBY

OBJECT-RECOGNITION PARADIGMS

Object-recognition memory is the ability to discriminate the familiarity of previously encountered objects.[1] People with normal memory may engage this ability hundreds of times each day, but impaired recognition occurs in many memory disorders, including those resulting from Alzheimer's disease, stroke, chronic alcoholism, encephalitis, and traumatic brain injury. The ability to distinguish between an object one has encountered previously and one that is new is so fundamental to normal memory function that understanding its neural bases seems necessary to develop a comprehensive picture of how the brain remembers things. Such knowledge may also contribute to better methods of diagnosing and treating certain memory disorders.

Rats also distinguish between objects they have previously encountered and ones they have not. The extent to which object-recognition memory involves similar processes in rats and humans is not entirely clear. Standardized tasks for assessing object recognition in rats, the effects of various brain lesions, and effects of drugs on this ability all suggest similar processes.

Two paradigms are most often used to assess object recognition in rats: delayed nonmatching-to-sample (DNMS) and novel-object-preference (NOP). On DNMS tasks, a sample object is briefly presented, and after a retention delay, it is presented again, along with a novel object (i.e., one the rat has not previously encountered on the current ses-

sion). The rat is rewarded for selecting the novel object. Reliably accurate performance requires, among other things, that the rat can recognize the sample object. Memory demands are controlled by varying duration of the delay or number of objects to remember on each trial. There are several trials per session, each using a different sample and novel object, so rats are consistently rewarded for selecting an unfamiliar object. Most DNMS procedures use pseudo-trial-unique objects–particular objects never recur within a session but may recur across multiple sessions widely separated in time. Rats also learn delayed *matching*-to-sample with objects, but they master DNMS more quickly because the nonmatching response is consistent with their natural bias for selecting novel objects.

The NOP task takes advantage of the tendency of rats to investigate novel objects more than familiar objects (Berlyne, 1950). Conventional procedures are similar to those described by Ennaceur and Delacour (1988): A rat is placed in an open-field arena and allowed to explore two identical sample objects for a few minutes. The rat is then removed for a delay, after which it returns to the arena with two new objects—one is identical to the sample and the other is novel. Normal rats spend more time exploring the novel object during the test, indicating that they recognize the sample object. With conventional procedures, rats can show a novel-object preference after delays of up to 24 hours and, with modified procedures, up to several weeks (see "Retrograde Object Recognition").

SENSORY SYSTEMS

Investigators often make the tacit assumption that rats perform DNMS and NOP tasks by discriminating the familiar *visual* features of sample objects. Rats have the opportunity to see, feel, and smell the objects on conventional versions of both tasks, however. Careful observation of a rat's behavior may give few clues to the type of sensory information used in a particular trial. A rat may sniff and palpate an object while investigating it, but this does not mean the rat is using olfactory or tactile information to make its choice. If, however, a rat performing a DNMS task consistently veers toward the correct object while approaching it and is still several centimeters away, it is likely using vision.

DELAYED NONMATCHING-TO-SAMPLE

TASK VARIANTS

Figure 36–1 illustrates a Y-maze procedure in which distinctive goal boxes containing ob-

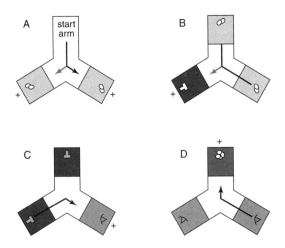

Figure 36–1. Y-maze DNMS task. Pluses indicate which arms the rat will be rewarded for entering. *A*, The first trial begins with the rat in one arm, from which it can enter either of the other two arms, both containing identical goal boxes. *B* to *D*, Each successive choice is between a box identical to the one containing the rat and a box the rat has not encountered during the current session. (From Aggleton, 1985.)

jects are inserted into the arms (Aggleton, 1985). The rat is confined for 20 seconds in a sample box and then placed in a "featureless" box for the retention delay, after which it receives access to the other two maze arms. Each arm contains a distinctive box with objects—one matches the sample box, and the other is novel. The rat is rewarded if it enters the novel box, which then serves as the sample for the next trial.

Other DNMS versions use discrete trials consisting of a sample phase and a choice phase. One method (Fig. 36–2A–D) uses a runway with a start area separated from a goal area by an experimenter-controlled door (Rothblat and Hayes, 1987). The goal area contains food wells over which objects are positioned. For the sample phase, the door to the start area is raised, and the rat approaches and displaces a sample object, for which it receives food reward. The rat is picked up and placed back in the start area for the delay, after which it again has access to the goal area, which now contains a duplicate of the sample and a novel object. A version in which the sample phase is located at one end of the apparatus and the choice phase is located at the other end, and rats are not handled between or within trials (Kesner et al., 1993) (Fig. 36–2E–H). The version shown in Figure 36–3 is similar but has a central start compartment, a pair of food wells at either end, and the end of the apparatus where the sample and choice phases occur changes randomly across trials (Mumby et al., 1990).

Before DNMS training, rats are familiarized with the apparatus and shaped to displace objects from food wells. This can be accomplished by training on a simple object-discrimination task, on which the same two objects are repeatedly presented together and selection of one of them is rewarded (Mumby et al., 1990). Rats learn object discriminations quickly, probably because manipulation of small objects is a natural behavior of rats seeking food (Barnett, 1956). The acquisition phase of DNMS training ensues, using a brief

Figure 36–2. *A to D*, DNMS procedures similar to those described by Rothblat and Hayes (1987). The rat is picked up and returned to the start area between the sample and choice phases of each trial. *E to H*, Procedures used by Kesner et al. (1993). A central door is raised and lowered by the experimenter to control the retention delay and the rat's opportunity to shuttle back and forth between the sample and choice ends of the runway. The rat is not handled between or within trials.

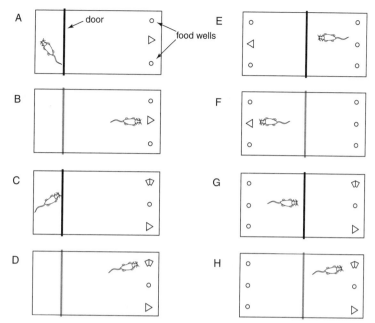

retention delay of only a few seconds, until the rat reaches some criterion level of accuracy, establishing that is has learned the nonmatching rule. Thereafter, recognition is assessed with longer delays, up to several minutes. Acquisition is measured by the number of trials or errors required to reach the performance criterion at the brief delay, and performance at longer delays is expressed as percentage of trials that are correct. Procedures that vary the number of objects to remember on each trial can be found for Y-maze DNMS (Steele and Rawlins, 1993) and discrete trial versions (Mumby et al., 1995).

MINIMIZING EXPERIMENTER EFFECTS

Rats likely perceive humans as large, noisy, smelly potential predators. The experimenter plays an interactive role in DNMS testing, so it is essential that rats are first well-tamed. A rat that perceives the experimenter as the most interesting thing in the room will pay more attention to the experimenter than to the task. This is often the most difficult and frustrating aspect of DNMS testing for persons inexperienced with rats and with how they react to movements and sounds.

Experimenters should also monitor their own behavior for inadvertent cues that rats could use to solve the task, such as unconscious movements the experimenter reliably makes as the rat approaches the correct object. If either the sample or the novel object is consistently handled more than the other, rats may learn to solve the task by discriminating the relative strength of the experimenter's scent on the objects (Mumby, Kornecook et al., 1995). The easiest solution is to use two identical copies of an object as the sample on each trial—one for the sample phase and the other for the test phase—so that all of the objects can be positioned before the trial begins.

FOOD DEPRIVATION

A common practice when using tasks motivated by food reward is to restrict the daily food intake of rats. Typically, rats are maintained at body weights around 80% to 85% of the weight of age-matched, free-feeding rats.

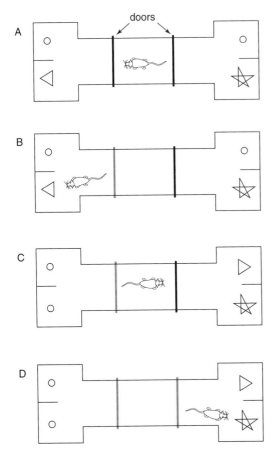

Figure 36–3. DNMS procedures described by Mumby et al. (1990). *A*, Before each trial, the rat is enclosed in the central area, and two objects are positioned over food wells, one at each end. *B*, A door is opened, and the rat approaches, and displaces the sample, and receives a food pellet. *C*, The sample is placed over the vacant food well at the other end. The second door remains closed for the retention delay. *D*, The rat displaces an object, and receives a food pellet if it is the novel object. The rat returns to the central area until the next trial; it is not handled between or within trials.

This level of deprivation is unnecessary and may be counterproductive when training rats on DNMS. Hyperactivity and problems with response inhibition interfere with accuracy, and both increase with increasing food deprivation. Deprivation to 85% of free-feeding weight may help initially, when rats are learning basic task procedures, but once they are readily approaching and displacing objects, they can be maintained at 90% to 100% of free-feeding weights.

Success in training rats on DNMS requires that they move about the apparatus at a pace that is neither too slow nor too fast and that they do not show persistent side biases (e.g., always picking the object on the left side). Rats that run too fast make frequent errors and may not take time to examine the sample at the beginning of a trial. Inexperienced experimenters may be pleased that their rats run quickly and waste no time while frantically leaping into the goal area and knocking over the first object they see, but such behavior significantly prolongs the acquisition phase of DNMS training. Fast-running and side biases often appear together. Slowing down a hasty rat is often all that is needed to eliminate a side bias, but additional measures may be required, such as remedial training on a simple object-discrimination task.

Slow rats are easily distracted between trials and during retention delays. This can be frustrating for impatient experimenters and may lead to unwise improvisations in protocol in attempts to get a rat back on task. Optimal performance requires adjustment of each rat's daily food ration in accordance with its current pace—rats that are too slow receive slightly less food and rats that are too frenzied receive slightly more. The objective is to optimize and equalize the pace of the rats. This is more important for successful DNMS training than making sure all rats eat the same amount of food.

INTERPRETING TREATMENT EFFECTS ON DNMS TASKS

Before concluding that a DNMS deficit is due to failure of object-recognition memory, several other possibilities must be considered. Some treatments produce nonspecific effects that interfere with successful DNMS performance, such as hyperactivity, an altered stress response, or problems with response inhibition. Nonspecific effects can last several days after highly invasive treatments, such as brain surgery, but often they subside eventually. If rats are not given a chance to overcome transient nonspecific effects, DNMS performance may be impaired even if recognition memory is unaffected. One solution is to train them first on an object-discrimination task before commencing with DNMS testing. Rats learn object-discrimination problems relatively easily, even after some treatments that impair DNMS performance, and they can overcome various nonspecific treatment effects while performing an object-discrimination task.

It is often assumed that after learning the nonmatching rule, rats will apply it on every trial, so accuracy will depend mostly on whether they recognize the sample object. A classic pattern of treatment effect on memory tests is the *delay-dependent deficit*, in which magnitude of the deficit increases as retention delay increases. The usual interpretation is that performance is normal at brief delays because the conditions do not sufficiently tax memory processes affected by the treatment. But delay-dependent effects can also occur for nonmnemonic reasons. Even after reaching asymptotic performance at short delays, certain skills must be mastered to achieve good scores at longer delays. With practice at long delays, rats may become better at attending to object features, avoiding distraction and frustration during the retention delay, withholding hasty choices, or discriminating the sample phase from the choice phase of a trial. Deficiencies in these skills diminish accuracy, even in animals with normal object-recognition abilities.

Pretreatment training can reduce the likelihood of seeing a treatment effect on DNMS performance (Mumby, 2001) but may make it easier to interpret an effect when one occurs. If a treatment produces DNMS deficits only when rats are not pretrained, it is possible that the treatment interferes with learning how to perform well on the task but may actually have no effect on object-recognition memory. To understand why this is so, keep in mind what rats learn during DNMS training: They do not learn how to recognize

objects—they already possess that capacity before the experiment begins. During training, rats learn the reward contingency (the nonmatching rule) and various skills necessary for good performance. Pretreatment training does not guarantee that acquired skills will be unaffected by a treatment, but if rats readily approach and displace objects and perform with better-than-chance accuracy on the first few posttreatment trials, they likely remember the nonmatching rule and general task procedures.

NOVEL-OBJECT PREFERENCE

BASIC TEST PROCEDURES

The NOP task has become a popular method for assessing object recognition in rats, mainly for practical reasons: it does not require food deprivation or prolonged training, so experiments can be completed in considerably less time than with DNMS tasks. The NOP task is just one of many procedures to assess how rats respond to various types of novelty. During the 1950s, while most Russian and Western psychologists were preoccupied with understanding how associations are learned through Pavlovian and operant conditioning procedures, Berlyne and some of his contemporaries were interested in how animals respond to certain features of the environment simply because they are new.

In one influential study, Berlyne (1950) allowed rats to explore three copies of an object in an open field arena. When one object was replaced with a novel object, the rats showed an exploratory preference for the novel object. It is not clear to what extent this occurred because that object was unfamiliar or because it was the odd object of the triplet, but the phenomenon inspired many other studies that examined conditions affecting how rats respond to novelty. Most of the relevant work of this period has been reviewed by Berlyne (1960) and Fowler (1965). There was much theorizing about emotions and drive states, such as surprise and curiosity, none of which is particularly relevant to understanding the processes of object-recognition memory. In many studies, however, behaviors of interest included object exploration, and much was learned about the variables that affect this behavior. Knowledge of some of these factors is essential to successful planning and interpretation of NOP experiments. Varieties of exploratory behavior and procedures to evoke and measure it are described by Berlyne (1960) and Fowler (1965) and in a recent analysis by Hughes (1997).

There is no standard operational definition of *object exploration*, but most investigators find it sufficient to include criteria that are likely to capture most of a rat's actual object investigation, such as having the head oriented toward the object and within a certain distance, usually a few centimeters. Renner and his colleagues provided useful analyses of the structure and organization of exploration and object investigation in rats (Renner, 1987; Renner and Seltzer, 1991, 1994). When using the NOP test to assess object recognition, it is not necessary to analyze the fine details of how rats interact with the objects. It is essential, however, to have an operational definition that can be applied easily and consistently.

Time spent exploring each object during the test phase is used to calculate a measure of relative exploratory preference for the sample and novel objects. For example, an *exploration ratio* reflects the proportion of total object exploration that was spent exploring the novel object $[T_{novel}/(T_{novel} + T_{sample})]$. Ancillary measures include time spent investigating the sample during the familiarization phase and the difference in time spent exploring the novel object and the sample on the test phase. Some investigators use difference scores during the test as the primary dependent measure, but a proportional or ratio measure has the advantage of controlling for individual differences in total object exploration.

ADDITIONAL FACTORS THAT AFFECT OBJECT EXPLORATION

Whether rats approach or avoid a novel object depends on the familiarity of the environment in which it is encountered. When the environment is familiar, novel objects evoke approach and investigation; when the environment is unfamiliar, rats avoid novel objects (Besheer and Bevins, 2000; Montgomery, 1955; Sheldon, 1969). Accordingly, rats should be familiarized to the apparatus before NOP testing, by allowing them to explore it for a short period (10 to 15 minutes) on two or three occasions. Subsequently, during NOP tests, responses evoked by background stimuli are less likely to overshadow how the rats respond to the objects.

The complexity of objects affects how much investigation they evoke (Berlyne, 1955). Before NOP tests are performed, objects should be screened by measuring how much investigation each one evokes in a non-choice situation. Objects that evoke either too much or too little investigation should be excluded. This reduces variance in undiscriminated object exploration and produces cleaner results on NOP tests.

The duration of the preference test can influence the likelihood of detecting a novel-object preference. The preference tends to be robust during only the first 1 or 2 minutes of a test session and diminishes thereafter, presumably because both objects become equally familiar as they are explored (Dix and Aggleton, 1999). A minute-by-minute assessment of differential exploration shows how preferences change over the test phase, allowing one to focus on the most sensitive portion (Mumby et al., 2002). The inclusion of portions of the test that come after rats stop discriminating merely adds noise to the data, which may obscure a significant preference that occurs in the early portions of the test.

INTERPRETING TREATMENT EFFECTS ON NOP TASKS

Despite the simplicity of NOP procedures, it is a challenge to discern when a treatment effect occurs because of impaired object recognition or because of some other reason. When a treatment has no effect and rats spend more time exploring novel objects than sample objects, it is likely that object-recognition abilities are not grossly impaired. When the preference is disrupted, however, and even if perceptual impairments can be ruled out, there are several potential reasons for a failure to spend more time exploring the novel object. One possibility is that a rat does not recognize the sample; another possibility is that the instinctive bias for exploring novel objects has been overshadowed or abolished.

A fundamental question that should be answered *before* comparing the scores of two groups is whether performance *within* each group indicates a significant preference. A one-sample *t* test can determine whether the mean exploration ratio of a group is significantly different from the chance level of 0.5. (Most studies report ratios in control rats between 0.6 and 0.7.) The first comparison *between* two groups, therefore, is on the appropriate *nominal* scale (i.e., preference versus no preference). Comparisons of group means should follow, but it is important to note that those comparisons merely concern the degree of preference. It is not clear to what extent stronger preference for the novel object in one rat relative to another rat should be taken as indicating superior recognition abilities.

RETROGRADE OBJECT RECOGNITION

Retrograde object recognition is the ability to discriminate the familiarity of objects that were encountered before a treatment. When the treatment is a brain lesion, tasks are needed on which control animals show significant retention after the period of post-surgery recovery. Neither the DNMS nor conventional NOP procedures are suitable. However, rats can show a novel-object preference after delays of several weeks if they are given repeated sample exposures (Mumby, Glenn, Nesbitt, & Kyriazis, 2002). To increase the amount of object exploration during the test, after such long retention delays, rats should first be rehabituated to the chamber before the test.

A NEW NOVEL-OBJECT PREFERENCE TASK

Evolutionary perspectives emphasize the importance of considering the behavior of rats in their natural environment when designing laboratory paradigms (Timberlake, 1984). The NOP task takes advantage of an innate exploratory bias, but the conventional procedures may constrain certain niche-related exploratory responses. For example, when exploring their natural environment, rats travel from place to place, investigating objects along the way, with a tendency to move to new places and objects rather than to return to ones they have just investigated. These aspects of natural exploratory behavior are continuously thwarted in the confined space of a standard open field with two objects.

A modified NOP procedure circumvents these limitations. (Mumby et al., submitted). The apparatus is a circular track that can be divided into different sectors (Fig. 36–4). Each divider has a door that can be set to open in only one direction, so when rats pass into a new sector, they cannot return to the previous sector. With all doors set to open in the same direction, rats travel around the track in that direction only, spending as much time as they choose in each sector and becoming familiar with several objects concurrently, as they encounter different pairs of identical objects in each sector. For the test, each sector

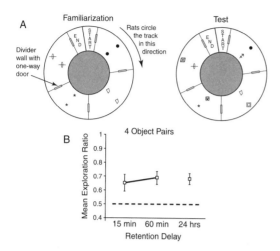

Figure 36–4. A, Circular-track for assessing novel-object preference. (See text for details.) B, Data from rats tested with four object pairs, and retention delays of 15 and 60 minutes (n = 12) or 24 hours (n = 12). Exploration ratio is the proportion of total object exploration during the test directed toward novel objects.

contains a copy of the sample and a novel object, and exploration of each object is recorded as the rat travels around the track. Because rats cannot return to a sector after they leave, several measurements of novel-object preference can be taken on each trial. Different manipulations can be made in different sectors, enabling simultaneous assessment of a rat's response to different types of novelty.

CONCLUSION

Despite superficial similarities, DNMS and NOP tests engage different behavioral and motivational systems and therefore entail different procedural pitfalls and interpretational challenges. Still, findings from lesion experiments using either task have so far been fairly consistent, suggesting that the hippocampus is not critical for object-recognition memory (Mumby, 2001), whereas the perirhinal cortex plays a more significant role. These conclusions are consistent with findings of changes in single-unit responses and c-*fos* expression

within the perirhinal cortex produced by repeatedly presented visual stimuli (Aggleton and Brown, 1999). Studies of the neural bases of object-recognition memory are beginning to focus on other structures and circuits, but it is too early to ascribe a particular role in object-recognition memory to any of them.

NOTE

1. This chapter discusses behavior pertaining to tests that require discriminating the familiarity of three-dimensional objects. Not considered are tasks that require learning relationships among individual objects and other events (e.g., reward).

REFERENCES

Aggleton JP (1985) One-trial object recognition by rats. Quarterly Journal of Experimental Psychology 37B:279–294.

Aggleton JP and Brown MW (1999) Episodic memory, amnesia, and the hippocampal-anterior thalamic axis. Behavioral Brain Science 22:425–444.

Barnett SA (1956) Behavior components in the feeding of wild and laboratory rats. Behaviour 9:24–43.

Besheer J and Bevins RA (2000) The role of environmental familiarization in novel-object preference. Behavioural Processes 50:19–29.

Berlyne DE (1950) Novelty and curiosity as determinants of exploratory behaviour. British Journal of Psychology 41:68–80.

Berlyne DE (1960) Conflict, arousal, and curiosity (Harlow HF, ed.). New York: McGraw-Hill.

Berlyne DE (1955) The arousal and satiation of perceptual curiosity in the rat. Journal of Comparative and Physiological Psychology 48:238–246.

Dix SL and Aggleton JP (1999) Extending the spontaneous preference test of recognition: evidence of object-location and object-context recognition. Behavioural Brain Research 99:191–200.

Ennaceur A and Delacour J (1988) A new one-trial test for neurobiological studies of memory in rats: I. behavioural data. Behavioural Brain Research 31:47–59.

Fowler H (1965) Curiosity and exploratory behavior. New York: Macmillan.

Hughes RN (1997) Intrinsic exploration in animals: motives and measurement. Behavioural Processes 41:213–226.

Kesner RP, Bolland BL, Dakis M (1993) Memory for spatial locations, motor responses, and objects: triple dissociation among the hippocampus, caudate nucleus, and extrastriate visual cortex. Experimental Brain Research 93:462–470.

Montgomery KC (1955) The relation between fear induced by novel stimulation and exploratory behavior. Journal of Comparative and Physiological Psychology 48:254–260.

Mumby DG (0000) Reducing constraints on exploratory behavior enhances novel-object preference in rats. (submitted to *Learning and Memory*).

Mumby DG (2001) Perspectives on object-recognition memory following hippocampal damage: lessons from studies in rats. Behavioural Brain Research 127:159–181.

Mumby DG, Kornecook TJ, Wood ER, Pinel JPJ (1995) The role of experimenter-odor cues in the performance of object-memory tasks by rats. Animal Learning and Behavior 23:447–453.

Mumby DG, Gaskin S, Glenn MJ, Schramek TE, Lehmann H (2002) Hippocampal damage and exploratory preferences in rats: memory for objects, places, and contexts. Learning and Memory 9:49–57.

Mumby DG, Pinel JPJ, Kornecook TJ, Shen MJ, Redila VA (1995) Memory deficits following lesions of hippocampus or amygdala in rats: assessment by an object-memory test battery. Psychobiology 23: 26–36.

Mumby DG, Pinel JPJ, Wood ER (1990) Nonrecurring-items delayed nonmatching-to-sample in rats: a new paradigm for testing nonspatial working memory. Psychobiology 18:321–326.

Mumby DG, Glenn MJ, Nesbitt C, Kyriazis DA (2002) Dissociation in retrograde memory for object discriminations and object recognition in rats with perirhinal cortex damage. Behavioural Brain Research 132:215–226.

Renner MJ (1987) Experience-dependent changes in exploratory behavior in the adult rat (Rattus norvegicus): overall activity level and interactions with objects. Journal of Comparative Psychology 101:94–100.

Renner MJ and Seltzer CP (1991) Molar characteristics of exploratory and investigatory behavior in the rat (Rattus norvegicus). Journal of Comparative Psychology 105:326–339.

Renner MJ and Seltzer CP (1994) Sequential structure in behavioral components of object investigation by Long-Evans rats. Journal of Comparative Psychology 108:335–343.

Rothblat LA and Hayes LL (1987) Short-term object recognition memory in the rat: nonmatching with trial-unique stimuli. Behavioral Neuroscience 101: 587–590.

Sheldon AB (1969) Preference for familiar versus novel stimuli as a function of the familiarity of the environment. Journal of Comparative and Physiological Psychology 67:516–521.

Steele K and Rawlins JNP (1993) The effects of hippocampectomy on performance by rats of a running recognition task using long lists of non-spatial items. Brain Research 54:1–10.

Timberlake W (1984) An ecological approach to learning. Learning and Motivation 15:321–333.

Piloting

37

ETIENNE SAVE AND BRUNO POUCET

THE PERCEPTION OF SPACE

In the wild, rats restrict their activity to a *home range*, which represents only a minor portion of the potentially useful environment (Bovet, 1998). An important place within the home range is the *nest*, which provides protection against predators and climatic conditions. Nevertheless, rats must leave their nest to look for food, water, and conspecifics, so they spend a substantial amount of time in moving from one place to another. Their survival depends on their ability to memorize locations (places) and to use behavioral strategies to navigate efficiently between their home base and other places of interest.

The selection of appropriate navigational strategies is primarily determined by the perception of space, that is, by the nature of the cues that can be used for navigation. There are two main categories of spatial cues: allothetic and idiothetic. *Allothetic cues* are provided by the environment and include visual, olfactory, and auditory information. *Allothetic navigation* refers to the process of "determining and maintaining a course or trajectory from one place to another" (Gallistel, 1990) by using environmental cues. *Idiothetic cues* are derived from the animal's own movements, including information provided by the vestibular, proprioceptive, and somatosensory systems; efference copies of motor commands; and external motion-related information such as optic flow. Idiothetic cues support a form of navigation called *dead reckoning*, in which an animal locates its starting point relative to

its current position (see Chapter 38). In recent hypotheses, allothetic navigation and dead reckoning are assumed to be complementary systems. The purpose of the present chapter is to describe allothetic navigation and the methods for studying allothetic navigation in the laboratory. Because a remarkable array of tasks have been developed, this chapter reviews only those tasks that have received widespread use and are important for understanding the theoretical basis of allothetic navigation.

THEORIES OF SPATIAL LEARNING

Several theories have been developed to account for spatial learning. For stimulus–response (S-R) theorists, spatial learning consists of chaining together a number of motor responses that link relevant external stimuli. This view was challenged by Tolman (1948), who argued that rats can learn the location of a place where they have been rewarded independent of the specific movements necessary to reach it. Tolman explained this ability by introducing the notion of a cognitive map, a mental representation held by the rat in which it encodes the elements of a task and the spatial relationships between these elements.

The Tolman notion of a cognitive map was further developed by O'Keefe and Nadel (1978), who proposed a theoretical framework to account for spatial behavior and its neural bases. They made a distinction between two main processes used by rats to perform spatial tasks. They proposed that rats can use either a taxon system or a locale system. The *taxon system* is based on S-R associations, whereas the locale system is based on spatial relationships in the form of a cognitive map. O'Keefe and Nadel further proposed that the brain regions subserving these forms of spatial navigation differ and are at least in part independent.

In contrast to O'Keefe and Nadel's (1978) notion of a cognitive map, associationists such as Sutherland and Rudy (1989) propose that all spatial behaviors can be described in terms of learned associations but that it is the form of the association, simple or configural, that determines which brain structures will be used. Simple associations correspond to S-R associations in theories of conditioning. Configural associations combine the representations of elementary stimulus events to build global representations. The configural association system also stores associations between the elementary events and the global representation. According to this theory, place learning is assumed to involve the configural association system, which encodes and stores the configuration of spatial relationships between cues that specifically define a location.

SPATIAL STRATEGIES

The taxon and locale systems enable the rats to use different strategies, including cue navigation, guidance, and route strategies (taxon system), and place navigation strategy (locale system).

CUE NAVIGATION

The most elementary navigational strategy is to move toward or away from a directly perceived cue. The use of such a system requires a radial stimulation gradient field centered on the source (Benhamou and Bovet, 1992). The notion of *gradient field* refers to the variation

of intensity of the source as a function of the distance. In the simplest case, intensity varies monotonically with the distance. The animal can therefore reach the goal (i.e., the origin of the gradient) by navigating in the gradient direction or, conversely, avoid a place by navigating in the opposite direction. A gradient field may originate from auditory and olfactory cues. Navigation toward a conspicuous visual cue has similar characteristics. The visual cue is considered to be at the center of a radial gradient where the gradient direction corresponds to the direction in the light flux. Thus, reducing the distance from the visual cue (which leads to an increase in the apparent cue size) is equivalent to navigating along the axis of greatest intensity. The animal can also use a similar mechanism to reach a hidden goal that is closely associated to a salient landmark (called a beacon). A *landmark* is therefore considered to be an intermediate goal whose spatial contiguity with the real goal has been learned through an associative process.

GUIDANCE

If the goal cannot be directly perceived and if there is no landmark closely associated to the goal, the animal must rely on environmental landmarks to navigate. One strategy is to use landmarks as directional cues. Guidance, therefore, requires that the animal directs its attention to particular landmarks and maintain some egocentric (i.e., relative to the animal) spatial relationships with respect to these landmarks to reach the goal (O'Keefe and Nadel, 1978). A place-recognition process is then needed to identify the goal location. Rats may be guided by either individual or configurations of landmarks. For example, they may be able to memorize a view of the spatial arrangement of landmarks seen from the goal (snapshot) and so adjust movements until a current view matches a memorized view (Collett et al., 1986).

ROUTES

To reach a remote goal, cue navigation and guidance may be elaborated into route learning. In *route learning*, rats may memorize a sequence of associations coupling specific landmarks and motor responses, such as, "after reaching the rock, turn left." Each landmark is considered as an intermediary goal and the sequence of these goals constitutes a route. Routes allow rapid navigation at the expense of behavioral flexibility. Environmental changes that alter the sequence of landmarks may be fatal to successful navigation. In addition, retracing the route cannot be achieved by simply inverting the order of the landmarks but requires active learning of a new sequence.

PLACE NAVIGATION

Place navigation requires formation and use of a map-like representation of the environment (a cognitive map) that encodes the geometrical relationships between landmarks and places, independent of the animal's position. The use of a mapping strategy allows an animal to reach a goal from a variety of points, including over trajectories that it has not previously used. A rat thus is able to infer its location relative to particular places whose position is encoded in the representation and to compute optimal trajectories to reach these places. Place navigation is flexible because if a possible path is obstructed or a landmark is missing, the animal remains able to generate successful trajectories by performing detours or shortcuts and by adapting its navigational behavior to environmental changes.

PROCEDURAL, WORKING, AND REFERENCE MEMORY

Spatial learning requires different kinds of memory. *Procedural memory* is the memory of how to perform a task (e.g., visit an arm maze, get a reward, return to a place, etc.). *Working memory* is a short-term store that allows a rat

to remember that, for example, the goal is at a specific location during the current trial. Erasing the content of working memory after a trial enables the animal to learn a new goal location in the next trial. If the goal location remains constant across trials, it is encoded in the long-term *reference memory*.

BEHAVIORAL TESTS

In this section, we describe a number of behavioral tests that have been developed to evaluate and study allothetic navigation strategies. Nevertheless, it should be borne in mind that each of the tasks can be solved by one or more of the processes described earlier, and so special "probe" trials are needed to identify a particular strategy used.

THE T-MAZE

Rodents have a tendency to optimize their foraging behavior so that they avoid entering places that they have already visited, where food may have been depleted or is absent. In a T- or Y-maze, this results in *alternation behavior*. Two kinds of tasks can be conducted. First, the spontaneous alternation paradigm exploits the innate tendency of rodents to alternate arm choices on successive trials. Second, in the delayed forced-choice alternation task, alternation is rewarded whenever the animal enters an arm that is different from that specified by the experimenter during a sample trial (Fig. 37–1A). The nature of the strategy used by the animals to solve these tasks remains unclear. Because the goal arm differs from one trial to another, rats cannot learn a fixed association between a single response and a location. There is evidence that they use a variety of cues, including extramaze spatial cues and movement-related cues (Dudchenko, 2001). Performance in these tasks also depends on working memory, because a rat must keep track of which arm has been visited on the previous trial. In addition, in the

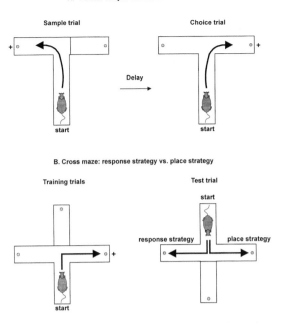

A. T-maze: delayed forced alternation task

Sample trial

Choice trial

Delay

start

start

B. Cross maze: response strategy vs. place strategy

Training trials

Test trial

start

response strategy

place strategy

start

Figure 37–1. *A,* Delayed forced alternation task in the T-maze. During the sample trial, the rewarded arm (+) is specified by the experimenter (the other arm is blocked). After a delay, the animal is required to choose between the two arms (choice trial) and is rewarded if it chooses the other arm. *B,* Response-versus-place strategy in the cross maze. During training, the rat is trained to enter the right arm. During the test trial, the animal is released from the opposite arm and is required to choose between the right arm and the left arm. A right turn indicates that the rat uses a response strategy based on stimulus–response association (association between a body turn and a place). A left turn indicates that the rat uses a place strategy based on a cognitive map.

delayed forced-choice alternation task version of the task, the introduction of a delay between the sample trial and the choice trial allows increased burden to be placed on the working memory demand.

THE CROSS-MAZE

The cross-maze has been developed to examine whether rats use a response strategy, that is, a learned association between a specific body turn and a place (taxon system) or a place strategy (locale system), to reach a target. The cross-maze consists of a sort of double T-maze constructed so that an animal can

reach the choice point connecting the two arms from two opposite starting arms. The apparatus is surrounded by numerous cues. During training, the animal is released from one starting arm and has to enter a specified arm, for instance, the right arm, to obtain a food reward. For the test trial, the animal is placed in the arm opposite the previous starting arm. If the rat performs a right turn to enter the goal arm, this suggests that it uses a response strategy based on learned S-R association. This outcome supports the S-R theory. In contrast, if the rat performs a left turn, this suggests that it can use a place strategy, and this supports the cognitive map theory (Fig. 37–1B). It is possible that rats are able to access either strategy (Packard and McGaugh, 1996).

THE RADIAL ARM MAZE

The radial arm maze is widely used to study spatial behavior, and many versions of the task have been developed (Foreman and Ermakova, 1998). The maze consists of a number of elevated runways (most often, eight) that radiate from a central platform, like the spokes of a wheel. In the continuous version of the task developed by Olton and Samuelson (1976), the animal is released on the central platform and allowed to freely explore the maze to collect small pieces of food from the ends of every arm. The optimal strategy consists of the rat visiting all of the arms without entering an already visited arm. A trial is complete when all food has been collected. Rats are extraordinarily good at performing the task; how they identify and remember the arm locations using the numerous cues surrounding the maze is not completely understood. Whether each arm location is encoded as a place in a cognitive map that guides the animal's choice or is considered to be an independent item in a list remains a matter of debate. The task also requires procedural memory (food can be found at the end of an arm) and working memory (a given arm has been exploited). Due to the regular geometry of the maze, the task may be successfully solved by making an egocentric response such as "keep turning left," although rats seldom use this strategy spontaneously. Nevertheless, to negate this strategy, only some arms may be baited. The radial arm task can also be used to differentiate the use of taxon strategies (some arms are marked by a local cue) and locale strategies (other arms can be identified only by distal cues).

THE PLACE PREFERENCE TASK

The place preference task combines random search and goal-directed navigation (Rossier et al., 2000). On a circular elevated platform, the rat must learn to locate an unmarked zone (the target zone) on the basis of room cues. Entering the target zone produces the release of a small (20 mg) food pellet from a feeder located above the arena. Because the pellet can land anywhere in the arena, the animal has to perform a pellet-chasing task (i.e., a random search) to get the reward while simultaneously learning that there is a target zone. This task involves place navigation abilities (locale system) but may be used to study dead reckoning if room cues are concealed.

THE MORRIS WATER TASK

The Morris water maze provides a versatile test to study navigation (Morris, 1981); unlike the tasks described earlier, food and water deprivation is not used to motivate the animals. In addition, the animals swim, a behavior at which rats are naturally adept. Briefly, the animal swims in a circular tank filled with opaque water to reach a submerged or visible platform, which is the only place to which it can escape from the water. In the place version of the task, there is no cue closely associated to the platform. Localization of the hidden platform can be performed and learned only by using distal room cues. To prevent

Place navigation task

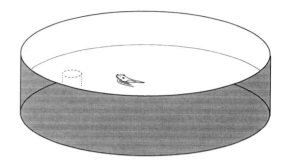

Cue navigation task

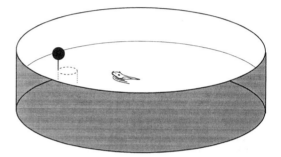

Figure 37–2. Two versions of the Morris water task. In the place version, the animal has to learn the location and navigate toward a submerged platform on the basis of distal room cues. This strategy is based on the locale system. In the cue version, the platform location is indicated by a beacon which guides the rat's navigation. This strategy is based on the taxon system.

the animal from using a local view for guidance, different starting places are used on successive trials. It is therefore assumed that learning results in elaboration of a cognitive map using the locale system (Fig. 37-2).

Acquisition of the Morris navigation task requires procedural memory and working memory. Procedural memory allows the rat to learn that it must swim, find an escape platform, climb on the platform, and so on. Working memory results in learning that the platform is at a specific location; rats can learn this in a single trial after mastering the procedural details of the task. The distinction between these two memories is important when considering the effects of brain damage on acquisition. For instance, poor performance may be wrongly attributed to spatial information processing impairment when it is due to procedural learning deficit (Whishaw et al., 1995). A matching-to-sample procedure can be used to study working spatial memory. During a sample trial, the animal has to locate the platform at a new location. During a matching trial, it can demonstrate that it knows that location. After one pair (or series) of trials, a new sample is presented.

Many variations of the Morris navigation task have been developed, allowing the study of a large range of navigation problems. For instance, if the platform surface is protruding above the water or if the submerged platform is closely associated with a beacon, rats are prone to use a cue navigation strategy rather than a place strategy. Guidance can also be used to solve the task. Animals can swim at a fixed distance of the wall corresponding to the distance of the platform until they come across the platform. In principle, normal rats do not use this strategy, but rats with brain damage may adopt it. Guidance may also be used if an animal is released from a constant start position. Navigation thus may be guided along an axis linking a prominent room cue to the starting point and including the platform.

The fact that an animal may use various strategies to solve water navigation problems requires the researcher to be aware of the animal's flexibility and allows the study of many different problems.

THE EXPLORATION TASK

Exploration and navigation are intimately related because navigation cannot occur without preliminary exploration of the environment (Renner, 1990). One of the major functions of exploration is information gathering, so it is necessary for the formation of spatial representations (O'Keefe and Nadel, 1978). Early studies have suggested that exploration, and, in particular, object exploration, is modulated by various factors such as the level of familiarity of the environment. More recently, Thinus-Blanc and colleagues (Poucet et al., 1986; Thinus-Blanc et al., 1987) developed a paradigm based on the exploration of a group of objects to study the nature of the spatial information processed during exploration. This paradigm consists of two successive phases: habituation and spatial change. First, repeated exposure to the same configuration of objects results in *habituation* of exploratory activity, that is, a decrement in the response with increasing familiarity. It is hypothesized that during this phase, rats acquire some knowledge of the spatial characteristics of the environment. Habituation thus reflects the elaboration of a spatial representation. Once habituation is achieved, the animals are exposed to a new configuration of the same familiar objects. A renewal of exploration following this change is assumed to reflect that the animal (1) encodes certain spatial relationships and (2) is able to compare the previous and current situations. Such renewal thus is an adaptive behavior aimed at updating the spatial representation of the environment.

The exploratory response to the spatial change depends on the nature of the change (Fig. 37–3). For instance, strong modification of the configuration by displacing one of four objects induces intense reexploration of all of the objects (Fig. 37–3B). In contrast, a moderate change yields reexploration specifically directed toward the displaced object (Fig. 37–3C). Topological modifications (two objects are switched [Fig. 37–3D]), also result in specific reexploration of the displaced objects (Poucet et al., 1986). Interestingly, when the metric properties of the configuration are modified without altering the affine properties, for example, by merely expanding a square arrangement (Fig. 37–3E), the rats do not exhibit any renewal of exploration (Thinus-Blanc et al., 1987). These results suggest that during exploration the animals encode some but not all geometric properties of the environment.

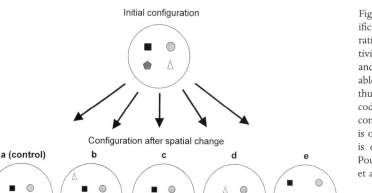

Initial configuration

Configuration after spatial change

| a (control) | b | c | d | e |

no re-exploration

re-exploration of all objects

re-exploration of the displaced object

re-exploration of the displaced objects

no re-exploration

Figure 37–3. Effects of various modifications (from *a* to *e*) of a configuration of objects on exploratory activity. A renewal of exploration (*b*, *c*, and *d*) indicates that the animals are able to detect the spatial change, thus suggesting that they have encoded some spatial relationships. In contrast, an absence of reexploration is observed when the configuration is only expanded. (Redrawn from Poucet et al., 1986, and Thinus-Blanc et al., 1987.)

WHICH TASK SHOULD BE USED?

The tasks that have been presented differ in many aspects. First, they differ in terms of motivation. The Morris water tasks are escape tasks that are based on aversive motivation. In contrast, the t-maze and radial maze tasks are appetitive tasks. This is a critical aspect that likely affects the rate of learning. Exploration tasks (including spontaneous alternation) are a special case because they are based on spontaneous motivation, providing an interesting alternative to other situations. Second, spatial tasks differ in terms of cognitive capacities. Open environments like the water maze or the place preference arena are more appropriate than structured mazes to study navigation per se because animals are free to select a strategy, allowing the animal to display its preferred navigational strategies. Third, the nature of the cues used by the animals differs from one task to another. In most tasks (water maze task, place preference task, radial arm-maze task, cross-maze task, etc.), the apparatus is located in a room that provides numerous cues and the rats have to use these cues to locate the platform, or the baited arms in the appetitive tasks. In these situations, it is somewhat difficult to study what kind of cues are used and how they are used

for navigation. One possibility therefore is to train rats in a controlled environment. The apparatus is isolated from the room by curtains, for example, and the experimenter provides a few cues that can be manipulated. Another aspect concerning the cues is that the processing of proximal cues (i.e., objects that are placed in the rat's locomotor space [exploration]), may be different from the processing of room cues (water maze).

BRAIN SUBSTRATES OF NAVIGATION

Damage to almost any portion of the brain can result in impairments in spatial navigation. Nevertheless, substantial evidence suggests that taxon navigation is subserved by basal forebrain structures, whereas locale navigation is subserved by temporal lobe structures, including the hippocampus (O'Keefe and Nadel, 1978). Hippocampal lesions produce place learning impairment in the water maze but do not affect navigation to a visible platform (Morris et al., 1982) or with a beacon closely associated with the platform (Save and Poucet, 2000). In the object exploration paradigm, hippocampal rats fail to detect the spatial change but are able to discriminate a novel object that replaces a familiar one (Save et al., 1992). An intact hippocampus is also required to solve the radial arm-maze task (Olton et al., 1979) and the T-maze (Dudchenko, 2001). However, animals with hippocampal damage that receive special training and testing can solve many spatial tasks, even place tasks. A number of other cortical structures (entorhinal, perihinal, postrhinal, retrosplenial, and parietal cortices; Aggleton et al., 2000) and subcortical structures (anterior thalamus, mammillary bodies, cerebellum) have been hypothesized to contribute to place navigation. The specific role of these structures in relation to the hippocampus remains to be elucidated. In contrast, the taxon system (cue, guidance, and routes strategies) does not seem to be dependent on the hippocampal system but involves the dorsal striatum and its allied structures (McDonald and White, 1994).

CONCLUSION

One major purpose of spatial laboratory work is to understand how rats "naturally" process spatial information and what brain structures they use to do so. Thus, a number of behavioral situations have been developed to test their spatial abilities and to study the involvement of the neural structures. It must be acknowledged that the apparatuses used and the environmental conditions (nature of cues, size of the environment, type of motivation, etc.) may be very different from the real conditions in which a rat navigation has evolved. In addition, laboratory animals themselves are different from wild animals and may exhibit behaviors or strategies that do not match their wild counterparts. Thus, researchers must be sensitive to the ecological validity of their results. On a positive note, however, behavioral models do exploit various aspects of natural behaviors such as exploratory activity, place learning, navigation, spatial memory, spontaneous alternation, and so on. Thus, it is reasonable to think that these main processes are common to

both laboratory and wild rats (e.g., Gaulin and Fitzgerald, 1989). In addition, it is reasonable to believe that the spatial strategies and the neural processes used by rats are those used by many other animal species, including humans.

REFERENCES

Aggleton JP, Vann SD, Oswald CJP, Good M (2000) Identifying cortical inputs to the rat hippocampus that subserve allocentric spatial processes: a simple problem with a complex answer. Hippocampus 10:466–474.

Benhamou S and Bovet P (1992) Distinguishing between elementary orientation mechanisms by means of path analysis. Animal Behaviour 43:371–377.

Bovet J (1998) Long-distance travels and homing: dispersal, migrations, excursions. In: Handbook of spatial research paradigms and methodologies, Volume 2: Clinical and Comparative studies (Foreman N and Gillett R, eds.), pp. 239–269. Hove, U.K.: Psychology Press.

Collett TS, Cartwright BA, Smith BA (1986) Landmark learning and visuo-spatial memories in gerbils. Journal of Comparative Physiology A 158:835–851.

Dudchenko PA (2001) How do animals actually solve the T maze? Behavioral Neuroscience 115:850–860.

Foreman N and Ermakova I (1998) The radial arm maze: twenty years on. In: Handbook of spatial research paradigms and methodologies, Volume 2: Clinical and Comparative studies (Foreman N and Gillett R, eds.), pp. 87–143. Hove, U.K.: Psychology Press.

Gallistel CR (1990) The organization of learning. Cambridge, MA: The MIT Press.

Gaulin SJC and Fitzgerald RW (1989) Sexual selection for spatial-learning ability. Animal Behaviour 37:322–331.

McDonald RJ and White NM (1994) Parallel information processing in the water maze: evidence for independent memory systems involving dorsal striatum and hippocampus. Behavioral and Neural Biology 61:260–270.

Morris RGM (1981) Spatial localization does not require the presence of local cues. Learning and Motivation 12:239–260.

Morris RGM, Garrud P, Rawlins JNP, O'Keefe J (1982) Place navigation impaired in rats with hippocampal lesions. Nature 297:681–683.

O'Keefe J and Nadel L (1978) The hippocampus as a cognitive map. Oxford: Clarendon Press.

Olton DS and Samuelson RJ (1976) Remembrance of places passed: spatial memory in rats. Journal of Experimental Psychology: Animal Behavior Processes 2:97–116.

Olton DS, Becker JT, Handelmann GE (1979) Hippocampus, space and memory. Behavioral and Brain Sciences 2:313–365.

Packard MG and McGaugh JL (1996). Inactivation of hippocampus or caudate nucleus with Lidocaine differentially affects expression of place and response learning. Neurobiology of Learning and Memory 65:65–72.

Poucet B, Chapuis N, Durup M, Thinus-Blanc C (1986) A study of exploratory behavior as an index of spatial knowledge in hamsters. Animal Learning and Behavior 14:93–100.

Renner MJ (1990) Neglected aspects of exploratory and investigatory behavior. Psychobiology 18:16–22.

Rossier J, Kaminsky Y, Schenk F, and Bures J (2000) The place preference task: a new tool for studying the relation between and place cell activity in rats. Behavioral Neuroscience 114:273–284.

Save E and Poucet B (2000) Involvement of the hippocampus and associative parietal cortex in the use of proximal and distal landmarks for navigation. Behavioural Brain Research 109:195–206.

Sutherland RJ and Rudy JW (1989) Configural association theory: the role of the hippocampa formation in learning, memory, and amnesia. Psychobiology 17:129–144.

Thinus-Blanc C, Bouzouba L, Chaix K, Chapuis N, Durup M, Poucet B (1987) A study of spatial parameters encoded during exploration in hamsters. Journal of Experimental Psychology: Animal Behavior Processes 13:418–427.

Tolman EC (1948) Cognitive maps in rats and men. Psychological bulletin 55:189–208.

Whishaw IQ, Cassel JC, Jarrard LE (1995) Rats with fimbria-fornix lesions display a place response in a swimming pool: a dissociation between getting there and knowing where. Journal of Neuroscience 15:5779–5788.

Dead Reckoning

38

DOUGLAS G. WALLACE AND
IAN Q. WHISHAW

Rats may use at least two navigational strategies that aid in the recovery of resources and protection from predation: allothetic and idiothetic (Gallistel, 1990). Allothetic navigation involves the use of external cues (visual, auditory, or olfactory cues). Idiothetic navigation involves the use of uses cues generated by self-movement (proprioceptive and vestibular cues, sensory flow, or efferent copies of movement commands). A prominent form of idiothetic navigation is dead reckoning, in which a rat locates its present position in relation to a starting position, to which it can return. A central point to be made in this chapter is that dead reckoning is an innate, online form of behavior that is central to the rat's survival. This chapter presents behavioral techniques used to investigate dead reckoning and describes a theoretical framework for dead reckoning.

FOOD HOARDING AND
DEAD RECKONING

The rat's proclivity to hoard large food items in a refuge has afforded researchers a technique to examine the cues that organize naturally occurring spatial behavior (Whishaw et al., 1995). The food-hoarding paradigm involves placing a hungry rat in a refuge that permits access to a large arena such that it can search the arena for a randomly located food pellet. We use a circular table with eight holes located around the perimeter of the table (Fig.

38–1A). Each hole has runners located below the table to which a small box, "a basement apartment," can be affixed. Rats placed in the basement apartment are free to jump up onto the table and to explore it. If a rat is hungry and if it finds a food item on the table, provided the food item takes more than a few seconds to eat, the rat will carry the food item back to the starting refuge. There, it eats the food before making another trip out onto the table. This procedure can be conducted under a variety of environmental conditions (Wallace et al., 2002b). This chapter discusses the rationale for manipulating testing conditions and the behaviors displayed by rats.

Methodologically, the simplest testing procedure involves using the foraging table and a home base made visible by adding a second "upstairs apartment" above the basement apartment (Fig. 38–1B). The upstairs apartment is a small box that has a hole in one side and in the floor so that a rat can climb down into the underlying refuge. Under this condition, rats can potentially use three different strategies to reach the home base: (1) a *cue response*, the proximal cue associated with the visible upstairs apartment; (2) a *place response*, two or more distal cues whose relationships can indicate the location of the apartments; or (3) *dead reckoning*, cues generated by self-movement on an outward trip that can be used to calculate a direct return to the starting point.

During the cued training, the outward searching behavior of rats is slow and circuitous (Fig. 38–2, top). Rats frequently stop

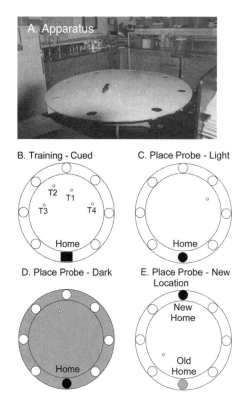

Figure 38–1. (A) Photograph of the room and circular table used to test rats. (B) Training, cued: a schematic of the table and a possible sequence of food pellet placements for one day of training with the cued home base. Place probes (light [C] and dark [D]): the arrangement of the table under place and dark probes. (E) Place probe, new location: the arrangement of the table and the home base when the rat was released from a novel location. The circle that is drawn tangent to the inner portion of the holes was used to code when a return trip was terminated.

training with the upstairs apartment removed (Fig. 38–1C). The home base is now indicated by the hole, which is indistinguishable from the other seven holes on the table, except for its relation to distal cues. Rats leave the "hidden" home base in search of a randomly located food pellet. The searching behavior is comparable in topography and kinematics to that observed during cued training (Fig. 38–2B). On finding the food pellet, the rat quickly orients toward the home base. Typi-

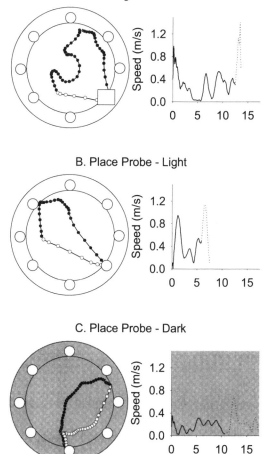

Figure 38–2. Each set of panels plot the kinematic and topographic profile of a representative trip. Black dots and solid lines represent the outward trip topography and kinematics, respectively, and white dots and dotted lines represent the homeward trip topography and kinematics, respectively. (A) Training, cued: One trip with a cued home base. (B) Place probe, light: one trip with a hidden home base. (C) Place probe, dark: one trip under dark conditions.

and make scanning movements or rear and then continue to search the table. On finding the food pellet, the animal quickly orients toward the home base. The path back to the home base is direct and is associated with a rapid increase in speed relative to the outward search behavior (Wallace et al., 2002b). The direct return may be guided the visible cue of the home base, but it could also be mediated by a place response or dead reckoning, as is described later.

PLACE PROBE

For the place probe, rats are released from the same location as experienced during cued

cally, the rat accelerates as it approaches the home base and then decelerates as it enters the home base (Wallace et al., 2002b). The place probe demonstrates the rat's ability to use distal environmental cues to locate the home base even though it had experienced the "cued" condition only during training.

DEAD RECKONING PROBE

The dead reckoning probe involves releasing the rat from the same location as experienced during cued training and the place probe, except that all light sources are eliminated (Fig. 38–1D). Under these conditions, infrared recording cameras and spotters are used to observe the rat's movements. Because there are no available distal or proximal cues, a rat can return directly to its starting point only through the use of self-movement cues generated on the outward trip. The topography of the searching and homing components is similar to that observed under cued training or the place probe (Fig. 38–2C). A rat's searching components are circuitous but, again, the homeward component is a direct path to the home base. Although movements on both outward and homeward components are slower under dark conditions, rats move faster on the homeward component. It is worth noting that it is unlikely that they are using other allothetic cues such as olfaction (the outward and homeward portion of the trip are different) or other olfactory or auditory cues because there are no other obvious allothetic cues in the test room. Nevertheless, a further control for the use of other surface or distal cues is embedded in the "new location" probe, which is described next.

NOVEL LOCATION PROBE

The way in which rats perform on the "place" and "dark" probe trials is relevant to the way in which rats solve a novel-location probe. This probe involves releasing the rat from an uncued home base that is in a novel location in the environment under light conditions.

For example, the home base is shifted 180° from the position experienced during cued training (Fig. 38–1E). This probe produces a conflict between distal environmental cues and self-movement cues. If the rat uses distal environmental cues to guide its movements, then it will attempt to return to the former location of the refuge. If the rat uses self-movement cues to organize its behavior, then the rat will return directly to the "new" home base.

After the rat leaves the home base in the new location, it searches the table for the food pellet (Fig. 38–3). After finding the food pel-

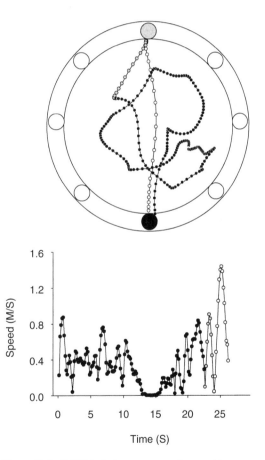

Figure 38–3. Solid black dots represent the searching segment; gray dots represent the segment after finding the food until the first hole was chosen; and white dots represent the segment of the trip after the first hole was chosen. *Top,* Topography of a representative rat when released from a novel location (large black circle) after receiving training from a different location (large gray circle). *Bottom,* Kinematics associated presented on top panel.

let, the rat quickly orients to the old home base location and makes a rapid direct path toward the old home base location. Thus, it initially uses a place response. On finding that the refuge is not available at the old home base location, the rat then orients to the new home base location. As the rat returns to the new home base location, it takes a direct path, and there is a symmetrical increase and decrease in speed. Thus, it now appears to use dead reckoning.

This pattern of responses suggests that rats have a hierarchy of navigational strategies they use as they are foraging for food items (Maaswinkel and Whishaw, 1999; Wallace et al., 2002b). First, they respond to familiar allothetic cues. If these cues are unreliable in predicting the location of the refuge, then the rats resort to the use of self-movement cues to guide navigation.

That rats switch to idiothetic cues to solve the "new" place problem can be confirmed by repeating the test in the dark. For example, if a rat were using diffuse olfactory cues present in the testing room to organize its spatial behavior, then one would anticipate a similar behavioral response as observed when the rat was released from a new position under light conditions. If, however, a rat were using self-movement cues to guide navigation, then one would predict that the rat should return directly back to the home base independent of its starting location. When rats are released from a novel refuge location under dark conditions, they return directly to the starting location and not to the previous location (Maaswinkel and Whishaw, 1999). This indicates that rats are not using nonvisual cues in the environment to organize their spatial behavior and are using dead reckoning.

NEURAL CONTROL OF DEAD RECKONING

Several studies have examined the neural basis of dead reckoning using the food-hoarding paradigm. Considerable evidence suggests a role for the limbic system, including the hippocam-

pus, in the processing of self-movement cues during dead reckoning–based navigation. Research has demonstrated that although impairments in dead reckoning were observed after hippocampal damage, other navigational strategies were intact (Maaswinkel et al., 1999). Whishaw et al. (2001b) also demonstrated a role for the posterior cingulate cortex in dead reckoning during food hoarding. In examining the role of the vestibular system in dead reckoning–based navigation Wallace et al. (2002b) also used the food-hoarding paradigm. Labyrinthectomies produced by intratympanic injections of arsenic acid disrupted performance on dead reckoning probes but did not impair performance on cued or place probes. This demonstrates an important role of the vestibular system and the hippocampus and other limbic structures in dead reckoning.

EXPLORATORY BEHAVIOR

Exploration is an obvious feature of the behavior of many animals (O'Keefe and Nadel, 1978). It is generally thought that this behavior is useful in that once an animal has explored an environment, it is subsequently able to use the information it has acquired to travel through that environment again (Whishaw and Brooks, 1999). A rat faces two problems when engaged in exploratory behavior. First, the information that it gathers on an outward trip may be of little value in guiding a return trip. Although it views and learns about various cues on its outward trip, it does not see those cues, or move in relation to them, from the vantage point of the journey home. Second, an animal may want to return directly home after a circuitous outward journey. How can it return home in a straight line after making a meandering outward trip? Rodents apparently solve this problem by using different strategies, one for outward behavior and one for homeward behavior. It is likely that a rat learns about the allothetic cues on the outward segment, whereas the homeward segment is guided by dead reckoning.

We have studied the exploratory behavior of rats in a testing apparatus that in many ways is designed to be a simplified analogy of the animal's natural world. We provide the animals with a visible home base, the upstairs apartment described earlier. From their home base, the animals can explore a large round table (on which there are no escape holes). There are no walls around the table because walls might limit an animal to displaying thigmotaxic (wall-following) behavior. Although an animal's only home base is the cage, its behavior is unconstrained with respect to opportunities to explore the table, to examine the surrounding room, and to return to the home base. In a featureless environment, rats set up virtual home bases where they turn, rear, and groom (Whishaw et al., 1983; Golani et al., 1993); from which they make slow excursions (marked by a number of pauses); and to which they return at a higher movement speed than when they left (Tchernichovski et al., 1998; Drai et al., 2000). Thus, the home base and open field used here assist the rat in displaying organized exploratory behavior.

THE OUTWARD TRIP

Our tests of rats' behavior on a tabletop confirm that an animal's exploratory movements can be divided into components, the most salient of which are *excursions*, *stops*, and *returns*. Initially, a rat makes excursions and returns to the vicinity of the home base entrance. Then the rat makes longer circuitous excursions on the table, and these excursions are marked by head scans, pauses, or stops. Periodically, the long excursions are also interrupted by direct shortcut returns to the home base. The elements of excursion (outward) and return (homeward) segments seem to be key components of the structure of exploratory movements (Whishaw and Brooks, 1999; Wallace et al., 2001c, 2002c).

We use the rat's final stop to behaviorally fractionate the outward trip segment from the homeward trip segment. A stop is defined as 2 seconds or more of movements in which speed does not exceed 0.1 m/sec. Several aspects of the exploratory data support this distinction. First, work examining the distribution of stops has demonstrated that the number of stops made during an exploratory trip has an upper limit that is not dependent on the size of the arena (Golani et al., 1993). In addition, the kinematics associated with movements before the last stop are qualitatively different from movements observed after the last stop. This latter point is considered in greater detail next.

The outward trip segment is characterized as a series of slow progressions punctuated by periods of stopping (Fig. 38–4). Speeds observed on the outward trip segment range from 0.2 to 0.6 m/sec. There is a high degree of variability in the topography of the outward trip segment between exploratory trips. Each outward trip segment has a unique kinematic

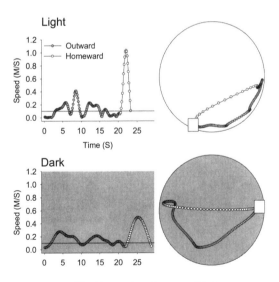

Figure 38–4. The kinematic and topographical representation of a single exploratory trip under light and dark conditions are plotted (top and bottom, respectively). The outward trip segments are represented by the gray dots, whereas the homeward trip segments are represented by the white dots. The solid black line in the kinematic plots represents the criterion speed for a stop.

profile (i.e., set of angular and linear speeds). Therefore, a rat gathers the maximal amount of information about an environment across exploratory trips.

THE HOMEWARD TRIP

In contrast to the outward trip segment, the homeward trip is a rapid, noncircuitous path that is directed toward the home base. The speeds observed on the homeward trip segment vary from 0.2 to 1.6 m/sec. As the rat progresses toward the home base, there is a symmetrical increase and decrease in speed in which the peak occurs at the midpoint independent of trip length (Fig. 38–4). The temporal pacing observed on the homeward trip segment is reminiscent of work examining the kinematics of planar reaching movements in humans that are executed independent of visual cues (Gordon et al., 1994). Such temporal pacing is consistent with the use of self-movement cues guiding behavior, rather than external guidance. The homeward trip segment originates from many different points in the environment across exploratory trips. This aspect of the homeward trip segment is further evidence that rats are not using a proximal cue such as olfactory guidance to organize their spatial behavior (for a description of odor tracking, see Wallace et al., 2002a).

EXPLORATION IN THE DARK

Spontaneous exploration has also been investigated under completely dark conditions in a novel environment. Spontaneous exploratory trips observed under dark conditions have a similar topographical organization to that observed under light conditions (Fig. 38–4, bottom). The outward trip segment is circuitous and varies in location between exploratory trips. The homeward trip segment is noncircuitous and is directed toward the home base. Although speeds observed on both trip segments are slower under dark conditions relative to light conditions; the temporal pacing

of the homeward trip segment is similar to that observed under light testing conditions. This is consistent with rats using self-movement cues, or dead reckoning–based navigation, to organize their homeward trips.

Because the light-versus-dark tests suggested that rats use dead reckoning in the dark, a parsimonious conclusion is that the animals also use dead reckoning to return home in the light. We found support for this idea with the following experiment. A rat was placed in a black box that was located on the edge of the table. This home base was clearly visible from all portions of the table. We hypothesized that if the rats were using allothetic cues to return to the starting location in the light test, the visibility of the home base might serve as a beacon. If the home was being used as a beacon, then removing it after an animal had left on an exploratory excursion should disrupt its homeward trip. Initially, the rat emerged from the home base box and explored its surface a number of times; therefore, it was clear that the rat was interested in the home base as an object. After this, it made some short excursions, followed by longer excursions and returns to the box. On removal of the home base as the rat was on a long outward excursion, thus eliminating the visible cue that marked the location of the home base, the rat still returned rapidly and directly to the previous location of the home base. After first sniffing around the vicinity of the previous location of the home base, it then made a number of rapid excursions and returned to that location, which seemed to indicate that it was attempting to find the missing home base. Thus, the behavior of the rats indicated that they could return accurately to the home base, even though it was no longer visible, and they expected to find the home base at that location. This result suggests that homeward trips in the light are not dependent on the ability to see the home base and thus they are likely to be produced by dead reckoning just as returns are in the dark.

NEURAL CONTROL OF EXPLORATION

The observation that hippocampal lesions disrupt dead reckoning in the food-hoarding paradigm has prompted researchers to evaluate the role of the hippocampus in the organization of exploration. Wallace et al. (2002c) examined the organization of exploratory trips subsequent to fimbria-fornix lesions, a major afferent/efferent pathway for the hippocampus proper. Although there were no observable differences between groups on the outward trip segment, the groups had different topographical and kinematic profiles on the homeward trip segment. Specifically, fimbria-fornix rats returned to the home base along a more circuitous path without a distinct difference in kinematics observed on the outward segment. Further evidence that the hippocampus is involved in dead reckoning comes from a similar study using rats with selective neurotoxic lesions of the hippocampus (Wallace and Whishaw, 2003). There is evidence for a role of the hippocampus in the organization of spontaneous exploration, but it is likely that other brain structures contribute to this naturally occurring behavior.

THEORETICAL FRAMEWORK FOR DEAD RECKONING

Dead reckoning observed during food hoarding and spontaneous exploration shares many components observed in nautical navigation (Fig. 38–5). We suggest four components that are critical for accuracy in dead reckoning–based navigation: (1) home base, (2) measurement of the rate of linear and angular movement, (3) some index of the amount of time that has passed, and (4) a central process that integrates linear and angular rates of movement with the associated temporal context to provide an internal representation of current distance and direction from the home base. Each of these components is discussed next.

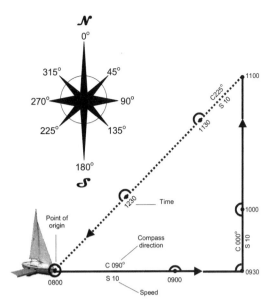

Figure 38–5. (A) Example of dead reckoning course for a ship. The ship starts from a known point of origin. During the course of the trip, the ship changes direction and then returns to the point of origin. The ship leaves the point of origin at 0800 hours, in an eastern direction (90°), and travels at 10 knots. The speed, direction, and current time are logged at regular intervals to provide the necessary information to return to the point of origin independent of external cues.

HOME BASE

The home base serves as a point of origin for rats as they explore or search for food. The home base is a critical component of a rat's environment. When a rat is placed in a novel environment, in the absence of a physical home base, it sets up one or more virtual home bases (Eilam and Golani, 1989). When a rat is provided with a physical shelter, it treats that location as a home base. If the shelter is removed, then the rats treat the former location as a home base (Whishaw et al., 2001a). One possibility is that the rat uses the home base as a discrete stimulus to reset or clear the dead reckoning process. Dead reckoning–based navigation is subject to an accumulation of errors (Maurer and Seguinot, 1995). One way to reduce the accumulation of errors across trips is to clear the dead reckoning process before the initiation of subsequent trips. It is possible that the importance of the home base to the rat reflects its function to clear the dead reckoning process.

VELOCITY

Changes in linear and angular velocity occur as rats move through an environment, and the vestibular system processes this information. Examining the speeds at which a rat travels through an environment, Drai et al. (2000) reported that animals typically travel at one of three speeds, or "gears." We observe that each of the gears is associated with specific components of exploration. For example, a rat's outward trips reflect movement within second gear, whereas third gear seems to be used for the return trip. The use of a limited number of speeds simplifies the calculations of travel velocities as a variable in estimating a homeward trip, thereby reducing the error present in estimating distance from the home base.

TEMPORAL CONTEXT

The temporal context in which the self-movement cues have occurred may play an important role in dead reckoning. Rats are able to discriminate differences in temporal intervals on the scale of seconds (Church and Gibbon, 1982), suggesting the rat has an adequate perception of time encountered during an exploratory trip. Little work has been conducted to examine the role of time perception in spatial navigation. We suggest that the scalar invariance property of interval time should be manifested in a rat's judgment of distance and direction from the home base. For example, a rat's ability to judge shorter distances will be highly reliable; however, as the distance increases, one will observe a systematic increase in the variability of distance judgments.

ONLINE PROCESSING

Dead reckoning involves online processing of changes in linear and angular velocities with respect to the temporal context in which they occur. The resulting internal representation provides information regarding the direction and distance of the home base from the rat's current location. The temporal pacing of the homeward trip is consistent with behavior guided by an internal representation rather than external cues. The process of dead reckoning may update the internal representation continually or just before returning to the home base. It seems that the former would minimize the amount of error in direction and distance estimates; however, the current data do not favor one method of updating over another.

CONCLUSION

Dead reckoning–based navigation has all of the appearances of an innate action pattern applied by rats to specific spatial problems. The rat processes online self-movement cues to return to the point of origin. The behavioral techniques discussed here provide a foundation for observing dead reckoning in the rat. Dead–reckoning based navigation provides a rich behavioral paradigm to evaluate the effect of manipulating the nervous system. In addition, dead reckoning may provide a foundation that enables other forms of spatial navigation and may serve as a temporal and spatial marker for significant spatially relevant events in a rat's life.

REFERENCES

Church RM and Gibbon J (1982) Temporal generalization. Journal of Experimental Psychology. Animal Behavior Processes 8:165–186.

Eilam D and Golani I (1989) Home base behaviour of rats (Rattus norvegicus) exploring a novel environment. Behavioural Brain Research 34:199–211.

Drai D, Benjamini Y, Golani I (2000) Statistical discrimination of natural modes of motion in rat exploratory behaviour. Journal of Neuroscience Methods 96:119–131.

Gallistel CR (1990) The organisation of learning. Cambridge, MA: The MIT Press.

Golani I, Benjamini Y, Eilam D (1993) Stopping behaviour: constraints on exploration in rats (Rattus norvegicus). Behavioural Brain Research 53:21–33.

Gordon J, Ghilardi MF, Cooper SE, Ghez C (1994) Accuracy of planar reaching movements. II. Systematic extent errors resulting from inertial anisotropy. Experimental Brain Research 99:112–130.

Maaswinkel H, Jarrard LE, Whishaw IQ (1999) Hippocampectomized rats are impaired in homing by path integration. Hippocampus 9:553–561.

Maaswinkel H and Whishaw IQ (1999) Homing with locale, taxon, and dead reckoning strategies by foraging rats: sensory hierarchy in spatial navigation. Behavioural Brain Research 99:143–152.

Maurer R and Seguinot V (1995) What is modeling for? A critical review of the models of path integration. Journal of Theoretical Biology 175:457–475.

Techernichovski O, Benjamini Y, Golani I (1998) The dynamics of long-term exploration in the rat. Part I. A phase-plane analysis of the relationship between location and velocity. Biological Cybernetics 78:423–432.

Wallace DG, Gorny B, Whishaw IQ (2002) Rats can track odors, other rats, and themselves: implications for the study of spatial behaviour. Behavioural Brain Research 131:185–192.

Wallace DG, Hines DJ, Pellis SM, Whishaw IQ (2002) Vestibular information is required for dead reckoning in the rat. Journal of Neuroscience 22:10009–10017.

Wallace DG, Hines DJ, Whishaw IQ (2002) Quantification of a single exploratory trip reveals hippocampal formation mediated dead reckoning. Journal of Neuroscience Methods 113:131–145.

Wallace DG and Whishaw IQ (2003) NMDA lesions of the Ammon's horn and the dentate gyrus disrupt the direct and temporally paced homing displayed by rats exploring a novel environment: Evidence for the role of the hippocampus in dead reckoning. European Journal of Neuroscience 18:513–523.

Whishaw IQ and Brooks BL (1999) Calibrating space: exploration is important for allothetic and idiothetic navigation. Hippocampus 9:659–667.

Whishaw IQ, Coles BL, Bellerive CH (1995) Food carrying: a new method for naturalistic studies of spontaneous and forced alternation. Journal of Neuroscience Methods 61:139–143.

Whishaw IQ, Hines DJ, Wallace DG (2001) Dead reckoning (path integration) requires the hippocampal formation: evidence from spontaneous exploration and spatial learning tasks in light (allothetic) and dark (idiothetic) tests. Behavioural Brain Research 127:49–69.

Whishaw IQ, Maaswinkel H, Gonzalez CLR, Kolb B (2001) Deficits in allothetic and idiothetic spatial behavior in rats with posterior cingulate cortex lesions. Behavioural Brain Research 118:67–76.

Whishaw IQ, Kolb B, Sutherland RJ (1983) The analysis of behaviour in the laboratory rat. In: Robinson TE, editor. Behavioural approaches to brain research, pp. 141–211. Oxford University Press: New York.

Whishaw IQ and Tomie JA (1997) Piloting and dead reckoning dissociated by fimbria-fornix lesions in a rat food carrying task. Behavioural Brain Research 89:87–97.

Fear

39

MATTHEW R. TINSLEY AND MICHAEL S. FANSELOW

During the past 30 years, there have been great advances in the conceptualization of defensive behavior. Behavioral and motivational approaches have stressed the organization of defensive behaviors into a functional behavior system tuned to the ecological and evolutionary requirements of the animals (Bolles, 1970; Fanselow, 1994). Advances in neurophysiological, neurochemical, and neuroanatomical techniques have allowed the mediating substrates of different defensive behaviors to be mapped and information on the behavioral and neural organization of defensive behavior to be integrated (Fanselow, 1994; Fendt and Fanselow, 1999). This synthesis has resulted in perhaps the best understanding of the processes involved in mediating a complex and functionally significant suite of behaviors in the rat. In the following chapter we present defensive behavior as a functional behavior system, describe the neural substrates of many of the various defensive behaviors described, and present a model that integrates the behavioral and neural organizations of defensive behavior.

THE ORGANIZATION OF DEFENSIVE BEHAVIOR

Current theories on the organization of the rat's defensive behavior are largely variations on Bolles' (1970) species-specific defensive reaction (SSDR) theory. This theory suggested that when confronted by a natural (e.g., a predator) or unnatural (e.g., a noxious stimulus) threat, the rat's behavior is restricted to a small number of innately determined defensive behaviors, the SSDRs. Determination of which behavior is produced and why is where these theories of defensive behavioral organization differ.

Bolles' (1970) original conception was that the behavior produced depended on prior experience, with SSDRs that had been previously unsuccessful being less likely than SSDRs that had not. However, attempts to reduce the production of SSDRs through operant punishment contingencies (Bolles, 1975) were not successful. In any event, learning via punishment requires the rat to be exposed to a number of predatory encounters before learning the correct behavioral strategy, which is unlikely in a natural situation (Fanselow et al., 1987).

A second theory suggests that the SSDR produced depends on features of the environment (Blanchard et al., 1976). This theory suggests that a rat will try to escape from a situation if escape is possible and will freeze if it knows that escape is not possible. One finding that seems to support this suggestion is that animals shocked immediately after being placed in a chamber do not show a freezing response. In contrast, animals shocked 3 minutes after placement, who have perhaps learned there is no exit, do freeze (Blanchard et al., 1976). However, this immediate shock deficit also occurs if other behaviors, such as defecation or conditional analgesia, are examined (Fanselow et al., 1994), suggesting that this procedure causes reduced freezing due to reduced fear conditioning. To further assess the influence of support stimuli, Fanselow and Lester (1988) presented a conditional fear stimulus in a variety of contexts that differed with respect to the support stimuli they contained. In each case, the stimulus elicited freezing, rather than some other SSDR.

These results led to the suggestion that what is important in determining the SSDR in which the animal is the level of fear it is experiencing (Fanselow and Lester, 1988). Fanselow and Lester (1998) and Fanselow (1994) suggested that defensive behavior is organized along a continuum of perceived threat or predatory imminence (Blanchard et al., 1989). When the animal perceives that predatory imminence is low, that the threat is temporally or spatially distal, it will engage in preencounter defensive behaviors such as meal pattern reorganization to reduce the risk of predation (Helmstetter and Fanselow, 1993) and stretched approach behavior to gain information (Blanchard et al., 1989). As predatory imminence increases, once a predator has been detected, behavior shifts to postencounter defensive behavior such as freezing. Finally, when contact with a predator is occurring or inevitable, the rat engages in circa-strike defensive behavior such as defensive fighting or jump attack. Conditional stimuli predictive of high predatory imminence, such as those paired with predator presentation or shock, cause the animal to engage in postencounter, not circa-strike, behavior as the rat attempts to avoid actual contact with the threat stimulus. This organization of defensive behavior along a predatory imminence continuum is illustrated in Figure 39–1.

BEHAVIORS DURING LOW PREDATORY IMMINENCE: FIELD STUDIES

The rat, being a heavily predated animal, is likely to spend most of its time in a state of at least some perceived threat. This suggests that most, if not all, of its behaviors outside the burrow may be modified by tonically employed, risk minimization strategies and overlapping phasically employed, preencounter defensive behaviors. Examples of the rat's tonically employed, risk minimization strategies have been described from naturalistic studies and include neophobia and trail making.

For a heavily predated animal, like the rat, nonneutral changes in its physical environment are more likely to be negative than positive. Hence, the appropriate risk minimization strategy is to treat all changes as negative until they are determined to be otherwise. The resulting avoidance of novelty has been termed neophobia. This effect was noted by Calhoun (1963) following the introduction of activity recorders in a naturalistic environment: "The rat would cautiously approach it, jump back, bypass it while keeping a foot or

	PREDATORY IMMINENCE $\longrightarrow$		
Defensive Mode	Pre-encounter Defense	Post-encounter Defense	Circa-Strike Defense
Natural Stimuli	Previous predatory encounter in the foraging area	Predator detected nearby	Physical contact, pain
Laboratory Stimuli	Very low density shock, exposure to bright light	Stimuli associated with aversive reinforcement	Physical contact, pain
Behavioral Response	Meal pattern re-organization, stretched approach, vigilance, potentiated startle	Freezing, conditional autonomic responses	Activity burst, defensive fighting, vocalization, unconditional analgesia

Figure 39–1. Defensive behavior is hypothesized as being organized along a psychological dimension of perceived predatory imminence. Increases in predatory imminence, corresponding to changes in external stimuli, shift the animal rightward along the continuum, resulting in qualitatively different defensive behaviors at differing levels of perceived threat.

two away" (Calhoun, 1963, p. 85). Importantly for laboratory studies on preencounter defensive behavior, exploratory behavior, and not defensive behavior, is induced if either a laboratory rat or a wild rat is placed into a completely novel environment, suggesting that defensive behavior "occurs . . . when there is a *change in an otherwise familiar situation*" (Barnett, 1963, p. 30 [author's italics]).

A second risk minimization strategy is trail making and use. Rats in both the wild and a controlled setting establish specific trails to resources such as sources of food or shelter (Calhoun, 1963). Trail following reduces the risk of predation by ensuring that the animal finds resources as reliably and efficiently as possible. These trails typically follow an approximation to the shortest distance between the two goals but deviate toward sources of overhead cover and vertical surfaces (thigmotaxis) (Calhoun, 1963). Trail-following behavior has been demonstrated in the laboratory by Timberlake and Roche (1998). In these studies, the radial-arm maze arrangement of arms and food cups was placed on the floor of a large room, allowing the animals to approach the food from all directions. Despite this, the rats continued to move along or beside the maze arms rather than using another, more efficient, search strategy.

BEHAVIORS DURING LOW PREDATORY IMMINENCE: LABORATORY STUDIES

At least two discrete behavioral responses to increased but low levels of threat have been described. Fanselow et al. (1988) described studies in which the foraging behavior of rats in a closed economy was responsive to low levels of threat. Following establishment of baseline performance, low densities of randomly timed shock were introduced and the effect on meal patterning and size was examined. Rats responded by reducing the number, but increasing the size, of their meals, result-

ing in the animals avoiding 50% of the shocks they would have experienced had they not altered their behavior. Subsequent studies have determined that rats will eat a given meal more quickly after tail shock (Dess and Vanderweele, 1994) and will eat a particular size of pellet more quickly in lighter and more exposed environments (Whishaw et al., 1992). Importantly, the animal's meal patterning returns toward normal after the cessation of shock (Fanselow et al., 1988), indicating that the change is a discrete behavior related to increases in perceived predatory imminence.

Light-enhanced startle is the second example in which exposure to conditions of mild threat or low predatory imminence lead to a phasic increase in defensive behavior. The startle response is a fast muscle contraction, most noticeable around the head and neck, that follows an unpredicted, intense stimulus (Fendt and Fanselow, 1999). The magnitude of the startle response is enhanced by manipulations that increase levels of fear and anxiety (Brown et al., 1951). In the light-enhanced startle procedure, the rat is first tested for its response to a number of short bursts of loud white noise in a darkened chamber before being tested in the same chamber with the same noise bursts under conditions of bright (700 footlamberts) illumination (Walker and Davis, 1997a). Startle amplitude is significantly enhanced during the illuminated period of the test (Walker and Davis, 1997a, 1997b).

BEHAVIORS DURING HIGH PREDATORY IMMINENCE: RESPONSES ELICITED BY PREDATOR CUES

Studies examining conditional defensive responses reinforced by predator exposure are rare. We have examined whether cat exposure conditions context fear by exposing rats to an inescapable cat for 5 minutes immediately, 15 seconds, or 120 seconds after placing the rats in a novel two-chamber context. Robust freezing, on the order of 40% to 50% for the delayed cat groups, was observed during cat exposure (Figure 39–2). Very little conditional freezing (5% to 10%) was seen during later context testing, indicating that cat exposure conditioned very little fear to the context. Adamec et al. (1998) found that cat odor–exposed rats later showed increased risk assessment behavior, whereas cat-exposed rats later showed reduced risk assessment behavior. Blanchard et al. (2001) found defensive behaviors conditioned with cat odor to a context included more crouch/freeze with sniffing behavior and significantly less risk assessment behavior than when behaviors were conditioned with cat odor on a discrete cue.

These results suggest that the conditional response to a context previously paired with a predator seems to be a suppression of behavior and an increase in freezing while the conditional response to a cue previously paired with a predator may be risk assessment behavior.

Studies have also examined unconditional responses to predator cues. The most common experimental method involves exposure to a cat (Blanchard and Blanchard, 1971), although studies have also demonstrated defensive behavior after exposure to cat odor (Blanchard et al., 1975), odiferous components of predator feces (Vernet-Maury et al., 1992), or the experimenter (Blanchard et al., 1981).

The most naturalistic of these studies have been those by Blanchard and Blanchard (1989) using their visible burrow system (VBS). Briefly placing a cat in the VBS leads the rats to flee into the burrows where they remain for several hours. Other behaviors include the production of 22 kHz ultrasonic vocalizations during and for 30 minutes subsequent to cat presentation (Blanchard et al., 1991) and the suppression of nondefensive behaviors (Blanchard et al., 1989). The rats begin to emerge from their burrows about 4 to 7 hours after the cat presentation and engage in risk assessment behaviors, although the rats' behavior after cat presentation does not return to pre–cat presentation baseline values for at least 24 hours (Blanchard et al., 1989).

Most studies of exposure to a predator use a procedure in which the rat and predator are placed in a chamber and the rat's behavior is assessed (Blanchard and Blanchard, 1971). These procedures typically involve a shorter period of more intense exposure, as the rat can continuously see, hear, and smell the cat and is unable to escape from the situation. Typical results show that rats react to the presence of a predator by engaging in *freezing*: the cessation of all body movement except that necessary for respiration (Bolles and Collier, 1976). Other responses include inhi-

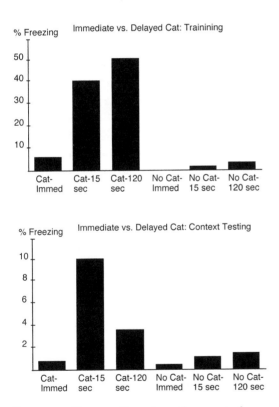

Figure 39–2. The upper frame shows the percent of samples during the 5 minute cat exposure during which the rats were freezing. Exposure to the cat after 15 or 120 seconds in the context causes robust freezing in rats. The lower frame shows the percent of samples during the 5 minute context reexposure 24 hours later. Despite the high levels of freezing during the cat exposure, freezing to the context is very low, indicating very little conditional fear of the context has been learned.

bition of locomotion, inhibition of exploratory behavior, and increased defecation (Satinder, 1976). Studies looking at the effects of cat exposure on endogenous antinociception have also shown that exposure to a cat leads to analgesia mediated by endogenous opioid systems (Lester and Fanselow, 1985).

Studies on the stimulus control of response to a predator have suggested that two main classes of stimuli are important: the motion of the predator and its odor. Blanchard et al. (1975) showed that rats do not freeze to the sound or the smell of a cat or to the sight of a dead cat but that either a moving cat or dog or the abrupt and rapid movement of an inanimate card causes freezing and inhibition of approach. In contrast, Griffith (1920) reports that presentation of a cat in the dark results in behavior we would now describe as freezing but that presentation of a cat in a glass jar does not. Presentation of a wooden block covered in a cloth impregnated with cat odor in the VBS elicits similar, but less intense, behavioral changes to those following cat presentation (Blanchard et al., 1989) and elicits risk assessment and crouch/freeze with sniffing behavior in a chamber (Blanchard et al., 2001). Rats show risk assessment behaviors when confined with a cat odor stimulus (Blanchard et al., 1991, 1993) but avoidance when the animal can retreat from the stimulus (Dielenberg and McGregor, 1999). This avoidance response habituates over repeated presentations of the stimulus (Dielenberg and McGregor, 1999).

VALIDITY OF STUDIES USING AVERSIVE STIMULI

The majority of studies of defensive behavior in rats have not examined responses to predators but have induced defensive behavior by using stimuli that are unconditionally aversive, such as loud noise, bright light, and electric shock. Just as studying laboratory rats to infer the behavior of wild rats provokes ques-

tions of external validity, so does using a stimulus such as electric shock. The most influential response to this criticism has come from the work of Robert Bolles (Bolles, 1970, 1975; Bolles and Fanselow, 1980). Bolles suggested that rats are equipped by evolution with preexisting defensive behaviors and that "the immediate and inevitable effect of severe and aversive stimulation on a[n] . . . animal is [of] . . . restricting its response repertoire to a narrow class of SSDRs" (Bolles, 1970, p. 34). Hence, using aversive stimulation to study defensive behavior is externally valid because the animal's reactions to both a predator and an aversive stimulus are governed by its SSDR organization. This is currently the dominant view in ethoexperimental approaches to defensive behavior.

BEHAVIORS DURING HIGH PREDATORY IMMINENCE: RESPONSES TO PREDICTORS OF AVERSIVE STIMULI

There are a number of conditional defensive behaviors to cues or contexts paired with an aversive stimulus, including freezing (Bolles and Collier, 1976), various forms of crouching (Blanchard and Blanchard, 1969), ultrasonic vocalization (Kaltwasser, 1991), and defecation (Fanselow, 1986), as well as autonomic measures such as hypertension (LeDoux et al., 1983), analgesia (Fanselow and Baackes, 1982), and both increases and decreases in heart rate (LeDoux et al., 1984). In addition, cues paired with shock have been shown to disrupt ongoing behavior (Estes and Skinner, 1941) and to potentiate unconditional startle behavior (Brown et al., 1951).

Of this variety of conditional responses to aversive stimulation, the most investigated is freezing. It is important to note that although freezing is an unconditional response to predator exposure (Blanchard and Blanchard, 1969, 1971), it is a conditional response to stimuli paired with aversive stimulation.

Fanselow (1980) compared freezing levels in the training context or a novel context 30 seconds or 24 hours after two levels of foot shock. Rats froze more in the training context than the novel context regardless of interval, indicating that freezing was a conditional response that required the presence of shock-associated cues to be expressed. In addition, there was no difference in the amount of freezing between rats tested 30 seconds after shock and 24 hours after shock, indicating that the shock itself is not required to elicit freezing.

The immediate shock deficit provides further evidence that freezing is a conditional response. Animals shocked immediately after being placed in a chamber do not later show a conditional freezing response to that chamber. In contrast, animals shocked 3 minutes after placement do freeze (Blanchard et al., 1976). This indicates that it is not merely the experience of being shocked that elicits freezing behavior. Subsequent investigation of this immediate shock deficit in freezing, and other measures of conditional fear to the context (Fanselow et al., 1994), indicates that it is due to the requirement for the animal to form a context representation before the shock delivery to be able to pair this representation, the to-be-conditioned stimulus, with the shock. Preexposure to the context alleviates the immediate shock deficit because it allows the animal to form the context representation (Fanselow, 1990).

BEHAVIORS DURING HIGH PREDATORY IMMINENCE: RESPONSES TO AVERSIVE STIMULI

By far the most common aversive stimulus used in experiments examining defensive behavior is electric shock. A rat's reaction to shock is a vigorous burst of activity that persists for a brief period after shock offset before it begins to engage in freezing behavior (Myer, 1971). Anisman and Waller (1973) determined that the duration of the activity burst is positively correlated with the intensity of the shock, and Fanselow (1980, 1982) characterized this reaction as the rat's unconditional response to the shock stimulus. Fanselow (1982, p. 453) provides a description of the activity burst, stating that it "was characterized by reflexive paw withdrawal, jumping and squealing . . . the animal moved rapidly, though in an uncoordinated manner, about the chamber."

Unconditional analgesia has also been demonstrated after footshock. Liebeskind and his colleagues (e.g., Lewis et al., 1980) demonstrated that prolonged, strong shock (30-minute intermittent or 3-minute continuous 3mA shock) induces opioid- and non–opioid-mediated analgesia in a tail-flick test. Subsequent studies at more commonly used intensities and durations (Fanselow et al., 1994) suggest that the expression of unconditional analgesia may depend on this strong shock.

THE NEURAL SUBSTRATES OF DEFENSIVE BEHAVIOR

THE AMYGDALA

Any discussion of the neural substrates of defensive behavior must begin with the amygdala. Early studies of the effects of temporal lobe lesions in monkeys (Brown and Schaffer, 1888) described a substantial loss of defensive behavior after extensive lesions of the medial temporal lobe that included the amygdala. More selective lesions determined this loss of defensive behavior was due to damage to the amygdala (Weiskrantz, 1956). Amygdala damage has been shown to affect unconditional freezing to a cat (Blanchard and Blanchard, 1972) and unconditional analgesia (Bellgowan and Helmstetter, 1996) and to attenuate unconditional autonomic responses (Iwata et al., 1986; Young and Leaton, 1996). Amygdala lesions also affect context freezing behavior conditioned by shock (Blanchard and Blanchard, 1972) and the acquisition of conditional bar press suppression (Kellicutt and Schwartzbaum, 1963).

Different regions within the amygdala play different roles in the coordination and expression of fear behavior. These nuclei can be roughly delineated into two major subsystems (Maren and Fanselow, 1996; Maren, 2001): the basolateral complex and the central nucleus. The basolateral complex includes the lateral, basolateral, and basomedial nuclei and forms the sensory input into the amygdala. Lesions to these structures disrupt the acquisition and expression of conditional defensive behaviors (LeDoux et al., 1990) by disrupting this sensory input. Selective lesions of specific input pathways from sensory processing regions to the basolateral complex affect conditioning of defensive behavior to CSs from the appropriate modality: lesions of the auditory cortex and auditory thalamus projections to the basolateral complex affect conditioning to auditory CSs (Campeau and Davis, 1995a), whereas lesions of the perirhinal cortex affect conditioning to visual CSs (Campeau and Davis, 1995a).

Although the basolateral complex is involved in the formation of CS–US associations, the central nucleus is involved in the expression of defensive behavior (Fanselow and Kim, 1994; Maren, 2001). Stimulation of the central nucleus produces autonomic responses similar to those caused by presentation of conditional fear stimuli (Kapp et al., 1982; Iwata et al., 1987). Although lesions of the central nucleus produce deficits in the acquisition and expression of fear conditioning these deficits seem to be caused by reduced performance of defensive behaviors rather than any inability to form the association between CS and US (Fanselow and Kim, 1994).

Additional studies have implicated regions efferent to the central nucleus in the production of specific autonomic and behavioral defensive responses. Regions involved in autonomic responses include the paraventricular nucleus of the hypothalamus and bed nucleus of the stria terminalis (BNST), which mediate glucocorticoid release; the lateral hypothalamus and dorsal motor nucleus of the

vagus (LeDoux et al., 1988; Kapp et al, 1991), which are involved in heart rate responses; and the parabrachial nucleus, which is involved in increased respiration. Regions involved in mediating behavioral responses include the periaqueductal gray, which is involved in expression of freezing behavior (Liebman et al., 1970; Kim et al., 1993) (see later), and endogenous opioid-mediated analgesia (Helmstetter and Landeira-Fernandez, 1990) and the nucleus reticularis pontis caudalis, which mediates the fear potentiation of acoustic startle (Davis et al., 1982).

THE BED NUCLEUS OF THE STRIA TERMINALIS

The BNST is a major output region of the basolateral amygdala, connecting it with the the hypothalamic-pituitary-adrenal axis (Alheid, 1995). Despite these similarities, studies have distinguished between the effects of lesions of the BNST and the central amygdala on defensive behavior. Pretraining and posttraining lesions of the BNST or infusions of the AMPA antagonist NBQX into the BNST do not affect fear-potentiated startle (Hitchcock and Davis, 1991; Walker and Davis, 1997b) or passive avoidance of a shock probe (Treit et al., 1998), whereas similar treatments directed at the central nucleus of the amygdala reduce expression of conditional fear. In contrast, lesions of the BNST, but not the central nucleus, reduce the expression of light-enhanced startle (Walker and Davis, 1997b) and CRH-enhanced startle (Lee and Davis, 1997). More recently. Fendt et al. (2003) demonstrated that unconditional freezing to an odor derived from predator feces was blocked by infusion of the GABA agonist muscimol into the BNST but not the amygdala. This pattern of results has given rise to the suggestion that the BNST is selectively involved in responses to unconditional fear stimuli. In a recent review, however, Walker et al. (2003) instead suggest that the BNST is involved in responding to long-duration aversive stimuli and can be characterized as part of an anxiety-mediating system, distinct from the amygdala-mediated fear system.

THE PERIAQUEDUCTAL GRAY

The ventrolateral regions of the periaqueductal gray (PAG) have long been implicated in the production of defensive behaviors (Liebman et al., 1970). Lesions of this region, particularly the caudalmost areas (LeDoux et al., 1988), reduce freezing (Kim et al., 1993) and suppress opioid-mediated conditioned analgesia (Helmstetter and Landeira-Fernandez, 1990). These effects occur after the presentation of shock, cues paired with shock, and after presentation of a cat and occur whether the lesion is performed before or after (DeOca et al., 1998). These results indicate that the ventrolateral PAG is involved in mediating the freezing response to predators and to predictors of aversive stimuli.

In contrast, lesions of the dorsolateral regions of the PAG have virtually no effect on freezing in response to either a cat (DeOca et al., 1998) or stimuli associated with shock (Fanselow, 1991a). Electrical stimulation of the dorsolateral PAG induces behaviors that are similar to those of the postshock activity burst. Chemical stimulation of this region shows similar effects, including the elicitation of defensive postures and flight in response to conspecifics (Bandler and Depaulis, 1988). Lesions of this region also dramatically reduce the postshock activity burst. These results indicate a double dissociation of function with the ventrolateral PAG involved in mediating freezing while the dorsolateral PAG is involved in mediating the postshock activity burst (Fanselow, 1991b).

INTEGRATING THE BEHAVIORAL AND NEURAL ORGANIZATION OF DEFENSIVE BEHAVIOR

Understanding the behavioral organization of defensive behavior and the neural substrates underlying particular behaviors has allowed the synthesis of this information into functional models of defensive behavior organization (Fanselow, 1994; Fendt and Fanselow, 1999). While differing authors may focus on particular parts of this organization—for example, Maren (2001) pays particular attention to the role of the amygdala, whereas Walker et al. (2003) review the role of the BNST, an archetypal model could be expected to resemble that presented in Figure 39–3.

A number of features of the organization of defensive behavior are easily discerned in Figure 39–3. The most obvious of these is the

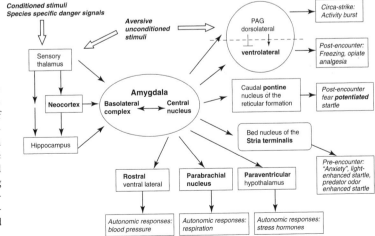

Figure 39–3. Sensory inputs and behavioral outputs (in italics) are connected via a functional network of brain structures. Central to this organization is the amygdala, which acts as an interface between these processes. Mapping the behavioral organization onto the underlying neural substrates allows a better appreciation for the relationship between the various structures involved in defensive behavior.

centrality of the amygdala in defensive behavior. Current conceptions of amygdala function in defensive behavior suggest it acts as an interface between sensory information and defensive behavior output (e.g., Maren, 2001) for both conditional and unconditional responses. Also clear is that the behavioral outputs related to differing levels of perceived predatory imminence are effected by differing structures. This is most clearly seen in the PAG, where circa-strike behaviors are mediated by the dorsolateral regions, whereas post-encounter freezing behavior is mediated by the ventrolateral regions. This model also highlights the possible relationship between preencounter defensive behaviors, which are hypothesized to be mediated by the BNST, and anxiety (Walker et al., 2003). Finally, this model also illustrates the distinction between defensive behavioral responses and autonomic responses such as bradycardia.

CONCLUSION

The understanding of defensive behavior at both a behavioral and neural level has been greatly enhanced over the past 30 years. The functional mapping of behavioral organization onto a network of brain structures has provided an invaluable and productive framework for generating testable hypotheses of the effects of experimental manipulations and may be beginning to inform studies of the substrates of clinical disorders related to dysfunction of defensive behavior in people (Walker et al., 2003). This synthesis of brain and behavior is much further advanced in the study of defensive behavior as a functional system and may provide a model for investigations into the neural substrates of behavior systems.

REFERENCES

Adamec R, Kent P, Anisman H, Shallow T, Merali Z (1998) Neural plasticity, neuropeptides and anxiety in animals—implications for understanding and treating affective disorder following traumatic stress in humans. Neuroscience and Biobehavioral Reviews 23:301–318.

Alheid GF, de Olmos JS, Beltramino CA (1995) Amygdala and extended amygdala. In: The rat nervous system, 2 ed. (Paxinos G, ed.), pp. 495–578. San Diego: Academic Press.

Anisman H and Waller TG (1973) Effects of inescapable shock on subsequent avoidance performance: role of response repertoire changes. Behavioral Biology 9:331–355.

Bandler R and Depaulis A (1988) Elicitation of intraspecific defence reactions in the rat from midbrain periaqueductal grey by microinjection of kainic acid, without neurotoxic effects. Neuroscience Letters 88:291–296.

Barnett SA (xxxx) A study in behaviour. London, UK: Methuen and Co.

Bellgowan PS and Helmstetter FJ (1996) Neural systems for the expression of hypoalgesia during nonassociative fear. Behavioral Neuroscience 110:727–736.

Blanchard DC, Williams G, Lee EM, Blanchard RJ (1981) Taming of wild Rattus norvegicus by lesions of the mesencephalic central gray. Physiological Psychology 9:157–163.

Blanchard RJ and Blanchard DC (1969) Crouching as an index of fear. Journal of Comparative and Physiological Psychology 67(3):370–375.

Blanchard RJ and Blanchard DC (1971) Defensive reactions in the albino rat. Learning and Motivation: 351–362.

Blanchard RJ and Blanchard DC (1972) Effects of hippocampal lesions on the rat's reaction to a cat. Journal of Comparative and Physiological Psychology 78(1):77–82.

Blanchard RJ and Blanchard DC (1989) Antipredator defensive behaviors in a visible burrow system. Journal of Comparative Psychology 103:70–82.

Blanchard RJ, Blanchard DC, Agullana R, Weiss SM (1991) Twenty-two kHz alarm cries to presentation of a predator, by laboratory rats living in visible burrow systems. Physiology and Behavior 50:967–972.

Blanchard RJ, Blanchard DC, Hori K (1989) Ethoexperimental approaches to the study of defensive behavior. In: Ethoexperimental approaches to the study of behavior. (Blanchard RJEB and F Paul, eds.), pp. 114–136. New York: Kluwer Academic/Plenum Publishers.

Blanchard RJ, Flannelly KJ, Blanchard DC (1986) Defensive behaviors of laboratory and wild Rattus norvegicus. Journal of Comparative Psychology 100:101–107.

Blanchard RJ, Fukunaga KK, Blanchard DC (1976) Environmental control of defensive reactions to footshock. Bulletin of the Psychonomic Society 8:129–130.

Blanchard RJ, Mast M, Blanchard DC (1975) Stimulus control of defensive reactions in the albino rat. Journal of Comparative Physiological Psychology 88:81–8.

Blanchard RJ, Yang M, Li CI, Gervacio A, Blanchard DC (2001) Cue and context conditioning of defensive behaviors to cat odor stimuli. Neuroscience and Biobehavioral Reviews Special Issue: Defensive Behavior and the Biology of Emotion 25:587–595.

Blanchard RJ, Yudko EB, Rodgers RJ, Blanchard DC (1993) Defense system psychopharmacology: an ethological approach to the pharmacology of fear and anxiety. Behavioural Brain Research 58:155–165.

Bolles RC (1970) Species-specific defense reactions and avoidance learning. Psychological Review 77(1):32–48.

Bolles RC and Collier AC (1976) The effect of predictive cues on freezing in rats. Animal Learning and Behavior 4:6–8.

Bolles RC, Uhl CN, Wolfe M, Chase PB (1975) Stimulus learning versus response learning in a discriminated punishment situation. Learning and Motivation 6:439–447.

Brown JS, Kalish HI, Farber IE (1951) Conditioned fear as revealed by magnitude of startle response to an auditory stimulus. Journal of Experimental Psychology 41:317–327.

Brown S and Schafer A (1888) An investigation into the functions of the occipetal and temporal lobes of the monkey's brain. Philosophical Transactions of the Royal Society of London Series B 179:303–327.

Calhoun JB (1963) The ecology and sociology of the Norway rat. US Dept of Health, Education and Welfare (PHS Monograph 1008). Bethesda, MD.

Campeau S and Davis M (1995) Involvement of the central nucleus and basolateral complex of the amygdala in fear conditioning measured with fear-potentiated startle in rats trained concurrently with auditory and visual conditioned stimuli. Journal of Neuroscience 15:2301–2311.

Davis M, Gendelman DS, Tischler MD, Gendelman PM (1982) A primary acoustic startle circuit: lesion and stimulation studies. Journal of Neuroscience 2:791–805.

De Oca BM, DeCola JP, Maren S, Fanselow MS (1998) Distinct regions of the periaqueductal gray are involved in the acquisition and expression of defensive responses. Journal of Neuroscience 18:3426–3432.

Dess NK and Vanderweele DA (1994) Lithium chloride and inescapable, unsignaled tail shock differentially affect meal patterns of rats. Physiology and Behavior 56:203–207.

Dielenberg RA and McGregor IS (1999) Habituation of the hiding response to cat odor in rats (Rattus norvegicus). Journal of Comparative Psychology 113:376–387.

Estes WK and Skinner BF (1941) Some quantitative properties of anxiety. Journal of Experimental Psychology 29:390–400.

Fanselow MS (1980) Conditioned and unconditional components of post-shock freezing. The Pavlovian Journal of Biological Science 15:177–182.

Fanselow MS (1982) The postshock activity burst. Animal Learning and Behavior 10:448–454.

Fanselow MS (1990) Factors governing one-trial contextual conditioning. Animal Learning and Behavior 18:264–270.

Fanselow MS (1991a) Analgesia as a response to aversive Pavlovian conditional stimuli: cognitive and emotional mediators. In: MR Denny (Ed.) Fear, avoidance, and phobias: a fundamental analysis. Hillsdale, NJ: Lawrence Erlbaum Associates, pp. 61–86.

Fanselow MS (1991b) The midbrain periaqueductal gray as a coordinator of action in response to fear and anxiety. In: The midbrain periacqueductal gray matter: Functional, anatomical and neuroschemical organization (Depaulis A and Bandler R, eds.). New York: Plenum Press.

Fanselow MS (1994) Neural organization of the defensive behavior system responsible for fear. Psychonomic Bulletin and Review 1:429–438.

Fanselow MS and Baackes M (1982) Conditioned fear-induced opiate analgesia on the formalin test: Evidence for two aversive motivational systems. Learning and Motivation 13:200–221.

Fanselow MS and Kim JJ (1994) Acquisition of contextual Pavlovian fear conditioning is blocked by application of an NMDA receptor antagonist D,L-2-amino-5-phosphonovaleric acid to the basolateral amygdala. Behavioural Neuroscience 108:210–212.

Fanselow MS and Kim JJ (1994) Acquisition of contextual Pavlovian fear conditioning is blocked by application of an NMDA receptor antagonist D,L-2-amino-5-phosphonovaleric acid to the basolateral amygdala. Behavioral Neuroscience 108:210–212.

Fanselow MS, Landeira-Fernandez J, DeCola JP, Kim JJ (1994). The immediate-shock deficit and postshock analgesia: Implications for the relationship between the analgesic CR and UR. Animal Learning and Behavior 22:72–76.

Fanselow MS and Lester LS (1988) A functional behavioristic approach to aversively motivated behavior: Predatory imminence as a determinant of the topography of defensive behavior. In: RC Bolles and MD Beecher (Eds) Evolution and learning. Hillsdale, NJ: Lawrence Erlbaum Associates, pp. 185–212.

Fanselow MS, Lester LS, and Helmstetter FJ (1988) Changes in feeding and foraging patterns as an antipredator defensive strategy: a laboratory simulation using aversive stimulation in a closed economy. Journal of Experimental and Analytical Behavior 50:361–374.

Fanselow MS, Sigmundi RA, and Williams JL (1987) Response selection and the hierarchical organization of species-specific defense reactions: The relationship between freezing, flight, and defensive burying. Psychological Record 37:381–386.

Fendt M, Endres T, Apfelbach R (2003) Temporary inactivation of the bed nucleus of the stria terminalis but not of the amygdala blocks freezing induced by trimethylthiazoline, a component of fox feces. Journal of Neuroscience 23:23–28.

Fendt M and Fanselow MS (1999) The neuroanatomical and neurochemical basis of conditioned fear. Neuroscience and Biobehavioral Reviews 23:743–760.

Griffith C (1920) The behavior of white rats in the presence of cats. Psychobiology 2:19–28.

Helmstetter FJ and Fanselow MS (1993) Aversively motivated changes in meal patterns of rats in a closed economy: The effects of shock density. Animal Learning and Behavior 21:168–175.

Helmstetter FJ and Landeira-Fernandez J (1990) Conditional hypoalgesia is attenuated by naltrexone applied to the periaqueductal gray. Brain Research 537:88–92.

Hitchcock JM and Davis M (1991) Efferent pathway of the amygdala involved in conditioned fear as measured with the fear-potentiated startle paradigm. Behavioral Neuroscience 105:826–842.

Iwata J, Chida K, LeDoux JE (1987) Cardiovascular responses elicited by stimulation of neurons in the central amygdaloid nucleus in awake but not anesthetized rats resemble conditioned emotional responses. Brain Research 418:183–188.

Iwata J, LeDoux JE, Meeley MP, Arneric S, Reis DJ (1986) Intrinsic neurons in the amygdaloid field projected to by the medial geniculate body mediate emotional responses conditioned to acoustic stimuli. Brain Research 383:195–214.

Kaltwasser MT (1991) Acoustic startle induced ultrasonic vocalization in the rat: a novel animal model of anxiety? Behavioural Brain Research 43:133–137.

Kapp BS, Gallagher M, Underwood MD, McNall CL, Whitehorn D (1982) Cardiovascular responses elicited by electrical stimulation of the amygdala central nucleus in the rabbit. Brain Research 234:251–262.

Kapp BS, Markgraf CG, Wilson A, Pascoe JP, Supple WF (1991) Contributions of the amygdala and anatomically-related structures to the acquisition and expression of aversively conditioned responses. In: L Da-

chowski and CF Flaherty (Eds) Current topics in animal learning: Brain, emotion and cognition. Hillsdale, NJ: Lawrence Erlbaum Associates, pp. 311–346.

Kellicutt MH and Schwartzbaum JS (1963) Formation of a conditioned emotional response (CER) following lesions of the amygdaloid complex in rats. Psychological Reports 12(2):351–358.

Kim JJ, Rison RA, Fanselow MS (1993) Effects of amygdala, hippocampus, and periaqueductal gray lesions on short- and long-term contextual fear. Behavioral Neuroscience 107:1093–1098.

LeDoux JE, Cicchetti P, Xagoraris A, Romanski LM (1990) The lateral amygdaloid nucleus: sensory interface of the amygdala in fear conditioning. Journal of Neuroscience 10:1062–1069.

LeDoux JE, Iwata J, Cicchetti P, Reis DJ (1988) Different projections of the central amygdaloid nucleus mediate autonomic and behavioral correlates of conditioned fear. Journal of Neuroscience 8:2517–2529.

LeDoux JE, Sakaguchi A, Reis DJ (1983) Alpha-methyl-DOPA dissociates hypertension, cardiovascular reactivity and emotional behavior in spontaneously hypertensive rats. Brain Research 259:69–76.

Lee Y and Davis M (1997) Role of the hippocampus, the bed nucleus of the stria terminalis, and the amygdala in the excitatory effect of corticotropin-releasing hormone on the acoustic startle reflex. Journal of Neuroscience 17:6434–6446.

Lester LS and Fanselow MS (1985) Exposure to a cat produces opioid analgesia in rats. Behavioral Neuroscience 99:756–759.

Lewis JW, Cannon JT, Liebeskind JC (1980) Opioid and nonopioid mechanisms of stress analgesia. Science 208:623–625.

Liebman JM, Mayer DJ, Liebeskind JC (1970) Mesencephalic central gray lesions and fear-motivated behavior in rats. Brain Research 23:353–370.

Maren S (2001) Neurobiology of Pavlovian fear conditioning. Annual Review of Neuroscience 24:897–931.

Maren S and Fanselow MS (1996) The amygdala and fear conditioning: has the nut been cracked? Neuron 16:237–240.

Myer JS (1971) Some effects of non-contingent aversive stimulation. In: Aversive conditioning and learning (Bursh FR, ed.). New York: Academic Press.

Roche JP and Timberlake W (1998) The influence of artificial paths and landmarks on the foraging behavior of Norway rats (Rattus norvegicus). Animal Learning and Behavior 26:76–84.

Treit D, Aujla H, Menard J (1998) Does the bed nucleus of the stria terminalis mediate fear behaviors? Behavioural Neuroscience 112:379–386.

Vernet-Maury E, Constant B, Chanel J (1992) Repellent

effects of trimethylthiazoline in the wild rat (Rattus norvegicus Berkenout). In: Chemical signals in vertebrates (Doty R and Muller-Schwarze D, eds.), pp. 305–310. New York: Plenum Press.

Walker DL and Davis M (1997) Anxiogenic effects of high illumination levels assessed with the acoustic startle response in rats. Biological Psychiatry 42:461–471.

Walker DL and Davis M (1997) Double dissociation between the involvement of the bed nucleus of the stria terminalis and the central nucleus of the amygdala in startle increases produced by conditioned versus unconditioned fear. Journal of Neuroscience 17:9375–9383.

Walker DL, Toufexis DJ, Davis M (2003) Role of the bed nucleus of the stria terminalis versus the amygdala in fear, stress, and anxiety. European Journal of Pharmacology 463:199–216.

Weiskrantz L (1956) Behavioral changes associated with ablation of the amygdaloid complex in monkeys. Journal of Comparative and Physiological Psychology 29:381–391.

Whishaw IQ, Dringenberg HC, Comery TA (1992) Rats (Rattus norvegicus) modulate eating speed and vigilance to optimize food consumption: Effects of cover, circadian rhythm, food deprivation, and individual differences. Journal of Comparative Psychology 106:411–419.

Young BJ and Leaton RN (1996) Amygdala central nucleus lesions attenuate acoustic startle stimulus-evoked heart rate changes in rats. Behavioural Neuroscience 110:228–237.

Cognitive Processes

ROBERT J. SUTHERLAND

40

The goals in this brief overview are two-fold: to provide the reader with an exposure to the kind of thinking about behavioral processes in rats that is characteristic of a cognitive approach and to provide some examples of experimental paradigms that have been used with some success to probe the nature of the rat's representations of its environment and its own actions. The examples are drawn mainly from recent work on attentional and memory processes, two of the traditional subject matters in cognitive science. At the outset, however, it is important to identify why an approach that is limited to studying just the rat's processing of current stimuli and details of on-going movements is inadequate—what motivates the study of cognition?

WHAT IS A COGNITIVE PROCESS?

The study of cognition in the rat and other nonhuman species has often triggered major controversies and misunderstandings. The main reason for this is that at the core what is "cognitive" about a cognitive process is control over behavior by an event or a relationship between events that are not now present, that is, not given by the immediate stimuli. We are often surprised by an animal when its behavior seems to reflect knowledge about future outcomes or provides an inference about the current situation or sensitivity to relationships that are "emergent." In this context, *emergent* means sensitivity to a relationship not explicitly trained. The traditional account

of such behavior involves the idea that through experience the rat's nervous system builds up a "representation" of certain, specific events, or relationships in its environments. The information contained in these representations guides appropriate behavior.

Cognitive neuroscience work with rats involves the study of how these representations are built, how and where they are stored in the brain, and in what ways they influence ongoing behavior.

STIMULUS–RESPONSE VIEWS UNDERSPECIFY BEHAVIOR

A key to understanding why cognition in the rat has generated heated controversy is an appreciation of the keen skepticism that animal behaviourists historically have shown toward complex, inferred, inner causes of behavior, especially when they involve conspicuous similarities to related human processes. There are many good reasons to be suspicious of anthropomorphic approaches. On the other hand, there are many clear examples of how embracing a narrow stimulus–response perspective has caused investigators to completely miss interesting neurobiological processes. In this regard, the field of circadian rhythm research offers a telling parallel to research on cognition in animals. Two contrasting views were played out in developing an understanding of how the daily activities of animals were organized into a daily cycle. One view, akin to the stimulus–response perspective, sought to discover the identity of the fluctuating cues in the environment that were the basis for rhythmic behavior. The alternative view held that there were biologically and chemically complex internal clocks, endogenous oscillating systems, in the brains of animals that control the daily activity and rest cycles in behavior and physiology. The former view yielded nothing of biological significance. This view always underpredicts the existing organization in the dynamics of behavior.

In contrast, the internal clock view spawned an understanding of a rich set of mechanisms involving "clock genes," hierarchically nested rhythmic, interacting neural systems, and circuitry to synchronize or entrain internally generating rhythms to environmental cues, not to mention interesting clinical insights about certain psychopathologies (Golombeck et al., 2003; Lowrey and Takahashi, 2000).

By analogy, in our subject domain, stimulus–response views always underpredict or underspecify complex behavioral processes controlled by neural representations of event relationships. A simple thought experiment should demonstrate this point. Consider a rat in a T-maze housed in a large room with abundant large visual cues around its perimeter. Imagine that the stem of the T-maze (the start arm) is always pointing southward during the rat's initial experiences. The rat is hungry, and on every trial it finds a small morsel of tasty food at the end of the arm pointing eastward. The rewarded arm is white and the other (west) arm is black; after several trials, we observe the rat efficiently running down the start arm and always making a right turn into the white goal arm. At this point, if we were good stimulus–response experimenters, we could easily conduct a "probe" test to determine if the rat is navigating to the food by approaching the white arm or by making a constant right turn. The probe should involve pitting the two sources of response control against each other, by having the white compete against the right turn. This can be done

by simply switching the white and black arms. When we conduct this "competitive test," we discover that the rat continues to make a right turn. We should be content that we have uncovered the simple stimulus–response rule, until some clever student carries out a second probe trial in which she rotates the T-maze so that the stem (the start arm) is pointing northward, leaving the white arm in its westward position. Now the rat beginning the probe trial running southward in the stem, reaches the point where it must turn right (always rewarded) into the white arm (always rewarded) and to our dismay it turns left into the black arm. We repeat this probe trial many times with other rats, always observing the same outcome. We note that by turning left into the black arm, the rats have arrived at the same location in the room at which they were always rewarded during all earlier experiences.

So far we have demonstrated that neither the white arm nor the right turn is a constancy in the rat's choice, but location within the room looks like it could be the key. One last probe trial should complete the picture. We notice that not only has the rat been going to the same spot in the room, even during the competitive tests, but also it has been heading toward the same direction in the room (eastward) on every trial. By shifting the T-maze eastward such that the end of the west arm is exactly where the end of the east arm had been and the east arm projects much farther eastward than it ever had before, we can complete our competitive testing. If the rat heads eastward on this trial to a point in the room it had never experienced, then we can rule out location in the room as the constancy. By the same token, if the rat turns into the westward arm, we can rule out directional constancy. The fact of the matter is that with real rats and real experiments it is likely that we could arrange the complex factors influencing choice such that we could have some rats choosing white arms, others making only right turns, another set going to a specific location in the room, and the rest always heading in a specific direction. A recent conceptually related series of experiments by Skinner et al. (2003) demonstrates this point.

Our stimulus–response perspective is here forlorn. The stimuli in the apparatus and in the room are invariant, it is always the same food reward, and no matter how fine-grained are the measurements of the movements the rat makes as it turns right, turns controlled by brightness, direction, or locations are the same turns.

IT IS THE REPRESENTATION

On a cognitive account of the rat in our thought experiment, information about brightness, turning and going straight, direction, and location relative to room cues form the grist for different types of representations. In addition, these can all be simultaneously built up in different neural networks. To extend our analogy from circadian rhythms, in which different fluctuating environmental cues such as lighting, temperature, noises, social activities, and feeding can synchronize or entrain various internal rhythms (they do not generate them), so too, in our experiments, different forms of cognitive representation are entrained by the cues in the environments. The problem of understanding complex behavior does not have a determinate solution unless it is known which of the various representational systems is engaged as the rat experiences an environment. At this point it is important to note one meaning of "representation" that is not intended here. If we say that a rat has built up representations of the rewards that are available in an environment, we do not mean that there are internalized copies of the reward items that from time to time can be regurgitated to be retasted or literally reexperienced. Here, *reward representation* means that there is a neural network in which a pattern of activity that corresponds to the reward item in such a way that neural operations involving that pattern can lead to

valid conclusions or inferences about what would be adaptive behavior in an environment that contains the specific reward item (see the section on reinforcer representations). An example from the spatial domain may be more concrete. By saying that a rat has a representation of direction and distance from its current location to its home, it is implied that some set of operations on the pattern of activity across the neural network representing that information can lead to a homeward trip with a direct trajectory having adaptive acceleration profiles (as has been shown by Wallace & Whishaw [Chapter 42], see also Gallistel, 1990). It is *not* meant that there is a literal Cartesian map that is called up, inspected with the mind's eye, and, based on mentally measured topographical coordinates on the map, a correct course plotted.

Cognitive processes in many ways resemble basic sensory or motor processes and in important ways they are different. One resemblance is that individual cognitive subsystems are specialized to enable specific types of action important in the rat's trafficking with its natural or social environment. Also, the processes are informed only by specific types of information. This allows for highly specialized and abstract processing, but this also imposes important constraints. Thus, the sort of cognitive systems that rats have do not permit all actions to be informed by all types of information. The functional organization of cognition confers the advantages of highly specialized processing that allows for flexible behavior in respect to cues and movements, as well as cognitive blind spots.

REPRESENTATIONS INTERACT

It was noted earlier that multiple different representations are being built up in different neural networks as a rat experiences an environment. A wide area of investigation that is relatively untapped involves the nature of the interactions among representational systems.

There are observations that suggest that under certain circumstances, different representations interfere with or inhibit one another and obviously they can be synergistic or supportive of each other. An especially clear example of the latter interaction involves the representations of head direction in a network including postsubiculum and anterior thalamus with the place field representation in the hippocampus. When the head direction system is disrupted by damage to either postsubiculum or anterior thalamus, hippocampal place fields are still intact, but aspects of their information content and stability between episodes in the same environment are degraded (Calton et al., 2003). The representation of head direction is clearly useful in building up and maintaining a representation of where the rat is in relation to visual and other environmental cues.

We expect that representations of cues or actions could support the building of representations in downstream systems. There is, however, evidence for the counterintuitive notion that acquiring one representation can block learning in a separate representational system and that this blocking is not due to the two systems merely driving competing movements. A nice example of this kind of interaction can be found in the work of McDonald and White, who studied rats learning the relationship between food reward and a particular cue in an arm of a simple maze. There was interference with this simple learning if the rats had had the opportunity to explore that maze and environment. The interference effect was shown to be due to the building up of a hippocampal system representation of the environment during initial exploration and the conditioning to the cue was shown to depend on amygdala circuitry (McDonald and White, 1995; White and McDonald, 1993, 2002).

In the rat, there are several nice examples of synergistic and antagonistic interactions between representations in different systems. What about different representations within

the same system? It is safe to say that we know little about how representations interact between systems, and we know almost nothing about within-system representational interactions beyond that they occur. The recent example comes from studying contextual avoidance learning (Fenton et al., 1998). Rats are placed on a circular tabletop with salient cues around the room. When they enter a specific region on the table, they receive a mild foot shock. They learn to quickly avoid entering that region. A moment's reflection reveals that, as in our preceding T-maze example, the identity of the region can be defined by more than one kind of information. To consider just two, the region can be represented by its relationship to the available cues around the room or by its relationship to the available cues on the table (which would be supported by self-motion information). We know from other work measuring the place field properties of hippocampal neurons that either of these types of information can serve as a frame of reference for the hippocampal representation (Gothard et al., 1996). Which does the rat use in this situation? The answer is both.

Fenton and co-workers (1998) demonstrated that both representations were "simultaneously" active by rotating the table in a slow and continuous manner. The rat, in the same episode, would avoid a region that rotated with the table frame and a region that was stable in the room frame. Furthermore, when one records from neurons in hippocampus in this situation, some have place fields relative to the table frame and others to the room frame (Zinyuk et al., 2000). Are these two representations of the environment simultaneously active, or does the hippocampal network switch quickly and coherently from one representation frame to the other? We do not know. It is possible that these two frameworks could be interleaved through the network, but equally one can imagine that as the rat attends to different features of the environment, these two representations are successively recalled. This latter possibility suggests that attention could be a critical process in the rat differentially allocating processing resources to different parts of the environment and hence to different representations.

We turn now to how it is that attention is studied in the rat.

ATTENTION

I would like to describe a limited set of behavioral tasks that have been used successfully to study somewhat different aspects of attention in the rat. Attention involves many separable processes but we will consider only two: (1) sustaining alert responsiveness in the face of occasional brief events and (2) selective attention.

SUSTAINED ATTENTION

Robbins and his co-workers have made extensive use of a serial reaction time task to map several components of the rat's attentional networks (Muir et al., 1996). The task, the five-choice serial reaction time task, requires the rat to detect a brief (500 milliseconds) visual stimulus. This visual target is presented through one of five small holes in a wall. Each of the five locations is used equally, often according to a random sequence. The rat's job is to quickly (within 5 seconds) poke its nose into the hole that had been lit, whereupon it is rewarded with a food pellet at a separate magazine. Five seconds later, the next trial begins. The task lends itself well to measuring several aspects of attentional performance. The typical measures include reaction time to nose-poke (from light target onset), response accuracy (nose-poking into the correct hole), reaction time to move to the food magazine, and number of trials during which no response is made. It is of importance that, in relation to attention, the rat must maintain an alert state to detect briefly presented visual targets, and because the target can unpredictably appear at one of five locations, good

performance requires that the rat actively scan across the relevant spatial layout. There are also opportunities to detect other kinds of behavioral effects such as nose-poking before target presentation (premature responding), repeated poking into the same hole (perseveration), and simply generally slow movement (Passetti et al., 2002).

SELECTIVE ATTENTION

Implicit in the five-choice serial reaction time task is the idea the rats can divide attention among multiple spatial locations, presumably by scanning across the layout, sampling successively, or simultaneously the target sites. The *covert orienting task* makes explicit that rats have the capacity to selectively attend to a region of space, such that their processing of information from that region is selectively enhanced. Posner (1980) designed this simple paradigm that permits measurement of the mechanisms that specifically underlie the selective spatial attention process. In brief his procedure involves subjects responding as quickly and as accurately as they can when they detect a briefly presented visual target. The target appears equally often to the left or right of head direction. For example, in an experiment by Stewart et al. (2001), a trial begins with the rat inserting its snout a specified distance into a hole to interrupt a photobeam. Just before the target appears, a cue is presented to the left or right. Usually the cue's location predicts where the target will appear, but sometimes the cue and target appear on opposite sides. The former case constitutes cue valid trials, and the latter cue invalid trials. There also are trials during which the target appears without prior cuing and still other trials when two cues, one left and one right, appear before the target. The rat's task is to detect the visual target and to quickly withdraw its snout from the nose-poke hole.

It is thought that the appearance of the brief cue on one side reflexively attracts attention to that region of space. If it did, then one should find that the reaction time (from target onset to snout withdrawal) would be faster than if no cue or two cues had appeared. Further, attention should be allocated away from target locations on the other side, leading to longer reaction times when the cue is invalid (i.e., when the cue and target appear on opposite sides). Each of these predictions has been confirmed in work with rats, strongly supporting the idea that rats do in fact have mechanisms for selectively and covertly directing attention to specific locations. The fact that reaction times are shorter on valid cue trials than on trials when two cues are presented rules out an explanation of the valid cue effect that is based on the cue simply alerting the rat to the impending target appearance.

If rats have a mechanism for selectively shifting attention to different points in space, do they also have mechanisms that enable shifting attention between cues (or objects) or cue types, independent of where these might be located in space? One approach to the study of nonspatial selective attention takes advantage of a cognitive phenomenon referred to as *attentional set*. Imagine that we present a subject with a variety of items that differ along several stimulus dimensions, say color, shape, and size, and require initially that the subject discriminate among the items based on color. After learning, we will find relatively easy transfer of discriminations to new colours that we introduce but more difficult transfer if we shift to requiring discriminations based on shape or size. If the stimulus dimension stays the same for a new discrimination, it is called an *intradimensional shift*, and if a different dimension is used, it is called an *extradimensional shift*. The finding that intradimensional shifts proceed more readily than extradimensional shifts is the basis for ascribing to the subject an attentional set. It is believed that the initial training with one dimension induces a selective scanning of objects for relevant perceptual features. This selective attention to the relevant dimension comes at the cost of diminished processing of values of irrelevant perceptual features.

Verity Brown and her colleagues designed a procedure for rats demonstrating that they can selectively allocate attention to perceptual features of objects, independent of spatial location. They have designed an *attentional set-shifting task* in which rats learn to dig in small bowls for a food reward. The bowls differ in odor, digging medium, and surface texture. Rats learn to discriminate readily using any of the three dimensions and readily learn new discriminations if the perceptual dimension of the new discrimination remains the same. They find it significantly more difficult to learn a new discrimination if it involves a shift to a previously irrelevant dimension. That is, rats have more difficulty with extradimensional compared with intradimensional shifts, a defining characteristic of an attentional set.

ATTENTIONAL NEURAL SYSTEMS

In several experiments, Sarter, Robbins, and others have shown that forebrain cholinergic systems projecting to neocortex are essential in supporting accurate sustained attention in the rat. The demonstrations of cholinergic involvement include reduced sustained attention performance in rats after selective elimination of forebrain cholinergic cells using the selective immunotoxin IgG-192 saporin as well as extracellular unit recordings and in vivo measurement of acetylcholine release in cortex (Dalley et al., 2001; Everitt and Robbins, 1997; McGaughy and Sarter, 1998, 1999; Sarter and Bruno, 1997; Sarter et al., 2001). The same cholinergic systems have been implicated in maintaining spatially selective attention in the rat, but most of this work has been conducted with systemically administered cholinergic drugs, limiting conclusions about locus of action (Phillips et al., 2000).

Medial prefrontal cortex damage in the rat has more complex effects in sustained attention tasks (Passetti et al., 2002); nonetheless, it is clear that circuits in this region are critical for sustained attention. Also in line with work from primates, medial prefrontal cortex damage dramatically disrupts the rat's ability to shift attentional set (Birrell and Brown, 2000).

MEMORY

One of the most interesting controversies bearing on cognitive processes in the rat is the extent to which memory representations enable rats to predict the future and to revisit the past.

CAN RATS PREDICT THE FUTURE?

Many cognitive representations, by their very nature, should confer on rats the ability to anticipate future outcomes of certain actions or events without actually experiencing them. We have seen one example of this involving a form of spatial navigation in which the representation of home base leads to the direction and acceleration characteristics of a trip home that are consistent with the rat having the expectation of finding its home at a certain spot—the rat begins to slow down at a particular part of the trajectory in anticipation of arriving home (Wallace and Whishaw, Chapter 42). How general is this property of memories? Are there other clear examples in the behavior of rats that show that they have representations of the properties of anticipated future events?

REINFORCER REPRESENTATIONS

In a typical Pavlovian conditioning experiment, a rat experiences some event (A) closely followed in time by some significant event (US), say food delivery, that has an ability reliably evoke a behavioral response (UR). In virtue, of the relationship between A and the food, A alone comes to evoke conditioned responses (CR), say, approaches to the site of food delivery. There are very clear examples

of the fact that during conditioning not only is an association formed between A and the CR but also A comes to activate a representation of the specific food. Further, that operations involving this food representation can change the way the rat subsequently responds to A. This is shown by the fact that if after conditioning we pair the food with illness (reinforcer devaluation) without involving A, we discover that when A subsequently occurs conditioned responding is diminished (Holland and Straub, 1979). This effect is specific to conditioned stimuli paired with the specific food type. Thus, without actually experiencing A together with a devalued US, the rat treated A differently, because of its associations with a modified representation of the US.

We find an exactly parallel mechanism in the case of instrumental learning. We have an environment that is structured such that if the rat performs action X, then it obtains a reward with one flavor (F1), and if it performs action Y, it obtains a reward with a second flavor (F2). Traditionally, it is thought that the rewards in this situation function primarily to increase the probability of occurrence of the two actions. More recent work, especially by Balleine and co-workers (Balleine and Dickinson, 1998a, 1998b), has shown the rat forms a connection between the actions and representations of the two outcomes. This is shown after initial training is complete by feeding the rat to satiation with one of the two flavors. The rat is then placed back into the training environment, and the two actions are measured in extinction (no further reward presentation). If F1 was devalued by satiation, then the probability of action X is selectively reduced; if F2 was devalued, then the probability of action Y is selectively reduced. Many experiments on this mechanism converge on the conclusions that the rat represents the action–outcome relationship and that the perceptual features of the outcome are included in this representation such that altering the motivation associated with the outcome can come to alter motivation to perform the associated action.

In both Pavlovian and instrumental conditioning, if the rat experiences a modification of the value attached to an outcome, it will automatically transfer a new value, without direct experience, to previously associated cues and actions.

CONFIGURAL/CONJUNCTIVE REPRESENTATIONS

Cues that rats encounter can occur alone or in conjunction with other cues. Sometimes the fact that cues occur together has an important predictive relationship to some outcome. Often, regardless of any relationship to other cues, a specific cue is unambiguously predictive. Social transmission of food preferences (see Chapter 23) offers us an example of a meaningful conjunction between cues (food flavor and essence of rat breath). There are often more arbitrary cue conjunctions that are used by the rat. An especially good example can be found in context fear conditioning. Rats that experience a shock after a discrete cue learn not only to fear the discrete cue but also to fear the constellation of cues comprising the context. In some interesting ways, the context conditioning is different from the discrete cue conditioning (Rudy and O'Reilly, 1999). For example, placing the rat into the context for the first time and almost immediately delivering the shock produces no fear of the context. The rat must explore the context for an extended interval before pairing it with shock if it is to acquire reliable context fear (Fanselow, 1990, 2000). If instead the rat explores individual features of the context to the same extent but never together, then pairing of the complete context with shock leads to no fear of the context (Rudy and O'Reilly, 1999). It appears to be the case that during the period of preshock exploration, the rat builds up a single configural or conjunctive representation of the elements of the context, a single reference frame. In principle rats could use just the individual, separate elements of the context to predict shock and generate fear—but they do not.

There are clear examples of experimental procedures in which resolving a discrimination requires that the animal represent the relationship among two or more cues. The simplest is the *negative patterning discrimination*, in which either cue A or cue B is associated with a reinforcer, but when A and B occur together, the reinforcer is never delivered (A+, B+, AB−). Rats do learn to respond readily to either cue but not to their co-occurrence. A simple account of this ability is that the rat not only has representations of the individual cues but also a configural or conjunctive representation of the two cues together. Each of these representations can enter into associations with outcomes. If the rat had only representations of the individual cues, negative patterning could not be solved. A second example may make this point even more clearly. The *transverse patterning problem* has the same formal structure as the "rock–paper–scissors" game. The rat experiences three cues, A, B, and C. They occur together in pairs. When A and B are together, choosing A is rewarded; when B and C are together, choosing B is rewarded; when C and A are together, choosing C is rewarded. Again, if the rat formed only representations of the individual cues, the problem cannot be solved. It is only through building up representations that include the co-occurrence or conjunctions of the pairs of cues that the discriminations can be resolved (Alvarado and Rudy, 1992). It is important to note that the *transitive inference problem* used by Eichenbaum and co-workers (see, for example, Dusek and Eichenbaum, 1997), unlike transverse patterning, does not require relational, configural, or conjunctive representations; instead, it can be solved by a series of simple or elemental representations (Frank et al., 2003; Van Elzakker et al., 2003). The transitive "inference" problem begins with training a rat on a series of discriminations problems, A+B−, B+C−, C+D−, D+E−. When presented with a novel combination BD, they choose B. Eichenbaum and co-workers assert that this must be due to formation of a complex ordered relational representation from which relative value of new combinations can be inferred. Rudy and O'Reilly (1999) have convincingly demonstrated that this is not necessary, that a solution exists based on the build up of associative strength of each cue's representation.

HIERARCHICAL REPRESENTATIONS

In the real world, cues that co-occur do not always appear simultaneously. When they do appear simultaneously, as described earlier, rats tend to come to represent them in a configural or conjunctive fashion. Configural representations likely participate in predicting outcomes in the same way as elemental cue representations (Rudy and O'Reilly, 1999; Sutherland and Rudy, 1989). This is often not the case if they appear in a sequence. Holland (1992) and others (Miller and Oberling, 1998; Swartzentruber, 1995) have made a convincing case that rats can represent cues in a special, hierarchical fashion, especially if they occur sequentially. The general name for experimental procedures to study this form of representation is *occasion-setting*.

The simplest occasion-setting paradigm, and one studied with great success by Holland and others, is the feature-negative discrimination. The rat is exposed to a cue A that always predicts a reinforcer, unless it is preceded by cue B (the occasion-setter). Thus the discrimination involves A+; B → A− and the meaning of A depends on prior occurrence of B. Rats readily resolve such discriminations responding differently to A depending on whether B preceded it. One could imagine that the rat could simply learn that B is directly inhibitory in respect to the reinforcer. However, it is clearly shown that the representation of B enters into a different kind of behavioral role. Instead of predicting a specific outcome, B predicts that specific relationships exist between two or more other events. That is, the representation of B predicts which outcomes other cue representations will predict. The specific cue predictions can thus be nested inside a higher-level representational structure. It appears that occasion-setters establish which of the possible event relationships that the rat has experienced are now likely to be operative. Interestingly, it has been shown that at least some aspects of occasion-setting functions are disrupted by selective damage to the hippocampal system (Holland et al., 1999).

FUNCTIONAL EQUIVALENCE

Honey and colleagues (Coutureau et al., 2002) have recently observed interesting behavioral examples showing that rats have even more complex, higher-level representations of event relationships. In their experiments, rats experience two cues, X and Y, that have a predictive relationship to food delivery. For half of the occasions, X predicts food and Y predicts no food (X+; Y−); for the other half of the occasions, the relationship is reversed (Y+; X−). When rats learned that a cue predicted food, they approached the food well and pushed open a small covering flap, so food–well inspection is the learned response. Honey et al. uses four different occasion-setters to signal for the rat, which of the two relationships between cues and food is operative. The occasion-setters never appear together. Two of the occasion-setters, A and B, signal X+; Y−, and the other two occasion-setters, C and D, signal Y+; X−. Thus the four component discrimination problems are A: X+,Y−; B: X+,Y−; C: X+,Y−; D: X+, Y−. After training, the rats receive additional experiences in the presence of A and C. With A (and not C), they receive delivery of many free food pellets. After this additional experience, the conditioned responding during B and then during D is measured. The interesting finding is that increased responding to B and not to D emerges during the test. The similarity between A and B and their distinctiveness from C and D does not reside in any perceptual similarity but rather in the fact that they predict the same relationships between other events. Thus, the new behavior to B emerges without any new experience with in virtue of its functional equivalence to A. Honey et al. have also shown two additional facts. If the occasion-setters are either static context cues or cues with discrete onsets and offsets, the new behavior transfers equally among them. Other methods of revaluing the occasion-setters, for example, shock in their presence, produce similar transfer through their equivalent signaling functions (Honey and Watt, 1999). Interestingly, the transfer does not occur in rats with selective entorhinal cortex damage (Coutureau et al., 2002).

This example demonstrates a further fact about the nature of the rat's representation of the elements of its environment. Not only do rats represent perceptual and motivational properties of events and outcomes and their predictive relationships, but also when an event signals that a certain predictive relationship obtains between other events, that signaling function becomes part of that event's representation. This permits interesting cognitive performance to emerge in behavioral responses to functionally related signals.

DO RATS REVISIT THE PAST?

It is clear that rats have much more rich representational processes than would be predicted by traditional stimulus–response accounts of their behavior. There are numerous examples, only some of them outlined earlier, indicating that rats use these representations built up through their experiences with an environment to predict the future—the outcomes and event relationships in that environment.

In terms of applying these cognitive paradigms to attempts to model human conditions, we should be heartened by many important convergences. Many of the same neural systems that have been established in humans as being critical for attentional processes (e.g., forebrain cholinergic projections, prefrontal and anterior cingulate cortex systems) have been convincingly shown to be involved in very similar, possibly the same, processes in rats. Likewise in the memory domains, we know from work with brain-injured humans or in neuroimaging studies that a ventral prefrontal cortex and amygdala network is involved in representing the affective and motivational value of significant events and cues that predict them (Bechara et al., 1999). The same system in the rat has been shown to be involved in the representational processes involved in cue–reinforcer associations and the behavioral effects of reinforcer revaluation (Balleine et al., 2003; Gallagher et al., 1999; Hatfield et al., 1998). Furthermore, some spatial and configural memory tasks (e.g., Morris water task or transverse patterning) that are affected by disrupting hippocampal system functions in rats are similarly affected by damage to the hippocampal system in humans (Astur et al., 2002; Reed and Squire, 1999; Rickard and Grafman, 1998) and cause reliable activations in this system during task performance in neuroimaging experiments (Ekstrom et al., 2003; Hanlon et al., 2002). Thus, the kinds of behavioral paradigms described earlier that are useful in clarifying cognitive processes in the rat are likely to be very valuable in work applied to human cognitive disorders.

What about episodic memories? Many believe that it is deficits in an episodic memory system that are at the core of human amnesia, most memory disorders, and certain dementias. Thus, the topic is of considerable applied importance. Tulving (1972) suggests that the key properties of episodic memories are that they represent *when* an episode occurred as well as *what* the event relationships were and *where* the episode took place. Furthermore, the reactivation of an episodic memory representation can occur outside of the relevant context. Clayton et al. (2003) suggest three related criteria for establishing episodic memory competence: (1) content, or what, where, and when; (2) an *integrated representation* of these three elements of content; and (3) *flexibility* in updating memory representations in light of new information gathered after the original episode. There is little doubt that humans have such memories, but do rats? Tulving and others doubt it, but they clearly cannot prove that no such episodic memory representations exist. It would be reasonable to conclude it is absent in the rat after numerous clever attempts to demonstrate it have failed (assuming that there is reason in principle that it cannot be demonstrated in the rat). I assert that there have probably never been ambitious, clever attempts to demonstrate episodic memories in rats using a paradigm that covers all of the relevant criteria. The trick in such a demonstration is to show that the rat's event memory includes an integrated representation of what, where, and when and that it can be updated in light of new relevant events.

One way to arrange such a demonstration has been devised for the experiments with the Western scrub jay (Clayton et al., 2001). Clayton has taken advantage of the fact that these jays naturally cache perishable food. It is well established that they remember where their caches are located and because they find some foods more palatable than others, their cache site preferences reveal that they remember what was cached. By varying when more palatable, but perishable, foods were cached in relation to the opportunities to retrieve cached foods, one can use the cache site retrieval preferences to learn if the jays remember when they cached the perishable but more palatable, foods. Clayton has a clear example of episodic-like memory in a bird species (Griffiths and Clayton, 2001). No similar attempt has been reported in rats.

We have seen that the rat's event representations contain information about where events took place and, based on the reinforcer devaluation experiments, it is clear that they remember what outcomes are expected. (See Day et al., 2003, for a very nice demonstration

of this in a novel learning procedure involving where–what paired associates.) What remains to be demonstrated is that the same event representation also contains information about when the event occurred.

I suggest that a simple way, given available data, to accomplish the demonstration of episodic memory in the rat is to use the revaluation procedure in Honey's functional equivalence experiments but with occasion-setters whose functions depend not only on their perceptual properties but on when they occur (e.g., time of day). Each occasion setter can have two time-stamps (say each cue occurs in the morning and the afternoon). We arrange the environment such that it is the conjunction of the cue identity and the time-stamp that signals where and what to do. We then revalue cue A at time 1 and test for transfer to cues that signal the same what–where relations as A when they are presented at the same time of day as the revalued cue A. If the rat "automatically" changes its behavior in the presence of a different cue with the same time-stamp and functional associations as the revalued A without any further direct experience with the different cue, we will have demonstrated that rats can show at least one form of time travel through their memories. Such a demonstration would bring us a step closer to measuring episodic memory in rats.

Memory researchers working with rats have only recently started to come to grips with the difficult conceptual and methodological challenges presented by demonstrating true episodic memory processes using nonlinguistic criteria.

CONCLUSION

The field of study of cognition in the rat has borne recent fruit. We now have a rich variety of behavioral tasks that enable reliable and valid measurements of several components of attention and of the characteristics of the rat's representations of important events and interevent relationships. We can be confident that there are striking similarities in the neural systems that subserve attentional and memory representational processes in rats and humans. We should be equally confident that we have not yet probed the full extent of these similarities, nor have we come to grips with what will likely be striking representational differences between rats and humans. It is an exciting time to be a student of rat cognition, as we continue to develop and refine behavioral tools for investigating the mind of the rat we will inevitably learn more about our own minds.

REFERENCES

Alvarado MC and Rudy JW (1992) Some properties of configural learning—An investigation of the transverse-patterning problem. Journal of Experimental Psychology Animal Behavior Processes 18:145–153.

Astur RS, Taylor LB, Mamelak AN, Philpott L, Sutherland RJ (2002) Humans with hippocampus damage display severe spatial memory impairments in a virtual Morris water task. Behavioural Brain Research 132:77–84.

Balleine BW and Dickinson A (1998a) The role of incentive learning in instrumental outcome revaluation by sensory-specific satiety. Animal Learning and Behavior 26:46–59.

Balleine BW and Dickinson A (1998b) Goal-directed instrumental action: contingency and incentive learning and their cortical substrates. Neuropharmacology 37:407–419.

Balleine BW, Killcross AS, Dickinson A (2003) The effect of lesions of the basolateral amygdala on instrumental conditioning. Journal of Neuroscience 23:666–675.

Bechara A, Damasio H, Damasio AR, Lee GP (1999) Different contributions of the human amygdala and ventromedial prefrontal cortex to decision-making. Journal of Neuroscience 19:5473–5481.

Birrell JM and Brown VJ (2000) Medial frontal cortex mediates perceptual attentional set shifting in the rat. Journal of Neuroscience 20:4320–4324.

Calton JL, Stackman RW, Goodridge JP, Archey WB, Dudchenko PA, Taube JS (2003) Hippocampal palce cell instability after lesions of the head direction cell network. Journal of Neuroscience 23:9719–9731.

Clayton NS, Bussey TJ, Dickinson A (2003) Can animals recall the past and plan for the future? Nature Reviews Neuroscience 4:685–691.

Clayton NS, Yu KS, Dickinson A (2001) Scrub jays (aphelocoma coerulescens) form integrated memories of the multiple features of caching episodes. Journal of Experimental Psychology Animal Behavior Processes 27:17–29.

Coutureau E, Killcross AS, Good M, Marshall VJ, Ward-Robinson J, Honey RC (2002) Acquired equivalence and distinctiveness of cues: II. Neural manipulations and their implications. Journal of Experimental Psychology Animal Behavior Processes 28:388–396.

Dalley JW, McGaughy J, O'Connell MT, Cardinal RN, Levita L, Robbins TW (2001) Distinct changes in cortical acetylcholine and noradrenaline efflux during contingent and noncontingent performance of a visual attentional task. Journal of Neuroscience 21:4908–4914.

Day M, Langston R, Morris RGM (2003) Glutamate-receptor-mediated encoding and retrieval of paired-associate learning. Nature 424:205–209.

Ekstrom AD, Kahana MJ, Caplan JB, Fields TA, Isham EA, Newman EL, Fried I (2003) Cellular networks underlying human spatial navigation. Nature 425:184–187.

Everitt BJ and Robbins TW (1997) Central cholinergic systems and cognition. Annual Review of Psychology 48:649–684.

Fanselow MS (1990) Factors governing one-trial contextual conditioning. Animal Learning and Behavior 18:264–270.

Fanselow MS (2000) Contextual fear gestalt memories, and the hippocampus. Behavioural Brain Research 110:73–81.

Fenton AA, Wesierska M, Kaminsky Y, Bures J (1998) Both here and there: Simultaneous expression of autonomous spatial memories in rats. Proceedings of the National Academy of Sciences U S A 95:11493–11498.

Frank MJ, Rudy JW, O'Reilly RC (2003) Transitivity, flexibility, conjunctive representations and the hippocampus. II. An computational analysis. Hippocampus 13:299–312.

Gallagher M, McMahan RW, Schoenbaum G (1999) Orbitofrontal cortex and representation of incentive value in associative learning. Journal of Neuroscience 19:6610–6614.

Gallistel CR (1990) The organisation of learning. Cambridge, MA: The MIT Press, 1990.

Golombek DA, Ferreyra GA, Agostino PV, Murad AD, Rubio MF, Pizzio GA, Katz ME, Marpegan L, Bekinschtein TA (2000) From light to genes: Moving the hands of the circadian clock. Frontiers in Bioscience 8:S285–S293.

Gothard KM, Skaggs WE, Moore KM, McNaughton BL (1996) Binding of hippocampal CA1 neural activity to multiple reference frames in a landmark-based navigation task. Journal of Neuroscience 16:823–835.

Griffiths DP and Clayton NS (2001) Testing episodic memory in animals. Physiology and Behavior 73:755–762.

Hanlon FM, Weisend MP, Huang MX, Astur RS, Moses SN, Lee RR (2002) Neural activation during performance of transverse patterning using magneto-encephalography. Journal of Cognitive Neuroscience Suppl S:163–163.

Hatfield T, Han J-S, Conley M, Gallagher M, Holland P (1996) Neurotoxic lesion of the basolateral, but not central, amygdala interfere with Pavlovian second-order conditioning and reinforcer-devaluation effects. Journal of Neuroscience 16:5256–5265.

Holland PC (1992) Occasion setting in Pavlovian conditioning. In: Medin D, editor. The psychology of learning and motivation. Vol. 28. San Diego: Academic Press; p. 69–125.

Holland PC and Straub JJ (1979) Differential effects of two ways of devaluing the unconditioned stimulus after Pavlovian appetitive conditioning. Journal of Experimental Psychology Animal Behavior Processes 5:65–78.

Honey RC and Watt A (1999) Acquired relational equivalence between contexts and features. Journal of Experimental Psychology Animal Behavior Processes 25:324–333.

Lowrey PL and Takahashi JS (2000) Genetics of the mammalian circadian system: Photic entrainment, circadian pacemaker mechanisms, and posttranslational regulation. Annual Review of Genetics 34:533–562.

McDonald RJ and White NM (1995) Information acquired by the hippocampus interferes with acquisition of amygdala-based conditioned cue preference (CCP) in the rat. Hippocampus 5:189–197.

McGaughy J and Sarter M (1998) Sustained attention performance in rats with intracortical infusions of 192 IgG-saporin-induced cortical cholinergic deafferentation: effects of physostigmine and FG 7142. Behavioral Neuroscience 112:1519–1525.

Miller RR and Oberling P (1998) Analogies between occasion setting and pavlovian conditioning. In: Schmajuk NA, Holland PC, editors. Occasion setting: Associative learning and cognition in animals. Washington, DC: American Psychological Association; p. 3–35.

Muir JL, Everitt BJ, Robbins TW (1996) The cerebral cortex of the rat and visual attentional function: dissociable effects of mediofrontal, cingulate,anterior, dorsolateral and parietal cortex lesions on a five

choice serial reaction time task. Cerebral Cortex 6:470–481.

Passetti F, Chudasama Y, Robbins TW (2002) The frontal cortex of the rat and visual attentional performance: Dissociable functions of distinct medial prefrontal subregions. Cerebral Cortex 12:1254–1268.

Phillips JM, McAlonan K, Robb WGK, Brown VJ (2000) Cholinergic neurotransmission influences covert orientation of visuospatial attention in the rat. Psychopharmacology 150:112–116.

Posner MI (1980) Orienting of attention. Quarterly Journal of Experimental Psychology 32:3–25.

Reed JM and Squire LR (1999) Impaired transverse patterning in human amnesia is a special case of impaired memory for two-choice discrimination tasks. Behavioral Neuroscience 113:3–9.

Rickard TC and Grafman J (1998) Losing their configural mind: Amnesic patients fail on transverse patterning. Journal of Cognitive Neuroscience, 10:509–524.

Rudy JW and O'Reilly RC (1999) Contextual fear conditioning, conjunctive representations, pattern completion, and the hippocampus. Behavioral Neuroscience 113:67–880.

Sarter M and Bruno JP (1997) Cognitive functions of cortical acetylcholine: toward a unifying hypothesis. Brain Research Reviews 23:28–46.

Sarter M, Givens B, Bruno JP (2001) The cognitive neuroscience of sustained attention: where top-down meets bottom-up. Brain Research Reviews 35:146–160.

Skinner DM, Etchegary CM, Ekert-Maret EC, Baker CJ, Harley CW, Evans JH, Martin GM (2003) An analysis of response, direction, and place learning in an open field and T maze. Journal of Experimental Psychology Animal Behavior Processes 29:3–13.

Stewart C, Burke S, Marrocco R (2001) Cholinergic modulation of covert attention in the rat. Psychopharmacology 155:210–218.

Sutherland RJ and Rudy JW (1989) Configural association theory: The role of the hippocampal formation in learning, memory, and amnesia. Psychobiology 17:129–144.

Swartzentruber D (1995) Modulatory mechanisms in Pavlovian conditioning. Animal Learning and Behavior 23:123–143.

Tulving E (1972) Episodic and semantic memory. In: Tulving E, Donaldson W, editors, Organization of memory New York: Academic Press, p. 381–403.

Van Elzakker M, O'Reilly RC, Rudy JW (2003) Transitivity, flexibility, conjunctive representations and the hippocampus. I. An empirical analysis. Hippocampus 13:292–298.

White NM and McDonald RJ (1993) Acquisition of a spatial conditioned place preference is impaired by amygdala lesions and improved by fornix lesions. Behavioural Brain Research 55:269–281.

White NM and McDonald RJ (2002) Multiple parallel memory systems in the brain of the rat. Neurobiology of Learning and Memory 77:125–184.

Zinyuk L, Kubik S, Kaminsky Y, Fenton AA, Bures J (2000) Understanding hippocampal activity by using purposeful behavior: Place navigation induces place cell discharge in both task-relevant and task-irrelevant spatial reference frames. Proceedings of the National Academy of Sciences USA 97:3771–3776.

Incentive Behavior

41

BERNARD W. BALLEINE

An *incentive* is any source of motivation that is acquired through learning; that is not innate. Incentive behavior is therefore any response that is demonstrably controlled by an incentive learning process of one kind or another. For the purposes of this chapter, the incentive behavior of rats is discussed in the context of evaluative, Pavlovian, and instrumental conditioning procedures productive of what we refer to here as evaluative, Pavlovian, and instrumental incentives, respectively (see Balleine, 2001; Dickinson and Balleine, 2002; Dayan and Balleine, 2002, for recent reviews). As described in this chapter, these incentive processes constitute a hierarchy: instrumental incentives involve in part processes engaged by Pavlovian incentives that in part involve processes engaged by evaluative incentives. Whether these incentive processes can in fact be fully dissociated structurally is still a matter of debate, and some current issues are discussed in the final section.

EVALUATIVE INCENTIVES

Since Pavlov (1927), it has become commonplace for students of learning to divide perceptual events or stimuli into those that are conditioned and those that are unconditioned. However, for Pavlov, this distinction was made solely on the basis of the behavioral response that they evoke on first presentation. Although a physiological disturbance may have an immediate and unlearned motivating effect (i.e., elicit an unconditioned response), the motivating effect of the perceptual features of the event productive of that disturbance (e.g., the taste, smell, and visual, auditory, or textural features) is established only once the animal has had some experience with these events. Thus, for Pavlov, "the effect of the sight and smell of food is not due to an inborn reflex, but to a reflex which has been acquired in the course of the animal's own individual existence" (1927; p. 23)—a process he referred to as "signalization."

In contrast, contemporary theories of learning acknowledge Pavlov's basic division between stimuli but obscure its basis by describing events like food, water, or shock as "unconditioned stimuli" (USs). In this way, these theories tend to conflate what is unconditioned about USs, the reflexive responses (URs) they elicit on first contact, with what is not; that is, the association between their sensory-perceptual features and, physiologically based, motivational systems activated by the detection of such things as calories, fluids, pain, and so on (c.f., Chapters 18–23). Though quite subtle, this distinction between the representation of an event and the response that it elicits is important. It identifies a fundamental form of learning that has largely been overlooked in contemporary accounts of learning and motivation in the rat and referred to here as evaluative incentive learning (Fig. 41–1). (De Houwer et al., 2001.)

References to evaluative incentives in rats have, occasionally, surfaced in the past, although cloaked in quite different terms, such as in the work of P. T. Young in 1949

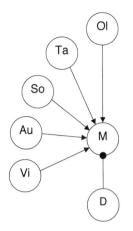

Figure 41–1. A model of the representational and motivational processes mediating evaluative incentive learning. The acquisition of biological significance by the sensory-perceptual features of events (i.e., olfactory [Ol], taste [Ta], somatosensory [So], auditory [Au], and visual [Vi] features) is shown here to involve the formation of an acquired connection (solid arrows) with physiologically based, motivational systems (M) modulated by visceral and humoral signals originating in peripheral regulatory systems (D, solid circle).

and in analyses of research in the 1940s and 1950s on what came to be called externalized or acquired drive (Bolles, 1975). Moll (1964), for example, reports evidence consistent with Pavlov's "signalization" process in young rats. On their first experience with food deprivation, Moll's rats ate substantially less than was required to make up their deficit or even to maintain them, although they rapidly learned to increase consumption over time and over presentations of the food. Similarly, Changizi, McGehee, and Hall (2002) observed that rat pups did not exhibit food-seeking behavior when food deprived unless they had previous experience with food deprivation and eating. Perhaps more surprisingly, Hall and colleagues also reported evidence that the same is true of water for thirsty rats (Hall et al., 2000; Changizi et al., 2002). Thus, for example, in preweanling rats or rats weaned onto a fluid diet, the induction of a strong, extracellular thirst was observed to have no immediate effect on water consumption relative to rats not made thirsty. After experience with water in the thirsty state had been allowed, however, sub-

sequent induction of thirst produced an immediate increase in water consumption. The representation of events such as food or water as biologically significant for hungry or thirst rats appears, therefore, to be acquired.

The procedures used to assess the acquisition of evaluative incentives have obvious similarities to those used to generate conditioned taste preferences and aversions. With regard to the former, rats are generally first deprived of some essential commodity or other (e.g., nutrients, fluids, or, more specifically, sodium or calcium), after which a stimulus (usually a taste) is paired with the delivery of the deprived commodity, presented either in solution with the taste or directly via intragastric, intraduodenal, hepatic portal, or intravenous routes. Evidence for evaluative incentive learning is established if, relative to rats given the taste and the infusion of the deprived commodity unpaired, the paired group significantly increases their willingness to contact and consume the taste (Sclafani, 1999). It has also been reported that treatments such as these increase the tendency of rats to show ingestive, orofacial fixed action patterns (FAPs) when the paired taste is contacted (Forestell and Lolordo, 2003). Deprivation of one or other commodity appears to be necessary to generate conditioned taste preferences of this kind (Harris et al., 2000), suggesting that evaluative conditioning is modulated by visceral and humoral signals originating in regulatory processes such as those that control feeding, drinking, and so on (Fig. 41–1) (Sudakov, 1990). Indeed, studies that have specifically manipulated deprivation state report, for example, that the acquisition of taste preferences by nutrient loads is strongly controlled by the degree of food deprivation (Harris et al., 2000). Studies of orofacial FAPs confirm that these reactions to taste stimuli are also modulated by motivational state. The taste reactivity patterns elicited by sugar solutions are augmented by hunger (e.g., Berridge, 1991) and those elicited by saline are enhanced by a sodium appetite (e.g., Berridge et al., 1984).

Garcia (1989; Garcia et al., 1989) has argued that conditioned taste aversions also are best viewed as an example of evaluative incentive learning (what he called "Darwinian conditioning"). This procedure involves the pairing of a (usually sweet) taste with the injection of an emetic agent, such as lithium chloride. Subsequently, both orofacial FAPs shift from acceptance to those associated with rejection (Berridge, 2000), and the consumption of substances that contain that taste is strongly and enduringly altered by this pairing. Garcia et al. (1989) argue that this effect reflects the formation of an association between taste afferents and brain stem autonomic centers that subsequent work has identified as the parabrachial nucleus for conditioned aversions (Reilly, 1999) as well as, interestingly enough, for conditioned preferences (Sclafani et al., 2001). The site of integration appears to differ for evaluative incentive learning involving olfactory, visual, auditory, and somatosensory features and likely involves the amygdala (Holland et al., 2002) along with its afferents in sensory cortex, brain stem, and hypothalamic nuclei.

PAVLOVIAN INCENTIVES

In contrast to evaluative conditioning, accounts of Pavlovian conditioning usually emphasize the formation of associations between stimulus representations rather than between stimuli and the activity of intrinsic motivational systems (Pearce and Bouton, 2001). It is quite possible, in fact, for Pavlovian associative processes to operate without a programmed motivational manipulation (e.g., in sensory preconditioning). Usually, however, evaluative incentives are heavily used in Pavlovian conditioning in the pairing of stimuli, usually called conditioned stimuli (CSs), that are relatively neutral with respect to a particular motivational state, with events that both elicit biologically potent responses (URs)

and are *represented* as biologically significant events (USs)—that have been established as evaluative incentives. It should come as no surprise, therefore, that numerous authors have suggested that Pavlovian CSs can acquire incentive properties (see Dickinson and Balleine, 2002, for a review).

One of the most sophisticated accounts of Pavlovian incentive learning is that developed by Konorski (1967). In Konorski's account, Pavlovian conditioning comes in two forms: *consummatory* and *preparatory*. Consummatory conditioning occurs when the form of the conditioned response reflects the specific sensory properties of the US, such as when a signal, or CS, predicting a food US elicits salivation, licking, or chewing (DeBold et al., 1965) or when predicting a shock to the eye elicits a blink response (Schmajuk and Christianson, 1990). Hence, Konorski assumed that consummatory conditioning reflects the formation of an association between the representation of the CS and the sensory and perceptual features of the US representation and that it is CS-related activation of the US representation via this connection that elicits the consummatory CR.

In contrast, preparatory conditioning reflects the acquisition of responses characteristic of the affective class to which that US belongs rather than its specific properties. These responses are quite general; for example, CSs paired with appetitive USs (e.g., food, water, etc.) often come to elicit approach, whereas those paired with aversive USs (e.g., foot shock or eye shock) elicit withdrawal. Konorski proposed that preparatory CRs are elicited by activation of affective processes by the CS that may take either of two routes— via the representation of the sensory features of the US or through a direct association, thereby producing purely preparatory conditioning. Hence, preparatory and consummatory conditioning can be dissociated. Ginn et al. (1983) established that although a 0.5 second tone simultaneously presented with leg shock elicited a leg flexion CR, a longer, 4 second, tone did not, although both CSs elicited a heart rate CR. A version of the associative structure thought to underlie the acquisition of Pavlovian incentives on this account is illustrated in Figure 41–2 (see Dickinson and Dearing, 1978; Dickinson and Balleine, 2002).

The affective processes engaged by preparatory conditioning can facilitate the performance of consummatory CRs based on the US as well as the performance of orienting responses (ORs) to the CS itself. Bombace et al. (1991) established a relatively long auditory CS as a signal for a shock to the rear leg and a second, short visual stimulus as a CS for eye shock. They found that the visual CS elicited

a conditioned eye-blink response of greater amplitude when presented during the auditory stimulus. This finding suggests that the sensory representation of the US both activates and can be activated by the general affective state conditioned to the CS. Furthermore, ORs to CSs (such as head jerking to tones or rearing to lights) usually habituate quite quickly but when the CS is paired with a US, their incidence increases (Holland, 1980), suggesting that the motivational support for sensory-specific responses that is provided by affective states conditioned to the CS may be quite general.

In Figure 41–2, different USs from the same affective class are proposed to activate a common affective system. Further evidence for this claim can be drawn from transreinforcer blocking studies. *Blocking* refers to the observation that pretraining one CS often reduces conditioning to a second CS when a compound of these two stimuli is paired with the US (Kamin, 1969); the pretrained CS is said to block conditioning to the added CS. Although the standard blocking procedure uses the same US during pretraining and compound training, Bakal et al. (1974) observed that a CS pretrained with a footshock US blocked conditioning to an added CS when the compound was paired with loud auditory US, even though the sensory properties of these USs differ substantially. What they have in common, however, is that they are aversive, and, as such, transreinforcer blocking is usually taken as evidence that the two USs activate a common affective process. Transreinforcer blocking also provides the best evidence for a common appetitive affective process. Ganesen and Pearce (1988) pretrained a CS with a water US before giving compound training with the added CS with a food US (or vice versa). Conditioned approach to the site of food delivery during the added CS was attenuated by pretraining with water, indicating that pretraining with the water US blocked conditioning to the added CS when paired with food.

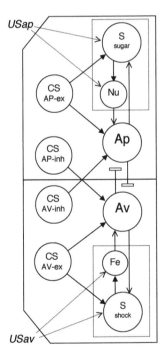

Figure 41–2. The Konorskian model of the representational and motivational processes mediating Pavlovian incentive learning (based on Dickinson and Dearing, 1979; Dickinson and Balleine, 2002). Here the model is elaborated for appetitive conditioning involving sugar and aversive conditioning involving shock. Evaluative processes are included in the dashed rectangle. Solid arrows represent acquired connections; open arrows, fixed connections; AP: appetitive; AV: aversive; ex: excitatory; inh: inhibitory; Nu: nutritive system; Fe: fear system.

APPETITIVE AND AVERSIVE INTERACTIONS

Although transreinforcer facilitation and blocking establish motivational commonalities, it is equally clear that there is a distinction, at least between appetitive and aversive processes. There is a wealth of evidence that CSs of one affective class inhibit responses controlled by CSs of the other affective class (Dickinson and Pearce, 1977). This inhibitory interrelationship is most clearly illustrated by counterconditioning experiments in which a previously established predictor of an aversive US is subsequently paired with an appetitive US or vice versa. Generally, rather than enhancing the performance of the previously conditioned CR, this treatment strongly attenuates it. This evidence has been reviewed extensively (Dickinson and Pearce, 1977; Dickinson and Dearing, 1979; Dickinson and Balleine, 2002) and favors the view that appetitive and aversive affective systems mutually inhibit one another.

In addition to counterconditioning effects, this opponent relationship also offers a straight-forward account of the properties of conditioned inhibitors. A conditioned inhibitor is a stimulus that acquires the capacity to inhibit the CR elicited by an excitatory CS as a result of being paired with the omission of an otherwise predicted US. It has long been known that an inhibitory CS of one affective class has properties in common with an excitatory CS of the opposite affective class (Fig. 41–2). Thus, a CS paired with the omission of expected food is aversive; rats will learned to escape from it (Daly, 1974). Moreover, using a transreinforcer blocking assay, conditioned excitors and inhibitors of opposite affective classes have been found to engage a common incentive process; for example, a CS that predicts the omission of a food US has been reported to block aversive conditioning with a shock US in rats (see Dickinson and Dearing, 1979; Dickinson and Balleine, 2002, for reviews).

MOTIVATIONAL CONTROL OF PAVLOVIAN CONDITIONING

Ramachandran and Pearce (1987) observed that the asymptotic level of magazine approach elicited by a CS paired with either food or water was reduced by the presence of the irrelevant motivational state—thirst in the case of the food reinforcer and hunger in the case of the water reinforcer. The suppression produced by the irrelevant deprivation state was motivational in origin because removal of the irrelevant state during an extinction test restored performance to the level observed in rats trained under the relevant state alone. Ramachandran and Pearce (1987) argue that this interaction between hunger and thirst does not occur at the level of the appetitive motivational systems within the Konorskian model but reflects the mechanism by which primary motivational states modulate the activation of specific US representations. This is an important point that raises a critical distinction between the interaction of motivational systems and the interaction between the mechanisms by which primary motivational states, such as hunger and thirst, modulate the capacity of relevant stimuli to activate these systems.

That some form of motivational modulation of this kind is required is well recognized; performance of the CR in appetitive Pavlovian conditioning has been found to depend directly on deprivation state (DeBold et al., 1965). Equally clearly, this modulation has, to some extent, to be US specific; it is hydration that modulates fluid representations and nutritional need that modulates food representations. The implication of the Ramachandran and Pearce (1987) results for the model, therefore, is not only that hunger enhances the activation of the appetitive motivational system by a CS paired with a nutritional US but also that thirst counteracts this enhancement thereby reducing facilitation of the US representation. Generally, therefore, and as suggested by Figure 41–2, motivational modulation of evaluative incentive processes serves to gate the ability of CSs to activate the affective system through the sensory representation of the US.

It is clear, however, that there should be cases where shifts in primary motivation do not affect conditioned responding. On the model presented in Figure 41–2, these cases should be those in which the CR is entirely preparatory, mediated by a direct connection with the affective system. A notable case of this kind is second-order conditioning in hungry rats, which is resistant to extinction of the first-order CS (Rizley and Rescorla, 1972) and is unaffected by devaluation of the US either by rotation or by satiation treatments (Holland and Rescorla, 1975).

INSTRUMENTAL INCENTIVES

Current evidence suggests that, in instrumental conditioning, rats encode the relationship between an action and its consequence or outcome and are extremely sensitive to the contingency between the performance of an action and the probability of outcome delivery (Balleine, 2001). Nevertheless, it has long been recognized that reference to the action–outcome association alone is not sufficient to determine the performance of an action; any learning that takes the form "action R leads to outcome O" can be used both to perform R and to avoid performing R. Of course, what is missing from this account is mention of the role that the incentive value of the outcome plays in controlling instrumental performance. It is now well established that the rats' experience of the affective and motivationally relevant properties of an outcome strongly controls the performance of actions instrumental to outcome delivery (Dickinson and Balleine, 1994; Balleine, 2001, for reviews). Evidence for this claim has mainly come from studies assessing the impact of shifts in primary motivation on instrumental performance.

One of the most striking properties of instrumental actions is that, in marked contrast to the Pavlovian CR, their performance is not *directly* sensitive to shifts in primary motivation. This was first observed when rats trained hungry to lever press and chain pull, with one action earning food pellets and the other a liquid sucrose solution, were subsequently shifted to water deprivation (Dickinson and Dawson, 1989). In this situation, the rats did not alter their performance of either action and, in fact, only increased their performance of the response trained with the liquid sucrose if they were first allowed to drink the sucrose when thirsty. The shift to water deprivation had no direct impact on performance. In subsequent studies, the same pattern of results has been found after a number of other posttraining shifts in motivation. For example, rats trained to lever press for food when food deprived do not immediately reduce their performance on the lever when they are suddenly shifted to an undeprived state (Balleine, 1992). Nor do they increase their performance immediately if they are trained undeprived and then given a test on the levers when food deprived (Balleine, 1992; Balleine et al., 1994). In both of these situations, rats modify their instrumental performance only after they have been allowed the opportunity to consume the specific food outcome in the test motivational state.

The need for consummatory contact with the instrumental outcome for a shift in primary motivation to affect instrumental performance has been found to be quite general and has been confirmed for a number of different motivational states and using a number of devaluation paradigms. For example, in addition to shifts from hunger to thirst and between hunger and satiety, incentive learning has been found to play a roll in (*1*) taste aversion–induced devaluation effects (Balleine and Dickinson, 1991, 1992), (*2*) specific satiety-induced devaluation (Balleine and Dickinson, 1998a), (*3*) posttraining shifts in water deprivation (Lopez et al., 1992), (*4*) changes in outcome value mediated by drug states (Balleine et al., 1994), and (5) changes in the value of thermoregulatory (Hendersen and Graham, 1979) and sexual (Everitt and Stacey, 1987) rewards (see Balleine, 2001, for a review). In all of these cases, it is clear that, after a shift in primary motivational state, rats have to learn about the effect of this shift on the incentive value of an instrumental outcome through consummatory contact before the shift will act to affect instrumental performance. This form of learning is referred to as *instrumental incentive learning*.

It is interesting to consider why Pavlovian CRs and goal-directed instrumental actions should differ in their sensitivity to the effects of shifts in primary motivation. In a recent review of the literature, Balleine (2001) argued that the primary distinction between Pavlovian and instrumental conditioning may lie in fact that, in instrumental conditioning, the representation of the outcome associated with an action is not directly connected with the motivational/affective structures typically directly activated by the CS. Rather it is only indirectly related to these structures via a connection with the emotional feedback induced by their activation. This account is based on an elemental model of outcome representation (Fig. 41–3), supposing that the most salient sensory features are directly connected with the motivational/affective processes whereas less salient features are not. It need only be assumed, on the basis of differential overshadowing, that CSs proximal to outcome delivery become associated with the more salient features, whereas actions (being more distal to outcome delivery) are associated with other, more diffuse, features. As a consequence, the performance of actions is not affected by shifts in motivational state until the diffuse elements of the outcome with which they are associated are revalued through consummatory contact with the outcome in the prevailing motivational state (Balleine, 2001).

On this account, therefore, instrumental incentive learning is mediated by an association

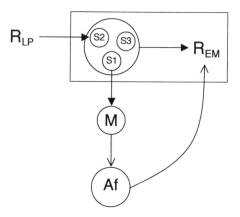

Figure 41–3. A model of instrumental incentive learning. Here the reward, or instrumental outcome, associated with an action such as lever pressing (i.e., LP) is illustrated as composed of several features (S1, S2, S3), the most salient of which is used in an evaluative connection with M that generates emotional feedback (EM) via fixed connections with the affective systems (Af). Incentive learning reflects the formation of an association between the various sensory-perceptual features of the outcome and the emotional response.

between the sensory features of the instrumental outcome and emotional feedback driven by the same motivational/affective processes that are engaged by evaluative and Pavlovian incentives. Instrumental incentive learning, although dependent on these processes, involves a distinct associative connection with this emotional response. By establishing the relative hedonic impact of the instrumental outcome, instrumental incentive learning allows rats (and, of course, other animals) to encode the goal, or incentive, value of the consequences of specific actions and so plays a critical role in action selection (Balleine and Dickinson, 1998b; Dickinson and Balleine, 2000).

CONCLUDING COMMENTS

The evaluative, Pavlovian, and instrumental incentive learning processes around which this brief review of incentive behavior has been organized can be readily distinguished descriptively and perhaps procedurally. They can also be distinguished functionally within a general analysis of the determinants of adaptive behavior as anchoring exploratory behavior, predictive learning, and behavioral choice, respectively, although this issue lies more in the province of evolutionary biology and is beyond the scope of this chapter (see Balleine and Dickinson, 1998b; Dickinson and Balleine, 2000, for further discussion). It remains to be established, however, whether these learning processes are mediated by distinct mechanisms, that is, whether their control over behavior can be dissociated one from the other. Unfortunately, to date, there has been relatively little systematic research of this issue, and what there has been has not always been focused on rats.

For example, arguably the best evidence that evaluative and Pavlovian incentive learning are mediated by distinct processes has come from studies of human learning where it has been reported that changes in the incentive value of a perceptual stimulus (like faces or flavors) can be produced by pairing them with noxious consequences in situations where subjects appear to be unaware of the contingency between these events (De Houwer et al., 2001). For obvious reasons, experiments of this kind are difficult to conduct in rats. Nevertheless, some dissociations have been reported that are suggestive of a similar effect. For example, lesions of the basolateral amygdala have been reported to have no effect on either sensory preconditioning or first-order Pavlovian conditioning using food (Holland et al., 2001; Blundell et al., 2003) or, in a conditioned suppression paradigm, shock USs (Killcross et al., 1997). These lesions do, however, strongly affect conditioned flavor aversions, aspects of fear conditioning (notably freezing behavior), and conditioned eating responses (Holland et al., 2001). In contrast, lesions of dorsal hippocampus can influence the association between stimuli in sensory preconditioning (Talk et al., 2002) but do not affect evaluative incentives produced either by conditioned taste preference or taste aversion procedures (Reilly et al., 1993).

Furthermore, Wyvell and Berridge (2000) reported that microinjection of a dopamine agonist into the nucleus accumbens of rats enhanced appetitive Pavlovian conditioning without affecting ingestion and rejection-related orofacial FAPs elicited by the intraoral infusion of a bitter-sweet sucrose-quinine solution. Conversely, dopamine antagonists attentuate Pavlovian conditioning for appetitive rewards (Berridge and Robinson, 1998) but do not affect ingestive orofacial FAPs (Pecina et al., 1997). Rather than affecting evaluative incentive learning, therefore, dopamine agonists and antagonists appear to act on the Pavlovian form of incentive learning. Indeed, Berridge and Robinson (1998; Robinson and Berridge, 1993) have consistently argued that the dopamine system mediates the motivational properties of CSs, something they refer to as its "incentive salience."

With regard to Pavlovian and instrumental incentive learning, numerous studies suggest that Pavlovian incentives can exert quite selective excitatory effects on instrumental performance. Thus, for example, there is good evidence that superimposing a Pavlovian excitor on an instrumental baseline can affect the performance of the instrumental response (Balleine, 1994), an effect referred to as Pavlovian-instrumental transfer. More impressive still is evidence of selective transfer effects; CSs paired with the outcome earned by an instrumental action have a greater excitatory effect on performance of that action than CSs paired with a different outcome (Colwill and Rescorla, 1988). Thus, and in contrast to the differential effects of shifts in primary motivation on Pavlovian and instrumental performance, this evidence appears to indicate that Pavlovian and instrumental conditioning share a common incentive process.

There is, however, good evidence against this claim. For example, a number of treatments have been found to affect transfer without affecting instrumental devaluation effects and vice versa. First, peripheral administration of the dopamine antagonists pimozide or

α-flupenthixol has been observed to attenuate the excitatory effects of a Pavlovian CS on instrumental performance without affecting instrumental outcome devaluation produced by a shift from a food deprived to a nondeprived state (Dickinson et al., 2000).

Second, Corbit and Balleine (2001) report that cell body lesions of the shell subregion of the nucleus accumbens profoundly attenuate selective transfer effects produced when a CS is paired with the same outcome as that earned by the instrumental action but have no effect on the sensitivity of rats to selective devaluation of the instrumental outcome induced by a specific satiety treatment. Conversely, lesions of the core subregion of the accumbens had no effect on the selective transfer effect abolished by the shell lesions but had a profound effect on the sensitivity of rats to the selective devaluation of the instrumental outcome. This study presents, then, a double dissociation between the neural processes that mediate Pavlovian and instrumental incentive learning processes.

Corbit and Balleine (2003) also report that Pavlovian CSs and instrumental incentive learning have dissociable effects on the performance of components of a heterogeneous chain of instrumental actions. Hungry, rats were trained to press first one lever (R1) and then a second lever (R2) to earn a food outcome (R1 → R2 → O). They were then given Pavlovian training in which a CS was paired with that same outcome. In a test of Pavlovian-instrumental transfer, the CS was found to elevate performance but only of the response proximal to food delivery (R2). The rats were then retrained on the chain after which they were shifted to an undeprived state and given an extinction test on the two actions. Half of the rats were allowed to eat the food outcome when sated before this test, whereas the remainder were not. On the test, performance on the chain was reduced but, importantly, only in the rats reexposed to the outcome when undeprived and then only on the distal response in the chain (R1). Perfor-

mance on R2 was not differentially affected by the instrumental incentive learning treatment.

The unavoidable conclusion from these studies is that Pavlovian and instrumental incentive processes not only are mediated by anatomically and neurochemically distinct systems but also are functionally independent. It is tempting to conclude that the same is true of evaluative incentive processes, although at present, and particularly in rats, there is too little in the way of systematic data on this issue to do so with any confidence.

ACKNOWLEGMENT

The preparation of this chapter was supported by a grant from the National Institute of Mental Health #MH56446 to the author.

REFERENCES

Bakal CW, Johnson RD, Rescorla RA (1974) The effect of change in US quality on the blocking effect. Pavlovian Journal of Biological Sciences 9:97–103.

Balleine B (1992) Instrumental performance following a shift in primary motivation depends upon incentive learning. Journal of Experimental Psychology: Animal Behavior Processes 18:236–250.

Balleine B (1994) Asymmetrical interactions between thirst and hunger in Pavlovian-instrumental transfer. Quarterly Journal of Experimental Psychology 47B:211–231.

Balleine BW (2001) Incentive processes in instrumental conditioning. In: Handbook of contemporary learning theories (Mowrer R and Klein S, eds.), pp. 307–366. Hillsdale, NJ: Erlbaum.

Balleine B, Ball J, Dickinson A (1994) Benzodiazepine-induced outcome revaluation and the motivational control of instrumental action in rats. Behavioral Neuroscience 108:573–589.

Balleine B and Dickinson A (1991) Instrumental performance following reinforcer devaluation depends upon incentive learning. Quarterly Journal of Experimental Psychology 43B:279–296.

Balleine B and Dickinson A (1992) Signalling and incentive processes in instrumental reinforcer devaluation. Quarterly Journal of Experimental Psychology 45B:285–301.

Balleine B and Dickinson A (1998a) The role of incentive learning in instrumental outcome revaluation by sensory-specific satiety. Animal Learning and Behavior 26:46–59.

Balleine BW and Dickinson A (1998b) Consciousness: the interface between affect and cognition. In: Consciousness and human identity (Cornwell J, ed.), pp. 57–85. Oxford: Oxford University Press.

Berridge, KC (1991) Modulation of taste affect by hunger, caloric satiety, and sensory-specific satiety in the rat. Appetite 16:103–120.

Berridge KC (2000) Reward learning: Reinforcement, incentives, and expectations. In: The psychology of learning and motivation (Medin DL, ed.), Vol. 40, pp. 223–278. New York: Academic Press.

Berridge KC, Flynn FW, Schulkin J, Grill HJ (1984) Sodium depletion enhances salt palatability in rats. Behavioral Neuroscience 98:652–660.

Berridge KC and Robinson TE (1998) What is the role of dopamine in reward: hedonic impact, reward learning, or incentive salience? Brain Research Reviews 28:309–369.

Blundell P, Hall G, Killcross S (2003) Preserved sensitivity to outcome value after lesions of the basolateral amygdala. Journal of Neuroscience 23:7702–7709.

Bolles RC (1975) Theory of motivation. New York: Harper & Row.

Bombace JC, Brandon SE, Wagner AR (1991) Modulation of a conditioned eyeblink response by a putative emotive stimulus conditioned with a hindleg shock. Journal of Experimental Psychology: Animal Behavior Processes 17:323–333.

Changizi MA, McGehee RM, Hall WG (2002) Evidence that appetitive responses for dehydration and food-deprivation are learned. Physiology and Behavior 75:295–304.

Colwill RM and Rescorla RA (1988) Associations between the discriminative stimulus and the reinforcer in instrumental learning. Journal of Experimental Psychology: Animal Behavior Processes 14:155–164.

Corbit L, Muir J, Balleine BW (2001) The role of the nucleus accumbens in instrumental conditioning: evidence for a functional dissociation between accumbens core and shell. Journal of Neuroscience 21:3251–3260.

Corbit LH and Balleine BW (2003) Pavlovian and instrumental incentive processes have dissociable effects on components of a heterogeneous instrumental chain. Journal of Experimental Psychology: Animal Behavior Processes 29:99–106.

Daly HB (1974) Reinforcing properties of escape from frustration aroused in various learning situations. In: The psychology of learning and motivation (Bower GH, ed.), Vol. 8, pp. 187–231. New York: Academic Press.

Dayan P and Balleine BW (2002) Reward, motivation and reinforcement learning. Neuron 36:285–298.

DeBold RC, Miller NE, Jensen DD (1965) Effect of strength of drive determined by a new technique for appetitive classical conditioning of rats. Journal of Comparative and Physiological Psychology 59:102–108.

De Houwer J, Thomas S, Baeyens F (2001) Association learning of likes and dislikes: A review of 25 years of research on human evaluative conditioning. Psychological Bulletin 127:853–869.

Dickinson A and Balleine BW (2002) The role of learning in the operation of motivational systems. In: Learning, motivation & emotion, Volume 3 of Steven's handbook of experimental psychology, 3rd ed. (Gallistel CR, ed.), pp. 497–533. New York: John Wiley & Sons.

Dickinson A and Balleine BW (2000) Causal cognition and goal-directed action. In: The evolution of cognition (Heyes C and Huber L, eds.), pp. 185–204. Cambridge, MA: MIT Press.

Dickinson A and Dawson GR (1989) Incentive learning and the motivational control of instrumental performance. Quarterly Journal of Experimental Psychology 41B:99–112.

Dickinson A and Dearing MF (1979) Appetitive-aversive interactions and inhibitory processes. In: Mechanism of learning and motivation (Dickinson A and Boakes RA, eds.), pp. 203–231. Hillsdale, NJ: Lawrence Erlbaum Associates.

Dickinson A and Pearce JM (1977) Inhibitory interactions between appetitive and aversive stimuli. Psychological Bulletin 84:690–711.

Dickinson A, Smith J, Mirenowicz J (2000) Dissociation of Pavlovian and instrumental incentive learning under dopamine antagonists. Behavioral Neuroscience 114:468–483.

Everitt BJ and Stacey P (1987) Studies of instrumental behavior with sexual reinforcement in male rats (Rattus novegicus): II. Effects of preoptic area lesions, castration, and testosterone. Journal of Comparative Psychology 101:407–419.

Forestell CA and Lolordo VM (2003) Palatability shifts in taste and flavour preference conditioning. Quarterly Journal of experimental Psychology 56B:140–160.

Ganesen R and Pearce JM (1988) Effects of changing the unconditioned stimulus on appetitive blocking. Journal of Experimental Psychology: Animal Behavior Processes 14:280–291.

Garcia J (1989) Food for Tolman: Cognition and cathexis in concert. In: Aversion, avoidance, and anxiety: Perspectives on aversively motivated behavior (Archer T and Nilsson L-G, eds.), pp. 45–85. Hillsdale, NJ: Lawrence Erlbaum Associates.

Garcia J, Brett L, Rusiniak KW (1989) Limits of Darwinian conditioning. In: Contemporary learning theories: Instrumental conditioning theory and the impact of biological constraints on learning (Klein SB and Mowrer RR, eds.), pp. 181–203. Hillsdale, NJ: Lawrence Erlbaum Associates.

Ginn SR, Valentine JD, Powell DA (1983) Concomitant Pavlovian conditioning of heart rate and leg flexion responses in the rat. Pavlov Journal of Biological Science 18:154–160.

Hall WG, Arnold HM, Myers KP (2000) The acquisition of an appetite. Psychological Science 11:101–105.

Harris JA, Gorissen MC, Bailey GK, Westbrook RF (2000) Motivational state regulates the content of learned flavor preferences. Journal of Experimental Psychology: Animal Behavior Processes 26:15–30.

Hendersen RW and Graham J (1979) Avoidance of heat by rats: Effects of thermal context on the rapidity of extinction. Learning and Motivation 10:351–363.

Holland PC (1980) CS-US interval as a determinant of the form of Pavlovian appetitive conditioned responses. Journal of Experimental Psychology: Animal Behavior Processes 6:155–174.

Holland PC and Rescorla RA (1975) The effect of two-ways of devaluing the unconditioned stimulus after first- and second-order appetitive conditioning. Journal of Experimental Psychology: Animal Behavior Processes 1:355–363.

Holland PC, Hatfield T, Gallagher M (2001) Rats with basolateral amygdala lesions show normal increases in conditioned stimulus processing but reduced conditioned potentiation of eating. Behavioral Neuroscience 115:945–950.

Holland PC, Petrovich GD, Gallagher M (2002) The effects of amygdala lesions on conditioned stimulus-potentiated eating in rats. Physiology & Behavior 76:117–129.

Kamin LJ (1969) Selective association and conditioning. In: Fundamental issues in associative learning (Mackintosh NJ and Honig WK, eds.), pp. 42–64. Halifax: Dalhousie University Press.

Killcross S, Robbins TW, Everitt BJ (1997) Different types of fear-conditioned behaviour mediated by separate nuclei within amygdala. Nature 388:377–380.

Konorski J (1967) Integrative activity of the brain: An interdisciplinary approach. Chicago: University of Chicago Press.

Lopez M, Balleine B, Dickinson A (1992) Incentive learning and the motivational control of instrumental performance by thirst. Animal Learning and Behavior 20:322–328.

Moll RP (1964) Drive and maturation effects in the development of consummatory behavior. Psychological Reports 15:295–302.

Pavlov IP (1927) Conditioned reflexes. Oxford University Press.

Pearce JM and Bouton ME (2001) Theories of associative learning in animals. Annual Review of Psychology 52:111–139.

Pecina S, Berridge KC, Parker LA (1997) Pimozide does not shift palatability: separation of anhedonia from sensorimotor effects. Pharmacology, Biochemistry and Behavior 58:801–811.

Ramachandran R and Pearce JM (1987) Pavlovian analysis of interactions between hunger and thirst. Journal of Experimental Psychology: Animal Behavior Processes 13:182–192.

Reilly S (1999) The parabrachial nucleus and conditioned taste aversion. Brain Research Bulletin 48:239–254.

Reilly S, Harley C, Revusky S (1993) Ibotenate lesions of the hippocampus enhance latent inhibition in conditioned taste aversion and increase resistance to extinction in conditioned taste preference. Behavioral Neuroscience 107:996–1004.

Rizley RC and Rescorla RA (1972) Associations in second-order conditioning and sensory preconditioning. Journal of Comparative and Physiological Psychology 81:1–11.

Schmajuk NA and Christiansen BA (1990) Eyeblink conditioning in rats. Physiology & Behavior 48(5):755–758.

Sclafani A (1999) Macronutrient-conditioned flavor preferences. In: Neural control of macronutrient selection (Bertoud H-R and Seeley RJ, eds.), pp. 93–106. Boca Raton, FL: CRC Press.

Sclafani A, Azzara AV, Touzani K, Grigson PS, Norgren R (2001) Parabrachial nucleus lesions block taste and attenuate flavor preference and aversion conditioning in rats. Behavioral Neuroscience 115:920–933.

Sudakov KV (1990) Oligopeptides in the organization of feeding motivation: a systemic approach. Biomedical Science 1:354–358.

Talk AC, Gandhi CC, Matzel LD (2002) Hippocampal function during behaviorally silent associative learning: dissociation of memory storage and expression. Hippocampus 12:648–656.

Wyvell CL and Berridge KC (2000) Intra-accumbens amphetamine increases the conditioned incentive salience of sucrose reward: Enhancement of reward "wanting" without enhanced liking or response reinforcement. Journal of Neuroscience 20:8122–8130.

Young PT (1949) Food seeking drive, affective process and learning. Psychological Review 56:98–121.

Models and Tests

Neurological Models

BRYAN KOLB

The *function* of any region of the brain is to produce behavior. It follows that if a brain region is *dysfunctioning*, then behavior will be altered in some way. The general presumption in neurology is that it ought to be possible to restore at least some normal functioning by pharmacological, behavioral, or surgical intervention. Consider Parkinson's disease as an example. Parkinson's patients have a wide range of symptoms, including two obvious ones: tremor and akinesia (i.e., the absence or poverty of movement). Although the cause of Parkinson's disease is the death of dopaminergic cells in the brain stem, the loss of those cells has a cascading effect on neurological functioning so that forebrain regions such as the basal ganglia and thalamus do not function properly, which in turn produces the observed behavioral symptoms. The simplest way to treat the disease would be to use drugs (such as L-DOPA) to increase the production of dopamine, but to date this treatment can only partially restore function. Other treatments thus are necessary, an example being the transplantation of embryonic dopaminergic cells into the affected brain. Again, although once believed to be a promising treatment, it has become clear that transplantation also has limitations. New treatments therefore must be developed.

The major problem in developing treatments for any neurological disorder, however, is that like most new treatments in medicine, they must be developed in nonhuman subjects first. The nervous system provides a unique problem for medical science, however, because unlike other body organs, such as the heart or pancreas, which appear to function similarly across a wide swath of animal species, the brain is different. The most obvious difference is in the relative size of the brain. The brain is more than twice a big relative to body size than that of our closest relatives, the chimpanzee, and about 15 times bigger than that of our most common laboratory animal, the rat. Thus, a fundamental problem for neurological science is the issue of whether nonhuman brains are similar enough to human brains to be useful in searching for cures to human neurological disorders. This problem is compounded further when we consider the issue of whether laboratory animals actually contract the same disorders as we do; further, there is the issue of how we would collect a sufficient number of animals with particular diseases to use as subjects in experiments.

The solution to these problems is actually quite simple. First, although species such as rats have brains that are much smaller than human brains, the overwhelming evidence is that the fundamental organization of the rat brain is not much different from the human brain. (For an extensive discussion of this issue, see Kolb and Whishaw, 1983, 2003.) Certainly rats do not have as complex a cognitive life as humans and obviously do not talk, but they do have sensory and motor systems that are sufficiently similar to ours to allow us to make generalizations across the species divide. Furthermore, rats have neurological systems controlling cognitive, emotional, and attentional processes, which are remarkably similar in general organization to human systems. Second, it is not necessary to wait until rats show symptoms of diseases. Rather, we can devise ways to induce different types of neurological disorders in otherwise healthy animals. The question remains, however, whether "artificially induced" disorders such as Parkinson's disease actually are good enough models of the "real," naturally occurring, disorder observed in humans. This is an empirical issue that requires careful analysis of behavior for each disorder under study. Of course, a disorder such as attention-deficit/hyperactivity disorder (ADHD) is going to provide special problems for animal modelers because the most obvious problem in children with hyperactivity syndromes is that they have problems in school. Rats obviously do not go to school! Nonetheless, it remains possible to study effectively other symptoms of such disorders, as we shall see.

The goal of this chapter is to examine some of the most well-developed models of human neurological disorders that involve the cerebral hemispheres. First, however, because so many neurological disorders involve cortical functioning, we must examine the usefulness of the rodent cerebral cortex as a model of human cortical organization.

CORTICAL ORGANIZATION IN RATS

One of the major obstacles in comparing the behavior of different species of mammals is that each species has a unique behavioral repertoire that permits the animal to survive in its particular environmental niche. There is, therefore, the danger that neocortical organization is uniquely patterned in different species in a way that reflects the unique behavioral adaptation of different species. One way to address this problem is to recognize that although the details of behavior may differ somewhat, mammals share many behavioral traits and capacities (e.g., Kolb and

Whishaw, 1983). For example, all mammals must detect and interpret sensory stimuli, relate this information to past experience, and act appropriately. Similarly, all mammals appear to be capable of learning complex tasks under various schedules of reinforcement. The details and complexity of these behaviors clearly vary, but the general capacities are common to all mammals. Warren and Kolb (1978) proposed that behaviors and behavioral capacities demonstrable in all mammals could be designated as *class-common* behaviors. In contrast, behaviors that are unique to a species and that have presumably been selected to promote survival in a particular niche are designated as *species-typical* behaviors. This distinction is important because it has implications for the organization of the cerebral cortex. We note that just because mammals have class-common behaviors does not prove that they have not independently evolved solutions to the class-common problems. There is little evidence in support of this notion, however. Neurophysiological, anatomical, and lesion studies reveal a similar topography in the motor, somatosensory, visual, and auditory cortices of the mammals, a topography that provides the basis for class-common neural organization of fundamental capacities of mammals.

Kaas (1987) has argued, for example, that all mammalian species have similar regions devoted to the analysis of basic sensory information (e.g., areas V1, A1, S1), the control of movement (M1), and a frontal region involved in the integration of sensory and motor information. We can extend Kaas's idea by suggesting that these regions have class-common functions. To be sure, there are large species differences in the details of the class-common behaviors. Monkeys (and humans) have chromatic vision and fine visual acuity compared with the largely achromatic vision and reduced acuity of cats or rats. Nevertheless, in all mammalian species studied, removal of visual cortex severely disrupts object recognition. Similarly, rats and cats have a large somatosensory representation of the whiskers, whereas monkeys and humans have no such representation, but in all species, the somatosensory cortex functions to represent skin-related receptors for tactile sensations. Finally, we have seen (see Chapter 15) that the motor systems of rats and primates are remarkably similar in general organization, although some details such as the use of an opposable thumb are obviously different. Thus, the basic operation of the sensory and motor systems represents class-common functions, even though the details of this recognition may vary in a species-typical manner.

But what about the so-called associative functions of the cerebral cortex? That is, are there parallel posterior parietal, anterior temporal, and prefrontal cortical systems in rats and primates? Pandya and Yeterian (1985) have noted that as the number of sensory representations increase in evolution, there is a corresponding increase in the size and number of associative regions that function to integrate this sensory information with motor output. It follows that although the complexity of each of the associative regions will vary across species with the nature of the basic sensory representations, there should be parallel associative regions across mammalian species. Here I consider each briefly.

As a gross generalization, we can argue that the posterior parietal region functions to use sensory information, and especially visual and tactile information, to guide movements in space. For primates much of the expansion in posterior parietal functioning is thus directed to visual and tactile guidance of limb movements to grasp and manipulate objects as well as to navigation in space. As noted in Chapter 15, the guidance of skilled limb movements in rats is largely under olfactory control, so it is hardly surprising that the posterior parietal region does not play a major role in guiding limbs of rats. The evidence from both lesion and electrophysiological studies is clear, however, in showing that as in primates, rats with posterior parietal lesions

have deficits in spatial navigation (Kolb et al., 1994).

The primary function of the temporal associative regions in humans is in the recognition of complex visual and auditory information, and especially in the recognition of meaningful patterns of such information such as in the recognition of faces, objects, words, and music. The temporal associative regions of rats have a much simpler organization than those of primates, but there now is little doubt that these regions are involved in complex pattern recognition of visual (e.g., objects) and auditory (e.g., species-typical vocalizations) inputs (Dean, 1990; Kolb et al., 1994).

Finally, the prefrontal region, which is presumed to control the somewhat mysterious "executive functions" in primates also has a parallel organization in rats. Recent studies have shown a striking parallel in anatomical organization and behavioral functioning in rats and primates (Uylings et al., 2003). Thus, for example, both rats and primates have at least two major subdivisions of the prefrontal cortex (a dorsal and medial division and an orbital division), each of which can be further subdivided into many subregions. Furthermore, there are a wide range of parallel symptoms after injury to the two regions in rats and monkeys (see a review, Uylings et al., 2003).

In sum, although obviously simpler, the cerebral cortical organization of rats provides a good approximation of what is observed in primates. This parallel provides a nice starting point for investigating the effects of behavioral disorders that involve cerebral dysfunctioning.

MODELS OF NEUROLOGICAL DISORDERS

A discussion of all neurological disorders or even a detailed discussion of a representative subset of disorders is beyond the scope of a short chapter. I therefore have selected the most common disorders, which I discuss in only general terms. (For a more detailed discussion of such models, see the two volumes edited by Boulton, Baker, and Butterworth, 1992.) One advantage of studying neurological, rather than psychiatric, disorders (see Szechtman and Eilam, Chapter 42) is that the former disorders are presumed to have a purely neurological cause, and thus it should be possible to induce the disease without particular concern about experiential factors. In parallel to psychiatric disorders, however, is the difficulty that many neurological disorders are characterized by disordered cognitive functioning. Thus, as discussed in detail by Szechtman and Eilam (Chapter 42) the analysis of behavior becomes critical as a means to making inferences about the nature of the diseases and the potential treatments that may be used to ameliorate the symptoms. Thus, depending on the location of the neurological disorder, there must be appropriate behavioral assays such as those described in the rest of this volume. I shall therefore not focus on the behavioral analysis of the different models so much as on the nature of neurological model itself.

In principle, it is possible to distinguish between disorders that are largely found in the adult brain and those that occur developmentally. These disorders present different challenges in developing models and are considered separately.

MODELS OF DISORDERS OF THE ADULT BRAIN

Stroke

Human ischemic stroke is diverse in its causes, location, and symptoms. A major advantage of rat models of stroke is that the injuries can be controlled to be reproducible, which allows for a systematic study of the behavioral sequelae and treatments of stroke. There are marked variations in details of cerebral blood flow in different strains so care must be taken in selecting a strain that can be compared with other literature (Ginsberg and Busto, 1989). Some strains show considerable variability

within the strain as well. For example, the Sprague-Dawley rat has at least six different branching patterns, which means that there will be considerable variance across strokes in different animals.

There are two general categories of stroke models: focal ischemia and global ischemia (Table 42–1) (Seta et al., 1992). We have found the Long-Evans rat to be a good model, particularly for behavioral studies of focal stroke, although they are not so commonly used for studies of ischemic stroke. The nature of the behavioral analysis is different in the two models because the pattern of injury is very different. Focal models that involve either the MCA or pial stripping involve motor cortex and variable amounts of striatum. The advantage is that there are excellent techniques for assessing motor recovery and compensation (see Chapter 15). One disadvantage of the MCA models is that large lesions may produce extreme motor deficits that are very difficult to assess, except with rather gross measures.

Ischemic models have been used extensively in studies of neuroprotective agents. The advantage of these models is that there is hippocampal cell death, thus allowing studies of cognitive behavior (see Chapter 39). The disadvantage is that the lesions tend to be variable and large group samples are often needed.

Cerebral Injury

People acquire cerebral injury in a variety of ways in addition to stroke, including traumatic head injury and surgically induced injury, such as when there is tumor removal, vascular surgery, or surgery for drug-intractable epilepsy. Cerebral injuries in rats are most commonly produced using either suction ablation or injections of selective neurotoxins or with a fluid percussion model of head trauma. The former methods produce focal injuries, whereas the head trauma model produces a much more diffuse injury. The behavior of animals with both types of injuries can be assessed using behavioral tasks sensitive to the functions of the dis-

Table 42–1. Models of Stroke in Rats

Model	Preparation
FOCAL MODELS	
Embolism models	Injection of blot clot fragments or microspheres into the carotid artery
Photothrombolitic model	Systemic injection of a chemical (e.g., Rose Bengal). Laser illumination of focal region of skull induces photochemical reaction, causing platelets and erythrocytes aggregation.
Endothelin-1	Local injection of endothelin-1 causes local degradation of blood vessels, causing cell death.
Middle cerebral artery (MCA) occlusion	Permanent occlusion by thermocoagulation of all, or part of, the MCA Reversible occlusion by using a snare ligature around the MCA
Pial stripping	The pia and blood vessels are stripped off of a localized region.
GLOBAL MODELS	
Bilateral carotid occlusion Two-vessel occlusion	Both common carotids are occluded, and blood pressure is reduced to 50 mm Hg.
Four-vessel occlusion	Vertebral arteries are cauterized; carotids are temporarily occluded with a clasp or snare.
Levine hypoxia-ischemia	Unilateral carotid occlusion followed by exposure to anoxic environment for about 45 minutes 24 hours later.

crete cortical areas injured (see Chapters 5–17, 39, 43), particularly in animals with focal injuries. Studies of head trauma typically also involve examination of cognitive functions using tests similar to those used for dementia (see Chapter 39).

Historically, the suction model has been most extensively used, but the major disadvantage of the suction model is that there must be a craniotomy to expose the tissue to be removed, although the advantage is that the tissue is visualized and by using stereotaxic coordinates it is possible to make consistent lesions. A second disadvantage is that the tissue is removed from the brain and thus does not slowly die as in a natural head injury. There are differences in the neuroanatomical and neurochemical reaction to injuries that do or do not have dying tissue, so this may be important in studies examining treatments for cerebral injury. The excitotoxic models only require that small bur holes be made to accommodate cannulae for infusion of the toxin,

but in the case of large lesions, such as removal of the entire motor cortex, there must be many bur holes and there is considerable variability in the lesion extent. One advantage of the excitotoxic lesions is that the lesions can be restricted to gray matter, thus reducing the diffuse effects of the injury. This may be a disadvantage in studies of recovery of function, however, because it would be atypical in humans to have lesions that spared the white matter.

The fluid percussion model of head trauma requires a craniotomy to expose the brain. A plunger strikes the brain with a specific amount of force, causing cell death and shearing and tearing of white matter. The lesions are quite variable in extent but do mimic closed head injuries in people.

Parkinson's Disease

Parkinson's disease has a complex and variable etiology and neuropathology, but the consistent common feature is a loss of dopamine neurons in the substantia nigra that leads to a variety of motor impairments. Although it was once believed that only primate models are useful in studying Parkinson's disease, there is no consensus nor current basis to suggest that one species is more predictive than another in the transition from research to clinical practice (Schallert and Tillerson, 2002). The most common rat model of Parkinson's disease is produced by injecting the neurotoxin 6-hydroxydopamine into either the medial forebrain bundle or the rostral striatum in one hemisphere. Such procedures normally produce a range of dopamine depletion (as measured by neurochemical assays such as high-performance liquid chromatography), and this variation can be correlated to behavioral impairments. For example, Tillerson et al. (1998) found a correlation of .92 between striatal dopamine depletion and the degree of forelimb motor impairment. The toxin injections are normally unilateral both because animals with extensive bilateral depletions will not eat and because unilateral lesions allow the possibility of comparisons between the motor performance of the two sides of the body.

Although the original behavioral measures used to reflect the extent of motor impairments were based on dopaminergic drug-induced asymmetries in behavior, these asymmetries do not predict the extent of motor impairment and may not be the best measure. A variety of tests have been devised that do not require drugs, including skilled forelimb reaching behavior (Miklyaeva and Whishaw, 1996), forelimb asymmetry during vertical exploration (Schallert and Lindner, 1990), tactile extinction (Schallert et al., 1982), and a variety of simple motor reflex tests described by Schallert and Tillerson (2002).

Huntington's Disease

Huntington's disease is an inherited, progressive, neurodegenerative disorder that is characterized by bizarre uncontrollable movements and postures. Anatomically, Huntington's disease is associated with gross generalized atrophy of the cerebral hemispheres and extensive cell death in the striatum. The striatal cell death is confined to a loss of medium spiny neurons that are the primary output neurons of the striatum. This cell death was originally believed to be the major cause of the behavioral abnormalities, but it has become clear that neural degeneration associated with the disease is quite widespread, involving the cerebral hemispheres, brain stem, and cerebellum (Emerich and Sanberg, 1992).

Rodent models of Huntington's disease have focused on the striatal pathology, but it must be remembered that the human disorder includes much broader pathology. There have been two general types of models developed: dyskinesia models based on striatal neurotransmitter imbalances and excitotoxin lesion models. The neurochemical models have been based on the assumption that it is not so much the striatal cell death but rather the effect of the striatal cell death on the balance of dopaminergic, GABAergic, and cholinergic systems in the striatum. Thus, various dopaminergic agonists and GABAergic and cholinergic antagonists have been infused into the striatum, and in each case dyskinetic movement patterns can be produced. The advantage of these models is that they are easy to prepare and the motor abnormalities do resemble those of the disease, but the major disadvantages are that the effects are not long lasting and it is difficult to see how such models would prove useful in developing clinically relevant treatments.

The excitotoxic models use selective toxic compounds (kainic acid, quinolinic acid) to kill striatal neurons. Both toxins produce pathology and behavioral effects that are reminiscent of Huntington's disease. The rat models have lead to the hypothesis that a fundamental deficit in Huntington's disease is a dysfunction of glutamatergic transmission, which results in the slow progressive cell death characteristic of the disease (Emerich and Sanberg, 1992). However, because the disease is genetic, these rat studies cannot provide information about the genetic defect (but see Viral Vector–Mediated Neurodegeneration). The models do provide a viable technique for examining clinical treatments.

Dementia of the Alzheimer's Type

Although there are many forms of dementia in humans, the most common and the most intensively studied is Alzheimer's disease. Like most dementias, Alzheimer's disease is a neurodegenerative disorder that can be reliably diagnosed only with postmortem examination of the cerebral pathology. The pathology in Alzheimer's is rather widespread, including death of magnocellular cholinergic neurons in the basal forebrain, loss of neurons in the brain stem monoaminergic projections from the raphe and locus ceruleus, and senile plaques within the amygdala, hippocampus, and cerebral cortex. The degeneration of the basal forebrain cholinergic neurons that innervate the cortex and hippocampus are believed to underlie some of the cognitive impairments associated with Alzheimer's disease; thus, animal models have tended to focus on the cholinergic loss. Like the models of Huntington's disease, although the pathology in Alzheimer's disease is rather widespread, the models have focused on a single aspect of the disease, which in Alzheimer's disease is the cholinergic aspect.

The three most common models of Alzheimer's are (1) aged rats, (2) lesions of the basal forebrain, and (3) pharmacological agents. The age-related models are based on the observation that many of the cognitive impairments in Alzheimer's disease also occur, although to a lesser degree, in normal aging. Thus, there is mild amnesia that progresses gradually over a period of years as well as a slow decline in general intellectual functions. Rats show age-related cell loss in the basal forebrain that are correlated with behavioral impairments in cognitive tasks. Surprisingly, however, there is no age-related loss in cortical markers of cholinergic activity, which may

limit the usefulness of the aged rat model. Furthermore, most aging studies in rats use a genetically homogeneous strain of rats, the Fisher 344, which is healthier in old age than other strains such as the Long-Evans, but there is some question over how representative this strain's spatial memory abilities might be.

Young animals do not develop the biochemical changes seen in Alzheimer's disease, but it is possible experimentally to produce such changes by making discrete basal forebrain lesions. Normally, studies use neurotoxins related to glutamate including kainic acid, ibotinic acid, and quinolinic acid. All are potent toxins that affect a slightly different population of neurons within the injection site. All produce mild, but consistent, deficits in memory in rats, and all produce a depletion in cortical and hippocampal markers for acetylcholine. One difficulty is that the memory impairments tend to recovery over time or with extensive postoperative training (Bartus et al., 1985), which means that the baseline performance will shift over time, thus making studies of therapeutic agents more difficult.

Finally, anticholinergic agents can be used to impair behavioral performance, the logic being that muscarinic antagonists can produce cognitive deficits in both humans and rats that resemble the impairments in Alzheimer's disease (but see Whishaw, 1985). The anticholinergic models have been used with some success to examine the effects of putative cognitive-enhancing drugs. The problem remains, however, that the effects of the anticholinergics are transient and are specific to brain cholinergic systems yet the disease is progressive and far more widespread.

Viral Vector–Mediated Neurodegeneration

The defective handling of proteins is a central feature of most major neurodegenerative diseases. Because the production of proteins is coded by genes, it is not surprising that it was recently discovered that neurodegenerative diseases can be caused by mutations in single cellular proteins. The genetic mutations can be induced in rats through the generation of transgenic animals carrying the disease-causing gene, but this model has the drawbacks that the mutated gene affects the entire brain and that the animal has the abnormal gene throughout its lifetime, rather than just in adulthood. A new rodent model has been developed that is based on advances in recombinant viral vector technology. Thus, a virus with a specific affinity for certain cell types can be injected into specific regions of the brain, such as the substantia nigra. Once inside the cell, the virus induces the expression of the abnormal gene (Kirik and Bjorklund, 2003). Although this technique is still in its infancy, vector systems have been used successfully to express either the mutated human huntingtin protein in striatal neurons as a model of Huntington's disease or mutated human alpha-synuclein in nigral dopamine neurons as a model of Parkinson's disease (de Almeida et al., 2002; Kirik et al., 2002). Viral vector–mediated gene transfer is expected to prove useful for producing other disease-causing proteins, the most likely candidate being amyloid precursor protein involved in Alzheimer's disease pathology. There are few reports of behavioral changes in animals infected with the viral vectors as the studies to date have been anatomical.

Seizure Disorders

One of the most common neurological conditions is the occurrence of seizures, either acutely in some disorders such as traumatic head injury or encephalitis or chronically as in epilepsy. The main reason for developing animal models is to understand the molecular mechanisms producing the seizures, as well as to screen potential anticonvulsant agents.

Most seizures are caused by an irritant of some sort in the nervous system, such as an injury, that results in abnormal electrical discharges. Seizures can be acute, however, resulting from electroshock or a chemical imbalance. Thus, there are two general types of seizure models in rats, one based on the development of an irritant in the brain and the other being chemically induced.

The most common focal model of seizures is known as *kindling*. Kindling was a metaphor proposed by Goddard to describe the observation that, in a manner similar to that in which burning small twigs ultimately produces a large fire, repeated subconvulsive stimulation of the brain by electrical current or pharmacological agents produces seizures that gradually increase in intensity. Kindling has been the object of intense study over the past 25 years, leading to the development of an enormous body of literature (e.g., Corcoran and Moshe, 1998; Teskey, 2001). One advantage of the kindling model is that the development of the seizures can be objectively measured both electrophysiologically and behaviorally, and thus it is possible to initiate anticonvulsant medications at many different time points in the development of the seizure disorder.

The other principal method of inducing seizures is to systemically or locally inject different drugs or other compounds, including penicillin, strychnine, tetanus toxin, aluminum, and bicucullin (see McCandless and FineSmith, 1992, for a review). These models have the advantage of ease of induction and reliability in the seizure disorder produced. The disadvantage is that chemically induced seizures are rare in humans and the mechanisms underlying the seizures are likely to be quite different in the different models.

Drug Addiction

Many people commonly take stimulant drugs like nicotine, cocaine, or heroin, all of which affect behavior and thus are said to be psychoactive. The long-term consequences of abusing psychoactive drugs are now well documented, but less is known about the how these drugs can chronically alter the nervous system. One experimental demonstration of drug-induced changes in the rat brain is known as *drug-induced behavioral sensitization*, often referred to just as *behavioral sensitization* (Robinson and Berridge, 2003). Behavioral sensitization is the progressive increase in the behavioral actions of a drug that occur after repeated administration of a constant dose of the drug. Behavioral sensitization occurs with most psychoactive drugs and is especially strong with psychomotor stimulants such as amphetamine. For example, when a rat is given a small dose of amphetamine, it may show a small increase in activity. When the rat is given the same dose on subsequent occasions, the increase in activity is progressively larger, showing behavioral sensitization. This drug-induced behavioral change persists for weeks or months so that if the drug is given in the same dose as before, the behavioral sensitization is still present. In a sense, the brain has some memory of the effects of the drug. Thus, addiction now is increasingly being viewed as a pathological process of learning (Berke and Hyman, 2000; Nestler, 2001).

Like the changes associated with learning different tasks (Kolb and Whishaw, 1999), behavioral sensitization is associated with changes in dendritic and spine morphology in both prefrontal cortex and the striatum (Robinson and Kolb, 1998), changes in striatal physiology (Gerdeman et al., 2003), and changes in the production of immediate-early genes and neurotrophic factors (Flores and Stewart, 2001). These various changes in brain organization and function are believed to underlie much of the pathological behavior of human drug addicts, and there is considerable interest in trying to relate changes in cognitive behaviors with morphological abnormalities.

MODELS OF DEVELOPMENTAL DISORDERS

One major advantage of using the rat as a model of developmental disorders is that the rat is born embryologically younger than laboratory primates or carnivores and thus it is possible to perform many manipulations, which would normally be prenatal in other laboratory species, postnatally in the rat. The rat's gestation period is about 22 days, with cerebral neurogenesis beginning around embryonic day 11 and finishing by birth (see Bayer and Altman, 1990, for details). Cell migration continues throughout the first postnatal week, at which time synaptogenesis begins and continues at a high rate until about postnatal day 25. The developing rat can thus be exposed to a variety of perturbations (e.g., injury, behavioral and pharmacological treatments) postnatally, at a time that would correspond to the third trimester of human development. Because the third trimester is so sensitive in human development, it is therefore possible to examine the effects of different brain manipulations and to explore possible ameliorative treatments in the developing rat model. I consider here two manipulations (injury, stress), although there is also extensive literature on the effects of alcohol and other drugs in both the prenatal and postnatal infant rat.

Perinatal Brain Injury

The perinatal rat brain can be disturbed using a variety of manipulations, including suction removal of discrete cortical areas, excitotoxic injuries, and hypoxia/ischemia. In general, these different manipulations show one consistent result. Thus, damage in the first week of life in infant rats produces more severe functional dysfunctions than similar damage in adulthood. In contrast, damage in the second week of life produces much less severe functional dysfunctions than similar injury in adulthood (for a review, see Kolb et al., 2001). These time points correspond roughly to injury in the third trimester and the first few postnatal months of human development, respectively. It is known that injury during the third trimester, including ischemia in premature infants, is particularly deleterious to the developing human brain whereas children with cerebral injuries in the first postnatal year show better outcomes, especially when the language zones are damaged (Kolb and Whishaw, 2003).

The markedly different behavioral outcomes from week 1 and week 2 injuries in the developing rat are correlated with distinctly different anatomical responses as well. For example, damage in the second week induces neurogenesis, gliogenesis, and dendritic hypertrophy, whereas damage in the first week generally has little effect on neurogenesis or gliogenesis and leads to dendritic atrophy (Kolb et al., 2001). In addition, my colleagues and I have shown recently that injury at the two timpanist produce differential effects on protein expression, changes that can be seen in some cases into adulthood. One example is the expression of a neurotrophic factor, basic fibroblast growth factor (bFGF). bFGF expression is markedly increased after injury in the second postnatal week but not after injury in the first postnatal week. In sum, the extensive literature on the effects of perinatal cortical injury in the rat provide an excellent model for understanding the effects of injury to the developing human brain. More important, however, the rat models have proved to be especially useful in understanding the effects of different treatments on stimulating functional compensation after early injuries (Kolb et al., 2001). Thus, the rat with cerebral injury in the first few days of life can show significant functional benefit from a variety of treatments, including tactile stimulation, dietary supplements, complex housing, and bFGF.

Perinatal Experience

Extensive literature shows that the developing rat is especially sensitive to postnatal manipulations. Two examples are neonatal handling and maternal separation. In most studies, handling involves maternal separation for about 15 minutes a day over the first 2 weeks of life (Levine, 1961; Caldji et al., 2000). This treatment enhances maternal behaviors such as licking and grooming of the infants, which in turn appears to alter the feedback regulation of stress-related systems in the nervous system. These changes have long-lasting benefits on cognitive functioning and reactivity to stress in adulthood. In contrast, maternal separation for longer periods, usually 3 to 4 hours per day, produces animals that are hyperresponsive to stress, including a hyperresponsive hypothalamic-pituitary response (Liu et al., 1997). Finally, other forms of perinatal experience, such as tactile stimulation during infancy or complex housing during the juvenile period, are also known to produce long-lasting changes in cortical organization (Kolb et al., 2003).

Taken together, these studies of perinatal experience are believed to provide models of how perinatal experience might chronically influence behavioral development in children. For example, it is believed that the studies of handling and maternal separation will provide insight into different developmental behavioral conditions (e.g., ADHD) and will provide models to investigate ways to treat such behavioral pathologies.

Attention-Deficit/Hyperactivity Disorder

Numerous recent reviews have described the behavioral and cognitive characteristics of ADHD (Barkley, 1997). It is now generally believed that ADHD results from a frontal-striatal abnormality, possibly lateralized to the right hemisphere (Heilman et al., 1991). Abnormalities in the dopaminergic systems projecting to the prefrontal cortex and basal ganglia are believed to underlie much of the frontal-striatal abnormality, and the most common treatment for ADHD is methylphenidate (Ritalin), which blocks dopamine reuptake. Correlated with the dopaminergic abnormality is an impairment in prefrontal functions, including in particular working memory and attentional functions. The cause of the dopaminergic abnormality has been difficult to determine, but predisposing factors, including genetics, premature birth, prenatal stress, and hypoxia/ischemia, among others, have been proposed (Sullivan and Brake, 2003).

ADHD has not proved to be easy to treat clinically and thus has led to considerable interest in developing an animal model. One way to proceed has been to take advantage of the normal variance in the performance of rats on various tests of working memory and cognitive functioning. Many studies have now shown that treating rats with methylphenidate can actually improve performance of poorly performing rats on tests of attentional processes. One strain of rats, the Kyoto SHR rat, has been proposed to be an especially good model, largely because there are known abnormalities in prefrontal dopaminergic innervation and this is correlated with behavioral abnormalities such as hyperactivity. The behavioral abnormalities can be reversed by dopaminergic agonists such as methylphenidate (Sullivan and Brake, 2003).

Other models of ADHD have focused on manipulating prefrontal development by perinatal anoxia (Brake et al., 1997). Interestingly, this treatment leads to prefrontal abnormalities that are lateralized to the right hemisphere, as is seen in humans (Brake et al., 2000). Anoxia is not the only manipulation to produce dopaminergic abnormalities; manipulations of the postnatal social environment between mothers and infants have shown abnormalities in dopaminergic systems in rats as well (Sullivan and Gratton, 2003).

In sum, there are now several different models in which dopaminergic innervation of the prefrontal/striatal circuitry in the rat is abnormal and is correlated with behavioral abnormalities including increased activity, poor working memory, and attentional deficits.

CONCLUSION

It is possible to produce brain dysfunctions in rats that can be shown to mimic a wide range of human neurological conditions. Although

it would be ideal to study neurological diseases in animals with brains very similar to ours, such as chimpanzees, this is impractical for both ethical and financial reasons. Rats provide an reasonable alternative, although their smaller brain and less complex cognitive processes place some constraints on the nature of the behavioral questions that can be addressed. There remains, however, an animal ethics issue that must be considered. There is no question that inducing neurological disorders in any animal means that there must be special consideration of animal care. Particular consideration must be given to animals used in models that might produce distress. Humans with neurological disorders often experience irritability, fear, anxiety, and pain. If the rat models successfully mimic the human condition, then we might expect that some of these symptoms would also be manifest in the rats. As Olfert (1992) has emphasized, any pain, suffering, distress, or deficits in function that negatively affect the animal's well-being and are not scientifically "necessary" for the study should be alleviated or minimized, regardless of the cost or convenience to the experimenter. Presumably the experiences of humans with disorders such as Parkinson's and Huntington's can serve as a guide to the distress that rats in models of these conditions might be enduring.

REFERENCES

de Almeida LP, et al (2002) Lentiviral-mediated delivery of mutant huntingtin in the striatum of rats induces a selective neuropathology modulated by polyglutamine repeat size, huntingtin expression levels, and protein length. Journal of Neuroscience 20:219–229.

Barkley RA (1997) Behavioral inhibition, sustained attention, and executive functions: constructing a unifying theory of ADHD. Psychological Bulletin 121:65–94.

Bartus RT, Flicker C, Dean RL, Pontecorvo M, Figuerdo JC, Fisher SK (1985) Selective memory loss following nucleus basalis lesions: long term behavioral recovery despite persistent cholinergic deficiencies. Pharmacology, Biochemistry, and Behavior 23:125–135.

Berke JD and Hyman SE (2000) Addiction, dopamine, and the molecular mechanisms of memory. Neuron 25:515–532.

Boulton AA, Baker GF, Butterworth RF (eds.) (1992) Animal models of neurological disease. Totowa, NJ: Human Press.

Brake WG, Noel MB, Boksa P, Gratton A (1997) Influence of perinatal factors on the nucleus accumbens dopamine response to repeated stress during adulthood: an electrochemical study in the rat. Neuroscience 77:1067–1076.

Brake WG, Sullivan RM, Gratton A (2000) Perinatal distress leads to lateralized medial prefrontal cortical dopamine hypofunction in adult rats. Journal of Neuroscience 20:5538–5543.

Caldji C, Diorio J, Meany MJ (2000) Variations in maternal care in infancy regulate the development of stress reactivity. Biological Psychiatry 48:1164–1174.

Corcoran ME and Moshe SL (1998) Kindling 5. New York: Plenum Press.

Dean P (1990) Sensory cortex: Visual perceptual factors. In: The cerebral cortex of the rat (Kolb B and Tees RC, eds.), pp. 275–308. Cambridge, MA: MIT Press.

Emerich DF and Sanberg PR (1992) Animal models of Huntington's disease. In: Animal models of neurological disease (Boulton AA, Baker GB, Butterworth RF, eds.), pp. 65–134. Totowa, NJ: Human Press.

Gerdeman GL, Partridge JG, Lupica CR, Lovinger DM (2003) It could be habit forming: drugs of abuse and striatal synaptic plasticity. Trends in Neuroscience 26:184–192.

Ginsberg MD and Busto R (1989) Rodent models of cerebral ischemia. Stroke 20:1627–1642.

Heilman KM, Voeller KK, Nadeau SE (1991) A possible pathophysiologic substrate of attention deficit hyperactivity disorder. Journal of Child Neurology 6(suppl):S76–S81.

Kaas JH (1987) The organization of neocortex in mammals: Implications for theories of brain function. Annual Review of Psychology 38:129–151.

Kirik D and Bjorklund A (2003) Modeling CNS neurodegeneration by overexpression of disease-causing proteins using viral vectors. Trends in Neurosciences 26:386–392.

Kirik D, et al (2002) Parkinson-like degeneration induced by targeted overexpression of alpha-synuclein in the nigrostriatal system. Journal of Neuroscience 22:2780–2791.

Kolb B, Gibb R, Gonzalez C (2001) Cortical injury and neuroplasticity during brain development. In: Toward a theory of neuroplasticity (Shaw CA and McEachern JC, eds.), pp. 223–243. New York: Elsevier.

Kolb B, Gibb R, Gorny G (2003) Experience-dependent changes in dendritic arbor and spine density in neocortex vary with age and sex. Neurobiology of Learning and Memory 79:1–10.

Kolb B, Buhrmann K, MacDonald R, Sutherland RJ (1994) Dissociation of the medial prefrontal, posterior parietal, and posterior temporal cortex for spatial navigation and recognition memory in the rat. Cerebral Cortex 4:15–34.

Kolb B and Whishaw IQ (1983) Generalizing in neuropsychology: problems and principles underlying cross-species comparisons. In: Behavioral contributions to brain research (Robinson TE, ed.). New York: Oxford University Press.

Kolb B and Whishaw IQ (2003) Fundamentals of human neuropsychology, 5th ed. New York: Worth.

Levine S (1961) Infantile stimulation and adaptation to stress. Research Publication of the Association of Nervous and Mental Disorders 43:280–291.

Liu D, Diorio J, Tannenbaum B, Caldji C, Francis D, Freedman A, et al (1997) Maternal care, hippocampal glucocorticoid receptors, and hypothalamic-pituitary-adrenal responses to stress. Science 277:1659–1662.

McCandless DW and FineSmith RB (1992) Chemically induced models of seizures. In: Animal models of neurological disease, II (Boulton AA, Baker GB, Butterworth RF, eds.), pp. 133–151. Totowa, NJ: Human Press.

Miklyaeva EI and Whishaw IQ (1996) Hemiparkinson analogue rats display active support in good limbs versus passive support in bad limbs on a skilled reaching task of variable height. Behavioral Neuroscience 110:117–125.

Nestler EJ (2001) Molecular basis of long-term plasticity underlying addiction. Nature Neuroscience Reviews 2:119–128.

Olfert ED (1992) Ethics of animal models of neurological diseases. In: Animal models of neurological disease, I (Boulton AA, Baker GB, Butterworth RF, eds.), pp. 1–29. Totowa, NJ: Human Press.

Pandya D and Yeterian EH (1985) Architecture and connections of cortical association areas. In: Cerebral cortex, 4: Association and auditory cortices (Peters A and Jones EG, eds.), pp. 3–61. New York: Plenum.

Robinson TE and Berridge KC (2003) Addiction. Annual Review of Psychology 54:25–53.

Robinson TE and Kolb B (1999) Alterations in the morphology of dendrites and dendritic spines in the nucleus accumbens and prefrontal cortex following repeated treatment with amphetamine or cocaine. European Journal of Neuroscience 11:1598–1604.

Schallert T and Lindner MD (1990) Rescuing neurons from trans-synaptic degeneration after brain damage: helpful, harmful or neutral in recovery of function? Canadian Journal of Psychology 44:276–292.

Schallert T and Tillerson JL (2002) Intervention strategies for degeneration of dopamine neurons in parkinsonism. In: Central nervous system diseases (Emerich DF, Dean RL, Sanberg PR, eds.), pp. 131–151. Totowa, NJ: Human Press.

Schallert T, Upchurch M, Lobaugh N, Farrar SB, Spiruso WW, Gilliam P, Vaughn D, Wilcox RE (1982) Tactile extinction: distinguishing between sensorimotor and motor asymmetries in rats with unilateral nigrostrial damage. Pharmacology, Biochemistry, and Behavior 18:753–759.

Seta KA, Crumrine RC, Whittingham TS, Lust WD, McCandless DW (1992) Experimental models of human stroke. In: Animal models of neurological disease, II (Boulton AA, Baker GB, Butterworth RF, eds.), pp. 1–50. Totowa, NJ: Human Press.

Sullivan RM and Brake WG (2003) What the rodent prefrontal cortex can teach us about attention-deficit/hyperactivity disorder: The critical role of early developmental events on prefrontal function. Behavioural Brain Research 146:43–55.

Sullivan RM and Gratton A (2003) Behavioral and neuroendocrine correlates of hemispheric asymmetries in benzodiazepine receptor binding induced by postnatal handling in the rat. Brain and Cognition 51:218–220.

Teskey GC (2001) Using kindling to model the neuroplastic changes associated with learning and memory, neuropsychiatric disorders, and epilepsy. In: Toward a theory of neuroplasticity (Shaw CA and McEachern JC, eds.), pp. 347–358. Philadelphia: Taylor & Francis.

Uylings HBM, Groenewegen HJ, Kolb B (2003) Do rats have a prefrontal cortex? Behavioural Brain Research 146:3–17.

Warren JM and Kolb B (1978) Generalizations in neuropsychology. In: Brain damage, behavior and the concept of recovery of function (Finger S, ed.). New York: Plenum Press.

Whishaw IQ (1985) Cholinergic receptor blockade in the rat impairs locale but not taxon strategies for place navigation in a swimming pool. Behavioral Neuroscience 99:979–1005.

Wenk GL (1992) Animal models of Alzheimer's disease. In: Animal models of neurological disease, I (Boulton AA, Baker GB, Butterworth RF, eds.), pp. 29–63. Totowa, NJ: Human Press.

Psychiatric Models

43

HENRY SZECHTMAN AND DAVID EILAM

Animal models have always played an integral part in the advancement of medical research, but their acceptance in psychiatry is both recent and instructive. The change in attitude from a past anathema to the present embracement of animal models reflects the profound shift of how today psychiatry views mental disorders and the successful efforts of basic science in showing the utility and the limits of animal studies of psychopathology. We first discuss the conflicts associated with a rise of a biological psychiatry perspective on mental illness and suggesting that the current preeminence of this viewpoint has aligned clinical psychiatry with behavioral neuroscience for the source of much of its basic science knowledge. We then elaborate on the methods of behavioral neuroscience and show how this methodology is in fact just what is needed to investigate mechanisms of mental disorders of interest to psychiatry. Next, we attempt to clear up some confusion revolving around the distinction between tests, models, and theories. Finally, we illustrate some of the discussed principles by examining a rat model of a psychiatric illness—obsessive-compulsive disorder.

PSYCHIATRY AND MENTAL ILLNESS

To understand what is asked of animal models in psychiatry, one must have an appreciation of the concerns and practices of psychiatry. Psychiatry is a medical specialty charged with the diagnosis, treatment, and prevention of mental disorders. This simple and straightforward textbook definition of psychiatry belies its true complexity. Unlike other medical specialties that deal with visible pathology of palpable bodily parts, the purviews of psychiatry are not physical entities but an intangible mental life. Thus, at the very heart of psychiatry lie theories about subjective experiences and human normality, theories that frame the psychiatric constructs of a mental disorder. Inevitably, notions what constitutes psychiatric illnesses intersect our most profound conceptions about the nature of man. A glimpse of this true essence of psychiatry can be obtained by considering the emergence of psychiatry in its historical context. As described by Berrios and Marková (2002):

> Born as a by-product of the nineteenth-century movement to organize society on a scientific basis, psychiatry was charged with the construction and enactment of normative views of madness. . . . Under the protection of medicine and the economic practices following the Industrial Revolution, psychiatrists developed representations of mental disease together with the professional and institutional apparatus to enjoin them. (p. 3)

Not surprisingly, a diversity of psychiatric constructions of mental disease arose over the years, attuned to major philosophical and social norms of humans as well as to the technological advances of the day. However, while with time clinical psychiatry had progressed in terms of successful patient management, no school of psychiatry can claim success in terms of cure or prevention of mental disease. Therefore, using this yardstick, an appropriate conceptualization of mental disease does not yet exist. Nevertheless, of relevance to the present discussion, we contrast the likelihood that a successful clinical science of mental illness will emerge from a biological versus a nonbiological psychiatry perspective.

BIOLOGICAL PSYCHIATRY PERSPECTIVE

Taking schizophrenia as an example of mental illness, we highlight two extreme positions on this disorder because the dichotomy that they represent illuminates the range of distinct conceptual approaches and attitudes in psychiatry. At one end of the pole are viewpoints that characterize schizophrenia not as an illness but rather as something extrinsic to the person—"a social fact and a political event" (Laing, 1967). At the other end is the perspective that schizophrenia is like any other medical disease caused by a distinct pathophysiology and as such is correctable by the appropriate medical intervention.

A reader of this book may be incredulous that schizophrenia could be conceived as anything but an illness. However, not only is there a long history of such viewpoints but one should expect that they will legitimately continue into the future until such time when the etiology of schizophrenia is ultimately identified. Before the advent of psychiatry, it was the philosophers who denied the mere possibility of mental disease "because the soul (mind) was a marker of the divine, [and therefore] it could become neither divided or diseased" (Berrios and Marková, 2002, p. 4). In recent times, psychiatrists such as Szasz (1961) and Laing (1967) similarly promulgated the position that schizophrenia is not a disease of the person but instead a societal construction and thus extrinsic to the individual.

In contrast to this, there is the perspective that the mental life and behaviors characteristic of schizophrenia are nothing but a disorder of brain function caused by a physical insult to normal brain biology. While the reader may resonate with the latter perspective, it too is an extreme position. This purely "medical" outlook seeks to explain all of the symptoms and all of the causes of schizophrenia solely in terms of biology, without allowing for any significant contribution from nonbiological factors. Ironically, probably in reaction to the extreme medical viewpoint, many scientists and mental health professionals fail to fully endorse, in turn, the role of biological factors in disorders such as schizophrenia.

A lack of consideration by the medical perspective of environmental and experiential factors in the causes, presentation, and treatment of schizophrenia is probably not the only reason for the skepticism by many vis-à-vis a biology of mental illness. For many scientists the opposition is no doubt also deeply philosophical as those scientists hold that biological explanations of schizophrenia will necessarily invalidate and eliminate explanations at a psychological level. This follows, they argue, from the supposition that the philosophy of biological reductionism that characterizes the scientific efforts of today's disciplines such as behavioral neuroscience or biological psychiatry leads to more fundamental accounts of mental life and disease by virtue of "reducing" psychological events to elements of biology. Consequently, the thinking goes, explanations of normal and abnormal behavior using a nonbiological language are at best tentative, awaiting replacement by an appropriate biological account in the future. Indeed, as pointed out by Teitelbaum and Pellis (1992), some have argued that a scientist would be wasting time by not directly proceeding to a biological account of behavior.

Biological reductionism is in fact the scientific philosophy that underlies the search in psychology and psychiatry for explanations of behavior and psychopathology, but it is also only an instance of a more general philosophy of reductionism that pervades all fields of science. Unfortunately, reductionism is much misinterpreted, as one can gleam from the opening sentence in *The Oxford Companion to Philosophy* entry for *reductionism*: "One of the most used and abused terms in the philosophical lexicon" (http://www.xrefer.com/entry/553365). Clark (1980) reviews many of the misconceptions associated with the term *reductionism*, especially as it relates to issues of explaining psychological phenomena in the language of biology (i.e., the brain). The main thesis advanced by Clark (1980) is that reductionism is simply a rational methodology with objective rules of procedure and evidence for mapping a *theory* in one field of science onto a *theory* in another field. Thus, with regard to phenomena at the level of behavior or mental life, *biological reductionism* refers to the methodology by which scientists attempt to find the correspondence (mapping) between components of a psychological theory of behavior or mental life and entities employed in a conceptual framework of another discipline, namely neuroscience (Clark used the older term, physiology, from which present-day neuroscience emerged). According to Clark (1980), psychological theories aim to provide an account for the observed relationships between environmental stimuli and behavior and are usually framed in terms of answers to two general questions: "What sort of machinery is required to produce the input-output relationships we observe? What jobs must parts of the system fulfill if the system as a whole is to manifest these relationships?" (Clark, 1980; p. 72). By specifying the "jobs" of the parts, psychological theory identifies the *function* of each component, that is, the nature of the transformation that each part performs on the inputs to it and the nature of the consequences that the output of each part has on the other components and on the system as a whole. In other words, psychological theories specify *what* different parts do but do not identify *how* it is physically done in the organism—how the function is implemented in the nervous system. Clark argues that reduc-

tionism of psychology to neuroscience is in essence the localization of function—the determination of the neural (structural) arrangement by which the psychologically identified function is implemented—and that functional and structural endeavors are complimentary:

> Psychological and physiological concepts can be related to one another in a unified model of the functioning of the nervous system. The various hypothetical states and processes invoked in psychological accounts are not abandoned in reduction; instead they are given a physiological identification. The presupposition of the psychological terms are not contradicted by increasing physiological knowledge; instead that knowledge allows us to explain how it is possible that neural states can be psychological states.
>
> (Clark, 1980, p. 184).

Before closing the present section, we consider the reason for our previous suggestion that biological psychiatry is more likely to yield a successful clinical science of mental illness than a nonbiological framework.

It should be first noted that a biological psychiatry perspective is not the extreme medical viewpoint described earlier but is confined only to the assertion that a mental disorder reflects some fault in the workings of the brain. In other words, using the language of biological reductionism, normal mental functions are carried out by particular arrangements of neural entities, and therefore mental disease or malfunction must correspond to a break in a required neural component(s) or to a breakdown in communication amongst the neural parts. How the neural failure is produced or can be repaired is an open question and biological psychiatry does not predetermine the nature of the answer. Thus, the conceptual framework of biological psychiatry allows for the possibility that neural parts can break because of psychological factors or because of physical insults or faulty genetics and that conversely both psychological and physical factors may yield neural repair effects. All that a biological psychiatry perspective specifies is that the roles of nonbiological and biological variables must be assessed at the proper levels of discourse. Hence, we suggest, the perspective of biological psychiatry is more likely to provide a successful clinical science of mental illness because it is a more comprehensive framework than either of the extreme positions described previously and because it includes the possible impact of psychological factors on the causes, presentation, and treatment of mental diseases. Moreover, by identifying mental disorders with a malfunction of the brain, the biological psychiatry perspective makes explicit that an understanding of these disorders can come about only through an understanding of the neural machinery underlying normal psychological functions. The pursuit of such basic knowledge is the purview of behavioral neuroscience, to which we now turn.

FROM PSYCHIATRY TO BEHAVIORAL NEUROSCIENCE: NEURAL MACHINERY OF MENTAL FUNCTIONS

Because the pathogenic mechanisms for most mental illnesses are unknown, psychiatrists make their diagnosis of mental disorder from evidence comprised primarily of behavioral data. Such data consist of (1) reports by patients on their subjective experiences and own behavior and (2) records by an external observer of patients' behavior and physical presentation. As mental disorders are classified on the basis of specific profiles in the behavioral data set, psychiatrists arrive at a diagnosis of a particular mental illness by matching the obtained behavioral findings to an identified data pattern of a disease. While in practice the process of diagnosis is more complicated, it remains the case that still today the primary source of psychiatric data comes from measures of behavior. This point is crucial—even though biological psychiatry holds that a fault in the workings of the brain underlies the pre-

sentation of a mental disorder, the classification of such a disease is not based on any identified brain pathology but on behavior. This is because no diagnostic brain pathology for psychiatric disorders has been yet identified. As is noted later, the demonstration of faulty brain workings in mental disorders is not likely to come from a clinical field but from a basic science discipline such as behavioral neuroscience.

Behavioral neuroscience is the discipline that seeks to map normal psychological functions onto specific arrangements of neurons. This task is doubly complex for it requires, on the one hand, that the entity to be localized is indeed a proper unit of psychological function and, on the other hand, that the mapping conforms to proper units of nervous system organization. Thus, the discipline requires expertise in the methods of both behavioral and neural sciences. This in principle is no different from requirements of biological psychiatry, where expertise in the evaluation of behavior and understanding of brain function is also needed. However, biological psychiatry, as an experimental science, can examine only *correlations* between mental life and brain function. In contrast, because its methods include manipulation of brain tissue, behavioral neuroscience can demonstrate the cause–effect mechanism by which a malfunction in neural machinery may result in abnormal mental experiences and behavior (for a discussion of behavioral neuroscience methods, see Teitelbaum and Pellis, 1992; Teitelbaum and Stricker, 1994). In other words, in the pursuit of localization of function, behavioral neuroscience simultaneously addresses two questions: What is the neural machinery of normal psychological functions? What is the nature of the behavioral abnormality that ensues from a perturbation of a particular set of neurons? In this respect, behavioral neuroscience is closely aligned with biological psychiatry by providing the basic science tools and knowledge required to identify what brain malfunction yields which type of mental disorder.

ON METHODS, TESTS, MODELS, AND THEORIES

Evidence that a specific neural dysfunction has a mechanistic role in a particular psychiatric disorder requires the experimental demonstration that the induction of this brain pathology yields the behavioral symptoms of the disease. Such experiments are ethically feasible using animals as subjects. Consequently, most laboratory studies use animal subjects, with rats and mice being the most common subjects. Of course, given the many obvious differences between rodents and humans, experiments on animals to understand phenomena in the human raise issues pertaining to the conduct and interpretation of such studies. The problem is especially vexing in relation to phenomena of interest to psychiatry with its focus on the mental life of the individual: How can one study and relate to the human the normal and abnormal mental life of so different an animal as a rat? Answers to such questions can be expected to be formalized in the use of animal models of psychiatric disorders.

Animal models of psychopathology are a relatively recent scientific phenomenon and the rationale underlying their use is still under debate. Nevertheless, several influential expositions have been published (McKinney and Bunney, 1969; Willner, 1984; Willner et al., 1992; Geyer and Markou, 1995). McKinney (1988), in *Models of Mental Disorders: A New Comparative Psychiatry*, reviewed the history of animal modeling in psychiatry and noted several pitfalls of early modeling in psychiatry. First, it was conceptually unreasonable of early workers to seek comprehensive animal models of psychiatric disorders because no model can be a miniature replica of the human illness. Rather, McKinney argued, one "should focus on the development of specific experimental systems to investigate selected aspects of the human syndromes" (p. 143). Second, and related to this point, there had been in the past a tendency to overgeneralize,

jumping too quickly to clinical conclusions from a given set of animal behavior experiments. Any generalization must consider not only the problem of cross-species comparisons but also the fact that animal experiments will not model the entire spectrum, but only aspects, of the disease. Third, early studies relied on descriptions, rather than on a quantitative analysis of animal behavior. Moreover, they did not consider the full richness of animal behavior, its evolution, and social structure. Consequently, comparisons to humans seemed strained considering the obvious complexity of human behavior and of psychiatric disorders. Finally, modeling of some disorders such as schizophrenia had been (and continues to be) especially problematic because diagnostic criteria keep changing, making it difficult to determine what should be modeled in animals.

A rationale for modeling psychopathology in the laboratory (which still seems appropriate today) had been proposed by Abramson and Seligman (1977). In accordance with the more general framework of what are models in psychology (Chapanis, 1961), Abramson and Seligman (1977) defined *modeling* as "the production, under controlled conditions, of phenomena analogous to naturally occurring mental disorders. Its goal is to understand the disorders" (p. 1). Furthermore, the authors viewed modeling as the attempt to bring the study of psychopathology under a "controlled setting in which particular symptom or constellation of symptoms is produced in miniature to test hypotheses about cause and cure. Confirmed hypotheses [would] then be further tested in situations outside the laboratory" (p. 1). The authors envisioned two types of paradigms. In one type, a standard test is administered under controlled conditions to subjects with mental disorder and compared with performance of normal controls. The purpose is to examine some a priori hypothesis about the determinants of the mental disorder and consequently the chosen test is one that measures the aspect of psychological function which the investigators hypothesize is disordered in the disease. For example, to examine hypotheses of disordered information processing in schizophrenia, investigators may compare performances of subjects with and without the disease on tests of sensorimotor gating such as prepulse inhibition of the startle response (Geyer et al., 1999). Using appropriate tests, other hypotheses of the disorder may be examined in the same manner.

The alternative paradigm is potentially much more powerful because it involves the use of the laboratory experimental model method. As argued by Abramson and Seligman (1977, p. 3),

> In this method, psychopathology is not merely brought into the laboratory for study; it is modeled and reproduced there. The advantage of studying a model over observing and comparing two groups is that because the disorder is produced in the lab, we can specify what causes it. The model is rarely an exact replica of the spontaneous psychopathology, and it is useful only to the degree that it mimics the phenomenon in question or suggests interesting experiments. . . .

The authors suggested that an ideal model would show *similarity* with the real-word disorder in terms of cause, symptoms, prevention and cures; they proposed the following 4 criteria in evaluating how good is a laboratory model of psychopathology (Abramson and Seligman, 1977; p. 3):

1. Is the experimental analysis of the laboratory phenomenon thorough enough to describe the essential features of its causes as well as its preventives and cures?
2. Is the similarity of symptoms between the model and naturally occurring psychopathology convincingly demonstrated?
3. To what extent is similarity of physiology, cause, cure and prevention found?
4. Does the laboratory model describe in all instances a naturally occurring psychopathology or only a subgroup? Is the laboratory phenomenon a model of a specific psychopathology,

or does it model general features of all psychopathologies?

While the meaning of the middle two criteria is self-evident, the authors elaborated that the first criterion regarding the "essential features" of the model refers to establishing how relevant is the procedure of inducing the disorder in the model to the real-world manner of precipitating the disease in the human (and likewise for the procedures of preventing and curing the disease). For instance, in a model of depression (Seligman, 1972; Miller and Seligman, 1973), the inducing procedure is the administration of electric shock but the authors show that the critical feature of this procedure for the induction of depression is not the shock itself but rather the production of an inescapable and uncontrollable situation, which the authors hold resembles the real-world inductor of depression. With regard to the last criterion, it refers to the need of establishing that the model is of a specific disease (or subtype) rather than of a general response to being ill.

It is important to note that, as outlined by Abramson and Seligman (1977), an animal model of a psychiatric disorder represents a *theory* of a naturally occurring psychopathology and as such is subject to the usual rules for confirmation of scientific theories. However, this terminology is often applied—especially in the psychopharmacology literature—to animal preparations that do not have the attributes of a theory (that is, the intention of describing in all its facets the true mechanism of the real-world psychopathology). Rather, the term "animal model" of a mental disorder is often used with reference to a preparation that is best described as a "behavioral test" that is sensitive, either directly or indirectly, to the presence of a particular psychopathology. For instance, apomorphine-induced vomiting is a behavior that responds to antipsychotic drugs (and was in fact one of the first "animal models" used in identifying potential drugs with antipsychotic activity

[Janssen et al., 1966]), but its status is more appropriately labeled as a "behavioral assay" of stimulation of dopaminergic receptors (Szechtman et al., 1988) and, for this reason, as a test that is indirectly sensitive to any psychopathology associated with dopaminergic hyperactivity. In essence, such "behavioral assays" conform to "standard tests" of psychological or neurobiological function as used in the first type of paradigm described earlier but do not constitute an animal preparation (model) of the psychopathology.

TOWARD AN ANIMAL MODEL OF OBSESSIVE-COMPULSIVE DISORDER

In this section we use a specific example of an animal model of obsessive-compulsive disorder (OCD) to evaluate its characteristics with reference to the criteria of Abramson and Seligman (1977) and to highlight the methods and principles relevant to the development of an animal model of a psychiatric disorder. The particular rat model seems strongest in addressing the *similarity* of symptoms and cure between the model and human OCD (criteria 2 and 3 of Abramson and Seligman, 1977; see earlier). Consequently, we discuss this model with reference to these criteria first.

The model of OCD (Szechtman et al., 1998) is produced by chronic treatment of rats with the D2/D3 dopamine agonist quinpirole. Rats receive repeated injections of the drug (0.5 mg/kg twice weekly for 10 doses) in a large, open field with four small objects (boxes) in it. On the last injection, the animals' behavior is evaluated and compared with the behavior of rats treated chronically with saline. The authors used the results of their behavioral analysis to stake the claim that quinpirole-treated rats show compulsive checking and that this is a symptom like the one shown by patients with OCD. The investigators provided two lines of evidence to show this similarity.

The first line of evidence came from a comparison of the structure of compulsive checking in the rat with that of the human. It was reasoned that whatever is the nature of the mental aberration in OCD, it must find expression in a characteristic pattern of behavior that is readily observable. Unfortunately, no ethological descriptions of human OCD compulsions were available. Consequently, the authors had to surmise what the spatiotemporal structure of compulsive checking is likely to be. For this they drew on the type of information solicited from patients by psychiatrists to identify OCD symptoms and on attempts in the psychiatric literature to characterize the essential features of clinical compulsions (Reed, 1985). Based on these sources, it became apparent that salient features of compulsions include (1) preoccupation with the performance of compulsions and a reluctance/resistance to engage in them, (2) a "ritual-like" quality in the performance of the compulsions, and (3) close coupling between OCD motor rituals and environmental stimuli/context. Accordingly, the authors suggested that a checking compulsion would reflect itself as (1) a preoccupation with and an exaggerated hesitancy to leave the item(s) of interest in one's territory, (2) a ritual-like motor activity pattern and (3) dependence of checking behavior on environmental context. Consequently, the authors proposed that the following five performance criteria identify the spatiotemporal structure of compulsive checking (Szechtman et al., 1998, p. 1477):

1. In the subject's territory, there would be one or two places/objects to which the subject returns excessively more often than to other places/objects in the environment.
2. The time to return to these preferred places/objects would be excessively shorter than the time to other places/objects.
3. Excessively fewer places would be visited between returns to the preferred places/objects.
4. A characteristic set of acts would be performed at the preferred place/object, which would dif-

fer from the acts performed at other locations/objects.

5. Activity would be altered when the environmental properties of the places/objects are changed.

Quinpirole rats met all of the five criteria for compulsive checking. As shown in Figure 43–1, their behavior indicated a preoccupation with and a reluctance to leave a particular location/object in the open field (the home base), as measured by the first three of the proposed criteria. Moreover, their behavior had a ritualistic quality in the preferred places/objects (Table 43–1). Finally, when an object was moved to a new location, their checking activity did shift to the new location as well as producing other changes in behavior (Szechtman et al., 1998).

The second line of evidence for similarity between quinpirole-induced compulsive checking and the symptom of OCD checking involved comparison of the possible motivational basis of this activity in the rat model and the human condition. In the human, compulsive checking appears motivated by concerns of safety and security and is seen as an exaggerated form of normal checking regarding one's well-being and security (Reed, 1985). The characteristics of the checking behavior in the rat model were consistent with having a similar motivational basis. In particular, rats normally do engage in checking behaviors ("risk assessment"; Blanchard, 1997) and the behavior of quinpirole-treated rats was directed at a likely stimulus for checking activity—the home base (Eilam and Golani, 1989). Thus, quinpirole checking was attached to stimuli with a plausible relationship to safety and security and in this regard could be called an exaggerated form of normal checking in the rat, similar to the human condition.

The similarity of symptoms in the rat model and human OCD was strengthened by the additional finding that quinpirole-treated rats, like OCD patients, may temporarily desist from engaging in compulsive checking (Szechtman et al., 2001). Criterion 3 of

Abramson and Seligman (1977)—similarity of "cure"—was shown by the effectiveness of clomipramine, a drug used in the treatment of human OCD, to partially attenuate quinpirole-induced checking behavior (Szechtman et al., 1998). Thus, in several respects, there is a strong similarity in the symptoms displayed by the quinpirole model and those found in human OCD psychopathology. However, this similarity would be strengthened by showing directly that the spatiotemporal structure of human OCD compulsions does have the same form as observed in the rat. We have begun to conduct such "ethological" studies and show the results of analyzing a compulsive ritual of one OCD patient (Table 43–2). As is evident, the structure of that ritual, even though it is of cleaning the nose rather than being a checking ritual, is remarkably similar in its form to the compulsive checking behavior of the quinpirole-treated rat (compare with Table 43–1).

With reference to criterion 1 of Abramson and Seligman (1977) regarding the "essential features" of the quinpirole model, this is a question that is currently not resolved. Quinpirole is a dopamine agonist, so clearly an important aspect of the "cause" of OCD symptoms in the model is chronic activation of dopamine receptors. However, it must be examined whether activation by only the drug produces compulsive checking (in which case the quinpirole preparation would be only a model of drug-induced OCD) or whether use of the drug represents a more general class of particular environmental conditions that hyperactivate the dopamine system(s). Likewise, how important is the fact that rats experience the drug injections in a large open field—what does this indicate about the induction of compulsive checking? From other studies in which the locomotor response to quinpirole was examined (Szechtman et al., 1993; Einat and Szechtman, 1993; Einat et al., 1996; Szumlinski et al., 1997), it is clear that the nature of the environmental context has a profound influence on the degree of locomotor sensitiza-

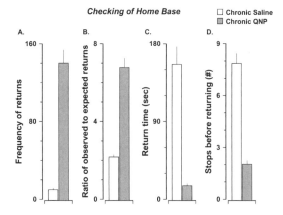

Figure 43–1. Induction of compulsive checking as identified by formal performance criteria. Performance measures are in reference to the home base established by each rat on the tenth open field test shown here and recognized as the locale with the longest total duration of stops. Generally, it is also the most frequently visited place in the open field. The open field is 160 × 160 cm without walls and subdivided into 25 locales (places). Locomotor activity is measured in relation to these 25 places. Quinpirole (QNP)-treated rats (*gray bars*) satisfied the first three criteria for compulsive checking because, compared with saline-treated rats (*open bars*), they showed (*A*) more frequent returns to the home base, (*B*) a higher-than-expected rate of returning to the home base, (*C*) more rapid returns to the home base, and (*D*) fewer visits to other places in the open field on trips from the home base. Values are given as mean and SEM. All differences between quinpirole- and saline-treated rats are significant. Adapted from Szechtman et al. (2001).

tion to quinpirole, with the induction of minimal sensitization when the drug is administered in the rats' home cage (Szechtman et al., 1993; Szumlinski et al., 1997). Therefore, it is likely that the type of environment in which the rat experiences chronic injections of quinpirole plays also an important role in whether the rat does or does not develop compulsive checking. We suspect that to develop OCD rituals, the environment must be such as to direct the rat's attention to events related to its safety and security, but this must be shown in formal studies.

As to whether the quinpirole model describes all instances or only a subtype of OCD (criterion 4 of Abramson and Seligman, 1977), it seems more likely that it is only a subtype of OCD that is being modeled. This opinion

is based on the observation that the effects of clomipramine on quinpirole checking were not lasting, as found for a subpopulation of OCD patients (Szechtman et al., 1998). Thus, there seem to be different populations of OCD patients, and the quinpirole model may represent the subtype that is less sensitive to the effects of clomipramine. However, this is another aspect of the model that needs investigation. Still other major aspects of the model that need examination concern how the neurobiology of checking in the quinpirole model relates to the neurobiology of the human disorder. In this endeavor, because knowledge about the neurobiology of the human psychopathology is relatively rudimentary, it is equally likely that the rat model will be used to generate hypotheses to be examined in human OCD. Indeed, the usefulness of an animal model lies not only in its convenience of providing a physical preparation to investigate the mechanisms of a disorder but also in its ability to generate novel hypotheses that can be examined in the clinic.

CONCLUDING REMARKS

Until quite recently, there was great skepticism in psychiatry regarding the usefulness of animal models of human psychiatric disorders. Now, however, the stage is set and wide open for the development of animal models for virtually any psychiatric disorder. This profound shift in attitude reflects largely the acceptance of the viewpoint that malfunction of mental life may be embodied in a breakdown of normal brain function, an attitude embraced by the rising field of biological psychiatry. Acceptance of animal models follows as a necessary corollary of this viewpoint because such models not only provide the physical preparation to investigate the mechanisms of normal and disordered mental life but also acknowledge the evolutionary roots of humans in the continuity of biological life. Of course, development of animal models of

Table 43-1. A Record of 2 Minutes of Activity of a Rat in a Large, Open Field, 40 Minutes After the Tenth Injection of Quinpirole

	Location in the open field		
Trip	Home Base (Corner Locale With Box)	Edge Locale	Center Locale With Box
1	A_3 V_u T_4 V_u V_d Rc_6 S $Trot_6$	A_2 Ra_1 $Trot_1$	A_5 Rc_3 $Trot_3$
2	A_3 V_u T_4 V_d Rc_6 S $Trot_6$	A_2 Ra_0 $Trot_0$	A_4 Rc_3 $Trot_3$
3	A_3 V_u T_4 V_d Rc_6 S Rc_2 T_2 Rc_6 S $Trot_6$	A_2 Ra_0 $Trot_0$	A_4 T_2 Rc_3 $Trot_3$
4	A_3 V_u T_2 Rc_5 S Ra_4 V_d T_2 Rc_6 S $Trot_6$	A_2 Ra_0 $Trot_0$	A_4 Rc_3 $Trot_3$
5	A_3 V_u T_2 Ra_2 Rc_6 V_u Rc_3 V_u T_0 Rc_4 V_d Ra_2 V_u T_2 Rc_6 S $Trot_6$	A_2 Ra_1 $Trot_1$	A_5 V_u T_0 Rc_3 $Trot_3$
6	A_3 V_u T_2 Ra_2 Rc_6 V_u Rc_7 S $Trot_7$		A_3 Rc_7 V_u Ra_6 V_u T_0 Rc_2 $Trot_3$
7	A_3 V_u T_2 Ra_4 Ra_2 V_u T_2 Rc_7 S $Trot_7$		A_3 V_u T_0 Ra_3 $Trot_3$
8	A_3 V_u T_0 V_d V_u Rc_7 S $Trot_7$		A_3 Rc_1 V_u T_0 Ra_6 T_4 V_u V_u T_4 Rc_3 $Trot_3$
9	A_3 V_u T_0 Rc_5 Ra_3 V_d V_u V_d Rc_6 S $Trot_6$	A_2 Ra_2	$Trot_2$
10	A_2 V_u T_0 V_u T_4 V_d Rc_4 V_d Rc_6 S $Trot_6$	A_2 Ra_1 $Trot_1$	A_5 V_u T_4 Rc_3 $Trot_3$
11	A_3 V_u T_2 Rc_6 S $Trot_6$	A_2 Rc_2 $Trot_2$	$Trot_3$
12	A_3 V_u T_2 Rc_6 S $Trot_6$	A_2 Ra_1 $Trot_1$	A_5 V_u T_4 Rc_3 $Trot_3$

Note: The record describes the sequence of movements performed by the rat during 12 successive round trips from the home base. A indicates arrival, with the direction of arrival indicated by the subscript numeral and specified in units of $1 = 45°$; V, vertical movement of the forequarters (u, upward; d, downward); T, establishing contact with the box (4, climbing on top of it with all four legs; 2, leaning on it with the forelegs; 0, only the snout makes contacts with the box); Rc, rotation (turning) clockwise to one of eight specified directions in intervals of $1 = 45°$; Ra, rotation (turning) anticlockwise to one of eight specified directions in intervals of $1 = 45°$; S, stepping down from the box; Trot, running in a trot gait to the specified direction, $1 = 45°$. The rat displayed a typical set of acts starting with arrival (A_3), vertical movement (V_u), and touching the box (T). The sequence of movements at this locale always ended with turning clockwise to direction 6 or 7 (Rc_6 or Rc_7), stepping down from the box (S), and trotting in that direction. On trips 1 to 5 and 9 to 12, the rat then stopped at an edge locale, where it made an anticlockwise turn (Ra, except for trip 11, when a clockwise turn occurred). It then ran to a center locale with an object in it and moved around the object, often making a vertical movement (V_u) and touching the object (T). It then trotted back to the home base, where it begun another trip. On trips 6 to 8, the rat did not stop at the edge locale but ran directly from the home base to the center locale, where it displayed the usual ritual and then returned to the home base. Thus, although there is a certain amount of flexibility within and across round trips, the behavior of this rat is a highly structured and repetitive ritual.

Table 43–2. Four Repetitions of a Ritual of Blowing and Wiping the Nose in an Obsessive-Compulsive Disorder Patient

Repeat	Act	Taking a Tissue Paper					Blowing or Wiping the Nose and/or Mouth			Handling the Used Tissue Paper				
1	Blow	H	C	Pr	F	Ra	Pl	$W_1 \times 4$	S	Rc			Pu	
	Wipe	H	C	Pr	F	Rc Ra	Pl	W_2	S		Fo			
							Pl	$W_3 \times 4$	S	Ra Rc		F		
							Pl	$W_4 \times 4$	S	Rc		F	Pu	Cleaning shirt and hands
2	Blow	H	C	Pr	F	Ra	Pl	$W_1 \times 4$	S	Rc F Ra		F	Pu	Cleaning shirt and hands
	Wipe	H	C	Pr	F	Rc Ra	Pl	$W_2 \times 2$	S		Fo			
							Pl	$W_3 \times 4$	S	Rc Rc		F		
							Pl	$W_4 \times 4$	S			F	Pu	Cleaning shirt and hands
3	Blow	Missing on tape												
	Wipe	H	C	Pr	F	Ra Rc	Pl	$W_2 \times 4$	S		Fo			
							Pl	$W_3 \times 4$	S	Rc Rc		F		
							Pl	$W_4 \times 4$	S			F	Pu	Cleaning shirt and hands
4	Blow	H	C	Pr	F	Ra	Pl	W_1	S	Rc Ra		F	Pu	
	Wipe	H	C	Pr	F	Ra Rc	Pl	$W_2 \times 2$	S		Fo			
							Pl	$W_3 \times 4$	S	Rc Ra		F		
							Pl	$W_4 \times 2$	S			F	Pu	Cleaning shirt and hands

Blow indicates the entire first row in each repetition and is a label for taking a piece of tissue paper, blowing the nose, and putting away the tissue; Wipe, the remaining rows in each repetition and a label for taking a piece of tissue paper, wiping the nose, upper lips, and under the nose, and putting away the tissue; H, holding horizontally in the left hand the roll of tissue paper; C, turning the roll so that two pieces of tissue hang down, holding them between the thumb and the other four fingers, and detaching the two pieces—the act takes place with what appears to be high concentration, with the eyes following every movement being performed; Pr, placing the tissue roll back on the dresser, with the left hand; F, holding vertically the tissue piece and folding it along a line perpendicular to the ground, and then smoothing the fold with two fingers from top to bottom; Fo, flipping over the folded piece of tissue; R, rotating the piece of the tissue by $90°$ clockwise (Rc) or anticlockwise (Ra); Pl, placing the piece of tissue on the body part to be wiped. The tissue covers the nose and the mouth, is held first with only one finger and then by four fingers at each side; W_1, blowing the nose strongly after a deep breadth and blowing the air intensely so that two holes are formed in the tissue—piping movements occur in the fingers starting with the median and ending with the lateral fingers; W_2, wiping the upper lip while moving the head to the right and opening the mouth, and then wiping the lower lip while moving the head to the left and closing the mouth; W_3, wiping the nose while moving the head up and down; W_4, wiping under the nose while moving the head up and down with closed eyes; S, staring at the tissue with high concentration for several seconds; Pu, placing the used piece of tissue, with the left hand, on top of the previously used tissues; $\times 2$ or $\times 4$, two or four repetitions, respectively.

Note: The record shows four repetitions of a ritual of blowing and wiping the nose observed during a 1 hour videotaped interview in the patient's home. Each repetition of the ritual took approximately 4 minutes. Although there is a certain amount of flexibility within and across repetitions of the ritual, behavior is highly structured and repetitive, as found in the quinpirole-treated rat in Table 43–1.

psychopathology is a complex task that at every stage must be evaluated with rigor, but the usefulness of such models is undisputable by providing both precise control over heredity and experience and the opportunity to constantly monitor the disorder (Abramson and Seligman, 1977). In developing the models, the challenge for behavioral neuroscientists lies, on the one hand, in applying their knowledge of psychology and skills in behavioral analysis to construct tests and methods of measurement that capture the expression of normal and disordered mental function. As illustrated in the rat model of OCD, an ethological approach that focuses on recording the spatiotemporal structure of behavior provides a natural window on mental life that lends itself readily to comparisons of the form of behavior across species as diverse as the human and the rat. The challenge is equally demanding, on the other hand, in using the available neuroscience tools to manipulate the nervous system in a meaningful fashion and thereby identify which perturbations of the brain yield disorders of psychological function and thereby psychiatric disorders.

ACKNOWLEDGMENTS

This work was supported by the Ontario Mental Health Foundation, the Canadian Institutes of Health Research, and the Israel Science Foundation. H.S. is a Senior Research Fellow of the Ontario Mental Health Foundation.

REFERENCES

Berrios GE and Marková IS (2002) Conceptual issues. In: Biological psychiatry (D'haenen HAH, Boer JA, Willner P, eds.), pp. 3–24. Chichester: Wiley.

Blanchard DC (1997) Stimulus and environmental control of defensive behaviors. In: The functional behaviorism of Robert C. Bolles: Learning, motivation and cognition (Bouton M and Fanselow MS, eds.), pp. 283–305. Washington, DC: American Psychological Association.

Chapanis A (1961) Men, machines, and models. American Psychologist 16:113–131.

Clark A (1980) Psychological models and neural mechanisms: An examination of reductionism in psychology. Oxford: Clarendon Press.

Eilam D and Golani I (1989) Home base behavior of rats (Rattus norvegicus) exploring a novel environment. Behavioural Brain Research 34:199–211.

Einat H, Einat D, Allan M, Talangbayan H, Tsafnat T, Szechtman H (1996) Associational and nonassociational mechanisms in locomotor sensitization to the dopamine agonist quinpirole. Psychopharmacology 127:95–101.

Einat H and Szechtman H (1993) Environmental modulation of both locomotor response and locomotor sensitization to the dopamine agonist quinpirole. Behavioural Pharmacology 4:399–403.

Geyer MA, Braff DL, Swerdlow NR (1999) Startle-response measures of information processing in animals: Relevance to schizophrenia. In: Animal models of human emotion and cognition (Haug M and Whalen RE, eds.), pp. 103–142.

Washington, DC: American Psychological Association.

Geyer MA and Markou A (1995) Animal models of psychiatric disorders. In: Psychopharmacology: The fourth generation of progress (Bloom FE and Kupfer DJ, eds.), pp. 787–798. New York: Raven Press.

Janssen PA, Niemegeers CJ, Schellekens KH (1966) Is it possible to predict the clinical effects of neuroleptic drugs (major tranquillizers) from animal data? Arzneimittel-Forschung 16:339–346.

Laing RD (1967) The politics of experience. Harmondsworth: Penguin.

McKinney WT Jr and Bunney WE Jr (1969) Animal model of depression. I. Review of evidence: Implications for research. Archives of General Psychiatry 21:240–248.

McKinney WT (1988) Models of mental disorders: A new comparative psychiatry. New York: Plenum Medical Book Co.

Miller WR and Seligman ME (1973) Depression and the perception of reinforcement. Journal of Abnormal Psychology 82:62–73.

Reed GF (1985) Obsessional experience and compulsive behaviour: A cognitive-structural approach. Orlando, FL: Academic Press, Inc.

Seligman ME (1972) Learned helplessness. Annual Review of Medicine 23:407–412.

Szasz TS (1961) The myth of mental illness: Foundations of a theory of personal conduct. New York: Hoeber-Harper.

Szechtman H, Eckert MJ, Tse WS, Boersma JT, Bonura CA, McClelland JZ, Culver KE, Eilam D (2001) Compulsive checking behavior of quinpirole-sensitized rats as an animal model of obsessive-compulsive disorder (OCD): Form and control. BMC Neuroscience 2:4.

Szechtman H, Eilam D, Ornstein K, Teitelbaum P, Golani I (1988) A different look at measurement and interpretation of drug-induced behavior. Psychobiology 16:164–173.

Szechtman H, Sulis W, Eilam D (1998) Quinpirole induces compulsive checking behavior in rats: A potential animal model of obsessive-compulsive disorder (OCD). Behavioral Neuroscience 112:1475–1485.

Szechtman H, Talangbayan H, Eilam D (1993) Environmental and behavioral components of sensitization induced by the dopamine agonist quinpirole. Behavioural Pharmacology 4:405–410.

Szumlinski KK, Allan M, Talangbayan H, Tracey A, Szechtman H (1997) Locomotor sensitization to quinpirole: Environment-modulated increase in efficacy and context-dependent increase in potency. Psychopharmacology 134:193–200.

Teitelbaum P and Pellis SM (1992) Toward a synthetic physiological psychology. Psychological Science 3:4–20.

Teitelbaum P and Stricker EM (1994) Compound complementarities in the study of motivated behavior. Psychological Review 101:312–317.

Willner P (1984) The validity of animal models of depression. Psychopharmacology 83:1–16.

Willner P, Muscat R, Papp M (1992) Chronic mild stress-induced anhedonia: A realistic animal model of depression. Neuroscience and Biobehavioral Reviews 16:525–534.

Neuropsychological Tests

GERLINDE A. METZ, BRYAN KOLB, AND IAN Q. WHISHAW

Having examined the behavior of the rat over 43 chapters focused on a wide range of behaviors, we must now try to present a simple, yet comprehensive battery of tests that could be recommended to provide a starting point for the description of the behavior of the rat. Obviously, particular chapters on specific topics such as sex or eating would be the place to start if there were a reason to suspect changes in such behaviors, but often that is not the case. For example, if one were interested in the effect of a genetic manipulation or perhaps in the effect of a drug to be used for the treatment of bowel cancer, the place to begin a behavioral analysis would be with a broad battery of tests that would serve to point the investigator to further examination with more specific tests. The behavioral repertoire of an animal, or its ethogram, is a summary of all of its innate and learned behaviors. Such a behavioral assessment combines principles of the neurological examination, developed for the assessment of the motor behavior of humans, and the neuropsychological assessment, developed for the assessment of human psychological functions. These assessment procedures have been tailored for the rat.

A neuropsychological examination of rat begins with a description of general appearance and sensorimotor responsiveness and continues by examining posture and immobility, locomotion, skilled movements, species-specific behavior, and learning (Table 44–1). It may not always be necessary to evaluate all of the behavioral categories, or to use all of the tests, the guiding principle being the answer to the question, "What do you want to know?" Thus, for example, studies of spinal cord function are unlikely to include measures of visual acuity. Nevertheless, one cannot always be certain about what one needs to know. Accordingly, it is often necessary to use the principles of differential diagnosis, in which both unimpaired and impaired functions are described. This chapter provides an overview of these seven main categories of behavioral assessment and describes some individual tests of behaviors in each category. The chapter is broken into seven sections that provide overviews on a selection of behavioral tests useful in assessing that behavioral category. Representative references are provided for each test, but the reader is directed to specific chapters earlier in the book for more extensive reference lists. The test descriptions are intended to present a summary of a given test procedure rather than presenting the apparatus and procedural details; these can be obtained in cited methods references.

It is important to note that there is no static test or test battery. Existing tests are constantly being revised and adapted to the demands and purpose of an experiment and to the type and severity of deficit predicted after a neurological insult (Whishaw et al., 1983, 1999). Furthermore, these tests are also being refined as understanding of the behavior of the rat and its neural substrate grows. The individual tests and assessment parameters may need to be modified to fit the specific research

Table 44–1. Categories of Behavior and
Neuropsychogical Assessment

Category	Tests Described
Appearance and responsiveness	General examination of health
	Tests of orienting
Sensory and sensorimotor behavior	Formalin pain test
	Hot and cold temperature test
	von Frey hair test
	Sticky dot test
	Limb-use asymmetry (cylinder) test
Posture and immobility	Posture
	Righting responses
Locomotion	Swimming
	Walking and running
	Circadian activity
	BBB scale
Skilled movements	Food handling
	Tray-reaching task
	Pellet-reaching task
	Rung and beam walking
Species-specific behavior	Grooming
	Food hoarding
	Exploration
	Play behavior
	Sexual behavior
Learning	Water task
	Radial arm maze
	Barnes maze
	Elevated-plus maze
	Context conditioning
	One-way avoidance
	Spontaneous alternation
	Two-choice association

protocol of the investigator to reveal parameters that best predict chronic outcome or long-term consequences of a manipulation. Furthermore, an optimal assessment would use a number of measurements from the test battery, each describing specific aspects of behavior. Therefore, a combination of different tests is used to gain a detailed and broad functional profile of an animal.

In addition to the selection of appropriate tests, the method of analysis determines the sensitivity of the individual measures. Parameters can include end point measures to quantify the frequency of occurrence of a specific event. Kinematic measures reconstruct a

movement to obtain information about its direction, velocity, or angle. Descriptive analysis of movement, which can include use of a formal language, describes the movement of body segments and the intersegmental relations. These methods of analysis will be mentioned in the description of the individual test procedures. It is important to remember that different measures may not necessarily correlate. A rat may perform a task in the correct way but have a terrible end point score or, alternatively, its performance may be terrible but its score excellent. Thus, single measures seldom capture function. As a rule of thumb, in initial screening, we favor observational descriptions along with end point measures. Obviously, because a test battery procedure involves the necessary use of multiple comparisons, some differences may occur by chance. Thus, we prefer to identify deficits only after we obtain a number of converging measures.

One of the problems in assessing behavior is that if a test procedure is laborious or time consuming, it will torpedo the usefulness of a test battery. Consequently, most of the tests described here are relatively simple and can be administered fairly quickly. Furthermore, the tests can be given concurrently so that an extensive battery of tests can be given in a relatively short time. In addition, there is a growing interest on the part of test designers to develop automated systems of analysis. So, for example, a photocell automated open field test can be administered in 10 minutes and can provide a good range of measures, including overall activity, habituation, rearing activity, thigmotactic tendencies, and turning biases. Such a test is sensitive to motor abnormality or the laterality of a motor impairment. Because rats are typically more active in the dark than in the light, an open field activity test repeated in the light and dark might also reveal something about a rat's visual sensitivity. Video-based automated monitoring systems can also reveal the organization of exploratory behavior in single animals or in animal pairs, thus providing a measure of social behavior.

Finally, we reiterate that the tests outlined in the following pages represent a general set of tests that are useful as a tool to generate a behavioral phenotype. Specific chapters earlier in this volume provide more detailed and specific measures of specific behavioral functions.

APPEARANCE AND RESPONSIVENESS

Much can be learned by simply examining the general appearance and responsiveness of animals (Table 44–2). This can be achieved by examining them in their home cage, by placing them in a novel environment, and by handling them. The analysis of appearance and responsiveness provides important clues on what other types of measures may be appropriate. For example, long claws may reflect a difficulty in making fine movements of the mouth that are necessary for claw trimming, and weeping eyes may indicate the accumulation of Hardarian material and thus poor grooming (Whishaw et al., 1983). The extent of a physical abnormality can be quantified by using a small ruler, such as to measure claw length. In the course of examining strain differences in spatial behavior of rats, we observed that male Dark Agouti rats, an inbred pigmented rat strain, were smaller than Long-Evans rats, tended not to grow as quickly, and had somewhat delicate heads and long snouts. Their appearance suggested that the process of

inbreeding had feminized the animals (Harker and Whishaw, 2002). Feminization may have contributed to their poorer spatial performance than that obtained in Long-Evans rats. Thus, the observation of physical appearance lead to hypotheses concerning function. This in turn may lead to experimentation in which males are given additional testosterone to determine if both general morphology and behavior might become more like that of the male Long-Evans rat.

It is useful to make up a checklist of appearance features and behaviors as a way of standardizing the appearance examination. We then recommend a five-step sequence for the assessment of appearance and responsiveness.

1. The animal is observed in its home environment. It is useful to have a test animal housed with, or beside, an experimental animal so direct comparisons can be made. Observations are made on the animals' fur, eyes, and feet to determine whether they are clean, a feature that indicates whether the animal is grooming.
2. The animal is given some simple tests of responsiveness in its home environment. Objects can be introduced into an animal's cage to observe its responsiveness to novel stimuli, and a substantial battery of sensorimotor tests can be given while the animal is in its home cage (see next section).
3. The animal is removed from its home cage and observed in a novel environment. Again it is useful to have a yoked control animal as a standard against which to compare an experimental subject. An animal's behavior can be quite different in a safe versus novel environment.

Table 44–2. Examination of Appearance and Responsiveness

Appearance	Inspect body shape, eyes, vibrissae, limbs, fur, tail, and coloring.
Cage side examination	Examine the animal's cage, including bedding material, nest, food storage, and droppings.
Handling	Remove the animal from its cage, and evaluate its response to handling, including movements and body tone, and vocalization. Lift the lips to examine teeth, especially the incisors, and inspect the digits and toenails. Inspect the genitals and rectum.
Body measurements	Weigh the animal and measure its body proportions (e.g., head, trunk, limbs, and tail). Measure core temperature with a rectal or aural thermometer.

4. The animal is given some simple tests in the novel environment (see next section). These include picking up the animal, assessing its muscle tone, examining its placing responses, dangling it by the base of the tail to observe its orientation in relation to gravity, and its response to the experimenter.
5. Specific morphological measures are made. The animal is weighed, and its eyes, ears, vibrissae, head shape, body shape, and tail shape are examined. Using a small plastic ruler, it is useful to directly measure physical features.

SENSORY AND SENSORIMOTOR BEHAVIOR

The object of sensory tests is to assess the general functional status of the sensory systems, including somatosensory, visual, and olfactory functions. Such measures may include a quick assessment both in the home cage and in an open space using the procedures outlined in Table 44–3, or they may involve specific tests outlined later, or more detailed tests provided in specific chapters elsewhere in the volume. It is important to note that informal observations, during which an animal is simply observed along with its control partner several times a day, can provide good observations that may later be refined into more formal tests. Many skilled behaviors can also be observed in the home cage. By presenting strips of paper each day, nest building behavior can be observed, especially in female rats. By presenting sunflower seeds, food handling can be examined, and by presenting tasty food

on a spatula, tongue use can be examined. Examination of the home cage can also provide clues about cognitive and motor status. For example, one can determine if food is chewed and eaten efficiently or if there are excessive food fragments and crumbs suggesting motor impairments in eating. Similarly, normal spatial behavior is suggested by compartmentalization of sites for elimination, eating, and sleeping. More formal tests involve removing the animal from its home cage; these include (1) somatosensory and pain testing using the von Frey hair test, or the formaline test (2) the formalin pain test, (3) thermal sensitivity using the hot and cold plate tests, (4) somatosensory detection and skilled movement using the sticky dot test, and (5) placing responses using the cylinder test.

THE VON FREY HAIR TEST

A vital modality in rats is their haptic sense, essential to their nocturnal life style. Rats use tactile receptors on their body, including their vibrissae, special receptors on their paws, and the many large "sinus" hairs on their body. When combined together, the information from these haptic receptors provide the animals with an acute sense of their surroundings. The responsiveness to tactile stimulation can be evaluated by mechanically stimulating the skin of the animal with various objects. A calibrated monofilament, the *von Frey hair*, can be used for threshold sensitivity evaluation.

In laboratory rodents, von Frey hair tests have been used to evaluate the effects of le-

Table 44–3. Examination of Sensory and Sensorimotor Behavior

Home cage	Response to auditory, olfactory, somatosensory, gustatory, vestibular, and visual stimuli. The home cage should provide easy viewing of an animal. Holes in the sides and bottom of the cage provide entry for probes to touch the animal or to present objects to the animal or to present food items. Animals are extremely responsive to inserted objects and treat capturing the objects as a "game." Slightly opening an animal's cage can attenuate its response to introduced stimuli, showing that it notices the change.
Open field	Response to auditory, olfactory, somatosensory, taste, vestibular, and visual stimuli. The same tests are administered. Generally, animals taken out of their home cage are more interested in exploring and so ignore objects that they responded to when in the home cage.

sions and maladaptive plasticity on sensory function as well as tactile sensitivity and pain sensitivity. Before testing, the animal is habituated to the testing environment. A good environment is a wire mesh cage or an elevated platform, either of which allows access to all parts of the animal's body. When the animal is still, specific locations on the skin can probed with a monofilament. von Frey monofilaments are calibrated to exert a force ranging from 5 to approximately 178 g/mm². The threshold force for each monofilament is the point at which it bends. The animal's responses of withdrawal or orienting toward the stimulus are recorded. Several scoring systems have been developed that allow categorizing the animals' reaction (e.g., see Marshall et al., 1971).

There are two ways to use von Frey hairs. One way is to present different strengths of monofilaments to one particular location of the skin to evaluate the threshold at which the animal starts responding to the stimulation of a part of the body. The other way is to present one monofilament to multiple regions of the body to topographically map the segments of the body (dermatomes) to tactile stimulation. This approach is useful when determining the severity and level of spinal lesions (Takahashi et al., 1995). The von Frey hair pinch test has been used in rat models of spinal cord injury, Parkinson's disease, and stroke and is sensitive to both spinal and supraspinal functions.

THE FORMALIN PAIN TEST

To assess analgesia (absence of pain perception) or hyperalgesia (increased pain perception) to mechanical, thermal, and chemical stimulation, several standard pain tests have been developed. Chemically induced pain is commonly assessed using formalin (Dubuisson and Dennis, 1977). In the formalin pain test, animals are injected with a 3% or 5% formalin solution subcutaneously into the dorsal surface of the hindpaw. Behavior is monitored

for a time interval of up to 60 minutes after the injection. There are two phases in the animals' response to the injection: an early phase and a late phase. The number of paw flinches per time interval and paw licking are easily quantifiable responses.

There are several confounding factors that can modify the response to formalin, including ambient temperature, stress, and number of treatments. Different strains of animals have a different stress response, thus restricting the validity of comparisons between groups of animals (Ramos et al., 2002). It is noteworthy that repeated exposure to formalin might influence the animals' responsivity to other behavioral manipulations and subsequent tests (Sorg et al., 2001).

HOT AND COLD TEMPERATURE TEST

Hot or cold plate tests are assessments of both loss of sensory perception and plasticity-induced changes in pain thresholds. Temperature and pain senses are mainly mediated via the spinothalamic, spinoreticular, and spinomesencephalic tracts, also known as the anterolateral system. These tracts decussate at the spinal level and ascend in the dorsal column to the contralateral thalamus and the somatic sensory cortex. Hot and cold tests make use of a range of temperatures to induce thermal sensations and induce a withdrawal response. A preferred apparatus is a heavy copper plate with heating or cooling coils distributed equally below its surface, so that the entire plate is at a constant temperature.

One form of the test involves placing an animal on a heated plate that is maintained at a temperature between 45° to 55° C or a cold plate of 4° C. Another form of the test involves briefly exposing the animal to a plate set at specific temperatures (e.g., an ascending or a descending series) to determine the "threshold" at which an animal makes a response. Responses include orientation to the site of the stimulation, flexion withdrawal reflexes, licking of the paw, and a generalized escape or attack response. These responses are transient and occur only during application of the stimulus. The latency for appearance of heat avoidance behavior (orientation, lifting and licking the paw, withdrawal) and the number of responses per time interval can be recorded. To avoid injury, the trial in the hot plate test should be terminated quickly after an animal responds.

Some commercial heat tests use infrared beams that can be selectively pointed to a specific part of the body, e.g., plantar foot or tail. Sensitivity thresholds to the heat are obtained by gradually increasing the temperature (Almasi et al., 2003).

The measures of withdrawal responses on a hot or cold plate can be used to screen analgesic effects of drugs and substances of abuse. Furthermore, they are part of standard test batteries used to evaluate the extent of spinal cord injury and supraspinal lesions. Structural rearrangements after a localized tissue injury or drug treatment might alter the pain threshold and excitability and thus change the latency of the initial withdrawal response. Using the hot or cold plate tests, malfunctional plasticity and abnormal processing that contribute to the development of neurogenic pain can be detected (e.g., Woolf et al., 1992).

STICKY DOT TEST

When adhesive-backed labels are attached to the paw of a normal rat, the animal orients toward the stimulus and removes it using its teeth. The latency and asymmetry in the behavioral response can be used to assess somatosensory function (Schallert and Whishaw, 1984). To determine the presence of a somatosensory asymmetry, an animal receives adhesive stimuli (usually round sticky dots of approximately 12 mm diameter) attached to the distal-radial aspect of each forelimb. The animal then is returned to a familiar cage and observed. After being replaced in its cage, a rat quickly contacts and removes each label from

the forelimbs by using the teeth. The dots are removed one at a time, and the latencies to contact and remove each stimulus, and the order of removal (left or right first), are recorded over several trials (Schallert et al., 2000). The order of contact and removal reflects whether the animal shows a bias. After a unilateral lesion affecting a forelimb, the animal may first contact and remove the dot from the good limb (the limb unaffected by the injury). The animal later orients toward the dot on the bad forelimb and may take more time to remove that dot. If an animal shows a somatosensory asymmetry in the majority of the trials, the magnitude of the asymmetry can be assessed in a follow-up test (Schallert et al., 2000). In this test, the size of the dot on the bad limb is progressively increased while the size of the dot on the good limb is decreased by an equal amount. If the stimulus on the bad limb becomes more salient than the stimulus on the good limb, the animal starts contacting and removing the stimulus from the bad limb first. Studies using this method have shown that the ratio sufficient to reverse the initial bias is related to the severity of brain damage (Barth et al., 1990).

The sticky dot test has been shown to reflect sensory impairments after unilateral lesions involving the sensorimotor cortex, lateral hypothalamus, and subcortical lesions such as pyramidal tract transection and striatal lesions (Schallert et al., 2000). A critical issue in lesion studies is that the damage of sensorimotor areas also results in impairments of motor function. Motor impairments are likely to lead to longer latency to remove the dot due to deficits in using the mouth and forelimb to remove the sticky dot. This criticism might be resolved by recording the contact time in addition to the removal time. The contact does not require the skill required to remove the dot.

LIMB USE ASYMMETRY (CYLINDER) TEST

Rats explore both horizontal and vertical surfaces (Gharbawie et al., 2004). When exploring a vertical surface, the animal lifts its forequarters to support itself as it investigates. When bracing against the wall, either one or both forelimbs are used. A normal animal typically uses both forelimbs equally often for support, but a unilateral lesion may bias the animal to prefer one limb (Schallert et al., 2000).

The cylinder apparatus can be placed either on a transparent bottom to allow video-recording from underneath the cylinder or on a table surface with a mirror arranged at an angle behind the cylinder to allow observation of the behavior from all directions. Three main categories of forelimb use are analyzed: lifting, moving along the wall, and landing. The forelimb used is recorded as the independent use of the left versus the right limb, or as the simultaneous use of both forelimbs. The standard measure is the number of left and right forelimb uses calculated as percentage of the total number of contacts (Schallert et al., 2000). Unilateral lesions that affect limb use reduce the use of the impaired forelimb. Interrater reliability for the test is found to be high (Schallert et al., 2000). Nevertheless, the sensitivity of this test to discrete motor disturbances might be reduced by the development of compensatory strategies, which could be misleadingly interpreted as recovery.

POSTURE AND IMMOBILITY

Animals spend a great deal of waking time partly or completely immobile. The posture of immobility is not always the same, however, and certain postures can be considered pathological. The initiation of movement from an immobile state can also reveal abnormalities in posture and body support. Table 44–4 summarizes several gross measures of posture and immobility. One test, the righting response, is described here.

Table 44–4. Examination Strategies of Posture and Immobility

Immobility and movement with posture	Animals usually have postural support when they move about, and they maintain posture when they stand still and remain still while rearing. Posture and movement can be dissociated: in states of catalepsy, postural support is retained while movement is lost.
Immobility and movement without posture	An animal has posture only with limb movement. When a limb is still, the animal collapses, unable to maintain posture when still. When still, the animal remains alert but has no posture, a condition termed *cataplexy*.
Movement and immobility of body parts	Mobility and immobility of body parts can be examined by placing a limb in an awkward posture or placing it on an object such as a bottle stopper and timing how long it takes an animal to move it.
Restraint-induced immobility	Restraint-induced immobility, also called tonic immobility or hypnosis, is induced by placing an animal in an awkward position, such as on its back. The time it remains in such a position is typically measured. Animals will maintain awkward positions while maintaining body tone or when body tone is absent. During tonic immobility, animals are usually awake.
Righting responses	Supporting, righting, placing, and hopping reactions are used to maintain a quadrupedal posture. When placed on side or back or dropped in a supine or prone position, adjustments are made to regain a quadrupedal position. Righting responses are mediated by tactile, proprioceptive, vestibular, and visual reflexes.
Environmental influences on immobility	Feeding fatigue potentiates immobility. Warning induces heat loss postures, such as sprawling, and thus potentiates immobility without tone. Cooling induces heat gain posture with shivering and thus potentiates immobility with muscle tone.

RIGHTING RESPONSES

Postural control is necessary for all types of motor performance. Postural adjustments depend on the position of the center of gravity in the body. In rats, the mechanics and control mechanisms for postural maintenance when standing or walking are relatively simple as compared to bipeds. When placed in a position of unstable equilibrium or when responding to a passive displacement of their limbs, rats attempt to maintain a prone quadrupedal body position in relation to the center of gravity of the body.

The magnitude and ability of righting responses in animals reflect their ability for sensorimotor integration. Sensory disruption of proprioceptive, vestibular, tactile, or visual input can disturb the position of the body and body parts in space. In turn, difficulties in responding to the sensory stimulus of limb or body displacement may be caused by the inability to recruit the appropriate musculature.

A number of tests for righting responses measure specific aspects of sensorimotor integration. The righting response most commonly tested requires that an animal quickly regain its quadrupedal position when placed on its side or back (stationary placing response). A normal animal first adjusts its head position by dorsiflexing the head and neck and then turns its forequarters of the body and later the hindquarters to right itself. When dropped facing upside down from the height of less than 1 meter onto a cushion, the animal will quickly right itself and adjust its posture to land on all four feet (acceleratory placing response; Pellis and Pellis, 1994). This righting response is mainly dependent on vestibular and visual cues. The sensory and motor aspects of the righting response are quantifiable by the use of limbs and axial musculature, completeness of righting, and the latency to regain a prone body position.

Because righting responses are easy to assess in laboratory rodents and do not require pretraining, they are often part of standard test batteries in neurotoxicology and pharmacology. Furthermore, it has been found that righting in the air occurs at a defined time point during development (Hard and Larsson, 1975), and so righting responses can be reliable indicators of maturation of the vestibular system. Righting responses also provide insight into processing mechanisms of motor control. For instance, an established phenomenon is an abnormality in regaining a prone body position after lesions of the dopaminergic system (Martens et al., 1996). Although righting responses are usually fast and require high-speed videorecording, they provide reliable measures of sensory and motor system function.

LOCOMOTION

Analysis of locomotion includes observation of rats as they walk, run, jump, turn, and swim (Table 44–5). Virtually any manipulation of the brain will affect at least one of these measures, although the effects are often subtle and may require analysis of slow motion videotapes. Locomotion can be measured with photocells, running wheels, or automated video tracking systems and by measures of ground reaction forces. Ultimately, a refined description can be obtained by using a movement notation system to document the movement of every joint and limb. One example is a system devised by Eshkol and Wachmann (1958) for classical ballet and later adapted for describing animal movements by Golani (1976). In addition, there are standardized tests of swimming, circadian rhythms, and there are composite tests such as the BBB scale, widely used for spinal cord injury.

SWIMMING

The swimming test takes advantage of the fact that rats are semiaquatic and thus are proficient swimmers. When rats swim, only the hind limbs are used for propulsion, while

Table 44–5. Tests of Locomotion

General activity	Included are videorecording, movement sensors, activity wheels, and open field tests.
Movement initiation	The warm-up effect: Movements are initiated in a rostral-caudal sequence, small movements precede large movements, and lateral movements precede forward movements, which precede vertical movements.
Turning and climbing	Components of movements can be captured by placing animals in cages, alleys, tunnels, etc.
Walking and swimming	Rodents have distinctive walking and swimming patterns. Rats and mice walk by moving limbs in diagonal couplets with a forelimb leading a contralateral hindlimb. They swim using the hindlimbs with the forelimbs held beneath the chin to assist in steering.
Exploratory activity	Rodents select a home base as their center of exploration, where they turn and groom, and make excursions of increasing distance from the home base. Outward trips are slow and involve numerous pauses and rears while return trips are more rapid.
Circadian activity	Most rodents are nocturnal and are more active in the night portion of their cycle. Peak activity typically occurs at the beginning and the end of the night portion of the cycle. Embedded within the circadian cycle are more rapid cycles of eating and drinking, especially during the night portion of the circadian cycle.

the forelimbs are held immobile underneath the chin or are tilted for steering. Although the ability to swim is rarely abolished by experimental manipulations, the performance of swimming movements changes during development and aging, after certain types of brain damage, and after some drug treatments. Most of these manipulations lead to disinhibition of the forelimb use when swimming in that animals make rhythmic strokes with their forepaws. Such changes in swimming have been described after lesions of the cortex, posterior hypothalamus, and cerebellum (Kolb and Whishaw, 1983; Whishaw et al., 1983).

The swim test requires a rectangular, transparent swimming pool with a water level of approximately 30 cm. A visible platform is placed on one end of the pool. The rat is introduced on the other end of the pool and quickly learns to swim to the platform (Stoltz et al., 1999). A few pretraining sessions before testing might be helpful to habituate the animal to the procedure. To analyze limb use, an animal's performance is videorecorded from a side view. The tapes can be analyzed on a frame-by-frame basis to count the number of forelimb strokes per limb. If the effects of uni-

lateral lesions are being assessed, the intact side serves as a control for the lesion side. Bilateral lesions, aging, or drug studies might impair both sides, so that data from control animals or preoperative test sessions will be necessary as a comparison.

CIRCADIAN ACTIVITY

The periodic changes of light intensity during night and day serve as a salient natural Zeitgeber in mammals. The light cycle regulates a variety of body functions, including cyclic behavior such as sleep and wakefulness, and physiological and biochemical processes, including body temperature and hormonal changes. The circadian clock assists in organizing the time course of these processes, thus optimizing the organism's performance in anticipating the rhythmic change of environmental conditions (Holzberg and Albrecht, 2003).

Circadian activity is usually tested by recording general activity in the home cage over one day and night cycle, but it can also be tested in constant light or in constant dark. Normal rats are active during the night portion of a cycle even if the lights are on or off during that portion of the cycle. Testing circadian activity requires a separate room with lighting regulated by a timer. The activity of animals over a 24-hour time interval is usually recorded by the interruption of photobeams mounted on the walls of the home cage or by the number of turns of a running wheel that is provided in the cage. The number of interruptions of photobeams or number of turns of the running wheel are summarized by a computer system that analyzes changes in the activity. Furthermore, the use of photobeams allows the distinction between locomotion and stereotyped movements. Successive interruptions of different beams reflect the animal's locomotion from one point to another, and successive interruptions of a specific beam indicate stereotyped movements such as grooming or circling.

The circadian cycle of rodents usually shows a burst of activity when the lights turn off followed by bouts of activity throughout the dark period, and another burst just before light turns on. During the light period, the animals are generally inactive. Animals that are placed in a new environment also show increased activity during the first hour when exploring the new environment. Detailed analyses of activity can reveal differences in experimental groups depending on environmental conditions such as external light cycles and feeding schedules, physiological conditions such as stress, aging, drugs, or brain dysfunctions (Weinert, 2000). Moreover, the computerized system makes it easy to record overall changes in activity after various manipulations and lesions to the brain.

BBB LOCOMOTOR RANKING SCALE

The BBB rating system, developed by Basso, Beattie, and Bresnahan (1995, 1996b), is used to rate the degree of paralysis in rat spinal cord injury models. The BBB scale is a widely used measure of spinal cord integrity. The integrity of spinal function is reflected in interlimb coordination, alternating activity of limb muscles and rotational paw position, and posture. Lesions of the spinal cord usually not only affect movement but also interrupt descending and ascending connections necessary to walk on difficult territory.

The 21-point BBB scale is based on a five-category score originally developed by Tarlov (1954) and has been refined to display discrete impairments of locomotion. The BBB scale rates the magnitude of limb movement, position of trunk and abdomen, paw placement and stepping positions, coordination, toe clearance, paw rotation, trunk stability, and tail position (Table 44–6). The 21 categories of the rating scale allow conclusions about trunk stability, weight support, and stepping and thus is tailored to the aim of preclinical studies in spinal cord injury.

Animals spontaneously explore a novel environment, and so no pretraining or other motivation is ordinarily required to test animals. There are, however, a few cautionary notes regarding the testing. For example, frequent repetition of test sessions leads to ha-

Table 44–6. Categories and Gait Features Assessed by the BBB Scale

Limb Movement	Trunk Position	Abdomen	Paw Placement	Stepping	Coordination	Toe Clearance	Paw Position	Trunk Stability	Tail
None	Side or mid	Drag	Sweep or	Dorsal or	Never	Never	Initial contact	Absent or	Up
Slight		Parallel	supported	plantar	Occasional	Occasional		Present	
					Frequent	Frequent			
Extensive		High			Consistent	Consistent	Lift off		Down

bituation of the animals to the testing environment and can reduce the exploratory activity. Stepping patterns of paralyzed animals also can be modified by textured surfaces, likely because of reflexive limb movements.

The BBB scale is a well-standardized scoring system that has been validated for spinal cord injury in rats in large-sample studies (Basso et al., 1996a); the results have been shown to correlate closely with results from other motor tasks (Metz et al., 2000). The behavioral abnormalities are directly related to the loss of spinal tissue (Basso et al., 1996b), and the scale was recently adapted for the use in mouse models (Ma et al., 2001), thus strengthening the applicability of the BBB scale as a multipurpose tool.

SKILLED MOVEMENTS

Skilled movements are voluntary movements that require irregular motor patterns, rotatory movement and movements that consist of a complex sequence of movements, or movements that counteract movements that normally move the body against gravity. These movements are likely mediated by neural mechanisms that are at least partly independent of those that support locomotion. Therefore the quantification of skilled movements can provide insights into neural function that are at least independent of those obtained from the study of locomotion. At its simplest, the use of skilled movements can be observed as

an animal eats laboratory chow, specialty food items such as sunflower seeds, which need to be shelled, and prey items, such as crickets, which have to be caught and prepared for eating. The movements used in these activities may be difficult to score objectively, and so many researchers prefer more formal tests (Table 44–7). Formal tests have the advantage in that the same movement is performed repeatedly and therefore can be scored with end point measures or qualitative measures. Tray reaching, single pellet reaching, and rung walking are examples of such skilled movement tasks that assess forelimb function. Each test assesses specific forelimb abilities, such as extending the forelimb, aiming and grasping, or applying force to retrieve an object (Whishaw et al., 1986; Montoya et al., 1991; Ballermann et al., 2001). The rung-walking task assesses hindlimb function as well as forelimb function.

TRAY-REACHING TASK

The tray-reaching task is a simple test of forelimb use (Whishaw et al., 1986). The reaching box consists of three solid walls and a front wall made of thin vertical metal bars to allow the animals to extend the forelimb through the full width of the front wall. Mounted on the outside of the front wall is a tray that is matched to the width of the wall. The tray can be filled with food, such as chicken feed or small food pellets, so animals can reach through the bars and retrieve food from any angle and position in the box. The box is

Table 44–7. Skilled Movements and Test Methods

Limb movements	Examples include bar pressing, reaching and retrieving food through a slot, spontaneous food handling of objects or nesting material, and limb movements used in fur grooming and social behavior. Rodents use limb movements that are order-typical and species-typical.
Climbing and jumping	Examples include movements of climbing up a screen, rope, ladder, etc. and jumping from one base of support to another.
Oral movements	Examples include mouth and tongue movements in acceptance or rejection of food such as spitting food out or grasping and ingesting food, as well as movements used in grooming, cleaning pups, nest building, and teeth and claw cutting.

mounted on a grid floor, so that lost and dropped food items fall through.

Training requires moderate food deprivation. The training procedure is relatively simple because there is no need to replace food items or handle the animals while they are being tested. Training sessions last about 30 minutes, but test sessions can be as brief as 5 minutes, during which performance can be videorecorded for analysis. Animals can also live in the test apparatus, should that be required. There are two main methods for analysis of limb use and reaching performance. First, animals can be allowed to use either forelimb and the preference of one limb over the other can be evaluated. Thus, when unilateral lesions are assessed, the ratio of the use of the good limb versus the bad limb reveals the degree of limb use asymmetry. Second, to measure success rates of one limb exclusively, the use of the other limb can be prevented with a cuff made from adhesive fabric tape wrapped around the distal aspect of the limb (Whishaw and Miklyaeva, 1996).

For end point measures of success, a *reach* is defined as the insertion of a forepaw through the bars of the cage, and a *successful reach* is defined as a movement that obtains food and brings it to the mouth for consumption. Performance is then described as hit percent: the number total reaches divided by successful reaches. Skilled forelimb movements in this task have been shown to be chronically affected by motor cortex lesion (Kolb et al., 1997) or corticospinal tract transection (Whishaw and Metz, 2002).

PELLET REACHING TASK

A sensitive test for fine motor control and bodily supporting adjustments in rats is the single pellet reaching task (Whishaw et al., 1991). The task is designed so that not only can measures of success be recorded but also behavior can be filmed and scored using standard scoring procedures. The pellet-reaching task requires that the animal extends its fore-limb through a slit to aim the forelimb to a single target, grasp a food pellet located on a shelf in front of the slit, for eating. Animals require 1 or 2 weeks of training to habituate to the apparatus and optimize the success of reaching movements. From videorecorded performance, a large variety of measures can be obtained from a single test session such as end point measures of reaching success and descriptive movement analysis of the limb and body movement components (Whishaw et al., 1991). The sequence and performance of the components are relatively fixed so that the ability to modify the elements of the reaching movement to adapt to context is limited. Animals with brain damage might return to baseline levels in their success rates, but the qualitative analysis of reaching movements will reveal whether the movement strategy is permanently changed.

In addition to the ample information about the organization of the motor system provided by the pellet reaching task, its relevance for investigations in animal models is established. Comparisons of rats with humans (Whishaw et al., 1992) demonstrates a homology between the reaching movements in the two species. Moreover, reaching movement abnormalities are similar between rodent models of human neurodegenerative disease and human patients (Whishaw et al., 2002). Performance of reaching movements can be compromised by even subtle brain damage that would spare coordinated movements such as locomotion, swimming, or grooming. Discrete lesions of the motor system, including the motor cortex and its efferent corticospinal tract, basal ganglia, and red nucleus cause recognizable disturbances of reaching movements (Whishaw et al., 1986; Metz and Whishaw, 2002a).

RUNG WALKING

Recent investigations also revealed that rats use skilled hindlimb movements to adapt their gait pattern to a difficult territory. A simple and sensitive test is the rung-walking task (Metz and Whishaw, 2002b). The rung-walking task resembles a horizontal ladder with rungs that can be adjusted individually. The rungs can be arranged in a regular pattern that allows the animals to anticipate the location of a rung or learn a specific sequence of patterns or in an irregular configuration to prevent animals from learning a specific pattern. Therefore, changing the pattern allows repeated testing sessions.

Animals can easily be trained to cross the horizontal ladder to reach their home cage. Performance is videorecorded for further analysis of end point measures and qualitative movement analysis. End point measures include the number of errors in placing a limb when crossing the beam (number of errors per step). The number of limb-placement errors also increases in the intact limbs, thus revealing compensatory adjustments. The error counts can be supplemented by describing the type of an error using a scoring system that allows a distinction between severe and mild errors and limb positioning on the rung (Table 44–8). Furthermore, the behavioral analysis can be supplemented by other techniques, such as electrophysiological recordings of muscle activity (Merkler et al., 2000).

The rung-walking test is sensitive to chronic movement deficits after adult and neonate lesions to the motor system and to physiological variables such as aging or stress (Metz et al., 2001).

SPECIES-TYPICAL BEHAVIORS

One of the primary functions of the brain is to produce behaviors that allow animals to adapt to their environment with a minimum of learning and its associated error. A number of behaviors that are mainly innate, relatively stereotyped, and characteristic of a species are called *species-typical behaviors*. Table 44–9 presents a catalogue of rat species-typical behaviors and their description. The following section provides some examples of tests of species-typical behaviors that can be used in a general analysis of behavioral phenotype.

FOOD HOARDING

Rats are in danger of predation whenever they leave their home base to search for food. A behavior that they use to minimize risk is to eat smaller pieces of food immediately and to carry larger pieces of food (i.e., food items that take more than a few seconds to consume) to a safe location for later consumption. Usually, this location is the nest or a home box in the laboratory. The rule that they use is that if food takes longer to consume than the round trip to the refuge, the exposure can be minimized by food carrying. Rats will not only carry food to a shelter to eat; they also carry and store food when they are not hungry. This behavior indicates that they place an incentive value on food.

Table 44–8. The Seven Categories of Limb Placement That Can Be Rated for Limb Placement

Category	Type of Foot Placement	Characteristics
0	Total miss	Deep fall after foot missed the rung
1	Deep slip	Deep fall after foot missed the rung OR slight fall after foot missed the rung
2	Slight slip	Slight fall after foot slipped off the rung
3	Replacement	Foot replaced from one rung to another
4	Correction	Foot aimed for one rung but was placed on another OR foot position on same rung was corrected
5	Wrist or digits	Foot placed on rung with either digits/toes or wrist/heel
6	Correct placement	Foot placed on rung in plantar fashion with full weight support

Table 44–9. Species-Typical Behaviors and Their Assessment

Grooming	Grooming movements are species-distinctive and are used for cleaning and temperature regulation. Begin with movements of paw cleaning and proceed through face washing, body cleaning, and limb and tail cleaning.
Food foraging/hoarding	Food carrying movements are species-distinctive and used for transporting food to shelter for eating, scattering food throughout a territory, or storing food in depots. Size of food, time required to eat, difficulty of terrain, and presence of predators influence carrying behavior. Both mouth-carrying or cheek-pouching are used by different species. Rodents also engage in food wrenching, in which food is stolen from a conspecific, and dodging, in which the victim protects food by evading the robber.
Eating	Incisors are used for grasping and biting, rear teeth are used for chewing, tongue is used for food manipulation and drinking.
Exploration/neophobia	Species vary in responses to novel territories and objects. Objects are explored visually or with olfaction, avoided, or buried. Spaces are explored by slow excursions into space and quick returns to a starting point. Spaces are subdivided into home bases, familiar territories, and boundaries.
Foraging and diet selection	Food preferences are based on size and eating time of food, nutritive value, taste, and familiarity. For colony species the colony is an information source with acceptable foods identified by smelling and licking snout of conspecifics.
Sleep	Rodents display all typical aspects of sleep including resting, napping, quiet sleep and rapid eye movement sleep. Most rodents are nocturnal, thus sleeping during the day with major activity periods occurring at sun up and sun down. Cycles in natural habitats vary widely with seasons.
Nest-building	Different species are nestbuilders, tunnel builders, and build nests for small family groups or large colonies. All kinds of objects are carried, manipulated, and shredded for nesting material.
Maternal behavior	Laboratory rodents typically have large litters that are immature when born. Pups are fed for the first 2 to 3 weeks of life and thereafter become independent.
Social behavior	Colony or family rodents have rich social relations including territorial defense, social hierarchies, family groupings, and greeting behaviors. Solitary rodents may have simplified social patterns. Defensive and attack behavior in males and females is distinctive.
Sexual behavior	Characteristic sexual behavior displayed by males and females. Males display territorial control or territory invasion, and engage in courtship and often group sexual behavior. Sexual behavior is often long-lasting with many bouts of chasing, mounting and intromission, and incidents of ejaculation. Mounting is followed by genital grooming and intromission is followed by immobility and high frequency vocalizations. Females engage in soliciting including approaches and darting, pauses and ear wiggling, and dodging and lordosis to facilitate male mounting.
Play behavior	Many rodents have rich play behavior with the highest incidence in the juvenile period. Play typically consists of attack in which snout-to-neck contact is the objective and defense in which the neck is protected.

Food-hoarding behavior can be tested in any apparatus in which an animal has a refuge and there is an open area that contains food. For example, a two-compartment box with one compartment illuminated more than the other or a home base attached to an alley provides a simple test enclosure. Food hoarding has also been used in structured test situations, including those in which an animal has to solve a maze puzzle or learn a spatial problem to find food.

The neural basis of food hoarding is not fully understood, but it is disrupted by frontal decortication, even though individual components of hoarding behavior may still be intact (Kolb, 1974). Food-hoarding behavior is disrupted after mesolimbic dopamine depletion (Stam et al., 1989), indicating that the dopaminergic projection to the prefrontal cortex plays a central role in structuring this behavior. Furthermore, rats with limbic system lesion, including hippocampal damage, show a disruption of food-carrying behavior unless stimulated by external signals (Whishaw, 1993). Other deficits found in food-hoarding tasks point toward a deficit in spatial navigation and acquiring problem-solving strategies. Thus, food-hoarding tasks are versatile tests in revealing sensory and cognitive aspects of natural behavior.

EXPLORATION

When rats are removed from their home cage and placed in an unfamiliar environment, they display structure in their exploratory behavior (Eilam and Golani, 1989). The rat will establish a home base at the location at which it was first placed or at a shelter provided. It will start exploring its environment and return to the home base to pause, rear, and groom. As it begins to explore, it will stretch its forequarter and head to examine the area surrounding the home base. It will then start to explore more remote areas by making trips away from the home base. The animal will travel mostly along a wall or the edge of the open field, and it will frequently return directly and quickly to the home base. Return trips are usually shorter than the outward-bound trips. The outward trips will become gradually longer until the entire area has been explored, and the animal might also adopt a new home base over the course of its exploration.

Although scoring the structure in exploratory behavior is laborious, computer-based programs are being developed. In addition, an interval of 5 or 10 minutes can provide ample data including the number of trips and stops, number of rears, the frequency of grooming, duration of trips, and kinematic measures of speed and path trajectories. Open field tests are used in test batteries because the behavior is easy to elicit and is reliable because pretraining or habituation is not necessary. Exploratory tests have proved useful in studies of the effects of drugs, brain damage, or physiological changes. Common neurological test batteries include the assessment of anxiety in the open field, which is indicated by exploratory activity, the time spent in peripheral compartments versus time spent in the center of the open field, the latency to enter a novel home base, and the habituation to the novel aspects of the field.

PLAY BEHAVIOR

Play is the most prominent form of social interaction during the juvenile stage of life; however, rats show play behavior at any stage of their development and in adulthood. Play consists of a ritualized sequence of movements. The individual components of play in rodents are often part of other behaviors such as aggression, predation, and sex. The complex movement sequences of play can be separated into the individual movement components, and each sequence is characteristic for a specific type of play. For instance, play fighting is marked by the attempt to repeatedly attack the nape area of a recipient with the snout, while the recipient attempts to avoid the contact. Because play fighting is the most reported form of social play and persists into adulthood, it is suitable as a measure of social interaction in rats.

Play fighting movements in juvenile and adult rats offer a variety of options for behavioral analysis. Although the defensive tactics undergo changes during development, the sequence of individual acts in play fighting is relatively fixed in both the initiator and the recipient. The major categories of play fighting are chasing, dodging, wrestling, and tumbling, all of which occur in play fighting in rats. The individual movement components of play fighting are easily recognizable. The relationship between the two interacting individuals or individual motor acts can be analyzed by the means of detailed descriptive analysis systems (Pellis and Pellis, 1983).

Play behavior has a cognitive component, and it is thought to serve gathering social and emotional information about conspecifics. In line with this theory is the finding that the complex sequences of play behavior can be disturbed or lost by neurologic conditions (Pellis et al., 1993). Furthermore, play behavior has been described as being a measure of the ability to initiate or respond to social contacts (Daenen et al., 2002). In summary, the analysis of play behavior represents a valuable measure of motor and cognitive function involved in social interaction.

LEARNING

Psychologists have been studying learning in rats for at least 100 years, and there are literally hundreds of different learning tests available in the literature (Table 44–10). The choice of learning tests will depend on the nature of the question being asked by the investigator, but a few simple tasks can provide considerable information about the cognitive status of an animal. Learning tests reveal at least four insights into behavior: (1) Can an animal master the procedures that the task requires? (2) What neural system is an animal using in performing/learning a task? (3) Is learning normal? (4) What is the structure of learning behavior? Investigators who screen animals for the effects of pharmacological manipulations or genetic influences will mainly be interested in the first three questions, whereas students of learning are likely to be interested in the fourth question as well.

Table 44–10. Learning and Measures of Learning

Classic conditioning	Unconditioned stimuli are paired with conditioned stimuli and the strength of an unconditioned response to the conditioned stimuli is measured. Almost any arrangement of stimuli, environments, treatment, or behavior can be used.
Instrumental conditioning	Animals are reinforced for performing motor acts such as running, jumping, sitting still, lever pressing, or opening puzzle latches.
Avoidance learning	Passive responses including avoiding preferred places or objects which have been associated with noxious stimuli such as electric shock. Active responses including moving away from noxious items or burying noxious items.
Object recognition	Including simple and recognition of one or more objects, matching to sample, and nonmatching to sample in any sensory modality. Tasks are formal in which an animal makes an instrumental response of inferential in which recognition is inferred from exploratory behavior.
Spatial learning	Dry land– and water-based tasks are used. Spatial tasks can be solved using *allothetic cues*, which are external and relatively independent to movements, or *idiothetic cues*, which include cues from vestibular or proprioceptive systems, reafference from movement commands, or sensory flow produced by movements themselves. Animals are required to move to/away from locations. *Cue tasks* require responding to a detectable cue. *Place tasks* require moving by using the relationships between a number of cues, no one of which is essential. *Matching tasks* require learning a response based on a single information trial.
Memory	Memory includes *procedural memory* in which response and cues remain constant from trial to trial. Tasks are constructed to measure one or both types of learning. Memory is typically divided into object, emotional, and spatial, and each category can be further subdivided into sensory and motor memory.

SWIMMING POOL TASKS

Some of the most popular tasks that provide information about place learning ability as well as procedural and working memory in animal studies are swimming pool spatial tasks (Morris et al., 1982). These tasks are especially useful in rats because they are semi-aquatic. Numerous modifications of this test have been developed, however, and the following will focus on the basic procedure that applies to all versions of the water task. A training or test session begins with introducing an animal to a round swimming pool that contains a hidden platform. The goal is for the rat to discover and to localize the platform and escape from the water. By using skim milk powder, the water is made opaque so that the animal is unable to see the platform submerged about 1 cm underneath the water surface. Over consecutive trials, the time to find the platform decreases as the animal learns to swim directly to the invisible platform with respect to distal cues surrounding the swimming pool. The animal's performance is measured as the time it takes to find the platform (escape latency), the distance traveled (swim distance), and the accuracy in targeting the platform over consecutive trials.

The procedural simplicity of the water task is opposed to the complex underlying processes that determine performance of this task. The task requires a variety of behavioral processes, including navigation strategy formation, place learning, memory, and the performance of visually guided behavior (Cain and Saucier, 1996; D'Hooge and De Deyn, 2001). Various aspects of spatial learning can be assessed using the following procedures:

1. *Procedural learning.* Procedural learning involves evaluating whether an animal can acquire the skills necessary to escape from the water. The water is tepid, and a platform is hidden at a fixed location with its surface 1 cm below the surface of the water. The cardinal compass points serve as starting points, and a rat swims until it finds the platform or until 60 seconds has elapsed, at which pint it is removed from the water. Two trials are administered each day for a total of 5 days. Latency to reach the platform, distance swum, and head direction can all be measured using commercial tracking systems.

2. *Matching-to-place learning.* If an animal can acquire the procedural aspects of the task, it can be tested for spatial learning ability using a matching-to-place version of the task. Each day the platform is moved to a new location, for a total of 5 to 7 days. Each day, the animal receives two swims from the same location—a sample swim in which it has to locate the platform at its new location and a matching swim—in which it demonstrates that it has learned the new location on the sample trial. Rats quickly become adept in learning new locations in a single trial. Typically, animals have long latencies on the first trial, because they display a win-stay behavior and search for the platform at its old location, and a short second trial latency, because they can learn the new location in a single trial.

3. *Cue learning.* If an animal cannot acquire the procedural aspects of the task, it can be presented with trials on which the platform is visible and so serves as a cue or beacon. A cue trial procedure is used to demonstrate that an animal can see, swim, and escape.

Many variations on these basic procedures are used, with major variations being the water temperature, size of the apparatus, number of swimming trials per day, and so on. The major advantages of the task are that no special deprivation is required to motivate the animals, so testing can proceed quickly. Consequently, a large number of studies have confirmed that acquisition, retention, and reversal of navigational strategies in the water maze involve a number of brain regions and neurochemical systems, each affecting specific parameters of performance in the water maze. Spatial learning has been related to the integration of hippocampal CA1 and CA3 regions, and hippocampal connectivity to nucleus accumbens and raphe nuclei via the fimbria fornix (Whishaw, 1987). Thus, interruption of hippocampal projections leads to deficits in the acquisition of new platform positions or new distal cues and to impairments in retention of previously learned information. Other brain structures influencing water task performance include the prefrontal cortex, striatum, cerebellum, and various neurochemical and neuromodulator systems (for reviews, see McNamara and Skelton, 1993; D'Hooge and De Deyn, 2001). It is worth noting that the number of structures sensitive to the task require caution in interpreting results. Animals may be impaired at the task for many reasons, only one of which is an impairment in learning or memory, per se.

Finally, it is noteworthy that a virtual water task for use in humans has been developed that could replicate the basic findings from animal studies (Hamilton et al., 2002). Thus, not only is the water maze a flexible test for animal studies but also the methodology is directly applicable to human spatial behavior.

RADIAL ARM MAZE

The radial arm maze analyzes spatial navigation strategies in rodents in that the animal is required to learn the location of food at one or more locations. The radial arm maze consists of a central platform from which a number of arms originate (Olton et al., 1979). Most commonly, eight arms are arranged around an octagonal central platform in equal spaces. Access to the arms can be controlled by guillotine doors that can be opened or closed for each trial. The maze is usually located in a room with salient visual cues such as posters on the walls, counters, cupboards, etc. The location of the arms is either fixed or flexible and can be marked by a cue on the arm such as a light or color. Food is located in an indentation at the farthest end of one or more arms. Before the start of an experiment, the animals must be food deprived and habituated to the food reward provided in the test. The task of the animal is to learn the location of the food over a number of trials. When using multiple locations of food, the animal has to

learn a sequential response strategy to pick up the food. By using doors to control access to the arms, the retention of a learned strategy can be tested by leaving the doors closed for a certain period of time before a new trial starts.

There are a number of standard training and testing procedure protocols available, with each focusing on specific aspects of procedural memory and retention (Jarrard, 1983). A common test strategy in the radial arm maze is to expose a rat to a habituation phase in which all arms are baited. In consecutive trials, the animal learns a strategy of entering the arms to collect all food items. After a few days, the training phase begins in which only four arms are baited. When the animal is released from the center platform, it is required to obtain all four food rewards within a certain time period. Over a number of test sessions, the same arms remain baited and the number of visits of every arm are recorded. In this arrangement, working memory is reflected by an animal visiting each of the four baited arms only once. Reference memory is reflected by an animal not visiting any of the four arms that has never been baited before.

The radial arm maze is especially suited to evaluate the ability to form a procedural memory of the task that might be interrupted by treatments or brain lesions. Like the water based tasks, the radial arm task is sensitive to damage of many brain structures. What distinguishes the swimming pool task and the radial arm task is that the former evaluates win-stay behavior, return to the previous escape location, whereas the latter evaluates win-shift behavior, because reward has been removed from that location, search at a new location.

BARNES TASK

The Barnes task is essentially a dry maze version of the swimming pool task, and it allows an animal to escape from an open area to a refuge. The Barnes task is a circular platform made of wood or stainless steel. In its original version, there are eight holes cut in the table that are spaced equidistant and arranged along the perimeter of the table (Barnes, 1979). A cage similar to the animal's home cage is placed underneath one of the holes and serves as a refuge. The common test procedure uses a fixed location for the refuge. The rat is re-leased in the center of the table, and its latency and accuracy in finding the refuge are de-pendent measures. Because the eight poten-tial refuge holes all appear the same to a viewer on the table, the correct hole can be quickly reached only by learning its location in relation to surrounding room cues.

The Barnes maze has been modified to evaluate search behavior. A food-deprived rat is placed in the refuge and can leave the refuge to find food on the table. When a rat is placed into its refuge, it will soon leave the refuge and start exploring the platform. Its outward-bound trips will become increasingly longer until the rat explored the entire surface of the platform. This version of the test allows for studying the structure of exploratory behav-ior (Whishaw et al., 2001). In addition, large food pellets can be placed on various or fixed locations on the table. The rat will then for-age for food, pick up the food item, and carry it to its refuge for consumption. This behavior requires that the animal memorize the location of its home base relative to the distant cues in the room and relative to self-movement cues. Interestingly, it was found that rats not only use these cues for path integration but also track olfactory information (Whishaw and Gorny, 1999). In their studies, Whishaw and Gorny (1999) described the use of strings that carry a new odor. Rats can be trained to track the odor and follow a string to locate a food reward on the platform. The animals also are able to fol-low their own scent or the scent of a conspe-cific to locate food or refuge locations.

To record the data, a camera can be mounted on the ceiling above the center of the platform to videorecord the behavior of the animals. When analyzing the video tapes, automated tracking systems might be useful to trace the path of the animal as it explores the platform.

ELEVATED-PLUS MAZE

When a rat is placed in a novel environment or it experiences a disturbance in its home cage, it can show an emotional response with signs of anxiety. *Anxiety* refers to an internal emotional state related to threat. The meas-ures of anxiety that have been established in animal research are based on observable be-havior that reflects the emotional state of fear and apprehension. One of these measures is the tendency of a rat to avoid lit open spaces, and the preference to move along walls when exploring a new environment.

The elevated-plus maze takes advantage of the fact that rats prefer to remain in en-closed compartments when they are placed in a novel environment. The elevated-plus maze is elevated above the floor. It has a small center area and four arms of equal length arranged in a plus-sign shape. Two of the arms that are facing each other are enclosed with side walls, and the other two arms are open without walls. When a rat is introduced into this test environment, it will spend more time exploring the enclosed arms than in the open arms. The typical measures of an animal's per-formance are the time spent in the enclosed arms versus the open arms and the number of entries of the two types of arms (Pellow et al., 1985).

The elevated-plus maze is a common test procedure to determine the effectiveness of drug treatments. For instance, it can be ex-pected that an anxiolytic drug would increase the number of entries in the open arms and also elevate the time spend in the open arms (Pellow and File, 1986). In contrast, an anxio-genic treatment or procedure would lead to the opposite effects by reducing the number of entries in the open arms and by promoting the animal's preference to stay in the enclosed area. Similar consequences can be drawn by deletion of genes that are involved in the control of these emotional responses. Thus, the elevated-plus maze is widely used not only in pharmacological research but also as a tool to test emotional behavior in genetic mouse models (Belzung and Griebel, 2001).

There are number of factors that can influence the behavior of an animal in the elevated-plus maze. First, manipulations that might arouse the animals and lead to stress-induced anxiety can be a disturbing confound that makes the interpretation of data difficult. Furthermore, the test apparatus itself influences the animals' responses. The size of the apparatus, the elevation above the ground, and the shape of the open arms might modify the fear to explore the open arms. Interestingly, it has been described that when the open arms are confined by a small ledge, the behavior changes over repeated test sessions in a different manner from trials that use no ledge on the open arms (Fernandez and File, 1996). These findings led to the conclusion that the elevated-plus maze in fact measures two distinct types of anxiety: the fear of the open space and the fear of elevation.

CONTEXT CONDITIONING

The rats' preference for some stimulus objects over others, its tendency to explore novel stimuli, and its avoidance to stimuli that signify threat or harm form the bases for "contextual" or "context conditioning" tasks (Otto and Giardino, 2001). The term *context* refers to the test situation prepared by the experiment, and the term *conditioning* refers to the fact that some previous experience in the situation will influence the rat's choice or preference behavior. In most of these tests, the contextual cues to which the rat responds are unspecified and are unexamined, as it is the animal's choice behavior that is of primary interest to the experimenter. Nevertheless, it is certainly the case that a rat's behavior in the task contains structure and that some cues are prepotent, although the measures of "prefer-

ence" made by the experimenter may not include the structure- or address-specific cues. The strengths of context conditioning tests are that they are simple to administer and require little pretraining of the animal and that end point scores of choice and preference are simple to collect. Many of the tests are also based on the facts that rats have a strong tendency to explore novel options and, when presented with a number of food sources, rats tend to alternate choices on successive trials. Among the many context conditioning tests that have been designed, those in greatest use are spontaneous alternation, forced alternation, and conditioned place preference.

Spontaneous Alternation

Spontaneous alternation tests are typically conducted in a T-maze or a Y-maze. The animal is placed in one arm of the maze to begin a trial and is allowed to move around in the maze until it has completely entered one of the other arms. At a specified time of minutes to as long as a day, it is given a second trial in which it is placed at the previously used starting location. If the animal chooses the previously unentered arm, it is scored as having performed an alternation. In principle, there is no limit to the number of additional trials that are administered, although for purposes of rapid screening, two trials may be sufficient. In some versions of the task, room cues are visible and it is assumed that the animal is making choices based on room cues. In other versions of the task, the alley is covered and it is assumed that the animal is making its choice using local cues, such as odors in the alley including those odors that the rat itself has left, or the animal is using self-movement cues, which is a record of its previous movements that is derived from vestibular or proprioceptive cues.

Forced Choice Alternation

A forced choice alternation test is a variation of the spontaneous alternation test in which on the first trial one of the two potential choice alleys is blocked so that the animal is forced to chose the only open alley. On the subsequent choice test, both choice alleys are open. If the animal chooses the previously blocked alley it is scored as having made an alternation.

Rewarded Alternation Tasks

In rewarded alternation tasks, animals are food or water deprived and reinforcement can be obtained by entering one of the choices. An animal may also be negatively reinforced with shock for entering an alley. The advantage of using reward is that many trials can be administered and the schedule and delay between choices can be controlled. In spontaneous-rewarded alternation tasks, reinforcement is always present in both choice alleys and the spontaneous alternations over successive trials are counted. In forced-choice rewarded alternation tasks, one of the alleys is blocked on a forced choice trial, and on a choice trial, both alleys are open. Reward may be removed from the forced choice alley for the choice trial to increase the probability of alternation behavior or reward may be present in both choice arms.

Conditioned Place Preference

Animals become conditioned to the location of an object or event that has been experienced as pleasant or noxious in a previous encounter (Jodogne et al., 1994). As a result of their previous experience, they may seek out or avoid places in which they have been reinforced previously. The apparatus for conditioned place preference is typically a box with two compartments, and the dependent measure is the time spent in either compartment. The box may allow the animal to view surrounding room cues and the compartments of the box may be distinctively marked with local cues (i.e., visual cues, odor cues, or bedding material). In the designs of the experiments, an animal is given some experience in one of the compartments of the box, the conditioning trial, and then later given an opportunity to chose which compartment it enters and subsequently spends its time, the place preference trial. Dependent measures are compartment choice and/or time spent in the compartments.

A typical experimental paradigm for conditioned place preference is conditioned reward trials using drugs. An animal is placed in one of the compartments after having received a drug, or the drug is administered while the animal is in that compartment. On a subsequent trial, the rat is allowed to choose compartments. If it chooses the previously rewarded compartment, it can be concluded that the drug is positively rewarding, whereas if it chooses the other compartment, it is concluded that the drug is negatively rewarding.

Shock-Induced Avoidance

A widely used variation of conditioned place preference test is a shock-induced avoidance or "passive-avoidance" test. Two compartments are used, and often the compartments vary in their reinforcing properties. The floor of the box is made of metal grids that can be electrified to present a mild shock to the feet of the animal. One compartment of the box may be painted black and may be in the dark, whereas the other arm is painted white and is illuminated. After one or two familiarization experiences with the apparatus, rats typically quickly enter the darkened compartment and remain there. On a subsequent test, the animal is placed in the nonpreferred compartment and then receives a foot shock as soon as it enters the preferred compartment. After this, the animal is then removed from the apparatus. In a subsequent avoidance phase of the test, the animal is replaced in the nonpreferred compartment, and its latency to enter the preferred compartment is measured.

The escape response can be used for classic conditioning experiments. In active avoidance conditioning, the animal can escape from one compartment to avoid a noxious stimulus such as gentle food shock in the other. A neutral stimulus (conditioned stimulus) such as a light flash or tone is presented before delivery of a footshock event (unconditioned stimulus). The animal will first learn to escape the footshock, but it will eventually associate the preceding neutral stimulus with the painful stimulus. When conditioning is complete, the escape response will occur to the onset of the conditioned stimulus and before the arrival of the unconditioned stimulus. Dependent measures are the number of trials for conditioning to occur and response latency.

CONCLUSION

In this chapter, we described a general battery of tests that can be used to describe the behavioral phenotype of the rat given a particular treatment. It was not possible to describe every behavior in detail, and many behaviors were not examined at all; we made no attempt to be inclusive in citing the relevant literature, only to provide examples. We noted in the introduction that the general test battery would normally be followed by more detailed analysis such as that described in the other chapters of the volume. The key point is that a great deal can be inferred about the functional state of the brain by careful behavioral analysis. Thus, although the analysis of molecular, chemical, anatomical, and physiological factors are critical to understanding brain function, in the end it is behavior that the brain is designed to produce. It thus could be argued that until you understand how the brain controls behavior, you really do not understand much at all about brain function.

REFERENCES

Almasi R, Petho G, Bolskei K, Szolcsanyi J (2003) Effect of resiniferatoxin on the noxious heat threshold temperature in the rat: a novel heat allodynia model sensitive to analgesics. British Journal of Pharmacology 139:49–58.

Ballermann M, Metz GA, McKenna JE, Klassen F, Whishaw IQ (2001) The pasta matrix reaching task: a simple test for measuring skilled reaching distance, direction, and dexterity in rats. Journal of Neuroscience Methods 106:39–45.

Barnes CA (1979) Memory deficits associated with senescence: a neurophysiological and behavioral study in the rat. Journal of Comparative and Physiological Psychology 93:74–104.

Barth TM, Jones TA, Schallert T (1990) Functional subdivisions of the rat somatic sensorimotor cortex. Behavioural Brain Research 39:73–95.

Basso DM, Beattie MS, Bresnahan JC (1995) A sensitive and reliable locomotor rating scale for open field testing in rats. Journal of Neurotrauma 12:1–21.

Basso DM, Beattie MS, Bresnahan DK, Anderson DK, Faden AI, Gruner JA, Holford TR, Hsu CY, Noble LJ, Nockels R, Perot PL, Salzman SK, Young W (1996a) MASCIS evaluation of open field locomotor scores: effects of experience and teamwork on reliability. Journal of Neurotrauma 13: 343–359.

Basso DM, Beattie MS, Bresnahan JC (1996b) Graded histological and locomotor outcomes after spinal cord contusion using the NYU weight-drop device versus transection. Experimental Neurology 139:244–256.

Belzung C and Griebel G (2001) Measuring normal and pathological anxiety-like behaviour in mice: a review. Behavioural Brain Research 125:141–149.

Daenen EW, Wolterink G, Gerrits MA, Van Ree JM (2002) The effects of neonatal lesions in the amygdala or ventral hippocampus on social behaviour later in life. Behavioral Brain Research 136: 571–582.

D'Hooge R and De Deyn PP (2001) Applications of the Morris water maze in the study of learning and memory. Brain Research and Brain Research Reviews 36:60–90.

Dubuisson D and Dennis SG (1977) The formalin test: a quantitative study of the analgesic effects of morphine, meperidine, and brain stem stimulation in rats and cats. Pain 4:161–174.

Eilam D and Golani I (1989) Home base behaviour of rats (Rattus norvegicus) exploring a novel environment. Behavioural Brain Research 34:199–211.

Eskhol N and Wachmann A (1958) Movement notation. London: Weidenfeld and Nicolson.

Fernandez C and File SE (1996) The influence of open arm ledges and maze experience in the elevated plusmaze. Pharmacology Biochemistry and Behavior 54:31–40.

Gharbawie O, Whishaw PA, Whishaw IQ (2004) The topography of three-dimensional exploration: a new quantification of vertical and horizontal exploration, postural support, and exploratory bouts in the cylinder test. Behavioural Brain Research, 151:125–35.

Golani I (1976) Homeostatic motor processes in mammalian interactions: a choreography of display. In

Perspectives in ethology, Vol II (Bateson PPG and Klopfer PH, eds.). New York: Plenum Press.

Hamilton DA, Driscoll I, Sutherland RJ (2002) Human place learning in a virtual Morris water task: some important constraints on the flexibility of place navigation. Behavioural Brain Research 129:159–170.

Hard E and Larsson K (1975) Development of air righting in rats. Brain Behavior and Evolution 11:53–59.

Harker TK and Whishaw IQ (2002) Place and matching-to-place spatial learning affected by rat inbreeding (Dark-Agouti, Fischer 344) and albinism (Wistar, Sprague-Dawley) but not domestication (wild rat vs. Long-Evans, Fischer-Norway). Behavioural Brain Research 134:467–477.

Holzberg D and Albrecht U (2003) The circadian clock: a manager of biochemical processes within the organism. Journal of Neuroendocrinology 15:339–343.

Jarrard LE (1983) Selective hippocampal lesions and behavior: effects of kainic acid lesions on performance of place and cue tasks. Behavioral Neuroscience 97:873–889.

Jodogne C, Marinelli M, Le Moal M, Piazza PV (1994) Animals predisposed to develop amphetamine self-administration show higher susceptibility to develop contextual conditioning of both amphetamine-induced hyperlocomotion and sensitization. Brain Research 657:236–244.

Kolb B (1974) Prefrontal lesions alter eating and hoarding behavior in rats. Physiology and Behavior 12:507–511.

Kolb B, Cote S, Ribeiro-da-Silva A, Cuello AC (1997) Nerve growth factor treatment prevents dendritic atrophy and promotes recovery of function after cortical injury. Neuroscience 76:1139–1151.

Kolb B and Whishaw IQ (1983) Dissociation of the contributions of the prefrontal, motor and parietal cortex to the control of movement in the rat. Canadian Journal of Psychology 37:211–232.

Ma M, Basso DM, Walters P, Stokes BT, Jakeman LB (2001) Behavioral and histological outcomes following graded spinal cord contusion injury in the C57Bl/6 mouse. Experimental Neurology 169:239–254.

Marshall JF, Turner BH, Teitelbaum P (1971) Sensory neglect produced by lateral hypothalamic damage. Science 221:389–391.

Martens DJ, Whishaw IQ, Miklyaeva EI, Pellis SM (1996) Spatio-temporal impairments in limb and body movements during righting in an hemiparkinsonian rat analogue: relevance to axial apraxia in humans. Brain Research 733:253–262.

McNamara RK and Skelton RW (1993) The neuropharmacological and neurochemical basis of place learning in the Morris water maze. Brain Research Reviews 18:33–49.

Merkler D, Metz GA, Raineteau O, Dietz V, Schwab ME, Fouad K (2001) Locomotor recovery in spinal cord-injured rats treated with an antibody neutralizing the myelin-associated neurite growth inhibitor Nogo-A. Journal of Neuroscience 21:3665–3673.

Metz GA and Whishaw IQ (2002a) Drug-induced rotation intensity in unilateral dopamine-depleted rats is not correlated with end point or qualitative measures of forelimb or hindlimb motor performance. Neuroscience 111:325–336.

Metz GA and Whishaw IQ (2002b) Cortical and subcortical lesions impair skilled walking in the variably spaced ladder rung walking task. Journal of Neuroscience Methods 115:169–179.

Metz GA, Dietz V, Schwab ME, van de Meent H (1998) The effects of unilateral pyramidal tract section on hindlimb motor performance in the rat. Behavioural Brain Research 96:37–46.

Metz GAS, Merkler D, Dietz V, Schwab ME, Fouad K (2000) Efficient testing of motor function in spinal cord injured rats. Brain Research 883:165–177.

Metz GA, Schwab ME, Welzl H (2001) The effects of acute and chronic stress on motor and sensory performance in male Lewis rats. Physiology and Behavior 72:29–35.

Montoya CP, Campbell-Hope LJ, Pemberton KD, Dunnett SB (1991) The "staircase test": a measure of independent forelimb reaching and grasping abilities in rats. Journal of Neuroscience Methods 36:219–228.

Morris RGM, Garrud P, Rawlins JN, O'Keefe J (1982) Place navigation impaired in rats with hippocampal lesions. Nature 297:681–683.

Olton DS, Becker JT, Handlemann GE (1979) Hippocampus, space and memory. Behavioral and Brain Sciences 2:313–365.

Otto T and Giardino ND (2001) Pavlovian conditioning of emotional responses to olfactory and contextual stimuli: a potential model for the development and expression of chemical intolerance. Annals of the New York Academy of Science 933:291–309.

Pellis SM and Pellis VC (1983) Locomotor-rotational movements in the ontogeny and play of the laboratory rat Rattus norvegicus. Developmental Psychobiology 16:269–286.

Pellis SM and Pellis VC (1994) Development of righting when falling from a bipedal standing posture: evidence for the dissociation of dynamic and static righting reflexes in rats. Physiology and Behavior 56:659–663.

Pellis SM, Castaneda E, McKenna MM, Tran-Nguyen LT, Whishaw IQ (1993) The role of the striatum in organizing sequences of play fighting in neonatally dopamine-depleted rats. Neuroscience Letters 158:13–15.

Pellow S and File SE (1986) Anxiolytic and anxiogenic drug effects on exploratory activity in an elevated plus-maze: a novel test of anxiety in the rat. Pharmacology Biochemistry and Behavior 24:525–529.

Pellow S, Chopin P, File SE, Briley M (1985) Validation of open:closed arm entries in an elevated plus-maze as a measure of anxiety in the rat. Journal of Neuroscience Methods 14:149–167.

Ramos A, Kangerski AL, Basso PF, Da Silva Santos JE, Assreuy J, Vendruscolo LF, Takahashi RN (2002) Evaluation of Lewis and SHR rat strains as a genetic model for the study of anxiety and pain. Behavioural Brain Research 129:113–123.

Schallert T and Whishaw IQ (1984) Bilateral cutaneous stimulation of the somatosensory system in hemidecorticate rats. Behavioural Neuroscience 98:518–540.

Schallert T, Fleming SM, Leasure JL, Tillerson JL, Bland ST (2000) CNS plasticity and assessment of forelimb sensorimotor outcome in unilateral rat models of stroke, cortical ablation, parkinsonism and spinal cord injury. Neuropharmacology 39:777–787.

Sorg BA, Tschirgi ML, Swindell S, Chen L, Fang J (2001) Repeated formaldehyde effects in an animal model for multiple chemical sensitivity. Annals of the New York Academy of Science 933:57–67.

Stam CJ, de Bruin JP, van Haelst AM, van der Gugten J, Kalsbeek A (1989) Influence of the mesocortical dopaminergic system on activity, food hoarding, social-agonistic behavior, and spatial delayed alternation in male rats. Behavioral Neuroscience 103:24–35.

Stoltz S, Humm JL, Schallert T (1999) Cortical injury impairs contralateral forelimb immobility during swimming: a simple test for loss of inhibitory motor control. Behavioural Brain Research 106:127–132.

Takahashi Y, Takahashi K, Moriya H (1995) Mapping of dermatomes of the lower extremities based on an animal model. Journal of Neurosurgery 82:1030–1034.

Tarlov IM (1954) Spinal cord compression studies. III. Time limits for recovery after gradual compression in dogs. Archives of Neurology and Psychiatry 71:588–597.

Weinert D (2000) Age-dependent changes of the circadian system. Chronobiology International 17:261–283.

Whishaw IQ (1987) Hippocampal, granule cell and CA3-4 lesions impair formation of a place learning-set in the rat and induce reflex epilepsy. Behavioural Brain Research 24:59–72.

Whishaw IQ (1993) Activation, travel distance, and environmental change influence food carrying in rats with hippocampal, medial thalamic and septal lesions: implications for studies on hoarding and theories of hippocampal function. Hippocampus 3:373–385.

Whishaw IQ and Gorny B (1999) Path integration absent in scent-tracking fimbria-fornix rats: evidence for hippocampal involvement in "sense of direction" and "sense of distance" using self-movement cues. Journal of Neuroscience 19:4662–4673.

Whishaw IQ, Haun F, Kolb B (1999) Analysis of behavior in laboratory rodents. In: Modern techniques in neuroscience research (Widhorst U, ed.). Heidelberg: Springer.

Whishaw IQ, Hines DJ, Wallace DG (2001) Dead reckoning (path integration) requires the hippocampal formation: evidence from spontaneous exploration and spatial learning tasks in light (allothetic) and dark (idiothetic) tests. Behavioural Brain Research 127:49–69.

Whishaw IQ, Kolb B, Sutherland RJ (1983) A neuropsychological study of behavior of the rat. In: Behavioral contributions to brain research (Robinson TE, ed.). New York: Oxford University Press.

Whishaw IQ and Metz GA (2002) Absence of impairments or recovery mediated by the uncrossed pyramidal tract in the rat versus enduring deficits produced by the crossed pyramidal tract. Behavioural Brain Research 134:323–336.

Whishaw IQ, Miklyaeva EI (1996) A rat's reach should exceed its grasp: analysis of independent limb and digit use in the laboratory rat. In: Measuring movement and locomotion: From invertebrates to humans (Ossenkopp KP, Kavaliers M, Sanberg PR, ed.), pp. 135–169. Austin, TX: RG Landes.

Whishaw IQ, O'Connor RB, Dunnett SB (1986) The contributions of motor cortex, nigrostriatal dopamine and caudate-putamen to skilled forelimb use in the rat. Brain 109:805–843.

Whishaw IQ, Pellis SM, Gorny BP (1992) Skilled reaching in rats and humans: evidence for parallel development or homology. Behavioural Brain Research 47:59–70.

Whishaw IQ, Pellis SM, Gorny BP, Pellis VC (1991) The impairments in reaching and the movements of compensation in rats with motor cortex lesions: an endpoint, videorecording, and movement notation analysis. Behavioural Brain Research 42:77–91.

Whishaw IQ, Suchowersky O, Davis L, Sarna J, Metz GA, Pellis SM (2002) Impairment of pronation, supination, and body co-ordination in reach-to-grasp tasks in human Parkinson's disease (PD) reveals homology to deficits in animal models. Behavioural Brain Research 133:165–176.

Woolf CJ, Shortland P, Coggeshall RE (1992) Peripheral nerve injury triggers central sprouting of myelinated afferents. Nature 355:75–78.

Index